pK_a Values for Various Compounds and Compound Classes

Compound or Class	pK_a
HI	~ −9.5
HBr	~ −9
HCl	~ −7
H_2SO_4	~ −5
H_3O^+	−1.7
$C_6H_5SO_3H$	~ −0.6
H_3PO_4	2.12
HF	3.45
CH_3CO_2H	4.75
RCO_2H	3 − 6
H_2CO_3	6.37
HCN	9.31
$CH_3\overset{O}{\overset{\|}{C}}CH_2\overset{O}{\overset{\|}{C}}CH_3$	9
NH_4^+	9.4
C_6H_5OH	10.00
CH_3NO_2	10.21
$CH_3\overset{O}{\overset{\|}{C}}CH_2\overset{O}{\overset{\|}{C}}OCH_2CH_3$	11
$CH_3CH_2O\overset{O}{\overset{\|}{C}}CH_2\overset{O}{\overset{\|}{C}}OCH_2CH_3$	13
H_2O	15.74
CH_3CH_2OH	15.9
ROH	15 − 19
$CH_3\overset{O}{\overset{\|}{C}}CH_3$	20
$CH_3\overset{O}{\overset{\|}{C}}OCH_2CH_3$	25
$RC \equiv CH$	~ 25
$CH \equiv CH$	~ 26
NH_3	~ 35
C_6H_6	~ 43
$CH_2 = CH_2$	~ 45
CH_3CH_3	~ 50

4

EDITION

Organic Chemistry

4

EDITION

Organic Chemistry

Ralph J. Fessenden
Joan S. Fessenden

University of Montana

Brooks/Cole Publishing Company
Pacific Grove, California

Brooks/Cole Publishing Company
A Division of Wadsworth, Inc.

Printed in the United States of America
10 9 8 7 6 5 4 3

Library of Congress Cataloging in Publication Data
Fessenden, Ralph J., [date]
    Organic chemistry/Ralph J. Fessenden, Joan S. Fessenden.—4th
ed.
      p. cm.
    ISBN 0-534-12252-3:
    1. Chemistry, Organic. I. Fessenden, Joan S. II. Title.
QD251.2.F49 1990
547—dc20

89-25353
CIP

Sponsoring Editor: *Harvey C. Pantzis*
Editorial Assistant: *Jennifer Kehr*
Production Editor: *Joan Marsh*
Manuscript Editor: *John Servideo*
Permissions Editor: *Carline Haga*
Cover Photo: *TIBWest, Steven Hunt*
Art Coordinator: *Michele Judge*
Interior Illustration: *Vantage Art*
Print Buyer: *Vena Dyer*
Typesetting: *Polyglot Compositors*
Cover Printing: *The Lehigh Press*
Printing and Binding: *Arcata Graphics/Hawkins*

Preface

When we wrote the first edition of *Organic Chemistry*, our goal was to produce a manageable text for the one-year introductory organic chemistry course—a text designed to guide students in their studies and not simply provide a compilation of organic chemical knowledge. For the most part, we think we were successful in achieving our goal. The response from users of the earlier editions has been gratifying.

Although we have made many minor changes and have rearranged the presentation in some places, the basic organization of the fourth edition is the same as that of previous editions. Most of the changes have been incorporated for improving the logical flow or for helping the student correlate the material.

The principal types of changes in the fourth edition are rearrangements of topics, in part to tighten the coverage by eliminating duplication. We have converted some of the longer chapters into two chapters. We have also added a few modern synthetic techniques such as asymmetric induction. These revisions and others are included in the following organizational discussion.

Organization

The text is organized into three parts: concepts of structure and bonding; organic reactions and mechanisms (sigma-bond chemistry followed by pi-bond chemistry); and topics of more specialized interest.

Introductory Material.

Chapters 1 and 2 are primarily reviews of atomic and molecular structure, electronegativity, hydrogen bonding, acid-base reactions, and molecular orbitals (presented for the most part in a pictorial way). Bonding in some simple nitrogen and oxygen compounds is included as a way of introducing functional groups. A brief introduction to resonance theory is also presented here.

In this edition, we have increased the coverage of acid-base chemistry to give some reasons for acidity differences. (Chapter 14 presents acidity in much more detail.) The thermodynamic control of reversible reactions is now defined here instead of later in the text. A table of classes of carbonyl compounds has been added for the student's reference.

The student's introduction to structural isomerism and nomenclature comes in Chapter 3. Besides discussing the nomenclature of alkanes, we introduce the naming of a few other compound classes that are used early in the book. In this edition, we have postponed the naming of esters and amines because these are not really necessary early in the course.

Chapter 4 on stereochemistry contains discussions of structure: geometric isomerism, conformation, chirality, and resolution. Fischer projections are also introduced in Chapter 4, although we do not use them again in the text until Chapter 22.

Organic Reactions.   Mechanisms are introduced in Chapter 5 with the substitution and elimination reactions of alkyl halides. We have several reasons for using the reactions of alkyl halides for our introduction to mechanisms. First, the typical S_N2 reaction path is a concerted reaction with a single transition state and is thus ideal for introducing transition-state diagrams and reaction kinetics. Second, the S_N1 path follows logically from the S_N2 mechanism and allows us to introduce steric hindrance and carbocations early in the course. Finally, ionic reactions allow us to apply the stereochemical principles just covered in Chapter 4.

A discussion of the E2 reaction path of alkyl halides follows the discussion of the substitution reactions. Although the E1 mechanism is mentioned here, the principal discussion of this reaction is postponed until Chapter 7, where the reactions of alcohols are presented.

Minor additions to Chapter 5 include a mention of chlorofluorocarbons and a definition of the kinetic control of reactions. The earlier definitions of thermodynamic and kinetic control in this edition clarify many mechanism discussions.

Chapter 6 covers free-radical chemistry, principally halogenation and polymerization. The topics of organometallic chemistry and Grignard reagents have been moved to Chapter 7 under the synthesis of alcohols. This placement is logical and eliminates duplication of coverage. Because of this reorganization, the section on solving synthesis problems has been moved from Chapter 6 to Chapter 7.

Chapter 7 (alcohols) now includes the introduction to Grignard reagents and lithium reagents. (Lithium dialkylcuprates have been moved to Chapter 15.) The discussion of phenols has been postponed to Chapter 12. The section on inorganic esters of alcohols has been shortened considerably.

The new Chapter 8 discusses the chemistry of ethers, epoxides, and sulfides, material that was formerly in Chapter 7.

Infrared and nmr spectroscopy, presented in Chapter 9, provides a break in the discussion of organic reactions. Spectroscopy is discussed later in this preface.

Chapter 10 (alkenes and alkynes) is an introduction to pi-bond chemistry. Minor additions to this chapter are examples of selective and chiral hydrogenation.

In this edition, the chemistry of benzene and substituted benzenes has been divided into two chapters. Chapter 11 covers benzene and electrophilic aromatic substitution. Chapter 12 includes the chemistry of alkylbenzenes, phenols, arylamines (including aryldiazonium salts), and aryl halides.

Carbonyl chemistry is introduced in Chapter 13 with the chemistry of aldehydes and ketones. The revisions in this chapter are minor. We have deleted reference to the use of organocadmium reagents for ketone synthesis here and in Chapter 15 (carboxylic acid derivatives). We have replaced these reagents with the more versatile lithium dialkylcuprates for ketone synthesis. Also in this chapter, we have added an example of asymmetric hydrogenation.

Chapter 14 continues carbonyl chemistry with the chemistry of carboxylic acids. Reasons for differences in acid strength of many types of compounds are discussed here. Chapter 15 describes the chemistry of carboxylic acid derivatives. Minor changes in this chapter include deletion of cadmium reagents and addition of amide dehydration for nitrile synthesis.

Chapter 16 introduces alkylation and condensation reactions. The Cannizzaro reaction has been deleted from the text but is used in a study problem. A subsection discussing the alkylation of less reactive carbonyl compounds has been added.

In Chapter 17 (amines), we have deleted a section on using amines in synthesis. Instead, we refer the student to other sections of the text where these reactions are discussed.

Topics of Specialized Interest. Chapters 18–21 cover the chemistry of polycyclic and heterocyclic compounds, natural products and some modern synthetic techniques (new to this edition), pericyclic reactions, and other spectroscopic techniques (uv and visible spectroscopy and mass spectrometry). Chapters 22–25 include bio-organic topics: carbohydrates, lipids, amino acids and proteins, and nucleic acids (new to this edition).

Chapter 18 has been shortened by the deletion of the nucleic acid and alkaloid discussions. New Chapter 19 includes an introduction to alkaloids, terpenes, and pheromones. Chapter 19 also describes three natural-product syntheses taken from the literature. These syntheses include the use of some innovative modern reagents as well as many types of reactions presented previously in the text.

Spectroscopy II has been relocated to precede the bio-organic chapters. The discussion of dyes and colored compounds has been shortened. An introduction to photochemical theory and reactions has been added.

Nomenclature

In keeping with current trends in nomenclature, we stress IUPAC names in this text; however, we also include some trivial names (such as acetone and *t*-butyl chloride) that are part of every organic chemist's vocabulary. As in previous editions, our presentation begins with a brief survey of systematic nomenclature in Chapter 3. The names presented there are those that the student will encounter again in chapters immediately following. The nomenclature for each class of compounds is then discussed in more detail in later chapters. An appendix is included for those who wish additional material or a quick source of reference.

Spectroscopy

Spectroscopy is discussed as early as we think feasible—Chapter 9. By this time, a student has a working knowledge of structure, a few functional groups, and a few reactions. However, those who wish to do so may cover the spectroscopy chapter right after Chapter 4, as soon as the students are familiar with organic structures.

We have included infrared and nuclear magnetic resonance spectroscopy in Chapter 9 because of their importance in structure determination. Sufficient background in the principles behind infrared and nmr spectroscopy is presented so that students can appreciate why spectra and structures are related, but the emphasis is on structure. Wherever appropriate after Chapter 9, we have included sections on the infrared and nmr characteristics of the compound classes being discussed. Structure-determination problems involving infrared and nmr spectra are included at the ends of many of these later chapters. In this edition, we have included a few additional ^{13}C nmr spectra here and in later problem sections.

Ultraviolet and mass spectra are described in Chapter 21. These topics are designed to stand alone; therefore, either or both can be presented along with infrared and nmr spectra if the instructor wishes. In this edition, Chapter 21 precedes the bio-organic chapters.

Synthesis

We have placed our formal discussions of synthesis in separate sections at the ends of appropriate chapters so that they may be emphasized or de-emphasized in the lecture presentation. The purposes of these sections are to provide additional review of material covered previously and to give students an opportunity to apply their knowledge.

Chapter 7 contains a discussion on approaching multistep synthesis problems and retrosynthetic analysis.

Bio-organic Material

Many students in the introductory organic course are majoring in medical or biological fields. Therefore, numerous sections and problems that are biological in nature have been included. We have selected material that is appropriate to the chemistry under discussion and that requires application of organic logic.

Problems

We are firm believers in problem solving as an important part of learning organic chemistry, and we have included more than 1150 unsolved problems in the text. Within the chapters, a number of worked-out sample problems are included to illustrate the approach to problem solving and to provide further information. Often these sample problems are followed directly by study problems with answers at the end of the book. Some of these study problems are designed to relate previous material to the present discussion. Others are designed to test students on their mastery of new material.

We have drawn heavily on recent literature for some of the more challenging problems. The supplement, *Drill Problems for Organic Chemistry* contains additional routine drill problems and their answers.

In the chapter-end problem sets, the first study problems follow the order of the chapter. The *Additional Problems* do not follow chapter order, but are in order of increasing difficulty. These problems vary from routine drill problems drawn from any previous section to challenging and thought-provoking problems.

Supplements

The student's *Solutions Manual* includes answers and solutions to chapter-end problems. The *Drill Problems for Organic Chemistry* supplement includes a brief discussion for each Chapter and additional drill problems with answers.

Many figures from the text, especially spectra, have been included in a set of *Overhead Transparencies*, available from the publisher.

Acknowledgments

We are indebted to users of previous editions for their suggestions. Many changes in this edition are the result of their comments. We also wish to thank reviewers of previous editions. Their suggestions helped us firm our ideas and helped us approach some topics in a new light.

The reviewers of the manuscript for this fourth edition have been very supportive and helpful. We are grateful to Professors Gordon S. Bates, University of British Columbia; Merle A. Battiste, University of Florida; Robert S. Bly, University of South Carolina; William Closson, SUNY (Albany); Michael T. Crimmins, University of North Carolina (Chapel Hill); Edwin M. Kaiser, University of Missouri (Columbia); Eberhard Kiehlmann, Simon Fraser University; Marek Majewski, University of Saskatchewan; James R. McKee, Philadelphia College of Pharmacy and Science; Chester W. Muth, West Virginia University; A. G. Pinkus, Baylor University; Frank P. Robinson, University of Victoria; Clay M. Sharts, San Diego State University; Robert D. Walkup, Texas Tech University; David R. Walt, Tufts University; Edgar W. Warnhoff, University of Western Ontario; Darrell J. Woodman, University of Washington; and Keith Yates, University of Toronto. We also are grateful to Professor Walter Hill, University of Montana, for his advice and suggestions for Chapter 25 (nucleic acids).

Finally, we thank the editorial and production staffs of Brooks/Cole Publishing Company, who have encouraged us and helped us in this edition. In particular, we are grateful to Harvey Pantzis, managing editor in chemistry, and Joan Marsh, production editor.

Ralph J. Fessenden
Joan S. Fessenden

Preface to the Student

Studying organic chemistry is similar to studying a foreign language. The vocabulary is new as are the "rules of grammar," such as predicting the course of an organic reaction.

When approaching new material, you may find it helpful to scan (1) the introductory paragraphs of a chapter or section, (2) the chapter summary, and (3) the chapter itself. If you do this *before the lecture* on that material, you will find the lecture more interesting and understandable. After the lecture, study the chapter sections and your lecture notes carefully, outlining the material wherever possible.

As a check on your understanding, you should work the problems within the chapter as you go along. (Cover the answers to Sample Problems with a sheet of paper, and try to work them as well.)

It is important to *write* your answers to the problems rather than simply to envision the answers in your mind. There are two reasons for writing out the answers: (1) for practice in the mechanical art of writing organic formulas, and (2) for reinforcing the subject material.

The chapter-end *Study Problems* are organized so that you can work them as you proceed through the text or you can work them when you have completed the entire chapter. The *Additional Problems* which draw on subject material more randomly should not be attempted until you have finished the chapter. These problems are gradated according to difficulty, with the early ones being drill problems and the later ones being thought problems.

See if you can answer the problems yourself, and *then* check the answers and the text. (If you try to take shortcuts, you are only deceiving yourself.) If your answers are correct, you have a reasonable mastery of the material. If you have several incorrect answers, go back and restudy the chapter, your lecture notes, and hints in the *Drill Problems for Organic Chemistry* supplement which accompanies this text.

Organic molecules are three dimensional. Besides a pencil and paper, you will need a molecular model kit. At the start of the course, especially in discussions of stereochemistry, you will find it helpful to construct models of organic molecules and compare them with their representations on paper.

For your convenience, this text contains an appendix covering the rules of organic nomenclature. (Nomenclature is also included where appropriate within the text.) A glossary of prefix symbols follows the appendix. The in-chapter problems are answered at the end of the text.

The *Drill Problems for Organic Chemistry* supplement contains additional hints and suggestions for studying organic chemistry. It also contains a brief chapter-by-chapter listing of important points; additional drill problems, whose answers are also contained in the supplement. The answers to the text's chapter-end problems are available in the *Solutions Manual*.

Brief Contents

Contents

1 ▽

Atoms and Molecules —A Review

Around 1850, organic chemistry was defined as **the chemistry of compounds that come from living beings**—hence the term *organic*. This definition was well-outgrown by about 1900. By that time, chemists were synthesizing new organic compounds in the laboratory, and many of these new compounds had no link with any living thing. Today, organic chemistry is defined as **the chemistry of the compounds of carbon**. This definition too is not entirely correct, because a few carbon compounds, such as carbon dioxide, sodium carbonate, and potassium cyanide, are considered to be inorganic. We accept this definition, however, because all organic compounds do contain carbon.

Carbon is but one element among many in the periodic table. What is so unique about carbon that its compounds justify a major subdivision in the study of chemistry? The answer is that carbon atoms can be covalently bonded to other carbon atoms and to atoms of other elements in a wide variety of ways, leading to an almost infinite number of different compounds. These compounds range in complexity from the simple compound methane (CH_4), the major component of natural gas and marsh gas, to the quite complex nucleic acids, the carriers of the genetic code in living systems.

A knowledge of organic chemistry is indispensable to many scientists. For example, because living systems are composed primarily of water and organic compounds, almost any area of study concerned with plants, animals, or microorganisms depends on the principles of organic chemistry. These areas of study include medicine and the medical sciences, biochemistry, microbiology, agriculture, and many others. However, these are not the only fields that depend on organic chemistry. Plastics and synthetic fibers are also organic compounds. Petroleum and natural gas consist mostly of compounds of carbon and hydrogen. Coal is a mixture of elemental carbon combined with compounds of carbon and hydrogen.

Where do we start? The cornerstone of organic chemistry is the covalent bond. Before we discuss the structure, nomenclature, and reactions of organic compounds in detail, we will first review some aspects of atomic structure and bonding (Chapter 1) and then molecular orbitals (Chapter 2) as these topics apply to organic compounds.

▼ Section 1.1
Electron Structure of the Atom

The most important elements to organic chemists are carbon, hydrogen, oxygen, and nitrogen. These four elements are in the first two periods of the periodic table, and their electrons are all found in the two electron shells closest to the nucleus. Consequently, our discussion of the electron structures of atoms will center mainly on elements with electrons only in these two electron shells.

Each electron shell is associated with a certain amount of energy. Electrons close to the nucleus are more attracted by the protons in the nucleus than are electrons farther away. Therefore, the closer an electron is to the nucleus, the lower its energy is and the more difficult it is to remove the electron in a chemical reaction.

The electron shell closest to the nucleus is the shell of lowest energy, and an electron in this shell is said to be at the **first energy level**. Electrons in the second shell, at the **second energy level**, are of higher energy than those in the first shell. Electrons in the third shell, at the **third energy level**, are of higher energy yet.

A. Atomic Orbitals

We cannot accurately determine the position of an electron relative to the nucleus of an atom. Instead, we must rely upon quantum theory to describe the most likely location of an electron. Each electron shell of an atom is subdivided into **atomic orbitals**, an atomic orbital being a region in space where the probability of finding an electron of a specific energy content is high (90–95%). **Electron density** is another term used to describe the probability of finding an electron in a particular spot; a higher electron density means a greater probability, while a lower electron density means a lesser probability.

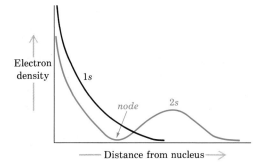

Figure 1.1 Graphic relationship between the 1s and 2s atomic orbitals.

The first electron shell contains only the spherical 1s orbital. The probability of finding a 1s electron is highest in this sphere. The second shell, which is slightly farther from the nucleus than the first shell, contains one 2s orbital and three 2p orbitals. The 2s orbital, like the 1s orbital, is spherical.

Figure 1.1 shows a graph of electron density in the 1s and 2s orbitals as a function of distance from the nucleus. It may be seen from the graph that the 1s and 2s orbitals do not have sharply defined surfaces, but rather the electron density increases and decreases over a range of distances from the nucleus. The result is that the 1s and 2s orbitals overlap each other.

The electron density–distance curve for the 2s orbital reveals two areas of high electron density separated by a zero point. This zero point, called a **node**, represents a region in space where the probability of finding an electron (the 2s electron in this case) is very small. All orbitals except the 1s orbital have nodes. Pictorial representations of the 1s and 2s orbitals are shown in Figure 1.2.

The second energy level also contains three 2p atomic orbitals. The 2p orbitals are at a slightly greater distance from the nucleus than the 2s orbital and are of slightly higher energy. The p orbitals are shaped rather like dumbbells; each p orbital has two lobes separated by a node (a nodal plane in this case) at the nucleus (see Figure 1.3).

A sphere (an s orbital) is nondirectional; that is, it appears the same when viewed from any direction. This is not the case with a p orbital, which can assume different orientations about the nucleus. The three 2p orbitals are at *right angles* to one another. This orientation allows maximum distance between the electrons in the three p orbitals and thus minimizes repulsions between electrons in different p orbitals. The mutually perpendicular p orbitals are

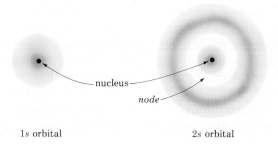

1s orbital   2s orbital

Figure 1.2 Pictorial representations of the 1s and 2s atomic orbitals.

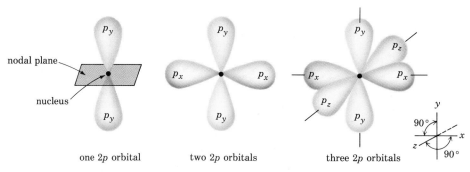

Figure 1.3 The shapes and orientations of the 2p orbitals.

Table 1.1 Atomic orbitals in the first three energy levels

Energy Level	*Atomic Orbitals*
1	$1s$
2	$2s$ $2p_x$ $2p_y$ $2p_z$
3	$3s$ $3p_x$ $3p_y$ $3p_z$ plus five $3d$

sometimes designated p_x, p_y, and p_z. The subscript letters refer to the x, y, and z axes that can be drawn through pictures of these p orbitals, as in Figure 1.3.

Since the three $2p$ orbitals are equivalent in shape and in distance from the nucleus, they have equal energies. Orbitals that have the same energy, such as the three $2p$ orbitals, are said to be **degenerate**.

The third electron shell contains one $3s$ orbital, three $3p$ orbitals, and also five $3d$ orbitals. The numbers of atomic orbitals at each of the first three energy levels are summarized in Table 1.1.

B. Filling the Orbitals

Electrons have spin, which can be either "clockwise" or "counterclockwise" ($+\frac{1}{2}$ or $-\frac{1}{2}$). The spin of a charged particle gives rise to a small magnetic field, or **magnetic moment**, and two electrons with opposite spin have *opposite magnetic moments*. The repulsion between the negative charges of two electrons with opposite spin is minimized by the opposite magnetic moments, allowing two such electrons to become *paired* within an orbital. For this reason, any orbital can hold a maximum of two electrons, and those electrons must be of opposite spin. Because of the number of orbitals at each energy level (one at the first energy level, four at the second, and nine at the third), the first three energy levels can hold up to two, eight, and 18 electrons, respectively.

The **aufbau principle** (German, "building up") states that as we progress from hydrogen (atomic number 1) to atoms of successively higher atomic number, orbitals become filled with electrons in such a way that the *lowest-energy orbitals are filled first*. A hydrogen atom has its single electron in a $1s$ orbital. The next element, helium (atomic number 2), has its second electron also in the $1s$ orbital. The two electrons in this orbital are paired.

A description of the electron structure for an element is called its **electron configuration**. The electron configuration for H is $1s^1$, which means one

Table 1.2 Electron configurations of the elements in periods 1 and 2

Element	Atomic Number	Electron Configuration
H	1	$1s^1$
He	2	$1s^2$
Li	3	$1s^2\ 2s^1$
Be	4	$1s^2\ 2s^2$
B	5	$1s^2\ 2s^2\ 2p^1$
C	6	$1s^2\ 2s^2\ 2p^2$
N	7	$1s^2\ 2s^2\ 2p^3$
O	8	$1s^2\ 2s^2\ 2p^4$
F	9	$1s^2\ 2s^2\ 2p^5$
Ne	10	$1s^2\ 2s^2\ 2p^6$

electron (superscript 1) in the $1s$ orbital. For He, the electron configuration is $1s^2$, meaning *two* electrons (superscript 2) in the $1s$ orbital. Lithium (atomic number 3) has two electrons in the $1s$ orbital and one electron in the $2s$ orbital; its electron configuration is $1s^2\ 2s^1$.

The electron configurations for the first- and second-period elements are shown in Table 1.2. In carbon and the succeeding elements, each $2p$ orbital receives one electron before any $2p$ orbital receives a second electron. This is an example of **Hund's rule**: In filling atomic orbitals, pairing of two electrons in degenerate orbitals does not occur until each degenerate orbital contains one electron. Therefore, an atom of carbon has an electron configuration of $1s^2\ 2s^2\ 2p_x{}^1\ 2p_y{}^1$, abbreviated $1s^2\ 2s^2\ 2p^2$ in Table 1.2.

▼ Section 1.2
Atomic Radius

The **radius of an atom** is the distance from the center of the nucleus to the outermost electrons, which are called the **valence electrons**. The atomic radius is determined by measuring the **bond length** (the distance between nuclei) in a symmetrical covalent compound such as Cl—Cl or H—H and dividing by two. Therefore, atomic radii are often called **covalent radii**. Values for atomic radii are usually given in nanometers, where 1 nm $= 10^{-9}$ m, or in angstroms, where 1 Å $= 10^{-10}$ m.

For H_2:

bond length $= 0.074$ nm or 0.74 Å

atomic radius $= 0.037$ nm or 0.37 Å

Atomic radii vary depending on the extent of attraction between the nucleus and its electrons. The greater the attraction, the smaller the atomic radius. What factors affect this attraction? The most important factors are *the number of protons in the nucleus* and *the number of shells containing electrons*.

H 0.37						
Li 1.225	Be 0.889	B 0.80	C 0.771	N 0.74	O 0.74	F 0.72
Na 1.572	Mg 1.364	Al 1.248	Si 1.173	P 1.10	S 1.04	Cl 0.994
						Br 1.142
						I 1.334

Figure 1.4 Atomic radii of some of the elements (in angstroms, Å, where $1 \text{ Å} = 10^{-8}$ cm).

A nucleus with a greater number of protons has a greater attraction for its electrons, including the outermost electrons. Consider the elements of the second row of the periodic table (lithium to fluorine). An atom of any of these elements has electrons in only the first two electron shells. As we progress stepwise from lithium to fluorine, a proton is added to the nucleus. At each step, the nucleus has a greater attraction for the electrons, and the atomic radius decreases (refer to Figure 1.4).

	Li	Be	B	C	N	O	F
atomic number:	3	4	5	6	7	8	9

decreasing atomic radius →

As we proceed from top to bottom within a group in the periodic table, the number of electron shells increases and, therefore, so does the atomic radius.

increasing atomic radius

H (1 shell)
Li (2 shells)
Na (3 shells)
K (4 shells)

In organic chemistry, atoms are bonded together in close proximity to one another by covalent bonds. We will find the concept of atomic radii useful in estimating the attractions and repulsions between atoms and in discussing covalent bond strengths.

Section 1.3
Electronegativity

Electronegativity is a measure of an atom's attraction for its outer bonding electrons, especially in covalently bonded compounds. The concept of electronegativity is used for predicting and explaining chemical reactivity. Like the atomic radius, electronegativity is affected by the number of protons in the

H 2.1						
Li 1.0	Be 1.5	B 2.0	C 2.5	N 3.0	O 3.5	F 4.0
Na 0.9	Mg 1.2	Al 1.5	Si 1.8	P 2.1	S 2.5	Cl 3.0
						Br 2.8
						I 2.5

Figure 1.5 Electronegativities of some elements (Pauling scale).

nucleus and by the number of shells containing electrons. A greater number of protons means a greater positive nuclear charge and thus an increased attraction for the bonding electrons. Therefore, electronegativity increases as we go from left to right in a given period of the periodic table.

Li   Be   B   C   N   O   F

increasing electronegativity

Attractions between oppositely charged particles increase with decreasing distance between the particles. Thus, electronegativity increases as we proceed from bottom to top in a given group of the periodic table because the valence electrons are closer to the nucleus.

increasing electronegativity
F
Cl
Br
I

The **Pauling scale** (Figure 1.5) is a numerical scale of electronegativities. This scale is derived from bond-energy calculations for different elements joined by covalent bonds. In the Pauling scale, fluorine, the most electronegative element, has an electronegativity value of 4. Lithium, which has a low electronegativity, has a value of 1. An element with a very low electronegativity (such as lithium) is sometimes called an **electropositive** element. Carbon has an intermediate electronegativity value of 2.5.

Section 1.4
Introduction to the Chemical Bond

Because of their different electron structures, atoms can become bonded together in molecules in different ways. In 1916, G. N. Lewis and W. Kössel advanced the following theories:

1. An **ionic bond** results from the transfer of electrons from one atom to another.
2. A **covalent bond** results from the sharing of a pair of electrons by two atoms.
3. Atoms transfer or share electrons so as to gain a **noble-gas electron configuration**. This configuration is usually eight electrons in the outer shell, corresponding to the electron configuration of neon and argon. This theory is called the **octet rule**.

An ionic bond is formed by electron transfer. One atom donates one or more of its outermost electrons to another atom or atoms. The atom that loses electrons becomes a positive ion, or **cation**. The atom that gains the electrons becomes a negative ion, or **anion**. The ionic bond results from the electrostatic attraction between these oppositely charged ions. We can illustrate electron transfer using dots to represent the valence electrons.

One electron is transferred from Na to Cl.

$$Na\cdot + \cdot \ddot{\underset{\cdot\cdot}{Cl}}\colon \longrightarrow Na^+ \; \colon\!\ddot{\underset{\cdot\cdot}{Cl}}\colon^- \quad \text{or} \quad Na^+ \; Cl^-$$

Each atom now has a complete octet in its outer shell. (The outer filled shell of Na^+ is not shown.)

A covalent bond is produced by the sharing of a pair of valence electrons between two atoms. Shared electrons result from the merging of the atomic orbitals into shared orbitals called **molecular orbitals**, a topic that we will discuss in Chapter 2. For now, we will use dots to represent valence electrons. With dot formulas, called **Lewis formulas** or **Lewis structures**, we can easily count electrons and see that the atoms attain noble-gas configurations: two electrons (helium configuration) for hydrogen and eight electrons for most other atoms. Note that both the bonding electrons and the unshared electrons of an outermost electron shell are shown as dots in Lewis formulas.

$$H\cdot + H\cdot \longrightarrow H\colon H$$

covalent bond

$$\colon\!\ddot{\underset{\cdot\cdot}{Cl}}\cdot + \cdot\ddot{\underset{\cdot\cdot}{Cl}}\colon \longrightarrow \colon\!\ddot{\underset{\cdot\cdot}{Cl}}\colon\!\ddot{\underset{\cdot\cdot}{Cl}}\colon$$

$$\cdot\dot{\underset{\cdot}{C}}\cdot + 4\,H\cdot \longrightarrow H\colon\!\overset{H}{\underset{H}{\ddot{C}}}\!\colon H \longleftarrow \textit{four covalent bonds}$$

The sharing of one pair of electrons between two atoms is called a **single bond**. Two atoms can share two pairs or even three pairs of electrons; these multiple bonds are called **double bonds** and **triple bonds**, respectively.

Sample Problems

Circle the eight bonding electrons associated with the carbon atom in each of the following structures:

$$\overset{\cdot\cdot}{\underset{\cdot\cdot}{O}}$$
$$H\!:\!\overset{..}{C}\!:\!H \qquad \overset{\cdot\cdot}{O}\!::\!C\!::\!\overset{\cdot\cdot}{O} \qquad H\!:\!C\!::\!N\!:$$

Solution:

For the structures in the preceding problem, circle the two electrons associated with each hydrogen atom and the eight electrons associated with each oxygen or nitrogen atom.

Solution:

When do atoms form ionic bonds and when do they form covalent bonds? Ionic bonds are formed when the electronegativity difference between two atoms is large (greater than about 1.7). For example, a sodium atom (electronegativity 0.9), with little attraction for its outer electron, readily loses this electron to a chlorine atom (electronegativity 3.0). On the other hand, the electronegativity difference between two carbon atoms is zero; between carbon and hydrogen, only 0.4; and between carbon and chlorine, 0.5. Because carbon has an electronegativity of 2.5, intermediate between the extremes of high and low electronegativity, it almost never forms ionic bonds with other elements. Instead, *carbon forms covalent bonds with other carbon atoms and with atoms of other elements.*

A. Number of Covalent Bonds

The number of covalent bonds that an atom forms depends upon the number of additional electrons needed for the atom to attain a noble-gas configuration. For example, a neutral atom of hydrogen needs to share one electron to attain the electron configuration of helium; therefore, hydrogen forms one covalent bond. A neutral atom of chlorine has seven electrons in its outer shell and needs to share one electron to attain the electron configuration of argon;

therefore, chlorine also forms one covalent bond. A neutral atom of carbon has four electrons in its outermost shell. Carbon needs to share four electrons to attain the electron configuration of neon; therefore, carbon forms four covalent bonds.

$$
\cdot \overset{\displaystyle \cdot}{\underset{\displaystyle \cdot}{C}} \cdot \quad + \quad 4\,H\cdot \quad \longrightarrow \quad H : \overset{\displaystyle H}{\underset{\displaystyle H}{\overset{\cdot\cdot}{\underset{\cdot\cdot}{C}}}} : H
$$

	forms 4	*forms 1*
	covalent bonds	*covalent bond*

$$
\cdot \overset{\displaystyle \cdot}{\underset{\displaystyle \cdot}{C}} \cdot \quad + \quad 4\,:\overset{\cdot\cdot}{\underset{\cdot\cdot}{Cl}}\cdot \quad \longrightarrow \quad :\overset{\cdot\cdot}{\underset{\cdot\cdot}{Cl}}:\overset{:\overset{\cdot\cdot}{Cl}:}{\underset{:\overset{\cdot\cdot}{Cl}:}{C}}:\overset{\cdot\cdot}{\underset{\cdot\cdot}{Cl}}:
$$

	forms 4	*forms 1*
	covalent bonds	*covalent bond*

Table 1.3 lists typical numbers of covalent bonds formed by elements commonly found in organic compounds. Table 1.3 also lists the groups in the periodic table where these elements are found. For the neutral atoms of elements in Groups IA–VIIA, the group number is the same as the number of outer electrons.

Table 1.3 Numbers of covalent bonds usually formed by some elements typically encountered in organic compounds

Element	*Number of Covalent Bonds*	*Group Number in Periodic Table*
H	1	IA
C	4	IVA
N	3^a	VA
O	2	VIA
Cl	1	VIIA
I	1	VIIA
Br	1	VIIA

[a] N can form four covalent bonds in some cases, such as in the NH_4^+ ion.

For simple structures, we can often deduce the Lewis formula for a compound of known composition from simply knowing the usual number of covalent bonds an element forms.

Sample Problems

Write the Lewis formulas for H_2O and C_2H_6.

Solution:

1. Determine the number of valence electrons of each atom: H = 1; O = 6; C = 4.
2. Draw the skeleton of the molecule, keeping in mind that H can form one covalent bond, O can form two, and C can form four.

H O H   H C C H with H above and below each C

3. Distribute the valence electrons in such a way that each H has two electrons and each other atom has an octet.

H:Ö:H   H:C:C:H with H above and below each C

Write the Lewis formula for C_3H_8 and two Lewis formulas for C_2H_6O.

Solution: H:C:C:C:H (with H above and below)   H:C:C:Ö:H (with H above and below)   H:C:Ö:C:H (with H above and below)

Study Problem

1.1 Each of the following structures contains a double or triple bond. Give the Lewis formulas.

(a) Cl_2CO **(b)** C_2Cl_4 **(c)** C_2H_3Br **(d)** C_2HF

B. Formal Charge

In some formulas, a few elements appear to form an unusual number of covalent bonds. We find that drawing completely correct Lewis formulas of these compounds is not possible unless we assign electrostatic charges, called **formal charges**, to some of the elements in these structures. For example, let us consider the Lewis structure of nitric acid.

:Ö:
H:Ö:N::Ö:

This oxygen has an octet but only one covalent bond.

This oxygen has an octet and two covalent bonds.

nitric acid, HNO_3

The formula for nitric acid shows three oxygens bonded to a nitrogen atom. Each oxygen and nitrogen atom has a complete octet, but one oxygen atom is bonded by only one covalent bond instead of by the usual two. If the two electrons in the N—O single bond were divided between the N and the O (one electron to each atom), the oxygen would have seven valence electrons—one more electron than a neutral oxygen atom has. Therefore, this oxygen is assigned an electrostatic charge, or formal charge, of -1. Similarly, the nitrogen atom has only four valence electrons—one fewer electron than a neutral nitrogen atom has. This electron-deficient nitrogen atom has a formal charge of $+1$.

The other atoms in nitric acid all have the same number of electrons as would their neutral atoms; their formal charges are all zero.

7 electrons "belong" to this O atom; it has a formal charge of −1.

Only 4 electrons "belong" to the N atom; it has a formal charge of +1.

In nitric acid, or any other electrically neutral molecule, the formal charges must cancel, or sum to zero.

$$\begin{array}{l} :\ddot{O}: \\ H:\ddot{O}:N^+::\ddot{O}: \end{array}$$

O has formal charge of −1 $\Big\}$ $+1 + (-1) = 0$
N has formal charge of +1

To determine if formal charges must be assigned to atoms within a Lewis structure, we use the following equation for each atom in the structure:

formal charge = (number of valence e^- in a neutral atom)
$\qquad - \frac{1}{2}$(number of shared e^-) − (number of unshared valence e^-)

Example

Using the formula, we can calculate the formal charges on N and each O in HNO_3. (H does not carry a formal charge in covalent molecules.)

$$\begin{array}{l} :\ddot{O}: \\ H:\ddot{O}:N::\ddot{O}: \end{array}$$

no. of valence e^- for neutral N = 5
$\frac{1}{2}$(no. of shared e^- for N) = 4
no. of unshared e^- for N = 0

formal charge for N = 5 − 4 − 0 = +1

For each O, the same technique is used.

formal charge:
6 − 1 − 6 = −1

$$\begin{array}{l} :\ddot{O}: \\ H:\ddot{O}:N::\ddot{O}: \end{array}$$

formal charge:
6 − 2 − 4 = 0

formal charge:
6 − 2 − 4 = 0

▼

Example

Let us calculate the formal charges, if any, in the following molecule.

$$\begin{array}{ll} \text{H} & \\ H:\ddot{C}:N::\ddot{O}: & \\ \text{H} & \end{array}$$

for C: 4 − 4 − 0 = 0
for N: 5 − 3 − 2 = 0
for O: 6 − 2 − 4 = 0

Note that none of the atoms in this molecule has a formal charge.   ▼

If the formal charges within a structure do not sum to zero, then the structure is an ion. The sum of the formal charges represents the ionic charge.

Example

Let us calculate the formal charges in the following formula to determine the ionic charge.

$$
\begin{bmatrix}
& :\ddot{\text{O}}: & \\
:\ddot{\text{O}}:\text{S}:\ddot{\text{O}}: & \\
& :\ddot{\text{O}}: &
\end{bmatrix}
\qquad
\begin{aligned}
&\text{for S:} & 6 - 4 - 0 &= +2 \\
&\text{for each O:} & 6 - 1 - 6 &= -1
\end{aligned}
$$

$$
\begin{aligned}
\text{ionic charge} &= (\text{formal charge of sulfur}) + 4(\text{formal charge of each oxygen}) \\
&= (+2) + 4(-1) \\
&= -2
\end{aligned}
$$

The formula, of course, represents the sulfate ion, $SO_4{}^{2-}$. ▼

Study Problems

1.2 Calculate the formal charges on C, N, and O in the following formulas and determine whether each represents a neutral molecule or an ion.

       H H   H H
(a) H:C:N:H   (b) H:C:N:H   (c) :C::O:
       H   H H

1.3 Calculate the formal charge on each atom in the following structures:

                  H   H :Ö:
(a) H:Ö:N::O: (b) H:C::N::N: (c) H:C:Ö:S:Ö:H
                                 H :Ö:

▼ ## Section 1.5
Chemical Formulas in Organic Chemistry

Lewis formulas are useful for keeping track of bonding electrons, but organic chemists rarely use true Lewis formulas. Let us consider the types of chemical formulas that are more frequently encountered.

An **empirical formula** tells us the types of atoms and their numerical ratio in a molecule. For example, a molecule of ethane contains carbon and hydrogen atoms in a ratio of 1 to 3; the empirical formula is CH_3. A **molecular formula** tells us the actual number of each type of atom in a molecule, not simply the ratio. The molecular formula for ethane is C_2H_6. A **structural formula** shows the *structure* of a molecule, that is, the order of attachment of the atoms. In order to explain or predict chemical reactivity, we need to know the structure of a molecule. Therefore, structural formulas are the most useful of the different types of formulas.

$$CH_3 \qquad C_2H_6$$

$$\begin{array}{c} \text{H} \quad \text{H} \\ | \quad | \\ \text{H--C--C--H} \\ | \quad | \\ \text{H} \quad \text{H} \end{array}$$

empirical formula   *molecular formula*   *structural formula*
for ethane   *for ethane*   *for ethane*

A. Structural Formulas

Lewis formulas are one type of structural formula. However, chemists usually represent a covalent structure by using a dash for each shared pair of electrons, and rarely show unshared pairs of valence electrons. Formulas with dashes for bonds are called **line-bond formulas**. In this text, we will also refer to them as **complete structural formulas**.

H:H   becomes   H—H

:C̈l:C̈l: becomes   :C̈l—C̈l: or   Cl—Cl

$$\begin{array}{c} \text{H} \\ \text{H:C:H} \\ \text{H} \end{array} \quad \text{becomes} \quad \begin{array}{c} \text{H} \\ | \\ \text{H--C--H} \\ | \\ \text{H} \end{array}$$

$$\begin{array}{c} \text{H} \qquad \text{H} \\ \text{:C::C:} \\ \text{H} \qquad \text{H} \end{array} \quad \text{becomes} \quad \begin{array}{c} \text{H} \qquad \text{H} \\ \diagdown \quad \diagup \\ \text{C=C} \\ \diagup \quad \diagdown \\ \text{H} \qquad \text{H} \end{array}$$

Although unshared pairs of valence electrons are not usually shown in line-bond formulas, we will sometimes show these electrons when we want to emphasize their role in a chemical reaction.

All represent the same molecule:

                                                        unshared pair of valence e⁻

$$\begin{array}{c} \text{H:N:H} \\ \text{H} \end{array} \quad \text{or} \quad \begin{array}{c} \text{H--N--H} \\ | \\ \text{H} \end{array} \quad \text{or} \quad \begin{array}{c} \text{H--N̈--H} \\ | \\ \text{H} \end{array}$$

Sample Problem

Write the complete structural formula, showing each bond as a line, for each of the following Lewis formulas:

$$\begin{array}{c} \text{H} \quad \text{H} \\ \text{H:C:O:C:H} \\ \text{H} \quad \text{H} \end{array} \qquad \begin{array}{c} \text{H H} \\ \text{H:C:C:O:H} \\ \text{H H} \end{array} \qquad \begin{array}{c} \text{:O:} \\ \text{H:C:H} \end{array}$$

Solution:

$$\begin{array}{c} \text{H} \qquad \text{H} \\ | \qquad | \\ \text{H--C--O--C--H} \\ | \qquad | \\ \text{H} \qquad \text{H} \end{array} \qquad \begin{array}{c} \text{H} \quad \text{H} \\ | \quad | \\ \text{H--C--C--O--H} \\ | \quad | \\ \text{H} \quad \text{H} \end{array} \qquad \begin{array}{c} \text{O} \\ \| \\ \text{H--C--H} \end{array}$$

B. Condensed Structural Formulas

Complete structural formulas are frequently condensed to shorter, more convenient formulas. In **condensed structural formulas**, bonds are not always shown and atoms of the same type bonded to one other atom are grouped together. The structure of a molecule is still evident from a condensed structural formula.

$$
CH_3CH_3 \quad \text{is a condensed structural formula for} \quad
\begin{array}{c}
\text{H}\ \ \text{H} \\
| \ \ \ | \\
\text{H---C---C---H} \\
| \ \ \ | \\
\text{H}\ \ \text{H}
\end{array}
$$

$$
CH_3CH_2OH \quad \text{is a condensed structural formula for} \quad
\begin{array}{c}
\text{H}\ \ \text{H} \\
| \ \ \ | \\
\text{H---C---C---O---H} \\
| \ \ \ | \\
\text{H}\ \ \text{H}
\end{array}
$$

Sample Problem

Write (a) the complete structural formula (showing all bonds as dashes), and (b) a condensed structural formula, for each of the following Lewis formulas:

```
    H H                  H   H   H
    .. ..                ..      ..
  H:C:C:Cl:            H:C : C : C:H
    .. ..                ..      ..
    H H                  H  :Cl:  H
                            ..
```

Solution:

$$
\textbf{(a)} \quad
\begin{array}{c}
\text{H}\ \ \text{H} \\
| \ \ \ | \\
\text{H---C---C---Cl} \\
| \ \ \ | \\
\text{H}\ \ \text{H}
\end{array}
\qquad
\begin{array}{c}
\text{H}\ \ \text{H}\ \ \text{H} \\
| \ \ \ | \ \ \ | \\
\text{H---C---C---C---H} \\
| \ \ \ | \ \ \ | \\
\text{H}\ \ \text{Cl}\ \ \text{H}
\end{array}
$$

(b) CH_3CH_2Cl $CH_3CHClCH_3$

Structural formulas may be condensed even further if a molecule has two or more identical groups of atoms. In these cases, parentheses are used to enclose a repetitive group of atoms. The subscript following the second parenthesis indicates the number of times the entire group is found at that position in the molecule.

$$
(CH_3)_2CHOH \quad \text{is the same as} \quad
\begin{array}{c}
CH_3 \\
| \\
CH_3\text{---C---OH} \\
| \\
\text{H}
\end{array}
$$

$$
(CH_3)_3CCl \quad \text{is the same as} \quad
\begin{array}{c}
CH_3 \\
| \\
CH_3\text{---C---Cl} \\
| \\
CH_3
\end{array}
$$

$$
CH_3(CH_2)_3CH_3 \quad \text{is the same as} \quad CH_3CH_2CH_2CH_2CH_3
$$

For the sake of clarity, double or triple bonds are usually shown in a condensed structural formula.

$CH_3CH{=}CH_2$ is the same as

$$\begin{array}{c} H \quad H \\ \backslash \quad / \\ C \\ H \diagup \quad \diagdown C{=}C \diagup H \\ \mid \qquad \backslash \\ H \qquad H \end{array}$$

$CH_3C{\equiv}CH$ is the same as $\quad H{-}\overset{\displaystyle H}{\underset{\displaystyle H}{\overset{\mid}{\underset{\mid}{C}}}}{-}C{\equiv}C{-}H$

$CH_3\overset{\displaystyle O}{\overset{\|}{C}}CH_2CH_3$ is the same as $\quad H{-}\overset{\displaystyle H}{\underset{\displaystyle H}{\overset{\mid}{\underset{\mid}{C}}}}{-}\overset{\displaystyle O}{\overset{\|}{C}}{-}\overset{\displaystyle H}{\underset{\displaystyle H}{\overset{\mid}{\underset{\mid}{C}}}}{-}\overset{\displaystyle H}{\underset{\displaystyle H}{\overset{\mid}{\underset{\mid}{C}}}}{-}H$

Study Problem

1.4 For each of the following formulas, write a more condensed formula:

(a) $CH_3\overset{\displaystyle CH_3}{\overset{\mid}{C}H}CH_2Cl$ (b) $CH_3\overset{\displaystyle Cl}{\overset{\mid}{C}H}Cl$ (c) $CH_3CH_2CH_2CH_2\overset{\displaystyle Cl}{\overset{\mid}{C}H}CH_2Cl$

(d) $\begin{array}{c} CH_3 \quad CH_3 \\ \diagdown \quad \diagup \\ C{=}C \\ \diagup \quad \diagdown \\ CH_3 \quad CH_3 \end{array}$ (e) $N{\equiv}C{-}CH_2{-}C{\equiv}N$

C. Cyclic Compounds and Polygon Formulas

A compound such as $CH_3CH_2CH_2CH_3$ is said to have its carbon atoms connected in a chain. Carbon atoms can be joined together in rings as well as in chains; a compound containing one or more rings is called a **cyclic compound**.

Cyclic structures are usually represented by **polygon formulas**, which are another type of condensed structural formula. For example, a triangle is used to represent a three-membered ring, while a hexagon is used for a six-membered ring.

$$\begin{array}{c} CH_2 \\ \diagup \quad \diagdown \\ H_2C{-}CH_2 \end{array} \quad \text{or} \quad \triangle \qquad \begin{array}{c} H_2C{-}CH_2 \\ \diagup \qquad \diagdown \\ H_2C \qquad CH_2 \\ \diagdown \qquad \diagup \\ H_2C{-}CH_2 \end{array} \quad \text{or} \quad \hexagon$$

In polygon formulas, a corner represents a carbon atom along with its hydrogens; the sides of the polygon represent the bonds joining the carbons. If an atom or group other than hydrogen is attached to a carbon of the ring, the number of hydrogens at that position is reduced accordingly.

C—C bond   *C with no H's*

C with two H's CH_3 Cl   Cl

C with one H

Rings can contain atoms other than carbon. These atoms and any hydrogens attached to them must be indicated in the polygon formula. Double bonds must also be indicated.

H₂C—CH₂
H₂C   NH   or   [hexagon] NH

HC=CH
H₂C   CH₂   or   [hexagon with double bond]   *double bond within ring*
H₂C—CH₂

H₂C—CH₂
H₂C   C=O   or   [pentagon] =O   *double bond from a ring carbon*
   CH₂

Study Problems

1.5 Draw complete structural formulas for the following structures, showing each C, each H, and each bond.

(a) [hexagon with double bond]   **(b)** [hexagon] NCH₃   **(c)** [hexagon with Cl and CH₃]

1.6 Draw polygon formulas for the following structures:

(a)
$$H_2C \overset{\overset{H_2}{C}}{} CH_2$$
H₂C——CH₂

(b)
H₂C   C(H₂)   CH   C(H₂)   CH₂
H₂C   C(H)   C(H)   CH₂
      H₂   H₂

(c)
H₂C——CH₂
H₂C O CH₂

(d)
CH₂—CHCH₃
CH₂—C(CH₃)₂

(e)
HC——CH
HC C CH
   H₂

Section 1.6
Bond Lengths and Bond Angles

We have discussed how chemists represent covalent compounds. Now let us consider some of the properties of covalent bonds. The distance that separates the nuclei of two covalently bonded atoms is called the **bond length**. Covalent bond lengths, which can be measured experimentally, range from 0.74 Å to 2 Å.

If there are more than two atoms in a molecule, the bonds form an angle, called the **bond angle**. Bond angles vary from about 60° up to 180°.

H   O   H   *bond lengths, 0.96 Å*
bond angle, 104.5°

N   H   *bond lengths, 1.008 Å*
H   H   *bond angles, 107.3°*

Most organic structures contain more than three atoms and are three-dimensional rather than two-dimensional. The preceding structural formula for ammonia (NH_3) illustrates one technique for representing a three-dimensional structure. A line bond (—) represents a bond in the plane of the paper. The solid wedge (◄) represents a bond coming out of the paper toward the viewer; the H at the wide end of the solid wedge is in front of the paper. The broken wedge (⸽⸽⸽) represents a bond pointing back into the paper; the H at the small end of the broken wedge is behind the paper.

in plane of paper ———   N ⸽⸽⸽⸽ H   *back, away from viewer*

H   H

forward, toward viewer

Section 1.7
Bond Dissociation Energy

There are two ways a bond can dissociate, or break apart. One way is by **heterolytic cleavage** (Greek *heteros*, "different"), in which both bonding electrons are retained by one of the atoms. Heterolytic cleavage usually results in the formation of ionic species.

Heterolytic cleavage: $H \overset{\frown}{-} H \longrightarrow H^+ + H:^-$

$H \overset{\frown}{-} \ddot{\underset{..}{C}}l: \longrightarrow H^+ + :\ddot{\underset{..}{C}}l:^-$

$H \overset{\frown}{-} \ddot{O}H \longrightarrow H^+ + {}^-:\ddot{O}H$

We use a curved arrow (⌢) in these equations to show the direction in which the pair of bonding electrons moves during bond breakage. In the heterolytic cleavage of HCl or H_2O, the bonding electrons are transferred to the more electronegative Cl or O.

The other process by which a bond may dissociate is **homolytic cleavage** (Greek *homos*, "same"). In this case, each atom involved in the covalent bond receives one electron from the original shared pair. Electrically neutral atoms or groups of atoms usually result.

Homolytic cleavage: $H \overset{\frown\frown}{-} H \longrightarrow H\cdot + H\cdot$

$H \overset{\frown\frown}{-} \ddot{\underset{..}{C}}l: \longrightarrow H\cdot + \cdot\ddot{\underset{..}{C}}l:$

$H_3C \overset{\frown\frown}{-} H \longrightarrow H_3C\cdot + H\cdot$

Note that the curved arrows in these equations have only half an arrowhead. This type of arrow (⌢), called a fishhook, is used to show the direction of shift of *one* electron, whereas the curved arrow with a complete head (⌢) is used to show the direction of shift of a *pair* of electrons.

Homolytic cleavage results in atoms or groups of atoms having unpaired electrons. An atom such as $H\cdot$ or a group of atoms such as $H_3C\cdot$ that contains an unpaired electron is called a **free radical** or, simply, a **radical**. Radicals are usually electrically neutral; therefore, there are no electrostatic interactions between radicals as there are between ions. Because a radical contains an atom with an incomplete octet, most radicals are unstable and very reactive. We usually represent a radical by including a single dot in its formula, without showing the remaining valence electrons. Thus, we represent $\cdot \ddot{\ddot{C}}l\colon$ as $\cdot Cl$ or $H\ddot{O}\cdot$ as $HO\cdot$.

Homolytic cleavage is more useful than heterolytic cleavage in determining the energies required for bond dissociations because calculations are not complicated by ionic attractions between the products. From measurements of the components of dissociating gases at high temperatures, the **change in enthalpy** ΔH (change in heat content, or energy) has been calculated for a large number of bond dissociations. For the reaction $CH_4 \rightarrow CH_3\cdot + H\cdot$, ΔH equals 104 kcal/mol. In other words, to cleave one hydrogen atom from each carbon atom in 1 mol of CH_4 requires 104 kcal. This value (104 kcal/mol) is the **bond dissociation energy** for the H_3C—H bond. The bond dissociation energies for several types of bonds are listed in Table 1.4.

To break a more stable bond requires a higher energy input. For example, cleavage of HF to $H\cdot$ and $F\cdot$ (135 kcal/mol) is difficult compared with cleavage of the O—O bond in hydrogen peroxide, HOOH (35 kcal/mol).

Note in Table 1.4 that atoms joined by multiple bonds require more energy for dissociation than the same atoms joined by single bonds ($CH\equiv CH$, 230 kcal/mol, versus CH_3—CH_3, 88 kcal/mol). Also note that other parts of

Table 1.4 Selected bond dissociation energies (in kcal/mol)

Miscellaneous Bonds		*C—H Bonds*		*C—X Bonds*[a]		*C—C Bonds*	
H—H	104	CH_3—H	104	CH_3—Cl	83.5	CH_3—CH_3	88
N≡N	226	CH_3CH_2—H	98	CH_3CH_2—Cl	81.5	CH_2=CH_2	163
F—F	37	$(CH_3)_2CH$—H	94.5	$(CH_3)_2CH$—Cl	81	$CH\equiv CH$	230
Cl—Cl	58	$(CH_3)_3C$—H	91	$(CH_3)_3C$—Cl	78.5		
Br—Br	46	CH_2=CH—H	108	CH_2=CH—Cl	84		
I—I	36						
H—F	135						
H—OH	110						
H—Cl	103			CH_3—Br	70		
H—Br	87			CH_3CH_2—Br	68		
H—I	71			$(CH_3)_2CH$—Br	68		
HO—OH	35			$(CH_3)_3C$—Br	67		

[a] X refers to a halogen.

a molecule may affect the bond dissociation energy:

$$H_3C\!\frown\!\frown\! H + 104 \text{ kcal/mol} \xrightarrow{\text{more difficult}} H_3C\cdot + H\cdot$$

$$(CH_3)_3C\!\frown\!\frown\! H + 91 \text{ kcal/mol} \xrightarrow{\text{easier}} (CH_3)_3C\cdot + H\cdot$$

Bond dissociation energies allow a chemist to calculate relative stabilities of compounds and to predict (to an extent) the courses of chemical reactions. For example, one reaction we will discuss later in this text is the chlorination of methane, CH_4:

$$CH_4 + Cl_2 \longrightarrow CH_3Cl + HCl$$

Will this reaction be **exothermic** (heat-releasing) or **endothermic** (heat-absorbing)? We can break down the reaction into its component parts and calculate from the individual bond dissociation energies whether energy will be liberated or required. In general, the greater the amount of energy liberated, the more favorable the reaction. (Note that in these equations, $+\Delta H$ indicates energy put into a reaction, while $-\Delta H$ indicates energy that is liberated.)

	ΔH
$Cl\!-\!Cl + 58 \text{ kcal/mol} \longrightarrow Cl\cdot + Cl\cdot$	$+58$ kcal/mol
$H_3C\!-\!H + 104 \text{ kcal/mol} \longrightarrow H_3C\cdot + H\cdot$	$+104$ kcal/mol
$H_3C\cdot + Cl\cdot \longrightarrow H_3C\!-\!Cl + 83.5 \text{ kcal/mol}$	-83.5 kcal/mol
$H\cdot + Cl\cdot \longrightarrow H\!-\!Cl + 103 \text{ kcal/mol}$	-103 kcal/mol

net reaction:
$$Cl_2 + CH_4 \longrightarrow CH_3Cl + HCl + 24.5 \text{ kcal/mol} \qquad \text{net } \Delta H = -24.5 \text{ kcal/mol}$$

We calculate that this reaction should be exothermic. When we run this reaction in the laboratory, we find that it is indeed exothermic.

Study Problem

1.7 Using bond dissociation energies from Table 1.4, predict which of the following reactions liberates more energy:

(a) $(CH_3)_3CH + Cl_2 \longrightarrow (CH_3)_3CCl + HCl$

(b) $CH_4 + Cl_2 \longrightarrow CH_3Cl + HCl$

▼ ## Section 1.8
Polar Covalent Bonds

Atoms with equal or nearly equal electronegativities form covalent bonds in which both atoms exert equal or nearly equal attractions for the bonding electrons. This type of covalent bond is called a **nonpolar bond**. In organic molecules, carbon–carbon bonds and carbon–hydrogen bonds are the most common types of nonpolar bonds.

Some compounds containing relatively nonpolar covalent bonds:

$$H\!-\!H \qquad N\!\equiv\!N \qquad H\!-\!\overset{\displaystyle H}{\underset{\displaystyle H}{C}}\!-\!H \qquad H\!-\!\overset{\displaystyle H}{\underset{\displaystyle H}{C}}\!-\!\overset{\displaystyle H}{\underset{\displaystyle H}{C}}\!-\!H \qquad \overset{H}{\underset{H}{{}}}\!C\!=\!C\!\overset{H}{\underset{H}{{}}}$$

In covalent compounds like H_2O, HCl, CH_3OH, or $H_2C=O$, one atom has a substantially greater electronegativity than the others. The more electronegative atom has a greater attraction for the bonding electrons—not enough of an attraction for the atom to break off as an ion, but enough so that this atom takes the larger share of electron density. The result is a **polar covalent bond**, a bond with an uneven distribution of electron density.

Besides electronegativity, another factor determining the degree of polarity of a bond is the *polarizability* of the atoms, which is the ability of the electron cloud to be distorted to induce polarity. The outer electrons of larger atoms are farther from the nucleus and are thus less tightly held than those of smaller atoms. Therefore, larger atoms are more polarizable than smaller ones. This means that the C—I bond, for example, can act as if it is polar even though the electronegativity difference between C and I is negligible.

The result of electronegativity differences and polarizability differences is a variety of bond types. We may think of chemical bonds as a continuum from nonpolar covalent bonds to ionic bonds. Within this continuum, we speak of the increasing **ionic character** of the bonds.

$$H—H \qquad CH_3—O—CH_3 \qquad H—O—H \qquad H—Cl \qquad Na^+ \; Cl^-$$

increasing ionic character of bonds

The distribution of electron density in a polar bond may be symbolized by **partial charges**: $\delta+$ (partial positive) and $\delta-$ (partial negative). Another way of representing different electron densities within a bond is by a crossed arrow ($\leftrightarrow$) that points from the partially positive end to the partially negative end of the bond.

$$\overset{\delta+}{H}—\overset{\delta-}{Cl} \quad \text{or} \quad H—Cl \qquad\qquad CH_3 \; CH_3 \quad \text{or} \quad CH_3 \; CH_3$$

Study Problem

1.8 Using partial charges, indicate the polarity of the bonds in the following compounds:

(a) CH_3Br **(b)** $CH_3\overset{O}{\underset{\|}{C}}OH$

A. Bond Moments

If a polar bond, such as an O—H bond, is subjected to an electric field, the bond feels a certain amount of "turning force." This force is simply the push of the electric field to align the bond in the field. A more polar bond feels more force than a less polar bond. The **bond moment**, a measure of the polarity of a bond, is calculated from the value of the force felt by that bond.

The bond moment is defined as $e \times d$, where e is the charge (in electrostatic units) and d is the distance between the charges (in Å), and is reported in units called *Debyes* (D). Bond moments range from 0 for a nonpolar

Table 1.5 Bond moments for selected covalent bonds

Bond[a]	Bond Moment, D	Bond[a]	Bond Moment, D
H—C	0.4	C—Cl	1.46
H—N	1.31	C—Br	1.38
H—O	1.51	C—I	1.19
C—N	0.22	C=O	2.3
C—O	0.74	C≡N	3.5
C—F	1.41		

[a] In each case, the more positive atom is on the left.

bond such as H—H up to 3–4 D for highly polar bonds such as C≡N (see Table 1.5). The bond moment for a particular bond is relatively constant from compound to compound.

Study Problems

1.9 Which of the indicated bonds in each pair of compounds is more polar?

(a) CH_3—NH_2 or CH_3—OH   (b) CH_3—OH or CH_3O—H
(c) CH_3—Cl or CH_3—OH

1.10 Use a crossed arrow to show the approximate direction of the bond moment (if any) of the double or triple bond in each of the following structures:

$$
\begin{array}{c}
O \\
\parallel
\end{array}
$$

(a) $CH_3C{\equiv}N$ (b) CH_3CCH_3 (c) CH=CH   (d) ⬡=O

B. Dipole Moments

The **dipole moment** μ is the *vector sum of the bond moments* in a molecule. Because vector addition takes into account the direction as well as the magnitude of the bond moments, the dipole moment is a measure of the polarity of the molecule as a whole.

The dipole moments for a few compounds are listed in Table 1.6. Note that the dipole moment of CCl_4 is zero, even though each C—Cl bond has a moment of 1.46 D. The reason for this apparent anomaly is that, in CCl_4 molecules, the vector sum of the bond moments is zero. We might say that the bond moments

Table 1.6 Dipole moments of selected compounds

Compound	Dipole Moment, D	Compound	Dipole Moment, D
H_2O	1.84	CH_3OCH_3	1.3
NH_3	1.46		
CH_3Cl	1.86	CH_3CH (with =O)	2.7
CCl_4	0		
CO_2	0		
		CH_3CCH_3 (with =O)	2.8

"cancel" one another. Carbon dioxide is another molecule with bond moments, but with a dipole moment of zero. Again, this is a case of a symmetrical molecule in which the bond moments cancel. On the other hand, the bond moments of a water molecule do not cancel, and water has a net dipole moment. From this observation, we can deduce that the water molecule is not symmetrical around the oxygen. Dipole moments can thus be used to help determine molecular geometry.

$\mu = 0$ $\mu = 0$ $\mu = 1.84$ D

▼ ■ Section 1.9
Attractions Between Molecules

A. Dipole–Dipole Interactions

Except in a dispersed gas, molecules attract and repel one another. These attractions and repulsions arise primarily from dipole–dipole interactions: attractions between opposite charges and repulsions between like charges. For example, in liquid CH_3Cl the partially negative chlorine atom in one molecule is attracted to the partially positive carbon in another molecule.

When the chlorine ends of two CH_3Cl molecules approach closely, the two molecules repel each other.

Nonpolar molecules are attracted to one another by weak dipole–dipole interactions called **London forces**. London forces arise from dipoles *induced* in one molecule by another. In this case, electrons of one molecule are weakly attracted to a nucleus of a second molecule; then the electrons of the second molecule are repelled by the electrons of the first. The result is an uneven distribution of electrons and an induced dipole. Figure 1.6 depicts how an induced dipole can arise when two molecules approach each other.

The various dipole–dipole interactions (attractive and repulsive) are collectively called **van der Waals forces**. The distance between molecules has an important effect on the strength of van der Waals forces. The distance at

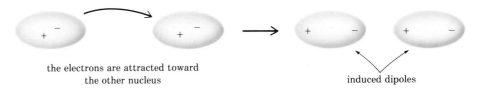

the electrons are attracted toward
the other nucleus

induced dipoles

Figure 1.6 Nonpolar molecules can induce dipoles in each other.

which attraction is greatest is called the **van der Waals radius**. If two atoms approach each other more closely than this distance, repulsions develop between the two nuclei and between the two sets of electrons. When the distance between two molecules becomes larger than the van der Waals radius, the attractive forces between the molecules decrease.

Continuous-chain molecules, such as $CH_3CH_2CH_2CH_2CH_3$, can align themselves in zigzag chains, enabling the atoms of different molecules to assume positions that match the van der Waals radii. Maximal van der Waals attractions can develop between such long-chain molecules. Branched molecules cannot approach one another closely enough for all the atoms to assume optimal van der Waals distances. Because more energy is necessary to overcome van der Waals attractions and to free molecules from the liquid state, continuous-chain compounds have higher boiling points than branched compounds of the same molecular weight and otherwise similar structures.

*branched chain;
lower boiling*

$$CH_3-\overset{\overset{\displaystyle CH_3}{|}}{\underset{\underset{\displaystyle CH_3}{|}}{C}}-CH_3$$

*continuous chain;
higher boiling*

bp 36°

$$CH_3-\overset{\overset{\displaystyle CH_3}{|}}{\underset{\underset{\displaystyle CH_3}{|}}{C}}-CH_3$$

bp 9.5°

B. Hydrogen Bonding

An especially strong type of dipole–dipole interaction occurs between molecules containing a hydrogen atom bonded to nitrogen, oxygen, or fluorine. Each of these latter elements is electronegative and has unshared valence electrons. Some typical compounds that contain an NH, OH, or FH bond are:

$$H-\overset{\cdot\cdot}{\underset{\displaystyle H}{O}}:\qquad CH_3-\overset{\cdot\cdot}{\underset{\displaystyle H}{O}}:\qquad H-\overset{}{\underset{\displaystyle H}{N}}-H\qquad CH_3-\overset{\cdot\cdot}{\underset{\displaystyle H}{N}}-H\qquad H-\overset{\cdot\cdot}{\underset{\cdot\cdot}{F}}:$$

In the liquid state, the molecules of these compounds have strong attractions for one another. A partially positive hydrogen atom of one molecule is attracted to the unshared pair of electrons of the electronegative atom of another molecule. This attraction is called a **hydrogen bond**.

$$H—\overset{\delta-}{\overset{..}{\underset{H}{O}}}:----\overset{\delta+}{H}—\overset{H}{\overset{|}{\underset{..}{O}}}: \qquad CH_3—\overset{\delta-}{\overset{..}{\underset{H}{O}}}:----\overset{\delta+}{H}—\overset{CH_3}{\overset{|}{\underset{..}{O}}}:$$

hydrogen bonds

Compounds or groups containing only carbon and hydrogen cannot undergo hydrogen bonding. As an example, consider methane, CH_4. Methane cannot undergo hydrogen bonding for two reasons:

1. Because the CH bond is relatively nonpolar, a CH_4 molecule does not have a partially positive H.
2. The carbon atom in CH_4 has no unshared electrons to attract a hydrogen atom.

The dissociation energy of a hydrogen bond is only 5–10 kcal/mol, much lower than the bond dissociation energy of a typical covalent bond (80–100 kcal/mol) but substantially stronger than most dipole–dipole attractions. The reason for this difference is the size of the atoms involved. A hydrogen atom is small compared to other atoms and can occupy a position very close to the unshared electrons of an electronegative atom. A strong electrostatic attraction results. Atoms larger than hydrogen cannot occupy positions so near each other; consequently, dipole–dipole attractions between other atoms are weaker.

closer   *farther apart*

Sample Problem

Show the hydrogen bonding between two molecules of $CH_3CH_2NH_2$.

Solution:

1. Look for one or more partially positive hydrogens, which are those bonded to a relatively electronegative atom.

$$CH_3CH_2—\overset{H}{\overset{|}{N}}—H$$

not these   *these*

2. Look for an electronegative atom (N, O, or F) with unshared electrons.

$$CH_3CH_2—\overset{H}{\overset{|}{\underset{..}{N}}}—H$$

3. Draw two molecules with a hydrogen bond between one partially positive H and the N of the other molecule.

$$CH_3CH_2-\overset{\overset{\displaystyle H}{|}}{\underset{\underset{\displaystyle H}{|}}{N}}\text{:}\text{-----}H-\overset{\overset{\displaystyle H}{|}}{N}-CH_2CH_3$$

Hydrogen bonds are not all the same strength. An O---HO hydrogen bond is stronger than an N---HN hydrogen bond. Why is this true? Oxygen is more electronegative than nitrogen; therefore, the O—H group is more polar and has a more positive H. This more positive H is more strongly attracted by a negative center.

$$CH_3-\overset{\overset{\displaystyle}{}}{\underset{\underset{\displaystyle H}{|}}{\ddot{O}}}\text{:}\text{-----}H-\overset{\displaystyle \ddot{O}}{}-CH_3 \qquad CH_3-\overset{\overset{\displaystyle H}{|}}{\underset{\underset{\displaystyle H}{|}}{N}}\text{:}\text{-----}H-\overset{\overset{\displaystyle H}{|}}{N}-CH_3$$

more positive H: *less positive H:*
stronger hydrogen bond   *weaker hydrogen bond*

Hydrogen bonds may form between two different compounds, such as between CH_3OH and H_2O or between CH_3NH_2 and H_2O. In these cases, there is often more than one possibility for hydrogen bonding. The following structures show two types of hydrogen bonds between CH_3NH_2 and H_2O. (In a mixture of these two compounds, hydrogen bonds can also form between two molecules of H_2O and between two molecules of CH_3NH_2.)

$$CH_3-\overset{\overset{\displaystyle H}{|}}{\underset{\underset{\displaystyle H}{|}}{N}}\text{:}\text{-----}H-\overset{\displaystyle \ddot{O}}{}\text{:} \qquad CH_3-\overset{\overset{\displaystyle H}{|}}{\underset{\underset{\displaystyle H}{|}}{N}}-H\text{-----}\text{:}\overset{\overset{\displaystyle H}{|}}{\ddot{O}}-H$$

more positive H: *less positive H:*
stronger hydrogen bond   *weaker hydrogen bond*

Table 1.7 shows the amount of energy needed to break some different types of hydrogen bonds. Note that the OH---N hydrogen bond is the strongest of the group. Because nitrogen is less electronegative than oxygen, its electrons are more loosely held and more easily attracted by another atom. The combination of nitrogen's loose electrons and the more positive hydrogen of an OH group leads to a quite strong hydrogen bond.

C. Effects of Hydrogen Bonding

Hydrogen bonds act rather like glue between molecules. Although a single hydrogen bond by itself is weak, all the molecules of a substance together may form a great many hydrogen bonds.

Boiling points generally increase with higher molecular weight because of increased van der Waals attractions. However, a hydrogen-bonded compound has a *higher boiling point* than would be predicted from molecular weight

Table 1.7 Approximate dissociation energies of some hydrogen bonds

Type of Hydrogen Bond	*Approximate Dissociation Energy (kcal/mol)*
—O—H---:N—	7
—O—H---:Ö—	5
—N—H---:N—	3
—N—H---:Ö—	2

considerations alone. For a hydrogen-bonded liquid to be volatilized, additional energy must be supplied for breaking all the intermolecular hydrogen bonds.

Ethanol (CH_3CH_2OH) and dimethyl ether (CH_3OCH_3) have the same molecular weight. However, ethanol has a much higher boiling point than dimethyl ether—ethanol is a liquid at room temperature while dimethyl ether is a gas. The difference in boiling points between these two compounds can be directly attributed to the fact that ethanol molecules form hydrogen bonds, while dimethyl ether molecules cannot form hydrogen bonds among themselves. Note that hydrogen bonding affects the boiling point to a much greater extent than branching does.

hydrogen bond   *no H to form a hydrogen bond*

$$CH_3CH_2\overset{..}{\underset{..}{O}}:\text{-----}H\text{---}\overset{..}{\underset{..}{O}}:$$

$$\underset{H}{\mid} \qquad \underset{CH_2CH_3}{\mid}$$

$$CH_3$$
$$\mid$$
$$:\underset{..}{O}\text{---}CH_3$$

ethanol
bp 78.5°

dimethyl ether
bp −23.6°

Solubility of covalent compounds in water is another property affected by hydrogen bonding. A compound that can form hydrogen bonds with water tends to be far more soluble in water than a compound that cannot. Sugars, such as glucose, contain many —OH groups and are quite soluble in water. Cyclohexane, however, cannot form hydrogen bonds and cannot break the existing hydrogen bonds in water; therefore, cyclohexane is water-insoluble.

glucose

cyclohexane

soluble in water   *insoluble in water*

The *shapes of large biomolecules* are also determined principally by hydrogen bonding. For example, proteins are held in specific orderly shapes by hydrogen bonds within each molecule or between molecules. Molecules of DNA—the genetic material in plant and animal cells—are held in double spirals partly because of hydrogen bonds between pairs of molecules.

Study Problem

1.11 Show all the types of hydrogen bonds (if any) that would be found in:

(a) liquid $CH_3CH_2CH_2NH_2$ (b) a solution of CH_3OH in H_2O
(c) liquid $CH_3CH_2OCH_2CH_3$ (d) a solution of CH_3OCH_3 in H_2O

▼ ## Section 1.10
Acids and Bases

According to the **Brønsted–Lowry concept** of acids and bases, an **acid** is a substance that can donate a positively charge hydrogen ion, or proton (H^+). Two examples of Brønsted–Lowry acids are HCl and HNO_3. A **base** is defined as a substance that can accept H^+; examples are OH^- and NH_3. (A typical source of hydroxide ions is $Na^+ OH^-$.) Although we speak of "proton donors" and "proton acceptors," curved arrows are used in organic chemistry to indicate the actions of *electrons*, not protons. Therefore, in the following equation, a curved arrow is drawn from the unshared electrons of the base to the proton that it is accepting.

$$HO:^- \quad + \quad H-Cl: \quad \longrightarrow \quad HO-H + :Cl:^-$$

a base   an acid
(<i>H acceptor</i>) (<i>H⁺ donor</i>)

A. Strong and Weak Acids and Bases

Recall from your general chemistry course that a **strong acid** is an acid that undergoes essentially complete ionization in water. Representative strong acids are HCl, HNO_3, and H_2SO_4. The ionization of these strong acids is a typical acid–base reaction. The acid (HCl, for example) loses a proton to the base (H_2O). The equilibrium lies far toward the right (complete ionization of HCl) because H_2O is a stronger base than Cl^- and HCl is a stronger acid than H_3O^+.

$$H-O-H \quad + \quad H-Cl: \quad \longrightarrow \quad H-\overset{\overset{\displaystyle H}{|}}{O}{}^+-H + :Cl:^-$$

stronger base   stronger acid
than Cl⁻   than H₃O⁺

A **weak acid**, by contrast, has only a small proportion of its molecules broken into ions in water. Carbonic acid is a typical weak acid. The equilibrium lies toward the nonionized side of the equation because H_3O^+ is the stronger acid and HCO_3^- is the stronger base.

$$H-\overset{..}{\underset{..}{O}}-H + H-\overset{..}{\underset{..}{O}}\overset{O}{\overset{||}{C}}OH \rightleftharpoons H-\overset{H}{\underset{|}{O^+}}-H + \;\; {}^-:\overset{..}{\underset{..}{O}}\overset{O}{\overset{||}{C}}OH$$

carbonic acid   bicarbonate ion

stronger acid   *stronger base*
than H_2CO_3   *than H_2O*

Also, recall that *bases* are classified as strong (such as OH^-) or weak (NH_3), depending on their affinity for protons.

$$:NH_3 \; + H-\overset{..}{\underset{..}{O}}-H \rightleftharpoons \overset{+}{N}H_4 \; + \;\; {}^-:\overset{..}{\underset{..}{O}}H$$

ammonia   ammonium ion   hydroxide ion

a weak base   *stronger acid*   *stronger base*
                                 than H_2O   *than NH_3*

Let us now consider some organic compounds that act as acids and bases. **Amines** are a class of organic compounds structurally similar to ammonia; an amine contains a nitrogen atom that is covalently bonded to one or more carbon atoms and that has an unshared pair of electrons.

Some common amines:

$$CH_3-\overset{..}{\underset{\underset{H}{|}}{N}}-H \quad CH_3-\overset{..}{\underset{\underset{CH_3}{|}}{N}}-H \quad CH_3-\overset{..}{\underset{\underset{CH_3}{|}}{N}}-CH_3$$

Amines, like ammonia, are weak bases and undergo reversible reactions with water or other weak acids. (The use of a strong acid drives the reaction to completion.)

$$CH_3\overset{..}{N}H_2 + H-\overset{..}{\underset{..}{O}}H \rightleftharpoons CH_3\overset{+}{N}H_3 + {}^-:\overset{..}{\underset{..}{O}}H$$

$$CH_3\overset{..}{N}H_2 + H-\overset{..}{\underset{..}{Cl}}: \longrightarrow CH_3\overset{+}{N}H_3 + :\overset{..}{\underset{..}{Cl}}:^-$$

An organic compound containing a **carboxyl group** ($-CO_2H$) is a weak acid. Compounds that contain carboxyl groups are called **carboxylic acids**. Acetic acid, CH_3CO_2H, is an example. One of the reasons for the acidity of carboxylic acids is the polarity of the O—H bond. (Another reason for this acidity will be discussed in Section 2.9.)

The carboxyl group:

$$-\overset{\overset{..}{\underset{..}{O}}}{\underset{}{\overset{||}{C}}}-\overset{\delta-}{\underset{..}{O}}-\overset{\delta+}{H} \quad \text{usually written} \quad -\overset{O}{\overset{||}{C}}OH, -COOH, \text{ or } -CO_2H$$

A proton is removed from the carboxyl group by a base. A carboxylate ion, an anion, is formed. Because carboxylic acids are only weakly acidic, these reactions do not proceed to completion unless a stronger base than water is used, as the reaction arrows in the following equations indicate.

$$CH_3\overset{\overset{\cdot\cdot}{\underset{\cdot\cdot}{O}}}{\underset{\Vert}{C}}\text{—}\overset{\cdot\cdot}{\underset{\cdot\cdot}{O}}\text{—}H \;+\; H\text{—}\overset{\cdot\cdot}{\underset{\cdot\cdot}{O}}\text{—}H \;\rightleftharpoons\; CH_3\overset{\overset{\cdot\cdot}{\underset{\cdot\cdot}{O}}}{\underset{\Vert}{C}}\text{—}\overset{\cdot\cdot}{\underset{\cdot\cdot}{O}}{:}^- \;+\; H\text{—}\overset{+}{\underset{\cdot\cdot}{O}}\text{—}H$$

acetic acid   acetate ion

a carboxylic acid   *a carboxylate ion*

$$CH_3\overset{\overset{\cdot\cdot}{\underset{\cdot\cdot}{O}}}{\underset{\Vert}{C}}\text{—}\overset{\cdot\cdot}{\underset{\cdot\cdot}{O}}\text{—}H \;+\; {:}NH_3 \;\rightleftharpoons\; CH_3\overset{\overset{\cdot\cdot}{\underset{\cdot\cdot}{O}}}{\underset{\Vert}{C}}\text{—}\overset{\cdot\cdot}{\underset{\cdot\cdot}{O}}{:}^- \;+\; \overset{+}{N}H_4$$

$$CH_3\overset{\overset{\cdot\cdot}{\underset{\cdot\cdot}{O}}}{\underset{\Vert}{C}}\text{—}\overset{\cdot\cdot}{\underset{\cdot\cdot}{O}}\text{—}H \;+\; {:}\overset{\cdot\cdot}{\underset{\cdot\cdot}{O}}H^- \;\longrightarrow\; CH_3\overset{\overset{\cdot\cdot}{\underset{\cdot\cdot}{O}}}{\underset{\Vert}{C}}\text{—}\overset{\cdot\cdot}{\underset{\cdot\cdot}{O}}{:}^- \;+\; H\overset{\cdot\cdot}{\underset{\cdot\cdot}{O}}H$$

Study Problems

1.12 Which of the following compounds or ions act as acids and which act as bases in H_2O?

(a) $^-{:}\overset{\cdot\cdot}{N}H_2$ (b) $CH_3CH_2CH_2\overset{\overset{\cdot\cdot}{\underset{\cdot\cdot}{O}}}{\underset{\Vert}{C}}\overset{\cdot\cdot}{O}H$ (c) $^-{:}\overset{\cdot\cdot}{\underset{\cdot\cdot}{O}}CH_2CH_3$

(d) [cyclohexane ring with $\overset{H}{\underset{H}{\overset{+}{N}}}$]   (e) [cyclohexane ring with $:NH$]

1.13 Rewrite and complete the following equations for acid–base reactions. Include in your answer: (1) electron dots; (2) curved electron-shift arrows; and (3) reaction arrows that show the direction of the equilibrium. (If the reaction proceeds essentially to completion, use a single reaction arrow.)

(a) $CH_3CH_2CO_2H + H_2O$ (b) [cyclohexane ring]$\text{—}CO_2H + OH^-$

(c) $(CH_3)_2NH + H_2O$ (d) [cyclohexane ring]$NH + CH_3CO_2H$

B. Conjugate Acids and Bases

The concept of conjugate acids and bases is useful for comparisons of acidities and basicities. The **conjugate base** of an acid is the ion or molecule that results after the loss of H^+ from the acid. For example, the chloride ion is the conjugate base of HCl. The **conjugate acid** of a base is the protonated form of the base. Thus, the conjugate acid of NH_3 is NH_4^+.

conjugate base   *conjugate acid*
of H_3O^+   *of H_2O*

$$HCl \;+\; H_2O \;\longrightarrow\; H_3O^+ \;+\; Cl^-$$

conjugate acid   *conjugate base*
of Cl^-   *of HCl*

$$\underset{\substack{\text{conjugate acid} \\ \text{of } CH_3CO_2^-}}{CH_3\overset{\overset{\displaystyle O}{\|}}{C}OH} + \underset{\substack{\text{conjugate base} \\ \text{of } ^+NH_4}}{NH_3} \rightleftharpoons \underset{\substack{\text{conjugate base} \\ \text{of } CH_3CO_2H}}{CH_3\overset{\overset{\displaystyle O}{\|}}{C}O^-} + \underset{\substack{\text{conjugate acid} \\ \text{of } NH_3}}{\overset{+}{N}H_4}$$

If an acid is a strong acid, its conjugate base is a weak base:

$$\underset{\substack{\text{strong acid} \\ \text{(loses } H^+ \text{ readily)}}}{HCl} + H_2O \longrightarrow H_3O^+ + \underset{\substack{\text{very weak base} \\ \text{(has little attraction for } H^+)}}{Cl^-}$$

On the other hand, if an acid is weak or very weak, its conjugate base is a moderately strong base or a strong base, depending on the affinity of the conjugate base for H^+.

$$\underset{\text{weak acid}}{CH_3\overset{\overset{\displaystyle O}{\|}}{C}OH} + H_2O \xleftarrow{\hspace{1cm}} \underset{\text{moderate base}}{CH_3\overset{\overset{\displaystyle O}{\|}}{C}O^-} + H_3O^+$$

$$\underset{\text{very weak acid}}{2\,H_2O} \xleftarrow{\hspace{1cm}} H_3O^+ + \underset{\text{strong base}}{OH^-}$$

Thus, as the acid strengths of a series of compounds increase, the base strengths of their conjugate bases decrease.

$$H_2O \qquad HCN \qquad CH_3CO_2H \qquad H_3PO_4 \qquad HCl$$

increasing acidity

$$^-OH \qquad ^-CN \qquad CH_3CO_2^- \qquad H_2PO_4^- \qquad Cl^-$$

decreasing basicity of conjugate base

C. Some Factors Affecting Acidity

The various factors that affect the strength of acids and bases are discussed in detail in Sections 14.5 and 17.6. Here, we will briefly describe only two of these factors: the size and the electronegativity of atoms.

In any equilibrium, such as the ionization of a weak acid, the thermodynamically more stable reactant or product predominates in the equilibrium mixture. We say that these reversible reactions are under **equilibrium control**, or **thermodynamic control**. The strength of an acid or base is thus a function of the stability of the protonated acidic form (H:B) versus that of the nonprotonated base (:B⁻), where H:B represents any Brønsted acid and :B⁻, its conjugate base.

$$H\!:\!B \rightleftharpoons H^+ + :B^-$$

A more stable protonated structure shifts the equilibrium toward the conjugate acid.   *A more stable base shifts the equilibrium toward the conjugate base.*

The ease with which a molecule or ion loses a proton depends on the strength of the bond to the proton. Two factors that affect bond strength are the size and the electronegativity of the atom bonded to the proton.

In the same group of the periodic table, the **size** of an atom bonded to a proton affects the acid strength of a compound. A larger atom can disperse the negative charge over a larger region and thus add stability to the conjugate base. As the stability of the conjugate base increases, the greater is the acid strength of the acid. For example,

$$\text{H—F} \qquad \text{H—Cl} \qquad \text{H—Br} \qquad \text{H—I}$$

increasing size of halogen; increasing acid strength

In the same row of the periodic table, the **electronegativity** of the atom bonded to the proton also affects the acidity. A more electronegative atom can carry a negative charge more readily than can a less electronegative atom. A proton is more readily removed from the more electronegative atom. For example,

$$\underset{\text{CH}_3}{\overset{\text{CH}_3}{\text{CH}_3\text{—C—H}}} \qquad \underset{\text{CH}_3}{\overset{\text{CH}_3}{\text{N—H}}} \qquad \overset{\text{CH}_3}{\text{O—H}} \qquad \text{F—H}$$

increasing electronegativity of circled atom; increasing acid strength

Study Problem

1.14 Predict which compound in each of the following pairs would be the stronger acid. Explain your choices.

(a) CH_4 or H_2O **(b)** CH_4 or NH_3 **(c)** H_2O or H_2S

D. Lewis Acids and Bases

Although many acid–base reactions involve the transfer of a proton from an acid to a base, some acid–base reactions do not involve proton transfer. For this reason, the more general Lewis concept of acids and bases was developed. A **Lewis acid** is a substance that can **accept a pair of electrons**. Any species with an electron-deficient atom can act as a Lewis acid; for example, H^+ is a Lewis acid. Many Lewis acids that we will encounter in this text are anhydrous metal salts (for example, $ZnCl_2$, $FeCl_3$, and $AlBr_3$).

$$\text{FeBr}_3 \quad + \quad :\!\ddot{\text{Br}}\!-\!\ddot{\text{Br}}\!: \quad \rightleftharpoons \quad \text{FeBr}_4^- + \ddot{\text{Br}}\!:^+$$

a Lewis acid
(electron acceptor)

A **Lewis base** is a substance that can **donate a pair of electrons**. Examples of Lewis bases are NH_3 and ^-OH, each of which has an unshared pair of valence electrons that it can donate to H^+ or to some other Lewis acid.

$$\ddot{N}H_3 \quad + \quad H\text{—}\overset{..}{\underset{..}{C}l}: \quad \longrightarrow \quad \overset{+}{N}H_4 \quad + \quad :\overset{..}{\underset{..}{C}}l:^-$$

a Lewis base
(electron donor)

Many organic reactions that we will consider in this text can be considered to be reactions of Lewis acids or bases. Two examples follow.

Lewis acid–base reactions:

$$H^+ + CH_3CH_2\overset{..}{\underset{..}{O}}H \rightleftharpoons CH_3CH_2\overset{H}{\underset{|}{O}}H$$

$$\underset{:O:}{\overset{\|}{CH_3CCH_3}} + \,^-:C\equiv N: \rightleftharpoons \underset{\underset{C\equiv N:}{|}}{\overset{:\overset{..}{O}:^-}{\underset{|}{CH_3CCH_3}}}$$

Sample Problem

Methylamine (CH_3NH_2) undergoes a Lewis acid–base reaction with boron trifluoride (BF_3) to yield $CH_3NH_2\text{—}BF_3$. (a) Write the equation for this reaction, showing the complete structural formula for the product and the formal charges on N and B. (b) Identify each reactant as a Lewis acid or Lewis base.

Solution:

$$CH_3\overset{..}{N}H_2 \quad + \quad BF_3 \quad \longrightarrow \quad \underset{H\ \ H\ \ F}{\overset{H\ \ H\ \ F}{H\text{—}C\text{—}\overset{+}{N}\text{—}\overset{-}{B}\text{—}F}}$$

Lewis base   Lewis acid

Study Problem

1.15 Identify the reactants in the following equations as Lewis acids or Lewis bases:

(a) $\underset{}{\overset{:\overset{..}{O}:}{\overset{\|}{CH_3CCH_3}}} + H^+ \rightleftharpoons CH_3\overset{:\overset{+}{O}H}{\underset{\|}{CCH_3}}$

(b) $(CH_3)_3C^+ + :\overset{..}{\underset{..}{C}}l:^- \longrightarrow (CH_3)_3C\overset{..}{\underset{..}{C}}l:$

(c) $\underset{}{\overset{O}{\overset{\|}{CH_3COCH_3}}} + \,^-:\overset{..}{\underset{..}{O}}CH_3 \rightleftharpoons \,^-:CH_2\overset{O}{\overset{\|}{COCH_3}} + H\overset{..}{\underset{..}{O}}CH_3$

E. Acidity Constants

A chemical reaction has an **equilibrium constant** K that reflects how far the reaction proceeds toward completion. For the ionization of an acid in water, this constant is called an **acidity constant** K_a. An equilibrium constant is determined by the following general equation, with concentration values given in molarity, M:

$$K = \frac{\text{concentrations of products in } M}{\text{concentrations of reactants in } M}$$

For acetic acid: $CH_3CO_2H \rightleftharpoons CH_3CO_2^- + H^+$

$$K_a = \frac{[CH_3CO_2^-][H^+]^*}{[CH_3CO_2H]}$$

where $[H^+]$ = molar concentration of H^+

$[CH_3CO_2^-]$ = molar concentration of $CH_3CO_2^-$

$[CH_3CO_2H]$ = molar concentration of CH_3CO_2H

The more ionized an acid is, the larger is the value for K_a because the values in the numerator are larger. *A stronger acid has a larger K_a value.* Any acid with a $K_a > 10$ is considered a strong acid. (For HCl, $K_a \cong 10^7$.) By contrast, typical carboxylic acids, such as acetic acid, have K_a values much smaller than 1. (For CH_3CO_2H, $K_a = 1.75 \times 10^{-5}$.)

$$K_a = \frac{[H^+][\text{anion}]}{[\text{un-ionized acid}]} \xleftarrow{\hspace{1cm}} \begin{array}{l} \textit{as numerator increases,} \\ K_a \textit{ increases} \end{array}$$

Just as pH is defined as the negative logarithm of hydrogen ion concentration, pK_a is defined as the negative logarithm of K_a. We will use pK_a values in this text for comparison of acid strengths. (The K_a values and pK_a values for some carboxylic acids are listed in Table 1.8. The pK_a table inside the front cover lists some other types of compounds as well.)

$$pH = -\log[H^+]$$

$$pK_a = -\log K_a$$

Examples

If $K_a = 10^{-3}$,   then $pK_a = 3$
If $K_a = 10^2$,   then $pK_a = -2$

---

* More correctly, the *activity*, or *effective concentration*, should be used instead of molarity. Because activities of ions approach their molarities in dilute solution, molarity is used for the sake of simplicity. In addition, the equilibrium expression should contain the hydrogen acceptor, water:

$$CH_3CO_2H + H_2O \rightleftharpoons CH_3CO_2^- + H_3O^+$$

$$K_a' = \frac{[CH_3CO_2^-][H_3O^+]}{[CH_3CO_2H][H_2O]}$$

In dilute solutions, the molar concentration of water is about 55.5. This constant factor is generally grouped with the equilibrium constant K_a, and the $[H_3O^+]$ term is simplified to $[H^+]$. Even though "naked" protons do not exist as such in solution, we will frequently use the symbol H^+ instead of H_3O^+ in this book to represent an aqueous acid.

$$K_a = K_a'[H_2O] = \frac{[CH_3CO_2^-][H_3O^+]}{[CH_3CO_2H]} = \frac{[CH_3CO_2^-][H^+]}{[CH_3CO_2H]}$$

Table 1.8 Acidity constants and pK_a values for some acids

Formula	K_a	pK_a
Strong:		
HCl	$\sim 10^7$	~ -7
H_2SO_4	$\sim 10^5$	~ -5
Moderately Strong:		
H_3PO_4	7.52×10^{-3}	2.12
Weak:		
HF	35.3×10^{-5}	3.45
HCO_2H	17.5×10^{-5}	3.75
CH_3CO_2H	1.75×10^{-5}	4.75
$CH_3CH_2CO_2H$	1.34×10^{-5}	4.87
Very Weak:		
HCN	4.93×10^{-10}	9.31
H_2O	1.80×10^{-16}	15.74

As K_a gets larger (stronger acid), pK_a gets smaller. *The smaller the value for* pK_a, *the stronger the acid.*

$$K_a: \quad 10^{-10} \qquad 10^{-5} \qquad 10^{-1} \qquad 10^2$$

$$\text{p}K_a: \quad 10 \qquad\quad 5 \qquad\quad 1 \qquad\quad -2$$

increasing acid strength

Sample Problem

Calculate the pK_a of an acid with K_a equal to 136×10^{-5}.

Solution:

$$\text{p}K_a = -\log K_a$$
$$= -\log(136 \times 10^{-5})$$
$$= -\log(1.36 \times 10^{-3})$$
$$= -(\log 1.36 - 3)$$
$$= 3 - \log 1.36$$
$$= 3 - 0.133 \longleftarrow \textit{from log table or calculator}$$
$$= 2.87$$

When the concentration of the undissociated carboxylic acid equals the concentration of the dissociated acid, the pH of the solution equals the pK_a of the carboxylic acid. This useful relationship can be expressed by the **Henderson–Hasselbalch equation**, a reexpression of the mass law equation.

Mass law equation:

$$K_a = \frac{[H^+][A^-]}{[HA]}$$

Henderson–Hasselbalch equation:

$$pH = pK_a + \log\frac{[A^-]}{[HA]}$$

When $[A^-] = [HA]$, $pH = pK_a$

Study Problem

1.16 Determine the pK_a of chloroacetic acid ($ClCH_2CO_2H$) if a solution containing 0.100 mol of undissociated acid and 0.100 mol of its sodium salt ($ClCH_2CO_2^- \ Na^+$) has a pH of 2.85.

F. Basicity Constants

The reversible reaction of a weak base with water, like the reaction of a weak acid with water, results in a small but constant concentration of ions at equilibrium. The **basicity constant** K_b is the equilibrium constant for this reaction. As in the case of K_a, the value for $[H_2O]$ is included in K_b in the equilibrium expression.

$$NH_3 + H_2O \rightleftharpoons NH_4^+ + OH^-$$

$$K_b = \frac{[NH_4^+][OH^-]}{[NH_3]}$$

$$pK_b = -\log K_b$$

With an increase in base strength, the value for K_b increases and the pK_b value decreases. *The smaller the value for pK_a, the stronger the base.*

K_b:	10^{-10}	10^{-7}	10^{-5}
pK_b:	10	7	5

increasing base strength →

The basicity of an amine (R_3N) is often expressed as the pK_a of its conjugate acid (R_3NH^+). Conversions between pK_b and pK_a values are discussed in Section 17.6.

Study Problems

1.17 List the following compounds in order of increasing basicity (weakest first). See Table 1.9 for pK_b values.

(a) NH_3 (b) CH_3NH_2 (c) $(CH_3)_2NH$

1.18 List the following anions in order of increasing basicity (weakest first):

(a) CH_3O^-, $pK_b = -1.5$ (b) $CH_3CO_2^-$, $pK_b = 9.25$ (c) Cl^-, $pK_b = 21$

Table 1.9 Basicity constants and pK_b values for ammonia and some amines

Formula	K_b	pK_b
NH_3	1.79×10^{-5}	4.75
CH_3NH_2	45×10^{-5}	3.34
$(CH_3)_2NH$	54×10^{-5}	3.27
$(CH_3)_3N$	6.5×10^{-5}	4.19

▼ Summary

The probable location (relative to the nucleus) of an electron with a particular energy is called an **atomic orbital**. The first electron shell (closest to the nucleus, lowest energy) contains only the spherical 1s orbital. The second shell (higher energy) contains a spherical 2s orbital and three mutually perpendicular, two-lobed 2p orbitals. Any orbital can hold a maximum of two paired (opposite spin) electrons.

The **atomic radius** equals half the distance between nuclei bonded by a nonpolar covalent bond, such as in H—H. The atomic radius increases as we go down any group in the periodic table and decreases as we go from left to right across a period. **Electronegativity** is a measure of the pull of the nucleus on the outer electrons of an atom in a covalent compound. Electronegativity decreases as we go down any group and increases as we go from left to right in the periodic table.

A chemical bond results from electron transfer (**ionic bond**) or electron sharing (**covalent bond**). The number of bonds an atom can form is determined by the number of valence electrons. Carbon has four valence electrons and forms four covalent bonds.

An **empirical formula** tells us the *relative number* of different atoms in a molecule, and the **molecular formula** tells us the *actual number* of each kind of atom in a molecule.

$$C_2H_5 \qquad\qquad C_4H_{10}$$

empirical   molecular
formula   formula

In **structural formulas**, which depict the structures of molecules, pairs of electrons may be represented by dots or by lines. Unshared valence electrons are not always shown in structural formulas.

$$\underset{\ddot{H}\ \ddot{H}}{\overset{H}{H:\ddot{C}:\ddot{N}:H}} \qquad \underset{H\quad H}{\overset{H}{H—C—\ddot{N}—H}} \qquad CH_3NH_2$$

Lewis formula   complete   condensed
*   structural formula   structural formula*

An atom may share two, four, or six electrons with another atom—two atoms may be joined by a **single bond**, a **double bond**, or a **triple bond**.

single bond ——————→

triple bond ——→

←—————— double bond

$$H\!-\!C\!\equiv\!C\!-\!\overset{\overset{\displaystyle H}{|}}{C}\!=\!CH_2$$

A **formal charge** is an electrostatic charge on an atom in a molecule.

$$CH_3\!-\!\overset{+}{N}\!\!\!\begin{array}{c}\ddot{O}\cdot\\ \\ \ddot{O}\cdot^-\end{array}$$

formal charges

The **bond length** is the distance between nuclei of covalently bonded atoms. The **bond angle** is the angle between two covalent bonds in a molecule. The **bond dissociation energy** (ΔH in the following equation) is the amount of energy needed to effect **homolytic cleavage** of a covalent bond. A greater bond dissociation energy means a more stable bond.

$$H_3C\!-\!H \longrightarrow H_3C\cdot + H\cdot \qquad \Delta H = +104 \text{ kcal/mol}$$

A **polar covalent bond** is a covalent bond with a charge separation. The **bond moment** is a measure of the polarity of the bond. The **dipole moment** is a measure of the polarity of the entire molecule.

Dipole–dipole attractions between molecules (**van der Waals attractions**) are generally well under 5 kcal/mol except for **hydrogen bonds** (attractions between a partially positive H and an unshared pair of electrons of N, O, or F), which require 5–10 kcal/mol for their dissociation. Hydrogen bonding leads to an increase in boiling point and water solubility of a compound.

A **Brønsted–Lowry acid** is a substance that can lose H^+; a **Brønsted–Lowry base** is a substance that can accept H^+. The strength of an acid or a base is reported as K_a (or pK_a) or as K_b (or pK_b), respectively. A stronger acid has a larger value for K_a (and a smaller pK_a); a stronger base has a larger K_b (and a smaller pK_b). (See Tables 1.8 and 1.9.)

anion

$$K_a = \frac{[H^+][A^-]}{[HA]} \quad \text{and} \quad pK_a = -\log K_a$$

$$K_b = \frac{[\overset{+}{B}H][OH^-]}{[B\!:\!]} \quad \text{and} \quad pK_b = -\log K_b$$

base

strong acids ($pK_a < -1$): HCl, HNO_3, H_2SO_4
weak acids ($pK_a > 3$): CH_3CO_2H, HCN, H_2O
strong bases: $^-OH, ^-OCH_3$
weak bases: NH_3, CH_3NH_2

The **conjugate base** of a strong acid is a weak base, while the conjugate base of a very weak acid is a strong base. In the following equation, as HA decreases in acid strength, A$^-$ increases in base strength.

$$H\overset{\frown}{-}A \rightleftharpoons H^+ + :A^-$$

conjugate acid of A$^-$ *conjugate base of HA*

A **Lewis acid** is a substance that can *accept* a pair of electrons, while a **Lewis base** is a substance that can *donate* a pair of electrons.

$$CH_3\overset{..}{N}H_2 + {}^{\frown}H^+ \longrightarrow CH_3\overset{\overset{\displaystyle H}{|}}{\overset{+}{N}}H_2$$

Lewis base   *Lewis acid*

Study Problems

Note to student: The following *Study Problems* follow the order of the chapter presentation. The *Additional Problems* that come after them *do not* follow the chapter order; these problems should be worked after you have covered the entire chapter. The *Additional Problems* are gradated in order of difficulty—the first problems are routine drill problems, while the last are more challenging. In each chapter in this book, the problems are similarly arranged. Solutions to all chapter-end problems are given in the *Solutions Manual.*

1.19 Without referring to the text, write the electron configuration (for example, $1s^2\,2s^1$) for a free, nonbonded atom of:

(**a**) carbon (atomic no. 6)   (**b**) silicon (14)   (**c**) phosphorus (15)   (**d**) sulfur (16)

1.20 What element corresponds with each of the following electron configurations?

(**a**) $1s^2\,2s^2\,2p^6\,3s^1$ (**b**) $1s^2\,2s^2\,2p^6\,3s^2\,3p^5$ (**c**) $1s^2\,2s^2\,2p^6$

1.21 From the electron configurations, which pairs of elements would you expect to exhibit similar chemical behavior? Why?

(**a**) $1s^2\,2s^1$ and $1s^2\,2s^2$ (**b**) $1s^2\,2s^1$ and $1s^2\,2s^2\,2p^6\,3s^2\,3p^6\,4s^1$
(**c**) $1s^2\,2s^1$ and $1s^2\,2s^2\,2p^6\,3s^1$

1.22 If an atom used two *p* atomic orbitals to form single covalent bonds with two hydrogen atoms, what would be the expected bond angle?

1.23 Give the *Lewis formula* (electron-dot formula) for each of the following structures:

(**a**)
$$\begin{array}{c} \text{H}\ \ \text{H}\ \ \text{H}\ \ \text{H} \\ |\ \ \ |\ \ \ |\ \ \ | \\ \text{H}-\text{C}-\text{C}-\text{C}-\text{C}-\text{H} \\ |\ \ \ |\ \ \ |\ \ \ | \\ \text{H}\ \ \text{H}\ \ \text{H}\ \ \text{H} \end{array}$$
(**b**)
$$\begin{array}{c} \text{H}\ \ \ \ \ \ \text{H}\ \ \ \ \ \ \text{H} \\ |\ \ \ \ \ \ \ |\ \ \ \ \ \ \ | \\ \text{H}-\text{C}\ \ \ \ \ \ \text{C}\ \ \ \ \ \ \text{C}-\text{H} \\ |\ \ \ \ \ \ \ |\ \ \ \ \ \ \ | \\ \text{H}\ \ \text{H}-\text{C}-\text{H}\ \ \text{H} \\ |\ \\ \text{H} \end{array}$$

(**c**) $CH_3\overset{\overset{\displaystyle Cl}{|}}{C}HCH(CH_3)_2$ (**d**) H_2O_2 (**e**) $H_2\overset{\displaystyle O}{\overset{\diagdown\diagup}{C}-CH_2}$

1.24 Calculate the formal charges of all atoms except H in each of the following structures:

(**a**) $CH_3-\overset{\overset{\displaystyle :\overset{..}{O}:}{|}}{\underset{\underset{\displaystyle :\overset{..}{O}:}{|}}{S}}-\overset{..}{\underset{..}{C}l}:$ (**b**) $CH_3-C\equiv N:$

(c) $CH_3\overset{\displaystyle :O:}{\overset{\|}{C}}-\ddot{\underset{..}{O}}:^-$ (d) $CH_3-\overset{\displaystyle :\ddot{O}:}{\underset{..}{S}}-CH_3$

(e) $CH_3-\overset{\displaystyle CH_3}{\underset{\displaystyle CH_3}{C^+}}$ (f) $\left[CH_3-\overset{\displaystyle H}{\underset{..}{O}}-H\right]^+$

1.25 Give the *complete structural formula* (showing each atom and using lines for bonds) for each of the following condensed formulas:

(a) $(CH_3)_2CHCHBrCH_3$ (b) $(CH_3)_2CH\overset{\displaystyle O}{\overset{\|}{C}}NHCH(CH_3)CH=CH_2$

(c) $(CH_3)_2C=C(CH_3)C\equiv CCH_3$

1.26 Write a *condensed structural formula* for each of the following structures:

(a) $H-\overset{\displaystyle H}{\underset{\displaystyle H}{C}}-\ddot{\underset{..}{O}}-H$ (b) $H-\overset{\displaystyle H}{\underset{\displaystyle H}{C}}-\overset{\displaystyle H}{\underset{\displaystyle H}{C}}-\overset{\displaystyle H}{N}:$ (c) $H:\overset{\displaystyle H}{\underset{\displaystyle H}{\ddot{C}}}:\overset{\displaystyle H}{\ddot{N}}:\overset{\displaystyle H}{\underset{\displaystyle H}{\ddot{C}}}:H$

1.27 Write the *molecular formula* for each of the following condensed structural formulas:

(a) $CH_3CH_2CH_2CH_2OH$ (b) $CH_3CH_2\overset{\displaystyle OH}{\overset{|}{C}}HCH_3$ (c) $(CH_3)_3COH$

1.28 Give the complete structural formula (showing each atom and using lines for bonds) for each of the following compounds:

(a) $H_2C=CHCH=CHCN$ (b) CH_3COCH_3
(c) CH_3CO_2H (d) $OHCCH_2CHO$

(*Hint:* Each structure contains at least one double or triple bond that is not shown.)

1.29 Show *the unshared pairs of valence electrons* (if any) in the following formulas:
(a) CH_3NH_2 (b) $(CH_3)_3N$ (c) $(CH_3)_3NH^+$ (d) CH_3OH
(e) $(CH_3)_3COH$ (f) $CH_2=CH_2$ (g) $H_2C=O$

1.30 Draw a polygon formula for each of the following cyclic structures:

(a)   (b)

(c)   (d)

1.31 Convert each of the following polygon formulas to a complete structural formula, showing each atom and each bond. Show also any unshared pairs of valence electrons.

(a)   **(b)**   **(c)**   **(d)**

1.32 Draw the polygon and the structural formulas for a carbon ring system containing:
(a) six ring carbons and a double bond;
(b) five ring carbons, one of which is part of a carbonyl group (C=O).

1.33 Which of the following reactions would be the most exothermic? By how much? (Use the bond dissociation energies given in Table 1.4, as well as the values for CH_3—F, 108 kcal/mol, and CH_3—I, 56 kcal/mol.)
(a) $CH_4 + F_2 \longrightarrow CH_3F + HF$
(b) $CH_4 + Cl_2 \longrightarrow CH_3Cl + HCl$
(c) $CH_4 + Br_2 \longrightarrow CH_3Br + HBr$
(d) $CH_4 + I_2 \longrightarrow CH_3I + HI$

1.34 Write chemical equations for (1) the *homolytic cleavage*, and (2) the *heterolytic cleavage*, of each of the following compounds at the indicated bond. (Apply your knowledge of electronegativities in the heterolytic cleavages and show by curved arrows the movement of the electrons.)
(a) CH_3CH_2—Cl   (b) H—OH   (c) H—NH_2
(d) CH_3—OH   (e) CH_3O—H

1.35 Which is the positive end and which is the negative end of the dipole in each of the following bonds? (Use $\delta+$ and $\delta-$.)
(a) C—Mg   (b) C—Br   (c) C—O
(d) C—Cl   (e) C—H   (f) C—B

1.36 Circle the most electronegative element in each of the following structures, and show the direction of polarization of its bond(s):

(a) CH_3OH (b) $CH_3\overset{\overset{\displaystyle O}{\|}}{C}CH_3$
(c) FCH_2CH_2OH (d) $(CH_3)_2NCH_2CH_2OH$

1.37 Arrange each of the following series of compounds in order of increasing electronegativity difference in the most polar bond:
(a) $CH_3CH_2CH_2NH_2$, $CH_3CH_2CH_3$, $CH_3CH_2CH_2OH$
(b) $CH_3CH_2CH_2Br$, $CH_3CH_2CH_2I$, $CH_3CH_2CH_2Cl$

1.38 Draw structures to show the hydrogen bonding (if any) you would expect in the following compounds in their pure liquid states:
(a) $(CH_3)_2NH$ (b) $CH_3CH_2OCH_3$ (c) CH_3CH_2F
(d) $(CH_3)_3N$ (e) $(CH_3)_2C=O$ (f) $CH_3OCH_2CH_2OH$

1.39 Show all types of hydrogen bonds in an aqueous solution of $(CH_3)_2NH$. Which is the strongest hydrogen bond?

1.40 Which compound would you expect to have the higher boiling point? Explain.
(a) CH_3Cl or CH_3I
(b) $CH_3CH_2CH_2Cl$ or $CH_3CH_2CH_2OH$
(c) $CH_3CH_2CH_2CH_2OH$ or $(CH_3)_3COH$
(d) $HOCH_2CH_2OH$ or $CH_3CH_2CH_2OH$

1.41 Complete the following equations for acid–base reactions:
(a) $CH_3O^- + H_2O \rightleftharpoons$ (b) $CH_3NH_2 + HCl \longrightarrow$

(c) HO_2CCO_2H + excess OH^- $\longrightarrow$ (d) ⬡NH + H^+ $\longrightarrow$

(e) $CH_3CO_2^-$ + H^+ $\rightleftharpoons$ (f) CH_3NH_2 + CH_3CO_2H $\rightleftharpoons$

(g) CH_3CO_2H + CH_3O^- $\longrightarrow$ (h) $CH_3\overset{+}{N}H_3$ + CH_3O^- $\longrightarrow$

(i) $HO\overset{\displaystyle O}{\overset{\|}{C}}OH$ + 2 CH_3O^- $\longrightarrow$

1.42 Write the formulas for **(a)** the conjugate acid of $(CH_3)_2NH$, and **(b)** the conjugate base of $CH_3CH_2CH_2CO_2H$.

1.43 From the pK_a values for the conjugate acids in Table 1.8 (page 35), list the following anions in order of increasing base strength:

(a) $CH_3CO_2^-$ **(b)** HCO_2^- **(c)** Cl^- **(d)** $H_2PO_4^-$

1.44 Calculate the pK_a of each of the following compounds, and arrange in order of increasing acidity (weakest acid first):

	Structure	K_a
(a)	CH_3CO_2H	1.75×10^{-5}
(b)	⬡$-OH$	1.0×10^{-10}
(c)	⬡$-CH_2CO_2H$	5.2×10^{-5}
(d)	CH_3CH_2OH	$\sim 10^{-16}$
(e)	CH_3CH_3	$\sim 10^{-50}$

1.45 Calculate the pK_b of the following bases, and arrange in order of increasing base strength:

	Structure	K_b
(a)	$CH_3\overset{\displaystyle O}{\overset{\|}{C}}NH_2$	4.3×10^{-14}
(b)	⬡$-NH_2$	4.3×10^{-10}
(c)	$(CH_3)_3CNH_2$	6.8×10^{-4}

1.46 Complete the following equations for acid–base reactions, showing by arrow lengths the direction of the equilibrium.

(a) $H\ddot{B}r\!: + :\!\ddot{F}\!:^-$ $\rightleftharpoons$

(b) $CH_3CH_2\ddot{N}H^- + CH_3\ddot{O}H$ $\rightleftharpoons$

1.47 What would be the approximate pH of a solution prepared from 0.0200 mol of acetic acid and 0.0100 mol of sodium hydroxide in 100.00 mL of water?

Additional Problems

1.48 Lead (IV) chloride, $PbCl_4$, is a liquid at room temperature (mp $-15°$), while lead(II) chloride, $PbCl_2$, is a high-melting solid (mp $501°$). What do these properties suggest about the bonding in these two compounds?

1.49 Assign the proper ionic charge to each of the following ions:

(a) $:C\!\!\equiv\!\!N\!:$ (b) $:C\!\!\equiv\!\!CH$ (c) $CH_3\overset{..}{O}H_2$

(d) $CH_3\overset{..}{\underset{..}{O}}:$ (e) $:\overset{..}{N}H_2$ (f) $CH_3\overset{\overset{\textstyle\cdot\,\cdot}{\parallel}}{\underset{}{C}}\overset{..}{\underset{..}{O}}:$

1.50 Assume that none of the following compounds contains double bonds. Can you devise a condensed structural formula for each? (More than one answer may be possible.)

(a) C_3H_6 (b) C_2H_4O (c) C_4H_8O

1.51 Assuming that each of the following compounds contains a double bond, write a condensed structural formula for each. (There may be more than one correct answer.)

(a) C_3H_6 (b) C_2H_4O (c) C_4H_8O

1.52 Using Table 1.4 (page 19), calculate the heats of reaction for:

(a) $CH_3CH_3 + Cl_2 \longrightarrow CH_3CH_2Cl + HCl$
(b) $CH_3CH_3 + Br_2 \longrightarrow CH_3CH_2Br + HBr$

Which reaction liberates more energy?

1.53 Write an equation to show the heterolytic cleavage (at the most likely position) of each of the following molecules. (Remember to take electronegativities into consideration.)

(a) $(CH_3)_2CHBr$ (b) CH_3CH_2Li

(c) $(CH_3)_2CHOCH(CH_3)_2$ (d) $\overset{\textstyle O}{\overset{\diagup\,\diagdown}{CH_2\!-\!CH_2}}$

1.54 BF_3 has a dipole moment of zero. Suggest a shape for the BF_3 molecule.

1.55 Arrange the following compounds according to increasing solubility in water, least soluble first:

(a) CH_3CO_2H (b) CH_3CH_3 (c) $CH_3CH_2OCH_2CH_3$

1.56 Why does compound **A** boil at a much lower temperature than compound **B**? (Use formulas in your answer.)

A   **B**

1.57 Diethyl ether, $CH_3CH_2OCH_2CH_3$, and 1-butanol, $CH_3CH_2CH_2CH_2OH$, are equally soluble in water, but the boiling point of 1-butanol is 83° higher than that of diethyl ether. What explanation can you give for these observations? (Use formulas in your answer.)

1.58 In the following reactions, which of the reactants is the Lewis acid and which is the Lewis base?

(a) $(CH_3)_3CCl + AlCl_3 \rightleftharpoons (CH_3)_3C^+ + AlCl_4^-$
(b) $(CH_3)_3C^+ + CH_2\!\!=\!\!CH_2 \rightleftharpoons (CH_3)_3CCH_2\overset{+}{C}H_2$
(c) $(CH_3)_3C^+ + H_2O \rightleftharpoons (CH_3)_3C\overset{+}{\underset{\textstyle\mid}{O}}H$
H

(d) $CH_2\!\!=\!\!CH_2 + Br_2 \rightleftharpoons \overset{\overset{\textstyle +}{Br}}{\overset{\diagup\,\diagdown}{CH_2\!-\!CH_2}} + Br^-$

1.59 Indicate the most likely position of attack by a Lewis acid on each of the following structures. Explain your answers.

(a) $HN \quad NH$ (b) (structure) $-O^-$ (c) CH_3OH (d) CH_3CH_2Cl

1.60 Like water, an alcohol can act as a weak acid or a weak base. Write equations for the reaction of methanol (CH_3OH) with   (a) concentrated sulfuric acid, and   (b) sodamide $(NaNH_2)$, an extremely strong base.

1.61 Complete the following equations:

(a) $\underset{\text{lactic acid}}{CH_3\overset{\displaystyle OH}{\underset{|}{C}}HCO_2H} + CH_3NH_2 \rightleftharpoons$

(b) $\underset{\text{limestone}}{CaCO_3} + \underset{\text{acetic acid}}{2\ CH_3CO_2H} \longrightarrow$

(c) $CH_3\overset{\displaystyle CH_3}{\underset{\displaystyle\underset{|}{CH_3}}{\underset{|}{C}}O^-} + H_2O \rightleftharpoons$

1.62 When a person is exercising strenuously, some of the muscles' energy comes from the anaerobic (without oxygen) conversion of glucose to lactic acid (see Problem 1.61). In the body, lactic acid undergoes reaction with either   (a) bicarbonate ions (HCO_3^-) or   (b) hydrogen phosphate ions (HPO_4^{2-}). Write equations for these reactions.

1.63 The pK_a of phenol is 10.00. Calculate the expected pH of a 0.010 M aqueous solution.

1.64 Calculate the approximate pH of a solution that is 0.100 M acetic acid and 0.200 M acetate ion. (For acetic acid, $pK_a = 4.75$.)

Orbitals and Their Role in Covalent Bonding

In Chapter 1, we took a brief look at atomic orbitals and at covalent bonding. In this chapter, we will discuss how covalent bonds are produced by the formation of molecular orbitals. Various approaches to molecular orbitals have been developed. The **molecular orbital (MO) theory** provides mathematical descriptions of orbitals, their energies, and their interactions. The **valence-shell electron-pair repulsion (VSEPR) theory** is based on the premise that valence electrons or electron pairs of an atom repel each other. These repulsions can be used to explain observed bond angles and molecular geometry. In the **valence-bond theory**, line-bond formulas are used to describe covalent bonds and their interactions.

Within their limitations, all of these theories are successful and are often in agreement. However, no single one of them is practical to use in all discussions of organic compounds and their reactions. For this reason, we will present some of the highlights of each theory without necessarily attempting to differentiate among them. You will find that a knowledge of molecular orbitals and molecular shapes is invaluable in a study of organic reactions—whether they occur in a flask in the laboratory or in the cells of an animal.

Section 2.1
Properties of Waves

Until 1923, chemists assumed that electrons were nothing more than negatively charged particles whirling about the atomic nuclei. In 1923, Louis de Broglie, a French graduate student, proposed the revolutionary idea that electrons have properties of waves as well as properties of particles. De Broglie's proposal met with skepticism at first, but his idea was the seed that grew into today's quantum-mechanical concept of electron motion and the molecular orbital theory.

Quantum mechanics is a mathematical subject. For our understanding of covalent bonds, we need only the results of quantum-mechanical studies, rather than the mathematical equations themselves. With this in mind, let us survey some of the basic concepts of wave motion as they pertain to the current theories of covalent bonds.

We will begin with some simple **standing waves** (Figure 2.1), the type of wave that results when you pluck a string, like a guitar string, that is fixed at both ends. This type of wave exhibits motion in only one dimension. By contrast, the standing waves caused by beating the head of a drum are two-dimensional, and the wave system of an electron is three-dimensional. The height of a standing wave is its **amplitude**, which may be up (positive value) or down (negative value) in relation to the resting position of the string. (Note that the + or − sign of amplitude is a mathematical sign, not an electrical charge.) A position on the wave at which the amplitude is zero is called a **node** and corresponds to a position on the guitar string that does not move as the string vibrates.

Two standing waves can be either **in phase** or **out of phase** in reference to each other. Intermediate states in which waves are only partially in phase are also possible. We can illustrate these terms with two wave systems on two identical vibrating strings. If the positive and negative amplitudes of the two waves correspond to each other, the two waves are *in phase*. If the mathematical signs of the amplitudes are opposite, the waves are *out of phase* (see Figure 2.2).

If two in-phase waves on the same string overlap, they **reinforce** each other. The reinforcement is expressed by addition of the mathematical functions of the same sign describing the waves. Conversely, a pair of overlapping waves that are out of phase **interfere** with each other. The process

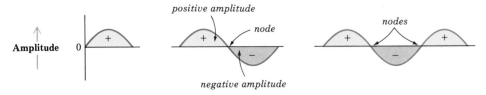

Figure 2.1 Some standing waves of vibrating strings with fixed ends (positive amplitudes shaded red).

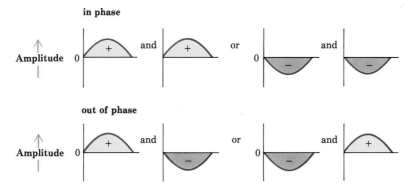

Figure 2.2 Two standing waves may be either in phase or out of phase.

of interference is represented by addition of two mathematical functions of opposite sign. Complete interference results in the cancelling of one wave by another. Partial overlap of two out-of-phase waves gives rise to a node. Figure 2.3 illustrates reinforcement and interference.

Although a three-dimensional electron wave system is more complicated than a one-dimensional string system, the principles are similar. Each atomic orbital of an atom behaves as a wave function and may have a positive or negative amplitude. If the orbital has both positive and negative amplitudes, it has a node. Figure 2.4 depicts the $1s$, $2s$, and $2p$ orbitals, including their signs of amplitude and their nodes.

One atomic orbital can overlap an atomic orbital of another atom. (Mathematically, the wave functions describing each overlapping orbital are added. These calculations are referred to as the **linear combination of atomic orbitals**, or **LCAO, theory**.) When the overlapping orbitals are in phase, the result is reinforcement and a **bonding molecular orbital**. On the other hand, interaction between atomic orbitals that are out of phase results in interference, creating a node between the two nuclei. Interference leads to an **antibonding molecular orbital**. We will expand upon these definitions of bonding and antibonding orbitals in Section 2.2B.

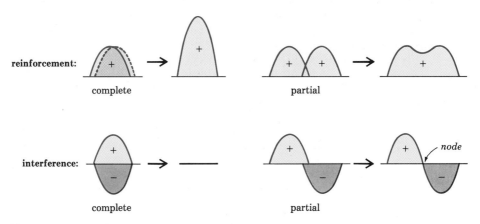

Figure 2.3 Reinforcement and interference of waves.

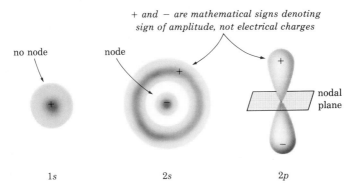

Figure 2.4 The 1*s*, 2*s*, and 2*p* orbitals with their signs of amplitude.

Section 2.2
Bonding in Hydrogen

Hydrogen (H_2) is the simplest molecule. We will look at the covalent bond of H_2 in some detail because many features of this bond are similar to those of more complex covalent bonds.

Let us consider two isolated hydrogen atoms, each with one electron in a 1*s* atomic orbital. As these two atoms begin bond formation, the electron of each atom becomes attracted by the nucleus of the other atom, as well as by its own nucleus. When the nuclei are at a certain distance from each other (the bond length, 0.74 Å for H_2), the atomic orbitals merge, or overlap, to reinforce each other and form a bonding molecular orbital. This molecular orbital encompasses both hydrogen nuclei and contains two paired electrons (one from each H). Both electrons are now equally attracted to both nuclei. Because a large portion of the negatively charged electron density of this new orbital is located between the two positively charged nuclei, repulsions between the nuclei are minimized. This molecular orbital results in the covalent bond between the two hydrogen atoms in H_2 (see Figure 2.5).

A. The Sigma Bond

The molecular orbital that bonds two hydrogen atoms together is *cylindrically symmetrical*—that is, symmetrical about a line, or axis, joining the two nuclei. Think of the axis as an axle, and rotate the orbital around this axis. If the

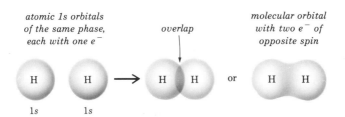

Figure 2.5 The formation of the bonding molecular orbital in H_2 from in-phase 1*s* orbitals.

Figure 2.6 The sigma bond of hydrogen is symmetrical about the axis joining the two nuclei.

axis

appearance of the orbital is not changed by the rotation, the orbital is symmetrical around that axis (see Figure 2.6).

Any molecular orbital that is symmetrical about the axis connecting the nuclei is called a **sigma (σ) molecular orbital**; the bond is a **sigma bond**. The bond in H_2 is only one of many sigma bonds we will encounter. (We will also encounter molecular orbitals that are not sigma orbitals—orbitals that are not symmetrical about their nuclear axes.)

B. The Bonding Orbital and the Antibonding Orbital

When a pair of waves overlap, they either reinforce each other or interfere with each other. Addition of two in-phase $1s$ atomic orbitals of two H atoms results in reinforcement and produces the σ bonding molecular orbital with a high electron density between the bonded nuclei.

If two waves are out of phase, they interfere with each other. Interference of two out-of-phase atomic orbitals of two hydrogen atoms gives a molecular orbital with a *node between the nuclei*. In this molecular orbital, the probability of finding an electron between the nuclei is *very low*. Therefore, this particular molecular orbital gives rise to a system in which the two nuclei are not shielded by the pair of electrons, and the nuclei repel each other. Principally because of the nuclear repulsion, this system is of *higher energy* than the system of two independent H atoms. This higher-energy orbital is the **antibonding orbital**, in this case, a "sigma star," or σ*, orbital (the * meaning "antibonding"). Figure 2.7 compares the shapes of the σ and σ* orbitals for H_2.

The energy of the H_2 molecule with two electrons in the σ bonding orbital is *lower* by 104 kcal/mol than the combined energy of two separate hydrogen atoms. The energy of the hydrogen molecule with electrons in the σ* antibonding orbital, on the other hand, is *higher* than that of two separate hydrogen atoms. These relative energies may be represented by the following diagram:

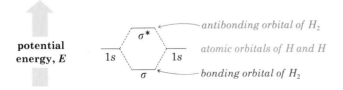

A molecular orbital, like an atomic orbital, can hold no electrons, one electron, or two paired electrons. The two electrons in a hydrogen molecule go into the lowest-energy orbital available, the σ bonding orbital. In the following

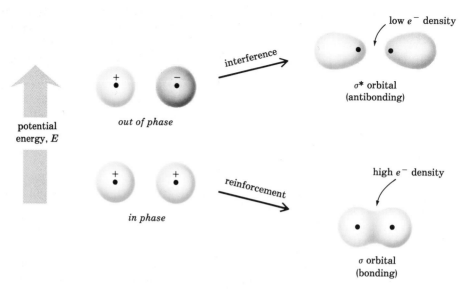

Figure 2.7 Reinforcement and interference of two 1s orbitals. (The + and − refer to phases of the wave functions, not electrical charges.)

diagram, we use a pair of arrows (one pointing up and one pointing down) to represent a pair of electrons of opposite spin.

We said in Chapter 1 that electrons in different atomic orbitals differ in energy because of the various distances of these electrons from the nucleus. The higher energy of a molecule with electrons in an antibonding orbital (compared to the energy of the molecule with electrons in a bonding orbital) does not arise from the electrons being farther from the nuclei. Instead, the higher energy arises from the presence of the node between the nuclei.

For the hydrogen molecule, the σ and σ^* orbitals are in the same general region of space. Although two particles of matter cannot occupy the same space at the same time, two orbitals may. Remember, orbitals are not matter, but are simply regions of space where the probability of finding an electron with a particular energy is high.

All bonding molecular orbitals have antibonding orbitals associated with them. In each case, a molecule with electrons in a bonding molecular orbital has a *lower energy* than the energy of the nonbonded atoms, and a molecule with electrons in an antibonding orbital has a *higher energy* than that of the nonbonded atoms. Because the antibonding orbitals are of high energy, the electrons are not generally found there. Almost all the chemistry in this text will deal with molecules in the **ground state**, the state in which the electrons are in the lowest-energy orbitals. However, we will encounter a few situations in which energy absorbed by a molecule is used to promote an electron from a low-energy orbital to a higher-energy orbital. A molecule is said to be in an **excited state** when one or more electrons are not in the orbital of lowest energy.

Section 2.3
Some General Features of Bonding and Antibonding Orbitals

Let us summarize some general rules that apply to all molecular orbitals, not only the molecular orbitals of H_2:

1. Any orbital (molecular or atomic) can hold a maximum of two electrons, which must be of opposite spin.
2. The number of molecular orbitals equals the number of atomic orbitals that went into their formation. (For H_2, two $1s$ atomic orbitals yield two molecular orbitals: σ and σ^*.)
3. In the filling of molecular orbitals with electrons, the lowest-energy orbitals are filled first. If two orbitals are degenerate (of equal energies), each gets one electron before either is filled.

Section 2.4
Hybrid Orbitals of Carbon

When a hydrogen atom becomes part of a molecule, it uses its $1s$ atomic orbital for bonding. The situation with the carbon atom is somewhat different. Carbon has two electrons in the $1s$ orbital; consequently, the $1s$ orbital is a filled orbital that is not used for bonding. The four electrons at the *second energy level* of carbon are the bonding electrons.

There are four atomic orbitals at the second energy level: one $2s$ and three $2p$ orbitals. However, carbon does not use these four orbitals in their pure states for bonding. Instead, carbon blends, or **hybridizes**, its four second-level atomic orbitals in one of three different ways for bonding:

1. *sp^3* **hybridization,**   used when carbon forms four single bonds.

2. *sp^2* **hybridization,**   used when carbon forms one double bond.

3. *sp* hybridization, used when carbon forms a triple bond or *cumulated* double bonds (two double bonds to a single carbon atom).

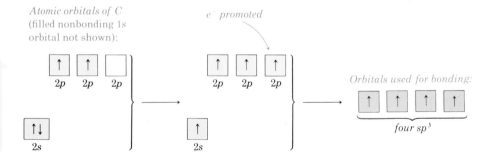

Why does a carbon atom form compounds with hybrid orbitals rather than with unhybridized atomic orbitals? The answer is that hybridization gives stronger bonds because of greater overlap and therefore results in more-stable, lower-energy molecules. As we discuss each type of hybridization, note that the *shape* of each hybrid orbital is favorable for maximum overlap with an orbital of another atom. Also note that the *geometries* of the three types of hybrid orbitals allow attached groups to be as far from each other as possible, thus minimizing their repulsions for each other.

A. *sp*³ Hybridization

In methane (CH_4), the carbon atom has four equivalent bonds to hydrogen. Each C—H bond has a bond length of 1.09 Å and a bond dissociation energy of 104 kcal/mol. The bond angle between each C—H bond is 109.5°. From this experimental evidence alone, it is evident that carbon does not form bonds by means of one *s* atomic orbital and three *p* atomic orbitals. If that were the case, the four C—H bonds would not all be equivalent.

According to present-day theory, these four equivalent bonds arise from complete hybridization of the four atomic orbitals (one 2*s* orbital and three 2*p* orbitals) to yield four equivalent *sp*³ orbitals. For this to be accomplished, one of the 2*s* electrons must be promoted to the empty 2*p* orbital. This promotion requires energy (about 96 kcal/mol), but this energy is more than regained by the concurrent formation of chemical bonds. The four *sp*³ orbitals have equal energies—slightly higher than that of the 2*s* orbital, but slightly lower than that of the 2*p* orbitals. Each of the *sp*³ orbitals contains one electron for bonding.

The preceding diagram is called an **orbital diagram**. Each box in the diagram represents an orbital. The relative energies of the various orbitals are signified by the vertical positions of the boxes within the diagram. Electrons are represented by arrows, and the direction of electron spin is indicated by the direction of the arrow.

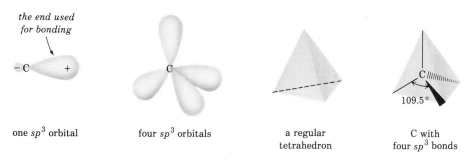

one sp^3 orbital   four sp^3 orbitals   a regular tetrahedron   C with four sp^3 bonds

Figure 2.8 The four sp^3-hybrid orbitals of carbon point toward the corners of a regular tetrahedron.

The sp^3 orbital, which results from a blend of the 2s and 2p orbitals, is shaped rather like a bowling pin: it has a large lobe and a small lobe (of opposite amplitude) with a node near the nucleus. Figure 2.8 shows one isolated sp^3 orbital. The small lobe of the hybrid orbital is not used for bonding because overlap of the large lobe with another orbital gives more complete overlap and results in a stronger bond.

Four sp^3-hybrid orbitals surround the carbon nucleus. Because of repulsions between electrons in different orbitals, these sp^3 orbitals lie as far apart from each other as possible while still extending away from the same carbon nucleus—the four orbitals point toward the corners of a regular tetrahedron (Figure 2.8). This geometry gives idealized bond angles of 109.5°. An sp^3-hybridized carbon atom is often referred to as a **tetrahedral carbon atom** because of the geometry of its bonds.

When an sp^3-hybridized carbon atom forms bonds, it does so by overlapping each of its four sp^3 orbitals (each with one electron) with orbitals from four other atoms (each orbital in turn containing one electron). In methane (Figures 2.9 and 2.10), each sp^3 orbital of carbon overlaps with a 1s orbital of hydrogen. Each of the resultant sp^3–s molecular orbitals is symmetrical around the axis passing through the nuclei of the carbon and the hydrogen. The covalent bonds between C and H in methane, like the H—H covalent bond, are sigma bonds.

Ethane (CH_3CH_3) contains two sp^3-hybridized carbon atoms. These two carbon atoms form a C—C sigma bond by the overlap of one sp^3 orbital from each carbon (sp^3–sp^3 sigma bond). Each carbon atom has three remaining sp^3

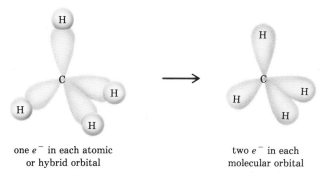

one e^- in each atomic or hybrid orbital   two e^- in each molecular orbital

Figure 2.9 Formation of C—H sigma bonds in methane, CH_4. (The small lobes of the sp^3 orbitals are not shown.)

CH$_4$			
condensed	complete	dimensional	molecular
structural formula	structural formula	formula	model

Figure 2.10 Some different ways of representing methane.

ethane

Figure 2.11 Formation of the sp^3–sp^3 sigma bond in ethane, CH$_3$CH$_3$.

CH$_3$CH$_3$			
condensed	complete	three-dimensional	molecular
structural formula	structural formula	formula	model

Figure 2.12 Some different ways of representing ethane.

orbitals, and each of these overlaps with a 1s orbital of a hydrogen atom to form a C—H sigma bond. Each carbon atom in ethane is tetrahedral (see Figures 2.11 and 2.12).

sp^3–sp^3 sigma bond

sp^3–s sigma bond

In any molecule, any carbon atom bonded to four other atoms is in the sp^3-hybrid state, and the four bonds from that carbon are sigma bonds. When carbon is bonded to four other atoms, the sp^3 hybridization allows maximal overlap and places the four attached atoms at the maximum distances from each other. If possible, the sp^3 bond angles are 109.5°. However, other factors, such as dipole–dipole repulsions or the geometry of a cyclic compound, can cause deviations from this ideal bond angle.

Examples of structures with sp^3-hybridized carbons (each C has four sigma bonds):

$$
\begin{array}{ccccc}
& \text{H} & \text{H}\quad\text{H} & \text{H} & \\
& | & |\quad| & | & \\
\text{H}-&\text{C}-\text{H} & \text{H}-\text{C}-\text{C}-\text{H} & \text{H}-\text{C}-\text{O}-\text{H} & \\
& | & |\quad| & | & \\
& \text{H} & \text{H}\quad\text{H} & \text{H} &
\end{array}
$$

Sample Problem

Give the complete structural formula (showing all atoms and bonds) for propane $(CH_3CH_2CH_3)$. Which types of orbitals overlap to form each bond?

Solution:

each C—H bond is sp^3–s

$$
\begin{array}{ccc}
\text{H} & \text{H} & \text{H} \\
| & | & | \\
\text{H}-\text{C}-&\text{C}-&\text{C}-\text{H} \\
| & | & | \\
\text{H} & \text{H} & \text{H}
\end{array}
$$

sp^3–sp^3

Study Problem

2.1 Write the complete structural formula for each of the following compounds, showing each bond. Which types of orbitals overlap to form each bond?

(a) $(\dot{C}H_3)_3CH$ **(b)**

B. sp^2 Hybridization

When carbon is bonded to another atom by one double bond, the carbon atom is in the sp^2-hybrid state.

Examples of compounds with sp^2-hybridized carbons:

$$
\begin{array}{cc}
\text{H}\diagdown\qquad\diagup\text{H} & \text{H}\diagdown \\
\quad\text{C}=\text{C} & \quad\text{C}=\text{O} \\
\text{H}\diagup\qquad\diagdown\text{H} & \text{H}\diagup
\end{array}
$$

    ethylene   formaldehyde

To form sp^2 bonding orbitals, carbon hybridizes its $2s$ orbital with only two of its $2p$ orbitals. One p orbital remains unhybridized on the carbon atom. Because three atomic orbitals are used to form the sp^2 orbitals, three sp^2-hybrid orbitals result. Each sp^2 orbital has a shape similar to that of an sp^3 orbital and contains one electron that can be used for bonding.

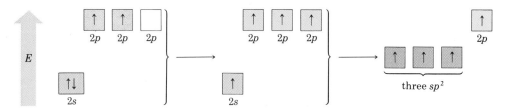

The three sp^2 orbitals around a carbon nucleus lie as far apart from one another as possible, that is, the sp^2 orbitals lie in a plane with angles of 120° (ideally) between them. An sp^2-hybridized carbon atom is said to be a **trigonal** (triangular) carbon. Figure 2.13 shows a carbon atom with three sp^2 orbitals and the one unhybridized p orbital, which is perpendicular to the sp^2 plane.

In ethylene ($CH_2{=}CH_2$), two sp^2-hybridized carbons are joined by a sigma bond formed by the overlap of one sp^2 orbital from each carbon atom. (This sigma bond is one of the bonds of the double bond.) Each carbon atom still has two sp^2 orbitals left for bonding with hydrogen. (Each carbon atom also has a p orbital, which is not shown in the following structure.)

Planar sigma-bond structure of ethylene (p orbitals not shown):

$$
\begin{array}{c}
\text{H} \qquad\quad \text{H} \\
\diagdown \;^{121°}\;\nearrow \\
\text{C}\!-\!\text{C}\;\Big\rangle\,118° \\
\diagup \qquad\quad \diagdown \\
\text{H} \qquad\qquad \text{H}
\end{array}
$$

What of the remaining p orbital on each carbon? Each p orbital has two lobes, one above the plane of the sigma bonds and the other (of equal amplitude, but opposite phase) below the plane. Each p orbital contains one electron. If these p electrons become paired in a bonding molecular orbital, then the energy of the system is lowered. Because the p orbitals lie side by side in the ethylene molecule, the *ends* of the orbitals cannot overlap, as they do in sigma-bond formation. Instead, the two p orbitals overlap their *sides* (see Figure 2.14). The result of this side-to-side overlap is the **pi (π) bond**—a bond joining the two carbons and located above *and* below the plane of the sigma bonds. The pi bond is the second bond of the double bond.

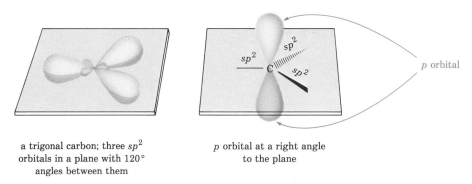

a trigonal carbon; three sp^2 orbitals in a plane with 120° angles between them

p orbital at a right angle to the plane

Figure 2.13 Carbon in the sp^2-hybrid state.

The π bond is a two–lobed molecular orbital containing one pair of e^- and having a node at the site of the sigma bond.

Figure 2.14 Formation of the sp^2–sp^2 sigma bond and the p–p pi bond in ethylene, CH_2=CH_2. (The + and − refer to phases of the wave functions, not electrical charges.)

Any carbon atom that is bonded to three other atoms is in the sp^2-hybrid state. In stable compounds, the p orbital on the sp^2-hybridized carbon must overlap with a p orbital of an adjacent atom, which can be another carbon atom or an atom of some other element.

C. Some Features of the Pi Bond

Each p orbital contributing to a pi bond has two lobes and has a node at the nucleus. It is not surprising that the pi orbital also is two-lobed and has a node. Unlike a sigma orbital, a pi orbital is not cylindrically symmetrical. However, just like any molecular orbital, a pi orbital can hold a maximum of two paired electrons.

A $2p$ orbital of carbon is of slightly higher energy than an sp^2 orbital. For this reason, a pi bond, which is formed from two $2p$ orbitals, has slightly higher energy and is slightly less stable than an sp^2–sp^2 sigma bond. The bond dissociation energy of the sigma bond of ethylene's carbon–carbon double bond is estimated to be 95 kcal/mol, while that of the pi bond is estimated to be only 68 kcal/mol.

The more-exposed pi electrons are more vulnerable to external effects than are electrons in sigma bonds. The pi bond is polarized more easily—we might say that the pi electrons are more mobile. The pi electrons are more easily promoted to a higher-energy (antibonding) orbital. Also, they are more readily attacked by an outside atom or molecule. What does this vulnerability mean in terms of the chemistry of pi-bonded compounds? In a molecule, the pi bond is a site of chemical reactivity.

Another property of the pi bond is that it holds the doubly bonded carbon atoms, and the atoms to which these carbons are bonded, in a rigid shape.

Figure 2.15 The portion of a molecule surrounding a pi bond is held in a planar structure unless enough energy is supplied to break the pi bond.

Rotation of the doubly bonded carbon atoms and their substituent groups around the double bond would require that the pi bond first be broken (see Figure 2.15). In chemical reactions, molecules may have sufficient energy (about 68 kcal/mol) for this bond to break. In a flask at room temperature, however, molecules do not have enough energy for this bond breakage to occur. (Approximately 20 kcal/mol is the maximum energy available to molecules at room temperature.) The significance of the rigidity of pi bonds will be discussed in Chapter 4.

In a structural formula, a double bond is indicated by two identical lines. Keep in mind that the double bond is not simply two identical bonds, but that the double line represents one relatively stable sigma bond and one relatively reactive pi bond.

Sample Problem

What type of overlap (sp^3–s, for example) is present in each bond of $CH_3CH{=}CH_2$? What is each bond angle (approximately)?

Solution:

Study Problems

2.2 Give the complete structural formula for each of the following compounds. Indicate which types of orbitals are used to form each bond.

(a) $CH_2 = C(CH_3)_2$ (b) $CH_2 = CHCH = CH_2$ (c)

2.3 For compound (b) in Problem 2.2, draw the structure showing the pi bonds with their correct geometry in reference to the sigma bonds. (Use lines to represent the sigma bonds.)

D. The Bonding and Antibonding Orbitals of Ethylene

The carbon-carbon sigma bond in ethylene results from overlap of two sp^2 orbitals. Because *two sp^2* orbitals form this bond, *two* molecular orbitals result. The other molecular orbital, arising from interference between the two sp^2 orbitals, is the antibonding σ^* orbital. This antibonding orbital is similar to the σ^* orbital of hydrogen: it has a node between the two carbon nuclei and is of high energy. The two electrons of the C—C sigma bond in ethylene are usually found in the lower-energy σ orbital.

In the pi bond joining the two carbons of ethylene, each carbon atom contributes one p orbital for a total of two p orbitals; therefore, two pi molecular orbitals result. One of these is the π bonding orbital that arises from the overlap of two in-phase p orbitals. The other orbital is the π^* antibonding orbital, which arises from interference between two p orbitals of opposite phase. These two orbitals are frequently designated π_1 for the bonding orbital and π_2^* for the antibonding orbital. Some molecules contain several π orbitals, which we number in order of increasing energy; the subscript numbers are useful for differentiating the π orbitals. (We do not usually use subscript numbers for σ orbitals because σ^* orbitals are usually of minor importance to the organic chemist.)

Figure 2.16 shows the orbital representations of the π_1 and π_2^* orbitals of ethylene. Note that besides the node at the σ-bond site, the π_2^* orbital has an additional node *between the two carbon nuclei*. A minimum of the pi electron density is located between the nuclei in this orbital; thus, the π_2^* orbital is of higher energy than the π_1 orbital. In the ground state of ethylene, the pi electrons are found in the lower-energy π_1 orbital.

The following diagram comparing energies of the σ, σ^*, π_1, and π_2^* orbitals shows that the σ^* orbital is of higher energy than the π_2^* orbital. The amount of energy required to promote an electron from the σ orbital to the σ^* orbital is therefore greater than the energy required to promote a π electron to the π^* orbital.

Ground state of C=C in ethylene:

E

σ^* _____

π_2^* _____

π_1 ⟰ (↑↓)

σ (↑↓)

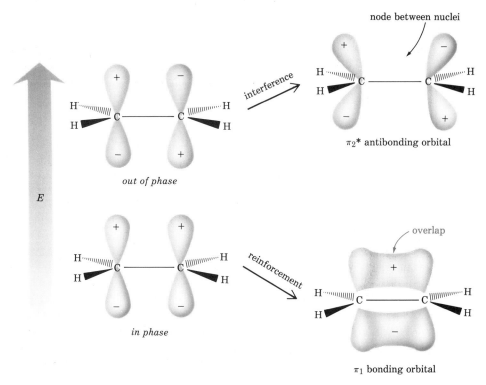

Figure 2.16 The π_1 bonding orbital and the $\pi_2{}^*$ antibonding orbital of ethylene.

Why is more energy required to promote a σ electron than a π electron? Sigma bonding electrons are found close to the nucleus and, for the most part, directly between the nuclei. Promotion of one of these sigma electrons results in severe nuclear repulsion. However, the π bonding electrons are farther from the nucleus. Promotion of one of these pi electrons does not result in such severe nuclear repulsions. Also, when a pi electron is promoted, the nuclei are still shielded from each other by σ bonding electrons.

Because of the large amount of energy required to promote a sigma electron, electron transitions of the $\sigma \rightarrow \sigma^*$ type are rare and relatively unimportant to the organic chemist. However, $\pi \rightarrow \pi^*$ electron transitions, which require less energy, are important. For example, $\pi \rightarrow \pi^*$ transitions are responsible for vision, a topic that will be mentioned in Chapter 21, and for the energy capture needed for photosynthesis.

Two possible excited states of $C{=}C$ *in ethylene:*

E. *sp* Hybridization

When a carbon atom is joined to only two other atoms, as in acetylene (CH≡CH), its hybridization state is *sp*. One 2*s* orbital blends with only one 2*p* orbital to form two *sp*-hybrid orbitals. Two unhybridized 2*p* orbitals remain, each with one electron.

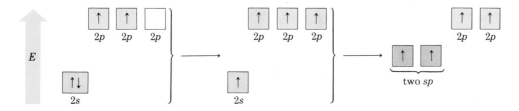

The two *sp* orbitals lie as far apart as possible, in a straight line with an angle of 180° between them. The *p* orbitals are perpendicular to each other and to the line of the *sp* orbitals (see Figure 2.17).

In CH≡CH, the two carbon atoms are joined by an *sp–sp* sigma bond. Each carbon is also bonded to a hydrogen atom by an *sp–s* sigma bond. The two *p* orbitals of one carbon then overlap with the two *p* orbitals of the other carbon to form *two* pi bonds. One pi bond is above and below the line of the sigma bonds, as shown in Figure 2.18; the other pi bond is located in front and back.

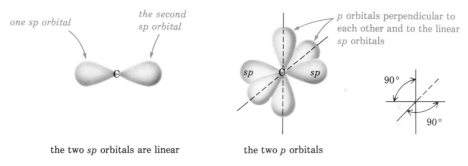

Figure 2.17 Carbon in the *sp*-hybrid state.

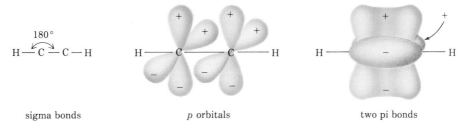

Figure 2.18 Bonding in acetylene, CH≡CH.

As you might guess, the chemical reactions of a compound containing a triple bond are not too different from those of a compound containing a double bond. Instead of one pi bond, there are two.

one σ bond, two pi bonds

$$H-C{\equiv}C-H \qquad CH_3-C{\equiv}C-H \qquad H-C{\equiv}N$$

Study Problem

2.4 What type of overlap is present in each carbon–carbon bond of $CH_3-C{\equiv}CH$? What are the approximate bond angles around each carbon?

F. Effects of Hybridization on Bond Lengths

A $2s$ orbital is of slightly lower energy than a $2p$ orbital. On the average, $2s$ electrons are found closer to the nucleus than $2p$ electrons. For this reason, a hybrid orbital with a greater proportion of s character is of lower energy and is closer to the nucleus than a hybrid orbital with less s character.

An sp-hybrid orbital is one-half s and one-half p; we may say that the sp orbital has 50% s character and 50% p character. At the other extreme is the sp^3 orbital, which has only one-fourth, or 25%, s character.

	hybridization of carbon	*percent s character*
$CH{\equiv}CH$	sp	50
$CH_2{=}CH_2$	sp^2	$33\frac{1}{3}$
CH_3CH_3	sp^3	25·

Because the sp orbital contains more s character, it is closer to its nucleus; it forms shorter and stronger bonds than the sp^3 orbital. The sp^2 orbital is intermediate between sp and sp^3 in its s character and in the length and strength of the bonds it forms.

Table 2.1 shows the differences in bond lengths among the three C—C and C—H bond types. Note that the sp–s CH bond in $CH{\equiv}CH$ is the shortest, while the sp^3–s CH bond is the longest. We see an even wider variation in the C—C bonds because these bond lengths are affected by the *number* of bonds joining the carbon atoms as well as by the hybridization of the carbon atoms.

Table 2.1 Effect of hybridization on bond length

	longest bonds from C		*shortest bonds from C*
	H H │ │ H—C—C—H │ │ H H	H H \\ / C=C / \\ H H	H—C≡C—H
percent s character:	25	$33\frac{1}{3}$	50
C—C bond length:	1.54 Å	1.34 Å	1.20 Å
C—H bond length:	1.09 Å	1.08 Å	1.06 Å

Study Problem

2.5 For each of the following structures, list the numbered bonds in order of increasing bond length (shortest bond first):

(a) $CH_3 \overset{①}{-} CH_2CH \overset{②}{=} CH_2$ **(b)** structure with H, C, CH_2C=CH_2 and numbered bonds ① and ② **(c)** ring structure with numbered bonds ①, ②, ③

G. Summary of the Hybrid Orbitals of Carbon

1. When a carbon atom is bonded to *four other atoms*, the bonds from the carbon atom are formed from four equivalent sp^3 orbitals. The sp^3-hybridized carbon is **tetrahedral**.

tetrahedral

Examples: CH_4, $CHCl_3$,

sp³

2. When carbon bonds to *three other atoms*, the bonds from the carbon atom are formed from three equivalent sp^2 orbitals with one p orbital remaining. The sp^2 orbitals form three sigma bonds; the p orbital forms a pi bond. The sp^2-hybridized carbon is **trigonal**.

σ and π

trigonal

Examples: $CH_3CH=CH_2$, $H_2C=O$,

sp²

3. When carbon bonds to *two other atoms*, the bonds from the carbon atom are formed from two equivalent sp orbitals, with two p orbitals remaining. The two p orbitals overlap with two p orbitals of another atom to form two pi bonds. The sp orbitals form two equivalent and **linear** sigma bonds.

one σ, two π

$H-C\equiv C-H$

linear

Examples: $CH_3C\equiv CH$, $HC\equiv N$

sp carbons

▼ ## Section 2.5
Functional Groups

Although sp^3–sp^3 carbon–carbon bonds and sp^3–s carbon–hydrogen bonds are common to almost all organic compounds, surprisingly, these bonds do not usually play a major role in organic reactions. For the most part, it is the presence of either pi bonds or other atoms in an organic structure that confers reactivity. A site of chemical reactivity in a molecule is called a **functional group**. A pi bond or an electronegative (or electropositive) atom in an organic molecule can lead to chemical reaction; either one of these is considered a functional group or part of a functional group.

Some functional groups (marked with color):

$$Br\ CH_2\quad CH{=}C\ H_2\qquad CH_3CH_2NH_2\qquad CH_3CH_2OH$$

Study Problem

2.6 Circle the functional groups in the following structures:

(a) $CH_2{=}CHCH_2\overset{\displaystyle O}{\overset{\|}{C}}H$ (b) ⬡—NH_2 (c)

Compounds with the same functional group tend to undergo similar chemical reactions. For example, each of the following series of compounds contains a **hydroxyl group** (—OH). These compounds all belong to the class of compounds called **alcohols**, and all undergo similar reactions.

Some alcohols:

$$CH_3CH_2OH\qquad (CH_3)_3COH\qquad ⬡{-}OH$$

Because of the similarities in reactivity among compounds with the same functional group, it is frequently convenient to use a general formula for a series of these compounds. We usually use R to represent an **alkyl group**, a group that contains only sp^3-hybridized carbon atoms plus hydrogens. By this technique, we can represent an alcohol as ROH. Table 2.2 shows some functional groups and some classes of compounds with generalized formulas. A more extensive table is shown inside the front cover of this book.

R—   means an alkyl group, such as CH_3—, CH_3CH_2—, or ⬡—

ROH   means an alcohol, such as CH_3OH, CH_3CH_2OH, or ⬡—OH

Table 2.2 Some functional groups and compound classes

Functional Group		Class of Compound	
Structure	*Name*	*General Formula*	*Class Name*
C=C	double bond	$R_2C{=}CR_2$	alkene
C≡C	triple bond	$RC{\equiv}CR'^{\,a}$	alkyne
—NH_2	amino group	RNH_2	amine
—OH	hydroxyl group	ROH	alcohol
—OR	alkoxyl group	$R'OR^{\,a}$	ether

^{a} R′ refers to an alkyl group that may be the same as or different from R.

Sample Problem

The following compounds are all **amines**. Write a general formula for an amine.

$$CH_3NH_2 \qquad CH_3CH_2NH_2 \qquad (CH_3)_2CHCH_2NH_2$$

Solution: RNH_2

Study Problem

2.7 The following compounds are all **carboxylic acids**. Write the general formula for a carboxylic acid.

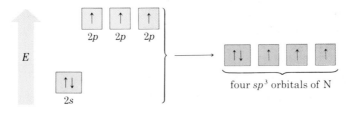

$$CH_3CO_2H \qquad CH_3CH_2CH_2CO_2H$$

▼ Section 2.6
Hybrid Orbitals of Nitrogen and Oxygen

A. Amines

Many important functional groups in organic compounds contain nitrogen or oxygen. Electronically, nitrogen is similar to carbon, and the atomic orbitals of nitrogen hybridize in a manner very similar to those of carbon:

As these orbital diagrams show, nitrogen can hybridize its four second-level atomic orbitals to four equivalent sp^3 bonding orbitals. However, note one important difference between nitrogen and carbon. While carbon has four electrons to distribute in four sp^3 orbitals, nitrogen has *five* electrons to distribute in four sp^3 orbitals. One of the sp^3 orbitals of nitrogen is filled with a

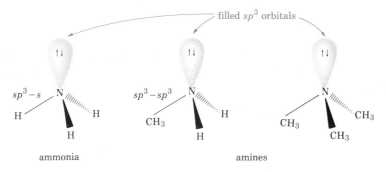

Figure 2.19 Bonding in ammonia and in two amines.

pair of electrons, and nitrogen can form compounds with only three covalent bonds to other atoms.

A molecule of ammonia contains an sp^3-hybridized nitrogen atom bonded to three hydrogen atoms. An **amine** molecule has a similar structure: an sp^3 nitrogen atom bonded to one or more carbon atoms. In either ammonia or an amine, the nitrogen has one orbital filled with a pair of unshared valence electrons. Figure 2.19 shows the geometry and the filled orbitals of ammonia and two amines: the similarities in structure are evident in this figure.

The unshared pair of electrons in the filled orbital on the nitrogen of ammonia and amines allows these compounds to act as bases. When an amine is treated with an acid, these unshared electrons are used to form a sigma bond with the acid. The product is an **amine salt**.

$$CH_3\text{—}\overset{\cdot\cdot}{N}\text{—}CH_3 + H\text{—}\overset{\cdot\cdot}{\underset{\cdot\cdot}{Cl}}: \longrightarrow CH_3\text{—}\overset{\overset{\textstyle H}{|}}{\underset{\underset{\textstyle CH_3}{|}}{N^+}}\text{—}CH_3 \quad :\overset{\cdot\cdot}{\underset{\cdot\cdot}{Cl}}:^-$$

an amine   *an amine salt*

Like carbon, nitrogen is also found in organic compounds in the sp^2- and sp-hybrid states. Again, the important difference between nitrogen and carbon is that one of the orbitals of nitrogen is filled with a pair of unshared electrons.

$$\underset{\underset{\textstyle CH_3}{|}}{CH_3CH\text{=}\overset{\cdot\cdot}{N}} \qquad\qquad CH_3CH_2C\text{≡}N:$$

sp² hybridized   *sp hybridized*

Study Problem

2.8 Give the complete structural formula for each of the following compounds and tell which types of orbitals overlap to form each bond:

(a) $(CH_3)_2NCH_2NH_2$ **(b)** $H_2N\overset{\overset{\textstyle NH}{\|}}{C}NHCH_2CN$ **(c)** ⬡NH

B. Water, Alcohols, and Ethers

Like carbon and nitrogen, oxygen forms bonds with sp^3-hybrid orbitals. Because oxygen has six bonding electrons, it forms two covalent bonds and has two filled orbitals.

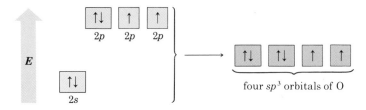

There are a number of classes of organic compounds that contain sp^3-hybridized oxygen atoms. For the present, let us consider just two, alcohols and ethers: ROH and ROR′. The bonding to the oxygen in alcohols and ethers is directly analogous to the bonding in water. In each case, the oxygen is sp^3 hybridized and has two pairs of unshared valence electrons, as is shown in Figure 2.20.

Figure 2.20 Bonding in water, in the alcohol CH_3OH, and in the ether CH_3OCH_3.

C. Carbonyl Compounds

The **carbonyl group** (C=O) contains an sp^2-hybridized carbon atom connected to an oxygen atom by a double bond. It is tempting to think that a carbonyl oxygen is in the sp^2-hybrid state just as the carbonyl carbon is; however, chemists are not truly sure of the hybridization of a carbonyl oxygen because there is no bond angle to measure.

The geometry of a carbonyl group is determined by the sp^2-hybridized carbon. The carbonyl group is *planar* around the trigonal sp^2 carbon. The carbon–oxygen bond contains a pair of *exposed pi electrons*. The oxygen also has *two pairs of unshared electrons*. Figure 2.21 shows the geometry of a carbonyl group.

The carbonyl group:

trigonal and planar   *unshared pairs of e⁻*   *polar*

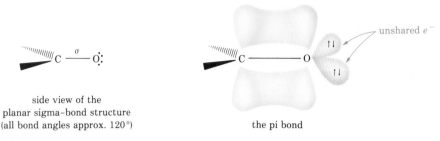

Figure 2.21 Bonding in the carbonyl group.

The carbonyl group is more polar than the C—O group in an alcohol or ether. The probable reason for this enhanced polarity is that the pi electrons are more easily drawn toward the electronegative oxygen than are the C—O sigma electrons.

The carbonyl group is part of a variety of functional groups. The functional group and class of compound are determined by the other atoms bonded to the carbonyl carbon. If one of the atoms bonded to the carbonyl carbon is a hydrogen, then the compound is an **aldehyde**. If two carbons are bonded to the carbonyl carbon, then the compound is a **ketone**. In Chapter 1 you already encountered **carboxylic acids**, in which an OH group is bonded to the carbonyl group. Table 2.3 lists a few classes of compounds containing carbonyl groups.

Sample Problem

Classify each of the following compounds as an aldehyde, a ketone, or a carboxylic acid:

$$(a)\ \langle\ \bigcirc\ \rangle -CO_2H \qquad (b)\ CH_3CH_2\overset{\displaystyle O}{\overset{\|}{C}}H$$

$$(c)\ \langle\ \bigcirc\ \rangle -\overset{\displaystyle O}{\overset{\|}{C}}CH_3 \qquad (d)\ HCHO$$

Solution:

(a) carboxylic acid
(b) aldehyde
(c) ketone
(d) aldehyde

Study Problem

2.9 Write condensed structural formulas for four-carbon compounds that exemplify **(a)** an aldehyde; **(b)** a ketone; **(c)** a carboxylic acid; and **(d)** a carboxylic acid that is also a ketone. (There may be more than one correct answer for each part.)

Table 2.3 Some compound classes that contain carbonyl groups

General Formula	*Name of Class*
$\overset{\displaystyle O}{\overset{\displaystyle \|}{\text{RCH}}}$ or RCHO	aldehyde
$\overset{\displaystyle O}{\overset{\displaystyle \|}{\text{RCR}'}}$	ketone
$\overset{\displaystyle O}{\overset{\displaystyle \|}{\text{RCOH}}}$, RCOOH, or RCO_2H	carboxylic acid
$\overset{\displaystyle O}{\overset{\displaystyle \|}{\text{RCOR}'}}$, RCOOR′, or RCO_2R'	ester
$\overset{\displaystyle O}{\overset{\displaystyle \|}{\text{RCNR}'_2}}$	amide
$\overset{\displaystyle O\ \ O}{\overset{\displaystyle \|\ \ \|}{\text{RCOCR}'}}$	anhydride

Section 2.7
Conjugated Double Bonds

An organic molecule may contain more than one functional group. In most polyfunctional compounds, each functional group is independent of another; however, this is not always the case. Let us consider some compounds with more than one carbon–carbon double bond.

There are two principal ways double bonds can be positioned in an organic molecule. Two double bonds originating at adjacent atoms are called **conjugated double bonds.**

Conjugated double bonds:

$$CH_2{=}CH{-}CH{=}CH_2$$

adjacent carbon atoms

Double bonds joining atoms that are not adjacent are called **isolated**, or **nonconjugated**, double bonds.

Isolated double bonds:

$$CH_2{=}CH{-}CH_2{-}CH{=}CH_2$$

carbon atoms are not adjacent

Study Problems

2.10 Vitamin A_1 has the structure that follows. How many conjugated double bonds does it contain? How many isolated double bonds?

$$H_3C \quad CH_3 \quad \underset{\displaystyle |}{C} \!\!=\!\! \underset{\displaystyle H}{C} \!\!-\!\! \underset{\displaystyle H}{\overset{\displaystyle CH_3}{C}} \!\!=\!\! \underset{\displaystyle H}{C} \!\!-\!\! \underset{\displaystyle H}{C} \!\!=\!\! \underset{\displaystyle H}{\overset{\displaystyle CH_3}{C}} \!\!-\!\! CH_2OH$$

2.11 Draw the condensed structural formula of an eight-carbon compound that has   **(a)** three conjugated double bonds;   **(b)** two conjugated double bonds and one isolated double bond;   **(c)** three isolated double bonds. (There may be more than one correct answer.)

Isolated double bonds behave independently; each double bond undergoes reaction as if the other were not present. Conjugated double bonds, on the other hand, are not independent of each other; there is electronic interaction between them. Let us choose the simplest of the conjugated systems, $CH_2\!\!=\!\!CH\!-\!CH\!\!=\!\!CH_2$, called 1,3-butadiene, to discuss this phenomenon. Figure 2.22 illustrates the *p*-orbital overlap in 1,3-butadiene.

We have numbered the carbon atoms of 1,3-butadiene in Figure 2.22 for reference. There are two pairs of *p* orbitals that form two pi bonds: one pi bond between carbons 1 and 2, and one pi bond between carbons 3 and 4. However, the *p* orbitals of carbons 2 and 3 are also adjacent, and *partial overlap* of these *p* orbitals occurs. Although most of the pi-electron density is located between carbons 1–2 and 3–4, some pi-electron density is also found between carbons 2–3. (Figure 2.22 shows a composite of the bonding orbitals in 1,3-butadiene. For a more-detailed discussion of the π molecular orbitals of this compound, see Section 20.1. Also, refer to Figure 20.1, page 829, for a representation of the four π molecular orbitals.)

We use a number of terms to describe this pi-bond interaction in conjugated systems. We may say that there is **partial *p*-orbital overlap** between the central carbons. We may also say that the bond between carbons 2 and 3 in 1,3-butadiene has **partial double-bond character.** Yet another way of describing the system is to say that the pi electrons are **delocalized,** which means that the pi-electron density is distributed over a somewhat larger region

each C is sp²

top view of planar σ bonds
(*p* orbitals not shown)

side view showing *p* orbitals
(hydrogens not shown)

Figure 2.22 The bonding in 1,3-butadiene, $CH_2\!\!=\!\!CH\!-\!CH\!\!=\!\!CH_2$.

within the molecule. By contrast, **localized electrons** are restricted to two nuclei; nonconjugated double bonds contain localized pi electrons.

Study Problems

2.12 Draw the *p* orbitals of each of the following compounds, using lines to show pi-bond overlap and dotted lines to show partial overlap, as we have done in Figure 2.22.

(a) CH_2=$CHCH_2CH$=$CHCH$=CH_2 (b)

2.13 Write the formula for an open-chain four-carbon ketone in which the carbonyl group is in conjugation with a carbon–carbon double bond.

▼ ## Section 2.8
Benzene

Benzene (C_6H_6) is a cyclic compound with six carbon atoms joined in a ring. Each carbon atom is sp^2 hybridized, and the ring is planar. Each carbon atom has one hydrogen atom attached to it, and each carbon atom also has an unhybridized *p* orbital perpendicular to the plane of the sigma bonds of the ring. Each of these six *p* orbitals can contribute one electron for pi bonding (see Figure 2.23).

With six *p* electrons, benzene should contain three pi bonds. We could draw three pi bonds in the ring one way (formula **A**), or we could draw them another way (formula **B**). However, we might also wonder—could this type of *p*-orbital system lead to *complete* delocalization of all six *p* electrons (formula **C**) instead of just partial delocalization?

sigma-bond structure
(Each C is sp^2 and
has one *p* electron.)

placement of pi bonds
(The circle represents complete delocalization.)

It is known that all carbon–carbon bond lengths in benzene are the same, 1.40 Å. All six bonds are longer than C—C double bonds, but shorter than C—C single bonds. If the benzene ring contained three localized double bonds separated by three single bonds, the bonds would be of different lengths. The fact that all the carbon–carbon bonds in the benzene ring have the same length suggests that the benzene ring does not contain alternate single and double bonds.

From the bond lengths plus a body of other evidence that will be presented in later chapters, chemists have concluded that benzene is a symmetrical molecule and that each of the six ring bonds is like each of the other ring bonds.

side view showing *p* orbitals
(all in phase)

aromatic pi cloud

Figure 2.23 The bonding in benzene, C_6H_6.

Instead of alternate double and single bonds, the six pi electrons are *completely delocalized* in a cloud of electronic charge shaped rather like a pair of donuts. This cloud of pi electrons is called the **aromatic pi cloud** of benzene. Figure 2.23 depicts the lowest-energy bonding orbital of benzene, the one commonly used to represent the aromatic pi cloud. (Figure 11.6, page 472, shows the six π orbitals of benzene.) For most purposes in this text, we will use a hexagon with a circle inside it to represent benzene; the circle represents the aromatic pi cloud.

Benzene is just one member of a class of **aromatic compounds**, compounds that contain aromatic pi clouds. Historically, the term "aromatic" derives from the fact that many of these compounds have distinctive odors. We will discuss aromatic compounds in detail in Chapters 11, 12, and 18.

▼ Section 2.9
Resonance

Methane (CH_4) and ethylene ($CH_2{=}CH_2$) are examples of organic compounds with structures that may reasonably be described using single line-bond formulas. In each case, a line joining two atomic symbols represents a covalent bond between two atoms.

Benzene is a compound that cannot be accurately represented by a single line-bond formula. In benzene, the electron delocalization causes the pi electrons to encompass more than two atoms. (The circle in the hexagon representing the aromatic pi cloud in benzene is a fairly recent addition to organic symbolism.) In order to describe the pi-electron distribution in benzene using classical line-bond formulas, we must use *two* formulas.

Neither structure actually exists—the real structure is a composite of these two resonance structures.

resonance structures for benzene
(Kekulé formulas)

These two line-bond formulas for benzene are called **Kekulé formulas** in honor of Friedrich August Kekulé, who first proposed them in 1872. Kekulé's original proposal was brilliant for its time, but unfortunately incorrect. His idea was that the two structures for benzene shifted back and forth so fast that neither could be isolated independently of the other. We now know that benzene does not shift between two different structures; the real structure of benzene is a composite of the two structures. We say that benzene is a **resonance hybrid** of the two structures, which are called **resonance structures**.

Whenever we can describe a molecular structure by two or more line-bond formulas that differ *only in the positions of the electrons* (usually pi electrons), none of these formulas will be in complete accord with the chemical and physical properties of the compound. If different resonance structures can be written for a compound, we can assume a delocalization of electron density. These statements are true for all aromatic structures, as well as for some other structures we will mention shortly.

The real structure of naphthalene is a composite of the resonance structures:

The real structure of pyridine is a composite of the resonance structures:

The important thing to keep in mind is that resonance symbols are not *true* structures; the true structure is a composite of all the resonance symbols. Many chemists have made the analogy that a rhinoceros (a real animal) might be described as a resonance hybrid of a unicorn (imaginary) and a dragon (imaginary). A rhinoceros does not shift back and forth from unicorn to dragon, but is simply an animal with characteristics of both a unicorn and a dragon.

To show that two or more formulas represent resonance structures (imaginary) and not real structures in equilibrium, we use a double-headed arrow ($\leftrightarrow$). By contrast, we indicate an equilibrium by *two* arrows ($\rightleftarrows$).

Resonance: rhinoceros: unicorn $\longleftrightarrow$ dragon
　　　　　　　　　(*real*)　　　　　(*imaginary*)

　　　　　　　　benzene:
　　　　　　　　(*real*)　　　　　(*imaginary*)

Equilibrium: $2 H_2O \rightleftarrows H_3O^+ + OH^-$
　　　　　　　　(*real*)　　　　(*real*)　　(*real*)

Aromatic compounds are not the only compounds for which single line-bond formulas are inadequate. The **nitro group** ($-NO_2$) is a group of atoms that we can best describe using resonance structures. A single line-bond structure for the nitro group shows two types of N—O bond. However, it is known that the two N—O bonds are the same length, intermediate between the lengths of an N—O single bond and an N=O double bond. We must use two line-bond resonance structures to describe this fact for a compound such as CH_3NO_2. To show that the bonds from nitrogen to the oxygen in the NO_2 group are the same, some chemists represent the nitro group with dotted lines for the partial double bonds.

The **carbonate ion** $(CO_3{}^{2-})$ is an ion that cannot be represented by a single line-bond structure. Each C—O bond in the carbonate ion is the same length. We must use *three* resonance structures to describe the real structure.

A. Electron Shifts

When writing resonance structures, keep in mind that nuclei of the atoms in a molecule cannot change positions—only electrons are delocalized. Also keep in mind that, with rare exceptions, only pi electrons or unshared valence electrons can be delocalized.

In a series of resonance symbols, we often show electron delocalization by small curved arrows that allow us to progress systematically from one resonance symbol to another. These electron shifts are purely artificial because the electrons do not truly shift, but are delocalized. Electron-shift arrows may be drawn only in the following ways:

from a pi bond to an adjacent bond position: =X⤻ ↔ —X=

from a pi bond to an adjacent atom: =Ẍ— ↔ —Ẍ—

from an atom to an adjacent bond position: —Ẍ— ↔ —X=

Check each arrow in the following series of resonance structures to see how these rules are applied.

When using electron-shift arrows, we must pay special attention to how many electrons an atom can accommodate. An atom of the period-2 elements can accommodate a maximum of eight valence electrons. For purposes of keeping track of these electrons, it is advisable to show the unshared electrons in resonance structures.

not possible: N cannot accommodate ten e⁻

In this book, we show the series of resonance structures to be read from left to right. For example, the curved arrows in the following left-hand resonance structure for nitromethane show how the electrons are "shifted" to get to the right-hand resonance structure.

Electron-shift arrows showing how to proceed from the left-hand structure to the right-hand structure:

Some instructors prefer that curved arrows be shown on all resonance structures so that the series can be read in either direction—from right to left or from left to right.

Electron-shift arrows showing how to proceed in either direction:

Sample Problem

Fill in the electron-shift arrows for the resonance structures of (a) pyrimidine and (b) the acetate ion.

Solution:

(a) Write the first structure. To arrive at the second structure above, shift each double bond to the adjacent bond position, just as for benzene. Draw the curved arrows one at a time. (It does not matter where you begin in this structure.)

(b) Write the first structure. Draw the arrows one at a time. Then draw the second structure.

Study Problem

2.14 The nitrate ion, NO_3^-, contains three equivalent N—O bonds. Write resonance structures for the nitrate ion.

B. Major and Minor Contributors

In each example we have shown so far, the bonding has been the same in each resonance symbol. When the bonding is the same, the resonance structures are of equal energy and are equivalent to each other. *Equivalent resonance structures contribute equally to the real structure.*

Dissimilar resonance structures for a compound or ion may not all contribute equally to the real structure. Consider the following example:

formaldehyde

This pair of resonance structures shows that electron shifts can lead to an atom with an incomplete octet (the C in the second resonance structure). Here, the more electronegative oxygen atom becomes electron-rich and the less electronegative carbon atom becomes electron-poor—the second resonance structure shows a charge separation.

In general, resonance structures in which all period-2 elements have octets of valence electrons make a greater contribution to the real composite structure than a resonance structure with incomplete octets. Also, a resonance structure with a lesser amount of charge separation makes a greater contribution to the real structure. Thus, we say that the left-hand structure shown for formaldehyde is a *more important resonance structure*, or a *major contributor* to the real structure, while the right-hand structure is a *less important resonance structure*, or a *minor contributor*. (This is similar to saying that a rhinoceros is like a dragon with just a little unicorn thrown in.)

Formaldehyde is like   H—C—H   with a small amount of   H—C—H thrown in.

C. Summary of Rules for Writing Resonance Structures

We will encounter resonance structures frequently in the coming chapters. Let us summarize the rules for writing resonance structures:

---

1. *Only pi electrons or unshared pairs of valence electrons (not sigma electrons or nuclei) may be shifted*, and they may be shifted only to adjacent atoms or bond positions.
2. Resonance structures in which an atom carries more than its quota of electrons (eight for the period-2 elements) are not contributors to the real structure.
3. The more important resonance structures show each atom with a complete octet and as little charge separation as possible.

     $CH_3\overset{+}{N}$ ⟷ $CH_3\overset{2+}{N}$

    more important because   *less important because N has*
    each atom has an octet   *only six electrons and because*
                        a greater charge separation exists

Sample Problem

Which of the following resonance structures is the major contributor to the real structure?

   $CH_3C—OCH_3$ ⟷ $CH_3C—OCH_3$ ⟷ $CH_3C=OCH_3$

Solution: The first structure is the major contributor because each C and O atom has an octet and no C or O atom has a formal charge. The second and third structures are less important contributors because they contain charge separations and the second structure contains a C atom with an incomplete octet.

Study Problem

2.15 Which of the following resonance structures is the major contributor to the real structure? Explain your answer.

$$:\overset{..}{\underset{..}{O}}:^- \qquad \qquad :\overset{..}{\underset{|}{O}}:^- \qquad \qquad :\overset{..}{\underset{||}{O}}$$

$$CH_3\overset{|}{C}=CH\overset{\frown}{}\overset{+}{C}H_2 \quad \longleftrightarrow \quad CH_3\overset{|}{\underset{+}{C}}-CH=CH_2 \quad \longleftrightarrow \quad CH_3\overset{||}{C}-CH=CH_2$$

D. Resonance Stabilization

If a structure is a resonance hybrid of two or more resonance structures, *the energy of the real structure is lower than that of any single resonance structure.* The real structure is said to be **resonance-stabilized**. In most cases, the energy difference is slight, but for aromatic systems, like benzene or naphthalene, the energy difference is substantial. (The reasons for aromatic stabilization will be discussed in Chapter 11.)

The hypothetical structure $\overset{O}{\underset{^-O \quad O^-}{||}}C$ is of higher energy than $\overset{O^{\delta-}}{\underset{\delta^- O \quad O^{\delta-}}{||}}C$

The hypothetical structure ⬡ is of substantially higher energy than ⬡

Resonance stabilization is most important if two or more resonance structures for a compound are *equivalent or nearly equivalent in energy.* A high-energy, minor-contributing resonance structure adds little stabilization.

$$\overset{\cdot\overset{..}{O}\cdot}{\underset{||}{H-C-H}} \quad \text{gains only a small amount of stabilization from} \quad \overset{:\overset{..}{O}:^-}{\underset{+}{H-C-H}}.$$

The reason for the energy differences between hypothetical resonance structures and the real structure of a compound is not entirely understood. Certainly, part of the reason is that a delocalized electron is attracted to more than one nucleus. It is generally true that a system with delocalization of electrons or of electronic charge has lower energy and greater stability than a similar system with localized electrons or electronic charge.

The principal reason that carboxylic acids are acidic is that the carboxylate ion is stabilized by delocalization of the negative charge.

$$\overset{\cdot\overset{..}{O}\cdot}{\underset{||}{CH_3\overset{\frown}{C}O}}-H + H_2O \quad \rightleftharpoons \quad \left[\overset{:\overset{..}{O}}{\underset{||}{CH_3\overset{}{C}}}\overset{\frown}{\underset{..}{O}}:^- \quad \longleftrightarrow \quad \overset{:\overset{..}{O}:^-}{\underset{|}{CH_3\overset{}{C}=\overset{..}{O}:}} \right] + H_3O^+$$

*In the carboxylate ion, the − charge
is shared by both O atoms
(two equivalent resonance structures).*

Sample Problem

Which of the following compounds are resonance-stabilized? (*Note*: Because the valence electrons of sulfur are in the third energy level, a sulfur atom can accommodate more than eight electrons.)

(a) $CH_3CH_2CH_3$　　(b)

$$\underset{H}{\overset{H}{\diagdown}} C = C \underset{H}{\overset{CH_3}{\diagup}}$$

(c)

$$CH_3O-\underset{O}{\overset{O}{\underset{\|}{\overset{\|}{S}}}}-O^-$$

Solution:

(a) is not resonance-stabilized because it contains no pi electrons.
(b) is not resonance-stabilized because there is no position in the molecule where the pi electrons can be delocalized.
(c) is resonance-stabilized:

$$CH_3OS\overset{\ddot{O}}{\underset{\ddot{O}\colon}{\|}}\ddot{O}\colon^- \longleftrightarrow CH_3OS\overset{\ddot{O}}{\underset{\colon\ddot{O}\colon^-}{\|}}=\ddot{O}\colon \longleftrightarrow CH_3OS\overset{\colon\ddot{O}\colon^-}{\underset{.\ddot{O}.}{\|}}=\ddot{O}\colon$$

Study Problem

2.16 (1) Write resonance structures for the following ions. (2) Indicate which, if any, of the resonance structures is the major contributor or whether the structures are of similar energy. [In **(b)**, **(e)**, and **(f)**, use Kekulé formulas.]

(a) $CH_3CH=CH\overset{+}{C}HCH_3$

(b) 〈 ◯ 〉$-\overset{+}{C}H_2$

(c) $CH_3\overset{\ddot{O}}{\underset{\|}{C}}\bar{C}H_2$

(d) $\colon\bar{C}H_2C\equiv N\colon$

(e) 〈 ◯ 〉$-\ddot{O}\colon^-$

(f) 〈 ◯ 〉$-\overset{\ddot{O}}{\underset{\|}{C}}\ddot{O}\colon^-$

▼ Summary

A covalent bond is the result of two atomic orbitals overlapping to form a lower-energy **bonding molecular orbital** in which two electrons are paired.

The four atomic orbitals of C undergo **hybridization** in bond formation:

sp^3 hybridization: four equivalent single bonds (*tetrahedral*).

sp^2 hybridization: three equivalent single bonds (*trigonal*) plus one unhybridized p orbital.

sp hybridization: two equivalent single bonds (*linear*) plus two unhybridized p orbitals.

Overlap of a hybrid orbital with an orbital of another atom (end-to-end overlap) results in a **sigma bond**. Overlap of a p orbital with a parallel p orbital of another atom (side-to-side overlap) results in a **pi bond**.

Each bonding molecular orbital has an **antibonding molecular orbital** associated with it. Antibonding orbitals are of higher energy than either bonding orbitals or the atomic orbitals that went into their formation. In the ground state, molecules have their electrons in the lowest-energy orbitals, usually the bonding orbitals.

A **functional group** is a *site of chemical reactivity* in a molecule and arises from a *pi bond* or from *differences in electronegativity* between bonded atoms. A double bond or a triple bond is a functional group. The following general formulas show examples of other functional groups:

Pi bonds extending from adjacent carbon atoms are called **conjugated double bonds**. In the case of benzene and other aromatic compounds, complete delocalization of pi electrons results in the **aromatic pi cloud**.

Resonance structures can be used to show delocalization of p electrons.

If all resonance structures are not equivalent, the lowest-energy resonance structure is the **major contributor**.

$$CH_3-\overset{\overset{\displaystyle \cdot\cdot}{\overset{\displaystyle O:}{\|}}}{C}-CH_3 \quad\longleftrightarrow\quad CH_3-\overset{\overset{\displaystyle :\overset{\cdot\cdot}{O}:^-}{|}}{\underset{+}{C}}-CH_3$$

major contributor

Delocalization of pi electrons in a molecule, or of electronic charge in an ion, results in a slight increase in the stability of the system. However, for aromatic compounds, the increase in stability is substantial.

Study Problems*

2.17 Do the following structures represent the same or different compounds? Explain.

$$\underset{H}{\overset{Br}{H-C-Br}} \qquad \underset{Br}{\overset{Br}{H-C-H}} \qquad \underset{H}{\overset{H}{Br-C-Br}}$$

2.18 Draw a structure that shows the *p*-orbital overlap forming the pi bonds in each of the following compounds. (Use lines for sigma bonds.)

(a) $(CH_3)_2C{=}C(CH_3)_2$ (b) $CH_3C{\equiv}CCH_3$

2.19 Indicate the hybridization of each carbon in the following structures:

(a) $CH_3OCH_2CH_3$ (b) $HC{\equiv}C\overset{\overset{\displaystyle CH_2}{\|}}{C}CH_3$ (c)

2.20 In each of the following structures, do the two indicated carbons lie in the same plane as each other and the double-bonded carbons?

(a) $\underset{H_3C}{\overset{H_3C}{>}}C{=}C\underset{H}{\overset{CH_3}{<}}$ (b) $\underset{H_3C}{\overset{H_3C}{>}}C{=}C\underset{H}{\overset{CH_2CH_3}{<}}$ (c)

2.21 What would be the expected *sigma-bond angles* in the following structures?

$$\overset{(a)}{\diagdown}C{\equiv}C\underset{(b)}{-}H \qquad CH_3\overset{(c)}{\underset{(f)\ }{\overset{H}{\diagup}}}\overset{(d)}{C}{=}CH-\overset{H}{\underset{H}{C}}\overset{(e)}{-}H$$

2.22 Show the structure of each of the following molecules or ions. Use lines to represent sigma bonds and dumbbell shapes to represent *p* orbitals. Show *p*-orbital overlap with

---

* For information concerning the organization of the *Study Problems* and *Additional Problems*, see the *Note to student* on page 39.

dashed lines. For example, CH_2=CH_2:

(a) H_3O^+ (b) HCN   (c) OH^- (d) NH_2NH_2
(e) $(CH_3)_4N^+$ (f) $^-OCH_3$ (g) CH_2=CHC≡CH

2.23 The molecular-orbital diagram that follows is for the carbon–carbon double bond in cyclohexene.

cyclohexene E σ^* ___ π_2^* ___ π_1 ___ σ ___

(a) Indicate by arrows which orbitals contain the double-bond electrons when cyclohexene is in the ground state.
(b) Absorption of ultraviolet light by a double bond results in the promotion of an electron to a higher-energy orbital. Which electron transition requires the least energy when cyclohexene goes to an excited state?

2.24 In each of the following structures or pairs of structures, which of the indicated bonds (1 or 2) is the shorter one?

(a) Cl—①—CH_3,   ②—Cl   (b) $\underset{H}{\overset{H}{C}}$=CH—C≡C—②—H

(c) —①—CH_3,   —②—CH_3 (d) CH_3—①—OCH_3, $CH_3\overset{O}{\overset{\|}{C}}CH_3$ ②

2.25 *Piperazine* is the drug of choice for treatment of patients with roundworms or pinworms. *Pyridine* is a pungent-smelling aromatic compound used as a solvent and reagent in organic laboratories.

(a) HN   NH   (b)

piperazine   pyridine

(1) Draw the complete structural formulas of these compounds, showing each atom and using a line for each covalent bond.
(2) Indicate unshared electron pairs by dots.
(3) What is the hybridization of each ring atom?

2.26 Circle and name the functional groups in the following structures:

(a) $(CH_3)_2C{=}CHCH_2CH_2\overset{\overset{\displaystyle CH_3}{|}}{C}HCH_2\overset{\overset{\displaystyle O}{\|}}{C}H$

citronellal

chief constituent of citronella oil;
used in perfumes and as a mosquito repellent

(b) $HOCH_2\overset{\overset{\displaystyle O}{\|}}{C}CH_2OH$

dihydroxyacetone

formed in the metabolism of carbohydrates;
also used in the tanning industry

(c) estradiol

a female hormone

(d) cortisone

an adrenal steroid
used to treat arthritis

2.27 Give a three-carbon open-chain structure to illustrate each of the following compound types: **(a)** alkene; **(b)** alkyne; **(c)** ether; **(d)** alcohol; **(e)** amine; **(f)** ketone; **(g)** aldehyde; **(h)** carboxylic acid. (There may be more than one correct answer in each case.)

2.28 Redraw each structure to emphasize the carbonyl group:

(a) $CH_3CH_2CH_2CO_2H$

butanoic acid

foul-smelling compound
found in rancid butter

(b) vanillin

vanilla flavoring

(c) $CH_3CO_2CH_2CH_2CH(CH_3)_2$

3-methylbutyl acetate

banana oil

2.29 Write a condensed structural formula for each of the following compounds. (There may be more than one answer.)
(a) a ketone, C_4H_8O **(b)** a cyclic alcohol, $C_6H_{12}O$
(c) an alkene, C_5H_{10} **(d)** an aldehyde, C_3H_6O

2.30 Indicate whether each of the following compounds contains *delocalized* or *localized* pi electrons:

(a) **(b)** $CH{\equiv}C\overset{\overset{\displaystyle O}{\|}}{C}H$ **(c)** $CH_2{=}CHCH_2\overset{\overset{\displaystyle O}{\|}}{C}H$

2.31 Give the structure of an open-chain five-carbon compound with **(a)** two double bonds in conjugation; **(b)** two isolated double bonds. (More than one answer may be possible.)

2.32 In the following structure for cyclohexanone,   (**a**) show the placement of a C—C double bond in conjugation with the carbonyl group, and   (**b**) indicate the hybridization of each carbon atom in this structure.

cyclohexanone

2.33 Which of the following pairs of structures represent resonance symbols for a molecule or ion?

(**a**) $(CH_3)_2CHC\overset{\overset{O}{\|}}{}O^-$, $(CH_3)_2CH\overset{\overset{O^-}{|}}{C}=O$ (**b**) $CH_3\overset{\overset{+OH}{\|}}{C}CH_3$, $CH_3\overset{\overset{OH}{|}}{C}CH_3$

(**c**) $CH_3\overset{\overset{O}{\|}}{C}CH_3$, $CH_3\overset{\overset{OH}{|}}{C}=CH_2$ (**d**)

(**e**) $CH_2=C=O$, $CH\equiv C-OH$

2.34 Give the important resonance contributors for each of the following structures:

(**a**) $H\overset{\overset{O}{\|}}{C}O^-$ (**b**) $HO\overset{\overset{O}{\|}}{C}O^-$ (**c**)

(**d**) $CH_3\overset{+}{C}HCH=CH_2$ (**e**) $CH_2=CHCH=CH\overset{+}{C}H_2$

(**f**) $CH_3\overset{\overset{O}{\|}}{C}\ddot{C}H_2^-$ (**g**) $:N\equiv C\overset{-}{\underset{\cdot\cdot}{C}}HC\equiv N:$

(**h**) $^-O-$$-\overset{\overset{O}{\|}}{C}H$ (**i**)

2.35 Which of the following resonance structures would be the major contributor to the real structure? Explain.

(**a**)

(**b**)

(**c**)

2.36 Write the important Kekulé resonance-contributing structures for each of the following aromatic compounds.

(a) **(b)** **(c)**

2.37 Which ion in each of the following pairs would be more resonance-stabilized? Explain your answers with structures.

(a) ⬡—$\overset{+}{\overset{\cdot\cdot}{C}}HCH_3$ or ⬡—$CH_2\overset{+}{C}H_2$

(b) $CH_3CH_2\overset{+}{C}HCH_3$ or $CH_2{=}CH\overset{+}{C}HCH_3$

(c) $:\overset{-}{C}H_2CH_3$ or $:\overset{-}{C}H_2NO_2$

Additional Problems

2.38 *Allene* has the following structure: $CH_2{=}C{=}CH_2$. **(a)** What is the hybridization of each carbon? **(b)** What would be the expected angle between the two pi bonds? **(c)** Are the two pi bonds conjugated? **(d)** Can electronic charge be delocalized throughout both pi bonds?

2.39 The O—C—O bond angle in CO_2 is 180°. Write the Lewis structure and draw the *p*-orbital picture. What is the hybridization of the carbon?

2.40 What are the electron distributions in the bonding and antibonding orbitals of H_2^+ and H_2^-?

2.41 *Triplet methylene* is a very unstable species with the formula $:CH_2$. Assume that the H—C—H bond angle is 180°. Draw the orbital picture (using lines for sigma bonds). Indicate the hybridization state of the carbon atom.

2.42 In solution, the amine **A** is not basic, as expected, but is practically neutral. One suggestion for the lack of basicity of **A** is that it exists in form **B** as well as in form **A**. **(a)** Are forms **A** and **B** resonance structures? Explain. **(b)** Are forms **A** and **B** interconvertible by electron shifts? If so, show these shifts with curved arrows.

A **B**

2.43 *Guanidines* and *amidines* are classes of compounds that react as bases. Draw the resonance structures for the conjugate acids—that is, the cations.

(a) $CH_3\overset{\cdot\cdot}{N}H-\overset{\overset{\displaystyle :NH}{\|}}{C}-\overset{\cdot\cdot}{N}H_2 + H^+ \rightleftharpoons CH_3\overset{\cdot\cdot}{N}H-\overset{\overset{\displaystyle +NH_2}{\|}}{C}-\overset{\cdot\cdot}{N}H_2$

    a guanidine

(b) $CH_3-\overset{\overset{\displaystyle :NH}{\|}}{C}-\overset{\cdot\cdot}{N}H_2 + H^+ \rightleftharpoons CH_3-\overset{\overset{\displaystyle +NH_2}{\|}}{C}-\overset{\cdot\cdot}{N}H_2$

    an amidine

3 ▼

Structural Isomerism, Nomenclature, and Alkanes

In Chapter 2, we discussed the bonding of several compounds that contain only carbon and hydrogen. A compound containing only these two elements is called a **hydrocarbon**. Methane (CH_4), ethylene ($CH_2{=}CH_2$), and benzene (C_6H_6) are all examples of hydrocarbons.

Hydrocarbons with only sp^3 carbon atoms (that is, only single bonds) are called **alkanes** (or **cycloalkanes** if the carbon atoms are joined in rings). Some typical alkanes are methane, ethane (CH_3CH_3), propane ($CH_3CH_2CH_3$), and butane ($CH_3CH_2CH_2CH_3$). All these alkanes are gases found in petroleum deposits and used as fuels. Gasoline is primarily a mixture of alkanes. At the end of this chapter, we will discuss some of these aspects of hydrocarbon chemistry.

Alkanes and cycloalkanes are said to be **saturated hydrocarbons**, meaning "saturated with hydrogen." These compounds do not undergo reaction with hydrogen. Compounds containing pi bonds are said to be **unsaturated**; under the proper reaction conditions, they undergo reaction with hydrogen to yield saturated products.

Saturated hydrocarbons:

$$CH_4 \; + H_2 \xrightarrow{\text{catalyst}} \text{no reaction}$$

methane
an alkane

$+ H_2 \xrightarrow{\text{catalyst}} \text{no reaction}$

cyclohexane
a cycloalkane

Unsaturated hydrocarbons:

$$CH_2{=}CH_2 + H_2 \xrightarrow{\text{Ni catalyst}} CH_3CH_3$$

ethylene   ethane

$$CH_3C{\equiv}CH + 2\,H_2 \xrightarrow{\text{Ni catalyst}} CH_3CH_2CH_3$$

propyne   propane

$+ 3\,H_2 \xrightarrow[\text{heat, pressure}]{\text{Ni catalyst}}$

benzene   cyclohexane

In this chapter, we will discuss primarily alkanes and cycloalkanes—the saturated hydrocarbons. Because saturated hydrocarbons lack a true functional group, their chemistry is not typical of most organic compounds; however, these compounds provide the carbon skeletons of organic compounds that do contain functional groups. Therefore, the discussion of saturated hydrocarbons provides an excellent opportunity for introducing the variations in the structures of organic compounds and the naming of organic compounds.

Section 3.1
Structural Isomers

Variations in the structures of organic compounds can arise from different numbers of atoms or types of atoms in molecules. However, variations in structure can also arise from *the order in which the atoms are bonded to one another in a molecule.* For example, we can write two different structural formulas for the molecular formula C_2H_6O. These two structural formulas represent two different compounds: dimethyl ether (bp $-23.6°$), a gas that has been used as a refrigerant and as an aerosol propellant, and ethanol (bp $78.5°$), a liquid that is used as a solvent and in alcoholic beverages.

$$CH_3OCH_3 \qquad\qquad CH_3CH_2OH$$

dimethyl ether   ethanol

a gas at room temperature   *a liquid at room temperature*

Two or more different compounds that have the same molecular formula are called **isomers** of each other. If the compounds with the same molecular formula have their atoms bonded in different orders, they have different structures and are said to be **structural isomers** of each other. (We will encounter other types of isomerism later.) Dimethyl ether and ethanol are examples of a pair of structural isomers.

Alkanes that contain three or fewer carbons have no isomers. In each case, there is only one possible way in which the atoms can be arranged.

No isomers: CH_4 CH_3CH_3 $CH_3CH_2CH_3$

methane   ethane   propane

The four-carbon alkane (C_4H_{10}) has two possibilities for arrangement of the carbon atoms. As the number of carbon atoms increases, so does the number of isomers. The molecular formula C_5H_{12} represents three structural isomers; C_6H_{14} represents five isomers; and $C_{10}H_{22}$ represents 75 isomers!

Structural isomers for C_4H_{10}:

$CH_3CH_2CH_2CH_3$

butane, bp $-0.5°$

a continuous-chain alkane

$$\overset{\displaystyle CH_3}{\overset{|}{CH_3CHCH_3}}$$

methylpropane, bp $-12°$

a branched-chain alkane

Sample Problem

Write formulas for the three structural isomers of C_5H_{12}.

Solution:

(a) $CH_3CH_2CH_2CH_2CH_3$ (b) $CH_3CH_2\underset{\underset{\displaystyle CH_3}{|}}{C}HCH_3$ (c) $CH_3\underset{\underset{\displaystyle CH_3}{|}}{\overset{\overset{\displaystyle CH_3}{|}}{C}}CH_3$

You might have listed $CH_3\underset{\underset{\displaystyle CH_3}{|}}{C}HCH_2CH_3$. This structure is the same as (b).

Different positions of a functional group in a molecule also lead to structural isomerism. The alcohols 1-propanol and 2-propanol are structural isomers with slightly different properties. The alkenes 1-butene and 2-butene are also structural isomers with different properties.

$CH_3CH_2CH_2OH$ $CH_3\overset{\overset{\displaystyle OH}{|}}{C}HCH_3$

1-propanol   2-propanol
bp 97°   bp 82°

$CH_2{=}CHCH_2CH_3$ $CH_3CH{=}CHCH_3$

1-butene   2-butene
bp $-6.3°$ bp 3.7°

Study Problems

3.1 Draw the formula of a structural isomer for each of the following compounds. (There is more than one answer for each.)

(a) $CH_3CH_2\overset{\displaystyle O}{\overset{\displaystyle \|}{C}}H\underset{\displaystyle CH_3}{|}CH$ (b) [cyclohexane ring with OH and CH_3 on same carbon]   (c) $CH_2{=}CHCH_2OH$

3.2 For each of the following compounds, write a structural formula of an isomer that has a *different functional group*. (There may be more than one correct answer.)

(a) $CH_3\overset{\displaystyle O}{\overset{\displaystyle \|}{C}}OH$ (b) [cyclohexane ring]$-OCH_3$ (c) $CH_3CH_2\overset{\displaystyle O}{\overset{\displaystyle \|}{C}}CH_2CH_3$

Example: An isomer of $CH_3CH_2CH_2OH$ that has a different functional group is $CH_3OCH_2CH_3$, but not $(CH_3)_2CHOH$, which is an isomer with the *same* functional group.

A. Isomers or Not?

Molecules can move around in space and twist and turn in "snake-like motion," as Kekulé once described it. (Kekulé, you will recall, was the chemist who proposed a structure for benzene.) We may write the same structure on paper in a number of ways. The *order of attachment* of the atoms is the factor that determines if two structural formulas represent isomers or the same compound. For example, all the following formulas show the same order of attachment of atoms; they all represent the same compound and do not represent isomers.

All represent the same compound:

$CH_3CH_2\overset{\displaystyle OH}{\overset{\displaystyle |}{C}}HCH_3$ $CH_3\overset{\displaystyle CH_3CH_2}{\overset{\displaystyle |}{C}}HOH$ $\overset{\displaystyle CH_3CH_2}{\underset{\displaystyle CH_3}{{\diagdown}CHOH{\diagup}}}$ $CH_3\overset{\displaystyle OH}{\overset{\displaystyle |}{C}}HCH_2CH_3$

Sample Problem

Which of the following pairs of formulas represent isomers and which represent the same compound?

(a) $CH_3\overset{\displaystyle Cl}{\overset{\displaystyle |}{C}}HCHCl_2$ and $Cl_2CH\overset{\displaystyle Cl}{\overset{\displaystyle |}{C}}HCH_3$ (b) [cyclohexene ring] and [cyclohexene ring]

(c) $Cl{-}$[benzene ring]$-Cl$ and [benzene ring with Cl at top and Cl at bottom]

$$\overset{\displaystyle OH}{|}$$

(d) $(CH_3)_3C\overset{|}{C}HCH_3$ and $(CH_3)_3CCH_2CH_2OH$

Solution: (c) and (d) represent pairs of structural isomers. In (a) and (b) the structures are oriented differently, but the order of attachment of the atoms is the same; therefore, the formulas in (a) and (b) represent the same compound.

B. A Ring or Unsaturation?

Given a molecular formula for a hydrocarbon, we can often deduce a reasonable amount of information about its structure. For example, all acyclic (noncyclic) alkanes have the general formula C_nH_{2n+2}, where n = the number of carbon atoms in the molecule. Propane ($CH_3CH_2CH_3$, or C_3H_8) has three carbon atoms ($n = 3$). The number of hydrogen atoms in propane is $2n + 2$, or eight. Try the formula on the following alkanes:

$$CH_4 \qquad CH_3\overset{\overset{\displaystyle CH_3}{|}}{C}HCH_3 \qquad CH_3CH_2CH_2\overset{\overset{\displaystyle CH_3}{|}}{C}HCH_3$$

The presence of a ring or a double bond reduces the number of hydrogens in the formula by two for each double bond or ring; that is, a compound with the general formula C_nH_{2n} contains one **double-bond equivalent**—either one double bond or one ring.

$$CH_3CH_2CH_2CH_3 \qquad CH_3CH_2CH=CH_2 \qquad \square$$
$$C_4H_{10} \qquad\qquad C_4H_8 \qquad\qquad C_4H_8$$
$$C_nH_{2n+2} \qquad\qquad C_nH_{2n} \qquad\qquad C_nH_{2n}$$

| *an alkane* | *an alkene* | *a cycloalkane* |

A compound with the general formula C_nH_{2n-2} might have one triple bond, two rings, two double bonds, or one ring plus one double bond. We can say that the formula C_nH_{2n-2} represents a structure with two double-bond equivalents.

Three of many possible structural isomers for C_8H_{14}:

$$CH_3(CH_2)_5C\equiv CH$$

one triple bond   *two rings*   *one double bond and one ring*

Study Problem

3.3 What can you deduce about the possible structures represented by each of the following molecular formulas?

(a) $C_{16}H_{34}$ (b) $C_{10}H_{20}$ (c) C_5H_8

Section 3.2
How Organic Nomenclature Developed

In the middle of the 19th century, the structures of many organic compounds were unknown. At that time, compounds were given names that were illustrative of their origins or their properties. Some compounds were named after friends or relatives of the chemists who first discovered them. For example, the name *barbituric acid* (and hence the common drug classification *barbiturates*) comes from the woman's name Barbara. At one time, the carboxylic acid HCO_2H was obtained from the distillation of red ants. This acid was given the name *formic acid* from the Latin *formica*, which means "ants." These names are both called **trivial names** or **common names**. In many respects, these trivial names are like nicknames; both these compounds have more formal (but seldom-used) names.

Even today, trivial names are coined for new compounds, especially for compounds with unwieldy formal names. Three examples of compounds with descriptive trivial names follow.

cubane   prismane   basketene

Faced with the specter of an unlimited number of organic compounds, each with its own quaint name, organic chemists in the late 19th century decided to systematize organic nomenclature to correlate the names of compounds with their structures. The system of nomenclature that has been developed is called the **IUPAC system**, for the *International Union of Pure and Applied Chemistry*, the organization responsible for the continued development of organic nomenclature. Other systems of nomenclature, related to the IUPAC system, are now being developed to ease the computer-indexing of organic names. However, all the formal systems of nomenclature are very similar.

In the next section, we will present a brief survey of the IUPAC nomenclature system, along with some frequently used trivial names. In future chapters we will cover the nomenclature of each class of compounds in more detail. A more complete outline of organic nomenclature is presented in the appendix.

Section 3.3
A Survey of Organic Nomenclature
A. Continuous-Chain Alkanes

The IUPAC system of nomenclature is based upon the idea that the structure of an organic compound can be used to derive its name and, in turn, that a unique structure can be drawn for each name. The foundations of the IUPAC system are the names of the continuous-chain alkanes. The structures and names of the first ten continuous-chain alkanes are shown in Table 3.1. (Alkanes with a greater number of carbons are listed in Table A1 in the appendix.)

Table 3.1 The first ten continuous-chain alkanes

Number of Carbons	Structure	Name
1	CH_4	methane
2	CH_3CH_3	ethane
3	$CH_3CH_2CH_3$	propane
4	$CH_3(CH_2)_2CH_3$	butane
5	$CH_3(CH_2)_3CH_3$	pentane
6	$CH_3(CH_2)_4CH_3$	hexane
7	$CH_3(CH_2)_5CH_3$	heptane
8	$CH_3(CH_2)_6CH_3$	octane
9	$CH_3(CH_2)_7CH_3$	nonane
10	$CH_3(CH_2)_8CH_3$	decane

The compounds in Table 3.1 are arranged so that each compound differs from its neighbors by only a methylene (CH_2) group. Such a grouping of compounds is called a **homologous series**, and the compounds in such a list are called **homologs**.

From Table 3.1, you can see that all the alkane names end in **-ane**, which is the IUPAC ending denoting a saturated hydrocarbon. The first parts of the names of the first four alkanes (methane through butane) are derived from the traditional trivial names. The higher alkane names are derived from Greek or Latin numbers; for example, pentane is from the Greek *penta*, "five."

Let us briefly consider the derivations of the names for the first four alkanes. *Methane* (CH_4) is named after methyl alcohol (CH_3OH). *Methyl*, in turn, is a combination of the Greek words *methy* (wine) and *hyle* (wood). Methyl alcohol can be prepared by heating wood in the absence of air. Even today, this alcohol is sometimes referred to as wood alcohol.

The name *ethane* (CH_3CH_3) is derived from the Greek word *aithein*, which means "to kindle or blaze." Ethane is quite flammable. The name for *propane* ($CH_3CH_2CH_3$) is taken from the trivial name for the three-carbon carboxylic acid, propionic acid ($CH_3CH_2CO_2H$). *Propionic* is a combination of the Greek *proto* (first) and *pion* (fat). Propionic acid is the first (or lowest-molecular-weight) carboxylic acid to exhibit properties of fatty acids, which are acids that can be obtained from fats. *Butane* ($CH_3CH_2CH_2CH_3$) is named after butyric acid—the odorous component of rancid butter (Latin *butyrum*, "butter").

B. Cycloalkanes

Cycloalkanes are named according to the number of carbon atoms in the ring, with the prefix **cyclo-** added.

cyclopentane   cyclohexane   cyclooctane

C. Side Chains

When alkyl groups or functional groups are attached to an alkane chain, the continuous chain is called the **root**, or **parent**. The groups are then designated in the name of the compound by prefixes and suffixes on the name of the parent.

A **side chain**, or **branch**, is an alkyl group branching from a parent chain. A continuous-chain alkyl group is named after its own alkane parent with the **-ane** ending changed to **-yl**. (CH_4 is methane; therefore, the CH_3- group is the *methyl group*. CH_3CH_3 is ethane; therefore, the CH_3CH_2- group is the *ethyl group*.) The names for the first five continuous-chain alkyl groups are listed in Table 3.2.

How is the name of a side chain incorporated into the name of a compound? To illustrate the technique, we will use hexane as the parent. If there is a methyl group on the second carbon of the hexane chain, the compound is named 2-methylhexane; *2-* for the position of attachment on the parent, *methyl* for the branch at this position, *hexane* for the parent. Methylhexane is one word.

Sample Problem

Why is the following compound *not* named 1-methylhexane? What is its correct name?

$$CH_2CH_2CH_2CH_2CH_2CH_3$$
$$|$$
$$CH_3$$

Solution: The structure contains a continuous chain of seven carbons; it is called *heptane*.

Table 3.2 The first five continuous-chain alkyl groups

Structure	Name
CH_3-	methyl
CH_3CH_2-	ethyl
$CH_3CH_2CH_2-$	propyl
$CH_3(CH_2)_2CH_2-$	butyl
$CH_3(CH_2)_3CH_2-$	pentyl

The general procedure to be followed in naming branched alkanes follows:

1. Find the longest continuous chain (the parent chain), which may or may not be shown in a straight line, and name this chain.
2. Number the parent chain, starting at the end closer to the branch.
3. Identify the branch and its position.
4. Attach the number and the name of the branch to the name of the parent.

Sample Problems

Name the following compound:

$$CH_3CH_2CHCH_2CH_2CH_2CH_3$$
$$\quad\quad\quad | $$
$$\quad\quad CH_2CH_2CH_3$$

Solution: The longest continuous chain is eight carbons. The parent is *octane*.

numbering:

$$
\begin{array}{ccccccc}
 & & ④ & ⑤ & ⑥ & ⑦ & ⑧ \\
C & C & C & C & C & C & C \\
 & & | & & & & \\
 & & C & C & C & & \\
 & & ③ & ② & ① & &
\end{array}
$$

start at the end closer to the branch

alkyl group: ethyl at carbon 4

name: 4-ethyloctane

Write the structure for 3-methylpentane.

Solution:

the parent chain: C—C—C—C—C   (pentane)

numbering: C—C—C—C—C
         ① ② ③ ④ ⑤

the alkyl group:
$$
\begin{array}{c}
CH_3 \\
| \\
C-C-C-C-C
\end{array}
$$
   (3-methyl)

add the hydrogens (each carbon must have four bonds):

$$
\begin{array}{c}
CH_3 \\
| \\
CH_3CH_2CHCH_2CH_3
\end{array}
$$

Name the following structure. In this name, a prefix number is not needed. Why not?

$$(CH_3)_2CHCH_2CH_3$$

Solution: Methylbutane. A prefix number is not needed because there is only one methylbutane:

$$
\begin{array}{ccc}
CH_3 & & CH_3 \\
| & & | \\
CH_3CHCH_2CH_3 & \text{is the same as} & CH_3CH_2CHCH_3
\end{array}
$$

The only other place to attach the methyl group would be on an end carbon; however, this compound is pentane, not methylbutane:

$$CH_2CH_2CH_2CH_3$$
$$|$$
$$CH_3$$

D. Branched Side Chains

An alkyl group may be branched, rather than a continuous chain. The following examples show branched side chains on a cyclohexane ring and on a heptane chain, respectively.

$$\overset{CH_3}{\underset{}{|}}$$
⬡—CHCH$_2$CH$_3$

$$\overset{CH_3CHCH_3}{\underset{}{|}}$$
$$CH_3CH_2CH_2CHCH_2CH_2CH_3$$

Common branched groups have specific trivial names. For example, the two propyl groups are called the **propyl group** and the **isopropyl group**.

$$CH_3CH_2CH_2—$$
propyl (or *n*-propyl)

$$\overset{CH_3}{\underset{}{|}}$$
$$CH_3CH—$$
isopropyl

To emphasize that a side chain is *not* branched, a prefix *n*- (for *normal*-) is often used. (The *n*- is redundant since the absence of a prefix also indicates a continuous chain.) The prefix iso- (from *isomeric*) is used to indicate a methyl branch at the end of the alkyl side chain.

A four-carbon side chain has four structural possibilities. The **butyl**, or ***n*-butyl**, group is the continuous-chain group. The **isobutyl group** has a methyl branch at the end of the chain. These two butyl groups are named similarly to the propyl groups.

methyl branch at end

$$CH_3CH_2CH_2CH_2—$$
butyl (or *n*-butyl)

$$\overset{CH_3}{\underset{}{|}}$$
$$CH_3CHCH_2—$$
isobutyl

The ***secondary*-butyl group** (abbreviated *sec*-butyl) has two carbons bonded to the head (attaching) carbon. The ***tertiary*-butyl group** (abbreviated *tert*-butyl or *t*-butyl) has three carbons attached to the head carbon.

*two C's on
head carbon*

$$\overset{CH_3}{\underset{}{|}}$$
$$CH_3CH_2CH—$$
sec-butyl

*three C's on
head carbon*

$$\overset{CH_3}{\underset{}{\overset{|}{CH_3—C—}}}$$
$$|$$
$$CH_3$$
t-butyl

Sample Problem

Name the following compounds:

$$\text{CH}_3 \qquad\qquad\qquad\qquad \text{CH}_3\text{CHCH}_3$$

(a) ⬡–CHCH$_2$CH$_3$ (b) CH$_3$CH$_2$CH$_2$CHCH$_2$CH$_2$CH$_3$

Solution: (a) *sec*-butylcyclohexane;   (b) 4-isopropylheptane.

Study Problem

3.4 Give structures for   **(a)** *n*-propylcyclohexane;   **(b)** isobutylcyclohexane;
(c) 4-*t*-butyloctane.

E. Multiple Branches

If two or more branches are attached to a parent chain, more prefixes are added
to the parent name. The prefixes are placed alphabetically, each with its number
indicating the position of attachment to the parent.

$$\text{CH}_3 \qquad\qquad\qquad\qquad \text{CH}_3$$

$$\text{CH}_3\text{CH}_2\text{CHCHCH}_2\text{CH}_2\text{CH}_3 \qquad \text{CH}_3\text{CH}_2\text{CCH}_2\text{CH}_2\text{CH}_3$$

$$\text{CH}_2\text{CH}_3 \qquad\qquad\qquad\qquad \text{CH}_2\text{CH}_3$$

4-ethyl-3-methylheptane   3-ethyl-3-methylhexane

If two or more substituents on a parent are the same (such as two methyl
groups or three ethyl groups), these groups are consolidated in the name. For
example, *dimethyl* means "two methyl groups" and *triethyl* means "three ethyl
groups." The prefixes (*di-*, *tri-*, etc.) that denote number are listed in Table 3.3. In
a name, the *di-* or *tri-* prefix is preceded by position numbers. If *di* is used, two
numbers are required; if *tri* is used, three numbers must be present. Note the use
of commas and hyphens in the following examples:

$$\text{CH}_2\text{CH}_3$$

$$\text{CH}_3$$

$$\text{CH}_3\text{CCH}_2\text{CH}_2\text{CH}_3$$

$$\text{CH}_3$$

CH$_3$CH$_2$ CH$_2$CH$_3$

2,2-dimethylpentane   1,3,5-triethylcyclohexane

Table 3.3 Prefixes for naming multiple substituents

Number	Prefix
2	di-
3	tri-
4	tetra-
5	penta-
6	hexa-

Sample Problem

Name the following compounds:

(a) [cyclohexane ring with two CH(CH$_3$)$_2$ groups on one carbon]   CH(CH$_3$)$_2$
CH(CH$_3$)$_2$

(b) H$_3$C— [cyclopentane ring with CH$_2$CH$_2$CH$_3$ at top and —CH$_2$CH$_2$CH$_3$ at right]

(c)
CH(CH$_3$)$_2$
|
CH$_3$CH$_2$CH$_2$CHCHCH$_2$CH$_2$CH$_3$
|
CH(CH$_3$)$_2$

Solution: (a) 1,1-diisopropylcyclohexane (b) 4-methyl-1,2-dipropylcyclopentane
(c) 4,5-diisopropylnonane. (We number a ring so that the prefix numbers are as low as
possible. We alphabetize *dipropyl* as *propyl*.)

F. Other Prefix Substituents

Like alkyl branches, some functional groups are named as prefixes to the parent
name. Some of these groups and their prefix names are listed in Table 3.4. The
rules governing the use of these prefixes are identical to those for the alkyl
groups except that the parent is the longest continuous chain *containing the
functional group*. The position of the functional group is specified by a number
(as low as possible), and identical groups are preceded by *di-* or *tri-*.

NO$_2$ Br
| |
CH$_3$CH$_2$CHCH$_2$CH$_2$

Cl Cl
| |
CH$_3$—C—C—CH$_3$
| |
Cl Cl

1-bromo-3-nitropentane   2,2,3,3-tetrachlorobutane

Study Problem

3.5 Give structures for:

(a) 1,1,2-trichloroethane   **(b)** 1,2-dichloro-4-nitrocyclohexane

Table 3.4 Some substituents named as prefixes

Substituent	Prefix Name
—NO$_2$	nitro-
—F	fluoro-
—Cl	chloro-
—Br	bromo-
—I	iodo-

G. Alkenes and Alkynes

In IUPAC nomenclature, carbon–carbon unsaturation is always designated by a change in the *ending* of the parent name. As we have indicated, if the parent hydrocarbon contains no double or triple bonds, the suffix **-ane** is used. If a double bond is present, the **-ane** ending is changed to **-ene**; the general name for a hydrocarbon with a double bond is **alkene**. A triple bond is indicated by **-yne**; a hydrocarbon containing this group is an **alkyne**.

$$CH_3CH_3 \qquad\qquad CH_2{=}CH_2 \qquad\qquad\qquad HC{\equiv}CH$$

      ethane   ethene   ethyne

an alkane   (*trivial name*: ethylene)   (*trivial name*: acetylene)
                       an alkene   *an alkyne*

When the parent contains four or more carbons, a prefix number must be used to indicate the position of the double or triple bond. The chain is numbered so as to *give the double or triple bond as low a number as possible*, even if a prefix group must then receive a higher number. Only a single number is used for each double and triple bond; it is understood that the double or triple bond *begins at this numbered position and goes to the carbon with the next higher number*. Thus, a prefix number 2 means the double or triple bond is between carbons 2 and 3, not between carbons 2 and 1.

C=C starts at carbon 1   *C=C starts at carbon 2*

$$CH_2{=}CHCH_2CH_3 \qquad\qquad CH_3CH{=}CHCH_3$$

         1-butene   2-butene

a branch precedes the name

$$CH_3$$
$$CH_3CHCH_2CH_2C{\equiv}CH$$

         5-methyl-1-hexyne

In some names, the prefix 1 is implied, but not written, for a carbon–carbon double bond or other principal functional group. The number can be omitted only in cases where there is no doubt about the position of a functional group.

carbon 1

       1-chlorocyclohexene
     instead of 1-chloro-1-cyclohexene

If a structure contains more than one double or triple bond, the name becomes slightly more complex. The following example of a **diene**, a compound with two double bonds, shows the placement of the numbers and the *di*. (Note that an *a* is inserted before *di* to ease the pronunciation.)

$$CH_2\!=\!CHCH\!=\!CH_2$$

1,3-butadiene

—*two double bonds*

four C's

The nomenclature of alkenes, as well as that of other compounds, is discussed in greater detail in later chapters, as well as in the appendix.

Sample Problem

Name the following compounds:

(a) $(CH_3)_2CHCH_2CH_2CH\!=\!CH_2$ (b) $CH_2\!=\!CHCH\!=\!CHCH\!=\!CH_2$
(c) $CH\!\equiv\!CC\!\equiv\!CH$

Solution: (a) 5-methyl-1-hexene (b) 1,3,5-hexatriene (c) 1,3-butadiyne, or simply butadiyne (the numbers are understood because there is only one possible way to position two triple bonds in a four-carbon chain)

Study Problems

3.6 Name the following compounds:

(a) $CH_3CH\!=\!CH_2$ (b) ⬡   (c) ⬡   (d) $CH_3CH\!=\!CHCl$

3.7 Give structures for:

(a) cyclopentene (b) 1,3-pentadiyne (c) 1-methylcyclobutene. (Note that the number 1 is implied for the carbon–carbon double bond.)

H. Alcohols

In the IUPAC system, the name of an alcohol (ROH) is the name of the parent hydrocarbon with the final **-e** changed to **-ol**. Prefix numbers are used when necessary; the hydroxyl group (—OH) receives the lowest prefix number possible.

$$CH_3CH_2CH_3 \qquad CH_3CH_2CH_2OH \qquad \underset{\displaystyle \overset{|}{OH}}{CH_3CHCH_3}$$

propane   1-propanol   2-propanol

$$\underset{\displaystyle \overset{|}{CH_3}}{CH_3CHCH_2CH_3} \qquad \underset{\displaystyle \overset{|}{CH_3}}{HOCH_2CHCH_2CH_3} \qquad \underset{\displaystyle \overset{|}{CH_3}}{CH_3CHCH_2CH_2OH}$$

methylbutane   2-methyl-1-butanol   3-methyl-1-butanol

⬠   ⬠—OH

cyclopentane   cyclopentanol

I. Aldehydes and Ketones

Because an aldehyde (RCHO) contains a carbonyl group bonded to a hydrogen atom, the aldehyde group must be at the beginning of a carbon chain. The aldehyde carbon is considered carbon 1; therefore, no number is used in the name to indicate the position. The ending of an aldehyde name is **-al**.

| *the aldehyde group* | propanal | 3-methylbutanal |

A keto group, by definition, cannot be at the beginning of a carbon chain. Therefore, except for propanone and a few other simple ketones, a prefix number is necessary. The chain should be numbered to give the carbonyl group the lowest possible number. The ending for a ketone name is **-one**.

$$
\underset{\text{the keto group}}{\overset{\displaystyle O}{-\!\overset{\|}{C}\!-}}
\qquad
\underset{\substack{\text{propanone}\\(\textit{trivial name: acetone})}}{\overset{\displaystyle O}{CH_3\overset{\|}{C}CH_3}}
\qquad
\underset{\text{3-methyl-2-hexanone}}{\overset{\displaystyle O}{CH_3\overset{\|}{C}\underset{\underset{\displaystyle CH_3}{|}}{CH}CH_2CH_2CH_3}}
$$

J. Carboxylic Acids

A carboxyl group, like an aldehyde group, must be at the beginning of a carbon chain and again contains the first carbon atom (carbon 1). Again, a number is not needed in the name. The ending for a carboxylic acid name is **-oic acid**.

| *the carboxyl group* | ethanoic acid
(*trivial name*: acetic acid) | 5,5-dimethylhexanoic acid |

Study Problem

3.8 Write structures and names for an aldehyde, a ketone, and a carboxylic acid, each containing five carbons in a continuous chain.

K. Benzene Compounds

The benzene ring is considered to be a parent in the same way that a continuous-chain alkane is. Alkyl groups, halogens, and the nitro group are named as prefixes to benzene.

isopropylbenzene   bromobenzene   nitrobenzene

When a benzene ring is attached to an alkane chain with a functional group or to an alkane chain of seven or more carbon atoms, benzene is considered a substituent instead of a parent. The name for benzene as a substituent is **phenyl**.

the phenyl group   2-phenylethanol   2-phenyloctane

$(C_6H_5—)$

L. Conflicts in Numbering

A structure that has more than one type of substituent can sometimes be numbered in more than one way. Should the name for $ClCH_2CH{=}CH_2$ be 1-chloro-2-propene or 3-chloro-1-propene? To cover such situations, a *system of nomenclature priorities* for prefix numbers has been developed (Table 3.5); the higher-priority substituent receives the lower number. (A more complete list of nomenclature priorities is found in Table A8 in the appendix.)

From Table 3.5, we can see that a double bond is higher in priority than Cl. When numbering a carbon chain, we give the double bond the lowest possible number. The name of $ClCH_2CH{=}CH_2$ is therefore 3-chloro-1-propene, and not 1-chloro-2-propene. Similarly, the following compound is numbered to give the higher-priority group the lower prefix number.

start at end closer to OH

Cl_2CHCH_2OH
② ①

2,2-dichloroethanol

In a molecule containing both a double bond and a triple bond, the longest continuous chain containing these groups is numbered to give the lower prefix numbers. If there is a choice, the double bond receives the lower number. In the name, the suffix for the double bond precedes that for the triple bond.

lower prefix numbers
than 2-penten-4-yne

$CH_2{=}CH{-}C{\equiv}C{-}CH_3$ $CH{\equiv}CCH{=}CHCH_3$

1-penten-3-yne   3-penten-1-yne

Other examples illustrating the use of nomenclature priority will be discussed in subsequent chapters.

Table 3.5 Nomenclature priorities of selected functional groups[a]

Partial Structure	Name
—CO₂H	-oic acid

$$-CO_2H \quad \text{-oic acid}$$

	Partial Structure	Name
	$-CO_2H$	-oic acid
	$\overset{\displaystyle O}{\underset{\displaystyle \|}{-CH}}$	-al
increasing priority	$\overset{\displaystyle O}{\underset{\displaystyle \|}{-C-}}$	-one
	$-OH$	-ol
	$-NR_2$	-amine
	$\diagup C{=}C \diagdown$	-ene
	$-C{\equiv}C-$	-yne
	$R{-}, C_6H_5{-}, Cl{-}, Br{-}, -NO_2$, etc.	prefix substituents

[a] A more complete list of functional groups by nomenclature priority is shown in Table A8 in the appendix.

Study Problem

3.9 Name each of the following compounds:

(a) $CH_3CH_2CH{=}CHCH_2CH_2CH_2NO_2$ (b)

$$\overset{\displaystyle OH}{\underset{\displaystyle |}{C_6H_5{-}CHCH_2CH_2Br}}$$

▼ ## Section 3.4
Alkanes

Most organic compounds have a portion of their structures composed of carbon atoms and hydrogen atoms. A fat is one example of an organic compound with ester groups and with long hydrocarbon chains, which may be alkyl or alkenyl (containing a carbon–carbon double bond).

$$CH_2O\overset{\displaystyle O}{\overset{\displaystyle \|}{C}}(CH_2)_{14}CH_3$$

long hydrocarbon chains

$$CHO\overset{\displaystyle O}{\overset{\displaystyle \|}{C}}(CH_2)_{14}CH_3$$

$$CH_2O\overset{\displaystyle O}{\overset{\displaystyle \|}{C}}(CH_2)_7CH{=}CH(CH_2)_7CH_3$$

a typical animal fat

Early chemists did not know the molecular structure of a fat, but they did know that many compounds containing long hydrocarbon chains behave similarly to fats—most of these compounds are water-insoluble and are less dense than water. For this reason, compounds with hydrocarbon chains are referred to as **aliphatic compounds** (Greek *aleiphatos,* "fat"). The term *aliphatic compound* is usually contrasted to *aromatic compound,* such as benzene or a substituted benzene.

Some aliphatic compounds:

$$CH_3CHCH_2CH_3 \quad CH_3CH_2CH_2OH \quad CH_3CH_2NHCH_2CH_3$$

(with CH_3 substituent on the second carbon)

Some of the physical and chemical properties of an aliphatic compound arise from the alkyl part of its molecules. Therefore, much of what we have to say about alkanes and cycloalkanes is true for other organic compounds as well. Of course, the properties of a compound are also greatly determined by any functional groups that may be present. For example, a hydroxyl group in a molecule leads to hydrogen bonding and a large change in physical properties. Ethane (CH_3CH_3) is a gas at room temperature, while ethanol (CH_3CH_2OH) is a liquid.

A. Physical Properties of Alkanes

Alkanes are nonpolar compounds. As a result, the attractive forces between molecules are weak. The continuous-chain alkanes through butane are gases at room temperature while the C_5 to C_{17} alkanes are liquids (see Table 3.6). The continuous-chain alkanes with 18 or more carbon atoms are solids.

The boiling point of a compound depends, in part, on the amount of energy needed by the molecules of that compound to escape from the liquid into the vapor phase. The boiling points of the compounds of a homologous series, such as the alkanes in Table 3.6, increase by about 30° for each additional methylene (CH_2) group. This increase in boiling point is due principally to an increase in the van der Waals attractions between molecules of increasing length. Homologous series of compounds other than alkanes show similar effects.

$$CH_3I \qquad CH_3CH_2I \qquad CH_3CH_2CH_2I$$

| iodomethane | iodoethane | 1-iodopropane |
| bp 43° | bp 72° | bp 102° |

As we mentioned in Section 1.9A, branching in the hydrocarbon portion of a molecule lowers the boiling point from the expected value because of

Table 3.6 Boiling points of some alkanes

Structure	Bp, °C	Structure	Bp, °C
CH_4	-162	$CH_3(CH_2)_4CH_3$	69
CH_3CH_3	-88.5	$CH_3(CH_2)_5CH_3$	98
$CH_3CH_2CH_3$	-42	$CH_3(CH_2)_6CH_3$	126
$CH_3(CH_2)_2CH_3$	0	$CH_3(CH_2)_7CH_3$	151
$CH_3(CH_2)_3CH_3$	36	$CH_3(CH_2)_8CH_3$	174

interference with van der Waals attractions between the molecules in the liquid state.

Because they are nonpolar, alkanes are soluble in nonpolar or slightly polar solvents such as other alkanes, diethyl ether ($CH_3CH_2OCH_2CH_3$), or benzene. The solubility arises from van der Waals attractions between the solvent and the solute. Alkanes are insoluble in water.

All the alkanes have densities less than that of water, a fact easy to remember because we know that gasoline and motor oil (which are principally alkanes) float on water.

B. Chemical Properties of Alkanes

Alkanes and cycloalkanes are relatively unreactive compared with organic compounds containing functional groups. For example, many organic compounds undergo reaction with strong acids, bases, oxidizing agents, or reducing agents. Alkanes and cycloalkanes generally do not react with these reagents. Because of their lack of reactivity, alkanes are sometimes referred to as **paraffins** (Latin *parum affinis*, "slight affinity").

We will discuss two principal reactions of alkanes in this text. One is the *reaction with halogens*, such as chlorine gas. We will present this reaction in detail in Chapter 6. The other important reaction of alkanes is *combustion*. The remainder of this chapter will be concerned primarily with the combustion of alkanes and their use as a source of energy.

Halogenation:

$$CH_4 \;+\; Cl_2 \xrightarrow{\text{light}} \quad CH_3Cl \quad + \text{ HCl} + \text{other products}$$

methane   chloromethane

$$CH_3CH_2CH_3 + Br_2 \xrightarrow{\text{light}} \quad CH_3\overset{\overset{\displaystyle Br}{|}}{C}HCH_3 \;+\; HBr + \text{other products}$$

propane   2-bromopropane

Combustion:

$$CH_4 + 2\,O_2 \xrightarrow{\text{spark}} \quad CO_2 \quad + 2\,H_2O$$

carbon dioxide

$$CH_3CH_2CH_3 + 5\,O_2 \xrightarrow{\text{spark}} 3\,CO_2 + 4\,H_2O$$

C. Combustion

Combustion is the process of burning—the rapid reaction of a compound with oxygen. Combustion is accompanied by the release of light and heat, two forms of energy that humans have sought since they first built a fire and found that it kept them warm. Although we will present the subject of combustion under the heading of alkanes, keep in mind that almost all organic compounds can burn.

Combustion of organic mixtures, such as wood, is not always a simple conversion to CO_2 and H_2O. Instead, combustion is the result of a large number

Table 3.7 Heats of combustion for some hydrocarbons

Name	$-\Delta H$, kcal/mol	Name	$-\Delta H$, kcal/mol
methane	213	cyclopropane	500
ethane	373	cyclobutane	656
propane	531	cyclopentane	794
butane	688	cyclohexane	944

of complex reactions. One type of reaction that occurs is **pyrolysis**, the thermal fragmentation of large molecules into smaller molecules in the absence of oxygen. Pyrolysis of large molecules in wood, for example, yields smaller gaseous molecules that then react with oxygen above the surface of the wood. This reaction with oxygen gives rise to the flames. On the surface of the wood, a slow, but very hot, oxidation of the carbonaceous residue takes place. Most of the heat from a wood or coal fire results from this slow oxidation, rather than from the actual flames.

    Complete combustion is the conversion of a compound to CO_2 and H_2O. If the oxygen supply is insufficient for complete combustion, **incomplete combustion** occurs. Incomplete combustion leads to carbon monoxide, or sometimes carbon as carbon black or soot.

Incomplete combustion:

$$2\ CH_3CH_2CH_3 + 7\ O_2 \longrightarrow 6\ CO + 8\ H_2O$$
$$\text{carbon monoxide}$$

$$CH_3CH_2CH_3 + 2\ O_2 \longrightarrow 3\ C + 4\ H_2O$$
$$\text{carbon}$$

D. Heat of Combustion

The energy released when a compound is oxidized completely to CO_2 and H_2O is called the **heat of combustion ΔH**. Under controlled laboratory conditions, ΔH may be measured. (As we described in Section 1.7, the value for ΔH is negative when energy is liberated because the molecules have lost energy.)

$$CH_4 + 2\ O_2 \longrightarrow CO_2 + 2\ H_2O + 213\ \text{kcal/mol}$$
$$\text{methane}$$

$$2\ CH_3CH_2CH_2CH_3 + 13\ O_2 \longrightarrow 8\ CO_2 + 10\ H_2O + 688\ \text{kcal/mol}$$
$$\text{butane}$$

    Values for heats of combustion (Table 3.7) depend primarily upon the number of carbon and hydrogen atoms in a molecule. In a homologous series, the energy liberated increases by about 157 kcal/mol for each additional methylene group.

Section 3.5
The Hydrocarbon Resources

A. Natural Gas and Petroleum

Natural gas is 60–90% methane depending on its source. The other components of natural gas are ethane and propane, along with nitrogen and carbon dioxide. Natural gas found in the Texas panhandle and in Oklahoma also is a source of helium. Deposits of natural gas are usually found with petroleum deposits.

Petroleum, also called *crude oil*, is a complex mixture of aliphatic and aromatic compounds, including sulfur and nitrogen compounds (1–6%). In fact, over 500 compounds have been detected in a single sample of petroleum. The actual composition varies from deposit to deposit.

Although the origin of petroleum is still subject to debate, it was probably formed by the anaerobic decay of microorganisms or other organic material.

Because crude oil is a mixture of many compounds, it is not very useful in the form removed from the ground. Separating the crude oil into useful components is called **refining**. The first step in refining is a fractional distillation, called **straight-run distillation**. The fractions that are collected are listed in Table 3.8.

The gasoline fraction of straight-run distillation is too scanty for the needs of our automobile-oriented society, and straight-run gasoline has poor burning characteristics in automobile engines. To increase both quantity and quality of the gasoline fraction, the higher-boiling fractions are subjected to *cracking* and *reforming*.

Catalytic cracking is the process of heating the high-boiling material under pressure in the presence of a catalyst (usually finely divided, acid-washed aluminum silicate clay). Under these conditions, large molecules are cracked, or broken, into smaller fragments, and continuous-chain alkanes are isomerized to branched-chain alkanes.

Steam cracking is a technique for converting alkanes to alkenes, and **catalytic reforming** converts aliphatic compounds to aromatic compounds. The alkenes and aromatics formed in these cracking and reforming procedures are used as starting materials for making plastics and other synthetic organic compounds.

In automobile engines, continuous-chain hydrocarbons burn unevenly and cause knocking, which is the ticking or rattling noise heard when a car

Table 3.8 Fractions of straight-run distillation

Boiling Range, °C	Number of Carbons	Name	Use
under 30	1–4	gas fraction	heating fuel
30–180	5–10	gasoline	automobile fuel
180–230	11–12	kerosene	jet fuel
230–305	13–17	gas oil	diesel fuel, heating fuel
305–405	18–25	heavy gas oil	heating fuel

Residue: (1) Volatile oils: lubricating oils, paraffin wax, and petroleum jelly. (2) Nonvolatile material: asphalt and petroleum coke.

accelerates uphill. To prevent knocking, branched alkanes and aromatic compounds, which burn more evenly than continuous-chain compounds, are added to automobile fuels. Fortunately, cracking procedures provide both branched alkanes and aromatics.

At one time, isooctane (a trivial name for 2,2,4-trimethylpentane) was the alkane with the best antiknock characteristics for automobile engines, and *n*-heptane was the poorest. These two compounds were used to develop an octane rating for petroleum fuels.

$$\begin{array}{c} \quad\; CH_3 \;\; CH_3 \\ \quad\; | \qquad | \\ CH_3CCH_2CHCH_3 \\ \quad\; | \\ \quad\; CH_3 \end{array} \qquad CH_3(CH_2)_5CH_3$$

2,2,4-trimethylpentane   heptane
   ("isooctane")

A gasoline's **octane number**, which is a rating of the gasoline's quality in automobile engines, is assigned by comparing the gasoline's burning characteristics with those of isooctane–heptane mixtures. An octane number of 100 means that the gasoline is equivalent in burning characteristics to pure isooctane. Gasoline with an octane number of 0 is equivalent to pure heptane. An octane number of 75 is given to gasoline that is equivalent to a mixture of 75% isooctane and 25% heptane. Straight-run gasoline has an octane number of about 70. Some compounds have better burning characteristics than isooctane; therefore, octane numbers of greater than 100 are possible.

Additives are also mixed with gasoline to increase octane ratings. Some common additives with octane numbers over 100 are benzene, ethanol, *t*-butyl alcohol [$(CH_3)_3COH$], and *t*-butyl methyl ether [$(CH_3)_3COCH_3$]. The additive mixture used in leaded gasoline, called *Ethyl fluid*, contains approximately 65% tetraethyllead, 25% 1,2-dibromoethane, and 10% 1,2-dichloroethane. The halogenated hydrocarbons are essential for conversion of the lead to volatile lead halides, which are removed from the cylinder in the exhaust.

$$(CH_3CH_2)_4Pb \qquad BrCH_2CH_2Br \qquad ClCH_2CH_2Cl$$

   tetraethyllead   1,2-dibromoethane   1,2-dichloroethane
                        (ethylene dibromide, EDB)   (ethylene dichloride)

A gasoline engine puts forth a variety of pollutants: unburned hydrocarbons, carbon monoxide, and nitrogen oxides. Catalytic converters are now installed in automobiles to convert nonoxidized and partly oxidized compounds to more highly oxidized and acceptable forms of exhaust. The platinum catalyst used in these converters is "poisoned" (made nonfunctional) by lead compounds; therefore, leaded gasoline cannot be used in cars equipped with catalytic converters without destroying the catalyst.

B. Coal

Coal is formed by the bacterial decomposition of plants under varying degrees of pressure. Coal is classified by its carbon content: *anthracite*, or hard coal, contains the highest carbon content, followed by *bituminous (soft) coal, lignite,* and, finally, *peat.* Because some coals also contain 2–6% sulfur, the burning of coal can lead to severe air pollution and acid rain.

When coal is subjected to heat and distillation in the absence of air, a process called **destructive distillation**, three crude products result: *coal gas* (primarily CH_4 and H_2), *coal tar* (the condensable distillate), and *coke* (the residue). Both coal gas and coke are useful fuels. (Today, coke is used primarily in the manufacture of steel.) Coal tar is rich in aromatic compounds, which are formed in the destructive distillation.

Until petroleum became plentiful and cheap in the 1940s, coal was a primary source of synthetic organic compounds. Currently, however, over 90% of the organic chemicals produced in the United States are synthesized from petroleum. The reasons for the switch to petrochemicals are that petroleum-refining processes are less costly and also that these processes give rise to less pollution than most coal-refining processes. Unfortunately, the world's petroleum resources are rapidly being depleted and becoming more expensive, while the coal reserves of the world are still very extensive. Economical ways to convert coal to useful fuels and chemicals (with a minimum of air pollution) are under investigation at the present time.

The conversion of coal to gaseous or liquid fuels, called synthetic fuels or "syn fuels," is referred to as **coal gasification** or **coal liquefaction**, respectively. Many gasification plants use the German-developed **Lurgi process**, or modifications of this technique, in which a bed of coal is treated with steam at high temperatures to yield *synthesis gas* ($CO + H_2$). Synthesis gas itself is only a moderately efficient fuel, and carbon monoxide is highly toxic. Thus, synthesis gas is treated with additional hydrogen to yield methane.

Coal gasification:

$$\underset{\text{coal}}{C} + \underset{\text{steam}}{H_2O} \xrightarrow{\text{heat}} \underset{\text{"synthesis gas"}}{\underline{CO + H_2}} \xrightarrow[\text{Ni catalyst}]{2\,H_2} \underset{\text{methane}}{CH_4} + H_2O$$

The liquefaction of coal is its conversion to liquid alkanes. The classical process for accomplishing this conversion is the **Fischer–Tropsch synthesis**, which was developed in Germany during World War II. The Republic of South Africa synthesizes most of its gasoline and organic chemicals by this process.

Coal liquefaction (Fischer–Tropsch synthesis):

$$\underset{\text{coal}}{C} + \underset{\text{steam}}{H_2O} \xrightarrow{\text{heat}} CO + H_2 \xrightarrow[\text{heat, pressure}]{\overset{H_2}{Fe\ catalyst}} \text{alkanes} + H_2O$$

Because coal reserves, too, are finite, other sources of hydrocarbons are being investigated and developed. These include oil shale in the United States; tar sands in the Athabasca basin in Alberta, Canada; and biological sources, such as agricultural wastes and plants of the *Euphorbia* genus (succulents with a high percentage of hydrocarbons in their sap).

▼ Summary

Compounds with the same molecular formula, but different structures (order of attachment of the atoms), are **structural isomers** of each other.

The **IUPAC system of nomenclature** is based on the names of the *continuous-chain alkanes* as parents. If a hydrocarbon chain forms a ring, the prefix *cyclo-* is added to the alkane name. Branches and functional groups are indicated in a name by prefixes or suffixes.

The longest continuous chain containing the functional group (if any) is the parent. The chain is numbered from the end nearer to the branches or functional groups. (The functional group of highest priority, as listed in Table 3.5, receives the lowest number.) Positions of substitution on the chain are then specified by these numbers.

1,4-dichloro-4-methyl-2-pentene: $\overset{\text{Cl}}{\underset{\underset{CH_3}{|}}{CH_3\overset{}{C}CH}}=CHCH_2Cl$

The principal reactions of alkanes are **halogenation** and **combustion**. The **heat of combustion** of an alkane (or other compound) is the heat released when the compound is burned and represents the amount of energy by which the products ($CO_2 + H_2O$) are more stable (lower energy) than the reactants (higher energy).

Petroleum is the world's principal source of gasoline and organic chemicals today. Alkanes, alkenes, and aromatic compounds are obtained by **refining**: straight-run distillation, cracking, and reforming. In the future, **coal-gasification** or **coal-liquefaction** processes may be the principal sources of methane and other alkanes.

Study Problems*

3.10 Which of the following compounds are unsaturated? (We have used trivial names for these compounds because these names are encountered in industry and medicine.)

(a) phenylephrine *a nasal decongestant*

(b) $CH_2=CHCN$ acrylonitrile *a carcinogenic compound used in the plastics industry*

(c) $(CH_3)_2CHOH$ isopropyl alcohol *rubbing alcohol*

(d) HCO_2H formic acid *used in tanning and dyeing*

* For information concerning the organization of the *Study Problems* and *Additional Problems*, see the *Note to student* on page 39.

3.11 Write a formula for one structural isomer of each of the following compounds, all of which are used as inhalation anesthetics.

(a)   △

cyclopropane

(b) $CH_3CH_2OCH_2CH_3$

diethyl ether

(c) $\overset{\overset{\text{Cl}}{|}}{CF_3CHBr}$

1-bromo-1-chloro-2,2,2,-trifluoroethane
(halothane)

3.12 Indicate whether the following pairs of formulas represent the same compound or isomers.

(a) $\overset{\overset{\text{Cl}}{|}}{CH_3CHCH_2CH_2CH_3}$, $\overset{\overset{\text{Cl}}{|}}{CH_3CH_2CH_2CHCH_3}$

(b) $(CH_3)_3CCl$, $\overset{\overset{\text{CH}_3}{|}}{\underset{\underset{\text{CH}_2\text{Cl}}{|}}{CH_3CH}}$

(c) $CH_3CH_2\overset{\overset{}{}}{\underset{\underset{\underset{\underset{\text{CH}_2\text{OH}}{|}}{\text{CH}_2}}{|}}{C}}HCH_3$, $CH_3CH_2\overset{\overset{\text{OH}}{|}}{\underset{\underset{\text{CH}_3}{|}}{CH}}CH_2CH_2$

(d) a benzene ring with —OH substituent;   a cyclohexene ring with HO substituent

3.13 Write condensed structural formulas for the indicated compounds:
(a) five structural isomers for C_6H_{14}
(b) all the isomeric alcohols for $C_4H_{10}O$
(c) all the isomeric amines for $C_4H_{11}N$
(d) all the structural isomers for C_3H_6BrCl

3.14 What is the molecular formula and the general formula (e.g., C_nH_{2n}) for each of the following compounds?

(a) $CH_2{=}CHCH{=}CHCH_2CH(CH_3)_2$ (b) (bicyclic ring structure)

3.15 Write the formula for one structural isomer of each of the following compounds:

(a) (decalone structure)   (b) (lactone structure)   (c) (cyclohexene structure)

3.16 For each of the following compounds, write the structural formula of an isomer that has a *different functional group* (see Problem 3.2, page 90).

(a) (cyclohexene structure)   (b) $HOCH_2\overset{\overset{\text{O}}{\|}}{C}H$ (c) $H\overset{\overset{\text{O}}{\|}}{C}CH_2CH_3$

3.17 The antimicrobial activity of a homologous series of 4-*n*-alkylresorcinols against *Eberthella typhosa* increases with increasing chain length up to six carbons and then falls off dramatically. Use formulas of three of the active compounds to illustrate the term "homologous series."

a 4-*n*-alkylresorcinol

3.18 Write structural formulas for a homologous series of alkyl bromides (monobromo-alkanes) from C_5 to C_{10}.

3.19 Give the condensed structural (or polygon) formula of each of the following compounds:

(**a**) 2,2-dimethyloctane   (**b**) 3,4-diethylheptane
(**c**) 4-ethyl-2,4-dimethylnonane   (**d**) 1,3-diisopropylcyclohexane
(**e**) *sec*-butylcyclopentane   (**f**) *t*-butylbenzene
(**g**) isobutylcycloheptane   (**h**) 1-methyl-3-pentylcyclohexane
(**i**) 4-isopropylheptane

3.20 Write the IUPAC name for each of the following compounds:

(**a**) $(CH_3CH_2)_2CHCH_2CH_3$ (**b**) $(CH_3)_2CHCH(CH_3)_2$

(**c**)   (**d**)

(**e**) $CH_3CH_2CH_2C(CH_3)_2CH_2C(CH_3)_3$ (**f**) $[(CH_3)_3C]_2CHCH_3$

3.21 Each of the following names is *incorrect*. Give a more suitable name in each case.
(**a**) 6-methyloctane; (**b**) *t*-butylmethane; (**c**) 3-methyl-3-isopropyl-4-methyldecane.

3.22 Circle and name the functional groups in the following structures:

(**a**)   (**b**)

lactic acid

in sour milk

α-irone

main perfume ingredient in violets

(**c**)   (**d**) $HCCH_2CH_2CCH_3$

3.23 Give structures for all the monochloro isomers of the following alkanes:
(**a**) *n*-pentane; (**b**) cyclopentane; (**c**) 2,2-dimethylbutane; (**d**) 2,2-dimethylpropane.

3.24 Write an IUPAC name for each of the following structures:

(a)

*One isomer is the
insecticide lindane.*

(b) $CH_2ClCHCl_2$

a cleaning solvent

(c) CH_3NO_2

a rocket fuel

3.25 Write the formula for each of the following names: **(a)** 1-bromo-1,2-diphenyl-propane; **(b)** hexachloroethane; **(c)** 2-iodo-1-octanol; **(d)** 1,1-dichloro-3-methyl-cyclohexane.

3.26 Write IUPAC names for the following compounds:

(a) $(CH_3)_2C=CHCH_3$ **(b)** $Cl_2C=CHCl$

(c)   **(d)** $CH_2=CHC=CHCH=CH_2$ with CH_3 above

(e) ⬡—Br   **(f)** ⬡—$(C≡C)_3$—⬡

3.27 Write the formulas and IUPAC names for **(a)** all the isomeric cyclopentanones that contain one methyl substituent on a ring carbon atom; **(b)** all the isomeric cyclopentenes with one isopropyl branch.

3.28 Write the IUPAC names for the following compounds:

(a) Cl_2CHCO_2H

*a corrosive compound used
as a topical astringent*

(b) $ClCH_2\overset{\displaystyle O}{\overset{\|}{C}}H$

*a strong irritant used to
facilitate tree bark removal*

(c) $BrCH_2\overset{\displaystyle O}{\overset{\|}{C}}CH_3$

*an irritant that has been used as
a chemical war gas*

3.29 Draw condensed structural formulas for the following compounds. (The IUPAC names are given in parentheses.)

(a) *erythritol* (1,2,3,4-butanetetrol), a naturally occurring alcohol isolated from algae and lichens;   **(b)** *ethohexadiol* (2-ethyl-1,3-hexanediol), an insect repellent that can cause irritation of the mucous membranes and eyes;   **(c)** *ethchlorvynol* (1-chloro-3-ethyl-1-penten-4-yn-3-ol), an addictive or habituating hypnotic with a pungent odor;   **(d)** *citral* (3,7-dimethyl-2,6-octadienal), which is found in lemongrass oil and is used as a citrus flavoring;   **(e)** *piperitone* (6-isopropyl-3-methyl-2-cyclohexenone), which is found in the oils of Sitka spruce, eucalyptus, and peppermint.

3.30 **(a)** From the data in Table 3.7 (page 106), predict the heat of combustion of octane.
   (b) From these data, could you also extrapolate the heat of combustion of 2-methylheptane? Explain.

3.31 Which one of each of the following pairs of structural isomers would you expect to have the higher boiling point: (a) hexane or 2,2-dimethylbutane?   (b) 2-butene or methylpropene?   (c) 1-pentanol or 2,2-dimethyl-1-propanol? Why?

Additional Problems

3.32 An alcohol with no carbon–carbon double bonds has the formula C_4H_8O. What are the possible structures?

3.33 A four-carbon compound contains an aldehyde group and a carboxyl group. What are two possible structures for this compound?

3.34 *Biacetyl*, $C_4H_6O_2$, is a ketone used to flavor margarine. This compound contains no rings or carbon–carbon double bonds. What is its structure?

3.35 *Oxirane*, also called ethylene oxide, C_2H_4O, is used as a sterilizing agent and a fumigant. It is extremely flammable and is also a strong irritant. This compound does not contain a double or triple bond. What is its structure?

3.36 (a) Complete, but do not balance, the following equation, showing all possible monochlorinated products:

$$CH_3CH_2CH_3 + Cl_2 \xrightarrow{\text{light}}$$

(b) For the equation in (a), show all possible *dichlorinated* products.
(c) If the substitution of Cl for H were completely random for the reaction in (a), what would be the ratio of isomeric products?

3.37 When a continuous-chain alkane is subjected to catalytic cracking, one reaction that occurs is the cleaving of the alkane into a smaller alkane and an alkene. Write a chemical equation showing the conversion of *n*-heptane to propene and an alkane.

3.38 When petroleum alkanes of six or more carbons are heated in the presence of platinum and rhenium on aluminosilicates, they can undergo cyclization and dehydrogenation (loss of H_2) to yield aromatic hydrocarbons. What aromatic hydrocarbons could be formed when a mixture of *n*-hexane, *n*-heptane, and *n*-octane is subjected to these conditions? (Use balanced equations in your answer.)

4 ▼

Stereochemistry

Stereochemistry is the study of molecules in space, that is, how atoms in a molecule are arranged in space relative to one another. The three aspects of stereochemistry that will be covered in this chapter are:

1. **Geometric isomers:** how rigidity in a molecule can lead to isomerism;
2. **Conformations of molecules:** the shapes of molecules and how they can change;
3. **Chirality of molecules:** how the right- or left-handed arrangement of atoms around a carbon atom can lead to isomerism.

It is often difficult to visualize a three-dimensional molecule from a two-dimensional illustration. Therefore, in our discussions of stereochemistry here and in subsequent chapters, we strongly urge you to use a set of molecular models.

▼ ## Section 4.1
Geometric Isomerism in Alkenes

In Chapter 3, we defined structural isomers as compounds with the same molecular formula but with different orders of attachment of their atoms. Structural isomerism is only one type of isomerism. A second type of isomerism is **geometric isomerism**, which results from rigidity in molecules and occurs in only two classes of compounds: *alkenes* and *cyclic compounds*.

Molecules are not quiet, static particles. They move, spin, rotate, and flex. Atoms and groups attached only by sigma bonds can rotate so that the overall shape of a molecule is in a state of continuous change. However, groups attached by a double bond cannot rotate around the double bond without the pi bond being broken. The amount of energy needed to break a carbon–carbon pi bond (68 kcal/mol in ethylene) is not available to molecules at room temperature. Because of the rigidity of a pi bond, groups attached to pi-bonded carbons are fixed in space relative to one another.

We usually write the structure for an alkene as if the sp^2-hybridized carbon atoms and the atoms attached to them are all in the plane of the paper. In this representation, we can visualize one lobe of the pi bond as being in front of the paper and the other lobe of the pi bond as being underneath the paper, behind the front lobe (see Figure 4.1).

Figure 4.1 **The groups attached to sp^2-hybridized carbons are fixed in relation to one another.**

In Figure 4.1, we show a structure with two Cl atoms (one on each sp^2-hybridized carbon) on one side of the pi bond and two H atoms on the other side. Because the double bond is rigid, this molecule is not readily interconvertible with the compound in which the Cl atoms are on opposite sides of the pi bond.

Two groups on the *same side of the pi bond* are said to be **cis** (Latin, "on the side"). Groups on the *opposite sides* are said to be **trans** (Latin, "across"). Note how the *cis* or *trans* designation is incorporated into the name.

cis-1,2-dichloroethene   *trans*-1,2-dichloroethene
bp 60°   bp 48°

The *cis*- and *trans*-1,2-dichloroethenes have different physical properties (such as boiling points); they are different compounds. However, these two compounds are not structural isomers because the sequence of atoms and the location of the double bond are the same in each compound. This pair of isomers falls into the general category of **stereoisomers**: different compounds that have the same structure, differing only in the *arrangement of the atoms in space*. This pair of isomers falls into a more specific category of **geometric isomers** (also called *cis–trans* **isomers**): stereoisomers that differ by groups being on the same side or on opposite sides of a site of rigidity in a molecule.

The requirement for geometric isomerism in alkenes is that each carbon atom involved in the pi bond have two different groups attached to it, such as H and Cl, or CH_3 and Cl. If one of the carbons of the double bond has two identical groups, such as two H atoms or two CH_3 groups, then geometric isomerism is not possible. (We urge you to make molecular models and verify for yourself this requirement of geometric isomerism.)

Sample Problem

Label each of the following pairs of structures as *structural isomers* of each other, as *geometric isomers* of each other, or as the *same compound*:

Solution:

(a) *same compound* (H's *trans* to each other in each);
(b) *structural isomers* (position of double bond is different; each structural isomer has a geometric isomer);
(c) *geometric isomers* (the first is *cis*; the second is *trans*).

A. (*E*) and (*Z*) System of Nomenclature

When there are three or four different groups attached to the carbon atoms of a double bond, a pair of geometric isomers exists, but it is sometimes difficult to assign *cis* or *trans* designations to the isomers.

$$Br \quad \backslash \quad C{=}C \quad / \quad F$$

I   Cl

cis or trans?

In our example, we can say that Br and Cl are *trans* to each other or that I and Cl are *cis* to each other. However, we cannot name the structure in its entirety as being either the *cis* or the *trans* isomer. Because of the ambiguity in cases of this type, a more general system of isomer assignment, called the **(*E*) and (*Z*) system**, has been devised.

The (*E*) and (*Z*) system is based on an assignment of priorities (not to be confused with nomenclature priorities) to the atoms or groups attached to each carbon of the double bond. If the higher-priority atoms or groups are on *opposite sides* of the pi bond, the isomer is (*E*). If the higher-priority groups are on the *same side*, the isomer is (*Z*). (The letter *E* is from the German *entgegen*, "across"; the letter *Z* is from the German *zusammen*, "together.")

If the two atoms on each double-bond carbon are different, priority is based on the atomic numbers of the single atoms directly attached to the double-bond carbons. *The atom with the higher atomic number receives a higher priority.* In our example, I has a larger atomic number than does Br; I is of higher priority. On the other carbon of the double bond, Cl is of higher priority than F.

	F	Cl	Br	I
atomic number:	9	17	35	53

increasing priority →

I higher priority than Br   *Cl higher priority than F*

(*Z*)-1-bromo-2-chloro-2-fluoro-1-iodoethene

(*E*)-1-bromo-2-chloro-2-fluoro-1-iodoethene

Sample Problem

Name the following compound by the (E) and (Z) system:

$$\begin{array}{c}\text{H}\qquad\quad\text{CH}_2\text{CH}_2\text{CH}_3\\[-4pt]\diagdown\qquad\diagup\\[-2pt]\text{C}=\text{C}\\[-2pt]\diagup\qquad\diagdown\\[-4pt]\text{H}_3\text{C}\qquad\quad\text{Cl}\end{array}$$

Solution: The left-hand carbon of the double bond has an H and a C attached to it; the C has the higher priority. The right-hand carbon has a Cl and a C attached to it; the Cl has a higher priority. (Look at the *single atoms* directly attached to the double-bond carbon: C, and not the entire —$\text{CH}_2\text{CH}_2\text{CH}_3$ group.) The higher-priority atoms are on the same side. The compound is named (Z)-3-chloro-2-hexene.

Study Problems

4.1 Name the following compounds by the (E) and (Z) system:

(a)
$$\begin{array}{c}\text{F}\qquad\quad\text{CH}_2\text{CH}_3\\[-4pt]\diagdown\qquad\diagup\\[-2pt]\text{C}=\text{C}\\[-2pt]\diagup\qquad\diagdown\\[-4pt]\text{H}\qquad\quad\text{Cl}\end{array}$$

(b)
$$\begin{array}{c}\text{H}_3\text{C}\qquad\quad\text{CH}_3\\[-4pt]\diagdown\qquad\diagup\\[-2pt]\text{C}=\text{C}\\[-2pt]\diagup\qquad\diagdown\\[-4pt]\text{Cl}\qquad\quad\text{Cl}\end{array}$$

4.2 The formula for maleic acid, a toxic dicarboxylic acid, is (Z)-$\text{HO}_2\text{CCH}=\text{CHCO}_2\text{H}$. Its (E) isomer, fumaric acid, is not toxic and is, in fact, an intermediate in some cellular reactions. Write the structures of maleic acid and fumaric acid, showing their stereochemistry.

B. Sequence Rules

Determination of priorities by atomic number alone cannot handle all cases. For example, how would we name the following compound by the (E) and (Z) system?

$$\begin{array}{c}\text{H}_3\text{C}\qquad\quad\text{CH}_2\text{CH}_3\\[-4pt]\diagdown\qquad\diagup\\[-2pt]\text{C}=\text{C}\\[-2pt]\diagup\qquad\diagdown\\[-4pt]\text{H}\qquad\quad\text{CH}_3\end{array}$$

To handle such a case, and others like it, a set of *sequence rules* to determine order of priority has been developed. These priority rules form the basis of the **Cahn–Ingold–Prelog nomenclature system**, named in honor of the chemists who developed the system.

Sequence Rules for Order of Priority:

1. If the atoms in question are different, the sequence order is by atomic number, with the atom of highest atomic number receiving the highest priority.

F   Cl   Br   I

increasing priority

2. If two isotopes of the same element are present, the *isotope of higher mass* receives the higher priority.

$^{1}_{1}$H, or H $^{2}_{1}$H, or D
hydrogen   deuterium

increasing priority

3. If two atoms are identical, the atomic numbers of the *next atoms* are used for priority assignment. If these atoms also have identical atoms attached to them, priority is determined at the first point of difference along the chain. The atom that has attached to it an atom of higher priority has the higher priority. (Do not use the sums of the atomic numbers, but look for the single atom of highest priority.)

three H's: lower priority

two H's and one C:
the C gives this group
higher priority

Cl gives this group
higher priority

(*E*)-3-methyl-2-pentene   (*Z*)-5-butyl-8-chloro-4-decene

4. Atoms attached by double or triple bonds are given single-bond *equivalencies* so that they can be treated like single-bonded groups in determining priority. Each doubly bonded atom is duplicated (or triplicated for triple bonds), a process better seen in examples.

Structure	*Equivalent for priority determination*
$\underset{\text{R}-\overset{\displaystyle\text{O}}{\overset{\|}{\text{C}}}-\text{R}}{}$	R—C—R with O and O—C
$\text{R}-\overset{\displaystyle\text{O}}{\overset{\|}{\text{C}}}-\text{OH}$	R—C—OH with O and O—C
R—C≡N	R—C—N with N, N and C, C
R—CH=CR$_2$	R—CH—CR$_2$ with C, C

Structure	Equivalent for priority determination

By this rule, we obtain the following priority sequence:

$$-CH=CR_2 \qquad \bigcirc \qquad -CN \qquad -CH_2OH \qquad \underset{\text{aldehyde}}{-\overset{\overset{\displaystyle O}{\|}}{C}H} \qquad \underset{\text{ketone}}{-\overset{\overset{\displaystyle O}{\|}}{C}-} \qquad -\overset{\overset{\displaystyle O}{\|}}{C}OH$$

increasing priority →

Sample Problems

List the following atoms or groups in order of increasing priority (lowest priority first):

$$-NH_2 \qquad -H \qquad -CH_3 \qquad -Cl$$

Solution: Increasing atomic number of the attached atom (N, H, C, Cl) gives increasing priority: H, CH_3, NH_2, Cl.

List the order of priority of the following groups (lowest to highest):

$$-CO_2H \qquad -CO_2CH_3 \qquad -CH_2OH \qquad -OH \qquad -H$$

Solution: Following the sequence rules:

$$-H \qquad -CH_2OH \qquad -CO_2H \qquad -CO_2CH_3 \qquad -OH$$

Which group is of higher priority, the *n*-butyl group or the isobutyl group?

Solution: The first carbon is the same in each (two H's and one C attached).

$$-\overset{\downarrow}{C}H_2CH_2CH_2CH_3 \qquad -\overset{\downarrow}{C}H_2CH(CH_3)_2$$

Therefore, we proceed to the second carbon, and we find that the isobutyl group is of higher priority.

$$-CH_2\overset{\overset{\displaystyle H}{|}}{\underset{\underset{\displaystyle H}{|}}{C}}CH_2CH_3 \qquad\qquad -CH_2\overset{\overset{\displaystyle H}{|}}{\underset{\underset{\displaystyle CH_3}{|}}{C}}CH_3$$

The *n*-butyl group has one C and two H's at the first point of difference.

The isobutyl group has two C's and one H at the first point of difference.

Study Problem

4.3 Tell whether each of the following compounds is (*E*) or (*Z*):

(a)
$$\begin{array}{c} D \\ \diagdown \\ H \end{array} C=C \begin{array}{c} H \\ \diagup \\ D \end{array}$$

(b)
$$\begin{array}{c} H_3C \\ \diagdown \\ H \end{array} C=C \begin{array}{c} CH_2CH_3 \\ \diagup \\ CH_3 \end{array}$$

(c)
$$\begin{array}{c} Cl \\ \diagdown \\ H \end{array} C=C \begin{array}{c} CH_2CH_3 \\ \diagup \\ CH(CH_3)_2 \end{array}$$

(d)
$$\begin{array}{cc} O & O \\ \parallel & \parallel \\ CH_3C & CCH_2Cl \\ \diagdown & \diagup \\ ClCH_2 \quad & \quad Cl \end{array} C=C$$

▼ ▬▬▬ ## Section 4.2
Geometric Isomerism in Cyclic Compounds

We have described how restricted rotation around a double bond can lead to geometric isomerism. Let us now consider restricted rotation in cyclic compounds.

Atoms joined in a small ring are not free to rotate around the sigma bonds of the ring. Rotation around the sigma bonds forming the ring would require that attached atoms or groups pass through the center of the ring. Van der Waals repulsions prevent this from happening unless the ring contains eight or more carbon atoms, depending on the size of the substituent. The most common rings in organic compounds are five- and six-membered rings. Therefore, we will concentrate our discussion on rings of six and fewer carbon atoms.

groups cannot rotate completely around ring bonds   CH₂CH₃   *a group can rotate completely around this bond*   H

For the moment, we will assume that the carbon atoms of a cyclic structure such as cyclohexane form a plane. (While this is not strictly correct, as we will show later in this chapter, it is often convenient to assume that they do lie in a plane.) For the present discussion, we will view the plane of the ring as being almost horizontal. The edge of the ring projected toward us is shaded more heavily.

away from viewer

toward viewer

Each carbon atom in the cyclohexane ring is joined to its neighboring ring carbon atoms and also to two other atoms or groups. The bonds to these two other groups are represented by vertical lines. A group attached to the top of a vertical line is said to be *above the plane of the ring*, and the group attached to the bottom of a vertical line is said to be *below the plane of the ring*.

H   H
— above the plane
H   CH₃
H   H
H   H
H   H

H   OH ←   — below the plane

In this symbolism, hydrogen atoms attached to the ring and their bonds are not always shown.

CH₃   is the same as

H   H
H   CH₃
H   H
H   H
H   H

H   OH

Another way of showing how groups are attached to the ring is by using a broken wedge to indicate a group below the plane of the ring and a solid line bond or a wedge to represent a group above the plane.

— above the plane —

—CH₃   H₃C—

ˈˈOH   HO

— below the plane —

Descriptions of substituents as being "above the plane" or "below the plane" are correct for only a particular representation of a structure. A molecule can be flipped over in space and the descriptions reversed.

CH₃
H   flip over   H
H   OH
OH   H₃C
                              H

The important point is that, in all the preceding formulas, the methyl group and the hydroxyl group are on *opposite sides* of the plane of the ring. When the two groups are on opposite sides of the ring, they are *trans*; when they are on the same side, they are *cis*. These designations are directly analogous to *cis* and *trans* in alkenes. The *cis*- and *trans*-compounds are geometric isomers of each other, just as *cis*- and *trans*-alkenes are.

trans-2-methylcyclohexanol   *cis*-2-methylcyclohexanol

Sample Problem

Tell whether each of the following compounds is *cis, trans,* or neither:

(a)

(b)

Solution: (a) *trans* because the hydroxyl groups are on opposite sides of the ring; (b) *cis* because the carboxyl groups are on the same side.

Study Problem

4.4 Draw formulas for the geometric isomers of 2-isopropyl-5-methylcyclohexanol. (One of these isomers, *menthol*, is used in cigarettes and throat lozenges.)

▼ ## Section 4.3
Conformations of Open-Chain Compounds

In open-chain compounds, groups attached by sigma bonds can rotate around these bonds. Therefore, the atoms in an open-chain molecule can assume an infinite number of positions in space relative to one another. Ethane is a small molecule, but even ethane can assume different arrangements in space, called **conformations**.

To represent conformations, we will use three types of formulas: **three-dimensional formulas** (which we will call simply **dimensional formulas**), **ball-and-stick formulas**, and **Newman projections**. (We suggest that you also use models to compare different conformations.) A ball-and-stick formula and a dimensional formula are three-dimensional representations of the molecular model of a compound (see Figure 4.2). A Newman projection is an end-on view of only *two carbon atoms* in the molecule. The bond joining these two carbons is hidden. The three bonds attached to the front carbon appear to go to the center of the projection, and the three bonds of the rear carbon are only partially shown.

three-dimensional formula   ball-and-stick formula   Newman projection
(*looking at the ball-and-stick formula end-on*)

Figure 4.2 A dimensional formula, a ball-and-stick formula, and a Newman projection of ethane.

bonds from the front carbon   bonds from the rear carbon   Newman projection for CH_3CH_3

Newman projections can be drawn for molecules with more than two carbon atoms. Because only two carbon atoms at a time can be shown in the projection, more than one Newman projection can be drawn for a molecule. For example, we can show two Newman projections for 3-chloro-1-propanol.

Looking at carbons 1 and 2:

OH

① CH_2

② CH_2

③ CH_2Cl

3-chloro-1-propanol   *dimensional*   *ball-and-stick*   *Newman*

Looking at carbons 2 and 3:

OH

① CH_2

② CH_2

③ CH_2Cl

dimensional   *ball-and-stick*   *Newman*

Because of rotations around sigma bonds, a molecule can assume any number of conformations. The different structures formed by the bond rotations are called **conformers** (from "conformational isomers"). Because conformers are easily interconverted, they cannot usually be isolated from one another, as can structural isomers.

In our formulas of ethane and 3-chloro-1-propanol, we have shown **staggered conformers**, in which the hydrogen atoms or the attached groups are as far apart from one another as possible. Because the C—C bond can undergo rotation, the hydrogen atoms might also be **eclipsed**, or as close as possible, one behind the other in the Newman projection. We will show them not quite eclipsed so they can be seen.

repulsion

staggered   *eclipsed*

The rotation around sigma bonds is often called **free rotation**, but it is not entirely free. The eclipsed conformation of ethane is about 3 kcal/mol less stable (of higher energy) than the staggered conformer because of minor repulsions between the bonding electrons to the hydrogen atoms. To undergo rotation from a staggered conformation to an eclipsed conformation, a mole of ethane molecules would require 3 kcal. Since this amount of energy is readily available to molecules at room temperature, the rotation can occur easily. However, even though the conformations of ethane are interconvertible at room temperature, at any given time we would expect a greater percentage of ethane molecules to be in the staggered conformation because of its lower energy. A diagram showing the changes in potential energy with rotation around the C—C bond in ethane is presented in Figure 4.3.

Butane ($CH_3CH_2CH_2CH_3$), like ethane, can exist in eclipsed and staggered conformations. In butane, there are two methyl groups, which are relatively large, attached to the center two carbons. If butane is viewed from the center two carbons, the presence of these methyl groups gives rise to two types of staggered conformations that differ in the positions of the methyl groups in relation to each other. The staggered conformation in which the methyl groups are at the maximum distance apart is called the *anti* conformer (Greek *anti*, "against"). The staggered conformations in which the methyl groups are closer are called *gauche* conformers (French *gauche*, "left" or "crooked"). Newman projections for one-half a complete rotation follow.

Partial rotation around the carbon 2–carbon 3 bond of butane (rear carbon rotating):

anti
(*lowest energy*)

eclipsed

gauche

eclipsed methyls
(*highest energy*)

The larger the groups attached to two carbon atoms, the greater is the energy difference between the molecule's conformations. It takes more energy to push together two bulky groups than two small groups. While it takes only 3 kcal/mol for ethane to rotate from staggered to eclipsed, it takes 4.5 kcal/mol for butane to rotate from *anti* to the conformation in which the methyls are eclipsed. The energy relationships of the complete rotation around the carbon 2–carbon 3 bond of butane are shown in Figure 4.4.

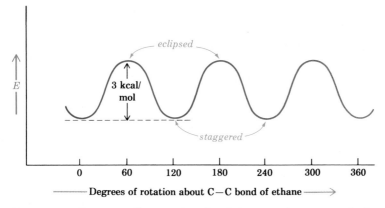

Figure 4.3 Energy changes involved in rotation around the carbon–carbon sigma bond of ethane.

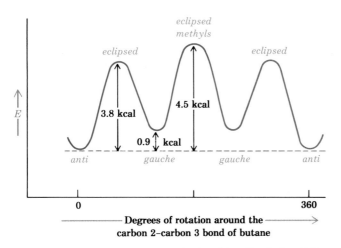

Figure 4.4 Energy relations (in kcal/mol) of the different conformations of butane.

Study Problem

4.5 Draw Newman projections for the *anti* and *gauche* conformations of   (**a**) 1-bromo-2-chloroethane, and   (**b**) 3-hydroxypropanoic acid, $HOCH_2CH_2CO_2H$.

▼ ## Section 4.4
Shapes of Cyclic Compounds

A. Ring Strain

In 1885, Adolf von Baeyer, a German chemist, theorized that cyclic compounds form planar rings. Baeyer further theorized that all cyclic compounds except for cyclopentane would be "strained" because their bond angles are not close to the tetrahedral angle of 109.5°. He proposed that, because of the abnormally small bond angles of the ring, cyclopropane and cyclobutane would be more reactive than an open-chain alkane. According to Baeyer, cyclopentane would be the most stable ring system (because its bond angles are closest to tetrahedral), and then reactivity would increase again starting with cyclohexane.

Bond angles according to Baeyer:

Baeyer's theory was not entirely correct. Cyclohexane and larger-sized rings are not more reactive than cyclopentane. We now know that cyclohexane is not a flat ring with bond angles of 120°, but rather a puckered ring with bond angles close to 109°, the normal sp^3 bond angles.

However, there is indeed what we call **ring strain** in the smaller ring systems. Cyclopropane is the most reactive of the cycloalkanes. Its heat of combustion is higher per CH_2 group than that of other alkanes (Table 4.1).

cyclohexane $+ 9 O_2 \longrightarrow 6 CO_2 + 6 H_2O + 944.5$ kcal/mol (157.4 kcal/mol for each CH_2)

$2 \triangle$ $+ 9 O_2 \longrightarrow 6 CO_2 + 6 H_2O + 499.8$ kcal/mol (166.6 kcal/mol for each CH_2)
cyclopropane

Table 4.1 Strain energies from heat of combustion data

	$-\Delta H$ (kcal/mol)	$-\Delta H$ per CH_2 (kcal/mol)[a]	Strain Energy per CH_2 (kcal/mol)[b]	Strain Energy, Total (kcal/mol)[c]
cyclopropane	499.8	166.6	9.2	27.6
cyclobutane	655.9	164.0	6.6	26.4
cyclopentane	793.5	158.7	1.3	6.5
cyclohexane	944.5	157.4	0	0

[a] $-\Delta H$/mol divided by number of CH_2 groups.
[b] The difference between (1) the value of $-\Delta H/CH_2$ for that compound, and (2) the value for cyclohexane, the assumption being that cyclohexane is not strained.
[c] Strain energy/CH_2 × number of CH_2 groups.

When treated with hydrogen gas at moderate temperatures, unstrained cyclo-alkanes such as cyclopentane do not react, but cyclopropane undergoes ring opening.

$$\text{cyclopentane} + H_2 \xrightarrow{\text{catalyst}} \text{no reaction}$$

$$\text{cyclopropane} + H_2 \xrightarrow{\text{catalyst}} CH_3CH_2CH_3$$
$$\text{propane}$$

Today we would say that the sp^3 orbitals of the carbon atoms in cyclo-propane cannot undergo complete overlap with each other because the angles between the carbon atoms of cyclopropane are geometrically required to be 60° (see Figure 4.5). The ring sigma bonds of cyclopropane are of higher energy than sp^3 sigma bonds that have the normal tetrahedral angle. The cyclopro-pane bonds are more easily broken than most other C—C sigma bonds and, in comparable reactions, more energy is released.

Cyclobutane is less reactive than cyclopropane but more reactive than cyclopentane. Following along reasonably with Baeyer's theory, the cyclopen-tane ring is stable and is far less reactive than the three- and four-membered rings.

With cyclohexane and larger rings, Baeyer's predictions fail. Cyclohexane and rings larger than cyclohexane are found in puckered conformations rather than as flat rings and are not particularly reactive. Larger rings are not as common in naturally occurring compounds as five- and six-membered rings. Baeyer felt the rarity was due to ring strain. We now realize that the rarity of larger rings is not due primarily to unusually high bond energies. Instead, the scarcity of these compounds arises from the decreasing probability that the ends of longer molecules will find each other to undergo a ring-forming reaction.

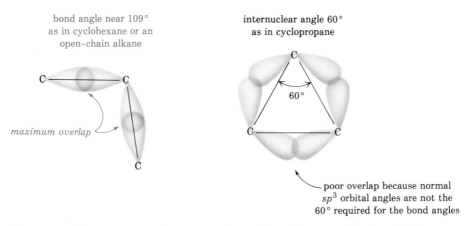

Figure 4.5 Maximum overlap cannot be achieved between the ring carbon atoms in cyclopropane.

Sample Problem

Would you expect the benzene ring to be flat or puckered?

Solution: The benzene ring is flat because the carbon atoms are sp^2 (not sp^3) hybridized. The normal positioning of sp^2 bonds is planar and with angles of 120° between them, corresponding to a regular planar hexagon.

Study Problem

4.6 Considering ring sizes, which of the following compounds would you expect to have substantial amounts of ring strain? [Note the arrangement of carbon atoms in (**b**); this type of bonding is also found in diamonds.]

(**a**)   (**b**)   (**c**)

decalin   adamantane

thujone

found in many plants

B. Ring Puckering and Hydrogen–Hydrogen Repulsions

If the cyclohexane ring were flat, all the hydrogen atoms on the ring carbons would be eclipsed. In the puckered conformer that we have shown, however, all the hydrogens are staggered. The energy of this puckered conformer of cyclohexane is lower than the energy of flat cyclohexane, both because of more-favorable sp^3 bond angles and also because of fewer hydrogen–hydrogen repulsions.

eclipsed   *staggered*

flat   *puckered*

What of the other cyclic compounds? Cyclopentane would have near-optimal bond angles (108°) if it were flat, but cyclopentane also is slightly puckered so that the hydrogen atoms attached to the ring carbons are staggered. Cyclobutane (flat bond angles of 90°) also is puckered, even though the puckering causes more-strained bond angles. Cyclopropane must be planar; geometrically, three points (or three carbon atoms) define a plane. The hydrogen atoms in cyclopropane necessarily are eclipsed.

*envelope form
of cyclopentane*

*butterfly form
of cyclobutane*

Section 4.5
The Conformers of Cyclohexane

The cyclohexane ring, either alone or in fused-ring systems (ring systems that jointly share carbon atoms), is the most important of all the ring systems. In this section, we will study the conformations of cyclohexane and substituted cyclohexanes. In Chapter 23, we will discuss the conformations of fused-ring systems.

There are many shapes that a cyclohexane ring can assume, and any single cyclohexane molecule is in a continuous state of flexing into different shapes. (Molecular models are invaluable for showing the relationships among the various conformations.) So far, we have shown the **chair form** of cyclohexane. This and some other shapes the cyclohexane molecule can assume are as follows:

chair   *half-chair*   *twist-boat*   *boat*

None of these other conformations has the favorable staggered-hydrogen structure of the chair form. The eclipsing of hydrogens, as in the boat form, adds to the energy of the molecule. Figure 4.6 shows models and Newman projections of the chair form and boat form; the staggered and eclipsed hydrogens are apparent in these representations.

The energy requirements for the interconversion of the different conformations of cyclohexane are shown in Figure 4.7. We can see that the chair form has the lowest energy, while the half-chair (which has an almost-planar structure) has the highest-energy. At any given time, we would expect most cyclohexane molecules to be in the chair form. Indeed, it has been calculated that about 99.9% of cyclohexane molecules are in the chair form at any one time.

A. Equatorial and Axial Substituents

The carbon atoms of the chair form of cyclohexane roughly form a plane. For purposes of discussion, an axis may be drawn perpendicularly to this plane. These operations are shown in Figure 4.8.

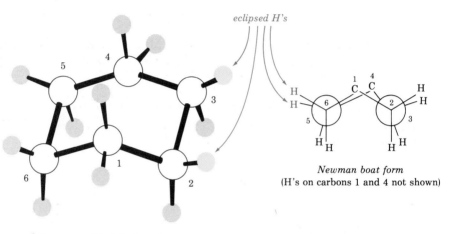

model of the chair form

Newman chair form
(H's on carbons 1 and 4 not shown)

model of the boat form

Newman boat form
(H's on carbons 1 and 4 not shown)

Figure 4.6 Molecular models and Newman projections of the chair and boat forms of cyclohexane.

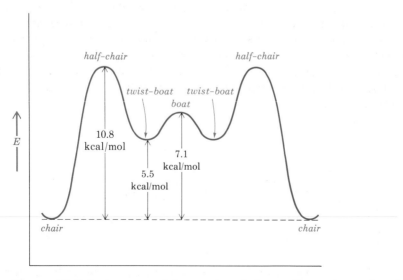

Figure 4.7 Relative potential energies of the conformations of cyclohexane.

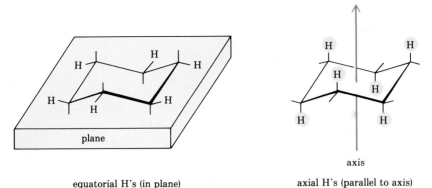

Figure 4.8 The equatorial and axial hydrogens of cyclohexane.

Each ring carbon of cyclohexane is bonded to two hydrogen atoms. The bond to one of these hydrogens is in the rough plane of the ring; this hydrogen atom is called an **equatorial hydrogen**. The bond to the other hydrogen atom is parallel to the axis; this is an **axial hydrogen**. Each of the six carbon atoms of cyclohexane has one equatorial and one axial hydrogen atom. (Again, refer to Figure 4.8.) In the flipping and reflipping between the conformers, axial becomes equatorial, while equatorial becomes axial.

A methyl group is larger than a hydrogen atom. When the methyl group in methylcyclohexane is in the axial position, the methyl group and axial hydrogens on the same side of the ring repel each other. Interactions between axial groups are called **1,3-diaxial interactions**. When the methyl group is in the equatorial position, the repulsions are minimized. Thus, the energy of the conformer with an equatorial methyl is lower. At room temperature, about 95% of methylcyclohexane molecules are in the conformation in which the methyl group is equatorial.

When the methyl group is in the axial position, the conformation of the ring is similar to that of butane with *gauche* methyl groups.

Gauche relationship in axial methylcyclohexane:

gauche butane

When the methyl group is in the equatorial position, the conformation of the ring is similar to that of butane with *anti* methyl groups.

Anti relationship in equatorial methylcyclohexane:

anti butane

Gauche butane is 0.9 kcal/mol less stable than *anti* butane. Methylcyclohexane with an axial methyl group is 1.8 kcal/mol less stable than the conformation with an equatorial methyl group. The energy is twice that for butane (2 × 0.9 kcal/mol) because the axial methyl group on the ring interacts with *two* axial hydrogen atoms (one at carbon 3 and one at carbon 5).

The bulkier the group, the greater is the energy difference between axial and equatorial conformers. In other words, a cyclohexane ring with a bulky substituent is more likely to have that group in the equatorial position. When the size of the substituent group reaches *t*-butyl, the difference in energies between the conformers becomes quite large. *t*-Butylcyclohexane is often said to be "frozen" in the conformation in which the *t*-butyl group is equatorial. The ring is not truly frozen, but the energy difference (5.6 kcal/mol) between the equatorial and axial positions of the *t*-butyl group means that only 1 in 10,000 molecules has the *t*-butyl group in an axial position at any given time.

favored

B. Disubstituted Cyclohexanes

Two groups substituted on a cyclohexane ring may be either *cis* or *trans*. The *cis*- and *trans*-disubstituted rings are geometric isomers and are not interconvertible at room temperature; however, either isomer may assume a variety of conformations. For example, consider some chair forms of *cis*-1,2-dimethylcyclohexane.

Some different representations of cis-1,2-dimethylcyclohexane,

both "down"   both "down"   both "up"

Because this is the *cis*-isomer, the methyl groups must both be on the same side of the ring regardless of the conformation. In each chair conformation that we can draw, one methyl is axial and the other is equatorial. For any *cis*-1,2-disubstituted cyclohexane, one substituent must be axial and the other substituent must be equatorial. (Refer back to Figure 4.8 or use molecular models and verify this statement for yourself.)

When *cis*-1,2-dimethylcyclohexane changes from one chair conformer to the other, the two methyl groups reverse their equatorial–axial status. The energies of the two conformers are equal because their structures and bonding are equivalent. Therefore, this compound exists primarily as a 50:50 mixture of these two chair-form conformers.

Conformers of cis-1,2-dimethylcyclohexane:

axial, equatorial (or *a,e*)   *equatorial, axial* (or *e,a*)

In *trans*-1,2-dimethylcyclohexane, the methyl groups are on *opposite* sides of the ring. In the chair form of the *trans*-isomer, one group must be attached to an "uppermost" bond, while the other is attached to a "lowermost" bond.

Some different representations of trans-1,2-dimethylcyclohexane,

No matter how the two adjacent *trans* groups are shown, they are *both axial* (*a*, *a*) or they are *both equatorial* (*e*, *e*). There is no way for two groups to be *trans* and 1,2 on the chair form of cyclohexane without assuming either the *a*,*a* or the *e*,*e* conformation.

Conformers of trans-1,2-dimethylcyclohexane:

$$a,a \qquad\qquad\qquad\qquad e,e$$
$$favored$$

A single methyl group on a cyclohexane ring assumes the equatorial position preferentially. *Two* methyl groups on a cyclohexane ring also assume the equatorial positions preferentially, if possible. In *trans*-1,2-dimethylcyclohexane, the *e*,*e* conformer is the preferred conformer and is of lower energy than the *a*,*a* conformation. The *trans e*,*e* conformer is also of lower energy (by 1.87 kcal/mol) than either conformer of the *cis*-compound, which must be *a*,*e* or *e*,*a*.

In summary,

For the 1,2-dimethylcyclohexanes:

cis: *a,e* ⇌ *e,a*   *equal energies*

trans: *a,a* → *e,e*   *more stable*

In the case of a 1,2-disubstituted cyclohexane, the *trans*-isomer is more stable than the *cis*-isomer because both substituents can be equatorial. However, when the two substituents are 1,3 to each other on a cyclohexane ring, the *cis*-isomer is more stable than the *trans*-isomer. The reason is that both substituents in the *cis*-1,3-isomer can be equatorial. In the *trans*-1,3-isomer, one group must be axial.

cis-1,3-dimethylcyclohexane:

$$a,a \qquad\qquad\qquad\qquad e,e$$
$$more\ stable\ and\ favored$$

trans-1,3-dimethylcyclohexane:

$$e,a \qquad\qquad\qquad\qquad a,e$$

Sample Problem

What are the possible equatorial–axial relationships for the *cis-* and *trans*-1,4-di-methylcyclohexanes? In each case, which conformer is of lower energy?

Solution: *cis*-1,4:   *a,e*;   *e,a*
            trans-1,4:   *a,a*;   *e,e*

The *cis*-conformers are of equal energy. The *e,e trans*-conformer is of lower energy than the *a,a trans*-conformer (and of lower energy than either *cis*-conformer).

Study Problem

4.7 Label each of the following disubstituted rings as *cis* or *trans* and as *a,a*; *e,e*; or *a,e*.

(a)

(b)

(c)

(d)

Section 4.6
Chirality

A. Chirality of Objects and Molecules

Consider your left hand. Your hand cannot be superposed on its mirror image, or placed over the mirror image so that all parts coincide. If you hold your left hand up to a mirror, the image looks like a right hand. If you do not have a mirror available, hold your hands together with the palms facing each other. You can see that they are mirror images. Try to superpose your hands (both palms down). You cannot do it (see Figure 4.9). This right- and left-handedness is also encountered in shoes and gloves. (Try wearing a left-handed glove on a right hand!)

Any object that *cannot be superposed on its mirror image* is said to be **chiral** (Greek *cheir*, "the hand"). Hands, gloves, and shoes are all chiral. Conversely, a plain cup or a cube is **achiral** (not chiral); these can be superposed on their mirror images. Figure 4.10 shows a cup being superposed on its mirror image. The same principles of right- and left-handedness also apply to molecules. A molecule that can be superposed upon its mirror image is *achiral*. A molecule that *cannot* be superposed on its mirror image is *chiral*. Figures 4.11 and 4.12 show an achiral molecule and a chiral molecule along with their mirror images.

An achiral molecule and its superposable mirror-image molecule are the same compound; they are not isomers. But a chiral molecule is *not* superposable

Figure 4.9 A *chiral* object *cannot* be superposed, or made to coincide completely, on its mirror image.

Figure 4.10 An *achiral* object *is* superposable on its mirror image.

on its mirror image; this molecule and its mirror-image molecule are different compounds, and represent a pair of stereoisomers called **enantiomers**. A pair of enantiomers is simply a pair of isomers that are *nonsuperposable mirror images*.

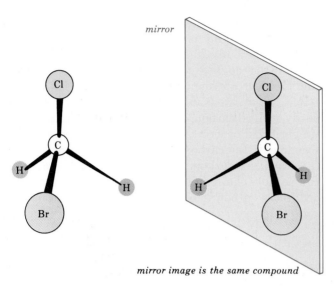

Figure 4.11 A molecule with a single carbon atom that has two identical sub-stituents (H in this case) is achiral and can be superposed upon its mirror image. (Try it with models.)

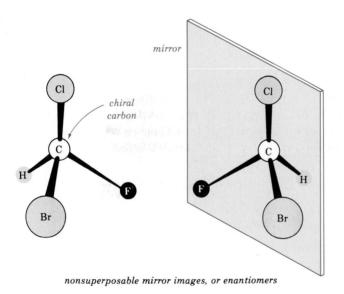

Figure 4.12 A molecule that has four different groups attached to one single carbon atom is chiral and is not superposable on its mirror image. (Try it with models.)

B. Chiral Carbon Atoms

The most common structural feature (but not the only one) that gives rise to chirality in molecules is that the molecule contains an sp^3-hybridized carbon atom with *four different groups* attached to it (Figure 4.12). Such a molecule is chiral and exists as a pair of enantiomers. A carbon atom with four different groups attached is usually called an **asymmetric carbon atom** or a **chiral carbon atom** (although, technically, it is the molecule and not the carbon atom that is chiral). By learning to recognize chiral carbons within a formula, we can greatly simplify the problem of identifying structures that can exist as enantiomers.

In order to identify a chiral carbon, we must determine that all four groups attached to the sp^3-hybridized carbon are different. In many cases, the problem is trivial; for example, if the carbon is attached to two or more H atoms ($-CH_2-$ or $-CH_3$), then the carbon cannot be chiral. However, in a few cases the problem can be more challenging. In these cases we must inspect each *entire group* attached to the carbon in question, not just the atoms bonded directly to that carbon.

$$CH_3CH_2CH_2CH_2 \overset{\displaystyle CH_2CH_3}{\underset{\displaystyle CH_3}{-C-}} CH_2CH_2CH_3$$

chiral carbon

Sample Problem

Chiral carbons are often starred for emphasis. Star the chiral carbons in the following structures:

$$CH_3\overset{Cl}{\underset{Br}{C}}CH_2CH{=}CHCH_3 \qquad CH_3\overset{O}{\overset{\|}{C}H}\overset{}{\underset{CH_2CH_3}{C}}OCH_3$$

Solution:

$$CH_3\overset{Cl}{\underset{Br}{\overset{*}{C}}}CH_2CH{=}CHCH_3 \qquad CH_3\overset{O}{\overset{\|}{\underset{CH_2CH_3}{\overset{*}{C}H}}}COCH_3$$

Study Problem

4.8 Star the chiral carbons (if any) in the following formulas:

$$\textbf{(a) } CH_3\overset{CH_3}{\underset{}{C}}HBr \qquad \textbf{(b) } CH_3CH_2\overset{CH_3}{\underset{}{C}}HBr \qquad \textbf{(c) } \bigcirc{-}CH_2\overset{F}{\underset{}{C}}H{-}\bigcirc$$

Drawing structures of a pair of enantiomers for a molecule with one chiral carbon is relatively easy. The *interchange of any two groups* around the chiral carbon results in the enantiomer. The following examples show two ways to convert the formula for one compound (**A**) to the formula for its enantiomer (**B**). (Use models to verify that the two formulas shown for **B** actually do represent the same molecule.)

interchange

$$\overset{CH_3}{\underset{CH_2CH_3}{H-C-Cl}}$$

A

$$\overset{CH_3}{\underset{CH_2CH_3}{Cl-C-H}}$$

B

$$\underbrace{}_{enantiomers}$$

interchange

$$\overset{CH_3}{\underset{CH_2CH_3}{H-C-Cl}}$$

A

$$\overset{CH_3}{\underset{Cl}{H-C-CH_2CH_3}}$$

B

$$\underbrace{}_{enantiomers}$$

Study Problem

4.9 Draw the formulas (either dimensional or ball-and-stick) for the two enantiomers of each of the following compounds:

(a) [benzene ring]—$\overset{Br}{\underset{}{CHCH_3}}$ **(b)** $CH_3CH_2\overset{OH}{\underset{}{CHCH_3}}$

C. Fischer Projections

In the late 1800s, the German chemist Emil Fischer introduced projection formulas for showing the spatial arrangement of groups around chiral carbon atoms. These projection formulas are called **Fischer projections**. Because Fischer devised these formulas for representing sugar molecules, we will illustrate the type of Fischer projection commonly used today with the simplest of sugars: 2,3-dihydroxypropanal (usually called glyceraldehyde) and 2,3,4-trihydroxybutanal (erythrose). Glyceraldehyde has one chiral carbon atom (carbon 2), while erythrose has *two* chiral carbons (carbons 2 and 3). As you can see from the following representations for these two compounds, a Fischer projection is simply a shorthand way to represent a ball-and-stick or dimensional formula.

For glyceraldehyde:

$$\overset{O}{\overset{\parallel}{①\,CH}}$$
$$H{\underset{②}{\diagup}}OH$$
$$③\,CH_2OH$$

or

$$\overset{O}{\overset{\parallel}{CH}}$$
$$H-C-OH$$
$$CH_2OH$$

becomes

$$\overset{O}{\overset{\parallel}{CH}}$$
$$H-\!\!\!-OH$$
$$CH_2OH$$

Fischer projection

For erythrose:

$$\overset{O}{\overset{\parallel}{①\,CH}}$$
$$H{\underset{②}{\diagup}}OH$$
$$H{\underset{③}{\diagup}}OH$$
$$④\,CH_2OH$$

or

$$\overset{O}{\overset{\parallel}{CH}}$$
$$H-C-OH$$
$$H-C-OH$$
$$CH_2OH$$

becomes

$$\overset{O}{\overset{\parallel}{CH}}$$
$$H-\!\!\!-OH$$
$$H-\!\!\!-OH$$
$$CH_2OH$$

Fischer projection

In drawing a Fischer projection, we assume that the molecule is *completely stretched out in the plane of the paper with all its substituents eclipsed*, regardless of any preferred conformation. The preceding formulas for erythrose show the conformation used for the Fischer projection. By convention, the carbonyl group (or group of highest nomenclature priority) is placed at or near the top. Thus, *the top carbon is carbon 1.* Each intersection of the horizontal and vertical lines represents a chiral carbon. *Each horizontal line represents a bond coming toward the viewer, while the vertical line represents bonds going back, away from the viewer.*

away from viewer

$$
\begin{array}{c}
\text{CHO} \quad\text{—— \textit{toward viewer}} \\
\text{H}\!\!-\!\!\!|\!\!-\!\!\text{OH} \\
\text{CH}_2\text{OH}
\end{array}
$$

A pair of enantiomers is easily recognized when Fischer projections are used.

$$
\begin{array}{c}
\overset{1}{\text{C}}\text{HO} \\
\text{H}\overset{2}{-\!\!|\!\!-}\text{OH} \\
\text{H}\overset{3}{-\!\!|\!\!-}\text{OH} \\
\text{HO}\overset{4}{-\!\!|\!\!-}\text{H} \\
\overset{5}{\text{C}}\text{H}_2\text{OH}
\end{array}
\qquad\qquad
\begin{array}{c}
\text{CHO} \\
\text{HO}-\!\!|\!\!-\text{H} \\
\text{HO}-\!\!|\!\!-\text{H} \\
\text{H}-\!\!|\!\!-\text{OH} \\
\text{CH}_2\text{OH}
\end{array}
$$

Carbons 2, 3, and 4   *The enantiomer shows all groups or atoms*
are chiral.   *on chiral carbons transposed*
                          from left to right.

A Fischer projection may be rotated 180° in the plane of the paper, but it may not be flipped over or rotated by any other angle. Either of these last two operations would take the formula out of the Fischer projection and lead to an incorrect structure.

Correct:

$$
\begin{array}{c}
\text{CHO} \\
\text{H}-\!\!|\!\!-\text{OH} \\
\text{CH}_2\text{OH}
\end{array}
\xrightarrow{\text{rotate 180°}}
\begin{array}{c}
\text{CH}_2\text{OH} \\
\text{HO}-\!\!|\!\!-\text{H} \\
\text{CHO}
\end{array}
$$

same as left-hand structure

Incorrect:

$$
\begin{array}{c}
\text{CHO} \\
\text{H}-\!\!|\!\!-\text{OH} \\
\text{CH}_2\text{OH}
\end{array}
\xrightarrow{\text{rotate 90°}}\!\!\!\!\times
\begin{array}{c}
\text{H} \\
\text{HOCH}_2-\!\!|\!\!-\text{CHO} \\
\text{OH}
\end{array}
$$

NOT THE SAME
(To verify the difference, draw
this projection as a dimensional
formula with the horizontal bonds
coming toward you.)

Fischer projections are a convenient shorthand way of representing chiral molecules. Because of the limitations placed upon them, such as the preceding one of rotation, Fischer projections must be used carefully. We suggest that you convert Fischer projections to ball-and-stick or dimensional formulas (or use models) when performing any spatial manipulations. In future chapters, we will stress dimensional formulas rather than Fischer projections until we discuss carbohydrates (Chapter 22).

Study Problems

4.10 The following Fischer projections represent two amino acids that are formed when proteins are hydrolyzed (cleaved with water). Convert these projections to dimensional formulas.

(a)
$$CO_2H$$
$$H_2N \!-\!\!\!\!+\!\!\!\!- H$$
$$CH(CH_3)_2$$

valine

(b)
$$CO_2H$$
$$H_2N \!-\!\!\!\!+\!\!\!\!- H$$
$$CH_2CO_2H$$

aspartic acid

4.11 Convert the following dimensional formulas for amino acids to Fischer projections.

(a)
$$NH_2$$
$$CH_3 \overset{C}{\diagdown} CO_2H$$
$$H$$

alanine

(b)
$$CH_2SH$$
$$H \overset{C}{\diagup} CO_2H$$
$$H_2N$$

cysteine

Section 4.7
Rotation of Plane-Polarized Light

With the exception of chirality, the structures of a pair of enantiomers are the same. Therefore, almost all of their physical and chemical properties are the same. For example, each pure enantiomer of a pair has the same melting point and the same boiling point. Only two sets of properties are different for a single pair of enantiomers:

1. interactions with other chiral substances, and
2. interactions with polarized light.

Ordinary light travels in waves, and the waves are at right angles to the direction of travel. **Plane-polarized light** is light in which all wave vibrations have been filtered out except for those in one plane. The plane polarization can be effected by passing ordinary light through a pair of calcite crystals ($CaCO_3$) or a polarizing lens. (The same principle is used in Polaroid sunglasses.) Figure 4.13 shows a simplified diagram of the plane polarization of light.

If plane-polarized light is passed through a solution containing a single enantiomer, the plane of polarization of the light is *rotated* either to the right or to the left (see Figure 4.14). The rotation of plane-polarized light is referred to as **optical rotation**. A compound that rotates the plane of polarization of plane-polarized light is said to be **optically active**. For this reason, enantiomers are sometimes referred to as **optical isomers**.

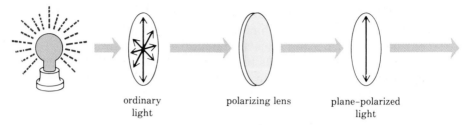

ordinary light   polarizing lens   plane-polarized light

Figure 4.13 The plane polarization of light.

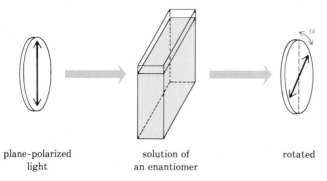

plane-polarized light   solution of an enantiomer   rotated

Figure 4.14 The plane of polarization of plane-polarized light is rotated by a single enantiomer.

A **polarimeter** is an instrument designed for polarizing light and then showing the angle of rotation of the plane of polarization of the light by an optically active compound. The amount of rotation depends on (1) the structure of the molecules; (2) the temperature; (3) the wavelength; (4) the number of molecules in the path of the light; and, in some cases, (5) the solvent.

Specific rotation is the amount of rotation by 1.00 gram of sample in 1.00 mL of solution in a tube with a path length of 1.00 decimeter (dm), at a specified temperature, wavelength, and solvent. The commonly used wavelength is 589.3 nm (the D line of sodium), where 1.0 nm $= 10^{-9}$ m. The specific rotation for a compound (at 20°, for example) may be calculated from the observed rotation by the following formula:

$$[\alpha]_{\text{D}}^{20} = \frac{\alpha}{lc}$$

where $[\alpha]_{\text{D}}^{20}$ = specific rotation of sodium D line at 20°

         α = observed rotation at 20°

         l = cell length in dm (1.0 dm = 10 cm)

         c = concentration of sample solution in g/mL

The fact that some compounds rotate the plane of polarization of plane-polarized light was discovered in 1815 by the French physicist Jean-Baptiste Biot. But it was Louis Pasteur who in 1848 made the momentous discovery that there are two types of sodium ammonium tartrate crystals and that these two types are mirror images of each other. (We will discuss the structure of tartaric acid in Section 4.9C.)

Pasteur painstakingly separated the "left-handed" crystals from the "right-handed" crystals with a pair of tweezers. Imagine his amazement when he found that (1) a solution of the original mixture of crystals did not rotate plane-polarized light; (2) a solution of left-handed crystals did rotate plane-polarized light; and (3) a solution of right-handed crystals also rotated plane-polarized light to exactly the same extent, but in the opposite direction. Pasteur's experiment and later experiments of innumerable other scientists lead us to the following conclusions:

1. A pair of pure enantiomers each rotates the plane of polarization of plane-polarized light the same number of degrees, but in the opposite directions (one to the left; one to the right).
2. A mixture of equal parts of a pair of enantiomers does not rotate plane-polarized light.

A. Some Terms Used in Discussing Optical Rotation

The enantiomer of any enantiomeric pair that rotates plane-polarized light to the *right* is said to be **dextrorotatory** (Latin *dexter*, "right"). Its mirror image, which rotates plane-polarized light to the *left*, is said to be **levorotatory** (Latin *laevus*, "left"). The direction of rotation is specified in the name by $(+)$ for dextrorotatory and $(-)$ for levorotatory. (In older literature, d for dextrorotatory and l for levorotatory are sometimes encountered.)

$$
\begin{array}{cccc}
\underset{\substack{| \\ \text{CH}_2\text{OH}}}{\overset{\substack{\text{O} \\ \| \\ \text{CH} \\ |}}{\text{H}\!-\!\text{C}\!\!-\!\text{OH}}} & \text{or} & \underset{\substack{| \\ \text{CH}_2\text{OH}}}{\overset{\substack{\text{O} \\ \| \\ \text{CH} \\ |}}{\text{H}\!-\!\!\!-\!\text{OH}}} & \underset{\substack{| \\ \text{CH}_2\text{OH}}}{\overset{\substack{\text{O} \\ \| \\ \text{CH} \\ |}}{\text{HO}\!-\!\text{C}\!\!-\!\text{H}}} & \text{or} & \underset{\substack{| \\ \text{CH}_2\text{OH}}}{\overset{\substack{\text{O} \\ \| \\ \text{CH} \\ |}}{\text{HO}\!-\!\!\!-\!\text{H}}}
\end{array}
$$

(+)-glyceraldehyde $\qquad\qquad$ (−)-glyceraldehyde

$[\alpha]_D^{20} = +8.7°$ $\qquad\qquad$ $[\alpha]_D^{20} = -8.7°$

A mixture of equal parts of any pair of enantiomers is called a **racemic mixture**, or **racemic modification**. A racemic mixture may be indicated in the name by the prefix $(\pm)$. (In older literature, *dl* is used to signify a racemic mixture.) Thus, racemic glyceraldehyde is called $(\pm)$-glyceraldehyde. (The term *racemic* comes from the Latin *racemis*, "a bunch of grapes." The reason for this unusual derivation is that *racemic* was first used to describe racemic tartaric acid, which was isolated as a by-product of wine-making.)

50% (+)-glyceraldehyde + 50% (−)-glyceraldehyde = $(\pm)$-glyceraldehyde

$[\alpha]_D^{20} = +8.7°$ $\qquad\qquad$ $[\alpha]_D^{20} = -8.7°$ $\qquad\qquad$ $[\alpha]_D^{20} = 0$

a racemic mixture

A racemic mixture does not rotate the plane of polarization of plane-polarized light because the rotation by each enantiomer is cancelled by the equal and opposite rotation by the other. A solution of either a racemic mixture or of an achiral compound is said to be **optically inactive**, but the causes of the optical inactivity are different.

Section 4.8
Assignment of Configuration: the (R) and (S) System

The order of arrangement of four groups around a chiral carbon atom is called the **absolute configuration** around that atom. (Do not confuse configurations with *conformations*, which are shapes arising from rotation around bonds.) A pair of enantiomers have opposite configurations. For example, (+)- and (−)-glyceraldehyde have opposite configurations. But which formula represents the dextrorotatory enantiomer and which represents the levorotatory one? Until 1951, chemists did not know! Prior to that time, it was known that (+)-glyceraldehyde and (−)-glyceric acid (2,3-dihydroxypropanoic acid) have the same configuration around carbon 2, even though they rotate plane-polarized light in opposite directions. But it was not known whether the OH at carbon 2 was to the right or to the left in the formulas as shown:

same configuration, but is the −OH on the right (as shown) or on the left?

(−)-glyceric acid   (+)-glyceraldehyde

To make formulas easier to work with, chemists decided in the late 19th century to assume that (+)-glyceraldehyde has the absolute configuration with the OH on carbon 2 to the right, as shown in the preceding formula. In 1951, x-ray diffraction studies by J. M. Bijvoet at the University of Utrecht in Holland showed that the original assumption was correct. (+)-Glyceraldehyde does indeed have the absolute configuration that chemists had been using for 60 years. If the early chemists had guessed wrong, the chemical literature would have been in a state of confusion—all pre-1951 articles would show configurations the *reverse* of those in more modern articles. The original assumption was indeed a lucky guess!

The direction of rotation of plane-polarized light by a particular enantiomer is a physical property. The absolute configuration of a particular enantiomer is a characteristic of its molecular structure. *There is no simple relationship between the absolute configuration of a particular enantiomer and its direction of rotation of plane-polarized light.* As we have said, the enantiomer of glyceric acid with the same absolute configuration as (+)-glyceraldehyde is levorotatory, not dextrorotatory.

We have shown how the direction of rotation of plane-polarized light can be indicated by (+) or (−). However, a system is also needed to indicate the absolute configuration—that is, the actual arrangement of groups around a chiral carbon. This system is the **(R) and (S) system**, or the **Cahn–Ingold–Prelog system**. The letter (R) comes from the Latin *rectus*, "straight," or "right," while (S) comes from the Latin *sinister*, "left." Any chiral carbon atom has either an (R) configuration or an (S) configuration; therefore, one enantiomer is (R) and the other is (S). A racemic mixture can be designated (R)(S), meaning a mixture of the two.

In the (R) and (S) system, groups are assigned a priority ranking using the same set of rules as are used in the (E) and (Z) system; however, the priority ranking is used in a slightly different manner. To assign an (R) or (S) configuration to a chiral carbon:

---

1. Rank the four groups (or atoms) attached to the chiral carbon in order of priority by the Cahn–Ingold–Prelog sequence rules (pages 119–121).
2. Project the molecule with the group of *lowest priority* to the rear.
3. Select the group of *highest priority* and *draw a curved arrow to the group of next highest priority*.
4. If this arrow is *clockwise*, the configuration is (R). If the arrow is *counterclockwise*, the configuration is (S).

Let us illustrate the use of this procedure by assigning (R) and (S) to the enantiomers of 1-bromo-1-chloroethane.

enantiomers of 1-bromo-1-chloroethane

1. Rank the four groups. In the case at hand, the order of priority of the four atoms by atomic number is Br (highest), Cl, C, H (lowest).
2. Draw projections with the lowest-priority atom (H) in the rear. (This atom is hidden behind the carbon atom in the following projections.)
3. Draw an arrow from highest priority (Br) to second highest priority (Cl).
4. Assign (R) and (S). Note how the (R) and (S) designation is incorporated into the name.

clockwise = (R)
(R)-1-bromo-1-chloroethane

counterclockwise = (S)
(S)-1-bromo-1-chloroethane

An easy way to remember which enantiomer is (R) and which is (S) is to compare the curved arrow to the steering wheel on an automobile. Turning the steering wheel to the *right* is equivalent to (R); turning it to the *left* is equivalent to (S).

right turn

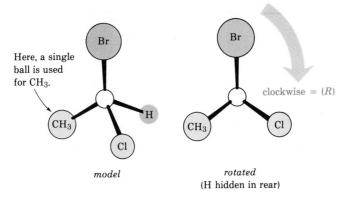

Figure 4.15 **Using a molecular model to determine (R) or (S) configuration.**

Placing a structure in the correct position for an (R) or (S) assignment is easily done with a molecular model. Construct the model, grasp the group of lowest priority with one hand, and turn the model so that the remaining three groups face you. Determine if the structure is (R) or (S) in the usual way. (See Figure 4.15.)

Sample Problems

Name the following compound, including an (R) or (S) designation.

HO   H   CH$_3$ or   HO—C—CH$_3$
         CH$_2$CH$_3$ CH$_2$CH$_3$

Solution:

1. order of priority: OH (highest), CH$_2$CH$_3$, CH$_3$, H (lowest)
2. projection with H in rear:

HO   CH$_3$ or   HO   CH$_3$
  CH$_2$CH$_3$ C
                   CH$_2$CH$_3$

3. counterclockwise = (S)
4. name: (S)-2-butanol

Draw the structure of (R)-(−)-2-butanol.

Solution:

1. Write the formula of the compound without regard to configuration of the chiral carbon.

OH
|
CH$_3$CHCH$_2$CH$_3$

2. Assign priorities around the chiral carbon: OH (highest), CH_2CH_3, CH_3, H (lowest).
3. Draw the projection with the lowest-priority group (H) in the rear. Place the remaining groups in such a way that from OH to CH_2CH_3 is clockwise for the (R) configuration.

OH   clockwise   OH

CH_3 CH_2CH_3 or CH_3 CH_2CH_3

4. Redraw the structure as a ball-and-stick or dimensional formula. (Models make this task easier.)

OH   OH

H_3C CH_2CH_3 or CH_3—C—CH_2CH_3

H   H

Note that in the solution of this problem, the sign of rotation in the name is not used. The (−) is the indication of direction of rotation of plane-polarized light, a physical property of (R)-2-butanol, and is not used in determining its configuration.

Study Problems

4.12 Assign each of the following molecules an (R) or (S) configuration:

(a)
CHO
H   OH
CH_2OH

(b)
Br
H_3C—D
H

(c) CH_3CH_2—C—CH_3
NH₂

(d)
CO_2H
H_2N—H
CH_2OH

4.13 Draw formulas for **(a)** (R)-3-bromoheptane and **(b)** (S)-2-pentanol that show the absolute configurations around the chiral carbons.

▼ ## Section 4.9
More Than One Chiral Carbon

Our discussion so far has concentrated primarily on compounds that contain only one chiral carbon, but compounds can have more than one chiral atom. Consider a compound with two different chiral carbons. Each of these two chiral carbons can be either (R) or (S); consequently, there are four different ways in which these configurations can be arranged in a molecule. A molecule with two different chiral carbons can therefore have four different stereoisomers.

chiral carbon 1	chiral carbon 2	total molecular configuration
(R)	(R)	(1R,2R)
(R)	(S)	(1R,2S)
(S)	(R)	(1S,2R)
(S)	(S)	(1S,2S)

Sample Problem

How many stereoisomers could exist for a compound that has three different chiral carbons?

Solution: eight: (1R,2R,3R); (1R,2R,3S); (1R,2S,3R); (1R,2S,3S); (1S,2R,3R); (1S,2R,3S); (1S,2S,3R); (1S,2S,3S).

The *maximum number of optical isomers* for a compound is 2^n, where n is the number of chiral atoms. If there are two chiral carbons, then there can be up to four stereoisomers ($2^2 = 4$); when there are three such carbon atoms, there can be up to eight stereoisomers ($2^3 = 8$).

Study Problem

4.14 What is the maximum number of stereoisomers for each of the following compounds?

(a) 1,2-dibromo-1-phenylpropane;   (b) 1,2-dibromo-2-methyl-1-phenylpropane;
(c) 2,3,4,5-tetrahydroxypentanal

A. (R) and (S) System for a Compound with Two Chiral Carbon Atoms

To assign (R) or (S) configurations to two chiral carbons in one molecule, we consider each chiral carbon in turn. We will use the sugar erythrose to illustrate the technique. (For convenience, we do not usually use broken wedges in a formula that shows more than one chiral carbon.)

carbons 2 and 3 are chiral

For carbon 2:

rotated so H is in rear

to go from OH (highest priority) to CHO (second) is clockwise; carbon 2 is (2R)

For carbon 3:

$$CHO$$
$$CHOH$$
$$H\text{—}C\text{◄}OH \xrightarrow{\text{rotated}}$$
$$CH_2OH$$

$$CHO$$
$$CHOH$$
$$C$$
$$HO \quad CH_2OH$$

> to go from OH to —CH(OH)CHO
> is clockwise; carbon 3 is (3*R*)

The IUPAC name for this stereoisomer is therefore (2*R*,3*R*)-2,3,4-tri-hydroxybutanal. Note that the numbers and letters in the single set of parentheses refer to the configurations around two different chiral carbons in one molecule. Contrast this notation with (*R*)(*S*), which means a racemic mixture.

B. Diastereomers

When a molecule has more than one chiral carbon, not all of the optical isomers are enantiomers. By definition, enantiomers (mirror images) come in pairs.

The four stereoisomers of 2,3,4-trihydroxybutanal:

CHO	CHO	CHO	CHO
H—C◄OH	HO►C◄H	H—C◄OH	HO►C◄H
H—C◄OH	HO►C◄H	HO►C◄H	H—C◄OH
CH₂OH	CH₂OH	CH₂OH	CH₂OH
(2*R*,3*R*)	(2*S*,3*S*)	(2*R*,3*S*)	(2*S*,3*R*)

enantiomers   *enantiomers*

Note that the (2*R*,3*R*)- and (2*S*,3*S*)-stereoisomers are enantiomers. The (2*R*,3*S*)- and (2*S*,3*R*)-stereoisomers are also enantiomers. However, the (2*S*,3*S*)- and (2*R*,3*S*)-stereoisomers are *not* enantiomers. (What other nonenantiomeric pairings can be arranged?) Any pair of stereoisomers that are not enantiomers are called **diastereomers**, or **diastereoisomers**. Thus the (2*S*,3*S*)- and (2*R*,3*S*)-stereoisomers are diastereomers. (*Geometric isomers* are also diastereomers by this definition.)

CHO	CHO
HO►C◄H	H—C◄OH
HO►C◄H	HO►C◄H
CH₂OH	CH₂OH

a pair of diastereomers:
stereoisomers that are not enantiomers

A pair of enantiomers have identical physical and chemical properties except for interactions with other chiral molecules and with polarized light.

Diastereomers, however, are chemically and physically different. They have different melting points and different solubilities and often undergo chemical reactions in a different fashion.

Sample Problem

Identify each pair of molecules as *structural isomers, enantiomers,* or *diastereomers.*

$$\text{(a)} \quad \underset{\underset{CH_2OH}{|}}{\overset{\overset{CO_2H}{|}}{H_2N\!-\!C\!-\!H}} \qquad \underset{\underset{CH_2OH}{|}}{\overset{\overset{CO_2H}{|}}{H\!-\!C\!-\!NH_2}}$$

$$\text{(b)} \quad \underset{\overset{|}{H} \qquad \overset{|}{H}}{\overset{CH_3 \qquad CH_3}{C\!=\!C}} \qquad \underset{H \qquad CH_3}{\overset{CH_3 \qquad H}{C\!=\!C}}$$

$$\text{(c)} \quad \begin{matrix} CHO \\ H\diagdown\;\diagup OH \\ H\diagup\;\diagdown OH \\ CH_2OH \end{matrix} \qquad \begin{matrix} CHO \\ H\diagdown\;\diagup OH \\ HO\diagup\;\diagdown H \\ CH_2OH \end{matrix}$$

$$\text{(d)} \quad \underset{\underset{CH_2OH}{|}}{\overset{\overset{CHO}{|}}{\underset{H-C-OH}{\overset{H-C-OH}{\underset{H-C-OH}{}}}}} \qquad \underset{\underset{CH_2OH}{|}}{\overset{\overset{CHO}{|}}{\underset{HO-C-H}{\overset{HO-C-H}{\underset{HO-C-H}{}}}}}$$

Solution: (a) and (d) are enantiomers. (b) and (c) are diastereomers.

C. *Meso* Compounds

A compound with n chiral carbon atoms can have a maximum of 2^n stereoisomers, but it may not have that many. Consider a pair of structures (**A** and **B**) with two chiral carbon atoms. At first glance, we might assume that **A** and **B** are enantiomers.

$$\underset{A}{\underset{\underset{CO_2H}{|}}{\overset{\overset{CO_2H}{|}}{\underset{H-C-OH}{H-C-OH}}}} \qquad \underset{B}{\underset{\underset{CO_2H}{|}}{\overset{\overset{CO_2H}{|}}{\underset{HO-C-H}{HO-C-H}}}}$$

Let us take compound **B** and rotate it 180° in the plane of the paper. We can see that compound **B** is identical to compound **A**! Indeed, **A** and **B** are mirror images, but the mirror images are superposable; therefore, **A** and **B** are the same compound.

$$\underset{B}{\underset{\underset{CO_2H}{|}}{\overset{\overset{CO_2H}{|}}{\underset{HO-C-H}{HO-C-H}}}} \quad \xrightarrow{\text{rotate }180°} \quad \underset{B\ superposable\ on\ A}{\underset{\underset{CO_2H}{|}}{\overset{\overset{CO_2H}{|}}{\underset{H-C-OH}{H-C-OH}}}}$$

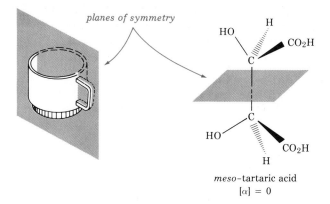

meso–tartaric acid
$[\alpha] = 0$

Figure 4.16 An internal plane of symmetry is a hypothetical plane that bisects an object into mirror-reflective halves. An object with an internal plane of symmetry is achiral (can be superposed on its mirror image).

How is it possible that a molecule with two chiral carbons is superposable on its mirror image? The answer is that, in at least one conformation, this molecule has an **internal plane of symmetry**. The "top" half of the molecule is the mirror image of the "bottom" half. We might say that the two halves of the molecule cancel each other so far as chirality is concerned. Therefore, the molecule as a whole is achiral and does not cause a rotation of plane-polarized light. Figure 4.16 illustrates an internal plane of symmetry.

A stereoisomer that contains chiral carbons but can be superposed on its mirror image is called a ***meso* form**. The compound we have been discussing is *meso*-tartaric acid.

A compound with two chiral carbons can have up to four stereoisomers. For tartaric acid, we have looked at two possibilities, and these two added up to only one isomer, ***meso***-tartaric acid. What about the other two stereoisomers of tartaric acid? Do they have an internal plane of symmetry? No, they do not. The top halves are not mirror images of the bottom halves. Rotation of either structure by 180° does not result in the other structure or in the *meso*-isomer. The following two stereoisomers of tartaric acid are enantiomers. They are both optically active and rotate the plane of polarization of plane-polarized light equally, but in opposite directions.

$$
\begin{array}{cc}
\mathrm{CO_2H} & \mathrm{CO_2H} \\
\mathrm{H-C-OH} & \mathrm{HO-C-H} \\
\mathrm{HO-C-H} & \mathrm{H-C-OH} \\
\mathrm{CO_2H} & \mathrm{CO_2H}
\end{array}
$$

NOT a plane of symmetry

$(2R,3R)$-$(+)$-tartaric acid $(2S,3S)$-$(-)$-tartaric acid

What are our conclusions about tartaric acid? Because of the symmetry in ***meso***-tartaric acid, there are only *three* stereoisomers for tartaric acid, rather than the four stereoisomers predicted by the 2^n rule. These three stereoisomers are a pair of enantiomers and the diastereomeric *meso* form.

The *meso* forms of some other compounds follow:

$$
\begin{array}{ccc}
& & CH_2OH \\
& CHO & | \\
Cl & | & H\!\!\rightarrow\!\!C\!\!\leftarrow\!\!OH \\
| & H\!\!\rightarrow\!\!C\!\!\leftarrow\!\!OH & | \\
H_3C\!\!\rightarrow\!\!C\!\!\leftarrow\!\!H & | & H\!\!\rightarrow\!\!C\!\!\leftarrow\!\!OH \\
\text{------} & \text{---}HO\!\!\rightarrow\!\!C\!\!\leftarrow\!\!H\text{---} & \text{------}\ plane\ of\ symmetry \\
H_3C\!\!\rightarrow\!\!C\!\!\leftarrow\!\!H & | & H\!\!\rightarrow\!\!C\!\!\leftarrow\!\!OH \\
| & H\!\!\rightarrow\!\!C\!\!\leftarrow\!\!OH & | \\
Cl & | & H\!\!\rightarrow\!\!C\!\!\leftarrow\!\!OH \\
& CHO & | \\
& & CH_2OH
\end{array}
$$

Study Problem

4.15 Draw all possible stereoisomers of 2,3-dichlorobutane. Indicate any enantiomeric pairs.

D. Chiral Cyclic Compounds

Some substituted cycloalkanes are chiral. For example, either *cis-* or *trans-*dimethylcyclopropane contains two chiral carbons.

Carbon 1 has four different groups attached:

$-H, -CH_3, -CH_2-$, and $-CH-$
                              $\quad\quad\quad\quad\quad\quad\quad |$
                              $\quad\quad\quad\quad\quad\quad CH_3$

Carbon 2 also has four different groups.

The *trans* isomer has no internal plane of symmetry and therefore exists as a pair of enantiomers—(1*R*,2*R*) and (1*S*,2*S*). The *cis* isomer does have an internal plane of symmetry; therefore, the *cis* isomer is a *meso* form.

The three stereoisomers of 1,2-dimethylcyclopropane:

trans: (1*R*,2*R*)   *trans*: (1*S*,2*S*)   *cis*: *meso*

enantiomers

Substituted cycloalkanes other than the substituted cyclopropanes are more complex. For example, the dimethylcyclohexanes include *cis* and *trans* isomers of 1,2-; 1,3-; and 1,4-disubstituted rings. Unlike the planar cyclopropane ring, these cyclohexane rings can exist in a variety of conformations, primarily chair forms. Fortunately, when considering the chirality of substituted cyclohexanes, we can generally use the simple polygon formulas. (If you wish to analyze the actual conformers of these compounds, use models.)

Neither the *cis*- nor the *trans*-1,4-dimethylcyclohexane is chiral because they lack chiral carbons. Note that both possess an internal plane of symmetry. Thus, neither isomer exists as a pair of enantiomers.

The two stereoisomers of 1,4-dimethylcyclohexane:

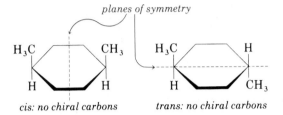

cis: no chiral carbons   *trans: no chiral carbons*

The *cis*- and *trans*-1,3-dimethylcyclohexanes have two chiral carbons each. The *cis* isomer has an internal plane of symmetry (in either the polygon formula, as shown, or in its chair form) and is a *meso* form. The *trans* isomer has no plane of symmetry and exists as a pair of enantiomers.

The three stereoisomers of 1,3-dimethylcyclohexane:

cis: meso   *trans: (1R,3R)*   *trans: (1S,3S)*

enantiomers

Sample Problem

Draw formulas for the chair–chair equilibrium for *cis*-1,3-dimethylcyclohexane and identify the internal plane of symmetry in each conformation.

Solution:

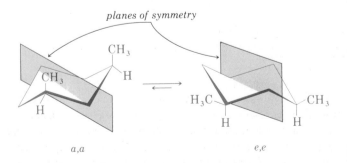

a,a   *e,e*

Study Problem

4.16 Draw formulas for the chair–chair equilibria for the two enantiomers of *trans*-1,3-dimethylcyclohexane and verify the mirror-image relationship of the enantiomers. (*Hint*: Make a model for each enantiomer.)

The 1,2-dimethylcyclohexanes also contain two chiral carbons. The *trans*-1,2-dimethylcyclohexane exists as a pair of enantiomers, while the *cis* isomer is a *meso* form.

The three stereoisomers of 1,2-dimethylcyclohexane:

cis: meso   *trans:* (1*R*,2*R*)   *trans:* (1*S*,2*S*)

enantiomers

Study Problems

4.17 *cis*-1,2-Dimethylcyclohexane actually exists as a pair of chair conformations that are nonsuperposable mirror images. These two chair conformations, however, are interconvertible by way of a *meso* conformation.

   (a) Verify that the 1*e*,2*a* chair conformation is indeed the enantiomer of the 1*a*,2*e* chair conformation. (Use models.)

   (b) What conformation must the molecule assume in order to have an internal plane of symmetry?

4.18 Draw the stereoisomers of the following compounds. Indicate enantiomeric pairs and *meso* forms, if any.

(a)

(b)

proline

an amino acid

Section 4.10
Resolution of a Racemic Mixture

In most laboratory reactions, a chemist uses achiral or racemic starting materials and obtains achiral or racemic products. Therefore, often we will ignore the chirality (or lack of it) of reactants or products in future chapters. On the other hand, we will also discuss many reactions in which stereochemistry is quite important.

Unlike most laboratory reactions, most biological reactions start with chiral or achiral reactants and lead to chiral products. These biological reactions are possible because of biological catalysts called **enzymes**, which are chiral themselves. Recall that a pair of enantiomers have the same chemical properties *except* for their interactions with other chiral substances (Section 4.7). Because enzymes are chiral, they can act very selectively in their catalytic action. For example, when an organism ingests a racemic mixture of alanine, only the (S)-alanine becomes incorporated into protein structures. The (R)-alanine is not used in proteins; instead, it is oxidized to a keto acid with the aid of other enzymes and enters other metabolic schemes.

$$CO_2H \qquad\qquad CO_2H$$
$$H_2N-C-H \qquad H-C-NH_2$$
$$CH_3 \qquad\qquad CH_3$$

(S)-alanine   (R)-alanine

Some enzymes can be separated from their living organisms and still retain their unique activity. In certain reactions, enzymes can be used to generate chiral products from achiral starting materials.

$$CH_3 \qquad\qquad\qquad CH_3$$
$$C=O \quad \xrightarrow{\text{enzymatic reduction}} \quad HO-C-H$$
$$CH_2CH_3 \qquad\qquad\qquad CH_2CH_3$$

2-butanone   (R)-2-butanol
                              $[\alpha]_D^{25} = -13°$
achiral

In the laboratory, the physical separation of a racemic mixture into its pure enantiomers is called the **resolution**, or **resolving**, of the racemic mixture. Pasteur's separation of racemic sodium ammonium tartrate was a resolution of that mixture. It is a very rare occurrence for enantiomers to crystallize separately. Therefore, the method that Pasteur used cannot be considered a general technique. Because a pair of enantiomers exhibit the same chemical and physical properties, they cannot be separated by ordinary chemical or physical means. Instead, chemists must rely on chiral reagents or chiral catalysts (which, one way or another, almost always have their origins in living organisms).

One way to resolve a racemic mixture, or at least to isolate one pure enantiomer, is to treat the mixture with a microorganism that will consume only one of the two enantiomers. For example, pure (R)-nicotine can be obtained from (R)(S)-nicotine by incubating the racemic mixture with the bacterium *Pseudomonas putida*, which oxidizes (S)-nicotine, but not the (R)-enantiomer.

(R)(S)-nicotine $\xrightarrow[\text{metabolizes (S)-nicotine}]{\textit{P. putida}}$ (R)-(+)-nicotine (78%)

racemic

The most general technique for resolving a pair of enantiomers is to subject them to a reaction with a chiral reagent to obtain a pair of **diastereomeric products**. Diastereomers, you will remember, are different compounds with different physical properties. Thus, a pair of diastereomers may be separated by ordinary physical means, such as crystallization.

Let us illustrate a general procedure for the laboratory resolution of $(R)(S)$-RCO_2H, a racemic mixture of a carboxylic acid, where (R)-RCO_2H and (S)-RCO_2H represent the two enantiomers. A carboxylic acid will react with an amine to yield a salt (Section 1.10):

| a carboxylic acid | an amine | | a salt |

Reaction of the $(R)(S)$ carboxylic acid with an amine that is a pure enantiomer results in a pair of diastereomeric salts: the amine salt of the (R) acid and the amine salt of the (S) acid.

In this reaction, the only possible products are the (R,S)-salt and the (S,S)-salt, which are not enantiomers of each other. The enantiomers of these two salts are the (S,R)-salt and the (R,R)-salt, respectively. Neither of these latter enantiomers can be formed in this reaction because only the (S)-amine was used as a reactant.

After separation, each diastereomeric salt is treated with a strong base to regenerate the amine. The amine and the carboxylate ion can be separated by extraction with a solvent such as diethyl ether (in which the amine is soluble, but the carboxylate salt is not). Acidification of the aqueous phase yields the free enantiomeric carboxylic acid.

This resolution of a racemic acid depends on salt formation with a single enantiomer of a chiral amine. Commonly used amines are amphetamine, which is commercially available as pure enantiomers, and the naturally occurring strychnine (page 738).

amphetamine

Study Problem

4.19 **(a)** Write Fischer projections for the enantiomers of amphetamine, and assign the (R) and (S) configurations.

(b) Write an equation to show the products of the reaction of (S)-amphetamine with $(R)(S)$-lactic acid.

$$
\begin{array}{c}
\text{OH} \\
| \\
\text{CH}_3\text{CHCO}_2\text{H}
\end{array}
$$

2-hydroxypropanoic acid
(lactic acid)

▼ Summary

Stereoisomerism is isomerism resulting from different spatial arrangements of atoms in molecules. **Geometric isomerism**, one form of stereoisomerism, results from groups being *cis* (same side) or *trans* (opposite sides) around a pi bond or on a ring. Geometric isomers of alkenes may also be differentiated by the letter (E), opposite sides, or (Z), same side.

Rotation of groups around sigma bonds results in different **conformations**, such as the **eclipsed**, *gauche*, **staggered**, and *anti* conformations. Lower-energy conformers predominate. Conformers are interconvertible at room temperature and therefore are not usually isolable isomers. A cyclic compound assumes puckered conformations to relieve strain of unfavorable bond angles and, more important, to minimize repulsions of substituents. For the cyclohexane ring, the chair-form conformer with substituents **equatorial** instead of **axial** is favored.

A **chiral molecule** is a molecule that is nonsuperposable on its mirror image. A pair of nonsuperposable mirror images are called **enantiomers** and represent another type of stereoisomerism. Each member of a pair of enantiomers rotates the plane of polarization of plane-polarized light an equal amount, but in opposite directions. An equimolar mixture of enantiomers, called a **racemic mixture**, is optically inactive.

Chirality usually arises from the presence of a carbon with four different atoms or groups attached to it. The arrangement of these groups around the chiral carbon is called the **absolute configuration** and may be described as (R) or (S). **Fischer projections** are often used to depict chiral molecules.

A molecule with more than one chiral carbon has more stereoisomers than a single enantiomeric pair. Stereoisomers that are not enantiomers are **diastereomers**. If a molecule has more than one chiral carbon and has a plane of symmetry in at least one conformation, then it can be superposed on its mirror image. Such a compound is optically inactive and is called a *meso* **form**.

The different types of isomerism may be summarized:

A. Structural isomers differ in order of attachment of atoms.

$$(\text{CH}_3)_2\text{CHCH}_3 \quad \text{and} \quad \text{CH}_3\text{CH}_2\text{CH}_2\text{CH}_3$$

B. Stereoisomers differ in arrangement of atoms in space.

1. Enantiomers: nonsuperposable mirror images

$$
\begin{array}{cc}
\text{CO}_2\text{H} & \text{CO}_2\text{H} \\
\text{H}-\text{C}-\text{OH} & \text{HO}-\text{C}-\text{H} \\
\text{HO}-\text{C}-\text{H} & \text{H}-\text{C}-\text{OH} \\
\text{CO}_2\text{H} & \text{CO}_2\text{H} \\
(2R,3R) & (2S,3S)
\end{array}
$$

2. Diastereomers: nonenantiomeric stereoisomers

Containing chiral carbons:

$$
\begin{array}{cc}
\text{CO}_2\text{H} & \text{CO}_2\text{H} \\
\text{H}-\text{C}-\text{OH} & \text{H}-\text{C}-\text{OH} \\
\text{HO}-\text{C}-\text{H} & \text{H}-\text{C}-\text{OH} \\
\text{CO}_2\text{H} & \text{CO}_2\text{H} \\
(2R,3R) & meso
\end{array}
$$

Achiral:

cis, or (*Z*)   *trans*, or (*E*)   *also called geometric isomers*

A pair of enantiomers have the same physical and chemical properties except for their interactions with other chiral substances and with polarized light. Enantiomers can be resolved by (1) treatment with a chiral reagent to yield a pair of diastereomers; (2) separation of the diastereomers, which do *not* have the same properties; and (3) regeneration of the separated enantiomers.

Study Problems*

4.20 *Cinnamic acid,* $C_6H_5CH{=}CHCO_2H$ (where C_6H_5 means a phenyl group), exists as a pair of stereoisomers. The *cis* isomer is a plant growth stimulant, while the *trans* isomer is not. Write formulas for the two isomers of cinnamic acid and label them.

4.21 Write structural formulas for the alkenes of molecular formula C_5H_{10} that exhibit geometric isomerism. Indicate the *cis* and *trans* structures.

4.22 Assign (*E*) or (*Z*) to each of the following alkenes.

(a)   **(b)**

* For information concerning the organization of the *Study Problems* and *Additional Problems*, see the *Note to student* on page 39.

(c)

$$\underset{H_3C}{\overset{H}{\diagdown}}C=C\underset{CO_2H}{\overset{CHO}{\diagup}}$$

(d)

$$\underset{Br}{\overset{Cl}{\diagdown}}C=C\underset{CH_3}{\overset{H}{\diagup}}$$

(e)

$$\underset{H_3C}{\overset{H}{\diagdown}}C=C\underset{CH_3}{\overset{N(CH_3)_2}{\diagup}}$$

(f)

$$\underset{C_6H_5}{\overset{H_3C}{\diagdown}}C=C\underset{H}{\overset{CO_2H}{\diagup}}$$

4.23 (a) One sex attractant compound of the female Douglas fir tussock moth has been identified as (*Z*)-1,6-heneicosadien-11-one. The (*E*) isomer shows little or no attractant power. Write formulas for these two compounds, identifying (*E*) and (*Z*) isomers. (*Hint*: Heneicosane is the 21-carbon continuous-chain alkane.)

(b) The principal component of the sex attractant secretion of the female arctiid moth (*Utetheisa ornatrix*) is named (3*Z*,6*Z*,9*Z*)-heneicosatriene. What is the structure of this compound?

(c) The principal component of the navel orangeworm (*Pamyelois transitella*) sex attractant is (11*Z*,13*Z*)-hexadecadienal. This sex attractant is unusual in that it contains a conjugated (*Z*,*Z*) diene system. Draw the structure of this compound. (*Hint*: Hexadecane is the name of the 16-carbon continuous-chain alkane.)

4.24 Draw the formulas and label the geometric isomers (if any) for the following compounds:

(a) CH_3—⬡—CH_3 **(b)** CH_3—⬡—CH_3

(c)

$$\text{CH}_2\text{OH}$$

(cyclohexane ring with O) —OH

4.25 Draw the Newman projections for the *anti* conformation and two types of eclipsed conformations of 1,2-diiodoethane. Of the two eclipsed conformations, which is of higher energy?

4.26 Draw the Newman projection for an *anti* conformer (if any) of each of the following compounds. Use the circled carbons as the center of the Newman projection.

(a) $HO_2CCH(CH_2CH_2)CO_2H$
$\quad\quad\quad\; |$
$\quad\quad\quad CH_3$

(b) $HO_2C(CHCH_2)CH_2CO_2H$
$\quad\quad\quad\;\; |$
$\quad\quad\quad CH_3$

(c) $HO_2C(CH)CH_2CH_2CO_2H$
$\quad\quad\quad\; |$
$\quad\quad\quad (CH_3)$

4.27 Indicate any strained rings in the following formulas.

(a)

α-pinene

in turpentine

(b)

⬡—$\underset{CH_2OH}{\overset{\overset{O}{\|}}{CHCO}}$— (bicyclic ring with NCH_3 and O)

scopolamine

a preoperative anesthetic

4.28 Label each of the indicated positions as axial or equatorial:

4.29 Which conformation or isomer in each pair is more stable? Explain.

(a)

(b)

(c) *cis*-1,4-dimethylcyclohexane   or   *trans*-1,4-dimethylcyclohexane

4.30 Which of the following conformations is more stable? Explain.

(a)

(b)

4.31 (a) Why is a *cis*-1,3-disubstituted cyclohexane more stable than the corresponding *trans*-structure?

(b) Is the *cis*-1,2-isomer more stable than the *trans*-1,2-disubstituted cyclohexane?

4.32 Draw the following compound in the chair form with all of the ring hydrogen atoms in axial positions. This compound is *glucose* (blood sugar). Suggest a reason why glucose is much more common in nature than any of its stereoisomers.

4.33 Draw the structure of the preferred conformation of   (a) 1-methyl-1-propylcyclohexane; (b) *cis*-1-methyl-2-propylcyclohexane;   (c) *trans*-1-methyl-3-propylcyclohexane.

4.34 Star the chiral carbons (if any):

(a) $(CH_3)_2CHCHBrCH_3$ (b) $CH_3CH_2CH_2\overset{\displaystyle CH_3}{\overset{\displaystyle |}{C}}HOH$

(c) $H_2NCH_2CO_2H$ (d) $H_2N\overset{}{\underset{\displaystyle \underset{CH_3}{|}}{C}HCO_2H}$

(e)

CH$_3$

carvone

spearmint oil

(f)

CHCO$_2$H

NH$_2$

ibotenic acid

a narcosis-potentiating substance
from poisonous Amanita *mushrooms*

4.35 Show the configuration of the enantiomer of each of the following compounds:

(a)
CHO
H—C—OH
H—C—OH
H—C—OH
CH$_2$OH

(b)
CO$_2$H
H$_2$N—C—H
CH$_2$⬡—OH

4.36 Give the structures and configurations for all the stereoisomers of saturated five-carbon alcohols (with one OH) that have stereoisomers, and indicate any enantiomeric pairs.

4.37 A carboxylic acid of the formula C$_3$H$_5$O$_2$Br is optically active. What is its structure?

4.38 The compound 1,2,3-decanetriol contains two chiral carbon atoms. Using these chiral carbons as the center of the Newman projections, draw the Newman projections for all stereoisomers of this triol.

4.39 Convert the following dimensional formulas to Fischer projections. (Molecular models may be helpful.)

(a)
CO$_2$H
H$_2$N—C—H
CH$_2$CH(CH$_3$)$_2$

(b)
CO$_2$H
C
H   OH
CH$_3$

(c)
CH$_3$
C'''Br
CH$_3$CH$_2$ CH
‖
O

(d)
H$_3$C   H
C—C
    OH
H   CH
NH$_2$ ‖
O

(e)
CO$_2$H Br
C—C
      H
H   CO$_2$H
Br

4.40 Convert the following Fischer projections to dimensional formulas (not necessarily the lowest-energy conformers):

(a)
CO$_2$H
H$_2$N——H
CH$_2$OH

(b)
CHO
H——OH
HO——H
CH$_2$OH

(c)
CH$_2$OH
C=O
H——OH
CH$_2$OH

4.41 For each Fischer projection in Problem 4.40, draw the Fischer projection of the enantiomer.

4.42 Which of the following formulas represents a chiral molecule? Star each chiral carbon.

(a)

(b)

(c)

(d)

4.43 Which of the following pairs of formulas represent enantiomeric pairs?

(a)   and

(b)   and

(c) $H_2N-\overset{CO_2H}{\underset{CH_2CH(CH_3)_2}{|}}-H$ and $H_2N-\overset{CO_2H}{\underset{(CH_3)_2CHCH_2}{|}}-H$

(d)   and

4.44 Which of the following formulas represent *meso* compounds?

(a)

CH_2OH
$H-C-OH$
$HO-C-H$
$H-C-OH$
$H-C-OH$
CH_2OH

sorbitol

an artificial sweetener

(b)

CHO
$H-C-OH$
$HO-C-H$
$HO-C-H$
$H-C-OH$
CH_2OH

galactose

a sugar

(c)

CO_2H
$H-C-OH$
$HO-C-H$
$HO-C-H$
$H-C-OH$
CO_2H

galactaric acid

4.45 Draw dimensional or ball-and-stick formulas for the following compounds. (Use models.)

(a) (R)-2-bromopropanoic acid

(b) ethyl (2R,3S)-3-amino-2-iodo-3-phenylpropanoate, $C_6H_5\overset{\underset{|}{NH_2}}{C}H\overset{}{C}HCH\overset{\underset{||}{O}}{I}COCH_2CH_3$

4.46 What is the *maximum number* of stereoisomers possible for (a) 3-hydroxy-2-methyl-butanoic acid, and (b) 2,4-dimethyl-1-pentanol? Draw and correctly name each.

4.47 (a) Which of the following Fischer projections represent enantiomers?
(b) Which are diastereomers?
(c) Which is a *meso* form?

4.48 Write Fischer projections for all possible configurations of 2,3,4-pentanetriol. Indicate enantiomeric pairs. Are there any *meso* forms?

4.49 Draw dimensional or ball-and-stick formulas for

(a) any enantiomeric pairs,
(b) a *meso* form, and
(c) any diastereomeric pairs for $HO_2CCHBrCH_2CHBrCO_2H$.

4.50 Assign an (R) or (S) designation to each of the following structures:

4.51 Which of the following formulas represents a *meso* form?

myo-inositol

a growth factor in animals

4.52 **(a)** Draw formulas for all the geometric isomers of 1,2,3,4-cyclobutanetetracarboxylic
acid.   **(b)** Which of these geometric isomers contain no chiral carbons?   **(c)** Which are
meso forms?   **(d)** Which can exist as enantiomers?

1,2,3,4-cyclobutanetetracarboxylic acid

Additional Problems

4.53 Red algae of the genus *Laurencia* that is collected near Sri Lanka contains as its
principal metabolite srilankenyne. Draw the most stable chair form of this compound.

srilankenyne

4.54 **(a)** Draw the structure for the most stable conformer of the following menthyl chloride:

(b) Draw the structure of the chair-form conformer in which Cl is *anti* to an H on one of
the adjacent carbon atoms. Would you expect a substantial percentage of the
menthyl chloride molecules to be in this latter conformation? Explain.

4.55 *Inositols* are 1,2,3,4,5,6,-hexahydroxycyclohexanes that are found in all cells.

(a) *scyllo*-Inositol is the most stable of all the inositols. Draw its structure in the chair
form.

(b) Only one stereoisomeric pair of inositols is a pair of enantiomers. Draw the
structures of this pair of enantiomers as polygon formulas.

4.56 Draw Newman projections for the most stable and the least stable conformations of
(a) (1*S*,2*R*)-1,2-dibromo-1,2-diphenylethane and   **(b)** *meso*-2,3-dichlorobutane.

4.57 Draw Newman projections for all stereoisomers of 1-bromo-1,2-diphenylpropane in
which the H on carbon 2 and the Br are *anti*.

4.58 Assign (*R*) and (*S*) configurations to all stereoisomers of 1-bromo-1,2-diphenylpropane.
Label enantiomeric pairs and diastereomers.

4.59 Write formulas for all the stereoisomers of 2,4-hexadiene. Name these isomers using the
(*E*) and (*Z*) system.

4.60 (S)-2-Iodobutane has an $[\alpha]_D^{24}$ of $+15.9°$.   **(a)** What is the observed rotation at 24°C of an equimolar mixture of (R)- and (S)-2-iodobutane?   **(b)** What is the observed rotation (at 24°C, 1-dm sample tube) of a solution (1.0 g/mL) of a mixture that is 25% (R)- and 75% (S)-2-iodobutane?

4.61 What is the specific rotation of each of the following solutions at 20°, sodium D-line?

   (a) 1.00 g of sample is diluted to 5.00 mL. A 3.00-mL aliquot is placed in a tube that is 1.0 cm long. The observed rotation is $+0.45°$.

   (b) A 0.20-g sample is diluted to 2.0 mL and placed in a 10-cm tube. The observed rotation is $-3.2°$.

4.62 Each of the following compounds is dissolved in an optically inactive solvent. Which would cause a rotation of the plane of polarization of plane-polarized light?

$$\underset{\text{NH}_2}{\text{CH}_3\text{CH}} - \underset{\text{NH}_2}{\text{CHCH}_3}$$

   (a) (2S,3R)-butanediamine, $\text{CH}_3\text{CH}-\text{CHCH}_3$
   (b) (2S,3S)-butanediamine
   (c) an equimolar mixture of (a) and (b)
   (d) an equimolar mixture of (b) and (2R,3R)-butanediamine
   (e) the principal constituent of oil of balsam (page 824).

4.63 Calculate the specific rotation of   **(a)** an equimolar mixture of (−)-tartaric acid and *meso*-tartaric acid, and   **(b)** an equimolar mixture of (−)- and (+)-tartaric acid. (The $[\alpha]_D^{20}$ for (+)-tartaric acid is $+12.0°$.)

4.64 A racemic mixture of *trans*-1,2-cyclohexanediamine is a by-product of the manufacture of nylon. Outline a route for the resolution of this compound using (+)-tartaric acid (page 153), which is a product of wine making.

4.65 The following compound is the sex attractant produced by the male dried-bean weevil. A molecule of this compound does not contain a chiral carbon atom with four different groups, yet it is optically active. Explain. (*Hint*: Consider the geometry of the molecule. Use models.)

(−)-methyl (E)-2,4,5-tetradecatrienoate

4.66 Which of the following compounds has an enantiomer? [Use models for **(b)** and **(c)**.]

   (a) α-terpineol   *found in many plant oils*

   (b) *a spiroheptadiene*

   (c)

   (d)

4.67 **(a)** Draw a pair of spirals that are mirror images of each other.   **(b)** Can a wood screw be achiral? Explain.

4.68 Draw polygon formulas for all the stereoisomers of   **(a)** 1-chloro-2-methylcyclopentane, and   **(b)** 1-chloro-3-methylcyclopentane.

4.69 Each of the following compounds can exist as a pair of enantiomers. Suggest why.

(a)

one enantiomer of
trans-cyclooctene

(b)

hexahelicene

Alkyl Halides; Substitution and Elimination Reactions

Organohalogen compounds are used extensively in modern society. Some are used as solvents, some as insecticides, and some as intermediates for the synthesis of other organic compounds. Naturally occurring organohalogen compounds are rather rare in land plants and animals. *Thyroxine*, a component of the human thyroid hormone thyroglobulin, is an example of a naturally occurring iodine compound.

thyroxine

Halogen compounds are somewhat more common in marine organisms, such as algae and seaweeds. The dye Tyrian purple is a bromine compound obtained in tiny amounts from a rare species of snail found in Crete. Tyrian purple was used as a dye by the Phoenician royalty and later by the Romans. (One may still hear the phrases "royal purple" and "born to the purple.") The structure of this compound is shown in Section 21.1.

Many organohalogen compounds are toxic and thus should be used with caution. For example, the solvents carbon tetrachloride (CCl_4) and chloroform ($CHCl_3$) both cause liver damage when inhaled in excess. Insecticides that contain halogens (such as DDT) have been widely used in agriculture. However, their use has declined in recent years because of their detrimental effects upon the environment. On the other hand, some halogen compounds seem to be quite safe and some are used as inhalation anesthetics. Examples of these anesthetics are halothane ($CF_3CHBrCl$) and methoxyflurane ($CH_3OCF_2CHCl_2$).

Chlorofluorocarbons (CFCs), also called Freons (a trademark registered by the DuPont Company), are inert nontoxic gases used as refrigerants in air conditioners and refrigerators. Freon-12, CF_2Cl_2, is the most commonly used refrigerant. CFCs have also been used as aerosol propellants for such products as hair sprays, underarm deodorants, and pesticides.

Unfortunately, CFCs have worked their way to the stratosphere where strong ultraviolet radiation causes them to decompose and release chlorine. Chlorine, in turn, catalyzes the decomposition of ozone and thus can destroy the protective ozone layer that surrounds the earth. For this reason, the United States banned the use of CFCs as aerosol propellants in 1978.

▼ Section 5.1
Types of Organohalogen Compounds

Compounds containing only carbon, hydrogen, and a halogen atom generally fall into one of three categories: **alkyl halides**; **aryl halides** (in which a halogen is bonded to a carbon of an aromatic ring); and **vinylic halides** (in which a halogen is bonded to a double-bonded carbon). A few examples follow:

Alkyl halides (RX):

CH_3I CH_3CH_2Cl

iodomethane   chloroethane

a topical anesthetic (bp 12)
that numbs by chilling

Aryl halides (ArX):

bromobenzene   a polychlorinated biphenyl (PCB)

one of a number of
toxic compounds that
have been used as cooling
fluids in transformers

Vinylic halides:

$$CH_2{=}CHCl \qquad\qquad CH_3CH{=}\overset{\displaystyle Br}{\overset{|}{C}}CH_3$$

chloroethene   2-bromo-2-butene
(vinyl chloride)

the starting material for
polyvinyl chloride (PVC),
a plastic used for piping,
house siding, phonograph
records, and trash bags

We have already defined R as the general symbol for an alkyl group. In a similar manner, Ar is the general symbol for an aromatic, or **aryl**, group. A halogen atom (F, Cl, Br, or I) may be represented by X. Using these general symbols, an alkyl halide is RX, and an aryl halide such as bromobenzene (C_6H_5Br) is ArX.

The carbon–halogen sigma bond is formed by the overlap of an orbital of the halogen atom and a hybrid orbital of the carbon atom. We cannot be sure of the hybridization (or lack of it) of the halogen atom in an organic halide because a halogen forms only one covalent bond and therefore has no bond angle around it. However, carbon uses the same type of hybrid orbital to bond to a halogen atom as it does to bond to a hydrogen atom or to another carbon atom.

An atom of F, Cl, or Br is electronegative with respect to a carbon atom. Although the electronegativity of iodine is close to that of carbon, iodine atoms are larger than carbon atoms or other halogens and are more readily polarized. Therefore, the C—X bonds of all the alkyl halides are polar. The dipole moments of the methyl halides follow. (Dipole moments were discussed in Section 1.8B.)

Dipole moments of the methyl halides:

CH_3F	CH_3Cl	CH_3Br	CH_3I
$\mu = 1.81\,D$	$1.86\,D$	$1.78\,D$	$1.64\,D$

A halogen atom in an organic compound is a functional group, and the C—X bond is a site of chemical reactivity. In this chapter, we will discuss reactions of only the alkyl halides. *Aryl halides and vinylic halides do not undergo the reactions that we will present in this chapter*, partly because a bond from an sp^2-hybridized carbon is stronger than a bond from an sp^3-hybridized carbon (Section 2.4F). Because this is the first chapter devoted to compounds containing functional groups, we will also use this chapter as an introduction to organic chemical reactions.

▼ ## Section 5.2
Physical Properties of Halogenated Alkanes

The names, boiling points, and densities of several halogenated alkanes are listed in Table 5.1. Except for fluorine, halogen atoms are heavy compared to carbon or hydrogen atoms. The increase in molecular weight and increase in polarizability (leading to increased van der Waals attractions) as halogen atoms are substituted into hydrocarbon molecules cause an increase in the boiling points of a series of compounds. Compare, for example, the boiling points of CH_3Cl, CH_2Cl_2, $CHCl_3$, and CCl_4.

Again, because of the mass of a halogen atom, the densities of liquid alkyl halides are often greater than those of other comparable organic compounds. Whereas most organic compounds are less dense than water, many common halogenated solvents, such as chloroform or dichloromethane, are more dense than water (densities greater than 1.0 g/mL). These compounds sink to the bottom of a container of water rather than float on top, as do most organic compounds. Halogenated hydrocarbons do not form strong hydrogen bonds with water and thus are insoluble in water.

Table 5.1 Physical properties of some halogenated alkanes

IUPAC Name	Trivial Name	Formula	Bp, °C	Density, g/mL at 20°
chloromethane	methyl chloride	CH_3Cl	-24	gas
dichloromethane	methylene chloride	CH_2Cl_2	40	1.34
trichloromethane	chloroform	$CHCl_3$	61	1.49
tetrachloromethane	carbon tetrachloride	CCl_4	77	1.60
bromomethane	methyl bromide	CH_3Br	5	gas
iodomethane	methyl iodide	CH_3I	43	2.28

▼ ## Section 5.3
Nomenclature and Classification of Alkyl Halides

In the IUPAC system, an alkyl halide is named with a **halo-** prefix. Many common alkyl halides also have trivial **functional-group names**. In these names, the name of the alkyl group is given, followed by the name of the halide.

$$\overset{\displaystyle Cl}{\underset{\displaystyle |}{CH_3CHCH_2CH_3}}$$

—Br

IUPAC: 2-chlorobutane   bromocyclohexane

trivial: *sec*-butyl chloride   cyclohexyl bromide

In chemical reactions, the structure of the alkyl portion of an alkyl halide is important. Therefore, we need to differentiate the four types of alkyl halides: *methyl*, *primary*, *secondary*, and *tertiary*.

A **methyl halide** is a structure in which one hydrogen of methane has been replaced by a halogen.

The methyl halides:

CH_3F	CH_3Cl	CH_3Br	CH_3I
fluoromethane	chloromethane	bromomethane	iodomethane

The *"head"* *carbon* of an alkyl halide is the carbon atom bonded to the halogen. A **primary (1°) alkyl halide** (RCH_2X) has *one* alkyl group bonded to the head carbon. In the following examples, the head carbons and their hydrogens are circled.

Primary alkyl halides (one alkyl group attached to head):

CH_3—$CH_2 Br$ $(CH_3)_3C$—$CH_2 Cl$ ⬡—$CH_2 I$

bromoethane   1-chloro-2,2-dimethylpropane   (iodomethyl)cyclohexane
(ethyl bromide)   (neopentyl chloride)

A **secondary (2°) alkyl halide** (R_2CHX) has *two* alkyl groups attached to the head carbon, and a **tertiary (3°) alkyl halide** (R_3CX) has *three* alkyl groups attached to the head carbon. (Note that a halogen attached to a cycloalkane ring must be either secondary or tertiary.)

Secondary alkyl halides (two alkyl groups attached to head):

CH_3
CH_3CH_2CH—Br

H_2C—CH_2
H_2C CH—Cl or   ⬠—Cl
C
H_2

2-bromobutane   chlorocyclopentane
(*sec*-butyl bromide)   (cyclopentyl chloride)

Tertiary alkyl halides (three alkyl groups attached to head):

CH_3
CH_3—C—Cl ⬠$<^{Br}_{CH_3}$
CH_3

2-chloro-2-methylpropane   1-bromo-1-methylcyclopentane
(*t*-butyl chloride)

Sample Problem

Classify each of the following alkyl halides as 1°, 2°, or 3°:

CH_3
(a) CH_3CHCH_2Cl (b) ⬡$<^{CH_2CH_3}_{Cl}$ (c) CH_3CHCl
                                                                        $CH_2CH_2CH_3$

Solution: (a) 1°;   (b) 3°;   (c) 2°.

Study Problems

5.1 Write two names for   **(a)** $(CH_3)_3CI$;   **(b)** $(CH_3)_2CHCl$;   **(c)** $(CH_3)_2CHCH_2I$.

5.2 Write structural formulas for   **(a)** 1,1-dibromobutane;   **(b)** 3-chloro-1-butene;   and **(c)** 2-fluoroethanol.

▼ ## Section 5.4
A Preview of Substitution and Elimination Reactions

A. Substitution Reactions

A carbon–halogen bond is polar, and the halogenated carbon of an alkyl halide has a partial positive charge. This carbon is particularly susceptible to attack by an anion or by any other species that carries an unshared pair of electrons in the outer shell. The result is a **substitution reaction**—a reaction in which one atom, ion, or group is substituted for another.

the σ-bond electrons
leave with the halogen

$$HO\overset{..}{:}^- + CH_3CH_2 \overset{\delta+}{\frown} \overset{\delta-}{Br:} \longrightarrow CH_3CH_2-\overset{..}{O}H + :\overset{..}{Br}:^-$$

hydroxide ion   bromoethane   ethanol

$$CH_3\overset{..}{O}:^- + CH_3CH_2CH_2 \frown \overset{..}{C}l: \longrightarrow CH_3CH_2CH_2-\overset{..}{O}CH_3 + :\overset{..}{C}l:^-$$

methoxide ion   1-chloropropane   methyl *n*-propyl ether

In substitution reactions of the alkyl halides, the halide is called the **leaving group**, a term meaning any group that can be displaced from a carbon atom. Halide ions are *good* leaving groups because they are very weak bases. Strong bases, such as ^-OH, are very poor leaving groups. In this chapter, we will discuss only the halides as leaving groups; we will introduce other leaving groups as we encounter them in subsequent chapters.

In substitution reactions of alkyl halides, the iodide ion is the halide most easily displaced, followed by the bromide ion and then the chloride ion. Because F^- is a stronger base than the other halide ions and because the C—F bond is distinctly stronger than other C—X bonds, fluorine is not as good a leaving group. From a practical standpoint, only Cl, Br, and I are good enough leaving groups to be useful in substitution reactions. For these reasons, when we refer to RX, we usually mean alkyl chlorides, bromides, and iodides.

RF   RCl   RBr   RI

increasing reactivity ➤

The species that attacks an alkyl halide in a substitution reaction is called a **nucleophile** (literally, "nucleus lover"), often abbreviated Nu:$^-$. In the preceding equations, OH^- and CH_3O^- are the nucleophiles. Generally, a nucleophile is any species that is attracted to a positive center. Thus, a nucleophile is a Lewis base. Most nucleophiles are anions; however, some neutral polar molecules, such as H_2O, CH_3OH, and CH_3NH_2, can also act as nucleophiles. These neutral molecules all contain unshared electrons that can be used to form sigma bonds. Substitutions by nucleophiles are called **nucleophilic substitutions**, or **nucleophilic displacements**.

The opposite of a nucleophile is an **electrophile** ("electron lover"), often abbreviated E^+. An electrophile is any species that is attracted toward a negative center; that is, an electrophile is a Lewis acid, such as H^+ or $ZnCl_2$. Electrophilic reactions are common in organic chemistry. You will encounter many of these reactions in later chapters.

B. Elimination Reactions

When an alkyl halide is treated with a *strong base*, an **elimination reaction** can occur. An elimination reaction is one in which a molecule loses atoms or ions from its structure. The organic product of an elimination reaction of an alkyl halide is an alkene. In this type of elimination reaction, the elements H and X are lost from the alkyl halide. Therefore, these reactions are also called **dehydrohalogenation reactions**. (The prefix **de-** means "minus" or "loss of.")

$$CH_3CH-CH_2 \ + \ ^-:\ddot{O}H \ \longrightarrow \ CH_3CH=CH_2 \ + \ H_2\ddot{O} + \ :\ddot{B}r:^-$$

2-bromopropane	propene
(isopropyl bromide)	(propylene)

$$CH_3-\underset{\underset{CH_3}{|}}{\overset{\overset{CH_3}{|}}{C}}-Cl \ + \ ^-OH \ \longrightarrow \ CH_3-\underset{\underset{CH_3}{|}}{\overset{\overset{CH_2}{||}}{C}} \ + \ H_2O \ + \ Cl^-$$

2-chloro-2-methylpropane	methylpropene
(*t*-butyl chloride)	(isobutylene)

C. Competing Reactions

A hydroxide ion or an alkoxide ion (RO^-) can react as a *nucleophile* in a substitution reaction or as a *base* in an elimination reaction. Which type of reaction actually occurs depends on a number of factors, such as the structure of the alkyl halide (1°, 2°, or 3°), the strength of the base, the nature of the solvent, and the temperature.

Methyl and primary alkyl halides tend to yield substitution products, not elimination products. Under equivalent conditions, *tertiary alkyl halides* yield principally elimination products, and not substitution products. *Secondary alkyl halides* are intermediate in their behavior. The relative proportion of the substitution product to the elimination product depends to a large extent upon the experimental conditions.

$$1°: \ CH_3CH_2Br \ + \ CH_3CH_2O^- \ \xrightarrow[25°]{CH_3CH_2OH} \ CH_3CH_2OCH_2CH_3$$
$$\text{(almost 100\%)}$$

$$2°: \ (CH_3)_2CHBr \ + \ CH_3CH_2O^- \ \xrightarrow[25°]{CH_3CH_2OH} \ (CH_3)_2CHOCH_2CH_3 \ + \ CH_2=CHCH_3$$
$$\text{(20\%)} \qquad\qquad \text{(80\%)}$$

$$3°: \ (CH_3)_3CBr \ + \ CH_3CH_2O^- \ \xrightarrow[25°]{CH_3CH_2OH} \ (CH_3)_3COCH_2CH_3 \ + \ CH_2=C(CH_3)_2$$
$$\text{(5\%)} \qquad\qquad \text{(95\%)}$$

Because more than one reaction can occur between an alkyl halide and a nucleophile or base, substitution reactions and elimination reactions are said to be **competing reactions**. Competing reactions are common in organic chemistry. Because mixtures of products are the rule rather than the exception when competing reactions occur, we will not balance most of the organic equations in this book.

In Sections 5.5–5.10, we will discuss two different types of substitution reactions (called S_N1 and S_N2 reactions) and two types of elimination reactions (E1 and E2). We will discuss each type of reaction individually. Then, we will summarize the factors that can help us predict which of these reactions will predominate in a given case.

D. Nucleophilicity Versus Basicity

Before proceeding with the details of substitution and elimination reactions, let us briefly consider the similarities and the differences between bases and nucleophiles. Under the proper circumstances, all bases can act as nucleophiles. Conversely, all nucleophiles can act as bases. In either case, the reagent reacts by donating a pair of electrons to form a new sigma bond.

Basicity is a measure of a reagent's ability to accept a proton in an acid–base reaction. Therefore, the relative base strengths of a series of reagents are determined by comparing the relative positions of their equilibria in an acid–base reaction, such as the degree of ionization in water.

OH⁻ acting as a base:

$$HO\overset{..}{:}\overset{\frown}{}H\overset{\frown}{}\overset{..}{\underset{..}{Cl}}: \longrightarrow H_2\overset{..}{\underset{.}{O}}: + :\overset{..}{\underset{..}{Cl}}:^-$$

$$\underset{CH_3CH—CH_2}{\overset{:\overset{..}{Br}:\ \ H}{}} + \ ^-:\overset{..}{O}H \longrightarrow CH_3CH{=}CH_2 + H_2\overset{..}{\underset{.}{O}}: + :\overset{..}{\underset{..}{Br}}:^-$$

strong bases

$$:\overset{..}{\underset{..}{I}}:^- \quad :\overset{..}{\underset{..}{Br}}:^- \quad :\overset{..}{\underset{..}{Cl}}:^- \quad R\overset{..}{O}H \quad H_2\overset{..}{\underset{.}{O}}: \quad ^-:C{\equiv}N: \quad ^-:\overset{..}{O}H \quad ^-:\overset{..}{O}R$$

increasing basicity →

In contrast to basicity, **nucleophilicity** is a measure of a reagent's ability to cause a substitution reaction. The relative nucleophilicities of a series of reagents are determined by their relative rates of reaction in a substitution reaction, such as a substitution reaction with bromoethane.

OH⁻ acting as a nucleophile:

$$H\overset{..}{O}:^- \quad CH_3\overset{\frown}{C}H_2\overset{..}{\underset{..}{Br}}: \longrightarrow CH_3CH_2{-}\overset{..}{O}H + :\overset{..}{\underset{..}{Br}}:^-$$

$$H_2\overset{..}{\underset{.}{O}}: \quad R\overset{..}{O}H \quad :\overset{..}{\underset{..}{Cl}}:^- \quad :\overset{..}{\underset{..}{Br}}:^- \quad ^-:\overset{..}{O}H \quad ^-:\overset{..}{O}R \quad :\overset{..}{\underset{..}{I}}:^- \quad ^-:C{\equiv}N:$$

increasing nucleophilicity →

A list of relative nucleophilicities does not exactly parallel a list of base strengths; however, a stronger base is usually a better nucleophile than a weaker base. For example, OH⁻ (a strong base) is a better nucleophile than Cl⁻

or H_2O (weak bases). We will discuss the factors affecting nucleophilicity in greater detail in Section 5.10.

Because some alkyl halides can undergo competing substitution and elimination reactions, a reagent such as OH^- can act both as a nucleophile and as a base in one reaction vessel.

$$(CH_3)_2CHBr + \ ^-OH \xrightarrow{\ H_2O\ } (CH_3)_2CHOH + CH_2{=}CHCH_3$$

formed by ^-OH *formed by* ^-OH
acting as a   *acting as a base*
nucleophile

Section 5.5
The S$_N$2 Reaction

The reaction of bromoethane with hydroxide ion to yield ethanol and bromide ion (page 174) is a typical **S$_N$2 reaction**. (S$_N$2 means "substitution, nucleophilic, bimolecular." The term *bimolecular* will be defined in Section 5.5B.) Virtually any methyl or primary alkyl halide undergoes an S$_N$2 reaction with any relatively strong nucleophile: ^-OH, ^-OR, ^-CN, and others that we have not yet mentioned. Methyl or primary alkyl halides also undergo reaction with weak nucleophiles, such as H_2O, but these reactions are too slow to be of practical value. Secondary alkyl halides can also undergo S$_N$2 reactions. However, tertiary alkyl halides do not.

A. Reaction Mechanism

The detailed description of how a reaction occurs is called a **reaction mechanism**. A reaction mechanism must take into account all known facts. For some reactions, the number of facts known is considerable, and the particular reaction mechanisms are accepted by most chemists. The mechanisms of some other reactions are still quite speculative. The S$_N$2 reaction is one that has been studied extensively. There is a large amount of experimental data supporting the mechanism that we will present.

For molecules to undergo a chemical reaction, they must first collide. Most collisions between molecules do not result in a reaction. Instead, the molecules simply rebound. To undergo reaction, the colliding molecules must contain enough *potential energy* for bond breakage to occur. Also, the *orientation* of the molecules relative to each other is often an important factor in determining whether a reaction will occur. This is particularly true in an S$_N$2 reaction. In this section, we will first discuss the stereochemistry of the S$_N$2 reaction, then we will discuss the energy requirements.

B. Stereochemistry of an S$_N$2 Reaction

In the S$_N$2 reaction between bromoethane and hydroxide ion, the oxygen of the hydroxide ion collides with the rear of the head carbon relative to the bromine and displaces the bromide ion.

Overall S_N2 reaction:

rear attack

When a nucleophile collides with the back side of a tetrahedral carbon atom bonded to a halogen, two things occur simultaneously: (1) a new bond begins to form, and (2) the C—X bond begins to break. The process is said to be a one-step, or **concerted**, process. If the potential energy of the two colliding species is high enough, a point is reached where it is energetically more favorable for the new bond to form and the old C—X bond to break. As the reactants are converted to products, they must pass through an in-between state that has a high potential energy relative to the reactants and the products. This state is called the **transition state**, or the **activated complex**. Because the transition state involves two particles (Nu:$^-$ and RX), the S_N2 reaction is said to be **bimolecular**. (The "2" in S_N2 indicates bimolecular.)

reactants

transition state:
high potential energy,
equally able to go to reactants
or products

products

A transition state in any reaction is the fleeting high-energy arrangement of the reactants as they go to products. We cannot isolate a transition state and put it in a flask. The transition state is simply a description of "molecules in a state of transition." We will often use square brackets in an equation to show any temporary, nonisolable structure in a reaction. Here, we use brackets to enclose the structure of a transition state. Later, we will sometimes use brackets to indicate unstable products that undergo further reaction.

For the S_N2 reaction, the transition state involves a temporary rehybridization of the head carbon from sp^3 to sp^2 and finally back to sp^3 again. In the transition state, the carbon atom has three planar sp^2 bonds, plus two half-bonds using the p orbital.

As the nucleophile attacks the rear of the molecule, relative to the halogen atom, the other three groups attached to the carbon flatten out in the transition state, then flip to the other side of the carbon atom, much as an umbrella being blown inside out. (Models would be useful to help visualize this.) The flipping is called **inversion of configuration**, or **Walden inversion**, after the Latvian and German chemist Paul Walden who first reported it in 1896.

The existence of inversion as part of the mechanism of an S_N2 reaction has been beautifully demonstrated by reactions of pure enantiomers of chiral secondary alkyl halides. For example, the S_N2 reaction of (*R*)-2-bromooctane with ⁻OH yields almost exclusively (*S*)-2-octanol.

HO⁻ +

$$\text{HO}^- + \underset{\text{(R)-2-bromooctane}}{\overset{\text{H } CH_2(CH_2)_4CH_3}{\underset{CH_3}{\mid}}{C-Br}} \xrightarrow{S_N2} \underset{\text{(S)-2-octanol}}{\text{HO}-\overset{CH_2(CH_2)_4CH_3}{\underset{CH_3}{\mid}}{C}} + Br^-$$

Most reactions involving chiral molecules are carried out with racemic mixtures, that is, equal mixtures of (*R*) and (*S*) reactants. In these cases, the products also are racemic mixtures. Even though inversion occurs, we cannot observe the effects because half the molecules go one way and half go the other way.

$$\underset{racemic}{(R)\text{-RX} + (S)\text{-RX}} \xrightarrow[S_N2]{OH^-} \underset{racemic}{(S)\text{-ROH} + (R)\text{-ROH}}$$

Study Problem

5.3 Write an equation (showing the stereochemistry by using dimensional formulas) for the S_N2 reaction of (*S*)-2-bromobutane with ⁻CN and designate the configuration of the product by the (*R*) and (*S*) system.

C. Energy in an S_N2 Reaction

We have mentioned that colliding molecules need energy to undergo reaction. We will now look at these energy requirements in more detail.

Molecules moving around in a solution contain a certain amount of potential energy in their bonds and a certain amount of kinetic energy from their movement. Not all molecules in solution have exactly the same amount of potential or kinetic energy. However, we may speak of the *average energy* of the molecules. The total energy of the reaction mixture may be increased, usually by heating the solution. When heated, the molecules gain kinetic energy, collide more frequently and more energetically, and exchange some kinetic energy for potential energy.

Before a reaction can begin to occur, some of the colliding molecules and ions in the flask must contain enough energy to reach the transition state upon collision. Reaching the potential-energy level of the transition state is rather like driving an old car to a mountain pass. Does the car have enough energy to make the top? Or, will it stall and slide back down the mountain? Once you reach the top, which way do you go—back the way you came or on down the

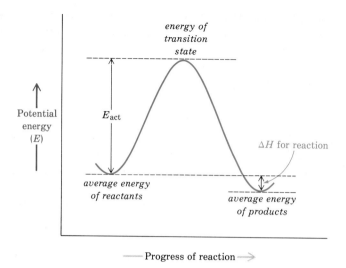

Figure 5.1 Energy diagram of an S_N2 reaction.

other side? Once you are descending the far side, the choice is easy; you can relax and let the car roll to the bottom.

Figure 5.1 shows an energy diagram for the progress of an S_N2 reaction. The potential energy required to reach the transition state forms an energy barrier; it is the point of maximum energy on the graph. For a colliding alkyl halide and nucleophile to reach the transition state, they need a certain minimum amount of energy called the **energy of activation E_{act}**. At the transition state, the molecules find it just as easy to go back to reactants or on to products. But, once over the top, the path of least resistance is that of going to products. The difference between the average potential energy of the reactants and that of the products is the change in enthalpy ΔH for the reaction (see Section 1.7).

D. Rate of an S_N2 Reaction

Each molecule that undergoes reaction to yield product must pass through the transition state, both structurally and energetically. Since the energies of all the molecules are not the same, a certain amount of time is required for all the molecules present to react. This time requirement gives rise to the **rate of a reaction**. The rate of a chemical reaction is a measure of how fast the reaction proceeds, that is, how fast reactants are consumed and products are formed. **Reaction kinetics** is the term used to describe the study and measurement of reaction rates.

The rate of a reaction depends on many variables, some of which may be held constant for a given experiment (temperature and solvent, for example). In this chapter, we will be concerned primarily with two variables: (1) the concentrations of the reactants, and (2) the structures of the reactants.

Increasing the concentration of reactants undergoing an S_N2 reaction increases the rate at which products are formed because it increases the frequency of molecular collisions. Typically, the rate of an S_N2 reaction is proportional to the concentrations of both reactants. If all other variables are held constant and the concentration of either the alkyl halide or the nucleophile is doubled, the rate of product formation is doubled. If either concentration is tripled, the rate is tripled.

$$Nu\!:^- + R\ddot{\underset{..}{X}}\!: \longrightarrow RNu + :\ddot{\underset{..}{X}}\!:^-$$

S$_N$2 rate $= k[RX][Nu\!:^-]$

In this equation, [RX] and [Nu$:^-$] represent the concentrations in moles per liter of the alkyl halide and the nucleophile, respectively. The term k is the proportionality constant, called the **rate constant**, between these concentrations and the measured rate of product formation. The value for k is constant for the same reaction under identical experimental conditions (solvent, temperature, etc.).

Sample Problem

What would be the effect on the rate of the S$_N$2 reaction of CH$_3$I with CH$_3$O$^-$ if the concentrations of *both* reactants were doubled and all other variables were held constant?

Solution: If the concentrations of both CH$_3$I and CH$_3$O$^-$ were doubled, the rate would quadruple—the reaction would proceed four times as fast.

Because the rate of an S$_N$2 reaction depends on the concentrations of two particles (RX and Nu$:^-$), the rate is said to be **second order**. The S$_N$2 reaction is said to follow **second-order kinetics**. (Although the S$_N$2 reaction is also bimolecular, not all bimolecular reactions are second-order and not all second-order reactions are bimolecular. For example, see Problem 5.62, page 224).

E. Effect of E_{act} on Rate and on Products

The effect of the energies of activation on the relative rates of reaction may be stated simply: *Under the same conditions, the reaction with the lower E_{act} has a faster rate.* The reason for this relationship is that, if less energy is required for reaction, a greater number of molecules have enough energy to react.

Let us consider a case in which one starting material can undergo two different irreversible reactions leading to two different products. (A reaction in which the E_{act} of the reverse reaction is substantially greater than the E_{act} of the forward reaction is exothermic and essentially irreversible.) When the starting material can undergo two such reactions, *the product of the faster reaction (the one with the lower E_{act}) predominates*. We say that the irreversible reaction is under **kinetic control** (contrasted to the equilibrium control of a reversible reaction). Figure 5.2 shows energy curves for two such reactions of one starting material.

The E_{act} is the energy of the transition state relative to that of the reactants. Therefore, relative rates of reaction are related to the energies of the transition states. In competing reactions of the same starting material, *the*

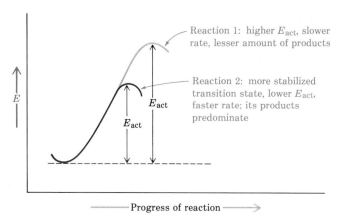

Figure 5.2. In competing reactions of a single starting material, the reaction with the lower E_{act} is the faster one. If the reactions are essentially irreversible, the products of the faster reaction predominate.

reaction with the lower-energy transition state is the faster reaction. From Figure 5.2, it is evident that the reaction with the lower-energy transition state has the lower E_{act}.

A species of low potential energy is more stable than one of high energy. Therefore, we can also say that *the reaction with a more stabilized transition-state structure is the faster reaction.* This concept is useful in analyzing competing reactions to determine which reaction predominates.

F. Effect of Structure on Rate

Reaction kinetics provide a valuable tool for exploring the effects of structure upon reactivity. Consider the following two reactions:

(1) $OH^- +$ CH_3Br $\longrightarrow$ $CH_3OH + Br^-$

bromomethane   methanol

a methyl halide

(2) $OH^- +$ CH_3CH_2Br $\longrightarrow$ $CH_3CH_2OH + Br^-$

bromoethane   ethanol

a 1° alkyl halide

Both are S_N2 reactions and both yield alcohols. The two reactions differ only in the alkyl portion of the alkyl halides. We may ask the following question: "Does this difference in the alkyl group have an effect upon the rate of the S_N2 reaction?" To answer such a question, the rates of the two reactions are measured under the same reaction conditions (the same solvent, the same concentrations, and the same temperature). Then, either the two rate constants (k_1 and k_2) are determined or, more commonly, the *relative rates* are determined.

$$CH_3Br \xrightarrow{\text{OH}^-} CH_3OH \qquad rate_1 = k_1[CH_3Br][OH^-]$$

$$CH_3CH_2Br \xrightarrow{\text{OH}^-} CH_3CH_2OH \qquad rate_2 = k_2[CH_3CH_2Br][OH^-]$$

relative rates of reaction of CH_3Br compared to $CH_3CH_2Br = \dfrac{rate_1}{rate_2}$

Under the experimental conditions used in one study, bromomethane undergoes reaction 30 times faster than bromoethane. (If it takes one hour for the bromoethane reaction to reach 50% completion, the bromomethane reaction would take about 1/30 as long, or only two minutes, to reach 50% completion!) We conclude that there is indeed a big difference in how the methyl and ethyl groups affect the rate of the reaction.

In a similar fashion, the relative rates of a variety of S_N2 reactions of alkyl halides have been determined. Table 5.2 shows some average relative rates (compared with the rates of ethyl halides) of S_N2 reactions of alkyl halides.

Table 5.2 Average relative reaction rates of some alkyl halides under typical S_N2 conditions

Alkyl Halide	Relative Rate
CH_3X	30
CH_3CH_2X	1
$CH_3CH_2CH_2X$	0.4
$CH_3CH_2CH_2CH_2X$	0.4
$(CH_3)_2CHX$	0.025
$(CH_3)_3CX$	~ 0

G. Steric Hindrance in S_N2 Reactions

In S_N2 reactions of the alkyl halides listed in Table 5.2, methyl halides show the fastest rate, followed by primary alkyl halides and then secondary alkyl halides. Tertiary alkyl halides do not undergo S_N2 reactions.

$$3°RX \qquad 2°RX \qquad 1°RX \qquad CH_3X$$

increasing rate of S_N2 reaction

As the number of alkyl groups attached to the head carbon increases ($CH_3X \rightarrow 1° \rightarrow 2° \rightarrow 3°$), the transition state becomes increasingly crowded with atoms. Consider the example of reactions of alkyl bromides with the methoxide ion (CH_3O^-) as the nucleophile ($CH_3O^- + RBr \rightarrow CH_3OR + Br^-$), shown in Figure 5.3.

Spatial crowding in structures is called **steric hindrance**. When large groups are crowded in a small space, repulsions between groups become severe and therefore the energy of the system is high. In an S_N2 reaction, the energy of a crowded transition state is higher than that of a transition state with less steric hindrance. For this reason, the rates of reaction become progressively slower in the series methyl, primary, secondary, and tertiary (see Figure 5.4). The energy of the S_N2 transition state of a tertiary alkyl halide is so high relative to other possible reaction paths that the S_N2 reaction does not proceed.

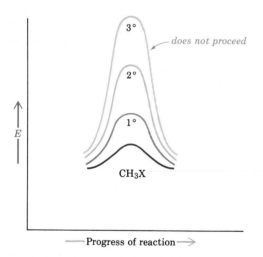

Figure 5.3 Steric hindrance in S_N2 reactions.

Figure 5.4 Energy diagram for S_N2 reactions of different types of alkyl halides.

Sample Problem

The S_N2 reaction of neopentyl bromide, $(CH_3)_3CCH_2Br$, with sodium ethoxide, $Na^+ \ ^-OCH_2CH_3$, proceeds about 0.00001 times as fast as the reaction of bromoethane. Explain.

Solution: Although neopentyl bromide is a primary alkyl halide, the alkyl group attached to the head carbon atom is very bulky. The steric hindrance in the transition state is considerable. Therefore, the E_{act} is high and the rate is slow.

Study Problems

5.4 Which member of each of the following pairs would undergo the faster S$_N$2 reaction? Explain your choices.

(a) ⬡—CH$_2$Cl   or   CH$_3$—⬡—Cl

(b) (CH$_3$)$_2$CHĊHCH$_2$CH$_3$ or   (CH$_3$)$_2$CHCH$_2$ĊHCH$_3$
        Cl   Cl

(c) ⬡—CH$_2$Cl   or   CH$_3$—⬡—Cl

5.5 Write an equation for the reaction of NaCN with each of your answers to Problem 5.4.

Section 5.6
The S$_N$1 Reaction

Because of steric hindrance, *t*-butyl bromide and other tertiary alkyl halides do not undergo S$_N$2 reactions. Yet, if *t*-butyl bromide is treated with a nucleophile that is a *very weak base* (such as H$_2$O or CH$_3$CH$_2$OH), substitution products are formed, along with elimination products. Because H$_2$O or CH$_3$CH$_2$OH is also used as the solvent, this type of substitution reaction is sometimes called a **solvolysis reaction** (from *solvent* and *-lysis*, "breaking down" or "loosening").

$$\xrightarrow[25°]{CH_3CH_2OH} (CH_3)_3COCH_2CH_3 + CH_2{=}C(CH_3)_2$$

                 t-butyl ethyl ether   methylpropene
                    (80%)   (20%)

(CH$_3$)$_3$CBr

$$\xrightarrow[25°]{H_2O} (CH_3)_3COH + CH_2{=}C(CH_3)_2$$

         t-butyl alcohol   methylpropene
           (70%)   (30%)

If tertiary alkyl halides cannot undergo S$_N$2 reactions, how are the substitution products formed? The answer is that tertiary alkyl halides undergo substitution by a different mechanism, called the **S$_N$1 reaction** (substitution, nucleophilic, unimolecular). The experimental results obtained in S$_N$1 reactions are considerably different from those obtained in S$_N$2 reactions. Typically, if a pure enantiomer of an alkyl halide containing a chiral C—X carbon undergoes an S$_N$1 reaction, predominantly *racemic* substitution products are obtained (not the inverted products observed in an S$_N$2 reaction). It has also been determined that the concentration of nucleophile generally has little effect upon the overall rate of an S$_N$1 reaction. (In contrast, the rate of an S$_N$2 reaction is directly proportional to the concentration of the nucleophile.) To explain these experimental results, we will discuss the mechanism of the S$_N$1 reaction of *t*-butyl bromide and water. For the moment, we will ignore the elimination product.

as the hydronium ion, H_3O^+

$$(CH_3)_3CBr \ + \ H_2O \longrightarrow (CH_3)_3COH \ + \ H^+ \ + \ Br^-$$

t-butyl bromide   *t*-butyl alcohol

A. The S_N1 Mechanism

An S_N1 reaction is an ionic reaction. The mechanism is complex because of interactions among solvent molecules, RX molecules, and the intermediate ions formed. For this reason, the mechanism we present here is somewhat idealized.

The S_N1 reaction of a tertiary alkyl halide is a *stepwise reaction*. Step 1 is the cleavage of the alkyl halide into a pair of ions: the halide ion and a **carbocation**, an ion in which a carbon atom carries a positive charge. Because S_N1 reactions involve ionization, these reactions are aided by polar solvents, such as H_2O, that can solvate, and thus stabilize, ions.

Step 1:

$$(CH_3)_3C\text{—}\overset{..}{\underset{..}{Br}}: \longrightarrow \left[(CH_3)_3\overset{\delta+}{C}\text{---}\overset{\delta-}{\overset{..}{\underset{..}{Br}}}: \right] \longrightarrow [(CH_3)_3C^+] \ + \ :\overset{..}{\underset{..}{Br}}:^-$$

transition state 1   *unstable carbocation intermediate*

Step 2 is the combining of the carbocation with the nucleophile (H_2O) to yield the initial product, a protonated alcohol.

Step 2:

$$[(CH_3)_3\overset{+}{C}] + \ H_2\overset{..}{\underset{..}{O}}: \longrightarrow \left[(CH_3)_3\overset{\delta+}{C}\text{---}\overset{\delta+}{\underset{\overset{|}{\underset{..}{O}H}}{}} \right] \longrightarrow (CH_3)_3C\text{—}\overset{\overset{H}{|}}{\underset{..}{\overset{+}{O}}H}$$

nucleophile   *transition state 2*   *protonated t*-butyl alcohol*

The final step in the sequence is the loss of H^+ by the protonated alcohol in a rapid, reversible acid–base reaction with the solvent.

Step 3:

$$(CH_3)_3C\overset{\overset{H}{\underset{|}{\overset{..}{\underset{..}{O}}}}}{\underset{..}{O}}H + H_2\overset{..}{\underset{..}{O}}: \rightleftharpoons (CH_3)_3C\overset{..}{\underset{..}{O}}H \ + \ H_3\overset{+}{O}:$$

excess   *t*-butyl alcohol*

The overall reaction of *t*-butyl bromide with water is thus actually composed of two separate reactions: the S_N1 reaction (ionization followed by combination with the nucleophile) and an acid–base reaction. The steps can be summarized:

means loss of Br^-

$$(CH_3)_3CBr \xrightarrow[\text{slow}]{-Br^-} [(CH_3)_3C^+] \xrightarrow[\text{fast}]{H_2O} (CH_3)_3\overset{+}{C}OH_2 \xrightarrow[\text{fast}]{-H^+} (CH_3)_3COH$$

$\underbrace{\hphantom{(CH_3)_3CBr \xrightarrow{-Br^-} [(CH_3)_3C^+] \xrightarrow{H_2O}}}_{S_N1 \ reaction}$ $\underbrace{\hphantom{(CH_3)_3\overset{+}{C}OH_2 \xrightarrow{-H^+} (CH_3)_3COH}}_{acid-base \ reaction}$

t.s. 1

t.s. 2

carbocation
intermediate

E

$(CH_3)_3CBr + H_2O$

$(CH_3)_3COH_2 + Br^-$

———— Progress of reaction ————→

Figure 5.5 Energy diagram for a typical S$_N$1 reaction.

Let us consider the energy diagram for an S$_N$1 reaction (Figure 5.5). Typically, Step 1 (the ionization) has a high E_{act}. It is the *slow step* in the overall process. Enough energy must be supplied to the tertiary alkyl halide to break the C—X sigma bond and yield the carbocation and the halide ion.

The carbocation is an **intermediate** in this reaction, a structure that is formed during the reaction and then undergoes further reaction to products. An intermediate is not a transition state. An intermediate has a finite lifetime; a transition state does not. At the transition state, molecules are undergoing bond-breaking and bond-making. The potential energy of a transition state is a high point on a potential-energy curve. By contrast, an intermediate is a short-lived, reactive product. No bond-breaking or bond-making is occurring in an intermediate. An intermediate is of lower energy than the transition states that surround it (and is at a lower point on the potential-energy curve), although it is of higher energy than the final products. The energy diagram in Figure 5.5 shows a dip for the carbocation formation. The dip is not a large one because the carbocation is a high-energy, reactive species.

Step 2 in the sequence of an S$_N$1 reaction is the reaction of the carbocation with a nucleophile. The two combine in a reaction having a low E_{act} (a fast reaction).

Sample Problem

Write equations for the steps in the S$_N$1 reaction of 2-chloro-2-methylbutane with methanol. (Include the deprotonation step in your answer.)

Solution:

Step 1 (ionization):

$$(CH_3)_2\overset{\overset{\displaystyle :\ddot{C}l:}{|}}{C}CH_2CH_3 \longrightarrow [(CH_3)_2\overset{+}{C}CH_2CH_3] + :\ddot{C}l:^-$$

a carbocation

Step 2 (combination with Nu:⁻):

$$HOCH_3$$
$$|$$
$$[(CH_3)_2\overset{+}{C}CH_2CH_3] + CH_3\ddot{O}H \longrightarrow (CH_3)_2\overset{+}{C}CH_2CH_3$$

a protonated ether

Step 3 (loss of H⁺ to solvent):

$$H-\overset{+}{\underset{|}{O}}CH_3 \qquad\qquad :\ddot{O}CH_3$$
$$CH_3\ddot{O}H + (CH_3)_2\overset{+}{C}CH_2CH_3 \longrightarrow (CH_3)_2\overset{|}{C}CH_2CH_3 + CH_3\overset{+}{\ddot{O}}H_2$$

an ether

Study Problem

5.6 Complete the following equations to show the expected solvolysis products.

(a) $(CH_3CH_2)_3CI + H_2O \longrightarrow$ **(b)** [cyclohexane ring with CH_3 and Cl substituents] $+ CH_3OH \longrightarrow$

B. Stereochemistry of an S$_N$1 Reaction

A carbocation (or **carbonium ion**, as it is also called) is a carbon atom with only three groups attached, instead of the usual four. Because there are only three groups, the bonds to these groups lie in a plane, and the angles between the bonds from the positive carbon are approximately 120°. To attain this geometry, the positive carbon is sp^2-hybridized and has an empty p orbital.

empty p orbital →

$$R_{\text{...}}\overset{+}{C}-R$$
$$R$$

sigma-bond angles
around C⁺ are
approximately 120°

Let us consider the S$_N$1 reaction of a chiral alkyl halide. When one enantiomer of 3-chloro-3,7-dimethyloctane is warmed with aqueous acetone, the halide undergoes solvolysis to 3,7-dimethyl-3-octanol that has been mostly racemized. The S$_N$1 mechanism can be used to explain this racemization. The first step is ionization to a carbocation and a halide ion.

Step 1:

$$\overset{Cl}{\underset{|}{C}}$$
$$CH_3\text{...}C$$
$$CH_3CH_2 \;(CH_2)_3CH(CH_3)_2 \qquad \xrightarrow{-Cl^-} \qquad \left[\begin{array}{c} CH_3\text{...}\overset{+}{C}-(CH_2)_3CH(CH_3)_2 \\ CH_3CH_2 \end{array} \right]$$

(R)-3-chloro-3,7-dimethyloctane   *planar carbocation*

In the second step, H$_2$O attacks the carbocation to form *two* protonated alcohols. (We show the alcohols *after* deprotonation in the following equations.)

If the ionization of the alkyl halide molecule is complete, the H$_2$O molecule can attack the empty p orbital from the top side, as we have shown the structure, and the carbocation yields the (R)-enantiomer of the alcohol. However, the H$_2$O molecule can also attack from the bottom side, yielding the (S)-enantiomer. Because attack can be from either side, racemization can occur.

Steps 2 and 3 (attack of H$_2$O to yield ROH$_2^+$, followed by loss of H$^+$)

attack from top

(*R*)-3,7-dimethyl-3-octanol

attack from bottom

(*S*)-3,7-dimethyl-3-octanol

C. Rate of an S$_N$1 Reaction

We mentioned earlier that the rate of a typical S$_N$1 reaction does not depend on the concentration of the nucleophile, but depends only on the concentration of alkyl halide.

$$S_N1 \text{ rate} = k[RX]$$

The reason for this behavior is that the reaction between R$^+$ and Nu:$^-$ is very fast, but the concentration of R$^+$ is very low. The fast combination of R$^+$ and Nu:$^-$ occurs only when a carbocation is formed. Therefore, the rate of the overall reaction is determined entirely by how fast RX can ionize and form R$^+$ carbocations. This ionization step (Step 1 of the overall reaction) is called the **rate-determining step**, or the **rate-limiting step**. In most stepwise reactions, the slowest step in the entire sequence is the rate-determining step because a reaction cannot ordinarily proceed faster than its slowest step does.

An S$_N$1 reaction is characteristically **first order** in rate because the rate is proportional to the concentration of only one reactant (RX). It is a **unimolecular reaction** because only one particle (RX) is involved in the *transition state of the rate-determining step*. (The "1" in S$_N$1 refers to unimolecular.)

Rate-determining step:

$$R{-}X \longrightarrow [\overset{\delta+}{R}{-}{-}\overset{\delta-}{X}] \longrightarrow R^+ + X^-$$

transition state 1
(*from one particle*)

D. Relative Reactivities in S$_N$1 Reactions

Table 5.3 lists the relative rates of reaction of some alkyl bromides under typical S$_N$1 conditions (solvolysis in water). Note that a secondary alkyl halide undergoes substitution 11.6 times faster than a primary alkyl halide under these

Table 5.3 Relative reaction rates of some alkyl bromides under typical S_N1 conditions

CH_3Br	1.00^a
CH_3CH_2Br	1.00^a
$(CH_3)_2CHBr$	11.6
$(CH_3)_3CBr$	1.2×10^6

[a] The observed reaction of the methyl or primary bromide probably occurs by a different path (S_N2, not S_N1).

conditions, while a tertiary alkyl halide undergoes reaction a million times faster than a primary halide!

Methyl:

$$CH_3Br \quad + H_2O \xrightarrow[S_N1]{\text{negligible rate}} CH_3OH \quad + Br^- + H^+$$

1°:

$$CH_3CH_2Br \quad + H_2O \xrightarrow[S_N1]{\text{negligible rate}} CH_3CH_2OH \quad + Br^- + H^+$$

2°:

$$(CH_3)_2CHBr + H_2O \xrightarrow[S_N1]{\text{very slow}} (CH_3)_2CHOH + Br^- + H^+$$

3°:

$$(CH_3)_3CBr \quad + H_2O \xrightarrow[S_N1]{\text{fast}} (CH_3)_3COH \quad + Br^- + H^+$$

The rates at which different alkyl halides undergo S_N1 reaction depend on the relative energies of activation leading to the different carbocations. In this reaction, the energy of the transition state leading to the carbocation is largely determined by the stability of the carbocation, which is partially formed in the transition state. We say that the transition state has **carbocation character**. Therefore, the reaction leading to a more stabilized, lower-energy carbocation has the faster rate. A tertiary alkyl halide yields a carbocation that is more stabilized than the carbocation from a methyl halide, a primary alkyl halide, or a secondary alkyl halide. Consequently, the reaction of a tertiary halide has the fastest rate.

E. Stability of Carbocations

A carbocation is unstable and quickly undergoes further reaction. However, we may still speak of *relative stabilities* of carbocations. The different types of carbocations that concern us here are the **methyl cation** (the carbocation resulting from ionization of a methyl halide), **primary carbocations** (from 1° alkyl halides), **secondary carbocations** (from 2° alkyl halides), and **tertiary carbocations** (from 3° alkyl halides). Some examples follow:

$^+CH_3$	$CH_3\overset{+}{C}H_2$	$(CH_3)_2\overset{+}{C}H$	$(CH_3)_3C^+$
methyl	1°	2°	3°

What increases the stability of a positively charged carbon atom? The answer is: anything that can *disperse the positive charge*. In alkyl cations, the principal phenomenon that disperses the positive charge is the **inductive effect**, a term used to describe the polarization of a bond by a nearby electronegative or electropositive atom. In a carbocation, the positively charged carbon is an electropositive center. The electron density of the sigma bonds is shifted toward the positive carbon. We will use arrows in place of line bonds to show the direction of this attraction.

Sigma-bond e⁻ are drawn toward the positive charge.

This shift of electron density creates a partial positive charge on the adjacent atoms. These partial positive charges, in turn, polarize the next sigma bonds. In this way, the positive charge of the carbocation is somewhat dispersed, and the carbocation is stabilized to some extent.

All the atoms help disperse the positive charge.

An alkyl group contains more atoms and electrons than does a hydrogen atom. When there are more alkyl groups attached to a positively charged carbon atom, there are more atoms that can help share the positive charge and help stabilize the carbocation. Because of the lack of stabilization, methyl and primary halides do not normally form carbocations.

$$^+CH_3 \qquad CH_3\overset{+}{C}H_2 \qquad (CH_3)_2\overset{+}{C}H \qquad (CH_3)_3C^+$$

methyl   1°   2°   3°

increasing carbocation stability; increasing S_N1 rate of RX

Study Problem

5.7 List the following carbocations in order of increasing stability (least stable first):

(a)   (b)   (c)

Another theory proposed to explain the relative stabilities of carbocations is **hyperconjugation**, the partial overlap of an sp^3–s orbital (a C—H bond) with the empty p orbital of the positively charged carbon.

one lobe of empty p orbital

An ethyl cation has only three C—H bonds that can overlap the empty *p* orbital, but the *t*-butyl cation has *nine* C—H bonds that can help disperse the charge in this fashion. Therefore, a tertiary carbocation is stabilized by a greater dispersal of the positive charge.

F. Rearrangements of Carbocations

The following secondary alkyl bromide can undergo solvolysis with methanol. However, in addition to the expected product, a *second* substitution product is observed. (The total yield of substitution products is not 100% because alkenes are also formed.)

$$\overset{\text{Br}}{\underset{\text{2-bromo-3,3-dimethylbutane}}{(CH_3)_3CCHCH_3}} \quad \xrightarrow[-Br^-, -H^+]{CH_3OH} \quad$$

$$\overset{OCH_3}{(CH_3)_3CCHCH_3}$$

2-methoxy-3,3-dimethylbutane (4%)

expected product

$$\overset{OCH_3}{(CH_3)_2CCH(CH_3)_2}$$

2-methoxy-2,3-dimethylbutane (41%)

unexpected product

Let us examine the intermediate carbocation in this reaction more carefully. The expected carbocation is a *secondary* carbocation.

$$CH_3-\overset{\overset{\displaystyle CH_3}{|}}{\underset{\underset{\displaystyle CH_3}{|}}{C}}-\overset{\overset{\displaystyle Br}{|}}{CH}CH_3 \quad \xrightarrow{-Br^-} \quad \left[CH_3-\overset{\overset{\displaystyle CH_3}{|}}{\underset{\underset{\displaystyle CH_3}{|}}{C}}-\overset{+}{C}HCH_3 \right]$$

expected carbocation: 2°

A secondary carbocation is of higher energy than a tertiary carbocation by about 11 kcal/mol. The energy of this particular carbocation can be lowered by the shift of a methyl group, with its bonding electrons, from the adjacent carbon atom. The result is the rearrangement of the secondary carbocation to a more stable tertiary carbocation.

1,2-Shift of a methyl group with its bonding electrons:

$$CH_3-\overset{\overset{\displaystyle CH_3}{|}}{\underset{\underset{\displaystyle CH_3}{|}}{C}}\overset{+}{-}CHCH_3 \quad \longrightarrow \quad CH_3-\overset{+}{\underset{\underset{\displaystyle CH_3}{|}}{C}}-\overset{\overset{\displaystyle CH_3}{|}}{C}HCH_3$$

a 2° carbocation   *a more stable 3° carbocation*

The shift of an atom or of a group from an adjacent carbon is called a **1,2-shift**. (The numbers 1,2 used in this context have nothing to do with nomenclature numbers, but refer to the positive carbon and the adjacent atom.) The

1,2-shift of a methyl group is called a **methyl shift**, or a **methide shift**. (The *-ide* suffix is sometimes used because $^-\!:CH_3$ is an anion. However, the 1,2-shift is a concerted reaction step, and no methide anion is actually formed.)

The presence of both secondary and tertiary carbocations in solution leads to the two observed products, the so-called "normal" product and the **rearrangement product**, a product in which the skeleton or the position of the functional group is different from that of the starting material.

$$[(CH_3)_3C\overset{+}{C}HCH_3] \xrightarrow[-H^+]{CH_3OH} (CH_3)_3CCHCH_3$$

2° *carbocation*   2-methoxy-3,3-dimethylbutane

with OCH$_3$ group on the CH.

$$[(CH_3)_2\overset{+}{C}CH(CH_3)_2] \xrightarrow[-H^+]{CH_3OH} (CH_3)_2CCH(CH_3)_2$$

3° *carbocation*   2-methoxy-2,3-dimethylbutane

with OCH$_3$ group on the C.

If an alkyl group, an aryl group, or a hydrogen atom (each with its bonding electrons) on an adjacent carbon atom can shift and thereby create a more stable carbocation, rearrangement can occur. Rearrangement can also occur when a pair of carbocations are of equivalent stabilities. The extent of rearrangement that will be observed in a reaction is often hard to predict and depends on a number of factors, including the relative stabilities of the carbocations in question and the reaction conditions (solvent, etc.). The following rearrangements exemplify 1,2-shifts and the formation of more-stable carbocations.

A methide shift:

CH$_3$—C(CH$_3$)$_2$—$\overset{+}{C}$HCH$_2$CH$_3$ $\longrightarrow$ CH$_3$—$\overset{+}{C}$(CH$_3$)$_2$—CHCH$_2$CH$_3$

a 2° carbocation   *a more stable 3° carbocation*

A hydride (H:⁻) shift:

CH$_3$—C(CH$_3$)(H)—$\overset{+}{C}$HCH$_3$ $\longrightarrow$ CH$_3$—$\overset{+}{C}$(CH$_3$)—CH$_2$CH$_3$

a 2° carbocation   *a more stable 3° carbocation*

Study Problems

5.8 Although many carbocations can form more-stable carbocations by 1,2-shifts, not all carbocations have structures that can yield more-stable carbocations by rearrangement. Indicate which of the following cations is likely to undergo rearrangement. Show by an arrow the shift of an alkyl group or a hydrogen, and give the structure of the rearranged carbocation.

(a) CH$_3$CH$_2$$\overset{+}{C}HCH_3$ **(b)** (CH$_3$)$_2$CH$\overset{+}{C}$HCH$_2$CH$_3$

5.9 What S_N1 products could be formed by the following reactants?

$$\underset{\displaystyle \quad}{(CH_3)_2CHCHCH_2CH_3} + CH_3CH_2OH \longrightarrow$$

$$\overset{\displaystyle Cl}{\underset{\displaystyle |}{}}$$

G. Summary of S_N1 and S_N2 Mechanisms

Substitution reactions of alkyl halides with nucleophiles can occur by an S_N1 path or by an S_N2 path. *Methyl, primary,* and *secondary alkyl halides* react principally by an S_N2 path, as shown in Figure 5.6. The rate of an S_N2 reaction is enhanced by increasing nucleophilicity of the attacking species. Typical good nucleophiles are ^-OH, ^-OR, and ^-CN.

Increasing hindrance around the halogenated carbon decreases the rate of S_N2 reaction. *Tertiary alkyl halides* are too hindered to undergo reaction by an S_N2 path, but can undergo reaction by an S_N1 path (by way of a carbocation intermediate) with a nucleophile such as H_2O or ROH (see Figure 5.6). Methyl and primary alkyl halides do not undergo S_N1 reactions at all. Secondary alkyl halides react slowly by this pathway.

S_N2: A concerted back-side attack
A *bimolecular reaction*: $Nu:^-$ and R_2CHX both involved in the transition state
Second order in rate: rate is proportional to the concentrations of both $Nu:^-$ and R_2CHX
Stereochemistry: inversion of configuration
Relative rates: $CH_3X > 1° RX > 2° RX$

S_N1: An *ionization reaction*, sometimes accompanied by rearrangement, followed by combination with $Nu:^-$
A *unimolecular reaction*: only R_3CX is involved in the transition state of the rate-determining step
First order in rate: rate is proportional to the concentration of R_3CX only, because its ionization is the slow, rate-determining step
Stereochemistry: racemization
Relative rates: $3° RX > 2° RX$

Figure 5.6 Summary of S_N1 and S_N2 mechanisms for substitution reactions of alkyl halides.

Section 5.7
Substitution Reactions of Allylic Halides and Benzylic Halides

Two special types of halides behave differently in S_N1 and S_N2 reactions from the alkyl halides we have been discussing. These are the **allylic halides** and the **benzylic halides**.

$$CH_2 = CH - CH_2 - $$
the allyl group

$$CH_2 = CHCH_2Cl$$
3-chloro-1-propene
(allyl chloride)

the benzyl group $- CH_2 - $

$- CH_2Br$
benzyl bromide

An atom or group that is attached to the carbon atom *adjacent to one of the sp²-hybridized carbon atoms* is said to be in the **allylic position** or the **benzylic position**. The halogen atoms in our previous examples and also in the following examples are in allylic or benzylic positions.

$$CH_3 \ CH = CHCH \ CH_3$$
$$\quad\quad\quad\quad\quad |$$
$$\quad\quad\quad\quad Cl$$

4-chloro-2-pentene

an allylic chloride

$$- CH \ CH_2CH_2CH(CH_3)_2$$
$$\quad\ |$$
$$\quad Br$$

1-bromo-4-methyl-1-phenylpentane

a benzylic bromide

Study Problem

5.10 Classify each of the following compounds as a vinylic halide, an allylic halide, a benzylic halide, or an aryl halide:

(a) $CH_3 - \bigcirc - CH = CHCH_2Br$

(b) $CH_3 - \bigcirc - CH = \overset{\overset{\textstyle Br}{\textstyle |}}{C}CH_3$

(c) $CH_3 - \bigcirc - CH = CHCH_3$
               $|$
              Br

(d) $BrCH_2 - \bigcirc - CH = CHCH_3$

A. S_N1 Reactions

Most primary alkyl halides undergo substitution by the S_N2 path exclusively and do not undergo S_N1 reactions. However, a primary allylic halide or benzylic halide is very reactive in both S_N1 and S_N2 reactions. Table 5.4 lists the relative reactivities of some halides under typical S_N1 conditions. You can see from the

Table 5.4 Relative reaction rates of some organic halides under typical S_N1 conditions

Halide	Relative Rate
CH_3CH_2X	1.0^a
$CH_2{=}CHCH_2X$	33
$C_6H_5CH_2X$	380
$(C_6H_5)_2CHX$	$\sim 10^5$

^{a} The observed reaction probably proceeds by an S_N2 path.

table that an allyl halide is more than 30 times more reactive than an ethyl halide, and a benzyl halide is almost 400 times as reactive. If *two* phenyl groups are present, the halide is 100,000 times as reactive!

$$CH_2{=}CHCH_2Cl + H_2O \longrightarrow CH_2{=}CHCH_2OH + Cl^- + H^+$$

allyl chloride   2-propen-1-ol
                                    (allyl alcohol)

increasing rate of S_N1 reaction

benzyl chloride   benzyl alcohol

a lachrymator:
a compound that causes
tears to flow

The reason for the enhanced reactivity of these two types of halides in an S_N1 reaction lies in the *resonance-stabilization of the carbocation and of the transition state leading to the carbocation.* Carbocations are stabilized by dispersal of the positive charge. *Inductive stabilization* involves dispersal of the positive charge through *sigma bonds.* We have used the inductive effect to explain the relative stabilities of 1°, 2°, and 3° carbocations. *Resonance-stabilization* involves the dispersal of the positive charge by *pi bonds.*

Let us consider the S_N1 reaction of allyl chloride with H_2O:

$$CH_2{=}CHCH_2{-}\overset{..}{\underset{..}{Cl}}: \xrightarrow[S_N1]{-:\overset{..}{\underset{..}{Cl}}:^-} [CH_2{=}CH\overset{+}{C}H_2] \xrightarrow[-H^+]{H_2\overset{..}{O}:} CH_2{=}CHCH_2\overset{..}{\underset{..}{O}}H$$

allyl cation

Recall from Section 2.9 that structures differing only in the position of pi electrons are resonance structures. If resonance structures can be drawn for a molecule or ion, the resonance hybrid (the real structure) is more stable than if delocalization of electrons or electrical charges could not occur. The two resonance structures for the allyl cation are equivalent in structure and bonding; therefore, they have the same energy content and contribute equally to the structure of the actual allyl cation. Because the allyl cation is resonance-stabilized, the energy of the transition state leading to its formation is

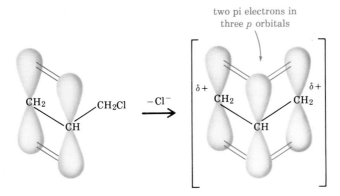

Figure 5.7 Formation of the allyl cation from allyl chloride in an S_N1 reaction.

relatively low. Consequently, the rate of its S_N1 reaction is fairly fast. Figure 5.7 shows the formation of the allyl cation and its bonding.

$$[CH_2\!\!=\!\!CH\!\!-\!\!\overset{+}{C}H_2 \longleftrightarrow \overset{+}{C}H_2\!\!-\!\!CH\!\!=\!\!CH_2]$$

resonance structures for the allyl cation
(equal contributors)

Both terminal (end) carbons in the allyl cation have an equal amount of positive charge. Which atom is attacked by the nucleophile? The answer is: either of them! Let us use another allylic system to illustrate this fact. The S_N1 reaction of 1-chloro-2-butene with H_2O in aqueous acetone leads to two products. These two products arise from attack of H_2O at either of the two partially positive carbon atoms. We say that the rearranged product is the result of **allylic rearrangement**.

$$CH_3CH\!\!=\!\!CHCH_2Cl \xrightarrow[S_N1]{-Cl^-} [CH_3CH\!\!=\!\!CH\!\!-\!\!\overset{+}{C}H_2 \longleftrightarrow CH_3\overset{+}{C}H\!\!-\!\!CH\!\!=\!\!CH_2]$$

1-chloro-2-butene

$$\Big\downarrow {{H_2O}\atop{-H^+}}$$

$$CH_3CH\!\!=\!\!CHCH_2 \qquad\qquad CH_3CHCH\!\!=\!\!CH_2$$
$$\overset{|}{OH} \qquad\qquad\qquad\quad \overset{|}{OH}$$

2-buten-1-ol (56%)   3-buten-2-ol (44%)

Benzyl halides also show enhancement of the S_N1 rate because of resonance-stabilization of the transition state leading to the carbocation. In this case, the pi electrons in the aromatic pi cloud of the benzene ring help disperse the charge.

$$\text{⬡}\!\!-\!\!CH_2Cl \xrightarrow[S_N1]{-Cl^-} \Big[\text{⬡}\!\!-\!\!\overset{+}{C}H_2\Big] \xrightarrow[-H^+]{H_2O} \text{⬡}\!\!-\!\!CH_2OH$$

benzyl chloride   benzyl cation   benzyl alcohol

We usually symbolize the aromatic pi cloud of benzene by a circle in the ring. However, in discussions of delocalization of pi electrons, Kekulé formulas are more convenient. With Kekulé formulas, we can count the pi electrons in the ring and readily see which atoms are electron-deficient. Note the similarity between the resonance structures for the benzyl cation and those for the allyl cation. The benzyl cation has five resonance structures similar to allylic resonance structures.

The first two resonance structures shown are the major contributors because these structures have aromatic stabilization. Therefore, the most positive carbon in the intermediate is the benzyl carbon. This is the carbon attacked by the nucleophile.

Study Problem

5.11 Predict the products of the solvolysis in water of:

(a)   (b)

B. S$_N$2 Reactions

Allylic halides and benzylic halides also undergo S$_N$2 reactions at faster rates than primary alkyl halides or even methyl halides. Table 5.5 lists the average relative rates of some halides in a typical S$_N$2 reaction.

The reason for the greater S$_N$2 reactivity of allylic and benzylic halides is that the allylic pi bond or the aromatic pi cloud reduces the energy of the

Table 5.5 Average relative S$_N$2 rates for some organic halides

Halide	Relative Rate
CH$_3$X	30
CH$_3$CH$_2$X	1
(CH$_3$)$_2$CHX	0.025
CH$_2$=CHCH$_2$X	40
C$_6$H$_5$CH$_2$X	120

S_N2 *transition state*
showing p orbitals

Figure 5.8 Stabilization of the transition state in an S_N2 reaction of allyl chloride.

transition state of an S_N2 reaction. In the transition state, the carbon undergoing reaction changes from the sp^3-hybrid state to the sp^2-hybrid state and has a *p* orbital. This *p* orbital forms partial bonds with both the incoming nucleophile and the leaving group. The entire grouping of atoms carries a negative charge. Adjacent *p* orbitals, as in an allylic or benzylic group, undergo partial overlap with the transitional *p* orbital. In this way, adjacent *p* orbitals help delocalize the negative charge and thus lower the energy of the transition state. Figure 5.8 shows the *p* orbitals of the allylic case; the benzylic case is similar.

For increased stabilization to occur in either S_N1 or S_N2 reactions of compounds with pi systems, the pi system must be *adjacent* to the reacting carbon. If it is farther away, it cannot overlap and cannot help stabilize the transition state.

pi bond too far
away to overlap
in transition state

too far away

Sample Problem

Which of the following compounds exhibit enhanced reactivities in S_N1 and S_N2 reactions because of resonance-stabilization or partial *p*-orbital overlap?

(a) $CH_3CH=CHCHBrCH=CH_2$ (b) $CH_3CH=CHCH_2CHBrCH_2CH_3$
(c) $C_6H_5CH=CHCH_2I$ (d) $CH_2=CHCH_2CHBrCH=CH_2$

Solution: (a), (c), and (d), in which the halogen atom is allylic. In (a), the Br atom is allylic with respect to *two* carbon–carbon double bonds; consequently, this halide is particularly reactive.

▼ ## Section 5.8
The E1 Reaction

A carbocation is a high-energy, unstable intermediate that quickly undergoes further reaction. One way a carbocation can reach a stable product is by combining with a nucleophile. This, of course, is the S_N1 reaction. However, there is an alternative: the carbocation can *lose a proton to a base in an elimination reaction*, an **E1 reaction** in this case, and become an alkene.

Substitution (S_N1):

$$(CH_3)_3CBr \xrightarrow{-Br^-} [(CH_3)_3C^+] \xrightarrow[\text{$-H^+$}]{\text{H_2O}} (CH_3)_3COH$$

t-butyl bromide   *t*-butyl cation   *t*-butyl alcohol

Elimination (E1):

$$(CH_3)_3CBr \xrightarrow{-Br^-} \left[(CH_3)_2\overset{+}{C}-CH_2 \overset{H}{\underset{}{}} \right] \longrightarrow (CH_3)_2C=CH_2 + H_3\overset{+}{O}:$$

t-butyl cation   methylpropene

The first step in an E1 reaction is identical to the first step in an S_N1 reaction: the ionization of the alkyl halide. This is the slow step and thus the rate-determining step of the overall reaction. Like an S_N1 reaction, a typical E1 reaction shows first-order kinetics, with the rate of the reaction dependent on the concentration of only the alkyl halide. Since only one reactant is involved in the transition state of the rate-determining step, the E1 reaction, like the S_N1 reaction, is unimolecular.

Step 1 (slow):

$$(CH_3)_3C\overset{..}{\underset{..}{Br}}: \longrightarrow \left[(CH_3)_3\overset{\delta+}{C}\text{----}\overset{\delta-}{\underset{..}{Br}}: \right] \longrightarrow \left[\begin{matrix} H_3C \\ H_3C \end{matrix} \overset{+}{C}-CH_3 \right] + :\overset{..}{\underset{..}{Br}}:^-$$

transition state 1   *carbocation intermediate*

In the second step of an elimination reaction, the base removes a proton from a carbon that is *adjacent to the positive carbon*. The electrons of that carbon–hydrogen sigma bond shift toward the positive charge, the adjacent carbon rehybridizes from the sp^3 state to the sp^2 state, and an alkene is formed.

Step 2 (fast):

$$\left[\begin{matrix} H_3C \\ H_3C \end{matrix} \overset{+}{C}-\overset{H}{\underset{}{C}}H_2 \right] + H_2\overset{..}{O}: \longrightarrow \left[\begin{matrix} H_3C \\ H_3C \end{matrix} \overset{\delta+}{C}\text{---}CH_2 \overset{\overset{\delta+}{H}\text{---}\overset{..}{O}H_2}{\underset{}{}} \right] \longrightarrow \begin{matrix} H_3C \\ H_3C \end{matrix} C=CH_2 + H_3\overset{+}{O}:$$

base   *transition state 2*   *an alkene*

Because an E1 reaction, like an S_N1 reaction, proceeds through a carbocation intermediate, it is not surprising that tertiary alkyl halides undergo this reaction far more rapidly than other alkyl halides. E1 reactions of alkyl halides occur under the same conditions as S_N1 reactions (polar solvent, very weak base, etc.). Therefore, S_N1 and E1 reactions are competing reactions. For this reason, E1 reactions of alkyl halides are relatively unimportant. You will learn in Chapter 7 that E1 reactions of alcohols, however, are quite important. We will discuss these reactions in greater detail at that time.

Study Problem

5.12 Some tertiary alkyl halides yield mixtures of alkenes as well as a substitution product when they are subjected to S_N1 conditions. Predict all likely products from the reactions of 2-bromo-2-methylbutane with ethanol under S_N1 conditions.

Section 5.9
The E2 Reaction

The most useful elimination reaction of alkyl halides is the **E2 reaction** (bimolecular elimination). E2 reactions of alkyl halides are favored by the use of strong bases, such as ^-OH or ^-OR, and high temperatures. Typically, an E2 reaction is carried out by heating the alkyl halide with K^+ ^-OH or Na^+ $^-OCH_2CH_3$ in ethanol.

$$\underset{\substack{\text{2-bromopropane}\\\text{(isopropyl bromide)}}}{CH_3\overset{\overset{\displaystyle Br}{|}}{C}HCH_3} + CH_3CH_2O^- \xrightarrow[\text{E2}]{\overset{CH_3CH_2OH}{\text{heat}}} \underset{\text{propene}}{CH_3CH=CH_2} + CH_3CH_2OH + Br^-$$

The E2 reaction does not proceed by way of a carbocation intermediate, but is a **concerted reaction**—that is, it occurs in one step, just as an S_N2 reaction does.

$$\overset{(1)}{\overset{\frown}{RO:^-}} + \overset{(2)}{H-CH_2-\overset{\overset{\displaystyle :\ddot{Br}:}{|\nearrow(3)}}{C}HCH_3} \longrightarrow R\ddot{O}H + CH_2=CHCH_3 + :\ddot{Br}:^-$$

(1) The base is forming a bond with the hydrogen.
(2) The C—H electrons are forming the pi bond.
(3) The bromine is departing with the pair of electrons
     from the carbon–bromine sigma bond.

The preceding equation shows the mechanism with arrows representing "electron-pushing." The structure of the transition state in this one-step reaction follows:

$$\begin{array}{c}
\ddot{R}\ddot{O}:^{\delta-} \\
| \\
H \\
| \\
CH_2=\!=\!=CHCH_3 \\
| \\
:\ddot{Br}:^{\delta-}
\end{array}$$

E2 transition state

In E2 reactions, as in E1 reactions, tertiary alkyl halides undergo reaction fastest, and primary alkyl halides react slowest. (When treated with a base, primary alkyl halides usually undergo substitution so readily that little alkene is formed unless the base is very strong and bulky.)

$$1°RX \quad 2°RX \quad 3°RX$$

increasing rate of E2

A. Kinetic Isotope Effect in E2 Reactions

One piece of experimental evidence that supports our understanding of the E2 mechanism is the difference in the rates of elimination of deuterated and nondeuteriated alkyl halides. A difference in the rates of reaction between compounds containing different isotopes is called a **kinetic isotope effect**.

Deuterium ($_1^2H$, or D) is an isotope of hydrogen with a nucleus containing one proton and one neutron. More energy (1.2 kcal/mol) is required to break a C—D bond than to break a C—H bond. We have postulated that the breaking of the C—H bond is an integral part of the rate-determining step (the *only* step) of an E2 reaction. What happens when the H that is eliminated is replaced by D? The stronger CD bond requires more energy to be broken. For this reason, the E_{act} is greater and the rate for the elimination reaction is slower.

When the following 2-bromopropanes are subjected to an E2 reaction with $CH_3CH_2O^-$ as the base, it has been observed that the deuterated compound undergoes reaction at *one-seventh* the rate that ordinary 2-bromopropane does, a fact that supports the E2 reaction mechanism we have described.

$$CH_3CH_2O^- + CH_3\overset{\overset{\displaystyle Br}{|}}{C}HCH_3 \xrightarrow[\text{fast}]{} CH_3CH_2OH + CH_2{=}CHCH_3 + Br^-$$

$$CH_3CH_2O^- + CD_3\overset{\overset{\displaystyle Br}{|}}{C}HCD_3 \xrightarrow[\text{slow}]{} CH_3CH_2OD + CD_2{=}CHCD_3 + Br^-$$

B. Mixtures of Alkenes

Often, E1 and E2 reactions are referred to as **beta (β) eliminations**. This term reflects which hydrogen atom is lost in the reaction. Different types of carbon and hydrogen atoms in a molecule may be labeled as α, β, and so forth, according to the Greek alphabet. The carbon atom *bonded to the principal functional group* in a molecule is called the **alpha (α) carbon**, and the adjacent carbon is the **beta (β) carbon**. The hydrogens bonded to the α carbon are called α hydrogens, while those bonded to the β carbon are β hydrogens. In a β elimination, a β hydrogen is lost when the alkene is formed. (Of course, an alkyl halide with no β hydrogen cannot undergo a β elimination.)

β carbons and hydrogens circled:

$$CH_3 \ CH \ CH_3 \qquad CH_3{-}\overset{\overset{\displaystyle CH_3}{|}}{\underset{\underset{\displaystyle CH_3}{|}}{C}}{-}Br \qquad CH_3 \ CH_2{-}\overset{\overset{\displaystyle CH_3}{|}}{\underset{\underset{\displaystyle CH_3}{|}}{C}}{-}Br$$

If 2-bromopropane or *t*-butyl bromide undergoes elimination, there is only one possible alkene product. However, if the alkyl groups around the α carbon are different and there is more than one type of β hydrogen, then more than one alkene can result. The E2 reaction of 2-bromobutane yields two alkenes because two types of hydrogen atoms can be eliminated: a hydrogen from a CH_3 group or a hydrogen from a CH_2 group.

$$HO^- \quad CH_3$$
$$CH_3-\overset{\displaystyle CH_3}{\underset{\displaystyle CH_3}{C}}-Br \longrightarrow CH_2=C\overset{\displaystyle CH_3}{\underset{\displaystyle CH_3}{}} + H_2O + Br^-$$

t-butyl bromide

only one type of β H

methylpropene

only possible alkene

$$HO^-$$
$$\overset{H \quad H}{CH_3\overset{|}{C}H\overset{|}{C}HCH_2}$$
$$\underset{Br}{|}$$

$-H^+$ $CH_3CH_2CH=CH_2$ 1-butene

$-H^+$ $CH_3CH=CHCH_3$ 2-butene

⎫ *two possible alkenes* ⎬ ⎭

2-bromobutane

two types of β H

Sample Problems

Circle the β carbons and hydrogens in the following structures:

(a) $CH_3CH_2\underset{\underset{Br}{|}}{CH}CH_2CH_2CH_3$ (b) [hexane ring with CH_3 and Br substituents]

Solution:

(a) $CH_3\ \text{CH}_2\ \text{CH}\ \text{CH}_2\ CH_2CH_3$ (b) [hexane ring with CH_3 and Br substituents]
$\quad\quad\quad\quad\underset{Br}{|}$

In the preceding problem, tell how many different *types* of β hydrogens are in each structure.

Solution: (a) two types; (b) two types (the ring CH_2 groups are equivalent to each other; the CH_3 is different).

Study Problem

5.13 Write the structures of the alkenes that could result from the E2 reaction of each of the alkyl bromides in the preceding Sample Problem.

C. Which Alkene Is Formed?

In 1875, the Russian chemist Alexander Saytzeff formulated the following rule, now called the **Saytzeff rule**. *In elimination reactions, the alkene with the greatest number of alkyl groups on the doubly bonded carbon atoms predominates in the product mixture.* We will refer to this alkene as the *more highly substituted alkene.*

The Saytzeff rule predicts that 2-butene would predominate over 1-butene as a product in the E2 reaction of 2-bromobutane. This indeed is what occurs. In the following reaction, the mixture of alkene products consists of 80% 2-butene and only 20% 1-butene.

$$
\underset{\text{2-bromobutane}}{CH_3CH_2\overset{\overset{\displaystyle Br}{|}}{C}HCH_3} \xrightarrow[CH_3CH_2OH]{Na^+ \; ^-OCH_2CH_3} \underset{\substack{\text{2-butene}\\(80\%)}}{CH_3CH=CHCH_3} + \underset{\substack{\text{1-butene}\\(20\%)}}{CH_3CH_2CH=CH_2}
$$

two R's on C=C, more highly substituted

one R on C=C

It has been determined that *more highly substituted alkenes are more stable than less substituted alkenes* (this will be discussed further in Chapter 10). Therefore, an E2 elimination leads to the *more stable alkene.* (In Chapter 7, we will describe how the E1 reaction also leads to the more stable alkene.)

$$
CH_2=CH_2 \qquad CH_3CH=CH_2 \qquad CH_3CH=CHCH_3 \qquad (CH_3)_2C=C(CH_3)_2
$$

increasing stability

To see why the more stable alkene (2-butene) is formed in preference to the less stable alkene (1-butene), let us consider the transition states leading to these two butenes. In either transition state, the base is removing a proton, and a double bond is being formed. We say that this transition state has some **double-bond character**, which we represent as a dotted line in the formula.

$$
\underset{\text{2-bromobutane}}{CH_3CH_2\overset{\overset{\displaystyle Br}{|}}{C}HCH_3} \xrightarrow{OR^-}
$$

$$
\left[\begin{array}{c} \overset{\displaystyle H}{\vdots}\overset{\delta-}{---OR} \\ CH_3CH_2CH\!=\!\!=\!CH_2 \\ \overset{\displaystyle |}{Br}^{\delta-} \end{array} \right] \longrightarrow \underset{\text{1-butene}}{CH_3CH_2CH=CH_2}
$$

transition state

$$
\left[\begin{array}{c} \overset{\delta-}{RO}---H \\ CH_3CH\!=\!\!=\!CHCH_3 \\ \overset{\displaystyle |}{Br}^{\delta-} \end{array} \right] \longrightarrow \underset{\text{2-butene}}{CH_3CH=CHCH_3}
$$

transition state

Because both transition states leading to the alkenes have some double-bond character, the transition state leading to the more stable alkene is itself more stabilized and of lower energy. The reaction with the lower-energy

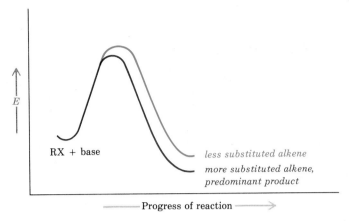

Figure 5.9 Energy diagram for a typical E2 reaction, showing why the more substituted alkene predominates.

transition state proceeds at a faster rate. Therefore, the more stable alkene is the predominant product (see Figure 5.9).

Sample Problems

Which is the more stable alkene,   (a) $(CH_3)_3CCH=CHCH_3$, or   (b) $CH_3CH=C(CH_3)_2$?

Solution: (b) with three R's is more stable than (a) with only two R's.

Predict the major alkene product of the E2 dehydrohalogenation of $CH_3CH_2C(CH_3)_2Cl$.

Solution: The two possible alkenes are $CH_3CH=C(CH_3)_2$ and $CH_3CH_2\underset{\underset{CH_3}{|}}{C}=CH_2$.

The first predominates (3 R's versus 2 R's).

Study Problem

5.14 Predict the major alkene product of the E2 dehydrohalogenation of 1-chloro-1-methylcyclohexane.

Part of our question about which alkene would be formed in dehydrohalogenation has been answered: the most highly substituted alkene predominates in the product mixture. The most highly substituted alkene can often exist as *cis* and *trans* diastereomers (geometric isomers). Is there a selectivity as to which diastereomer is formed? Experimentally, it has been determined that *trans* alkenes are generally more stable than their *cis* isomers, presumably because of less steric hindrance in the *trans* isomers. Therefore, it is not surprising that *trans* alkenes predominate as products of E2 reactions. Again, the reason is a more stabilized transition state. The following equation shows the results of an E2 reaction of 2-bromopentane.

$$CH_3CH_2CH_2CH=CH_2$$

1-pentene (31%)

$$CH_3CH_2 \quad\quad CH_3$$
$$C=C$$
$$H \quad\quad\quad H$$

cis-2-pentene (18%)

$$CH_3CH_2 \quad\quad H$$
$$C=C$$
$$H \quad\quad\quad CH_3$$

trans-2-pentene (51%)

$$CH_3CH_2CH_2\overset{\displaystyle Br}{\underset{}{C}}HCH_3 \xrightarrow[\text{heat}]{\substack{Na^+ \ ^-OCH_2CH_3 \\ CH_3CH_2OH}}$$

2-bromopentane

most stable alkene: major product

D. Stereochemistry of an E2 Reaction

In the transition state of an E2 elimination, the attacking base and the leaving group are generally as far apart as possible, or *anti*. For this reason, the E2 elimination is often referred to as **anti-elimination**.

anti-Elimination:

dimensional   *ball-and-stick*   *Newman*

The interesting feature about *anti*-elimination is that the *anti* positioning of the H and Br that are lost determines the stereochemistry of the product alkene. To see how this happens, let us look at the E2 reactions of some stereoisomeric halides. The compound 1-bromo-1,2-diphenylpropane has two chiral carbon atoms (carbons 1 and 2) and four stereoisomers.

The four stereoisomers of $\overset{③}{C}H_3\overset{②}{C}H-\overset{①}{C}HBr:$
$$\underset{C_6H_5 \ C_6H_5}{}$$

(1R,2R)   (1S,2S)   (1R,2S)   (1S,2R)

enantiomers   enantiomers

Because there is only one β hydrogen in the starting halide, any of these stereoisomers yields $C_6H_5(CH_3)C=CHC_6H_5$. However, geometric isomerism is possible in this product.

only one β hydrogen

$$CH_3\overset{|}{\underset{|}{C}}H—\overset{|}{\underset{|}{C}}HBr + {}^-OR \xrightarrow{E2} CH_3\overset{|}{\underset{|}{C}}=\overset{|}{\underset{|}{C}}H + ROH + Br^-$$
$$\quad\quad C_6H_5\ \ C_6H_5 \quad\quad\quad\quad\quad C_6H_5\ \ C_6H_5$$

When either (1*R*,2*R*)-1-bromo-1,2-diphenylpropane or its (1*S*,2*S*)-enantiomer undergoes E2 reaction, the (*Z*)-alkene is formed exclusively; no (*E*)-alkene is formed.

(1*R*,2*R*)   (*Z*)-1,2-diphenyl-1-propene

The reason for all (*Z*) and no (*E*) product is that there is only one conformation of either of these enantiomers in which the Br and the beta hydrogen are *anti*. In either the (1*R*,2*R*)- or the (1*S*,2*S*)-enantiomer, the *anti* alignment of H and Br puts the phenyl groups on the same side of the molecule, and the (*Z*)-alkene results. If the elimination could occur regardless of the conformation of the enantiomers, then some (*E*)-alkene would also be observed.

anti H and Br
means
cis-phenyls

(1*R*,2*R*)   (*Z*)-*alkene*

Study Problem

5.15 Write equations for the *anti*-elimination of the (1*S*,2*S*)-enantiomer, as we have done for the (1*R*,2*R*)-enantiomer.

Just the opposite situation prevails with the (1*R*,2*S*)- or (1*S*,2*R*)-enantiomers. Either of these isomers yields the (*E*)-alkene and not the (*Z*)-alkene. The reason, once again, is that there is only one conformation for each of these enantiomers in which the Br and the single beta hydrogen are in an *anti* relationship. In these conformations, the phenyl groups are on opposite sides of the molecule.

anti H and Br
means
trans-phenyls

(1*R*,2*S*)   (1*S*,2*R*)   (*E*)-*alkene*

A reaction in which different stereoisomers of the reactant yield stereoisomerically different products is said to be a **stereospecific reaction**. The E2 reaction is an example of a stereospecific reaction.

Halocycloalkanes, such as chlorocyclohexane, can also undergo E2 reactions. In these cases, the conformations of the ring play an important role in the course of the reaction. In order to be *anti* on a cyclohexane ring, the leaving group (such as chlorine) and a β hydrogen must be 1,2-*trans* and *diaxial*. No other conformation places the H and Cl *anti* to each other. (Try it with models.) Even though this conformation is not the favored one, a certain percentage of halocycloalkane molecules are in this conformation at any given time and can thus undergo elimination.

Cl is equatorial and	*Cl is axial and*
not anti to any β hydrogens	*anti to two β hydrogens*

OH⁻ could attack   cyclohexene
either H shown
                          (from attack at either H)

Study Problems

5.16 Draw dimensional formulas for the conformers of (1*R*,2*S*)-1-bromo-1,2-diphenylpropane and (1*S*,2*R*)-1-bromo-1,2-diphenylpropane that undergo E2 reaction.

5.17 1,2-Dibromo-1,2-diphenylethane contains two chiral carbon atoms and has a pair of enantiomers plus a *meso* diastereomer. Any of the stereoisomers of this compound can undergo an E2 reaction to yield 1-bromo-1,2-diphenylethene. The *meso* form yields one geometric isomer of the alkene, while the racemic mixture of enantiomers yields the other geometric isomer. Predict the stereochemistry of the products of these two reactions.

5.18 Write an equation using conformational formulas that illustrates the E2 reaction of *trans*-1-chloro-2-methylcyclohexane in aqueous sodium hydroxide. (Be sure to show the required conformation of the ring for *anti*-elimination.)

E. Hofmann Products

Most dehydrohalogenations follow the Saytzeff rule, and the more substituted alkene predominates. However, under some circumstances, the major product of an E2 dehydrohalogenation is the *less substituted, less stable alkene*. When the less substituted alkene is the predominant product, we say that the reaction yields the **Hofmann product**.

When is the less substituted alkene likely to be the predominant product? A common phenomenon leading to the less substituted alkene is **steric hindrance** in that transition state leading to the most substituted alkene. Steric hindrance can raise the energy of this transition state so substantially that the reaction follows a different course and yields the less substituted alkene. The steric hindrance may be caused by any one of three factors. First of all, it may be caused by the **size of the attacking base**. In the elimination reaction of 2-bromobutane with the small ethoxide ion in ethanol, the more substituted alkene predominates. With the bulky *t*-butoxide ion in *t*-butyl alcohol, however, the 1- and 2-butenes are formed in equal amounts.

$$CH_3CH_2O^- \text{ (small)} \rightarrow CH_3CH=CHCH_3 + CH_3CH_2CH=CH_2$$
2-butene (80%)   1-butene (20%)

Br
|
$CH_3CH_2CHCH_3$

2-bromobutane

$$(CH_3)_3CO^- \text{ (bulky)} \rightarrow CH_3CH=CHCH_3 + CH_3CH_2CH=CH_2$$
2-butene (50%)   1-butene (50%)

Reaction with the t-butoxide ion:

attack at carbon 3:
more steric hindrance
than with attack
at carbon 1

$(CH_3)_3CO^-$
$$H-\overset{\displaystyle H}{\underset{\displaystyle H}{C}}-\overset{\displaystyle H}{\underset{\displaystyle Br}{CH}}CHCH_3 \longrightarrow CH_3CH=CHCH_3$$
2-butene

attack at carbon 1:
less steric hindrance
than with attack
at carbon 3

$^-OC(CH_3)_3$
$$CH_3CH_2CH\overset{\displaystyle H}{\underset{\displaystyle Br}{C}}\overset{}{\underset{\displaystyle H}{-H}} \longrightarrow CH_3CH_2CH=CH_2$$
1-butene

Second, steric hindrance might be caused by the **bulkiness of groups surrounding the leaving group** in the alkyl halide. The hindered 2-bromo-2,4,4-trimethylpentane yields the less substituted alkene in an E2 reaction, even with a small base like the ethoxide ion.

more crowded β H   less crowded β H

$$\underset{\underset{\displaystyle CH_3 \; CH_3}{|\quad\quad|}}{CH_3CCH_2\overset{\overset{\displaystyle CH_3 \;\; Br}{|\quad\;\;|}}{C}-CH_3} \xrightarrow{CH_3CH_2O^-} (CH_3)_3CCH_2\underset{\underset{\displaystyle CH_3}{|}}{C}=CH_2$$

2-bromo-2,4,4-trimethylpentane   2,4,4-trimethyl-1-pentene

Third, if the **leaving group itself is large and bulky**, the Hofmann product may predominate. This type of reaction is discussed in Section 17.10.

Study Problems

5.19 Write formulas for both the Hofmann and Saytzeff products of the E2 reactions of
(a) 3-bromo-2-methylpentane and   (b) 1-chloro-1-methylcyclohexane.

5.20 Predict the major alkene products of the following E2 reactions:

(a) $CH_3CH_2CH_2CHBrCH_3 + {}^-OC(CH_3)_3 \longrightarrow$

(b) $CH_3CH_2CHBrCH_3 + OH^- \longrightarrow$

F. Summary of E1 and E2 Mechanisms

Elimination reactions of alkyl halides can proceed by an E1 path or by an E2
path. *Tertiary alkyl halides* can undergo E1 elimination as a side reaction in
solvolysis when water or an alcohol acts as a very weak base. Figure 5.10
summarizes some important features of the E1 pathway.

When a strong base is used to effect elimination, *tertiary alkyl halides,
secondary alkyl halides,* and, in some cases, *primary alkyl halides* undergo
reaction by an E2 path, as shown in Figure 5.10.

An ionization reaction is sometimes accompanied by rearrangement,
followed by loss of H^+ to the base
Generally produces the most stable alkene (Saytzeff rule)
Stereochemistry: not stereospecific
A unimolecular reaction
First order in rate
Relative rates: $3°$ RX > $2°$ RX

A concerted, bimolecular reaction
Produces the most stable alkene (with exceptions)
Stereochemistry: anti elimination of H and X
Second order in rate
Relative rates: $3°$ RX > $2°$ RX > $1°$ RX

Figure 5.10 Summary of E1 and E2 mechanisms for elimination reactions of
alkyl halides. (The symbol $B:^-$ represents a weak base for E1 reactions or a
strong base for E2 reactions of alkyl halides.)

Section 5.10
Factors Governing Substitution and Elimination Reactions

At the start of this chapter, we mentioned that S_N1, S_N2, E1, and E2 are *competing reactions*. A single alkyl halide could be undergoing substitution, elimination, and rearrangements all in the same reaction flask. If this happens, a mixture of a large number of products can result. However, a chemist can control the products of the reaction to a certain extent by a proper choice of the reagents and reaction conditions. We have already mentioned some of the factors that affect the course of substitution and elimination reactions. These and other major factors are:

1. the structure of the alkyl halide;
2. the nature of the nucleophile or base;
3. the nature of the solvent;
4. the concentration of the nucleophile or base;
5. the temperature.

A. The Alkyl Halide

We have mentioned that the type of alkyl halide affects the mechanism of the reaction. Now that we have looked at the four principal mechanisms by which an alkyl halide can undergo reaction with a nucleophile or base, we can summarize how the different alkyl halides act. (See also Figure 5.11.)

Methyl and 1°: $RCH_2X \xrightarrow[S_N2]{Nu:^-} RCH_2Nu$

2°: R_2CHX
- $\xrightarrow[S_N2 + E2]{Nu:^-} R_2CHNu +$ alkenes
- $\xrightarrow[E2]{strong\ B:^-}$ alkenes

3°: R_3CX
- $\xrightarrow[S_N1\ and\ E1]{weak\ Nu:^-} R_3CNu +$ alkenes
- $\xrightarrow[E2]{strong\ B:^-}$ alkenes

Allylic and benzylic: $RX \xrightarrow[S_N1\ and\ S_N2]{Nu:^-} RNu$

Figure 5.11 The principal reactions that different classes of alkyl halides undergo. Typical nucleophiles ($Nu:^-$) are ^-OH and ^-CN. Typical weak nucleophiles are H_2O and ROH. Typical strong bases ($B:^-$) are ^-OH and $^-OCH_2CH_3$.

Methyl and primary alkyl halides tend to undergo S_N2 reactions. They do not form carbocations and thus cannot undergo S_N1 or E1 reactions. Primary alkyl halides undergo E2 reactions less readily than other alkyl halides.

Secondary alkyl halides can undergo reaction by any path, but S_N2 and E2 are more common than E1 or S_N1. The reactions of secondary alkyl halides are more subject to control by conditions in the reaction flask (concentration of nucleophile, solvent, etc.) than are reactions of other alkyl halides.

Tertiary alkyl halides undergo primarily E2 reactions with a strong base (such as ⁻OH or ⁻OR), but undergo the S_N1 reaction and some E1 reaction with a very weak base (such as H_2O or ROH).

Allylic and benzylic alkyl halides undergo substitution reactions readily: generally S_N1 with weak nucleophiles and S_N2 with moderately strong nucleophiles.

B. The Nucleophile or Base

The difference between nucleophilicity and basicity was discussed in Section 5.4D. As we mentioned in that section, a strong base is generally also a good nucleophile. Two other factors can affect the relative nucleophilicities of reactants, sometimes dramatically. One of these factors is the *solvent* used for the reaction. Solvent effects will be discussed in Section 5.10C.

The *polarizability* of an ion or molecule is another factor that affects its nucleophilicity. The outer electrons of larger atoms are farther from the nucleus and less tightly held than those of smaller atoms. The outer electrons of larger atoms are therefore more easily distorted by attraction to a positive center and can attack a partially positive carbon atom more readily. For example, an iodide ion is usually a better nucleophile than a chloride ion.

The degree of nucleophilicity versus basicity can affect the course of a reaction. The reaction of a *primary alkyl halide* with a strong nucleophile (see page 175) follows an S_N2 path, even if the nucleophile is also a strong base. However, for a *tertiary alkyl halide*, any moderately strong base favors E2 reactions. Only the weakest bases (H_2O, ROH) result in substitution (by an S_N1 path).

For *secondary alkyl halides*, strong nucleophiles (such as ⁻CN) favor S_N2 reactions, while weak nucleophiles (such as H_2O) favor carbocation reactions, primarily S_N1 with some E1. Strong bases (such as ⁻OH or ⁻OR) favor E2 reactions.

strong nucleophile: S_N2
weak nucleophile: S_N1
strong base: E2

Study Problem

5.21 Predict which is generally the better nucleophile: CH_3S^- or CH_3O^-. Explain the reason for your answer.

Table 5.6 Relative rates of typical S_N1 reactions in various solvents

Solvent	Formula	Dielectric Constant	Approximate Relative Rate
formic acid	HCO_2H	58	15,000
water	H_2O	78.5	4,000
80% aqueous ethanol	$CH_3CH_2OH-H_2O$	67	185
ethanol	CH_3CH_2OH	24	37
acetone	$CH_3\overset{\displaystyle O}{\overset{\|}{C}}CH_3$	21	0.5
tetrahydrofuran (THF)	[pentagon ring with O]	7	0.05

C. The Solvent

The solvent exerts its influence on substitution and elimination reactions by its ability or inability to solvate ions: carbocations, nucleophiles or bases, and leaving groups. The ability of a solvent to solvate ions is determined by its polarity, which is usually reported as a **dielectric constant**. Whereas a dipole moment is a measure of the polarity of a single molecule, the dielectric constant is a measure of the polarity of a liquid (many molecules with interactions between them). A highly polar solvent has a high dielectric constant. Table 5.6 lists some common organic solvents, their dielectric constants, and the relative rates of a typical S_N1 reaction in that solvent.

While dielectric constants can provide a guide for solvent selection, there are no firm rules about how to predict the best solvent for a given reaction. (The solubilities of the reactants must be considered too!) In general, a *very polar solvent* (such as water) encourages S_N1 reactions by helping stabilize the carbocation through solvation.

S_N2 reactions are also favored by polar solvents, but only by solvents that cannot solvate the nucleophile. A solvent that can solvate (and thus stabilize) an anion reduces its nucleophilicity. By contrast, a solvent that cannot solvate an anion enhances its nucleophilicity. The chloride ion is a far better nucleophile in dimethylformamide (DMF) or dimethyl sulfoxide (DMSO), where it is not solvated, than in ethanol, where it is solvated.

ethanol can solvate a negative ion

DMF and DMSO contain no H capable of solvating a negative ion

$$CH_3CH_2\overset{\delta+}{O}H\text{---}Cl^-\text{---}\overset{\delta+}{H}OCH_2CH_3 \qquad HCN(CH_3)_2 \qquad CH_3-\overset{\displaystyle O}{\overset{\|}{S}}-CH_3$$

DMF   DMSO

D. Concentration of the Nucleophile or Base

By controlling the concentration of nucleophile or base, a chemist has direct control over the rates of S_N2 and E2 reactions. Increasing the concentration of nucleophile generally has no effect on the rates of S_N1 or E1 reactions, but increases S_N2 or E2 reaction rates proportionally.

high concentration of Nu:⁻ or base: S_N2 or E2
low concentration of Nu:⁻: S_N1 or E1

E. Temperature

An increase in temperature increases the rates of all substitution and elimination reactions. However, an increase in temperature usually leads to a greater increase in elimination products. (The reason for this is that elimination reactions usually have higher E_{act}'s than do substitution reactions, and higher temperatures enable a greater number of molecules to reach the elimination transition state.)

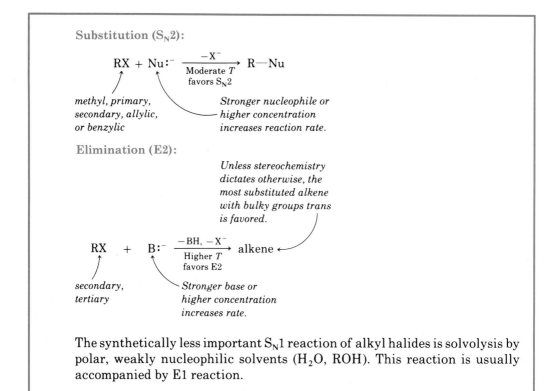

Figure 5.12 Summary of principal factors affecting important substitution and elimination reactions of alkyl halides.

Sample Problem

Predict the products of the following reactions, all of which generate one substitution or elimination product in good yield, and tell which mechanism(s) would be followed.

(a) $BrCH_2CH_2CH_2Br + 2\ Na^+\ {}^-CN \xrightarrow[\text{heat}]{\text{aqueous } CH_3CH_2OH}$

(b) $C_6H_5S^-\ Na^+ + CH_3\overset{\overset{\displaystyle Br}{|}}{CH}-\overset{\overset{\displaystyle O}{\|}}{C}OCH_2CH_3 \xrightarrow[25°]{THF}$

(c) $O_2N-\!\!\!\bigcirc\!\!\!-CH_2Cl + Na^+\ {}^-O\overset{\overset{\displaystyle O}{\|}}{C}CH_3 \xrightarrow[heat]{CH_3CO_2H}$

(d) $CH_3CH_2CH_2\overset{\overset{\displaystyle CH_3}{|}}{C}=CHCH_2Cl + Na^+\ {}^-OCH_3 \xrightarrow[25-50°]{CH_3OH}$

(e)

$+ K^+\ {}^-OC(CH_3)_3 \xrightarrow[90°]{(CH_3)_2S=O}$

Solution:

(a) Even though a polar solvent mixture is used (to dissolve the NaCN), the combination of primary halide and strong nucleophile yields the S_N2 products:

$NCCH_2CH_2CH_2CN + 2Na^+\ Br^-$

(b) A strong nucleophile, a secondary halide, moderately low temperature, and nonpolar solvent lead to S_N2 products:

$CH_3\overset{\overset{\displaystyle C_6H_5S}{|}}{CH}-\overset{\overset{\displaystyle O}{\|}}{C}OCH_2CH_3 + Na^+\ Br^-$

(c) The acetate ion is a weak nucleophile, but a benzylic halide is a reactive halide. From the material presented in this chapter, we would predict the substitution products to arise by either an S_N1 path or an S_N2 path.

$O_2N-\!\!\!\bigcirc\!\!\!-CH_2-O\overset{\overset{\displaystyle O}{\|}}{C}CH_3 + Na^+\ Cl^-$

(d) This is a reaction of an allylic halide. The conditions (strong nucleophile) indicate an S_N2 mechanism, not an S_N1 mechanism. For this reason, and because of the steric hindrance at carbon 3, we would not expect an allylic rearrangement. (In fact, only one substitution product could be isolated when this reaction was carried out.)

$CH_3CH_2CH_2\overset{\overset{\displaystyle CH_3}{|}}{C}=CHCH_2OCH_3 + Na^+\ Cl^-$

(e) Do not worry about the complexity of a structure. This compound, which contains a tertiary halogen, is being subjected to E2 conditions—strong bulky base and heat. We would expect the Hofmann product because of the bulky base and because of the geometry of the substrate molecule. (The formation of the most substituted alkene here would be geometrically impossible.)

$$+ \ K^+ \ Cl^- \ + \ HOC(CH_3)_3$$

Study Problem

Predict the products and probable mechanistic paths of the following reactions. Explain your answers.

(a) $CH_3(CH_2)_5\overset{\displaystyle I}{C}HCH_3 \xrightarrow[100°]{\text{aqueous } CH_3CH_2OH}$

(b) $C_6H_5O^- \ Na^+ \ + \ HOCH_2\overset{\displaystyle OH}{C}HCH_2Cl \xrightarrow[\text{heat}]{CH_3CH_2OH}$

(c) $CH_3CH_2OCH_2CH_2Br \ + \ Na^+ \ ^-CN \longrightarrow$

(d) $CH_3CHClCH{=}CH_2 \ + \ CH_3CH_2OH \xrightarrow{\text{heat}}$

▼ ## Section 5.11
Synthesizing Other Compounds from Alkyl Halides

From a practical standpoint, only S_N2 and E2 reactions are useful for synthesizing other compounds from alkyl halides. S_N1 and E1 reactions usually yield mixtures of products.

A large number of compound classes can be prepared from alkyl halides and other halogen compounds by S_N2 reactions. We have presented only a few nucleophiles in the chapter discussions, but a variety of nucleophilic reagents are useful (see Table 5.7). Of special interest are the reactions of RX with an alkoxide (RO^-) or a phenoxide (ArO^-) to yield ethers. You will encounter these ether preparations in more detail in Section 8.3.

$$RCH_2X \ + \ R'O^- \longrightarrow RCH_2OR' \ + \ X^-$$

$$HOCH_2(CH_2)_{10}CH_2Br \ + \ Na^+ \ ^-OCH_3 \xrightarrow{CH_3OH} HOCH_2(CH_2)_{10}CH_2OCH_3 \ + \ NaBr$$

$$(\sim 100\%)$$

$$(96\%)$$

The cyanide ion is a useful strong nucleophile that leads to nitriles (RCN). Nitriles can, in turn, be converted to amines or carboxylic acids. (See Section 15.11D.)

$$RCH_2X \quad or \quad R_2CHX + \; ^-CN \longrightarrow RCH_2CN \quad or \quad R_2CHCN + X^-$$

$$\text{(90\%)}$$

Ammonia and amines, too, can be used as nucleophiles. These reactions are discussed in Section 17.5A.

$$RCH_2X \quad or \quad R_2CHX + R_3N\colon \longrightarrow RCH_2\overset{+}{N}R_3 \; X^- \quad or \quad R_2CH\overset{+}{N}R_3 \; X^-$$

$$CH_3(CH_2)_3Br + excess \colon NH_3 \longrightarrow CH_3(CH_2)_3\overset{+}{N}H_3 \; Br^-$$

$$\text{(45\%)}$$

Reactive alkyl halides (CH_3I, an allylic halide, or a benzylic halide), but not ordinary primary or secondary alkyl halides, react with the weakly nucleophilic carboxylate ion ($RCO_2{}^-$) to yield esters.

an allylic bromide   (96%)

An *alkene* can be prepared by heating a secondary or tertiary alkyl halide with a *strong base* such as potassium hydroxide or the alkali metal salt of an alcohol in an alcohol solvent. Generally, the more highly substituted, *trans* alkene is the product. The less substituted alkene can sometimes be prepared if a bulky base, such as $K^+ \; ^-OC(CH_3)_3$, is used.

An example of an alkene preparation follows. Note that the least substituted alkene is formed here, both because of the location of the only *anti*

Table 5.7 Some types of compounds that can be synthesized from alkyl halides

Reactants[a]			Principal Product	Typical Reagents
1° RX	+ ⁻OR′	⟶ ROR′	**an ether**	$Na^+ \; ^-OCH_2CH_3$, $Na^+ \; ^-OC_6H_5$
1° RX	+ ⁻OH	⟶ ROH	**an alcohol**	$Na^+ \; ^-OH$, $K^+ \; ^-OH$
1° or 2° RX	+ ⁻CN	⟶ RCN	**a nitrile**	$Na^+ \; ^-CN$
1° or 2° RX	+ ⁻SR′	⟶ RSR′	**a sulfide, or thioether**	$Na^+ \; ^-SCH_2CH_3$
1° or 2° RX	+ ⁻OCR′ (O)	⟶ ROCR′ (O)	**an ester**[b]	$Na^+ \; ^-O_2CCH_3$
1° or 2° RX[c]	+ I⁻	⟶ RI	**an alkyl iodide**	$Na^+ \; ^-I$
1° or 2° RX	+ NR′₃	⟶ R$\overset{+}{N}$R′₃X⁻	**an ammonium salt**	$(CH_3)_3N$
2° or 3° R₂CHCXR₂	+ ⁻OR′	⟶ R₂C=CR₂	**an alkene**	$K^+ \; ^-OH$, $Na^+ \; ^-OCH_2CH_3$, $K^+ \; ^-OC(CH_3)_3$

[a] Where 1° RX is specified, methyl halides, allylic halides, and benzylic halides may also be used.
[b] A reactive halide must be used.
[c] Where X = Cl or Br.

hydrogen atoms and because of the geometric impossibility of forming a double bond at the bridgehead carbon.

Table 5.7 summarizes the types of products that can be obtained by S_N2 and E2 reactions of alkyl halides.

Study Problem

5.23 Write equations to show how you would synthesize the following compounds from organic halides and other appropriate reagents. If there are two routes to the compound, choose the better one. If two routes are equivalent, show both.

(a) $CH_3CH_2CH_2O-$⬡

(b) ⬡$-CH_2SCH_3$

(c) $CH_3CH_2OCH_2CH_2CH_3$

(d) ⬡$-\overset{\overset{O}{\|}}{C}OCH_2CH=CH_2$

(e) ⬡$-OCH_2CH_3$

(f) $(CH_3)_2CHCH=CHCH_3$

▼ Summary

An alkyl halide contains a good **leaving group** (X^-) and is readily attacked by **nucleophiles** ($Nu:^-$). Reaction occurs by one or more of four possible paths: S_N1, S_N2, **E1**, **E2**.

An S_N1 or E1 reaction proceeds through a **carbocation intermediate**:

$$RX \xrightarrow{-X^-} [R^+] \begin{cases} \xrightarrow[S_N1]{Nu:^-} RNu \\ \xrightarrow[E1]{-H^+} \text{alkene} \end{cases}$$

A carbocation intermediate usually leads to a mixture of products: *a substitution product, an alkene,* and also *rearrangement products.* Rearrangement products occur if the carbocation can form a more stable carbocation by a 1,2-shift of H, Ar, or R. If RX is optically active, racemization can occur in an S_N1 reaction.

The rate of a typical S_N1 or E1 reaction depends on the concentration of only RX. Thus, these reactions are said to be **first order**. The **rate-determining step** (slow step) in an S_N1 or E1 reaction is the formation of R^+. The stability of R^+ determines the **energy of the transition state** (E_{act}) in this step because the transition state has carbocation character. The *order of stability* of carbocations is $3° > 2° \gg 1° \gg CH_3^+$. For this reason, the likelihood of RX to undergo S_N1 or E1 reaction is $3° > 2° \gg 1° \gg CH_3X$. Allylic and benzylic halides undergo S_N1 reactions readily because of resonance-stabilization of the intermediate carbocation.

An S_N2 reaction is a **concerted reaction** that leads to *inversion of configuration*. Inversion can be observed if RX is optically active. An E2 reaction is also a concerted reaction that results by *anti*-elimination of H^+ and X^-.

$$Nu\!:^- + R\!-\!X \xrightarrow{\ S_N2\ } NuR + X^-$$

$$RO^- \quad \underset{\underset{X}{|}}{R_2C}\!-\!\overset{\overset{H}{|}}{C}R_2 \xrightarrow{\ E2\ } ROH + R_2C\!=\!CR_2 + X^-$$

Both S_N2 and E2 reactions follow **second-order kinetics**. The rate is dependent on the concentrations of both RX and $Nu\!:^-$ because both are involved in the transition state. Because of steric hindrance, the order of reactivity of RX in S_N2 reactions is $CH_3X > 1° > 2° \gg 3°$.

Because the transition state has double-bond character, the order of reactivity of RX in E2 reactions is $3° > 2° \gg 1°$, the same order as in the E1 reaction. The *most substituted alkene* usually predominates in E2 reactions (**Saytzeff rule**). The *trans* alkene usually predominates over the *cis* alkene. If steric hindrance inhibits the formation of the most substituted alkene, then the *least substituted alkene* predominates (**Hofmann product**).

Study Problems*

5.24 Name each of the following compounds by the IUPAC system:

(a) $CCl_3CH\!=\!CH_2$ (b) $CH_3\overset{\overset{Br}{|}}{C}HCH_2CH_2Br$ (c)

(d)   (e)

*For information concerning the organization of the *Study Problems* and *Additional Problems*, see the *Note to student* on page 39.

5.25 Give the structure for each of the following compounds:

(a) isobutyl iodide

(b) 1-iodo-2-methylpropane

(c) *cis*-1,3-dichlorocyclohexane

(d) 2-bromo-3-methyl-1-butanol

(e) (2R,3R)-2-bromo-3-chlorobutane

5.26 Classify the following organohalogen compounds as methyl, 1°, 2°, or 3°:

(a) $(CH_3)_3CCH_2Cl$ (b) $(CH_3CH_2)_3CCl$ (c)

5.27 Which compound in each of the following pairs would undergo more rapid S_N2 reaction?

(a) $(CH_3)_3CI$ or $(CH_3CH_2)_2CHI$

(b) $(CH_3)_2CHI$ or $(CH_3)_2CHCl$

(c)

(d)

5.28 If Reaction 1 and Reaction 2 are two different irreversible reactions of a single starting material, and if Reaction 1 has a higher-energy transition state than Reaction 2,

(a) which reaction has the higher E_{act}?

(b) which reaction has the faster rate?

(c) which products will predominate in the product mixture?

5.29 What is the effect on the rate of the S_N2 reaction of CH_3I and OH^- when:

(a) The concentration of CH_3I is tripled and that of OH^- is doubled?

(b) The concentration of OH^- is halved?

(c) The temperature is increased?

(d) The ratio of solvent to reactants is doubled?

5.30 Which of the following two syntheses for *t*-butyl ethyl ether would be preferred and why?

(a) $(CH_3)_3CO^- + CH_3CH_2Br$ (b) $(CH_3)_3CBr + CH_3CH_2O^-$

5.31 Predict the likely substitution products when the following compounds are treated with NaCN:

(a) $BrCH_2\overset{\overset{O}{\|}}{C}CH_3$ (b) $CH_3\overset{\overset{CHCl}{\|}}{C}CH_2Cl$ (c) $HOCH_2CH_2Cl$

bromoacetone   *a fumigant*   2-chloroethanol

5.32 Write equations showing the structures of the transition state and the product for the S_N2 reaction of $^-OCH_3$ with each of the following alkyl halides:

(a) (R)-2-bromobutane

(b) *trans*-1-chloromethyl-4-methylcyclohexane

(c) (R)-2-bromo-3-methylbutane

(d) (S)-2-bromo-3-methylbutane

5.33 Predict the product of an S_N2 reaction of each of the following nucleophiles with (S)-2-iodohexane:

(a) $^-SCH_2CH_3$ (b) $CH_3C\equiv C:^-$ (c) $(R)\text{-}CH_3CH_2\overset{\overset{O^-}{|}}{C}HCH_2CH_2CH_3$

5.34 Complete the following equations for S_N2 reactions.

(a)

(b)

(c) $(CH_3CH_2O\overset{\overset{O}{\|}}{C})_2\overset{..}{C}H^- + CH_3I \longrightarrow$

(d) $+ OH^- \longrightarrow$

5.35 Suggest reagents for the preparation of each of the following compounds by an S_N2-type reaction of an organohalogen compound with a nucleophile:

(a) $(CH_3)_2CHOC_6H_5$ **(b)** **(c)** $CH_3\overset{\overset{O}{\|}}{C}O-$

5.36 Which of the following carbocations is the most stable? Which is the least stable? Explain.

(a) $-CH_2^+$ **(b)** $(CH_3CH_2)_2CH^+$ **(c)** $(CH_3CH_2)_3C^+$

5.37 Label the indicated parts of the following energy diagram for an S_N1 reaction:

5.38 (1) Show the products of the solvolysis of the following compounds in *aqueous ethanol*. (2) Which halide would react the fastest?

(a) $(CH_3)_2CHBr$ **(b)** $(CH_3)_3CBr$ **(c)** $(CH_3)_3CI$

5.39 Predict the products of the aqueous solvolysis of the following halides:

(a) *cis*-1-iodo-3-methylcyclohexane **(b)** (R)-2-iodooctane

(c) $(2R,4S)$-2-iodo-4-methylhexane **(d)**

5.40 The reaction between sodium phenylselenide ($Na^+ C_6H_5Se^-$) and allyl chloride in ethanol gives an 84% yield of one product. What is this product?

5.41 Each of the following carbocations is capable of undergoing rearrangement to a more stable carbocation. Suggest a structure for the rearranged carbocation.

(a) $(CH_3)_3C\overset{+}{C}HCH_2CH_3$ **(b)** $CH_2=CHCH_2\overset{+}{C}HCH_3$

(c) $(CH_3)_2CH\overset{+}{C}HCH_2CH(CH_3)_2$ (d) ⟨cyclohexyl⟩$-\overset{+}{C}HCH_3$

5.42   The following reactions can proceed with rearrangement. Show the initial carbocation, the rearranged carbocation, and the major rearrangement product in each case.

(a) $(CH_3)_3CCHICH_3 + H_2O \longrightarrow$
(b) $(CH_3)_2CHCHICH_2CH_3 + CH_3CH_2OH \longrightarrow$

5.43   Give the important resonance structures of each of the following cations:

(a) $CH_2{=}CHCH{=}CH\overset{+}{C}H_2$ (b) ⟨benzene⟩$-\overset{+}{C}H-$⟨benzene⟩

(c) ⟨cyclohexene⟩$\overset{+}{}-CH{=}CH_2$ (d) ⟨cyclohexadiene⟩$-\overset{+}{C}HCH_3$

5.44   Complete the following equations, showing structures of all likely S_N1 substitution products:

(a)
$$\overset{Br}{\underset{}{\text{⟨cyclohexane ring⟩}}}{=}CH_2 + CN^- \longrightarrow$$

(b) $CH_3CH{=}CHCH{=}CHCH_2Cl + HOCH_3 \longrightarrow$

(c)
$$\text{⟨phthalimide⟩}N\overset{|}{\underset{CH_2OH}{CHCO_2^-}} + O_2N{-}\text{⟨benzene⟩}{-}CH_2Br \longrightarrow$$

5.45   (a) Write the equations for the steps of the E1 reaction of 2-iodohexane with H_2O, showing only the major alkene product.
(b) Which step determines the rate of reaction?
(c) What other alkene products could be formed?

5.46   Each of the following alkyl halides can undergo rearrangement in an E1 reaction. For each, show the initial carbocation, the rearranged carbocation, and the anticipated major rearrangement product.

(a) $(CH_3)_2CHCHCl\overset{+}{C}H(CH_3)_2$ (b) $(CH_3)_3CCHBrCH_2CH_2CH_3$

5.47   Which compound in each of the following pairs undergoes E2 reaction more rapidly? Why?

(a) $(CH_3)_2CHCHBrCH_2CH_3$ or $(CH_3)_2CBrCH_2CH_2CH_3$
(b) $(CH_3)_2CHCHICH_3$ or $(CH_3)_2CHCH_2CH_2I$
(c) CH_3CHICH_3 or $CH_3CHBrCH_3$

5.48   Each of the following compounds yields two alkenes upon dehydrohalogenation. Predict the structures of the alkenes and predict which alkene would predominate in each product mixture.

(a) $CH_3(CH_2)_8\overset{\overset{CH_3}{|}}{\underset{\underset{Br}{|}}{C}}CO_2H$ (b) ⟨cyclohexanone ring with CH_3 and Cl substituents⟩

5.49   Which hydrogen(s) must be replaced by deuterium if a maximum kinetic isotope effect is to be observed in an E2 reaction of each of the following compounds?

(a) $(CH_3)_2CClCH_3$ **(b)**   **(c)**

5.50 For each of the following reactions, indicate whether the product shown is the Saytzeff product or the Hofmann product:

(a) $CH_3CHClCH_2CH_2CH_3 \longrightarrow CH_2{=}CHCH_2CH_2CH_3$
(b) $CH_3CHClCH_2CH_2CH_3 \longrightarrow CH_3CH{=}CHCH_2CH_3$
(c) $CH_3CH_2CHICH(CH_3)_2 \longrightarrow CH_3CH_2CH{=}C(CH_3)_2$

5.51 Predict the principal E2 product of the reaction of each of the following compounds with $Na^+\ ^-OCH_3$:

(a) (*S*)-2-bromopentane   **(b)** 2,6-dichloroheptane
(c) (1*S*,2*S*)-1-bromo-1,2-diphenylbutane   **(d)** (*S*)-1-chloro-1-cyclohexylethane

5.52 Complete the following equations, showing only the major organic product, and predict which reaction mechanism (S_N1, S_N2, E1, E2) is the most likely:

(a) $(CH_3)_2CHBr + KI \xrightarrow{\text{acetone}}$

(b) $(CH_3)_2CHBr + KOH \xrightarrow[\text{heat}]{CH_3CH_2OH}$

(c) $(CH_3)_2CHBr + CH_3CH_2OH \xrightarrow{\text{heat}}$

(d) $(CH_3)_2CClCH_2CH_3 + Na^+\ ^-OCH_3 \xrightarrow{CH_3OH}$

(e) $CH_3CHBrCH_2CH_2CH_3 + KOH \xrightarrow[\text{heat}]{CH_3CH_2OH}$

(f) $(CH_3)_2CBrCH_2CH_3 + K^+\ ^-OC(CH_3)_3 \xrightarrow{(CH_3)_3COH}$

(g) $ClCH_2CO_2H\ + 2\,Na^+\ ^-OCH_2CH_3 \longrightarrow$

    chloroacetic acid

     a herbicide

5.53 Which reaction would be more likely to give a rearrangement product: **(a)** 2-bromo-3-methylbutane with 10% aqueous KOH, or **(b)** 2-bromo-3-methylbutane with 0.1% aqueous KOH? Explain.

Additional Problems

5.54 Write equations to show how you could prepare the following compounds from organic halides:

(a) *trans*-$C_6H_5CH{=}CHC_6H_5$ **(b)**

(c) (2*S*,3*R*)-$C_6H_5CHCHCH_3$ **(d)** $ClCH_2CH_2CH_2CN$

(e)

(f)

5.55 Herbicides can be selective or nonselective. 2,4-D (the sodium salt of 2,4-dichlorophen-oxyacetic acid) is an example of a selective herbicide. It overstimulates the growth of broadleaf plants (but not grasses), causing them to die. *Paraquat* is an example of a nonselective herbicide. Suggest the starting materials that would be needed to prepare each herbicide by an S_N2 reaction.

2,4-D   paraquat diiodide

5.56 2,3-Dibromopropene reacts with an excess of ethylamine ($CH_3CH_2NH_2$) to yield $CH_3CH_2\overset{+}{N}H_3 \, Br^-$ and another product.

(a) What is the structure of the initial substitution product?

(b) What is the structure of the final product, which is the product of an acid–base reaction?

5.57 The reaction of 2-chloro-2-methyl-1-phenylpropane with methanol yields a mixture composed of a substitution product A (54%) and two elimination products, B (27%) and C (19%). What are the structures of A, B, and C?

5.58 Explain the following observations:

(a) The following halide is inert toward nucleophiles under either S_N1 or S_N2 conditions.

(b) *Quinuclidine* undergoes reaction with iodomethane about 50 times faster than does triethylamine, even though these two amines have almost equal basicities.

quinuclidine

(c) When heated in a solvent, (S)-4-bromo-*trans*-2-pentene undergoes racemization.

(d) Chloromethyl methyl ether ($ClCH_2OCH_3$) undergoes rapid S_N1 reaction.

5.59 A compound with the formula C_4H_9Cl, upon treatment with a strong base, yields three isomeric alkenes. What is the structure of this alkyl halide?

5.60 (2R,3S)-2-Bromo-3-deuteriobutane undergoes an E2 reaction when treated with $NaOCH_2CH_3$ in ethanol. What would be the principal product(s)?

5.61 (2S,3S)-3-Bromo-2-methoxybutane undergoes an S_N2 reaction with CH_3O^- to yield an optically inactive product. Explain. (Use an equation in your answer.)

5.62 Consider the following hypothetical two-step reaction:

$$A + B \underset{\text{fast}}{\rightleftharpoons} [AB] \xrightarrow{\text{slow}} \text{products}$$

(a) Would this reaction show overall *first-order* or *second-order* kinetics? Explain.

(b) Is this reaction *unimolecular* or *bimolecular*? Explain.

5.63 1-Bromo-2-phenylethene is added to molten KOH. The product (67%) is distilled from the mixture. What is the product? Write the equation for a mechanism showing its formation.

5.64 Suggest a mechanism for each of the following reactions:

(a) $(CH_3)_2CHCl + AgNO_3 + H_2O \longrightarrow (CH_3)_2CHOH + AgCl$

(b) $CH_2=CHCH(Br)CH_3 \xrightarrow[H_2O]{LiBr} CH_3CH=CHCH_2Br$

(c) $CH_3CH_2CH_2Cl + NO_2^- \longrightarrow CH_3CH_2CH_2NO_2 + CH_3CH_2CH_2ONO + Cl^-$

(d) $CH_2=CHC(Cl)(CH_3)_2 + H_2O \xrightarrow{Ag_2O} CH_2=CHC(OH)(CH_3)_2 + HOCH_2CH=C(CH_3)_2$
(86%)   (14%)

(e) $BrCH_2(CH_2)_8CH_2CO_2H \xrightarrow[100°]{\substack{K_2CO_3 \\ (CH_3)_2S=O}}$

(83%)

5.65 Explain the following observation:

(76%)

5.66 Formulas for two of the stereoisomers of menthyl chloride follow. One of these menthyl chlorides undergoes rapid E2 reaction when treated with a base to yield two alkenes: 75% A and 25% B. The other menthyl chloride undergoes a slow E2 reaction and yields only one alkene.

(a) Which menthyl chloride undergoes the more rapid reaction? Why?

(b) What two alkenes are formed in this reaction, and which one predominates? Explain.

(c) Which menthyl chloride undergoes the slow reaction, and what is the structure of the single alkene product? Explain why the reaction is slow and yields only one alkene.

6 ▼

Free-Radical
Reactions

Many organohalogen compounds are prepared industrially by the reaction of hydrocarbons and halogens, two relatively inexpensive starting
materials. Direct halogenation reactions often proceed explosively and, as a
general rule, give mixtures of products. For these reasons, direct halogenation
is used only occasionally in the laboratory.

$$CH_3CH_3 + Cl_2 \xrightarrow{\text{light}} CH_3CH_2Cl + ClCH_2CH_2Cl + HCl + \text{other products}$$

ethane

Direct halogenation reactions proceed by a **free-radical**, or **radical**,
mechanism, which is different from the mechanisms discussed in Chapter 5 for
nucleophilic substitutions and eliminations. Radical reactions have biological
and practical importance. For example, organisms utilize atmospheric oxygen
by a sequence of reactions that begins with a radical oxidation–reduction.
Butter and other fats become rancid partly by radical reactions with oxygen.

Polymerization reactions are reactions that yield giant molecules from the joining of small molecules. Many polymerization reactions are also radical reactions.

$$\text{many units of } CH_2{=}CH_2 \xrightarrow{\text{initiator}} \text{?}{-}CH_2CH_2{-}CH_2CH_2{-}CH_2CH_2{-}\text{?}$$

ethylene   *a portion of a polyethylene molecule,*
                                   a polymer used for plastic bags and
                                   many other items

▼ ## Section 6.1
A Typical Radical Reaction: Chlorination of Methane

The term **free radical** or **radical** refers to any atom or group of atoms that has *one or more unpaired electrons*. Although a radical usually has no positive or negative charge, such a species is highly reactive because of its unpaired electron and incomplete octet. A radical is usually encountered as a high-energy, highly reactive, short-lived, nonisolable reaction intermediate.

We usually symbolize a radical with a single dot representing the unpaired electron.

Lewis formulas for typical radicals:

$$:\!\overset{..}{\underset{..}{Cl}}\!\cdot \qquad :\!\overset{..}{\underset{..}{Br}}\!\cdot \qquad H\!:\!\overset{\overset{\textstyle H}{|}}{\underset{\underset{\textstyle H}{..}}{C}}\!\cdot$$

Usual formulas for radicals (note the omission of the paired electrons):

$$Cl\cdot \qquad Br\cdot \qquad H_3C\cdot \quad \text{or} \quad CH_3\cdot \quad \text{or} \quad \cdot CH_3$$

Study Problem

6.1 Write Lewis formulas for the following radicals:
(a) $HO\cdot$ (b) $CH_3O\cdot$ (c) $CH_3CH_2\cdot$

The chlorination of methane in the presence of light (symbolized hv; see page 325) is a classical example of a radical reaction. The result of the reaction of Cl_2 with CH_4 is the *substitution* of one or more chlorine atoms for hydrogen atoms on the carbon.

$$CH_4 \;+\; Cl_2 \;\xrightarrow{hv}\;$$

methane

$$CH_3Cl \quad + \quad CH_2Cl_2 \quad + \quad CHCl_3 \quad + \quad CCl_4 \quad + HCl$$

chloromethane   dichloromethane   trichloromethane   tetrachloromethane
(methyl chloride) (methylene chloride)   (chloroform)   (carbon tetrachloride)

Although methane is the simplest alkane, four organic products can be formed in its chlorination. Small amounts of higher alkanes, such as ethane, and their chlorinated products may also be formed. First, we will discuss the reactions leading to CH_3Cl. Then, we will expand the discussion to the formation of other products.

The mechanism of this type of radical reaction is best thought of as a series of stepwise reactions. Each step falls into one of the following categories: (1) **initiation** of the radical reaction; (2) **propagation** of the radical reaction; and (3) **termination** of the radical reaction.

A. Initiation

As the term implies, the initiation step is the initial formation of radicals. In the chlorination of methane, the initiation step is the homolytic cleavage of Cl_2 into two chlorine radicals. (You may wish to review the discussion of bond dissociation in Section 1.7.) The energy for this reaction step is provided by light or by heating the mixture to a very high temperature.

Step 1 (initiation):

$$Cl—Cl + 58 \text{ kcal/mol} \xrightarrow{h\nu \text{ or heat}} 2 \text{ Cl} \cdot$$

B. Propagation

After its formation, the chlorine radical starts a series of reactions in which new radicals are formed. Collectively, these reactions are called the **propagation steps** of the radical reaction. In effect, the initial formation of a few radicals results in the propagation of new radicals in a self-perpetuating reaction called a **chain reaction**.

As the first propagation step, the reactive chlorine radical abstracts a hydrogen atom from methane to yield a methyl radical and HCl.

$$Cl \cdot \quad H \colon CH_3 + 1 \text{ kcal/mol} \longrightarrow H \colon Cl + \cdot CH_3$$

The methyl radical is also reactive. In the second propagation step, the methyl radical abstracts a chlorine atom from Cl_2.

$$\cdot CH_3 \quad Cl \colon Cl \longrightarrow CH_3Cl + Cl \cdot + 25.5 \text{ kcal/mol}$$
$$\text{chloromethane}$$

This step yields one of the products of the overall reaction, chloromethane. This step also regenerates a chlorine radical that can abstract a hydrogen atom from another methane molecule and start the propagation sequence over again.

The overall sequence so far is:

Initiation:

$$Cl_2 \xrightarrow{h\nu \text{ or heat}} 2 \text{ Cl} \cdot$$

Propagation:

$$CH_4 + Cl\cdot \longrightarrow \cdot CH_3 + HCl$$

$$\cdot CH_3 + Cl_2 \longrightarrow CH_3Cl + Cl\cdot \overset{\text{\textit{can undergo}}}{\underset{\text{\textit{reaction with } CH_4}}{}}$$

The reaction is self-perpetuating because a single Cl· starts a reaction sequence in which a new Cl· (along with products) is formed. However, as you might imagine, the reaction does not continue indefinitely. The *number of cycles* (that is, the number of passes through the propagation steps) is called the **chain length**. The chain length of a radical reaction depends partly upon the energies of the radicals involved in the propagation. (We will discuss this subject shortly.) For chlorination of a hydrocarbon, the chain length is about 10,000.

C. Termination

The propagation cycle is broken by **termination reactions**. Any reaction that results in the destruction of radicals or in the formation of stable, nonreactive radicals can terminate the propagation cycle. The chlorination of methane is terminated principally by radicals combining with other radicals. This is a process of destruction of radicals. In Section 6.7, we will mention termination by formation of stable, nonreactive radicals.

Termination steps:

$$Cl\cdot + \cdot CH_3 \longrightarrow CH_3Cl$$

$$\cdot CH_3 + \cdot CH_3 \longrightarrow CH_3CH_3$$

The second termination step shown is an example of a **coupling reaction**: the joining together of two alkyl groups.

In summary,

(1) **initiation:** increases the net number of radicals;
(2) **propagation:** leaves the net number of radicals unchanged;
(3) **termination:** decreases the net number of radicals.

Study Problem

6.2 Write equations for the initiation, propagation, and termination reactions leading to the formation of chlorocyclohexane from cyclohexane and chlorine.

D. Why Radical Reactions Yield Mixtures of Products

Radical reactions are often characterized by a multitude of products. For example, the chlorination of methane can yield four organic products. The reason for the formation of these mixtures is that the high-energy chlorine radical is not particularly selective about which hydrogen it abstracts during the propagation step.

While chlorine is undergoing reaction with methane, chloromethane is being formed. In time, the chlorine free radicals are more likely to collide with chloromethane molecules than with methane molecules, and a new propagation cycle is started. In this new cycle, chloromethyl radicals ($\cdot CH_2Cl$) are formed. These undergo reaction with chlorine molecules to yield dichloromethane (CH_2Cl_2). As in the previous cycle leading to CH_3Cl, another chlorine radical is regenerated in the process.

Propagation steps leading to dichloromethane:

$$Cl\cdot + CH_3Cl \longrightarrow HCl + \cdot CH_2Cl$$

$$\cdot CH_2Cl + Cl_2 \longrightarrow \underset{\text{dichloromethane}}{CH_2Cl_2} + Cl\cdot$$

Sample Problem

Write the propagation steps leading to the formation of trichloromethane (chloroform) from dichloromethane.

Solution:

$$\cdot Cl + CH_2Cl_2 \longrightarrow HCl + \cdot CHCl_2$$

$$\cdot CHCl_2 + Cl_2 \longrightarrow Cl\cdot + CHCl_3$$

Study Problem

6.3 Write the propagation steps leading to the formation of tetrachloromethane (carbon tetrachloride) from trichloromethane.

The radical chlorination of methane can yield four organic products (or more, if coupling products are considered). Higher alkanes can produce even larger numbers of products because there are more hydrogens available that can enter into propagation reactions.

Study Problem

6.4 How many chloroalkanes could be produced in the chlorination of ethane? Write formulas for their structures.

Sample Problem

A chemist wishes to make chloroethane from chlorine and ethane. If he wants to avoid higher chlorination products, would he use: (a) an equimolar mixture of CH_3CH_3 and Cl_2; (b) an excess of Cl_2; or (c) an excess of CH_3CH_3?

Solution: (c) By using an excess of CH_3CH_3, the chemist increases the probability of collisions between $Cl\cdot$ and CH_3CH_3 and decreases the probability of collisions between $Cl\cdot$ and CH_3CH_2Cl.

▼ ## Section 6.2
Relative Reactivities of the Halogens

The halogens vary dramatically in their reactivity toward alkanes in radical reactions. Fluorine undergoes explosive reactions with hydrocarbons. Chlorine is next in terms of reactivity, followed by bromine. Iodine is nonreactive toward alkanes.

$$I_2 \quad Br_2 \quad Cl_2 \quad F_2$$

increasing reactivity as free-radical agents

The relative reactivity of the halogens toward alkanes is *not* due to the ease with which X_2 molecules are cleaved into radicals. From the bond dissociation energies for the halogens, we can see that the relative ease of homolytic cleavage is almost the reverse of their reactivity in halogenation reactions.

	F_2	Cl_2	Br_2	I_2
bond dissociation energy (kcal/mol):	37	58	46	36

The order of reactivity of the halogens in radical halogenations (relative rates of reaction under similar conditions) depends on the energy of activation for the rate-determining step. In radical halogenations, we can approximate the order of reactivity by examining the heats of reaction. The propagation steps of fluorination are highly exothermic. Because fluorination has a low E_{act}, the heat produced by the exothermic reaction increases the rate of reaction by increasing the temperature. The result is an extremely rapid, explosive reaction.

$$
\begin{array}{llr}
F\cdot + CH_4 \longrightarrow HF + CH_3\cdot + 31\,kcal/mol & \Delta H = -31\,kcal/mol \\
CH_3\cdot + F_2 \longrightarrow CH_3F + F\cdot + 71\,kcal/mol & \Delta H = -71\,kcal/mol \\
\hline
CH_4 + F_2 \longrightarrow CH_3F + HF + 102\,kcal/mol & net\ \Delta H = -102\,kcal/mol
\end{array}
$$

Just the reverse situation is encountered with iodine: the propagation steps are *endothermic*—that is, the products are of higher energy than the reactants. Most important, the energy required by $I\cdot$ to abstract hydrogen from a C—H bond is *substantially* endothermic. The result is that the iodine radical does not enter into a chain reaction. $I\cdot$ is an example of a *stable radical*, a radical that does not abstract hydrogens.

$$
\begin{array}{llr}
I\cdot + CH_4 + 33\,kcal/mol \longrightarrow HI + CH_3\cdot & \Delta H = +33\,kcal/mol \\
CH_3\cdot + I_2 \longrightarrow CH_3I + I\cdot + 20\,kcal/mol & \Delta H = -20\,kcal/mol \\
\hline
CH_4 + I_2 + 13\,kcal/mol \longrightarrow CH_3I + HI & net\ \Delta H = +13\,kcal/mol
\end{array}
$$

Chlorine and bromine are intermediate between fluorine and iodine in ΔH of the propagation steps and therefore are also intermediate in reactivity. Figure 6.1 shows energy diagrams for the reactions of Cl_2 and Br_2 with methane.

$$
\begin{array}{llr}
Cl\cdot + CH_4 + 1\,kcal/mol \longrightarrow HCl + CH_3\cdot & \Delta H = +1\ \ kcal/mol \\
CH_3\cdot + Cl_2 \longrightarrow CH_3Cl + Cl\cdot + 25.5\,kcal/mol & \Delta H = -25.5\,kcal/mol \\
\hline
CH_4 + Cl_2 \longrightarrow CH_3Cl + HCl + 24.5\,kcal/mol & net\ \Delta H = -24.5\,kcal/mol
\end{array}
$$

$$
\begin{array}{llr}
Br\cdot + CH_4 + 17\,kcal/mol \longrightarrow HBr + CH_3\cdot & \Delta H = +17\,kcal/mol \\
CH_3\cdot + Br_2 \longrightarrow CH_3Br + Br\cdot + 24\,kcal/mol & \Delta H = -24\,kcal/mol \\
\hline
CH_4 + Br_2 \longrightarrow CH_3Br + HBr + 7\,kcal/mol & net\ \Delta H = -7\,kcal/mol
\end{array}
$$

Figure 6.1 Energy diagrams for the radical chlorination and bromination of methane.

In summary, we find that chlorine and bromine are the most useful radical halogenating agents.

Section 6.3
Stereochemistry of Radical Halogenation

An alkyl radical is a species in which a carbon atom has three atoms or groups bonded to it and a single, unpaired electron. We will look at the structure of the methyl radical. Other alkyl radicals have similar bonding around the radical carbon.

A radical carbon is in the sp^2-hybrid state. The three sp^2 orbitals are planar in the methyl radical and very nearly planar in most other radicals. The unpaired electron of the radical is in the p orbital. The structure is very similar to that of a carbocation, except that the p orbital of a carbocation is empty.

methyl radical, CH_3·

When a pure enantiomer of a chiral alkyl halide undergoes S_N1 reaction at the chiral carbon, racemization is observed. As we discussed in Chapter 5, racemization arises from the nucleophile being able to attack either lobe of the empty p orbital of the carbocation. If a hydrogen is abstracted from the chiral carbon of a pure enantiomer in a radical reaction, racemization also occurs.

$$CH_3CH_2{-}\overset{H}{\underset{CH_3}{C}}{-}CH_2Cl + Cl_2 \xrightarrow{\ h\nu\ } CH_3CH_2\overset{Cl}{\underset{CH_3}{C}}CH_2Cl \quad + \ HCl$$

(S)-1-chloro-2-methylbutane   (R)(S)-1,2-dichloro-2-methylbutane

(racemic)

The preceding reaction can lead to a number of products. There are five carbon atoms in the molecule that can lose a hydrogen and gain a chlorine. We would also expect to find as products trichlorinated alkanes, tetrachlorinated alkanes, and so forth. But we are interested only in the one product that has been chlorinated at the chiral carbon. When we isolate this specific product, we find that it is a racemic mixture of the (R) and (S) enantiomers. Just as in an S_N1 reaction, this evidence leads us to believe that the bonds around the radical carbon are planar (sp^2-hybrid orbitals) and that a chlorine atom can add to either lobe of the p orbital.

Study Problem

6.5 Another dichloroalkane formed in the chlorination of (S)-1-chloro-2-methylbutane is $CH_3CHClCH(CH_3)CH_2Cl$. Is this dichloroalkane racemic or not?

Section 6.4
Hydrogen Abstraction: The Rate-Determining Step

Unlike the kinetics of substitution and elimination reactions, the kinetics of a radical reaction are quite complex. Simple rate expressions, such as first-order or second-order, are not encountered in radical chemistry. The reason for this complexity is that the steps in a radical reaction are enmeshed in a cyclical process of varying chain lengths. However, evidence does point to the **hydrogen-abstraction step** as the step governing the overall rate at which products are formed. For example, methane (CH_4) undergoes chlorination 12 times faster than perdeuteriomethane (CD_4), indicating that the CH bond is broken in the rate-determining step of the reaction.

H (or D) abstraction is the rate-determining step:

$$CH_4 + Cl\cdot \xrightarrow{\text{faster}} CH_3\cdot + HCl$$

$$CD_4 + Cl\cdot \xrightarrow{\text{slower}} CD_3\cdot + DCl$$

A. Which Hydrogen Is Abstracted?

The hydrogen atoms in organic compounds can be classified as **methyl** (CH_4), **primary** (bonded to a 1° carbon), **secondary** (bonded to a 2° carbon), **tertiary** (bonded to a 3° carbon), **allylic** (on a carbon adjacent to a double bond), or **benzylic** (on a carbon adjacent to an aromatic ring).

2° hydrogens

$CH_3CH_2CH_3$ $CH_2{=}CHCH_3$ ⬡—CH_3

1° hydrogens   *allylic hydrogens*   *benzylic hydrogens*

Sample Problem

Classify each circled H as 1°, 2°, 3°, allylic, or benzylic:

(a) $(CH_3)_3CC$ Ⓗ$_3$ (b) $(CH_3)_3C$ Ⓗ   (c)

(d)

(e)

Solution: (a) 1°; (b) 3°; (c) 2°; (d) allylic and 3°; (e) benzylic and 2°.

These different types of hydrogen atoms are not abstracted at identical rates by radicals. Instead, there is a degree of selectivity in hydrogen abstraction. The reaction of propane with a small amount of chlorine under radical conditions yields two monochlorinated products, 1-chloropropane and 2-chloropropane, with the 2-chloropropane predominating.

$$CH_3CH_2CH_3 + Cl_2 \xrightarrow{hv} CH_3\overset{\underset{\displaystyle |}{Cl}}{C}HCH_3 + CH_3CH_2CH_2Cl$$

propane	2-chloropropane (55%)	1-chloropropane (45%)
	(isopropyl chloride)	(*n*-propyl chloride)

There are *six* primary hydrogens and *two* secondary hydrogens in propane. The ratio of primary to secondary hydrogens in propane is 6/2, or 3/1. If all the hydrogens underwent abstraction at equal rates, we would observe three times more 1-chloropropane than 2-chloropropane in the product mixture. This is *not* what is observed when the reaction is carried out. Instead, slightly more 2-chloropropane is formed. We conclude that secondary hydrogens are abstracted faster than primary hydrogens.

Another example of how the relative rates of hydrogen abstraction affect the product ratio follows. (We will discuss the greater selectivity of Br_2 in Section 6.5.)

$$CH_3\overset{\underset{\displaystyle |}{CH_3}}{C}HCH_3$$

methylpropane
(isobutane)

$\xrightarrow[hv]{Cl_2}$ $(CH_3)_2CHCH_2Cl$ + $(CH_3)_3CCl$ + other products

1-chloro-2-methylpropane   *t*-butyl chloride
(isobutyl chloride)
(50%)   (30%)   (20%)

$\xrightarrow[hv]{Br_2}$ $(CH_3)_3CBr$

t-butyl bromide
(almost 100%)

Study Problem

6.6 In the chlorination of methylpropane, what would be the expected product ratio of isobutyl chloride to *t*-butyl chloride if all the hydrogens were abstracted at equal rates?

Table 6.1 Average relative rates of hydrogen abstraction

Hydrocarbon	Reagent[a]	
	Br_2	Cl_2
CH_3—H	0.0007	0.004
CH_3CH_2—H	1	1
$(CH_3)_2CH$—H	220	4.3
$(CH_3)_3C$—H	19,400	6.0
$C_6H_5CH_2$—H	64,000	1.3
$(C_6H_5)_2CH$—H	6.2×10^5	2.6
$(C_6H_5)_3C$—H	1.14×10^6	9.5

[a] The two columns contain data from two separate studies of relative rates. The chlorination of ethane proceeds much more rapidly than bromination under the same conditions.

Through these and similar experiments, the order of reactivity of hydrogens toward radical halogenation has been determined. The relative rates for halogenation reactions of a few compounds are given in Table 6.1.

$$
\underset{\text{methyl}}{\overset{\overset{\displaystyle H}{|}}{H_3C}} \quad \underset{1°}{\overset{\overset{\displaystyle H}{|}}{CH_3CH_2}} \quad \underset{2°}{\overset{\overset{\displaystyle H}{|}}{(CH_3)_2CH}} \quad \underset{3°}{\overset{\overset{\displaystyle H}{|}}{(CH_3)_3C-H}} \quad \text{allylic and benzylic}
$$

increasing rate of reaction toward Br_2 ⟶

B. Relative Stabilities of Alkyl Radicals

To understand why some hydrogens are abstracted more easily than others, we must look at the transition states of the hydrogen-abstraction steps. The following equations show the hydrogen-abstraction steps in the chlorination of methane and methylpropane. (The symbol $\delta\cdot$ is used to show that both the chlorine atoms and the carbon atoms have partial radical character in the transition states.)

methane · · · transition state · · · planar methyl radical

methylpropane · · · transition state · · · planar *t*-butyl radical

The reactivity sequence $3° > 2° > 1° > CH_4$ arises from the stabilities of the transition states leading to the radicals. Because the transition states have radical character, their stabilities parallel the stabilities of the radicals themselves. These stabilities can be related to the bond dissociation energies of the C—H bonds being broken. A bond breakage leading to a more stable radical requires less energy than one leading to a less stable (higher-energy) radical.

bond dissociation energy (kcal/mol):	CH_3—H	CH_3CH_2—H	$(CH_3)_2CH$—H	$(CH_3)_3C$—H
	104	98	94.5	91

decreasing bond strength ⟶

The order of radical stability, just like that of carbocation stability, increases as we proceed from methyl to tertiary. It is thought that the radical intermediates are stabilized by interaction with neighboring sigma bonds, possibly by hyperconjugation (see Section 5.6E),

$\cdot CH_3$	$\cdot CH_2CH_3$	$(CH_3)_2\overset{\cdot}{C}H$	$(CH_3)_3C\cdot$	allylic and benzylic
methyl	1°	2°	3°	

increasing stability ⟶

As in the case of carbocation reactions, we find enhanced radical reactivity at allylic and benzylic positions because of resonance-stabilization of the intermediate.

allylic

$$CH_2=CH\overset{\cdot}{C}H_3 \xrightarrow[500°]{Cl_2} CH_2=CHCH_2Cl$$

propene   3-chloropropene (allyl chloride) (90%)

benzylic

$$\text{(ring)}-CH_2CH_3 \xrightarrow[h\nu]{Cl_2} \text{(ring)}-\overset{Cl}{\underset{|}{CH}}CH_3 \;+\; \text{(ring)}-CH_2CH_2Cl$$

ethylbenzene   (1-chloroethyl)benzene (56%)   (2-chloroethyl)benzene (44%)

Study Problems

6.7 Write resonance structures for the following free radicals:

(a) $CH_2=CH\overset{\cdot}{C}H_2$ **(b)** (ring)—$\overset{\cdot}{C}H_2$

6.8 List the following free radicals in order of increasing stability (least stable first):

(a) (ring)·   **(b)** (ring)·   **(c)** (ring)—CH_3 **(d)** (ring)—$\overset{\cdot}{C}H$—(ring)

C. Rearrangements of Radicals

Alkyl radicals are, in many respects, similar to carbocations. Both are sp^2 hybrids; both undergo racemization if reaction occurs at a resolved chiral carbon; and both show the same order of stability in terms of structure. Carbocations tend to undergo rearrangement to more-stable carbocations. Does the same hold true for radicals? No, this is one of the differences between free radicals and carbocations. While radical rearrangements are not unknown, they are not common.

Rearrangement:

$$
\begin{array}{ccc}
\underset{\begin{array}{c}|\\ CH_3\end{array}}{\overset{\begin{array}{c}CH_3\\ |\end{array}}{CH_3\overset{+}{C}-\overset{\curvearrowleft}{C}HCH_3}} & \longrightarrow & \underset{\begin{array}{c}|\\ CH_3\end{array}}{\overset{\begin{array}{c}CH_3\\ |\end{array}}{CH_3\overset{+}{C}-CHCH_3}} \xrightarrow[-H^+]{H_2O} \underset{\begin{array}{c}|\\ CH_3\end{array}}{\overset{\begin{array}{c}HO\quad CH_3\\ |\quad\ \ |\end{array}}{CH_3C-CHCH_3}}
\end{array}
$$

   a 2° carbocation   *a 3° carbocation*   *a 3° alcohol*

No rearrangement:

$$
\underset{\begin{array}{c}|\\ CH_3\end{array}}{\overset{\begin{array}{c}CH_3\\ |\end{array}}{CH_3C-\overset{\cdot}{C}HCH_3}} \xrightarrow[-Cl\cdot]{Cl_2} \underset{\begin{array}{c}|\\ CH_3\end{array}}{\overset{\begin{array}{c}CH_3\quad Cl\\ |\quad\ \ |\end{array}}{CH_3C-CHCH_3}}
$$

   a 2° radical   *a 2° alkyl halide*

▼ Section 6.5
Selective Radical Halogenations

A. Bromine versus Chlorine

Although radical halogenations often lead to mixtures of products, good yields of single products can be obtained in some cases. Compare the product ratios of the chlorination and the bromination of propane:

$$
CH_3CH_2CH_3
\begin{cases}
\xrightarrow[hv]{Cl_2} & \underset{\begin{array}{c}|\\ \end{array}}{\overset{\begin{array}{c}Cl\\ |\end{array}}{CH_3CHCH_3}} + CH_3CH_2CH_2Cl \\[2em]
& \text{2-chloropropane}\quad\ \text{1-chloropropane} \\
& \quad\ (55\%)\qquad\qquad\ (45\%) \\[2em]
\xrightarrow[hv]{Br_2} & \underset{\begin{array}{c}|\\ \end{array}}{\overset{\begin{array}{c}Br\\ |\end{array}}{CH_3CHCH_3}} + CH_3CH_2CH_2Br \\[1em]
& \text{2-bromopropane}\quad\ \text{1-bromopropane} \\
& \quad\ (98\%)\qquad\qquad\ (2\%)
\end{cases}
$$

We can see that bromine, which yields 98% 2-bromopropane, is more selective about abstracting a secondary hydrogen than is chlorine. The selectivity of bromine arises from the fact that bromine is less reactive than chlorine in radical halogenations. To see why this is so, we will consider a pair of hypothetical energy diagrams (Figure 6.2).

Reaction 1 in Figure 6.2 is an *exothermic reaction with a low* E_{act}. The structure of the transition state in Reaction 1 is *very close to that of the reactants.* Like Reaction 1, the hydrogen-abstraction step in the chlorination of propane is exothermic and has a low E_{act}. Therefore, the transition state in this reaction

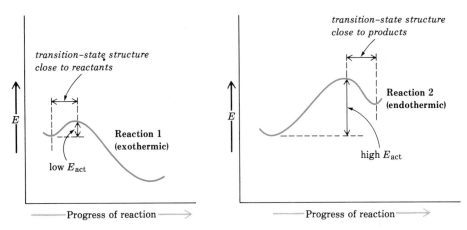

Figure 6.2 Energy diagrams showing the relationship of transition-state structure to the exothermic or endothermic nature of a reaction.

step resembles the reactants more than it does the products. In this transition state, very little bond breaking or bond forming has occurred.

Cl· + H—C(H)(CH₃)(CH₃) ⟶ [transition state] ⟶ Cl—H + ·C(H)(CH₃)(CH₃)

propane   transition state   isopropyl
                    resembles reactants   free radical

Now let us look at Reaction 2 in Figure 6.2. This is an *endothermic reaction with a high E_{act}*. The structure of the transition state in Reaction 2 is *close to that of the products*. The hydrogen-abstraction step in radical bromination is more endothermic and has a higher E_{act} than in chlorination. The structure of the transition state in bromination has a greater resemblance to the product alkyl radical because bond breaking and bond forming are practically complete.

Br· + H—C(H)(CH₃)(CH₃) ⟶ [transition state] ⟶ Br—H + ·C(H)(CH₃)(CH₃)

*transition state
resembles products*

Because this transition state in bromination resembles the alkyl radical, it is highly influenced by the stability of the alkyl radical. The reaction proceeds through the lower-energy transition state to yield the lower-energy, more stable radical. $CH_3\dot{C}HCH_3$ is highly favored over $CH_3CH_2\dot{C}H_2$. By contrast, the transition state in chlorination is less influenced by the stability of the alkyl radical. $CH_3\dot{C}HCH_3$ is only slightly favored over $CH_3CH_2\dot{C}H_2$. Therefore, chlorine is more likely to yield product mixtures.

In 1955, the American chemist George F. Hammond first proposed what we now call the **Hammond postulate**, a working rule for the relationships of different states in a reaction. According to Hammond, in a single reaction path, reaction states that differ little in energy also differ little in geometry. What we have just described concerning chlorination and bromination is part of this working rule. In an exothermic reaction with a low E_{act}, the geometry of the transition state resembles that of the reactant because their energies are similar. In the endothermic reaction with a high E_{act}, the energy difference between the reactants and the transition state is large. Therefore, the geometry of the transition state is more similar to that of the product than to that of the reactants.

B. Benzylic and Allylic Halogenations

Toluene can be selectively halogenated at the benzylic position with either chlorine or bromine. If more than one alkyl position on a benzene side chain is open to attack, as in ethylbenzene, the more selective bromine is the reagent of choice for halogenation at the benzylic position.

toluene   *a benzyl halide* (100%)

(1-chloroethyl)benzene (56%)   (2-chloroethyl)benzene (44%)

ethylbenzene

(1-bromoethyl)benzene (100%)

Alkenes can be directly halogenated in the allylic position, but very high temperatures and a low concentration of halogen must be used to prevent reaction at the double bond. A reagent that is more specific than Br_2 for allylic and benzylic halogenations is **N-bromosuccinimide (NBS)**. An NBS reaction can be initiated by light or by some source of radicals, such as a peroxide (ROOR).

General:

NBS   succinimide

cyclohexene + NBS $\xrightarrow[\text{initiator}]{\text{CCl}_4}$ 3-bromocyclohexene (87%)

—CH$_2$CH$_2$CH$_2$Br + NBS $\xrightarrow[\text{initiator}]{\text{CCl}_4}$ —CHCH$_2$CH$_2$Br

3-bromo-1-phenylpropane   1,3-dibromo-1-phenylpropane (~100%)

NBS acts as a brominating agent by providing a low, but constant, concentration of Br$_2$, which is the halogenating agent. The Br$_2$ is generated by the reaction of HBr (a product of the halogenation) and NBS. Therefore, as Br$_2$ is consumed, more is formed.

Br$_2$ consumed:

$$\text{CH}_2\text{=CHCH}_3 + \text{Br}_2 \longrightarrow \text{CH}_2\text{=CHCH}_2\text{Br} + \text{HBr}$$

Br$_2$ generated:

NBr + HBr $\longrightarrow$ NH + Br$_2$

Study Problem

6.9 Predict the major monohalogenation products of NBS halogenation of the following compounds: **(a)** *n*-butylbenzene;   **(b)** 1-phenylpropene.

▼ # Section 6.6
Other Radical Reactions

As we have mentioned, radical reactions are not limited to the halogenation of hydrocarbons, but are encountered in many areas of organic chemistry. Here, we will limit our discussion to sigma-bond radical reactions and consider only a few of the many processes known to proceed by radical mechanisms.

A. Pyrolysis

In Section 3.4C, we defined **pyrolysis** as the *thermal decomposition of organic compounds in the absence of oxygen.* When organic molecules are heated to high temperatures, carbon–carbon sigma bonds rupture and the molecules are broken into radical fragments. (The temperature required depends on the bond dissociation energies.) This fragmentation step, called **thermally induced homolysis** (homolytic cleavage caused by heat), is the initiation step for a series of radical reactions. The following equations illustrate some possible pyrolysis reactions of pentane. (There are other possible positions of cleavage and subsequent reaction.)

Initiation (homolysis):

$$CH_3CH_2\overset{\frown}{C}H_2\overset{\curvearrowleft}{}CH_2CH_3 \xrightarrow{\text{heat}} CH_3CH_2CH_2\cdot + CH_3CH_2\cdot$$

Propagation (hydrogen abstraction):

$$CH_3CH_2\cdot \quad CH_3CH_2\overset{..}{C}HCH_2CH_3 \longrightarrow CH_3CH_3 + CH_3CH_2\dot{C}HCH_2CH_3$$

Termination by coupling (joining together):

$$CH_3CH_2\cdot \quad CH_3CH_2\cdot \longrightarrow CH_3CH_2CH_2CH_3$$

Termination by disproportionation (oxidation–reduction of two equivalent species):

$$CH_3CH_2\cdot \quad \overset{..}{C}H_2\overset{\frown}{}CH_2 \longrightarrow CH_3CH_3 + CH_2{=}CH_2$$

two ethyl radicals   ethane   ethylene

                                   reduced   *oxidized*

Controlled pyrolysis has been used industrially for the cracking of high-molecular-weight compounds into lower-molecular-weight compounds. Until about 1925, pyrolysis of wood was the major source of methanol (wood alcohol). Thermal cracking was once the only petroleum cracking technique available. Now, the cracking of petroleum is accomplished using catalysts (Section 3.5), and methanol is largely produced by the catalytic hydrogenation of carbon monoxide ($CO + 2H_2 \rightarrow CH_3OH$).

B. Oxygen as a Radical Reagent

Molecular oxygen is different from the compounds we have been studying so far because a stable molecule of O_2 in the ground state has two unpaired electrons. Oxygen is said to be a **diradical**. The structure of O_2 cannot be adequately explained by line-bond formulas because two $2p$ electrons are in antibonding orbitals. For our purposes, we will represent molecular oxygen as $\cdot O{-}O\cdot$ or simply O_2.

Oxygen is a stable diradical and therefore is a selective radical agent. A compound that contains double bonds, allylic or benzylic hydrogens, or tertiary hydrogens is susceptible to **air oxidation**, also called **auto-oxidation** or **autoxidation**. Compounds with only primary and secondary hydrogens are not as susceptible. (From our discussion of radical halogenation reactions, the relative reactivities of these hydrogens should not be surprising.)

(57%)

Fats and vegetable oils often contain double bonds. Auto-oxidation of a fat yields a mixture of products that includes low-molecular-weight (and foul-smelling) carboxylic acids. For example, rancid butter contains the odorous butanoic acid.

Linseed oil and other vegetable oils, which contain many double bonds, are used as drying oils in paint and varnish. These compounds are purposely allowed to undergo air oxidation because the molecules join together, or *polymerize*, into a tough film on the painted surface.

Auto-oxidation initially leads to **hydroperoxides**, compounds containing the —OOH group, which are readily converted to mixtures of alcohols, ketones, and other products. Because mixtures are the usual result, auto-oxidation is rarely used as an organic synthetic technique in the laboratory.

Auto-oxidation:

$$R_3CH + \cdot O{-}O\cdot \longrightarrow R_3COOH \longrightarrow \text{mixture of products}$$

a hydroperoxide

In the laboratory, a chemist is most likely to encounter auto-oxidation products as undesirable impurities in ethers and aldehydes. In ethers, the carbon adjacent to the oxygen is the position of attack. Peroxides of ethers explode when heated. For example, diethyl ether is a common laboratory solvent that is purified by distillation. Unless the peroxides have been removed prior to distillation (by a reducing agent, for example), they will become concentrated in the distilling flask as the ether is boiled away. The result could easily be an explosion.

$$CH_3CH_2OCH_2CH_3 + O_2 \longrightarrow CH_3CH_2O\overset{\displaystyle OOH}{\underset{\displaystyle |}{C}}HCH_3$$

diethyl ether   *a hydroperoxide*

The product of aldehyde auto-oxidation is a carboxylic acid, which is formed by reaction of the intermediate peroxy acid with the aldehyde.

$$CH_3\overset{O}{\overset{\|}{C}}H \xrightarrow{O_2} CH_3\overset{O}{\overset{\|}{C}}OOH \xrightarrow{CH_3CH} 2\ CH_3\overset{O}{\overset{\|}{C}}OH$$

ethanal   peroxyacetic acid   acetic acid
(acetaldehyde)

Study Problem

6.10 When exposed to air, diisopropyl ether forms peroxides at a more rapid rate than diethyl ether does. Suggest a reason for this behavior.

The **cumene process** is an important industrial auto-oxidation reaction because *two* valuable chemicals—phenol and acetone—are obtained from one starting material.

In this series of reactions, isopropylbenzene (trivial name, cumene) is heated in air to yield a hydroperoxide. Oxygen attacks at the reactive benzylic position, just as we would predict. Treatment with aqueous sulfuric acid causes the hydroperoxide to undergo rearrangement to yield the products.

▼ Section 6.7
Free-Radical Initiators and Inhibitors

A **free-radical initiator** is anything that can initiate a radical reaction. The action of light to bring about radical halogenation is the action of an initiator. There are several types of compounds that may be added to a reaction mixture to initiate radical reactions. These compounds are sometimes erroneously called *free-radical catalysts*. They are not truly catalysts because they are often consumed in the reaction.

Any compound that can easily decompose into free radicals can act as an initiator. **Peroxides** (ROOR) are one example. They form radicals easily because the RO—OR bond dissociation energy is only about 35 kcal/mol, lower than that for most other bonds. Benzoyl peroxide and peroxybenzoic acid are two peroxides that are commonly used in conjunction with NBS brominations.

benzoyl peroxide

peroxybenzoic acid

As the name implies, a **free-radical inhibitor** inhibits a radical reaction. An inhibitor is sometimes referred to as a **free-radical "trap."** The usual action of a free-radical inhibitor is to undergo reaction with reactive radicals to form relatively stable and nonreactive radicals.

An inhibitor used to control auto-oxidation is called an **antioxidant** or, in the food industry, a **preservative**. *Phenols*, compounds with an —OH group attached to an aromatic ring carbon, are effective antioxidants because the radical products of these compounds are resonance-stabilized and thus nonreactive compared with most other radicals.

phenol   "trapped"

the simplest of
the phenols

Resonance structures of the phenol radical:

The food preservative BHT is a synthetic phenol. (BHA, another preservative, is closely related to BHT. Instead of a methyl group on the ring, BHA has an —OCH_3 group.) A naturally occurring preservative found in vegetable oils, especially wheat germ oil, is α-tocopherol, or vitamin E.

phenol hydroxyl groups

"butylated hydroxytoluene"   α-tocopherol
BHT   (vitamin E)

Study Problems

6.11 *Azobisisobutyronitrile* (AIBN) is often used as a free-radical initiator because it yields radicals (along with N_2) readily upon heating.   **(a)** Write an equation with fishhook arrows ($\frown$) for this thermal decomposition.   **(b)** Suggest a reason why AIBN yields radicals readily.

AIBN

6.12 Arylamines, like phenols, can act as antioxidants. For example, *N*-phenyl-2-naphthylamine is added to rubber articles to prevent radical degradation of the rubber. Write an equation that shows how this amine can function as an antioxidant.

N-phenyl-2-naphthylamine

▼ ## Section 6.8
Polymers

Polymers are giant molecules, or **macromolecules**. Natural polymers include proteins (such as silk, muscle fibers, and enzymes), polysaccharides (starch and cellulose), and nucleic acids. Synthetic polymers are about as diverse as nature's polymers. The original synthetic polymers were devised to mimic natural polymers. For example, the DuPont Company introduced synthetic rubber and nylon ("artificial silk") in the 1930s. Today, we wear polyester clothes, sit on vinyl chairs, and write on Formica table tops. Our rugs may be made of polyester, polyacrylic, or polypropylene. Sky divers use nylon parachutes. We paint walls with latex paint and protect wood floors with polyurethane. Automobiles may have synthetic rubber tires and vinyl upholstery. Dishes may be melamine. Other common polymeric products include food wrap, Teflon coating for frying pans, hairbrushes, toothbrushes, epoxy glue, electrical insulators, plastic jugs, heart valves, airplane windshields—the list could continue! The technology of macromolecules has become a giant in the world of industry.

Polymers fall into three general classifications. *Elastomers* are those polymers with elastic properties, like rubber. *Fibers* are threadlike polymers, such as cotton, silk, or nylon. *Plastics* can be thin sheets (kitchen wrap); hard and moldable solids (piping, children's toys), or coatings (car finishes, varnishes). The multiplicity of properties depends on the variety of structures that are possible in polymers.

Many useful polymers can be prepared by radical reactions, and we will discuss some of these here. Other types of polymers will be covered in appropriate sections elsewhere in this text.

A polymer (Greek *poly-* and *meros*, "many parts") is made up of thousands of repeating units of small parts, the **monomers** ("one part"). In a polymerization reaction, the first products are **dimers** ("two parts"), then **trimers**, **tetramers**, and finally, after a series of reaction steps, the polymer molecules. The polymers that we will discuss here are called **addition polymers** because they are formed by the addition of monomers to each other without the loss of atoms or groups.

A synthetic polymer is usually named from the name of its monomer prefixed with **poly-**. For example, ethylene forms the simple polymer *polyethylene*, which is used for things like cleaners' bags and plastic piping.

$$CH_2{=}CH_2 \xrightarrow[\text{or catalyst}]{O_2,\ \text{heat, pressure}} \overset{\overbrace{\qquad\qquad\qquad}^{\textit{repeating monomeric units}}}{-CH_2CH_2-CH_2CH_2-CH_2CH_2-}$$

ethylene   polyethylene

the monomer   *the polymer*

The equations for polymerization are conveniently represented in the following format, where x is used to mean "a large number."

$$x\ CH_2{=}CH_2 \xrightarrow{\text{catalyst}} +CH_2CH_2+_x$$

Frequently, the end groups of macromolecules are unknown—they may arise from impurities in the reaction mixture. In some cases, the end groups can be controlled. The properties of a polymer are governed almost entirely by the bulk of the polymer molecule rather than by the end groups. To empha- size the basic structure of the polymer, it is customary not to include the end groups in the formula unless they are specifically known.

A. Mechanism of Radical Polymerization

A radical polymerization is started by a catalyst or an initiator such as O_2 or a peroxide. The resulting polymer is formed by a chain-propagation process. Let us use the polymerization of styrene as our example.

Overall reaction:

$$x \ CH_2{=}CH \xrightarrow{\text{initiator}} {+}CH_2{-}CH{+}_x$$

styrene   polystyrene

used in toothbrush handles and in the manufacture of Styrofoam

In this reaction, the initiation step is the cleavage of the initiator into radicals. An initiator radical attacks the pi electrons of the carbon–carbon double bond. The results are a new sigma bond (between RO and C) and a new radical, as shown in the following equations. The new radical can then attack another carbon–carbon double bond to add a second monomer to the growing chain.

Initiation:

$$ROOR \longrightarrow 2 \ RO\cdot$$

Propagation:

$$RO{\cdot}\overset{\frown}{CH_2}{=}\overset{\frown}{CH} \longrightarrow RO{-}CH_2\dot{C}H$$
$$\underset{C_6H_5}{|} \qquad\qquad \underset{C_6H_5}{|}$$

$$ROCH_2\dot{C}H \ \overset{\frown}{CH_2}{=}\overset{\frown}{CH} \longrightarrow ROCH_2CH{-}CH_2\dot{C}H \quad \text{etc.}$$
$$\underset{C_6H_5}{|} \quad \underset{C_6H_5}{|} \qquad\qquad \underset{C_6H_5}{|} \quad \underset{C_6H_5}{|}$$

Theoretically, the chain growth could go on indefinitely, which of course does not happen. The termination steps for polymerization are typical radical termination steps. Two radicals may meet and join, or two radicals might undergo **disproportionation**, one being oxidized to an alkene and one being reduced to an alkane. (The squiggles in the following formulas are used to indicate that a much larger molecule than shown is involved in the reaction.)

Coupling:

$$\wideparen{}{-}CH_2\dot{C}H \ + \ \dot{C}HCH_2{-}\wideparen{} \longrightarrow \wideparen{}{-}CH_2CH{-}CHCH_2{-}\wideparen{}$$
$$\underset{C_6H_5}{|} \quad \underset{C_6H_5}{|} \qquad\qquad \underset{C_6H_5}{|} \ \underset{C_6H_5}{|}$$

terminated

Disproportionation:

$$\xi-CH_2\dot{C}H \;\; + \;\; \dot{C}HCH_2-\xi \;\; \longrightarrow \;\; \xi-CH=CH \;\; + \;\; CH_2CH_2-\xi$$
$$\quad\quad\;\; \underset{C_6H_5}{|} \quad\quad \underset{C_6H_5}{|} \quad\quad\quad\quad\quad\quad \underset{C_6H_5}{|} \quad\quad \underset{C_6H_5}{|}$$

terminated

There are two ways in which styrene molecules could join together to form polystyrene: (1) *head-to-tail*, or (2) *head-to-head and tail-to-tail*.

tail   *head*

$$CH_2=CH$$
$$\quad\quad\;\; \underset{C_6H_5}{|}$$

head-to-tail   *head-to-head*   *tail-to-tail*

$$\xi-CH_2CH-CH_2CH-CH_2CH-\xi \quad\quad\quad \xi-CH_2CH-CHCH_2-CH_2CH-\xi$$
$$\quad\quad \underset{C_6H_5}{|} \quad\; \underset{C_6H_5}{|} \quad\; \underset{C_6H_5}{|} \quad\quad\quad\quad\quad \underset{C_6H_5}{|} \;\; \underset{C_6H_5}{|} \quad\quad\quad \underset{C_6H_5}{|}$$

Polystyrene is an example of a *head-to-tail polymer*. Let us discuss the reason for this orientation of the monomers. A more stable radical intermediate means a lower-energy transition state and a faster rate of reaction. Of the two possible modes of radical attack, one leads to a less stable primary radical, while the other leads to the more stable secondary benzylic radical. The repetitive formation of benzylic radicals leads to a head-to-tail joining of styrene monomers.

$$R\cdot + CH_2=CH \quad\longrightarrow\quad RCH_2-\dot{C}H \quad\quad \text{and not} \quad\quad RCH-\dot{C}H_2$$
$$\quad\quad\quad\quad \underset{C_6H_5}{|} \quad\quad\quad\quad\quad \underset{C_6H_5}{|} \quad\quad\quad\quad\quad\quad\quad\quad \underset{C_6H_5}{|}$$

benzylic,   *1°, less stable*
more stable

again benzylic,
more stable

$$\xrightarrow{CH_2=CHC_6H_5} \; RCH_2CH-CH_2\dot{C}H \;\; \text{etc.}$$
$$\quad\quad\quad\quad\quad\quad \underset{C_6H_5}{|} \quad\;\; \underset{C_6H_5}{|}$$

Resonance structures of a benzylic radical:

The polymerization of vinyl chloride and methyl methacrylate are other examples of radical reactions leading to head-to-tail products.

$$x\ CH_2{=}\underset{\underset{\textstyle Cl}{|}}{CH} \xrightarrow{\text{initiator}} \left(CH_2{-}\underset{\underset{\textstyle Cl}{|}}{CH}\right)_{x}$$

vinyl chloride   polyvinyl chloride (PVC)

used for flooring, piping, siding,
phonograph records, and garbage bags

$$x\ CH_2{=}\underset{\underset{\textstyle CO_2CH_3}{|}}{\overset{\overset{\textstyle CH_3}{|}}{C}} \xrightarrow{\text{initiator}} \left(CH_2{-}\underset{\underset{\textstyle CO_2CH_3}{|}}{\overset{\overset{\textstyle CH_3}{|}}{C}}\right)_{x}$$

methyl methacrylate   polymethyl methacrylate

Plexiglas and Lucite

A chemist is not limited to using a single monomer in the production of a polymer. To achieve desired properties, a chemist might use a mixture of two, three, or even more monomers. A mixture of two different monomers results in a **copolymer**, such as Saran (used in kitchen wrap).

$$x\ CH_2{=}CHCl\ +\ \ x\ CH_2{=}CCl_2 \xrightarrow{\text{initiator}} (CH_2CHCl{-}CH_2CCl_2)_x$$

vinyl chloride   1,1-dichloroethene   Saran
         (vinylidene chloride)

a copolymer
(not necessarily alternating)

Study Problems

6.13 The monomer for *Teflon* is $CF_2{=}CF_2$. What is the structure of Teflon?

6.14 *Orlon* has the formula $\left(CH_2\underset{\underset{\textstyle CN}{|}}{CH}\right)_x$. What is the structure of its monomer?

B. Structure and Stereochemistry of Polymers

Polymers, like any other organic compound, may have functional groups and chiral carbons. They can undergo hydrogen bonding and dipole–dipole interactions. The chemical composition of a polymer chain is referred to as its **primary structure**. How the chain is arranged in relation to itself and to other chains is called the **secondary structure**. This secondary structure may be as important to the properties of a polymer as its chemical composition.

A polymer may be a tangled mass of continuous chains or branched chains. The result is a soft amorphous solid such as soft rubber. On the other hand, a polymer may be composed of continuous chains held together by hydrogen bonds or by other dipole–dipole attractions. This type of polymer structure lends itself to fibers or hard, moldable plastics. A more-ordered polymer is said to have a higher degree of **crystallinity** than the amorphous, or noncrystalline, polymer.

a noncrystalline polymer   *a crystalline polymer*

Let us consider the polymerization of propylene.

$$x \ \ CH_2\!\!=\!\!\overset{\overset{\displaystyle CH_3}{|}}{CH} \ \ \xrightarrow{\ \ catalyst\ \ } \ \ \left(\!\!CH_2\!-\!\!\overset{\overset{\displaystyle CH_3}{|}}{CH}\!\!\right)_{\!\!x}$$

polypropylene

used for rugs and upholstery

Three types of products could result from the head-to-tail polymerization of propylene. (1) The methyl groups at the newly formed chiral carbons could be protruding from the chain in a random fashion. This is an **atactic polymer** (a soft, amorphous product). (2) The methyl groups could alternate from one side of the chain to the other. This is a **syndiotactic polymer**. (3) The methyl groups could all be on the same side. Then, the polymer is said to be **isotactic** (see Figure 6.3). Because of their orderly arrangements, the chains of the latter two polymers can lie closer together and the polymers are more crystalline.

Until 1955, most addition polymers were made by radical paths, which produce atactic polymers. In that year, however, Karl Ziegler and Giulio Natta introduced a new technique for polymerization. These two chemists received the 1963 Nobel prize for their discovery, a type of catalyst that permits control

Figure 6.3 The three types of polypropylene molecules.

of the stereochemistry of a polymer during its formation. [A commonly used Ziegler–Natta catalyst is $(CH_3CH_2)_3Al$ complexed with $TiCl_4$.] Ziegler–Natta catalysts function by undergoing reaction with the monomeric alkene. Then, new monomers are *inserted* between the catalyst and the growing polymer.

$$\boxed{\text{catalyst}} \diagup CH_2CH_3 + CH_2{=}CH_2 \longrightarrow \boxed{\text{catalyst}}{-}CH_2CH_2{-}CH_2CH_3 \quad \text{etc.}$$

In the commercial synthesis of polypropylene and other polymers, Ziegler–Natta catalysts are used in order to produce the more desirable syndiotactic or isotactic polymers.

▼ Summary

A **free radical**, or **radical**, is an atom or group of atoms with an unpaired electron. Radical reactions are **chain reactions** that involve **initiation** (formation of radicals); **propagation** (reactions in which new radicals are consumed and formed); and **termination** (coupling, disproportionation, or the formation of stable radicals). Racemization occurs at a reacting, resolved, chiral carbon during a radical reaction.

The order of reactivity of H toward radical abstraction is $CH_4 < 1° < 2° < 3° <$ allylic or benzylic. The order of reactivity is the result of the relative stabilities of the radical intermediates.

Cl_2 is more reactive and less selective than Br_2 in radical halogenations. **N-Bromosuccinimide** (NBS) is a selective brominating agent for *allylic* and *benzylic positions*.

$$CH_3CH_2CH_3 \quad \underset{h\nu}{\overset{Cl_2}{\longrightarrow}} \quad CH_3CHClCH_3 + CH_3CH_2CH_2Cl + HCl$$

$$\underset{h\nu}{\overset{Br_2}{\longrightarrow}} \quad CH_3CHBrCH_3 + HBr$$

$$CH_2{=}CHCH_3 \quad \overset{NBS}{\longrightarrow} \quad CH_2{=}CHCH_2Br$$

$$\bigcirc{-}CH_2R \quad \overset{NBS}{\longrightarrow} \quad \bigcirc{-}CHBrR$$

Other radical reactions besides halogenation include **pyrolysis**, a thermal free-radical decomposition of organic compounds, and **auto-oxidation**, a free-radical oxidation by O_2 that results in decomposition of fats, oils, rubber, ethers, and aldehydes. The **cumene process** is a useful auto-oxidation.

Free-radical initiators are substances that cause the formation of radicals. Light and peroxides (which contain the easily broken $-O-O-$ bond) are examples. **Free-radical inhibitors**, such as phenols, are substances that form nonreactive radicals.

Polymers are giant molecules composed of repetitive units of *monomers*. Some polymers can be formed by radical addition reactions of alkenes.

$$x\ CH_2{=}CH \xrightarrow[\text{catalyst}]{\substack{\text{initiator}\\ \text{or}}} \left(CH_2CH\right)_x$$

with R substituent

The stereochemistry of a polymer can sometimes be controlled by the use of an appropriate catalyst in the polymerization reaction.

Study Problems*

6.15 Write Lewis formulas for   (a) $CH_3CH_2O\cdot$ and   (b) $CH_2{=}CH{-}\dot{C}H_2$.

6.16 Label the following reactions as *initiation, propagation,* or *termination* steps:

(a) $(CH_3)_3C\cdot + CH_2{=}CH_2 \longrightarrow (CH_3)_3C{-}CH_2CH_2\cdot$

(b) $C_6H_5\overset{\text{O}}{\overset{\|}{C}}OOH \longrightarrow C_6H_5\overset{\text{O}}{\overset{\|}{C}}O\cdot + \cdot OH$

(c) $2\ CH_3CH_2CH_2\cdot \longrightarrow CH_3CH_2CH_3 + CH_3CH{=}CH_2$

(d) $2\ CH_3CH_2\cdot \longrightarrow CH_3CH_2CH_2CH_3$

(e) $Br\cdot + CH_2{=}CH_2 \longrightarrow \cdot CH_2CH_2Br$

(f) $(CH_3)_2\overset{\text{CN}}{\overset{|}{C}}{-}N{=}N{-}\overset{\text{CN}}{\overset{|}{C}}(CH_3)_2 \longrightarrow 2\ (CH_3)_2\overset{\text{CN}}{\overset{|}{C}}\cdot + N_2$

6.17 Write equations for the steps in the radical dichlorination of cyclopentane to yield 1,2-dichlorocyclopentane.

6.18 If all H's were abstracted at equivalent rates, what would be the relative percentages of monochlorination products of an equimolar mixture of $CH_3CH_2CH_3$ and cyclohexane?

6.19 List the products that would be obtained from the radical monochlorination of each of the following compounds. (Do not forget to indicate stereoisomers.)

(a) (*R*)-1,2-dichloropropane   (b) (*R*)-2-chlorobutane

6.20 Only one monochlorination product is obtained from an alkane with the molecular formula C_5H_{12}. What is the structure of the alkane?

6.21 Rank the following radicals in order of increasing stability (least stable first):

(a) $CH_3CH_2\dot{C}(CH_3)_2$ (b) $CH_3\dot{C}HCH_3$ (c) $C_6H_5\dot{C}HCH_3$

(d) $CH_3CH{=}CHCH_2\dot{C}H_2$ (e) $C_6H_5\dot{C}HCH{=}CH_2$

6.22 Rank the following hydrocarbons in order of increasing ease of radical bromination:

(a) $CH_3CH_2CH(CH_3)_2$ (b) $CH_3CH_2CH_3$ (c) $C_6H_5CH_2CH_3$

6.23 Draw all the important resonance structures for the following radicals:

(a) cyclohexane ring with $=CH_2$ and radical   (b) benzene ring with $\dot{C}HCH_3$ (c) cyclohexane ring with $\overset{\text{O}}{\overset{\|}{C}}O\cdot$

---

*For information concerning the organization of the *Study Problems* and *Additional Problems*, see the *Note to student* on page 39.

6.24 Each of the following compounds is treated with NBS and a peroxide in CCl_4. Predict the monohalogenation product.

(a) $(CH_3)_2C{=}CHCOCH_3$ (b)   (c)

6.25 Suggest a mechanism that explains the following observation. (The * represents an isotopic label: ^{14}C.)

6.26 Predict the coupling and disproportionation products of the following radicals:

(a) $(CH_3)_2\dot{C}CH_2CH_2CH_3$ (b)

6.27 In 1900, the Russian-born chemist Moses Gomberg, then at the University of Michigan, attempted to synthesize hexaphenylethane. Gomberg's synthesis failed. Instead of synthesizing the still unknown hexaphenylethane, Gomberg had accomplished something far more important—the first synthesis of a relatively stable radical, the triphenylmethyl radical. **(a)** Suggest a reason for the difficulty of synthesizing hexaphenylethane. **(b)** Suggest a reason why the triphenylmethyl radical is more stable than most other radicals. Use formulas in your answer.

6.28 Which of the following compounds would form a hydroperoxide (ROOH) readily upon exposure to air?

(a)   (b)   (c)

p-cymene

found in plant oils

vanillin

dioxane

a useful solvent, but carcinogenic

6.29 What alkene monomers would be needed to prepare the following polymers?

(a) $+CH_2CH+_x$

polyvinylpyrrolidone

used as a blood plasma expander and in cosmetics

(b) $+CH_2CH+_x$

polyvinyl acetate

used in latex paints and paper coatings

(c) $+CH_2CH+_x$

polyvinyl ethyl ether

a viscous gum used on pressure-sensitive tape

6.30 Styrene ($C_6H_5CH{=}CH_2$) can be copolymerized with maleic anhydride. Write an equation for this reaction.

maleic anhydride

6.31 Suggest a reaction mechanism for the preparation of polyisobutylene from an appropriate alkene using benzoyl peroxide (page 244) as an initiator. In your mechanism, show the initiation step, the propagation steps, and two possible termination steps.

polyisobutylene

Additional Problems

6.32 The radical bromination of $C_6H_5CH_2D$ with Br_2 shows a kinetic isotope effect. Suggest a reason for the fact that, if HBr is removed from the reaction mixture as soon as it is formed, the isotope effect is even greater.

6.33 The aqueous electrolysis of salts of carboxylic acids (called the **Kolbe electrolysis**) yields carbon dioxide and hydrocarbons by a radical path. What would be the products of the electrolysis of sodium acetate (CH_3CO_2Na)?

6.34 Upon radical bromination, *n*-pentane yields almost exclusively two monobromo compounds, *A* and *B*. Upon treatment with $NaOCH_3$ under E2 conditions, *A* and *B* yield predominantly the same product, *C*. What are the structures of *A*, *B*, and *C*?

6.35 If not prepared with an antioxidant, polypropylene undergoes rapid deterioration when exposed to the air. Polyethylene does not exhibit the same rate of deterioration. Explain why these two polymers differ in their air stabilities.

6.36 When an excess of radical initiator is used in the polymerization of an alkene, the average polymeric molecular weight is often less than that when just a trace of initiator is used. Explain why this is true.

6.37 Instrumental studies (electron spin resonance, or ESR) have shown that the anion-radical that would be obtained from the one-electron reduction of 2,4-cyclopentadienone is a resonance hybrid of the following two formulas. Rewrite these formulas as Kekulé formulas and use electron-shift arrows to show the resonance structures.

6.38 Suggest a mechanism for the following reaction:

$$(CH_3)_3COH \xrightarrow{\text{initiator}} \underset{\underset{CH_3}{|}}{HOCCH_2} - \underset{\underset{CH_3}{|}}{CH_2COH}$$

with CH_3 groups above each central carbon.

6.39 2-Heptene undergoes reaction with NBS to yield a monobrominated product in 58–64% yield.   (a) What two possible monobrominated products could be formed?   (b) Why is one of these two products formed in greater yield?

6.40 Some addition polymers can be depolymerized by heat.   (a) Write equations for the depolymerization of (1) polystyrene and (2) Teflon (polytetrafluoroethene).   (b) What danger would reactions such as these pose to electronic technicians soldering insulated wiring or to firefighters?

7 ▼

Alcohols

Alcohols (ROH) are compounds containing a hydroxyl group (—OH) bonded to an sp^3-hybridized carbon. Many alcohols are commonly encountered in everyday life. Ethanol (ethyl alcohol, grain alcohol, or just alcohol), CH_3CH_2OH, is used in beverages. 2-Propanol (isopropyl alcohol, or rubbing alcohol), $(CH_3)_2CHOH$, is used as a bacteriocidal agent. Methanol (methyl alcohol, or wood alcohol), CH_3OH, is used as an automobile gas-line antifreeze. In the laboratory and in industry, all these compounds are used as solvents and reagents.

Contrast the structure of an alcohol to those of phenols (hydroxyl group on an aryl sp^2-hybridized carbon), enols (hydroxyl group on an alkenyl sp^2-hybridized carbon), or ethers (no hydroxyl group).

CH_3OH	—OH	$R_2C\!=\!CR$ (OH)	ROR
an alcohol	a phenol	an enol	an ether

As described in Chapter 2, alcohols are similar in structure to water. The oxygen is in the sp^3-hybrid state and has two unshared pairs of valence electrons. Like the hydroxyl groups in water, the hydroxyl group in an alcohol is polar.

methanol
an alcohol

water

It is not surprising that alcohols can undergo hydrogen bonding in either the pure state or in an aqueous medium.

hydrogen bonds

Section 7.1
Physical Properties of Alcohols

A. Boiling Points

Because alcohols can form hydrogen bonds with other alcohol molecules, they have higher boiling points than alkyl halides of comparable molecular weights. Table 7.1 compares the boiling points of some alcohols and chloro compounds with the same carbon skeletons.

B. Solubility in Water

Alcohols of low molecular weight are miscible with water, while the corresponding alkyl halides are water-insoluble. This water solubility is directly attributable to hydrogen bonding between alcohols and water.

The hydrocarbon portion of an alcohol is **hydrophobic**, that is, it repels water molecules. As the length of the hydrocarbon portion of an alcohol molecule increases, the water solubility of the alcohol decreases. When the hydrocarbon chain is long enough, it overcomes the **hydrophilic** (water-loving)

Table 7.1 Comparison of the boiling points of some alcohols and chloroalkanes

Alcohol	Bp, °C	Chloroalkane	Bp, °C
CH_3OH	64.5	CH_3Cl	−24
CH_3CH_2OH	78.3	CH_3CH_2Cl	13
$CH_3CH_2CH_2OH$	97.2	$CH_3CH_2CH_2Cl$	46
$HOCH_2CH_2OH$	197	$ClCH_2CH_2Cl$	83.5

Table 7.2 Physical properties of some alcohols

IUPAC Name	Trivial Name	Formula	Bp, °C	Density, g/mL at 20 °C	Solubility in H_2O
methanol	methyl alcohol	CH_3OH	64.5	0.79	∞
ethanol	ethyl alcohol	CH_3CH_2OH	78.3	0.79	∞
1-propanol	propyl alcohol	$CH_3CH_2CH_2OH$	97.2	0.80	∞
2-propanol	isopropyl alcohol	$(CH_3)_2CHOH$	82.3	0.79	∞
1-butanol	butyl alcohol	$CH_3(CH_2)_3OH$	117	0.81	8.3 g/100 mL

properties of the hydroxyl group. The three-carbon alcohols, 1- and 2-propanol, are miscible in water, while only 8.3 grams of 1-butanol dissolves in 100 mL of water. (These solubilities are summarized in Table 7.2.)

Branching increases water solubility. Although 1-butanol is only slightly soluble, *t*-butyl alcohol, $(CH_3)_3COH$, is miscible with water. The reason for this is that the *t*-butyl group is more compact and less hydrophobic than the *n*-butyl group. An increase in the number of OH groups also increases hydrophilicity and solubility. Sucrose (table sugar, page 925) has twelve carbons, but it also has eight hydroxyl groups and is readily soluble in water.

Section 7.2
Basicity and Acidity of Alcohols

In acidic solution, alcohols are protonated. This reaction is an acid–base equilibrium with the alcohol acting as a base. It is the same type of reaction that occurs between water and a proton.

hydronium ion

an alcohol   *an oxonium ion*

In each case, an empty 1*s* orbital of H^+ overlaps with one of the filled valence orbitals of the oxygen, and an O—H sigma bond is formed (see

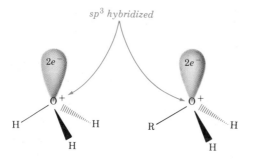

Figure 7.1 Bonding in the hydronium ion (H_3O^+) and in an oxonium ion (ROH_2^+).

Figure 7.1). The product of the reaction with water is the protonated water molecule, or the **hydronium ion**. A protonated alcohol molecule is called an **oxonium ion**.

An alcohol can also *lose* a proton to a strong base to yield an **alkoxide ion**, RO^-. Alkoxides are *strong bases*, generally stronger than hydroxides. To prepare an alkoxide from an alcohol, we need a base stronger than the alkoxide itself, such as an alkali metal hydride (NaH, KH).

$$\langle \rangle\text{—OH} + Na^+\ H^- \longrightarrow \langle \rangle\text{—O}^-\ Na^+ + H_2\uparrow$$

Another convenient method for the preparation of alkoxides is the treatment of an alcohol with an alkali metal such as sodium or potassium. The alkali metal is oxidized to a cation, and the hydrogens of the OH groups are reduced to hydrogen gas.

$$CH_3CH_2OH + Na \longrightarrow CH_3CH_2O^-\ Na^+ + \tfrac{1}{2}H_2\uparrow$$

$$(CH_3)_3COH + K \longrightarrow (CH_3)_3CO^-\ K^+ + \tfrac{1}{2}H_2\uparrow$$

Methanol and ethanol undergo fairly vigorous reaction with sodium metal. As the size of the R group is increased, the vigor of the reaction decreases. Sodium and water react explosively. Sodium and ethanol undergo reaction at a very controllable rate. Sodium and 1-butanol undergo a very sluggish reaction. With alcohols of four or more carbons, the more reactive potassium metal is generally used to prepare the alkoxide.

$$H_2O \quad CH_3OH \quad CH_3CH_2OH \quad CH_3CH_2CH_2OH$$

decreasing reactivity toward Na or K

Like water, a pure alcohol can undergo ionization. However, in their pure states, alcohols ionize to a lesser extent than does water.

Table 7.3 pK_a values for water and some alcohols in dilute aqueous solution

Compound	pK_a
H_2O	15.74
CH_3OH	15.5
CH_3CH_2OH	15.9
$(CH_3)_3COH$	~18

$$CH_3\ddot{O}\!\!-\!\!H + CH_3\ddot{O}H \;\Longleftrightarrow\; CH_3\ddot{O}\!:^- + CH_3\overset{+}{\ddot{O}}H_2$$

One reason for the lesser amount of ionization of pure alcohols is that alcohols have lower dielectric constants. Because they are less polar, alcohols are less able to support ions in solution than are water molecules.

In dilute aqueous solution, alcohols have approximately the same pK_a values as water (see Table 7.3). For example, the pK_a of methanol in water is 15.5, while that of pure water is 15.74 (not 14, which is the pK_w).

Water (pK_a = 15.74):

$$H\ddot{O}\!\!-\!\!H + H_2\ddot{O}\!: \;\Longleftrightarrow\; H\ddot{O}\!:^- + H_3\ddot{O}^+$$

Methanol in H_2O (pK_a = 15.5):

$$CH_3\ddot{O}\!\!-\!\!H + H_2\ddot{O}\!: \;\Longleftrightarrow\; CH_3\ddot{O}\!:^- + H_3\ddot{O}^+$$

▼ **Section 7.3**
Nomenclature and Classification of Alcohols

A. IUPAC Names

The IUPAC names of alcohols are taken from the names of the parent alkanes, but with the ending **-ol**. A prefix number, chosen to be as low as possible, is used if necessary.

			OH
	CH_3OH	$CH_3CH_2CH_2OH$	$CH_3\overset{\mid}{C}HCH_3$
IUPAC:	methanol	1-propanol	2-propanol

More than one hydroxyl group is designated by *di-*, *tri-*, etc., just before the -ol ending. The final -e of the parent alkane name is retained before the consonant "d" of the -diol ending.

OH
|
$CH_3\overset{\mid}{C}HCH_2CH_2OH$

1,3-butanediol

a diol

Study Problems

7.1 Name the following compounds:

(a) (CH₃)₂CHCHCH(CH₃)₂ $\overset{\text{OH}}{|}$ (b) (CH₃)₂C—C(CH₃)₂ $\overset{\text{OH OH}}{|\ |}$

OH OH
CH₃C — C - CH₃
CH₃ CH₃

7.2 Write the structures for   **(a)** 3-ethyl-3-methyl-2-pentanol, and   **(b)** 2,2-dimethyl-1,4-hexanediol.

A hydroxyl group is often found in a molecule that contains other functional groups. In the IUPAC system, the numbering and the suffix in the name of a multifunctional compound are determined by nomenclature priority. A more complete list of nomenclature priorities is given in the appendix.

$$-\text{R, } -\text{X, etc.}\qquad \underset{\diagup}{\overset{\diagdown}{\text{C}}}{=}\underset{\diagup}{\overset{\diagdown}{\text{C}}}\qquad -\text{OH}\qquad \overset{\text{O}}{\overset{\|}{-\text{C}-}}\qquad \overset{\text{O}}{\overset{\|}{-\text{CH}}}\qquad -\text{CO}_2\text{H}$$

increasing nomenclature priority

Carboxylic acids, aldehydes, and ketones have higher nomenclature priority than the hydroxyl group. One of these groups receives the lowest nomenclature number and is also given the suffix position in the name. The lower-priority OH group is then named by the prefix **hydroxy-**, as in the following examples:

$$\underset{\text{CH}_3\text{CHCOH}}{\overset{\text{HO O}}{\overset{|\ \|}{}}}\qquad\qquad \underset{\text{HOCH}_2\text{CH}_2\text{CH}}{\overset{\text{O}}{\overset{\|}{}}}\qquad\qquad \underset{\text{HOCH}_2\text{CH}_2\text{CCH}_3}{\overset{\text{O}}{\overset{\|}{}}}$$

2-hydroxypropanoic acid   3-hydroxypropanal   4-hydroxy-2-butanone
(lactic acid)

In a compound that contains an OH group and also has either a double bond or a group named as a prefix, the hydroxyl group has the higher nomenclature priority. In these cases, the OH receives the lowest prefix number and is given the *-ol* ending. Note in the following examples how a double-bond suffix is inserted into the name of an unsaturated alcohol.

3,3-dichlorocyclohexanol   CH₂=CHCH₂CH₂OH   4-methyl-2-cyclohexenol
                              3-buten-1-ol

Study Problem

7.3 Name the following compounds by the IUPAC system:

(a) ⬠—OH   (b) HO—⬡—Br   (c) BrCH₂CH₂OH

B. Trivial Names

Just as CH_3I may be called methyl iodide, CH_3OH may be called methyl alcohol. This type of name is a popular way of naming alcohols with common alkyl groups.

$(CH_3)_3COH$ $(CH_3)_2CHOH$

t-butyl alcohol   isopropyl alcohol

A diol (especially a 1,2-diol) is often referred to as a **glycol**. The trivial name for a 1,2-diol is that of the corresponding **alkene** followed by the word **glycol**. Epoxides and 1,2-dihalides are often named similarly. The naming of a saturated compound as a derivative of an alkene is unfortunate. However, the practice arose quite innocently in the early years of organic chemistry, because all these compounds can be prepared from alkenes.

		OH OH	Br Br	O
	$CH_2{=}CH_2$	$CH_2{-}CH_2$	$CH_2{-}CH_2$	$CH_2{-}CH_2$
IUPAC:	ethene	1,2-ethanediol	1,2-dibromoethane	oxirane
trivial:	ethylene	ethylene glycol	ethylene dibromide	ethylene oxide

an epoxide

C. Classification of Alcohols

Alcohols, like alkyl halides, can be classified as **methyl, primary, secondary,** or **tertiary**, as well as **allylic** or **benzylic**.

CH_3OH CH_3CH_2OH $(CH_3)_2CHOH$ $(CH_3)_3COH$

methyl   1°   2°   3°

OH
|
$CH_3CH{=}CHCH_2OH$ ⬡—$CHCH_3$

an allylic alcohol   *a benzylic alcohol*
(and 1°)   *(and 2°)*

Study Problem

7.4 Classify each of the following alcohols as 1°, 2°, 3°, allylic, or benzylic (all classifications that apply):

(a)   **(b)**   **(c)**

▼ ■ # Section 7.4
Preparation of Alcohols

The ethanol used in beverages is obtained by the enzyme-catalyzed **fermentation of carbohydrates** (sugars and starches). Fermentation is the decomposition of organic compounds into simpler compounds by the action of the biological catalysts known as enzymes. In the fermentation of carbohydrates, one type of enzyme converts carbohydrates to glucose and then to ethanol. Another type leads to vinegar (acetic acid), with ethanol as an intermediate.

$$C_6H_{12}O_6 \xrightarrow{\text{enzymes}} CH_3CH_2OH$$

glucose   ethanol

a sugar

The source of the carbohydrates used for fermentation depends on availability and on the purposes of the alcohol. In the U.S., carbohydrates are obtained primarily from corn and from the molasses residue of sugar refining. However, potatoes, rice, rye, or fruit (grapes, blackberries, etc.) may also be used.

Fermentation of any of these fruits, vegetables, or grains ceases when alcoholic content reaches 14 to 16 percent. If a higher concentration of alcohol is desired, the mixture is distilled. The distillate is an azeotrope of 95% ethanol −5% water. (An azeotrope is a mixture that boils at a constant boiling point as if it were a pure compound.) This distillate may then be used to fortify the fermentation mixture, or it may be diluted with water to the desired strength.

Because alcoholic beverages are taxed in almost all countries of the world, most ethanol sold for laboratory or industrial purposes (and not taxed as a liquor) is **denatured**, that is, small amounts of toxic impurities are added so that the ethanol cannot be diverted from the laboratory or factory into illegal beverages.

Fermentation is rarely used as a synthetic technique in the laboratory. We have already discussed one laboratory reaction that yields alcohols as products: $RX + {}^-OH$ (Chapter 5). In this section, we will present a survey of this reaction and a few others that are commonly used to synthesize alcohols.

A. Nucleophilic Substitution Reactions

The reaction of an alkyl halide with hydroxide ion is a nucleophilic substitution reaction. When primary alkyl halides are heated with aqueous sodium hydroxide, reaction occurs by an S_N2 path. Primary alcohols can be prepared in good yields by this technique. Because secondary and tertiary alkyl halides are likely to give elimination products, they are not generally as useful for preparing alcohols.

$$CH_3CH_2CH_2Br + OH^- \xrightarrow{\text{heat}} CH_3CH_2CH_2OH + Br^-$$

1-bromopropane   1-propanol

a 1° alkyl halide   *a 1° alcohol*

B. Reduction of Carbonyl Compounds

Alcohols can be prepared from carbonyl compounds by **reduction reactions** in which hydrogen atoms are added to the carbonyl group. For example, *reduction of a ketone* by catalytic hydrogenation or with a metal hydride yields a *secondary alcohol*. Yields are often 90–100%. These reactions will be discussed in more detail in Section 13.7.

$$CH_3\overset{O}{\overset{\|}{C}}CH_3 \xrightarrow[\text{(2) H}_2\text{O, H}^+]{\text{(1) NaBH}_4} CH_3\overset{OH}{\overset{|}{C}H}CH_3$$

acetone   2-propanol

cyclohexanone   cyclohexanol

C. Hydration of Alkenes

When an alkene is treated with water plus a strong acid, which acts as a catalyst, water (H^+ and OH^-) adds to the double bond in a **hydration reaction**. The product is an alcohol. Many alcohols, such as laboratory ethanol, are made commercially by the hydration of alkenes. Limitations and variations of hydration reactions, as well as the mechanism, will be discussed in Chapter 10.

$$CH_2{=}CH_2 + H_2O \xrightarrow{\text{H}^+} CH_3CH_2OH$$

ethylene   ethanol

cyclohexene   cyclohexanol

Study Problem

7.5 Write equations that show how each of the following alcohols can be prepared from (1) an alkene, and (2) a ketone:

(a) 2-butanol   **(b)** 2,4-dimethylcyclopentanol

D. Grignard Reactions

An **organometallic compound** is defined as a compound in which *carbon is bonded directly to a metallic atom* (such as mercury, zinc, lead, magnesium, or lithium) or to certain *metalloids* (such as silicon, arsenic, or selenium).

$CH_3CH_2CH_2CH_2Li$	$(CH_3)_4Si$	CH_3ONa
n-butyllithium	tetramethylsilane (TMS)	sodium methoxide
organometallic	*organometallic*	*not considered organometallic* (*no carbon–metal bond*)

One of the most useful classes of reagents in organic synthesis is that of the organomagnesium halides (RMgX). These compounds are called **Grignard reagents** after the French chemist Victor Grignard, who received the Nobel prize in 1912 for work in this area of organometallic chemistry. A Grignard reagent is the product of the reaction between magnesium metal and an organohalogen compound in an ether solvent.

$$R-X + Mg \xrightarrow{\text{diethyl ether}} R-Mg-X$$

a Grignard reagent

The reaction is general and does not depend to any great extent upon the nature of the R group. Primary, secondary, and tertiary alkyl halides, as well as allylic and benzylic halides, all form Grignard reagents.

$$(CH_3)_3CBr + Mg \longrightarrow (CH_3)_3CMgBr$$

t-butyl bromide   *t*-butylmagnesium bromide

$$\langle\!\bigcirc\!\rangle-CH_2Cl + Mg \longrightarrow \langle\!\bigcirc\!\rangle-CH_2MgCl$$

benzyl chloride   benzylmagnesium chloride

Aryl and vinylic halides (X on the doubly bonded carbon) are generally quite inert toward nucleophilic substitution and elimination. These compounds are not as reactive as alkyl halides toward magnesium, but their Grignard reagents can still be prepared.

$$\langle\!\bigcirc\!\rangle-Br + Mg \longrightarrow \langle\!\bigcirc\!\rangle-MgBr$$

bromobenzene   phenylmagnesium bromide

an aryl Grignard reagent

$$CH_2\!\!=\!\!CHI + Mg \longrightarrow CH_2\!\!=\!\!CHMgI$$

iodoethene   ethenylmagnesium iodide
(vinyl iodide)   (vinylmagnesium iodide)

a vinylic Grignard reagent

The most used solvent for a Grignard reagent is diethyl ether $(CH_3CH_2OCH_2CH_3)$, which is nonreactive toward Grignard reagents, but can donate unshared electrons to the empty orbitals of Mg and help dissolve the organomagnesium compound. The ethyl groups provide a hydrocarbon environment that acts as the solvent for the alkyl portion of a Grignard reagent.

$$CH_3CH_2\underset{\ddot{\underset{\ddot{}}{O}}}{\diagdown}CH_2CH_3$$

$$CH_3-Mg-I$$

$$CH_3CH_2\underset{\ddot{O}}{\diagup}\diagdown CH_2CH_3$$

In most organic compounds, carbon carries either no partial charge or a partial positive charge. In a Grignard reagent, carbon is bonded to an electropositive element and consequently carries a *partial negative charge*.

$$\overset{\delta+}{CH_3CH_2} \overset{\delta-}{-Br} + Mg \xrightarrow{\text{diethyl ether}} \overset{\delta-}{CH_3CH_2} \overset{\delta+}{-Mg} -Br \qquad \textit{carbon is } \delta-$$

Because a Grignard reagent has a partially negative carbon, (1) it is an *extremely strong base*, and (2) the alkyl or aryl portion of the Grignard reagent can act as a *nucleophile*.

Grignard reagents as bases. A hydrogen that can be abstracted from a compound by a Grignard reagent is said to be an *acidic hydrogen* in relation to the Grignard reagent. Grignard reagents undergo rapid reaction with compounds that have acidic hydrogens such as water, alcohols (ROH), amines (R_2NH), carboxylic acids (RCO_2H), and terminal alkynes ($RC\equiv CH$), to yield hydrocarbons and metal salts. For example,

$$\overset{\delta-}{R} -MgX + H-\ddot{O}H \longrightarrow RH + XMg\ddot{O}H$$

Because of the basicity of the carbon in a Grignard reagent, water and other compounds containing acidic hydrogens must be excluded from a Grignard reaction mixture unless a hydrocarbon is the desired product.

Grignard reagents as nucleophiles. The most important reactions of Grignard reagents are those with carbonyl compounds. In a carbonyl group (C=O), the electrons in the carbon–oxygen bonds (sigma and pi) are drawn toward the electronegative oxygen. The carbon of the carbonyl group, which has a partial positive charge, is attacked by the nucleophilic carbon of the Grignard reagent. The following equations show how Grignard reagents undergo reactions with ketones.

Reaction of RMgX with ketones:

General:

$$R-\overset{\overset{\ddot{O}:^{\delta-}}{\|}}{\underset{\delta+}{C}}-R + R'-MgX \longrightarrow R-\overset{\overset{:\ddot{O}:^- \ ^+MgX}{|}}{\underset{R'}{C}}-R$$

$$\overset{\overset{O}{\|}}{CH_3CCH_3} + CH_3CH_2CH_2CH_2-MgBr \longrightarrow \overset{\overset{O^- \ ^+MgBr}{|}}{\underset{CH_2CH_2CH_2CH_3}{CH_3CCH_3}}$$

The product of the reaction of RMgX with a ketone is the magnesium salt of an alcohol. When treated with water or aqueous acid, this magnesium salt yields the alcohol and a mixed inorganic magnesium salt.

$$\text{R}-\overset{\overset{\displaystyle :\ddot{\text{O}}:^- \ ^+\text{MgX}}{|}}{\underset{\underset{\displaystyle \text{R}'}{|}}{\text{C}}}-\text{R} + \text{H}^+ \longrightarrow \text{R}-\overset{\overset{\displaystyle :\ddot{\text{O}}\text{H}}{|}}{\underset{\underset{\displaystyle \text{R}'}{|}}{\text{C}}}-\text{R} + \text{Mg}^{2+} + \text{X}^-$$

a strong base   an acid   a 3° alcohol

Grignard reactions provide an excellent route to alcohols with complex carbon skeletons. A Grignard reaction:

1. with formaldehyde yields a *primary alcohol*;
2. with any other aldehyde yields a *secondary alcohol*; and
3. with a ketone yields a *tertiary alcohol*.

Reaction of RMgX with formaldehyde, $H_2C{=}O$:

$$\overset{\overset{\displaystyle O}{\|}}{\text{H}\text{C}\text{H}} \xrightarrow[\text{(2) } H_2O,\ H^+]{\text{(1) RMgX}} \text{RCH}_2\text{OH}$$

methanal   *a 1° alcohol*
(formaldehyde)

Reaction of R′MgX with other aldehydes:

$$\overset{\overset{\displaystyle O}{\|}}{\text{R}\text{C}\text{H}} \xrightarrow[\text{(2) } H_2O,\ H^+]{\text{(1) R′MgX}} \overset{\overset{\displaystyle \text{OH}}{|}}{\text{RCHR}'}$$

an aldehyde   *a 2° alcohol*

Some other Grignard reactions also lead to alcohols. The reaction of a Grignard reagent with *ethylene oxide* yields a *primary alcohol*. The reaction of a Grignard reagent with an *ester* leads to a *tertiary alcohol*. (If a formate ester is used, a secondary alcohol is the product.)

1° Alcohols from ethylene oxide (Section 8.5A):

$$\overset{\overset{\displaystyle O}{\diagup \ \diagdown}}{\text{CH}_2-\text{CH}_2} \xrightarrow[\text{(2) } H_2O,\ H^+]{\text{(1) } C_6H_5MgBr} \text{C}_6\text{H}_5-\text{CH}_2\text{CH}_2\text{OH}$$

ethylene oxide   2-phenylethanol

2° Alcohols from formate esters (Section 15.5C):

$$\overset{\overset{\displaystyle O}{\|}}{\text{H}\text{C}\text{O}\text{C}\text{H}_3} \xrightarrow[\text{(2) } H_2O,\ H^+]{\text{(1) 2 } CH_3CH_2MgBr} \overset{\overset{\displaystyle \text{OH}}{|}}{\underset{\underset{\displaystyle \text{CH}_2\text{CH}_3}{|}}{\text{H}\text{C}}}-\text{CH}_2\text{CH}_3}$$

methyl formate   *from RMgX*

3-pentanol

3° Alcohols from other esters (Section 15.5C):

$$\overset{\overset{\displaystyle O}{\|}}{\text{CH}_3\text{C}\text{O}\text{C}\text{H}_2\text{CH}_3} \xrightarrow[\text{(2) } H_2O,\ H^+]{\text{(1) 2 } CH_3CH_2MgBr} \overset{\overset{\displaystyle \text{OH}}{|}}{\underset{\underset{\displaystyle \text{CH}_2\text{CH}_3}{|}}{\text{CH}_3\text{C}}}-\text{CH}_2\text{CH}_3}$$

ethyl acetate   *from RMgX*

3-methyl-3-pentanol

Lithium reagents. Lithium reagents (RLi) are closely related to Grignard reagents. These reagents are prepared similarly to Grignard reagents, by the reaction of lithium metal with an organic halide. They undergo the same types of reactions as Grignard reagents do. Because Li is more electropositive than Mg, lithium reagents are more reactive than Grignard reagents. Therefore, lithium reagents can be used for the synthesis of even very hindered alcohols. Also, lithium reactions can be run at very low temperatures.

General:

$$RX + 2\,Li \longrightarrow RLi + LiX$$

$$\underset{\text{R}'-\overset{\displaystyle O}{\overset{\|}{C}}-\text{R}''}{}\ \xrightarrow[\text{(2) H}_2\text{O, H}^+]{\text{(1) RLi}}\ \underset{\text{R}'-\overset{\displaystyle OH}{\underset{\displaystyle |}{\overset{|}{C}}}-\text{R}''}{\underset{\text{R}}{}}$$

Specific example:

$$(CH_3)_3CCC(CH_3)_3 + (CH_3)_3C-Li \xrightarrow{-70°} (CH_3)_3C\overset{O^-\ {}^+Li}{\underset{C(CH_3)_3}{\overset{|}{C}}}C(CH_3)_3 \xrightarrow{H_2O,\ H^+} (CH_3)_3C\overset{OH}{\underset{C(CH_3)_3}{\overset{|}{C}}}C(CH_3)_3$$

$$(81\%)$$

Sample Problems

A chemist   (a) treats iodobenzene with magnesium metal and diethyl ether;   (b) adds acetone; and finally,   (c) adds a dilute solution of HCl. Write an equation to represent each reaction.

Solution:

(a) $C_6H_5I + Mg \xrightarrow{\text{diethyl ether}} C_6H_5MgI$

(b) $CH_3\overset{O}{\overset{\|}{C}}CH_3 + C_6H_5MgI \longrightarrow CH_3\overset{OMgI}{\underset{C_6H_5}{\overset{|}{C}}}CH_3$

(c) $(CH_3)_2\overset{OMgI}{\overset{|}{C}}C_6H_5 + H^+ \longrightarrow (CH_3)_2\overset{OH}{\overset{|}{C}}C_6H_5 + Mg^{2+} + I^-$

A student maintains that a Grignard reagent undergoes reaction with water to yield an alcohol. What is wrong with this statement?

Solution: The carbon of the Grignard reagent is partially *negative* and undergoes reaction with a *positive group* (such as H^+), not with a negative group (such as ^-OH).

$$\underset{\delta-}{R}MgX + \underset{\delta+}{H}-OH \begin{array}{l} \nearrow\ R-H + XMgOH \\ \searrow\!\!\!\!\times\ R-OH + HMgX \end{array}$$

does not occur

Table 7.4 summarizes the preparations of alcohols.

Table 7.4 Summary of laboratory syntheses of alcohols[a]

Reaction	Section Reference
Primary Alcohols:	
$RCH_2X + {}^-OH \xrightarrow{S_N2} RCH_2OH$	5.5
$\overset{\overset{\displaystyle O}{\|}}{HCH} \xrightarrow[\text{(2) } H_2O, H^+]{\text{(1) } RMgX} RCH_2OH$	7.4D, 13.5D
$\overset{\overset{\displaystyle O}{\diagup \diagdown}}{CH_2CH_2} \xrightarrow[\text{(2) } H_2O, H^+]{\text{(1) } RMgX} RCH_2CH_2OH$	8.5A
$R_2C{=}CH_2 \xrightarrow[\text{(2) } H_2O_2, OH^-]{\text{(1) } BH_3} R_2CHCH_2OH$	10.10
$\overset{\overset{\displaystyle O}{\|}}{RCH} \xrightarrow{[H]^b} RCH_2OH$	13.7
$RCO_2R' \xrightarrow{[H]^b} RCH_2OH + R'OH$	15.5C
$RCO_2H \xrightarrow[\text{(2) } H_2O, H^+]{\text{(1) } LiAlH_4} RCH_2OH$	14.8
Secondary Alcohols:	
$\overset{\overset{\displaystyle O}{\|}}{RCH} \xrightarrow[\text{(2) } H_2O, H^+]{\text{(1) } R'MgX} \overset{\overset{\displaystyle OH}{\|}}{RCHR'}$	7.4D, 13.5D
$\overset{\overset{\displaystyle O}{\|}}{HCOR} \xrightarrow[\text{(2) } H_2O, H^+]{\text{(1) } 2\,R'MgX} \overset{\overset{\displaystyle OH}{\|}}{R'CHR'}$	15.5C
$\overset{\overset{\displaystyle O}{\|}}{RCR'} \xrightarrow{[H]^b} \overset{\overset{\displaystyle OH}{\|}}{RCHR'}$	7.4D, 13.5D
$RCH{=}CHR^c \xrightarrow{H_2O, H^+} \overset{\overset{\displaystyle OH}{\|}}{RCHCH_2R}$	10.8
Tertiary Alcohols:	
$\overset{\overset{\displaystyle O}{\|}}{RCR'} \xrightarrow[\text{(2) } H_2O, H^+]{\text{(1) } R''MgX} \overset{\overset{\displaystyle OH}{\|}}{\underset{\underset{\displaystyle R''}{\|}}{RCR'}}$	7.4D, 13.5D
$\overset{\overset{\displaystyle O}{\|}}{RCCl} \text{ or } \overset{\overset{\displaystyle O}{\|}}{RCOR'} \xrightarrow[\text{(2) } H_2O, H^+]{\text{(1) } R''MgX} \overset{\overset{\displaystyle OH}{\|}}{RCR''_2}$	15.3C, 15.5C

[a] The synthesis of 1,2-diols is discussed in Section 8.3C.
[b] The symbol [H] means a reducing agent such as H_2 + catalyst or a metal hydride (see page 283).
[c] Other preparations of 2° alcohols from alkenes include oxymercuration–demercuration (Section 10.9) and hydroboration–oxidation (Section 10.10).

Study Problems

7.6 Which of the following compounds could *not* be used to prepare a Grignard reagent? Explain.

(a) $CH_3\overset{\overset{\displaystyle NH_2}{|}}{C}HCH_2CH_2Br$ (b) ⟨ ⟩—Br   (c) [naphthalene]—Br

(d) $BrCH_2\overset{\overset{\displaystyle O}{||}}{C}OH$ (e) $CH_3C{\equiv}CCH_2CH_2I$ (f) $HOCH_2CH_2CH_2Br$

7.7 Suggest reagents for the following conversion:

⟨ ⟩—Br $\longrightarrow$ ⟨ ⟩—D

7.8 Suggest a Grignard synthesis for 1-cyclohexylethanol starting with bromocyclohexane and an aldehyde.

7.9 Write equations to show how the following conversions could be made:

(a) ⟨◯⟩—Br $\longrightarrow$ ⟨◯⟩—CH_2OH

(b) $CH_3CH_2CH_2CH_2Cl \longrightarrow CH_3CH_2CH_2CH_2OH$

(c) $CH_3CH_2Br \longrightarrow$ ⟨◯⟩—$\overset{\overset{\displaystyle \,}{|}}{\underset{\underset{\displaystyle OH}{|}}{C}}HCH_2CH_3$

▼ ## Section 7.5
Substitution Reactions of Alcohols

In acidic solution, alcohols can undergo substitution reactions.

$$CH_3CH_2CH_2CH_2{-}OH + H{-}Br \xrightarrow[\text{heat}]{H_2SO_4} CH_3CH_2CH_2CH_2{-}Br + H_2O$$

1-butanol   1-bromobutane (95%)

$$CH_3CH_2\overset{\overset{\displaystyle CH_3}{|}}{C}H{-}OH + H{-}Cl \xrightarrow{ZnCl_2} CH_3CH_2\overset{\overset{\displaystyle CH_3}{|}}{C}H{-}Cl + H_2O$$

2-butanol   2-chlorobutane (66%)

$$(CH_3)_3C{-}OH + H{-}Cl \longrightarrow (CH_3)_3C{-}Cl + H_2O$$

t-butyl alcohol   *t*-butyl chloride (88%)

Unlike alkyl halides, alcohols do not undergo substitution in neutral or alkaline solution. Why not? The reason is that, in general, a leaving group must be a fairly weak base. In Chapter 5, we saw that Cl^-, Br^-, and I^- are good leaving groups and are readily displaced from alkyl halides. These ions are very weak bases. But ^-OH, which would be the leaving group of an alcohol in neutral or alkaline solution, is a *strong base* and thus is a very poor leaving group.

good leaving group

$$CH_3CH_2{-}Br + {}^-OH \longrightarrow CH_3CH_2OH + Br^-$$

poor leaving group

$$CH_3CH_2{-}OH + Br^- \longrightarrow \text{no reaction}$$

In acidic solution, alcohols are protonated. Although —OH is a poor leaving group, $-OH_2{}^+$ is a good leaving group because it is lost as water, a very weak base. A weak nucleophile such as a halide ion can displace the water molecule to yield an alkyl halide, as shown in the equations at the start of this section.

a good leaving group

$$R{-}\overset{..}{\underset{H}{\overset{..}{O}}}{:} \;\overset{+H^+}{\rightleftharpoons}\; R{-}\overset{+}{\underset{H}{\overset{..}{O}H}} \;\overset{:\overset{..}{\underset{..}{X}}{:}^-}{\longrightarrow}\; R{-}\overset{..}{\underset{..}{X}}{:} \;+\; {:}\overset{..}{\underset{H}{O}H}$$

an alcohol   *an oxonium ion*   *an alkyl halide*

A. Reactivity of the Hydrogen Halides

In alcohol substitution reactions, the reactivity of the hydrogen halides is as follows:

	HF	HCl	HBr	HI
pK_a:	3.45	-7	-9	-9.5

increasing acid strength and increasing nucleophilicity of anion; increasing reactivity toward ROH ➤

Although HI, HBr, and HCl are all considered strong acids (almost completely ionized in water), HI is the strongest acid of the group. HF is a weak acid (Section 1.10C). Also, I^- is the strongest nucleophile of the series and F^- is the weakest (see Section 5.4D). The order of reactivity of hydrogen halides toward alcohols parallels the relative acid strengths and the relative nucleophilicities.

increasing rate of reaction ⬆

$$ROH + HI \longrightarrow RI + H_2O$$

$$ROH + HBr \longrightarrow RBr + H_2O$$

$$ROH + HCl \longrightarrow RCl + H_2O$$

B. Reactivity of Alcohols Toward Hydrogen Halides

The order of reactivity of alcohols toward the hydrogen halides follows:

methyl $1°$ $2°$ $3°$ benzylic and allylic

increasing reactivity of ROH toward HX ➤

All alcohols undergo reaction readily with concentrated aqueous HBr or HI to yield alkyl bromides and iodides. Tertiary alcohols, benzylic alcohols, and allylic alcohols also undergo reaction with concentrated aqueous HCl. However, primary and secondary alcohols are less reactive and require the help of anhydrous $ZnCl_2$ or a similar catalyst before they can undergo reaction with the less reactive HCl in a reasonable period of time.

↑
increasing reactivity toward HX

$$3°:\ (CH_3)_3COH\ +\ HCl\ \xrightarrow{25°}\ (CH_3)_3CCl\ +\ H_2O$$

$$2°:\ (CH_3)_2CHOH\ +\ HCl\ \xrightarrow{ZnCl_2}\ (CH_3)_2CHCl\ +\ H_2O$$

$$1°:\ CH_3CH_2OH\ +\ HCl\ \xrightarrow[heat]{ZnCl_2}\ CH_3CH_2Cl\ +\ H_2O$$

The function of the zinc chloride is similar to that of H^+. Anhydrous zinc chloride is a powerful Lewis acid with empty orbitals that can accept electrons from the oxygen. The formation of a complex of $ZnCl_2$ with the alcohol oxygen weakens the C—O bond and thus enhances the leaving ability of the oxygen group.

C. S_N1 or S_N2?

It has been observed that secondary alcohols and tertiary alcohols sometimes undergo rearrangement when treated with HX. Most primary alcohols do not. The conclusion is that secondary and tertiary alcohols undergo reaction with hydrogen halides by the S_N1 path (through a carbocation), while primary alcohols undergo reaction by the S_N2 path (back-side displacement).

Methyl and primary alcohols, S_N2:

$$CH_3CH_2\overset{..}{\underset{..}{O}}H\ \underset{}{\overset{H^+}{\rightleftharpoons}}\ CH_3CH_2{-}\overset{+}{\underset{..}{O}}H_2\ \xrightarrow{X^-}$$

protonated

$$\left[\overset{\delta-}{:\overset{..}{\underset{..}{X}}}{-}\underset{\underset{CH_3}{|}}{CH_2}{-}\overset{\delta+}{\overset{..}{O}H_2} \right] \longrightarrow :\overset{..}{\underset{..}{X}}{-}\underset{\underset{CH_3}{|}}{CH_2}\ +\ H_2\overset{..}{\underset{..}{O}}:$$

S_N2 transition state

Other alcohols, S_N1:

$$(CH_3)_2CH\overset{..}{\underset{..}{O}}H\ \overset{H^+}{\rightleftharpoons}\ (CH_3)_2CH{\overset{+}{\underset{..}{O}}}H_2\ \underset{}{\overset{-H_2\overset{..}{O}:}{\rightleftharpoons}}\ [(CH_3)_2\overset{+}{C}H]\ \xrightarrow{X^-}\ (CH_3)_2CHX$$

protonated   *carbocation intermediate*

Sample Problem

The only alkyl halide formed in the reaction of 3-methyl-2-butanol with HBr is a product of rearrangement. Give the structure of the product and show how it is formed.

Solution:

$$CH_3\underset{\underset{H}{|}}{\overset{\overset{CH_3}{|}}{C}}\!-\!\underset{\underset{OH}{|}}{CH}CH_3 \xrightarrow[-H_2O]{H^+} \left[CH_3\underset{\underset{H}{|}}{\overset{\overset{CH_3}{|}}{C}}\!-\!\overset{+}{C}HCH_3 \right] \longrightarrow$$

a 2° carbocation

$$\left[CH_3\underset{+}{\overset{\overset{CH_3}{|}}{C}}\!-\!CH_2CH_3 \right] \xrightarrow{Br^-} CH_3\underset{\underset{Br}{|}}{\overset{\overset{CH_3}{|}}{C}}\!-\!CH_2CH_3$$

a 3° carbocation   2-bromo-2-methylbutane

Section 7.6
Other Reagents Used to Convert Alcohols to Alkyl Halides

Other halogenating reagents besides hydrogen halides can convert alcohols to alkyl halides. Two common halogenating reagents are thionyl chloride (SOCl$_2$) and phosphorus tribromide (PBr$_3$). Both these reagents undergo reaction with alcohols to form intermediate *inorganic esters*, discussed in Section 7.8. The resulting inorganic ester groups are good leaving groups that can be displaced by halide ions. These reactions do not proceed by way of carbocations and thus no rearrangement occurs.

The following reactions are excellent methods for the preparation of alkyl halides from primary and secondary alcohols. (Tertiary alcohols generally undergo elimination reactions under these conditions.)

$$ROH + \underset{\substack{\text{thionyl}\\\text{chloride}}}{\overset{\overset{O}{\|}}{Cl}SCl} \longrightarrow \left[R\!-\!\overset{\overset{O}{\|}}{O}SCl + HCl \right] \longrightarrow RCl + SO_2 + HCl$$

good leaving groups

$$ROH + \underset{\substack{\text{phosphorus}\\\text{tribromide}}}{PBr_3} \longrightarrow [R\!-\!OPBr_2 + HBr] \longrightarrow RBr + HOPBr_2$$

The reaction of SOCl$_2$ with an alcohol is often carried out in the presence of pyridine (page 73) or a tertiary amine (R$_3$N:), which removes H$^+$ from the reaction mixture. In the presence of these compounds, inversion of configuration is observed if the starting alcohol is chiral. The following mechanism explains why inversion occurs.

Step 1:

(S)-2-butanol

R₃N:⟋   (S)-2-butyl chlorosulfite

an inorganic ester

Step 2 in an amine solvent, an S_N2 reaction:

(R)-2-chlorobutane

Phosphorus tribromide reacts with alcohols by a similar path. However, each PBr₃ molecule can brominate three ROH molecules.

$$3 \text{ CH}_3\text{CH}_2\overset{\overset{\text{OH}}{|}}{\text{CH}}\text{CH}_3 + \text{PBr}_3 \longrightarrow 3 \text{ CH}_3\text{CH}_2\overset{\overset{\text{Br}}{|}}{\text{CH}}\text{CH}_3 + \text{H}_3\text{PO}_3$$

2-butanol   2-bromobutane (86%)   phosphorous acid

Study Problems

7.10 Write equations for the following conversions:

(a)

(b)

7.11 Would you expect deuterium and chlorine to be *cis* or *trans* in the product of the following reaction?

Figure 7.2 summarizes the reactions for converting alcohols to alkyl halides by substitution reactions.

Figure 7.2 Summary of substitution reactions of alcohols leading to alkyl halides.

Section 7.7
Elimination Reactions of Alcohols

Alcohols, like alkyl halides, undergo elimination reactions to yield alkenes. Because water is lost in the elimination, this reaction is called a **dehydration reaction**.

3°: $(CH_3)_3COH$ $\xrightarrow[60°]{\text{conc. } H_2SO_4}$ $(CH_3)_2C=CH_2 + H_2O$
t-butyl alcohol   methylpropene
                                     (isobutylene)

2°: $(CH_3)_2CHOH$ $\xrightarrow[100°]{\text{conc. } H_2SO_4}$ $CH_3CH=CH_2 + H_2O$
2-propanol   propene
                                     (propylene)

1°: CH_3CH_2OH $\xrightarrow[180°]{\text{conc. } H_2SO_4}$ $CH_2=CH_2 + H_2O$
ethanol   ethene
                                     (ethylene)

increasing ease of dehydration

Although sulfuric acid is often the acid of choice for a dehydration catalyst, any strong acid can cause dehydration of an alcohol. Note the comparative ease with which a tertiary alcohol undergoes elimination. Simply warming it with concentrated H_2SO_4 leads to the alkene. Elimination is a prevalent side reaction in substitution reactions of tertiary alcohols with HX.

For secondary and tertiary alcohols, dehydration follows an **E1 path**. The hydroxyl group is protonated, a carbocation is formed with loss of a water molecule, and then a proton is eliminated to yield the alkene. (Primary alcohols probably undergo dehydration by an E2 path.) A secondary or tertiary alcohol yields little, if any, substitution product in hot sulfuric acid. However, primary alcohols may yield ethers or sulfates. These competing reactions are discussed in Section 8.3.

Let us consider the dehydration of 2-pentanol, a secondary alcohol that undergoes a typical E1 reaction.

Step 1 (protonation and loss of water):

$$CH_3CH_2CH_2\overset{:\ddot{O}H}{\underset{|}{C}}HCH_3 \xrightleftharpoons{H^+} CH_3CH_2CH_2\overset{+\ddot{O}H_2}{\underset{|}{C}}HCH_3 \xrightleftharpoons{-H_2\ddot{O}:} [CH_3CH_2CH_2\overset{+}{C}HCH_3]$$

protonated   a carbocation

Step 2 (loss of H$^+$):

$$CH_3CH_2\overset{H}{\underset{|}{C}}H\overset{+}{-}\overset{|}{C}HCH_3 \rightleftharpoons \left[CH_3CH_2\overset{H^{\delta+}}{\underset{|}{C}}H\overset{\delta+}{=\!=}\overset{|}{C}HCH_3 \right] \xrightleftharpoons{-H^+} CH_3CH_2CH=CHCH_3$$

transition state   2-pentene

In the second step, the carbocation loses H$^+$ (to H$_2$O, HSO$_4^-$, or another molecule of alcohol). In this second step, the double bond is partially formed in the transition state. For this reason, when more than one alkene can be formed, a typical E1 reaction yields predominantly the *more substituted, more stable alkene* (Saytzeff rule). Recall from Chapter 5 that E2 reactions also usually yield the more stable alkene.

more stable alkene

$$CH_3CH_2CH_2\overset{OH}{\underset{|}{C}}HCH_3 \xrightarrow[\substack{\text{heat} \\ E1}]{H_2SO_4} CH_3CH_2CH=CHCH_3 + CH_3CH_2CH_2CH=CH_2$$

2-pentanol   2-pentene (80%)   1-pentene (5%)

Study Problem

7.12 Would you expect to find *cis*- or *trans*-2-pentene as the predominant product in the dehydration of 2-pentanol? Explain.

In any elimination reaction in which the double bond can be in *conjugation with a benzene ring*, the conjugated product is formed in preference to the nonconjugated product. The conjugated alkene is of lower energy, as is the transition state leading to its formation. In fact, it is often difficult to isolate alcohols in which the hydroxyl group is one or two carbons away from a benzene ring. Dehydration is usually spontaneous under acidic conditions.

$$\bigcirc\!\!-CH_2\overset{OH}{\underset{|}{C}}HCH_2CH_3 \xrightarrow[-H_2O]{H^+}$$

1-phenyl-2-butanol

$$\bigcirc\!\!-CH=CHCH_2CH_3 \quad \text{and not} \quad \bigcirc\!\!-CH_2CH=CHCH_3$$

1-phenyl-1-butene   1-phenyl-2-butene

Recall from Section 5.6F that carbocations can undergo rearrangements. Are rearrangements observed in the E1 reactions of alcohols? Yes, they are. If a carbocation can undergo a 1,2-shift to a more stable carbocation, rearrangement products will be observed.

Rearrangements:

$$
\begin{array}{ccc}
\underset{\substack{\text{3,3-dimethyl-2-butanol}}}{\overset{\displaystyle CH_3\ \ OH}{CH_3-\overset{\overset{\displaystyle |}{\underset{\displaystyle |}{C}}}{\underset{\displaystyle CH_3}{}}-CHCH_3}}
& \xrightarrow[95°]{H_2SO_4} &
\underset{\substack{\text{2,3-dimethyl-2-butene}}}{\overset{\displaystyle CH_3 \quad CH_3}{\underset{\displaystyle CH_3 \quad CH_3}{C=C}}}
\end{array}
$$

$$
\begin{array}{ccc}
\underset{\substack{\text{2,3-dimethyl-1-butanol}}}{\overset{\displaystyle CH_3\ \ H}{CH_3CH-\overset{\overset{\displaystyle |}{\underset{\displaystyle |}{C}}}{\underset{\displaystyle CH_3}{}}-CH_2OH}}
& \xrightarrow[140°]{H_2SO_4} &
\underset{\substack{\text{2,3-dimethyl-2-butene}}}{\overset{\displaystyle CH_3 \quad CH_3}{\underset{\displaystyle CH_3 \quad CH_3}{C=C}}}
\end{array}
$$

Study Problem

7.13 Write an equation showing the mechanism of the conversion of 3,3-dimethyl-2-butanol to 2,3-dimethyl-2-butene.

Because rearrangements can occur in the dehydration of alcohols and because the dehydrations of primary alcohols are slow, the dehydration of an alcohol is not always the best method for preparing an alkene. In many cases, it is preferable to convert the alcohol to the alkyl halide and then subject the alkyl halide to an E2 reaction.

Figure 7.3 summarizes the dehydration reactions of alcohols.

$$1°: \quad R_2CH-CH_2OH \quad \xrightarrow[\substack{\text{E2}}]{\substack{\text{conc. } H_2SO_4, \text{ heat} \\ -H_2O}} \quad R_2C=CH_2$$

$$2°: \quad R_2CH-\overset{\displaystyle OH}{\overset{\displaystyle |}{CHR}} \quad \xrightarrow[\substack{\text{E1}}]{\substack{\text{conc. } H_2SO_4, \text{ heat} \\ -H_2O}} \quad R_2C=CHR$$

most stable alkene predominates;
rearrangement possible

$$3°: \quad R_2CH-\overset{\displaystyle OH}{\overset{\displaystyle |}{CR_2}} \quad \xrightarrow[\substack{\text{E1}}]{\substack{\text{conc. } H_2SO_4 \\ -H_2O}} \quad R_2C=CR_2$$

most stable alkene predominates;
rearrangement possible

Figure 7.3 Summary of dehydration reactions of alcohols.

Sample Problem

Predict the major products of dehydration of   (a) 3-pentanol and   (b) neopentyl alcohol (2,2-dimethyl-1-propanol). (*Hint*: Neopentyl alcohol undergoes rearrangement when treated with H_2SO_4.)

Solution:

$$\text{(a)} \quad \underset{\underset{\text{OH}}{|}}{CH_3CH_2CHCH_2CH_3} \quad \xrightarrow{-H_2O} \quad CH_3CH_2CH{=}CHCH_3$$

trans-2-pentene

more stable alkene

$$\text{(b)} \quad \underset{\underset{CH_3}{|}}{(CH_3)_2C{-}CH_2}{-}\overset{..}{\underset{..}{O}}H_2 \quad \xrightarrow{-H_2\overset{..}{O}:} \quad \left[\underset{\underset{CH_3}{|}}{(CH_3)_2\overset{+}{C}{-}CH_2} \right] \quad \xrightarrow{-H^+} \quad (CH_3)_2C{=}CHCH_3$$

rearranged

2-methyl-2-butene

more stable alkene

Study Problem

7.14 Propose a mechanism for the following conversion:

Section 7.8
Esters of Alcohols

A. Carboxylic Esters

Alcohols undergo reaction with carboxylic acids and carboxylic acid derivatives to yield **esters of carboxylic acids**. These reactions, called *esterification reactions*, their mechanisms, and the product esters will be covered in detail in Chapters 14 and 15.

$$\underset{\substack{\text{acetic acid} \\ \textit{a carboxylic acid}}}{\overset{\overset{\displaystyle O}{\|}}{CH_3C\,OH}} \quad + \quad \underset{\text{ethanol}}{H\,OCH_2CH_3} \quad \underset{}{\overset{H^+,\,\text{heat}}{\rightleftharpoons}} \quad \underset{\substack{\text{ethyl acetate} \\ \textit{an ester}}}{\overset{\overset{\displaystyle O}{\|}}{CH_3COCH_2CH_3}} + H_2O$$

$$\underset{\substack{\text{benzoic acid} \\ \textit{a carboxylic acid}}}{\overset{\overset{\displaystyle O}{\|}}{\text{⬡}{-}C\,OH}} \quad + \quad \underset{\text{1-propanol}}{H\,OCH_2CH_2CH_3} \quad \underset{}{\overset{H^+,\,\text{heat}}{\rightleftharpoons}} \quad \underset{\substack{\textit{n}\text{-propyl benzoate} \\ \textit{an ester}}}{\overset{\overset{\displaystyle O}{\|}}{\text{⬡}{-}COCH_2CH_2CH_3}} + H_2O$$

B. Inorganic Esters

Inorganic esters of alcohols are compounds **prepared** by the reaction of alcohols and either mineral acids (such as HNO_3 or H_2SO_4) or acid halides of mineral acids (such as $SOCl_2$).

Nitrate esters. Nitrate esters, $RONO_2$ (for example, nitroglycerin and PETN), are explosives. When detonated, these compounds undergo fast intramolecular oxidation–reduction reactions to yield large volumes of gases (N_2, CO_2, H_2O, O_2). Organic nitrates and nitrites, RONO, are also used as *vasodilators* (substances that dilate blood vessels) in the treatment of certain types of heart disease.

$$\begin{array}{ccc}
CH_2ONO_2 & & CH_2ONO_2 \\
| & & | \\
CHONO_2 & & O_2NOCH_2CCH_2ONO_2 \\
| & & | \\
CH_2ONO_2 & & CH_2ONO_2 \\
\text{glyceryl trinitrate} & & \text{pentaerythritol tetranitrate} \\
\text{(nitroglycerin)} & & \text{(PETN)}
\end{array}$$

Phosphate esters. Phosphate esters of phosphoric acid and its anhydrides are extremely important in biochemistry. In living systems, diphosphate groups are common leaving groups in substitution reactions.

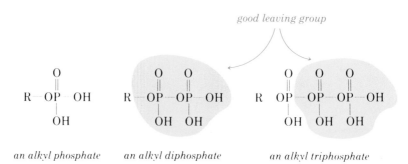

an alkyl phosphate   *an alkyl diphosphate*   *an alkyl triphosphate*

Sulfates. The reaction of concentrated sulfuric acid with alcohols can lead to monoalkyl or dialkyl sulfate esters. The monoesters are named as **alkyl hydrogen sulfates, alkylsulfuric acids**, or **alkyl bisulfates**. The three terms are synonymous. Alkyl hydrogen sulfates are strong acids, but dialkyl sulfates are not acidic.

$$\begin{array}{ccc}
O & O & O \\
\| \quad \text{—} acidic & \| & \| \\
CH_3OSOH & CH_3OSOCH_3 & CH_3OSOCH_2CH_3 \\
\| & \| & \| \\
O & O & O \\
\text{methyl hydrogen} & \text{dimethyl sulfate} & \text{methyl ethyl sulfate} \\
\text{sulfate}
\end{array}$$

Sulfonates. Sulfonates are inorganic esters with the general formula RSO_2OR. (Do not confuse the sulfonate structure with the sulfate structure. A sulfonate has an alkyl or aryl group attached *directly to the sulfur atom*.)

benzenesulfonic acid   methyl benzenesulfonate

a sulfonic acid   *a sulfonate*
(a strong acid)

We will concentrate our attention on only one class of sulfonates—the
p-toluenesulfonates (4-methylbenzenesulfonates), usually called **tosylates** and
abbreviated ROTs. The tosylates are prepared by the reaction of an alcohol
with *p*-toluenesulfonyl chloride (tosyl chloride). An amine, such as pyridine
(page 73), is often added to the reaction mixture to "trap" the HCl as it is
formed ($R_3N: + HCl \rightarrow R_3NH^+ Cl^-$).

an alcohol

tosyl chloride   *an alkyl tosylate*
(TsCl)   (ROTs)

In the formation of the tosylate, the carbon–oxygen bond of the alcohol is
not broken. If a tosylate is prepared using a single enantiomer of a chiral
alcohol, the tosylate retains the configuration of the starting alcohol.

(*S*)-2-butanol

(*S*)-2-butyl tosylate

Study Problem

7.15 How would you prepare   **(a)** *n*-propyl tosylate and   **(b)** (*R*)-2-hexyl tosylate?

The tosylate anion (as well as other sulfonate anions) is resonance-
stabilized and is a very weak base.

For this reason, the tosylate group is a far better leaving group than an OH group. The tosylate group can be displaced in S$_N$2 reactions by such weak nucleophiles as halide ions or alcohols. (No acidic catalyst is necessary.)

$$:\ddot{C}l:^- + \underset{\underset{CH_3}{|}}{CH_2}-\ddot{O}Ts \xrightarrow{S_N2} :\ddot{C}l-\underset{\underset{CH_3}{|}}{CH_2} + \quad ^-:\ddot{O}Ts$$

tosylate ion

$$CH_3\ddot{O}H + \underset{\underset{CH_3}{|}}{CH_2}-\ddot{O}Ts \xrightarrow{S_N2} CH_3\overset{+}{\underset{\underset{H}{|}}{\ddot{O}}}-\underset{\underset{CH_3}{|}}{CH_2} + \quad :\ddot{O}Ts$$

a protonated ether

$$\xrightarrow[-H^+]{} CH_3\ddot{O}CH_2CH_3$$

Figure 7.4 summarizes the types of esters of alcohols.

Study Problems

7.16 Write resonance structures for the methanesulfonate ion.

7.17 Predict the S$_N$2 product of the reaction of water with (*R*)-2-octyl tosylate.

Figure 7.4 Summary of the types of esters of alcohols. (Esters of carboxylic acids are discussed in more detail in Sections 14.7 and 15.5.) Tertiary alcohols are more likely to yield alkenes in these reactions. Secondary alcohols yield alkenes when treated with concentrated H$_2$SO$_4$.

Section 7.9
Oxidation of Alcohols

Oxidation is defined as the *loss of electrons* by an atom, while **reduction** is the *gain of electrons* by an atom.

Oxidation:

$$Na^0 \xrightarrow{-e^-} Na^+$$

$$Mg^0 \xrightarrow{-2e^-} Mg^{2+}$$

Reduction:

$$Fe^{3+} \xrightarrow{+e^-} Fe^{2+}$$

$$Cu^{2+} \xrightarrow{+2e^-} Cu^0$$

In organic reactions, it is not always easy to determine whether a carbon atom "gains" or "loses" electrons. However, oxidation and reduction of organic compounds are common reactions. Good rules of thumb to determine if an organic compound has been oxidized or reduced follow:

If a molecule gains oxygen or loses hydrogen, it is oxidized:

The symbol [O] represents an oxidizing agent.

$$CH_3CH_2OH \xrightarrow{[O]} CH_3CO_2H$$

$$\underset{\displaystyle CH_3\overset{\displaystyle OH}{\underset{|}{CH}}CH_3}{} \xrightarrow{[O]} CH_3\overset{\displaystyle O}{\overset{\|}{C}}CH_3$$

If a molecule loses oxygen or gains hydrogen, it is reduced:

The symbol [H] represents a reducing agent.

$$CH_3CO_2H \xrightarrow{[H]} CH_3CH_2OH$$

$$CH_3\overset{\displaystyle O}{\overset{\|}{C}}CH_3 \xrightarrow{[H]} CH_3\overset{\displaystyle OH}{\underset{|}{C}HCH_3}$$

We can list a series of compounds according to the increasing oxidation state of carbon:

	$CH_2{=}CH_2$	$CH{\equiv}CH$	$CH_3\overset{O}{\overset{\|}{C}}OH$	CO_2
CH_3CH_3	CH_3CH_2OH	CH_3CHO		
	CH_3CH_2Cl	CH_3CHCl_2	and its derivatives	

increasing oxidation state of C

Note that $CH_2{=}CH_2$ and CH_3CH_2OH are at the same oxidation level. This is not surprising because the difference between the two molecules is only a molecule of water. No oxidation–reduction reaction takes place in the interconversion of ethylene and ethanol.

$$CH_2{=}CH_2 \underset{-H_2O}{\overset{H_2O}{\rightleftarrows}} CH_3CH_2OH$$

Study Problems

7.18 List the following compounds in order of increasing oxidation state.

(a) $HO_2C(CH_2)_4CO_2H$ **(b)** ⟨⟩—OH   **(c)** ⟨⟩=O

7.19 List the following carboxylic acids in order of increasing oxidation state:

$$\text{(a)} \quad \underset{\text{hydroxyacetic acid}}{HOCH_2\overset{\displaystyle O}{\overset{\|}{C}}OH} \qquad \text{(b)} \quad \underset{\text{acetic acid}}{CH_3\overset{\displaystyle O}{\overset{\|}{C}}OH} \qquad \text{(c)} \quad \underset{\text{oxalic acid}}{HO\overset{\displaystyle O}{\overset{\|}{C}}-\overset{\displaystyle O}{\overset{\|}{C}}OH}$$

Alcohols can be oxidized to ketones, aldehydes, or carboxylic acids. These oxidations are widely used in the laboratory and in industry, and they also occur in biological systems.

$$RCH_2OH \xrightarrow{[O]} \begin{array}{l} \overset{\displaystyle O}{\overset{\|}{RCH}} \quad \textit{an aldehyde} \\[2mm] \xrightarrow{[O]} RCO_2H \quad \textit{a carboxylic acid} \end{array}$$

$$\underset{\text{a } 1° \text{ alcohol}}{RCH_2OH}$$

$$\underset{\textit{a 2° alcohol}}{\overset{OH}{\overset{|}{RCHR}}} \xrightarrow{[O]} \underset{\textit{a ketone}}{\overset{\displaystyle O}{\overset{\|}{RCR}}}$$

Direct reduction of an alcohol is not a common laboratory reaction. Instead, indirect methods of reduction are used. For example, dehydration of an alcohol leads to an alkene, which is readily reduced with H_2 to the alkane (Section 10.13).

$$\underset{\text{cyclohexanol}}{\text{⟨⟩—OH}} \xrightarrow[\text{heat}]{H_2SO_4} \underset{\text{cyclohexene}}{\text{⟨⟩}} \xrightarrow[\text{Pt catalyst}]{H_2} \underset{\text{cyclohexane}}{\text{⟨⟩}}$$

A. Combustion of Ethanol

Alcohols, like other organic compounds, can undergo combustion. Ethanol, usually from fermented cornstalks, can be mixed in gasoline as an octane booster. The mixture is popularly called "gasohol."

$$\underset{\text{ethanol}}{CH_3CH_2OH} + 3\,O_2 \longrightarrow 2\,CO_2 + 3\,H_2O + \text{energy}$$

The burning of ethanol has an interesting history. In the days of pirates and sailing ships, the alcoholic content of rum or whiskey was determined by pouring it on a small pile of gunpowder and igniting the vapor. If the flames died down and the gunpowder did not burn, the conclusion was that the rum had been watered down. If the gunpowder did burn, this was proof that the rum had not been diluted. The term "proof" derives from this custom of testing alcoholic beverages. Proof is twice the percentage of alcohol; 100-proof spirits are 50 percent ethanol.

B. Biological Oxidation of Ethanol

In a mammalian system, ingested ethanol is oxidized primarily in the liver with the aid of an enzyme called *alcohol dehydrogenase*. The product of this dehydrogenation is acetaldehyde, CH_3CHO. (The biological oxidation of methanol leads to formaldehyde, $HCHO$, which is toxic.) The acetaldehyde from ethanol is further oxidized enzymatically to the acetate ion, $CH_3CO_2^-$, which undergoes esterification with the thiol **coenzyme A** (often abbreviated HSCoA). The product of the esterification is **acetylcoenzyme A**. (The complete structure of acetylcoenzyme A is shown on page 712.) The acetyl group ($CH_3CO—$) in acetylcoenzyme A can be converted to CO_2, H_2O, and energy, or it can be converted to other compounds, such as fat.

$$CH_3CH_2OH \xrightarrow[\text{dehydrogenase}]{\text{alcohol}} \underset{\overset{\|}{O}}{CH_3CH} \xrightarrow{[O]} \underset{\overset{\|}{O}}{CH_3CO^-} \xrightarrow{HSCoA}$$

$$\underset{\text{acetylcoenzyme A}}{\underset{\overset{\|}{O}}{CH_3C}—SCoA} \begin{array}{l} \longrightarrow CO_2 + H_2O + \text{energy} \\ \longrightarrow \text{fat, etc.} \end{array}$$

C. Laboratory Oxidation of Alcohols

In general, laboratory oxidizing agents oxidize primary alcohols to carboxylic acids and secondary alcohols to ketones.

$$\underset{\text{a 1° alcohol}}{RCH_2OH} \xrightarrow{[O]} \underset{\text{a carboxylic acid}}{\underset{\overset{\|}{O}}{RCOH}}$$

$$\underset{\text{a 2° alcohol}}{\underset{\overset{|}{OH}}{RCHR}} \xrightarrow{[O]} \underset{\text{a ketone}}{\underset{\overset{\|}{O}}{RCR}}$$

Some typical oxidizing agents used for these oxidations are:

1. alkaline potassium permanganate solution: $KMnO_4 + {}^-OH$;
2. hot, concentrated HNO_3;
3. chromic acid: H_2CrO_4 prepared *in situ* from CrO_3 with aqueous H_2SO_4 (Jones' reagent) or $Na_2Cr_2O_7$ with aqueous H_2SO_4;

4. chromium trioxide (CrO_3) complexed with pyridine or with pyridine and HCl.

$$CrO_3 \cdot 2 : N \quad \overbrace{\text{pyridine}}$$

$$CrO_3Cl^-$$

$$\underset{H}{\overset{+}{N}}$$

chromium trioxide–   pyridinium chlorochromate (PCC)
pyridine complex

Primary alcohols are oxidized first to aldehydes. In aqueous solution, aldehydes are more easily oxidized than alcohols. Therefore, the oxidation usually continues until the carboxylic acid (or, in alkaline solution, its anion) is formed (see Section 13.5A).

$$CH_3(CH_2)_8CH_2OH \xrightarrow[H^+]{H_2CrO_4} CH_3(CH_2)_8CO_2H$$

1-decanol   decanoic acid (93%)

$$CH_3(CH_2)_3\overset{\overset{\displaystyle CH_2CH_3}{|}}{C}HCH_2OH \xrightarrow[(2)\ H^+]{(1)\ KMnO_4,\ OH^-} CH_3(CH_2)_3\overset{\overset{\displaystyle CH_2CH_3}{|}}{C}HCO_2H$$

2-ethyl-1-hexanol   2-ethylhexanoic acid (74%)

If the intermediate aldehyde has a low boiling point, it can be distilled from the reaction mixture before it is oxidized to the carboxylic acid. Yields of aldehydes by this method are usually low. Therefore, this technique is of limited synthetic value. Better reagents for oxidizing a primary alcohol to an aldehyde are the chromium trioxide–pyridine complex or pyridinium chlorochromate. These reagents are soluble in nonaqueous solvents, such as CH_2Cl_2, and thus do not oxidize aldehydes further to carboxylic acids.

$$CH_3(CH_2)_5CH_2OH + CrO_3 \cdot 2\ N \quad \text{or}\quad PCC \xrightarrow{CH_2Cl_2} CH_3(CH_2)_5\overset{\overset{\displaystyle O}{\|}}{C}H$$

1-heptanol   heptanal (72–84%)

Secondary alcohols are oxidized to ketones in excellent yields by standard oxidizing agents. (Acidic conditions are usually used because ketones can be oxidized further in alkaline solution.)

$$CH_3(CH_2)_5\overset{\overset{\displaystyle OH}{|}}{C}HCH_3 \xrightarrow[H^+]{H_2CrO_4} CH_3(CH_2)_5\overset{\overset{\displaystyle O}{\|}}{C}CH_3$$

2-octanol   2-octanone (96%)

$$\xrightarrow[H^+]{H_2CrO_4}$$

menthol   menthone (84%)

in mint

Tertiary alcohols are not oxidized under alkaline conditions. If the oxidation is attempted in acidic solution, the tertiary alcohol undergoes dehydration and then the alkene is oxidized. Alkene oxidation will be discussed in Chapter 10.

$$R_3COH \quad \begin{array}{l} \xrightarrow[\text{OH}^-]{\text{[O]}} \text{ no reaction} \\ \\ \xrightarrow[\text{H}^+]{} \text{alkenes} \xrightarrow{\text{[O]}} \text{alkene oxidation products} \end{array}$$

a 3° alcohol

The mechanisms of many oxidation reactions are not completely understood. Because of the large number of oxidation states of Mn and Cr, the mechanisms have the potential for being quite complicated. In some reactions, the oxidizing agent probably forms an inorganic ester with the alcohol and, by appropriate shifts of electrons and protons, the oxidized product results.

$$R_2CHOH + HCrO_4^- \xrightarrow{-H_2O} \left[\begin{array}{c} \overset{H}{\underset{R_2C}{|}} \overset{\ddot{O}:}{\underset{\ddot{O}}{\overset{||}{CrO^-}}} \\ O \end{array} \right] \longrightarrow R_2C=\ddot{O}: + \overset{H\ddot{O}:}{\underset{O}{\overset{|}{\underset{||}{CrO^-}}}}$$

Cr(VI)   *an inorganic ester*   Cr(IV)

Study Problem

7.20 Predict the organic products of H_2CrO_4 oxidation of: **(a)** cyclopentanol, and **(b)** benzyl alcohol. (*Hint*: The benzene ring is not affected by H_2CrO_4.)

Table 7.5 Types of compounds that can be obtained from alcohols

Reaction		Principal Product		Section Reference
Substitution:				
ROH + HX	⟶ RX	alkyl halide		7.5
ROH + PX$_3$ or SOCl$_2$	⟶ RX	alkyl halide		7.6
Elimination:				
OH				
$\overset{|}{R_2CCHR_2}$ + H$_2$SO$_4$	$\xrightarrow{\text{heat}}$ R$_2$C=CR$_2$	alkene		7.7
Alkoxide Formation:[a]				
ROH + Na	⟶ RO$^-$ Na$^+$	alkoxide		7.2
ROH + NaH	⟶ RO$^-$ Na$^+$	alkoxide		7.2
Esterification:				
ROH + R'CO$_2$H	$\xrightarrow{\text{H}^+}$ R'CO$_2$R	ester		7.8A, 14.7, 15.5
ROH + TsCl	⟶ ROTs	inorganic ester		7.8B
Oxidation:				
RCH$_2$OH + CrO$_3$·2 pyridine	⟶ RCHO	aldehyde		7.9
RCH$_2$OH + [O][b]	⟶ RCO$_2$H	carboxylic acid		7.9
R$_2$CHOH + [O]	⟶ R$_2$C=O	ketone		7.9

[a] RO$^-$ is used with methyl and 1° halides to synthesize ethers.
[b] Typical oxidizing agents used synthetically are KMnO$_4$ + $^-$OH and H$_2$CrO$_4$.

▼ Section 7.10
Use of Alcohols in Synthesis

Alcohols are versatile starting materials for the preparation of alkyl halides, alkenes, carbonyl compounds, and ethers. The types of compounds that can be obtained from alcohols are shown in Figure 7.5 and Table 7.5. From the reactions presented in this chapter, together with those presented previously, many types of compounds can be prepared from a variety of starting materials.

▼ Section 7.11
Synthesis Problems

Organic chemists often synthesize compounds in the laboratory. The syntheses may be simple and straightforward (for example, the preparation of a particular simple alcohol for a rate study), or they may be very involved (for example, the laboratory synthesis of a complex biological molecule). Even if you do not become a laboratory chemist, designing synthetic schemes on paper is a valuable way to learn to think in the language of organic chemistry.

In this text, you will encounter many **synthesis problems**: problems in which you are asked to show by equations how you would prepare a particular

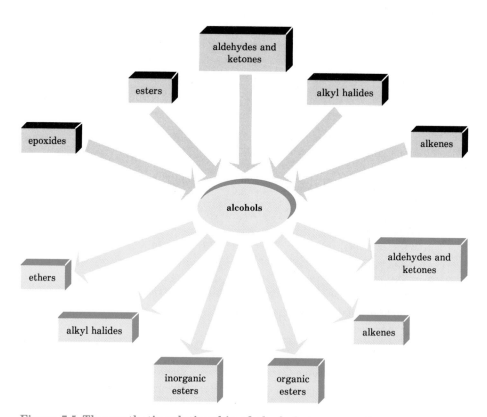

Figure 7.5 The synthetic relationship of alcohols to some other organic compounds.

compound. In some of these problems (but not all), a starting material will be specified. Often, there will be more than one correct solution to a synthesis problem. However, only one correct answer to each synthesis problem is usually given in this text or the accompanying study guide. If your answer is significantly different from the one given, you should verify its correctness with your instructor.

The synthesis problems posed in this text are not intended to be valid in a laboratory sense. The solution to a true laboratory synthesis problem includes a complete search of the chemical literature to see if a particular compound or sequence of reactions has already been studied by other chemists. Then, various possible pathways to the desired compound are drawn up. Each pathway is evaluated from a practical laboratory standpoint (likelihood of success, cost in terms of reagents and time, availability of starting materials, hazards, etc.). Of the various proposed pathways, one is selected. Finally, the synthetic sequence is tested in the laboratory. In solving synthesis problems in this text, you need consider only reactions or reaction sequences that are reasonable, based on the information previously presented in the text.

A. Solving Synthesis Problems

The following suggestions may help you answer synthesis problems correctly.

1. A typical study problem gives the reactants and asks for the product or products:

   $$A + B \longrightarrow ?$$

   In synthesis problems, the reverse question is asked: Given the product, what are the reactants?

   $$? + ? \longrightarrow C$$

   Thus, as you study organic reactions, you must learn them both ways. You must be able to answer such questions as "What reactions yield alcohols?" as well as "The reaction of a Grignard reagent with formaldehyde yields what?"

2. Use only reactions that give the product in reasonable yield. If one reaction gives 100% of the desired product, use that reaction. However, if no such reaction is available, use a reaction that gives 50 or 60% yield. Do not use reactions that give very low yields (under 25%).

3. It is acceptable to use flow equations, with reagents and reaction conditions above or below the arrows. This technique, while not necessary with a one-step synthesis, saves considerable time in writing the solutions to multistep syntheses.

   $$A \xrightarrow{\quad xy \quad} B \xrightarrow[100°]{\quad yz \quad} C$$

4. Unless you are asked to do so, you need not balance equations nor indicate minor products.

5. If the text shows a reaction of a simple compound, you can usually extrapolate the reaction to more-complex, but similar, structures.

Example.

$$CH_3\overset{\overset{\displaystyle O}{\|}}{C}CH_3 \xrightarrow[\text{(2) } H_2O,\, H^+]{\text{(1) } CH_3MgI} (CH_3)_3COH$$

This reaction can be extrapolated to other ketones and Grignard reagents.

$$CH_3CH_2\overset{\overset{\displaystyle O}{\|}}{C}CH_2CH_3 \xrightarrow[\text{(2) } H_2O,\, H^+]{\text{(1)} \langle \rangle -MgI} CH_3CH_2\overset{\overset{\displaystyle OH}{|}}{\underset{|}{C}}CH_2CH_3$$

6. In a synthesis problem, do not be intimidated by the complexity of a structure, but focus your attention on the functional groups. Inspect the structure for its important features and worry only about the small portion that undergoes reaction. When you have a larger vocabulary of organic reactions, you will want to inspect a complex structure for other functional groups that might also undergo reaction under the same conditions.

B. Multistep Synthetic Problems

Most syntheses in the laboratory (and many synthesis problems in this book) require more than a single step from commercially available starting materials to desired products. When you are confronted with a synthesis that will require two or more steps and a reaction sequence is not immediately evident, do *not* choose a likely starting material and try to convert it to the product. Instead, *start with the product and work backwards, one step at a time, to the starting material.* This procedure, called **retrosynthetic analysis**, is best explained with an example.

Example. Show by flow equations how you would prepare 3-deuteriopropene from a nondeuteriated hydrocarbon, standard inorganic reagents, and appropriate solvents.

1. Write the structure of the product.

$$CH_2=CHCH_2D$$

2. Consider not a possible starting hydrocarbon, but a reaction that leads directly to this given product. A Grignard reagent treated with D_2O is a way to introduce deuterium into a structure. Write the equation for this reaction.

$$CH_2=CHCH_2MgBr \xrightarrow{D_2O} CH_2=CHCH_2D$$

3. Next, what reagents are needed to prepare the allyl Grignard reagent? (Again, you are working "backwards.")

$$CH_2=CHCH_2Br \xrightarrow[\text{diethyl ether}]{Mg} CH_2=CHCH_2MgBr$$

4. Finally, what reaction could be used to prepare allyl bromide?

$$CH_2{=}CHCH_3 \xrightarrow{\text{NBS}} CH_2{=}CHCH_2Br$$

5. Because we have worked our way backwards to a nondeuterated hydrocarbon, we have solved the problem. The answer is now written forwards, rather than backwards.

$$CH_2{=}CHCH_3 \xrightarrow{\text{NBS}} CH_2{=}CHCH_2Br \xrightarrow[\text{diethyl ether}]{\text{Mg}}$$

$$CH_2{=}CHCH_2MgBr \xrightarrow{\text{D}_2\text{O}} CH_2{=}CHCH_2D$$

Sample Problem

Suggest a synthesis for 3-methyl-3-hexanol from alcohols of four carbons or fewer.

Solution:

1. Write the formula for 3-methyl-3-hexanol:

$$\underset{\underset{CH_3}{|}}{\overset{\overset{OH}{|}}{CH_3CH_2CH_2CCH_2CH_3}}$$

2. Decide on the reactants. This is a 3° alcohol, so it can be prepared by a Grignard reaction.

$$\overset{\overset{O}{\|}}{CH_3CCH_2CH_3} \xrightarrow[\text{(2) H}_2\text{O, H}^+]{\text{(1) CH}_3\text{CH}_2\text{CH}_2\text{MgBr}} \text{product}$$

3. The organic reactants in the preceding step can be obtained from alcohols.

$$\overset{\overset{OH}{|}}{CH_3CHCH_2CH_3} \xrightarrow{\text{H}_2\text{CrO}_4} \overset{\overset{O}{\|}}{CH_3CCH_2CH_3}$$

$$CH_3CH_2CH_2OH \xrightarrow{\text{HBr}} CH_3CH_2CH_2Br \xrightarrow[\text{diethyl ether}]{\text{Mg}} CH_3CH_2CH_2MgBr$$

4. Write the entire synthetic sequence.

$$CH_3CH_2CH_2OH \xrightarrow{\text{HBr}} CH_3CH_2CH_2Br \xrightarrow[\text{diethyl ether}]{\text{Mg}} CH_3CH_2CH_2MgBr$$

$$\overset{\overset{OH}{|}}{CH_3CHCH_2CH_3} \xrightarrow{\text{H}_2\text{CrO}_4} \overset{\overset{O}{\|}}{CH_3CCH_2CH_3} \xrightarrow[\text{(2) H}_2\text{O, H}^+]{\text{(1) CH}_3\text{CH}_2\text{CH}_2\text{MgBr}} \text{product}$$

Study Problems

7.21 Suggest a method for preparing ethyl acetate ($CH_3CO_2CH_2CH_3$) from ethanol and *no other organic reagent*.

7.22 Suggest syntheses for the following compounds from organic compounds containing six or fewer carbon atoms and any other required reagents:

(**a**) 3,5-dimethyl-3-hexanol   (**b**) cyclohexylmethanol

(**c**) (*E*)-5-methyl-1,5-heptadien-4-ol

▼ Summary

An **alcohol** (ROH) is a compound containing a hydroxyl group bonded to an sp^3-hybridized carbon. The molecules of an alcohol are polar and can form hydrogen bonds with other alcohol molecules, with water molecules, or with any molecules containing NH or OH.

An **alkoxide** (a compound containing RO^-) can be prepared from the reaction of an alcohol with a strong base, such as NaH, or with an alkali metal, such as Na.

In the laboratory, alcohols can be prepared by the S_N2 reaction of primary alkyl halides with OH^-; by the hydration of alkenes; by the reduction of carbonyl compounds; or by Grignard reactions of carbonyl compounds or epoxides. These reactions are summarized in Table 7.4.

Alcohols undergo *substitution reactions* with HX (1° alcohols, S_N2; 2° and 3° alcohols, S_N1). Alcohols undergo *elimination reactions* with H_2SO_4 or other strong acids. In either case, the order of reactivity of alcohols is 3° > 2° > 1°. Alkyl halides can be prepared from alcohols without rearrangement with $SOCl_2$ or PBr_3.

Substitution:

$$ROH + HX \longrightarrow RX + H_2O$$

Elimination:

$$R_2CH{-}\overset{\displaystyle OH}{\underset{\displaystyle |}{C}}R_2 \xrightarrow[\text{heat}]{H_2SO_4} R_2C{=}CR_2 + H_2O$$

An alcohol can undergo reaction with an acid or an acid derivative to yield an **ester of a carboxylic acid** (RCO_2R) or an **inorganic ester**, such as $RONO_2$ or $ROSO_3H$. Most inorganic ester groups are good leaving groups.

The **oxidation** of a primary alcohol results in a carboxylic acid (or aldehyde), while the oxidation of a secondary alcohol yields a ketone (see Table 7.5).

Study Problems*

7.23 Name all the oxygen-containing functional groups in the following compounds:

(**a**) the antihistamine *Benadryl*, $(CH_3)_2NCH_2CH_2OCH(C_6H_5)_2$

---

* For information concerning the organization of the *Study Problems* and *Additional Problems*, see the *Note to student* on page 39.

(b) the drug morphine (p. 820)
(c) tetrahydrocannabinol (the principal active ingredient in marijuana):

7.24 Draw formulas that show the hydrogen bonding in an aqueous solution of 2-methoxy-ethanol ($CH_3OCH_2CH_2OH$).

7.25 Complete the following equations for acid–base reactions:
(a) $CH_3CH_2OH + H^+ \rightleftharpoons$

(b) ⬡—$OH + K \longrightarrow$

(c) $(CH_3)_3CO^- + H_2O \rightleftharpoons$
(d) $CH_3CH_2CH_2CH_2OH + NaH \longrightarrow$
(e) $CH_3OH + conc. H_2SO_4 \rightleftharpoons$
(f) $CH_3CH_2\overset{+}{O}H_2 + H_2O \rightleftharpoons$

7.26 Draw formulas for the following compounds:
(a) 2-pentanol
(b) cyclopentanol
(c) 2,3-dimethyl-3-phenyl-2-butanol
(d) 1-phenylethanol
(e) 3-phenyl-1-propanol
(f) *t*-butyl alcohol
(g) isobutyl alcohol
(h) *cis*-2-buten-1-ol
(i) (*E*)-4-bromo-2,3-dimethyl-2-penten-1-ol
(j) (*R*)-2-butanol

7.27 Write an acceptable name for each of the following alcohols, and classify each as 1°, 2°, or 3°.

(a) $(CH_3)_2CHCH_2CH_2OH$ (b) ⬡····CH_3 HO

H₃C–CH–CH₂CH₂OH
　　|
　　CH₃

(c) $CH_3\overset{OH}{\underset{|}{C}}HCH_2CH_2\overset{OH}{\underset{|}{C}}HCH_2CH_3$ 2,5 hexanediol

(d) $\underset{H}{\overset{CH_3}{C}}=\underset{H}{\overset{CH_2CH_2OH}{C}}$ 2 penten-5-ol

7.28 Write a trivial name for each of the following compounds:
(a) $CH_3CH_2CH_2OH$ (b) $CH_2{=}CHCH_2OH$

(c) ⬡—CH_2OH (d) $CH_3\overset{OH}{\underset{|}{C}}HCH_3$ (e) ◇—OH

7.29 Write equations to show how the following alcohols could be prepared by S_N2 reactions.

(a) $\langle\bigcirc\rangle$—CH_2CH_2OH

(b) $CH_3CH_2CH_2\overset{\overset{\displaystyle OH}{|}}{C}HCH_3$

(c) (R)-$CH_3CH_2\overset{\overset{\displaystyle OH}{|}}{C}HCH_3$

(d)

$(CH_3)_3C$—[cyclohexane ring with H up, OH on opposite carbon, H down]

7.30 Write equations that show how each of the following alcohols can be prepared (1) by the reduction of a carbonyl compound, and (2) by a Grignard reaction:

(a) CH_3—$\langle\bigcirc\rangle$—CH_2OH

(b) $(CH_3)_2CHCH_2\overset{\overset{\displaystyle OH}{|}}{C}HCH_3$

7.31 Predict the organic products:

(a) [cyclohexane with substituent]—I + Mg $\xrightarrow{\triangle O}$

(b) [cyclohexane]=O $\xrightarrow[\text{(2) } H_2O,\ H^+]{\text{(1) } CH_3Li}$

(c) [cyclohexane]—Br $\xrightarrow[\substack{\text{(2) HCHO} \\ \text{(3) } H_2O,\ H^+}]{\text{(1) Mg, diethyl ether}}$

(d) $\langle\bigcirc\rangle$—$CH=CH\overset{\overset{\displaystyle O}{\|}}{C}H$ $\xrightarrow[\text{(2) } H_2O,\ H^+]{\text{(1) } HC\equiv CMgBr}$

7.32 (1) Which of the following compounds contain acidic hydrogens that would be removed by CH_3MgI? (2) Write equations showing the products (if any) of these reactions, assuming that an excess of CH_3MgI is used.

(a) $C_6H_5C\equiv CCH_3$ (b) $HOCH_2CH_2OH$

(c) $(CH_3CH_2)_2NH$ (d) $HO_2C\overset{\overset{\displaystyle CO_2H}{|}}{C}H-CO_2H$

7.33 How could you prepare each of the following compounds, starting with 2-bromopropane, magnesium, and other appropriate reagents?

(a) 2,3-dimethyl-2-butanol   (b) 3-methyl-2-butanol

7.34 Write equations showing how each of the following alcohols could be prepared by (1) a nucleophilic substitution reaction; (2) a Grignard synthesis; and (3) the reduction of a carbonyl compound.

(a) $CH_3CH_2\overset{\overset{\displaystyle OH}{|}}{C}HCH_3$

(b) $\langle\bigcirc\rangle$—$\overset{\overset{\displaystyle OH}{|}}{C}H$—$\langle\bigcirc\rangle$

(c) $CH_3CH_2CH_2CH_2OH$ (d) [cyclohexane]—CH_2OH

7.35 Write the equation for the reaction that occurs when each of the following alcohols is treated with HI. (Show the mechanisms.)

(a) 2-propanol   (b) 1-butanol

7.36 Complete the following equations:

(a) [cyclohexyl]—OH + HCl $\xrightarrow{ZnCl_2}$

(b) [aromatic ring with OCH$_3$ and CH$_2$OH substituents] + HCl $\xrightarrow{H_2O}$

(c) [cyclohexane ring with OCH$_3$, CH$_3$, CH$_3$, OH, CH(CH$_3$)$_2$ substituents] + HBr $\xrightarrow{H_2O}$

7.37 Complete the following equations for substitution reactions. Which reaction of the three has the fastest rate of reaction? Which has the slowest rate? Explain.

(a) [naphthalene ring with CH$_2$OH] + HBr $\longrightarrow$

(b) $CH_3CH_2CH_2OH$ + HBr $\longrightarrow$

(c) [cyclohexyl]—OH + HBr $\longrightarrow$

7.38 Give the structure of the expected rearranged halide from each of the following reactions:

(a) 3,3-dimethyl-2-butanol + HCl $\xrightarrow{ZnCl_2}$

(b) 2,2-diphenyl-1-ethanol + HI $\longrightarrow$

7.39 For each of the following alcohols, would you expect a substitution reaction with HBr to proceed by an S$_N$1 path or by an S$_N$2 path? Explain.

(a) CH_3OH (b) [cyclohexyl]—CHCH$_3$ with OH   (c) [decalin structure with OH]   (d) [phenyl]—C(CH$_3$)$_2$ with OH

7.40 Predict the organic products:

(a) $CH_3CH_2OCH_2CH_2OH$ $\xrightarrow{PBr_3}$

(b) [cyclohexene ring with H$_3$C, CH$_3$, CH$_2$OH, CH$_3$ substituents] $\xrightarrow{PBr_3}$

(c) $(CH_3)_2NCH_2CH_2OH$ $\xrightarrow{SOCl_2}$

7.41 Predict the major organic products of dehydration reactions of the following alcohols. Include the stereochemistry of the product if applicable.   (a) 1-phenyl-2-propanol;

(**b**) 1-butanol; (**c**) 2-butanol; (**d**) 4-methyl-1,4-pentanediol (elimination of 1.0 mol H_2O only); (**e**) citric acid.

$$CH_2CO_2H$$
$$|$$
$$HO-C-CO_2H$$
$$|$$
$$CH_2CO_2H$$

citric acid

7.42 For the following reaction, draw the structure of the initial intermediate inorganic ester; circle its leaving group; and draw the structure of the final product.

$$CH_2=CHCH_2OH + SOCl_2 \xrightarrow{R_3N}$$

7.43 Write equations for the following reactions: (**a**) (S)-2-pentanol and tosyl chloride; (**b**) (R)-2-butanol and H_2SO_4 at 180°; (**c**) (R)-2-butanol and $ClSO_3H$; (**d**) (R)-2-butyl tosylate and ethanol under S_N1 conditions.

7.44 Identify the inorganic ester groups in the following compounds:

(**a**) *a lecithin*, found in egg yolks (page 944)
(**b**) *sphingomyelin*, a component of nerve sheaves (page 944)
(**c**) a *nucleotide*, a hydrolysis product of nucleic acids (page 1009)

7.45 In which one of each of the following pairs of compounds is carbon in the higher oxidation state?

(**a**) $CH_3CH_2C{\equiv}CH$ or $CH_3CH_2CO_2H$
(**b**) CH_3CH_2CHO or $CH_3CH_2CH_2OH$

(**c**) —Cl or

7.46 For each of the following compounds, write the formula for a compound that would be at the next higher oxidation state and the formula for a compound that would be at the next lower oxidation state. (More than one answer may be correct.)

(**a**) —CH_2OH (**b**) $CH_3CH=CH_2$

(**c**) $CH_3CH_2\overset{\overset{\displaystyle O}{||}}{C}H$ (**d**) $CH_3CH_2CH-CHCH_3$
                                         | |
                                        Br   Br

7.47 For each of the following conversions, would an oxidizing agent, a reducing agent, or neither be required?

(**a**) $CH_4 \longrightarrow CH_3Cl$ (**b**)

(**c**) $CH_3CH_2Br \longrightarrow CH_3CH_2OH$ (**d**) —Br $\longrightarrow$ —MgBr

(**e**) $CH_3CH=CH_2 \longrightarrow CH_3CH_2CH_2Br$

$$\text{(f)} \quad CH_3CH{=}CH_2 \longrightarrow \underset{\underset{O}{\parallel}}{CH_3}COH + \underset{\underset{O}{\parallel}}{H}COH$$

7.48 Suggest an alcohol and an oxidizing agent for preparing

(a) 2-butanone; (b) butanal; (c) butanoic acid;
(d) 1,3-dichloropropanone; (e) cyclooctanone; (f) decanal.

7.49 Suggest a synthesis for each of the following compounds from an alcohol of four carbons or less and any other reagents needed:

(a) $CH_2\overset{\displaystyle CH_2}{\diagdown\!\!\diagup}CHCH_2CHClCH_2CH_3$ (b) $(CH_3)_2C{=}CHCH_3$

(c) $(CH_3)_2CHCH_2CH_2CH_2OH$ (d) $(CH_3)_2CHCH_2\underset{\underset{CN}{|}}{C}HCH_3$

Additional Problems

7.50 Predict the major organic products.

(a) $(CH_3)_2CHCH_2OH + HCl \xrightarrow{ZnCl_2}$

(b) $CH_3CH_2CH_2CH_2O^- K^+ + H_2O \longrightarrow$

(c) (S)- $\underset{\displaystyle OH}{\overset{\displaystyle H_3C}{\diagup}}$ $+ SOCl_2 \xrightarrow{pyridine}$

(d) (R)-2-butanol + SOCl₂ $\xrightarrow{pyridine}$

(e) $\bigcirc{-}CH_2CH_2OH + Na \longrightarrow$

(f) $CH_3CH_2CH_2CH_2OMgBr + H_2O \longrightarrow$

(g) $(C_6H_5)_2\underset{\underset{CH_2CH_3}{}}{C}\overset{\overset{OH}{|}}{}CH_2CH_3 \xrightarrow{H^+}$

7.51 Suggest reagents for the following conversions. Other organic reactants may also be used.

(a) $(CH_3)_2CHCH_2O\underset{\underset{CH_3}{|}}{C}HCH_2CH_3$ from alcohols containing four or fewer carbon atoms
(b) cycloheptanol from cycloheptanone
(c) cycloheptanone from cycloheptanol
(d) (S)-2-chlorobutane from (R)-2-butanol
(e) benzaldehyde from benzyl alcohol
(f) 1-butanol from 1-bromopropane
(g) benzyl alcohol from bromobenzene
(h) 1-phenylethanol from benzaldehyde (C_6H_5CHO)

(i) $CH_3\underset{\underset{O}{\parallel}}{C}CH_2CH_2\underset{\underset{O}{\parallel}}{C}H$ from $CH_3\underset{\overset{OH}{|}}{C}HCH_2CH_2\underset{\underset{O}{\parallel}}{C}H$

7.52 If 1-butanol is added to each of the following reagents, what would be the expected major products? (a) methylmagnesium iodide; (b) phenyllithium; (c) HBr; (d) potassium metal; (e) NaH.

7.53 Predict the major organic product of the reaction (if any) of (*R*)-2-heptanol with each of the following reagents: (**a**) H_2CrO_4;   (**b**) HI;   (**c**) Li metal;   (**d**) hot, conc. H_2SO_4;   (**e**) CH_3MgI;   (**f**) aqueous NaCl;   (**g**) aqueous NaOH;   (**h**) $SOCl_2$ in pyridine.

7.54 Suggest one (or more, if possible) synthetic route to each of the following alcohols, starting with an organic halide and other needed organic reagents:

(**a**) $(CH_3)_3CCHCH_3$ with OH

(**b**) structure with CH_2OH

(**c**) structure with HO and CH_3

(**d**) CH_3CH_2C— with OH and CH_3

(**e**) $(CH_3)_2CHCH$— with OH

(**f**) CH_2=$CHCC(CH_3)_2$ with CH_2 and OH

7.55 A chemist treated CH_3CH=$CHCH_2Cl$ with magnesium in anhydrous ether and then added acetone (propanone). After hydrolysis, instead of a single alkenyl alcohol as a product, the chemist obtained *two* alcohols. What are the structures of the two alcohols? Explain your answer.

7.56 Upon dehydration, 2,2-dimethylcyclohexanol yields two alkenes, both the result of rearrangements. One contains a five-membered ring. What are these alkenes? Write a mechanism for the formation of each alkene.

7.57 The compound *trans*-4-*t*-butyltosylcyclohexane was treated with KI under S_N2 conditions. When the product was isolated, it was found to be a mixture of *cis*- and *trans-t*-butyliodocyclohexanes. Suggest a reason for this and propose a mechanism.

7.58 When (2*S*,3*S*)-3-phenyl-2-tosylbutane is heated with $Na^+\ {}^-OCH_2CH_3$ in ethanol, a 91% yield of one product (a pure stereoisomer) is obtained. Write a flow equation to show the formation and structure of this product. Explain why only this one product is formed.

7.59 Inorganic reagents can often be complexed with an organic polymer to provide a convenient insoluble source of the reagent, which can be easily removed from the reaction mixture in the workup. For example, when poly(vinylpyridine) is treated with CrO_3/HCl, poly(vinylpyridinium chlorochromate) is obtained. In one study, this reagent was used to oxidize the following alcohols, resulting in product yields of 76–100%. What were the organic products?   (**a**) 2-octanol;   (**b**) 1-phenylethanol;   (**c**) 3-phenyl-2-propen-1-ol (cinnamyl alcohol).

7.60 Suggest a synthesis for each of the following compounds, starting with any organic compounds of three or fewer carbon atoms:

(**a**) 1-bromo-2-butene   (**b**) 3-ethyl-3-pentanol

7.61 Propose a reaction sequence for each of the following conversions. Use any inorganic or organic reagents you wish.

(**a**) any alcohol to CH_3O—⟨benzene ring⟩—CH_2CN

(**b**) benzyl bromide to stilbene (*trans*-1,2-diphenylethene).

(**c**) $C_6H_5CH_3$ to $C_6H_5CH_2CHCH_3$ with OH

(**d**) ⟨cyclopentyl⟩—CH_2CH_2OH to   ⟨cyclopentyl⟩—$CH_2CH_2CH_2CH$— with OH —⟨benzene ring⟩

7.62 Suggest reagents for the following conversions, which may require more than one step.

(a) cyclohexanone from bromocyclohexane

(b)

$$\underset{CH_2CH=CH_2}{\overset{OH}{\diagdown}}$$

from cyclohexanol

(c) 1-phenylcyclohexene from bromocyclohexane

7.63 When 1,2-diols are heated in acidic solution, they undergo a rearrangement called a **pinacol rearrangement**. The following conversion is just one example of this reaction. Suggest a mechanism.

$$\underset{\substack{\text{2,3-dimethyl-2,3-butanediol}\\ \text{(pinacol)}}}{\overset{OH \quad OH}{(CH_3)_2C-C(CH_3)_2}} \quad \xrightarrow{H^+} \quad \underset{\substack{\text{3,3-dimethyl-2-butanone (72\%)}\\ \text{(pinacolone)}}}{\overset{O}{(CH_3)_3CCCH_3}} \quad + H_2O$$

Ethers, Epoxides, and Sulfides

Like water and alcohols, an ether (ROR′) contains an sp^3-hybridized oxygen. But, in an ether the oxygen is bonded to two carbon atoms. The groups bonded to the ether oxygen can be alkyl, aryl, vinylic, or other carbon group. The ether oxygen, like other sp^3-hybridized oxygens, contains two pairs of unshared valence electrons.

$$CH_3CH_2 \overset{\cdot\cdot\overset{\cdot\cdot}{O}\cdot\cdot}{} CH_2CH_3$$

diethyl ether
("ether")

a solvent and
inhalation anesthetic

$$\bigcirc\!\!-\!\overset{\cdot\cdot}{\underset{\cdot\cdot}{O}}CH_3$$

methyl phenyl ether
(anisole)

used in perfumery

$$CH_2{=}CH \overset{\cdot\cdot\overset{\cdot\cdot}{O}\cdot\cdot}{} CH{=}CH_2$$

divinyl ether

an inhalation
anesthetic

Ethers can be either open-chain or cyclic. When the ring size (including the oxygen) is five or greater, the chemistry of the ether is similar to that of an open-chain ether. (There are some differences in rates of reactions because the oxygen atom in a cyclic ether is less sterically hindered than the oxygen atom in a comparable open-chain ether—the alkyl substituents are tied back in the ring.) **Epoxides** are three-membered ring ethers. Because of their small ring size, epoxides are more reactive than other ethers. Large ring systems with repeating $-OCH_2CH_2-$ units are called **crown ethers**. These compounds are valuable reagents that can be used to help dissolve inorganic salts in organic solvents.

Some cyclic ethers:

ethylene oxide   tetrahydrofuran   1,4-dioxane

an epoxide   THF

18-crown-6

a crown ether

We will discuss first the open-chain ethers and then the epoxides and crown ethers. At the end of the chapter, we will briefly discuss the sulfur analogs of alcohols and ethers—thiols (RSH) and sulfides (RSR′).

Section 8.1
Physical Properties of Ethers

Ethers are polar but not so polar as water and alcohols. The dipole moments of the following compounds show decreased polarity in the series H_2O, ROH, and ROR.

μ: 1.8 D   1.7 D   1.3 D

Ethers cannot form hydrogen bonds in the pure state because they have no hydrogen attached to the oxygen. However, ethers can form hydrogen bonds with water, alcohols, or phenols. Because of hydrogen bonding with water, the solubilities of the four-carbon compounds diethyl ether and 1-butanol are about the same, even though the boiling points are quite different.

$$CH_3CH_2OCH_2CH_3 \qquad CH_3CH_2CH_2CH_2OH$$

diethyl ether   1-butanol

	diethyl ether	1-butanol
MW:	74.1	74.1
bp:	34.6°	117°
approx. water solubility:	8 g/100 mL	8 g/100 mL

Table 8.1 Physical properties of some ethers and epoxides

Name	Formula	Bp, °C	Density, g/mL at 20°C	Solubility in H₂O
dimethyl ether	CH_3OCH_3	−24	gas	∞
diethyl ether	$CH_3CH_2OCH_2CH_3$	34.6	0.71	8 g/100 mL
tetrahydrofuran (THF)		66	0.89	∞
oxirane (ethylene oxide)	CH_2CH_2	13.5	0.88 (at 10°)	∞
methyloxirane (propylene oxide)	CH_3CHCH_2	34.3	0.86 (at 0°)	∞

Table 8.1 shows that other low-molecular-weight ethers are miscible with water. Of special note is tetrahydrofuran. Although it contains four carbons, as does diethyl ether, these alkyl carbons are held away from the oxygen by being bonded in a ring. The more exposed oxygen can form stronger hydrogen bonds than can the oxygen in diethyl ether. Therefore, tetrahydrofuran is more soluble.

diethyl ether   tetrahydrofuran

approx. water solubility: 8 g/100 mL   miscible

Section 8.2
Nomenclature of Ethers and Epoxides

Simple open-chain ethers are named almost exclusively by their trivial names, as **alkyl ethers**.

$CH_3CH_2OCH_2CH_3$ $(CH_3)_2CHOCH(CH_3)_2$ $CH_3OCH_2CH_3$

diethyl ether   diisopropyl ether   methyl ethyl ether
(or ethyl ether, or
simply "ether")

The names of more-complex ethers follow systematic nomenclature rules. An **alkoxy-** prefix is used when there is more than one alkoxyl (RO—) group or when there is a functional group of higher priority. (Note that a hydroxyl group has priority over an alkoxyl group.)

$$CH_3CH_2OCH_2CH_2CH_2\overset{\overset{\displaystyle OH}{|}}{C}HCH_3 \qquad CH_3OCH_2CH_2OCH_3$$

5-ethoxy-2-pentanol   1,2-dimethoxyethane
                                       (DME, or "glyme")

a common solvent

In the IUPAC system, epoxides are called **oxiranes**. In the numbering of these rings, the oxygen is always considered position 1.

$$CH_3\overset{③}{C}H\overset{①}{-}\overset{②}{C}HCH_2CH_3$$

2-ethyl-3-methyloxirane

Because an epoxide is cyclic, a substituted epoxide may be capable of geometric isomerism.

$$\underset{\textit{cis-2,3-dimethyloxirane}}{\overset{\displaystyle H_3C}{\underset{H}{\diagdown}}\overset{O}{\underset{}{\triangle}}\overset{CH_3}{\underset{H}{\diagup}}} \qquad \underset{\textit{trans-2,3-dimethyloxirane}}{\overset{\displaystyle H_3C}{\underset{H}{\diagdown}}\overset{O}{\underset{}{\triangle}}\overset{H}{\underset{CH_3}{\diagup}}}$$

Epoxide rings may be part of fused-ring systems. In such cases, the epoxide must be *cis* on the other ring unless the other ring is eight-membered or larger. (The required bond angles for the three-membered ring makes the *trans*-configuration impossible for smaller rings.) The prefix *cis* is not used in the names of these compounds.

cyclopentene oxide

a fused-ring system

Sample Problem

Name the following compounds:

(a) $(CH_3)_3COCH_3$ (b) [structure with H, OCH$_2$CH$_3$, OCH$_2$CH$_3$]   (c) [structure H_3C, O, H, C—C, H, $CH_2CH_2CH_3$]

Solution: (a) *t*-butyl methyl ether; (b) *cis*-1,2-diethoxycyclohexane; (c) *trans*-2-methyl-3-propyloxirane.

Study Problem

8.1 Write formulas for **(a)** ethyl phenyl ether; **(b)** 1,4-diisopropoxybenzene; **(c)** 2,2-di-methyloxirane.

▼ ## Section 8.3
Preparation of Ethers and Epoxides

A. Diethyl Ether

Diethyl ether is undoubtedly the most popular organic laboratory solvent. Of historical interest is its introduction in the 1800s, along with chloroform and nitrous oxide (N_2O, laughing gas), as a general anesthetic. Diethyl ether and nitrous oxide are still used as anesthetics. (Chloroform, however, has a narrow margin of safety and leads to liver damage, as do many of the chlorinated hydrocarbons.) Diethyl ether is volatile, its vapors are explosive, and it has a tendency to cause nausea. Despite these drawbacks, it is physiologically a relatively safe anesthetic. Other ethers that are used as anesthetics are methyl propyl ether ($CH_3OCH_2CH_2CH_3$) and ethyl vinyl ether ($CH_3CH_2OCH{=}CH_2$).

Industrially, diethyl ether is prepared from ethanol in the presence of sulfuric acid. This reaction was first reported in the 1500s! Until 1800, it was thought that diethyl ether contained sulfur in its structure when, in fact, sulfur was an impurity arising from the sulfuric acid.

$$CH_3CH_2OH \xrightarrow{\ H_2SO_4\ } CH_3CH_2OSO_3H \xrightarrow{\ CH_3CH_2OH\ } CH_3CH_2OCH_2CH_3$$

ethanol   ethyl hydrogen sulfate   diethyl ether

In Chapter 7, we described the conversion of alcohols to alkenes in the presence of sulfuric acid. The reactions that actually occur depend on the reaction conditions.

When an alcohol is mixed with H_2SO_4, a series of reversible reactions occurs. (The scheme that follows is simplified; alkenes can also go to sulfates, sulfates can go to ethers, etc.) Which reaction product predominates depends on the structure of the alcohol, the relative concentrations of reactants, and the temperature of the reaction mixture. In general, *primary alcohols* give sulfate esters at low temperatures, ethers at moderate temperatures, and alkenes at high temperatures. (In all cases, mixtures would be expected.) *Tertiary alcohols* and, to a large extent, *secondary alcohols* yield alkene products.

$$\xrightleftharpoons{0°} ROSO_2OH + ROSO_2OR + H_2O$$

$$1°{:}\ \ ROH + H_2SO_4 \xrightleftharpoons{140°} ROR + H_2O$$

$$\xrightleftharpoons{170°} \text{alkenes} + H_2O$$

$$2° \text{ and } 3°{:}\ \ ROH + H_2SO_4 \rightleftharpoons \text{alkenes} + H_2O$$

Because of the large numbers of possible products, this type of dehydration reaction is seldom used in the laboratory for the synthesis of ethers.

Study Problem

8.2 Predict the major organic products:

(a) ⬡—OH + H_2SO_4 $\xrightarrow{\text{heat}}$

(b) CH_3CH_2OH (excess) + H_2SO_4 $\xrightarrow{\text{warm}}$

(c) 1 CH_3CH_2OH + 1 H_2SO_4 $\xrightarrow{\text{cold}}$

(d) ⬡—OH + HNO_3 $\xrightarrow{\text{cold}}$

B. Williamson Ether Synthesis

The **Williamson ether synthesis** is a versatile laboratory procedure for synthesizing ethers. The Williamson synthesis is the S_N2 reaction of an alkyl halide with an alkoxide or phenoxide, a reaction we discussed in detail in Chapter 5.

$$\text{RO}^- + \text{R}'\text{X} \xrightarrow{S_N2} \text{ROR}' + \text{X}^-$$

R' = CH_3 or 1°

R = CH_3, 1°, 2°, 3°, or Ar

Best yields are obtained when the alkyl halide is methyl or primary. (Secondary and tertiary alkyl halides lead to alkenes, while aryl and vinyl halides do not undergo S_N2 reactions.) The alkoxide that can be used in a Williamson synthesis has fewer limitations. It may be methyl, primary, secondary, tertiary, allylic, or benzylic. Usually, either the sodium or potassium alkoxide or phenoxide is used.

Synthesis of dialkyl ethers:

$$\underset{\text{methoxide ion}}{CH_3O^-} + \underset{\text{1-chloropropane}}{CH_3CH_2CH_2-Cl} \xrightarrow{S_N2} \underset{\text{methyl propyl ether}}{CH_3OCH_2CH_2CH_3} + Cl^-$$

$$\underset{\text{propoxide ion}}{CH_3CH_2CH_2O^-} + \underset{\text{iodomethane}}{CH_3I} \xrightarrow{S_N2} \underset{\text{methyl propyl ether}}{CH_3OCH_2CH_2CH_3} + I^-$$

$$\underset{\text{t-butoxide ion}}{(CH_3)_3CO^-} + CH_3I \xrightarrow{S_N2} \underset{\text{t-butyl methyl ether}}{(CH_3)_3COCH_3} + I^-$$

but not $CH_3O^- + (CH_3)_3CI$ (Why not?)

Synthesis of an alkyl aryl ether:

$$\underset{\text{phenoxide ion}}{⬡-O^-} + \underset{\text{bromoethane}}{CH_3CH_2Br} \xrightarrow{S_N2} \underset{\text{ethyl phenyl ether}}{CH_3CH_2O-⬡} + Br^-$$

but not $CH_3CH_2O^- + ⬡-Br$ (Why not?)

Study Problem

8.3 Show by equations the best method for preparing each of the following ethers by a Williamson synthesis:

(a) [naphthalenyl-OCH$_3$ structure]

(b) $(CH_3)_2CHOCH_2CH_3$

C. Epoxides

Treatment of an alkene with a peroxycarboxylic acid (RCO_3H or $ArCO_3H$) in an inert solvent, such as $CHCl_3$ or CCl_4, yields an epoxide, or oxirane. Peroxybenzoic acid ($C_6H_5CO_3H$) and m-chloroperoxybenzoic acid are commonly used peroxy acids.

[Reaction scheme:]

$C_6H_5{-}CH{=}CH_2$ + m-chloroperoxybenzoic acid $\longrightarrow$ phenyloxirane (styrene oxide) + m-chlorobenzoic acid

styrene   *m*-chloroperoxybenzoic acid (*m*-CPBA)   phenyloxirane (95%) (styrene oxide)   *m*-chlorobenzoic acid

The reaction path involves transfer of an oxygen from the peroxy acid directly to the alkene in one step.

[Mechanism scheme]

A second way to synthesize an epoxide is by an intramolecular Williamson synthesis using a **1,2-halohydrin**, a compound with a halogen and a hydroxyl group on adjacent (1,2) carbons.

General:

[Reaction scheme showing halohydrin to epoxide]

an epoxide

Specific example:

[Reaction scheme]

(*S*)-2-chloro-1-propanol   (*R*)-methyloxirane

Study Problem

8.4 Write equations for two syntheses of *cis*-2,3-dimethyloxirane.

▼ ## Section 8.4
Substitution Reactions of Ethers

Ethers are quite unreactive and behave more like alkanes than like organic compounds containing functional groups. Ethers undergo auto-oxidation (Section 6.6B) and combustion (which occurs readily), but they are not oxidized by laboratory reagents; nor do ethers undergo reduction, elimination, or reactions with bases. When they are heated with strong acids, ethers do undergo substitution reactions. For example, when heated with HI or HBr, an ether yields an alcohol and an alkyl halide. Under these conditions, the alcohol can react further with the HI or HBr to yield additional alkyl iodide or bromide.

$$CH_3CH_2OCH_2CH_3 + HI \xrightarrow{\text{heat}} CH_3CH_2I + HOCH_2CH_3$$

diethyl ether   iodoethane   ethanol

$$\xrightarrow{HI} CH_3CH_2I$$

Ether cleavage with HI or HBr proceeds by almost the same path as the reaction of an alcohol with HX: protonation of the oxygen, followed by S_N1 or S_N2 reaction. (Protonation is necessary because RO^- is a poor leaving group, while ROH, like H_2O, is easily displaced.)

$$CH_3CH_2-\overset{..}{\underset{..}{O}}-CH_2CH_3 \underset{H^+}{\rightleftharpoons} CH_3CH_2-\overset{\overset{H}{|}}{\underset{..}{O}}{}^+-CH_2CH_3 \xrightarrow{I^-}$$

protonated

$$\left[\overset{CH_3}{\underset{\delta-}{:\overset{..}{\underset{..}{I}}----CH_2}}----\overset{\overset{H}{|}}{\underset{\delta+}{O}}-CH_2CH_3 \right] \longrightarrow CH_3CH_2\overset{..}{\underset{..}{I}}: + H\overset{..}{\underset{..}{O}}CH_2CH_3$$

S_N2 transition state

An alkyl phenyl ether, such as anisole, yields the alkyl iodide and phenol (not iodobenzene and methanol) because sp^2-hybridized carbons do not undergo reaction by an S_N1 or S_N2 path.

not cleaved

methyl phenyl ether   phenol   iodomethane
(anisole)

Sample Problem

Give the steps for the cleavage of diisopropyl ether by HI (an S_N1 reaction).

Solution:

$$(CH_3)_2CH\text{—}\ddot{O}\text{—}CH(CH_3)_2 \;\rightleftharpoons\; (CH_3)_2CH\overset{\overset{\displaystyle H}{|}}{\underset{}{\ddot{O}^+}}\text{—}CH(CH_3)_2 \;\xrightarrow{-\,HOCH(CH_3)_2}$$

protonated

$$[(CH_3)_2\overset{+}{C}H] \;\xrightarrow{\;\ddot{I}\!:^-\;}\; (CH_3)_2CH\text{—}\ddot{I}\!:$$

a carbocation

Study Problem

8.5 The **Zeisel procedure** for estimating the number of methoxyl ($CH_3O\text{—}$) or ethoxyl ($CH_3CH_2O\text{—}$) groups in an alkyl aryl ether consists of ether cleavage with excess HI, followed by distillation of the volatile methyl or ethyl iodide from the reaction mixture. The amount of volatile iodide is determined by reaction with aqueous silver nitrate. Write the equations for the reactions that would occur in the determination of the number of methoxyl groups in 1,2,3-trimethoxybenzene:

      OCH_3

      —OCH_3

      OCH_3

Section 8.5
Substitution Reactions of Epoxides

An epoxide ring, like a cyclopropane ring, cannot have normal sp^3 bond angles of 109°; instead, the internuclear angles are 60°, a geometric requirement of the three-membered ring. The orbitals forming the ring bonds are incapable of maximum overlap. Therefore, epoxide rings are strained. The polarity of the C—O bonds, along with this ring strain, contributes to the high reactivity of epoxides compared to the reactivity of other ethers.

    $\overset{\delta-}{\ddot{O}}$

       60°

$\underset{\delta+}{CH_2}\text{—}\underset{\delta+}{CH_2}$ *polar and strained*

    Opening of the strained three-membered ring results in a lower-energy, more stable product. The characteristic reaction of epoxides is ring opening, which can occur under either alkaline or acidic reaction conditions. These reactions of epoxides are referred to as **base-catalyzed** or **acid-catalyzed cleavage reactions**.

In base:

$$
\underset{\substack{\text{oxirane} \\ \text{(ethylene oxide)}}}{\overset{\displaystyle O}{CH_2-CH_2}} + H_2O \xrightarrow{OH^-} \underset{\substack{\text{1,2-ethanediol} \\ \text{(ethylene glycol)}}}{\overset{\displaystyle OH \quad OH}{CH_2-CH_2}}
$$

$$
\overset{\displaystyle O}{CH_2-CH_2} + CH_3OH \xrightarrow{^-OCH_3} \underset{\text{2-methoxyethanol}}{\overset{\displaystyle OH \quad OCH_3}{CH_2-CH_2}}
$$

In acid:

$$
\overset{\displaystyle O}{CH_2-CH_2} + H_2O \xrightarrow{H^+} \overset{\displaystyle OH \quad OH}{CH_2-CH_2}
$$

$$
\overset{\displaystyle O}{CH_2-CH_2} + HCl \longrightarrow \underset{\substack{\text{2-chloroethanol} \\ \text{(ethylene chlorohydrin)}}}{\overset{\displaystyle OH \quad Cl}{CH_2-CH_2}}
$$

A. Base-Catalyzed Cleavage

Epoxides undergo S_N2 attack by nucleophiles such as the hydroxide ion or alkoxides. The steps in the reactions of ethylene oxide with hydroxide ion (NaOH or KOH in water) and with methoxide ion (NaOCH$_3$ in methanol) follow:

$$
\overset{\displaystyle \ddot{O}}{CH_2-CH_2} + {}^-:\ddot{O}H \xrightarrow{S_N2} \overset{\displaystyle :\ddot{O}:^- \longrightarrow H-\ddot{O}H}{CH_2-CH_2OH} \rightleftharpoons \overset{\displaystyle :\ddot{O}H}{CH_2CH_2OH} + {}^-:\ddot{O}H
$$

abstracting a proton from H₂O   1,2-ethanediol

$$
\overset{\displaystyle \ddot{O}}{CH_2-CH_2} + {}^-:\ddot{O}CH_3 \xrightarrow{S_N2} \overset{\displaystyle :\ddot{O}:^- \longrightarrow H-\ddot{O}CH_3}{CH_2-CH_2OCH_3} \rightleftharpoons
$$

$$
\underset{\text{2-methoxyethanol}}{\overset{\displaystyle :\ddot{O}H}{CH_2CH_2OCH_3}} + CH_3\ddot{O}:^-
$$

In base-catalyzed cleavage, the nucleophile attacks the *less hindered carbon*, just as we would expect from an S_N2 attack (1° > 2° > 3°).

$$CH_3-\overset{\overset{\displaystyle O}{\diagup\diagdown}}{\underset{\underset{\displaystyle H}{|}}{C}}-CH_2$$

in base, attack at
less hindered carbon

:Nu$^-$

$$CH_3\overset{O}{\overset{\diagup\diagdown}{CH}}-CH_2 \quad \xrightarrow[CH_3CH_2OH]{Na^+ \ ^-OCH_2CH_3} \quad CH_3\overset{\overset{\displaystyle OH}{|}}{CH}CH_2OCH_2CH_3$$

methyloxirane
(propylene oxide)

1-ethoxy-2-propanol (83%)

A Grignard reagent contains a partially negative carbon atom and attacks an epoxide ring in the same manner as other nucleophiles. The product is the magnesium salt of an alcohol. The alcohol can be obtained by hydrolysis. The reaction of a Grignard reagent with ethylene oxide is a method by which the hydrocarbon chain of the Grignard reagent may be *extended by two carbons*.

$$CH_3(CH_2)_3MgBr + \overset{O}{\overset{\diagup\diagdown}{CH_2}}-CH_2 \quad \longrightarrow \quad CH_3(CH_2)_3CH_2\overset{\overset{\displaystyle O^- \ ^+MgBr}{|}}{CH_2}$$

n-butylmagnesium
bromide

two carbons added

$$\xrightarrow{H_2O, H^+} CH_3(CH_2)_3\overbrace{CH_2CH_2}OH$$

1-hexanol (62%)

Sample Problem

Suggest a synthesis for 2-phenylethanol starting with bromobenzene.

Solution:

1. Convert bromobenzene to a Grignard reagent.

$$C_6H_5Br + Mg \quad \xrightarrow{ether} \quad C_6H_5MgBr$$

2. Treat the Grignard reagent with ethylene oxide.

$$C_6H_5MgBr + \overset{O}{\overset{\diagup\diagdown}{CH_2}}-CH_2 \quad \longrightarrow \quad C_6H_5CH_2CH_2OMgBr$$

3. Hydrolyze the magnesium alkoxide.

$$C_6H_5CH_2CH_2OMgBr + H^+ \quad \xrightarrow{H_2O} \quad C_6H_5CH_2CH_2OH + Mg^{2+} + Br^-$$

B. Acid-Catalyzed Cleavage

In acidic solution, the oxygen of an epoxide is protonated. A protonated epoxide can be attacked by weak nucleophiles such as water, alcohols, or halide ions.

General:

protonated

As contrasted with base-catalyzed cleavage, attack in acid occurs at the *more hindered carbon.*

2-chloro-1-propanol (89%)

We must conclude that the protonated epoxide has a fair amount of carbocation character. If this is the case, there is a partial positive charge on the carbon with the greater number of alkyl groups (carbocation stability: $3° > 2° > 1°$). The subsequent nucleophilic attack is favored at the more positive carbon even though this carbon is more hindered.

In acid, attack occurs at the more hindered carbon because it has a greater + charge.

How do we know that a true carbocation is not formed? When the product of an epoxide-cleavage reaction is capable of geometric isomerism, only the *trans*-product is observed. If the reaction went through a true carbocation intermediate, we would observe both *cis*- and *trans*-products.

trans-1,2-cyclohexanediol (86%)

observed product

attack from either side
would give both cis and
trans, which does not occur

Figure 8.1 summarizes the substitution reactions of ethers.

ether cleavage in strong acid:

$$R-OR' \xrightarrow[\text{heat}]{\text{strong HX}} R-X + HOR' \xrightarrow{\text{HX}} RX + R'X$$

epoxide cleavage in acid:

$$R_2\overset{\displaystyle O}{C}-CHR + R'OH \xrightarrow{H^+} R_2C-\overset{\displaystyle OH}{\underset{\displaystyle OR'}{C}HR} \quad (R' = \text{alkyl, aryl, or H})$$

$$R_2\overset{\displaystyle O}{C}-CHR + X^- + H^+ \longrightarrow R_2C-\overset{\displaystyle OH}{\underset{\displaystyle X}{C}HR}$$

epoxide cleavage in base:

$$R_2\overset{\displaystyle O}{C}-CHR \xrightarrow[\text{Nu:H}]{\text{Nu:}^-} R_2C-\overset{\displaystyle OH}{\underset{\displaystyle Nu}{C}HR} \quad (Nu:^- = {}^-OH, {}^-SH, {}^-OR, {}^-CN, :NH_3, \text{etc.})$$

epoxide reaction with Grignard reagents:

$$\overset{\displaystyle O}{CH_2}-CH_2 \xrightarrow[\text{(2) H}_2\text{O, H}^+]{\text{(1) RMgX}} RCH_2CH_2OH$$

Figure 8.1 Summary of substitution reactions of ethers.

Study Problem

8.6 Predict the products of reaction of the following epoxide with   **(a)** sodium methoxide in methanol, and   **(b)** aqueous HCl:

CH$_3$

H

O

CH$_3$

CH(CH$_3$)$_2$

▼ ## Section 8.6
Crown Ethers

Crown ethers are cyclic ethers with structures consisting of repeating —OCH$_2$CH$_2$— units, derived from 1,2-ethanediol. These compounds are named as x-crown-y, where x = the total number of atoms in the ring and y = the number of oxygen atoms in the ring.

one unit of

O(H)

CH$_2$

CH$_2$

(OH)

18 atoms in the ring; 6 oxygen atoms in the ring

18-crown-6

Recall from general chemistry that a **chelate** (Greek *chele*, "claw") is a cyclic complex of a molecule and a metal ion. The unique feature of crown ethers is that they can chelate metal ions—different-sized cyclic ethers chelate different-sized metal ions. Thus, 18-crown-6 chelates K$^+$, 15-crown-5 chelates the smaller Na$^+$, and 12-crown-4 chelates Li$^+$. In these complexes, the crown ether is often referred to as the *host*, while the metal ion is called the *guest*. The ring closures leading to crown ethers are easily accomplished if the appropriate metal ion is in solution. The chelation occurs even before the ring is closed and thus the two end groups can find each other readily.

K$^+$ MnO$_4^-$

Na$^+$ $^-$CN

An ionic compound in which the metal ion is chelated by a crown ether is soluble in nonpolar organic solvents. For example, "purple benzene" is a reagent in which KMnO$_4$, complexed by 18-crown-6, is dissolved in benzene.

One advantage of having an ionic compound dissolved in a nonpolar solvent is fairly evident: an ionic reagent can be dissolved in an organic phase, where it can react with a water-insoluble organic compound. A second advantage is that the nucleophilicity of an anion such as ^-CN or $CH_3CO_2^-$ is greatly enhanced in nonpolar solvents, where the anion is poorly solvated, or "naked." An example of how a crown ether increases the rate of a substitution reaction in CH_3CN, a polar solvent that does not dissolve ionic compounds, follows:

$$\text{C}_6\text{H}_5-\text{CH}_2\text{Br} + \text{K}^+ \ ^-\text{OCCH}_3 \xrightarrow[\text{2 hr}]{\text{CH}_3\text{CN}} \text{C}_6\text{H}_5-\text{CH}_2\text{OCCH}_3 + \text{K}^+ \ \text{Br}^-$$

5% yield with no crown ether;
100% yield with 18-crown-6

▼ Section 8.7
Thiols and Sulfides

Sulfur is just below oxygen in the periodic table. Many organic compounds containing oxygen have sulfur analogs. The sulfur analog of an alcohol is called an **alkanethiol**, or simply **thiol**, or by its older name **mercaptan**. The —SH group is called a **thiol group** or a **sulfhydryl group**.

$$\text{CH}_3\ddot{\text{S}}\text{H} \qquad \overset{\overset{\textstyle :\ddot{\text{S}}\text{H}}{\displaystyle |}}{\text{CH}_3\text{CH}_2\text{CHCH}_3}$$

methanethiol   2-butanethiol

The most characteristic property of a thiol is its odor! The human nose is very sensitive to these compounds and can detect their presence at levels of about 0.02 parts thiol to one billion parts air. The odor of a skunk's spray is due primarily to a few simple thiols.

Hydrogen sulfide (H_2S, $pK_a = 7.04$) is a stronger acid than water ($pK_a = 15.7$). Thiols ($pK_a = \sim 8$) are also substantially stronger acids than alcohols ($pK_a = \sim 16$).

$$\text{CH}_3\text{CH}_2\text{SH} + \text{OH}^- \ \rightleftharpoons \ \text{CH}_3\text{CH}_2\text{S}^- + \text{H}_2\dot{\text{O}}$$

stronger acid   *stronger base*
than H_2O *than* RS^-

Sulfur is less electronegative than oxygen, and its outer electrons are more diffuse; therefore, sulfur atoms form weaker hydrogen bonds than oxygen atoms. For this reason, H_2S (bp $-61°$) is more volatile than water (bp $100°$), and thiols are more volatile than their analogous alcohols.

Treatment of an alkyl halide with the hydrogen sulfide ion (HS^-) leads to thiols. Good yields are obtained only if an excess of inorganic hydrogen sulfide is used because the resulting thiol (which is acidic) can ionize to form the RS^- ion, also a good nucleophile. The subsequent reaction of RS^- with the alkyl halide yields the **sulfide**, R_2S, the sulfur analog of an ether.

$$CH_3I + \ ^-SH \longrightarrow CH_3SH + I^-$$
$$\text{methanethiol}$$

$$CH_3SH \xrightarrow[\ ]{-H^+} CH_3S^- \xrightarrow[-I^-]{CH_3I} CH_3SCH_3$$
$$\text{dimethyl sulfide}$$

When a thiol is treated with a mild oxidizing agent such as I_2 or $K_3Fe(CN)_6$, it undergoes coupling to form a **disulfide**, a compound containing the S—S linkage. This reaction can be reversed by treatment of the disulfide with a reducing agent such as lithium metal in liquid NH_3.

$$2\ CH_3CH_2SH \xrightleftharpoons[[H]]{[O]} CH_3CH_2S-SCH_2CH_3$$
$$\text{ethanethiol} \qquad\qquad \text{diethyl disulfide}$$

This disulfide link is an important structural feature of some proteins (Section 24.1B). The disulfide bond helps hold protein chains together in their proper shapes. The locations of the disulfide bonds determine, for example, whether hair (a protein) is curly or straight.

A sulfide can be oxidized to a **sulfoxide** or a **sulfone**, depending on the reaction conditions. For example, 30% hydrogen peroxide in the presence of an acidic catalyst oxidizes a sulfide to a sulfoxide at 25° or to a sulfone at 100°.

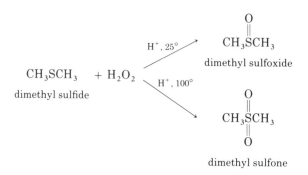

Dimethyl sulfoxide (DMSO) is prepared industrially by the air oxidation of dimethyl sulfide, which is a by-product of the paper industry. DMSO is a unique and versatile solvent. It has a high dielectric constant (49 D), but does not form hydrogen bonds in the pure state. (Why not?) It is a powerful solvent for both inorganic ions and organic compounds (see Section 5.10C). Reactants often have enhanced reactivities in DMSO compared to that in alcohol solvents. DMSO readily penetrates the skin and has been used to promote the dermal absorption of drugs; however, DMSO can also cause the absorption of dirt and poisons. A common complaint of people working with DMSO is that, when it is spilled on their hands, they can taste it!

Study Problem

8.7 Explain with formulas why dimethyl sulfoxide is miscible with water.

▼ Summary

Ethers can be prepared by the reaction of an alkoxide (RO⁻) or a phenoxide (ArO⁻), both good nucleophiles, with a methyl, primary, allylic, or benzylic halide (**Williamson ether synthesis**).

$$RO^- + R'X \longrightarrow ROR' + X^-$$

Epoxides $(R_2C{-}CR_2)$ are prepared by the reaction of a peroxy acid (RCO_3H) with an alkene or by a Williamson ether synthesis with a 1,2-halohydrin.

Ethers do not undergo elimination reactions, but can undergo *substitution reactions* when heated with HBr or HI.

$$ROR + HI \xrightarrow{\text{heat}} RI + ROH \xrightarrow{\text{HI}} RI$$

Epoxides are more reactive than other ethers and undergo ring opening with nucleophiles (in either alkaline or acidic solution) or with Grignard reagents. These reactions are summarized in Figure 8.1, page 313. **Crown ethers** are cyclic ethers that are used to chelate metal ions.

Thiols (RSH) and **sulfides** (RSR') are the sulfur analogs of alcohols and ethers. Thiols can be oxidized to disulfides (RSSR). Sulfides can be oxidized to sulfoxides and sulfones.

Study Problems*

8.8 Identify the ether, thiol, and sulfide groups in the following structures:

(a) methionine, an amino acid found in proteins (page 962)

(b) scopolamine (page 161), a preoperative anesthetic

(c)

papaverine

a component of opium

(d)

benzylpenicillin sodium

---

* For information concerning the organization of the *Study Problems* and *Additional Problems*, see the *Note to student* on page 39.

8.9 Draw formulas for the (R) and (S) enantiomers of (chloromethyl)oxirane (epichloro-hydrin).

$$CH_2-CHCH_2Cl$$

epichlorohydrin

8.10 Name the following ethers and epoxides:

(a) $(CH_3)_2CHOCH_2CH_3$ (b) $\bigcirc$—OCH_3

(c) CH_3OCHCH_2 (d)

8.11 Write formulas for the following ethers and epoxides:
(a) *t*-butyl phenyl ether
(b) 1-*sec*-butoxy-4-isobutoxybutane
(c) *cis*-2-isopropyl-3-cyclohexyloxirane
(d) styrene oxide (Styrene is the trivial name for phenylethene.)

8.12 2,6-Dichlorophenol is the sex attractant of two species of ticks. Write an equation to show how you would prepare the methyl ether of this compound.

2,6-dichlorophenol

8.13 Complete the following equations (C_6H_5— means a phenyl group).

(a) $HOCH_2CH_2Cl + (C_6H_5)_2CHOH \xrightarrow[\text{heat}]{H_2SO_4}$

(b) $CH_3CH_2\overset{\overset{\displaystyle OH}{|}}{C}HCH_2SCH_3 \xrightarrow[CH_3I]{NaH}$

(c) $+ CH_3CH_2CH_2CH_2Br \longrightarrow$

(d) $\bigcirc$—$CH_2CH_2CH_2Br + NaH \longrightarrow$

(e) $(C_6H_5)_2CHCl + Na^+ \ ^-OCH_2CH_2N(CH_3)_2 \longrightarrow$

8.14 Complete the following equations:

(a) $\xrightarrow[H_2O]{KOH}$

(b) (cyclooctene) + (benzene ring)—$\overset{\overset{\displaystyle O}{\|}}{C}$—COOH $\longrightarrow$

(c) (2S,3S)-CH$_3$CH$_2\overset{\overset{\displaystyle CH_3}{|}}{C}$HCHCH$_2$OH $\xrightarrow[\text{H}_2\text{O}]{\text{KOH}}$

with Cl substituent

8.15 Write equations showing how the following ethers or epoxides could be synthesized by a one-step reaction:

(a) (benzodioxane structure)

(b) (cyclohexyl—O—cyclohexyl structure)

(c) (bicyclic epoxide structure)　　　　(two methods)

(d) (dicyclohexyl epoxide structure)　　　　(two methods)

8.16 Complete the following equations:

(a) (tetrahydropyran) + excess HBr $\xrightarrow[\text{heat}]{\text{H}_2\text{SO}_4}$

(b) (naphthalene with OCH$_3$ groups) + excess HI $\xrightarrow{\text{heat}}$

(c) CH$_3$CH$_2$CH$_2$OCH$_3$ + excess HBr $\xrightarrow[\text{heat}]{\text{H}_2\text{SO}_4}$

(d) (oxolane/tetrahydrofuran) + 1 HCl $\xrightarrow{\text{heat}}$

(e) (7-oxabicyclo structure) + excess HBr $\xrightarrow[\text{heat}]{\text{H}_2\text{SO}_4}$

8.17 Predict the major organic products when 2,2,3-trimethyloxirane is treated with the following reagents: **(a)** 1-pentanol and HCl; **(b)** CH$_2$=CHCH$_2$MgBr, then dilute HCl; **(c)** phenyllithium, then dilute HCl; **(d)** a solution of phenol and NaOH.

8.18 Complete the following equations:

(a) CH$_2$CH$_2$ (epoxide) + HBr $\longrightarrow$

(b) CH$_2$CHCH$_2$N(CH$_2$CH$_3$)$_2$ (epoxide) + Na$^+$ $^-$SH $\longrightarrow$

(c) (epoxide) $\xrightarrow[\text{(2) H}_2\text{O, H}^+]{\text{(1) CH}_3\text{CH}_2\text{MgI, diethyl ether}}$

(d) CH$_2$CHCH$_2$Cl (epoxide) + excess CH$_3$SH $\xrightarrow[\text{CH}_3\text{OH}]{\text{OH}^-}$

(e) $\xrightarrow[\text{(2) H}_2\text{O, H}^+]{\text{(1) CH}_3\text{CH}_2\text{CH}_2\text{MgBr}}$

(f) $\overset{\text{O}}{\overset{\displaystyle\triangle}{\text{CH}_2\text{CHCH}_2\text{OH}}}$ + $\xrightarrow{25°}$

(g) $\xrightarrow{(\text{CH}_3)_2\text{CHNH}_2}$

8.19 Complete the following equations:

(a) $\text{CH}_3\text{CH}_2\text{SCH}_2\text{CH}_3 \xrightarrow[25°]{\text{H}_2\text{O}_2}$

(b) $-\text{SH} \xrightarrow[\text{(2) CH}_3\text{CH}_2\text{I}]{\text{(1) NaOH}}$

(c) $\text{CH}_3\text{CH}_2\text{CH}_2\text{SH} + \text{I}_2 \longrightarrow$

(d) $\underset{\overset{\displaystyle|}{\text{CH}_2\text{CH}_3}}{\overset{\overset{\displaystyle\text{CH}_3}{\displaystyle|}}{\text{CH}_3\text{CH}_2\text{SCSCH}_2\text{CH}_3}} \xrightarrow[\text{warm}]{\text{KMnO}_4}$

Additional Problems

8.20 When *trans*-2-chloro-1-cyclohexanol is treated with base, cyclohexene oxide is the product. However, when *cis*-2-chloro-1-cyclohexanol is treated with base, the product is cyclohexanone.

   (a) Why doesn't the *cis* isomer yield the oxide?
   (b) Write a mechanism for each reaction.

8.21 Allyl disulfide (CH_2=$\text{CHCH}_2\text{SSCH}_2\text{CH}$=$\text{CH}_2$) is a contributor to the odor of garlic. How would you prepare this compound from allyl alcohol?

8.22 The red fox (*Vulpes vulpes*) uses a chemical communication system based partly upon odorous compounds in its urine. The characteristic "skunky" odor of fox urine is due to two compounds, the structures of which are shown below. A synthetic mixture of these compounds induces characteristic marking behavior in wild red foxes. Suggest a synthesis for each compound, starting with a nonsulfur-containing alcohol.

(a) $\underset{}{\overset{\overset{\displaystyle\text{CH}_3}{\displaystyle|}}{\text{CH}_2\text{=CCH}_2\text{CH}_2\text{SCH}_3}}$ **(b)** $-\text{CH}_2\text{CH}_2\text{SCH}_3$

   3-methyl-3-butenyl methyl sulfide   2-phenylethyl methyl sulfide

8.23 Write flow equations to show how you would make the following conversions:

(a) to   **(b)** to

(c) $\overset{\text{O}}{\overset{\triangle}{CH_2-CH_2}}$ to $C_6H_5OCH_2CH_2OC_6H_5$

(d) (cyclopentyl)$-Br$ to (cyclopentyl)$-CH_2CH_2OH$

(e) $ClCH_2CH_2SCH_2CH_2Cl$ to (thiomorpholine ring with S at top and N–H at bottom)

(f) (phenyl)$-CH_2Br$ to (phenyl)$-CH_2\overset{\overset{\text{O}}{\|}}{\underset{\|}{\underset{\text{O}}{S}}}-$(phenyl)

8.24 Predict if the following equilibrium results in a racemic mixture. Explain your answer.

$$\overset{\text{O}}{\overset{\triangle}{CH_2}}-\overset{\overset{\text{O}^-}{\overset{\triangle}{}}}{\underset{\text{H}}{C}}\cdots CH_2 \rightleftharpoons \overset{\text{O}^-}{CH_2}-CH-\overset{\text{O}}{\overset{\triangle}{CH_2}}$$

(2R) *optically active?*

8.25 In a study of naphthalene metabolism in rats, the following diepoxide A was implicated as a biochemical intermediate:

A

When A was treated with methanol at room temperature, it was converted to compound B in 96% yield. Compound B was subsequently treated with an excess of sodium methoxide in methanol to yield compound C in 80% yield. Compound A was also converted to compound C in 76% yield by treatment with *p*-toluenesulfonic acid in methanol. Write an equation for each reaction, showing the structures and stereochemistry of B and C.

8.26 The following reaction is postulated to involve the displacement of the chlorine to form an epoxy group at that carbon. Write a plausible mechanism for the reaction with such an intermediate.

$$\overset{\text{O}}{\overset{\triangle}{CH_2CHCH_2Cl}} + KCN \xrightarrow{H_2O} NCCH_2\overset{\overset{\text{OH}}{|}}{CH}CH_2CN$$

epichlorohydrin (54–62%)

9 ▼

Spectroscopy I: Infrared and Nuclear Magnetic Resonance

Spectroscopy is the study of the interactions between radiant energy and matter. The colors that we see and the fact that we can see at all are consequences of energy absorption by organic and inorganic compounds. The capture of the sun's energy by plants in the process of photosynthesis is another aspect of the interaction of organic compounds with radiant energy. Of primary interest to the organic chemist is the fact that the wavelengths at which an organic compound absorbs radiant energy are *dependent upon the structure of the compound*. Therefore, we can use spectroscopic techniques to determine the structures of unknown compounds and to study the bonding characteristics of known compounds.

In this chapter, our emphasis will be on **infrared spectroscopy** and **nuclear magnetic resonance spectroscopy (nmr)**, both of which are used extensively in organic chemistry. In Chapter 21, we will broaden our discussion to include some other types of spectroscopy.

Section 9.1
Electromagnetic Radiation

Electromagnetic radiation is energy that is transmitted through space in the form of waves. Each type of electromagnetic radiation (radio waves, ultraviolet, infrared, visible, and so forth) is characterized by its **wavelength** (λ), the distance from the crest of one wave to the crest of the next wave. (See Figure 9.1.)

The entire spectrum of electromagnetic radiation is represented in Figure 9.2. The wavelengths that lead to vision range from 400 nm to 750 nm (1 nm = 10^{-9} m or 10^{-7} cm). However, the visible region is a very small part of the entire electromagnetic spectrum. Wavelengths slightly shorter than those of the visible region fall into the ultraviolet region, while slightly longer wavelengths fall into the infrared region.

In addition to being characterized by its wavelength, radiation can be characterized by its **frequency** (v), which is defined as the number of complete cycles per second (cps), also called *Hertz* (Hz). (See Figure 9.3.) Radiation of a higher frequency contains more waves per second; therefore, the wavelength must be shorter. By their definitions, wavelength and frequency are *inversely proportional*. This relationship can be expressed mathematically:

$$v = \frac{c}{\lambda}$$

where v = frequency in Hz,

   $c = 3 \times 10^{10}$ cm/sec (the speed of light), and

   λ = wavelength in cm.

In infrared spectroscopy, frequency is expressed as **wavenumbers**: the number of cycles per centimeter. Wavenumbers have units of *reciprocal centimeters* (1/cm, or cm^{-1}). The unit used for wavelength in infrared spectroscopy is the *micrometer*, μm (or *micron*, μ), where 1.0 μm = 10^{-6} m or 10^{-4} cm.

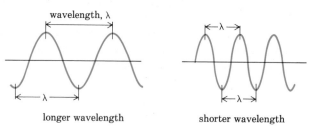

longer wavelength   shorter wavelength

Figure 9.1 Wavelength of electromagnetic radiation.

Figure 9.2 The electromagnetic spectrum.

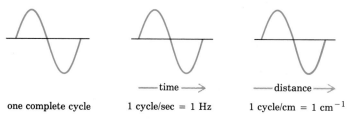

one complete cycle   1 cycle/sec = 1 Hz   1 cycle/cm = 1 cm^{-1}

Figure 9.3 Frequency of electromagnetic radiation.

Wavenumbers and wavelengths can be interconverted by the following equation:

$$\text{wavenumber in cm}^{-1} = \frac{1}{\lambda \text{ in cm}} = \frac{1}{\lambda \text{ in } \mu\text{m}} \times 10^4$$

A few common symbols and units used in spectroscopy are listed in Table 9.1.

Electromagnetic radiation is transmitted in particle-like packets of energy called **photons** or **quanta**. The energy of a photon is *inversely proportional to the wavelength*. (Mathematically, $E = hc/\lambda$, where h = Planck's constant.) Radiation of shorter wavelength has a higher energy. Therefore, a photon of ultraviolet light has more energy than a photon of visible light and has substantially more energy than a photon of radio waves.

Conversely, the energy of a photon of radiation is *directly proportional to the frequency* (more waves per unit time mean higher energy). The relationship is expressed in the equation $E = h\nu$. (The symbol $h\nu$ is often used in chemical equations to represent electromagnetic radiation.)

ultraviolet   visible   infrared   radio

increasing λ (or decreasing ν) means decreasing energy

Molecules absorb only specific wavelengths of electromagnetic radiation. Absorption of ultraviolet light (high-energy radiation) results in the promotion of an electron to a higher-energy orbital. Infrared radiation does not contain enough energy to promote an electron. Absorption of infrared radiation results in increased amplitudes of vibration of bonded atoms.

Table 9.1 Symbols commonly encountered in spectroscopy

Symbol or Unit	Definition
Frequency:	
ν	frequency in Hz (cycles per second)
cm^{-1}	wavenumber: frequency in reciprocal cm, or $1/\lambda$
Wavelength:	
λ	wavelength, usually expressed in units of meters (μm, nm, etc.)
μm	micrometer, same as micron (μ), 10^{-6} m
nm	nanometer, same as millimicron (mμ), 10^{-9} m
Å	angstrom, 10^{-10} m or 10^{-1} nm

The **intensity** of radiation is proportional to the number of photons, regardless of their energies. When a sample absorbs photons from a beam of transmitted radiation, whether of infrared or ultraviolet radiation, the number of transmitted photons in the beam must decrease. Thus, the intensity of the transmitted radiation decreases. It is this change in intensity that is measured in absorption spectroscopy.

Study Problems

9.1 Which has the higher energy:
  (a) Infrared radiation of 1500 cm^{-1} or of 1600 cm^{-1}?
  (b) Ultraviolet radiation of 200 nm or of 300 nm?
  (c) Radio waves of 60,000 Hz or of 60,004 Hz?

9.2 Make the following conversions:
  (a) 6.000 μm to cm^{-1} **(b)** 800 cm^{-1} to μm   **(c)** 1.5 μ to μm

Section 9.2
Features of a Spectrum

An infrared, visible, or ultraviolet spectrum of a compound is a graph of either *wavelength* or *frequency*, continuously changing over a small portion of the electromagnetic spectrum, versus either *percent transmission* (%T) or *absorbance* (A).

$$\%T = \frac{\text{intensity}}{\text{original intensity}} \times 100 \qquad A = \log\left(\frac{\text{original intensity}}{\text{intensity}}\right)$$

Most infrared spectra record wavelength or frequency versus %T. The absence of absorption by a compound at a particular wavelength is recorded as 100%T (ideally).When a compound absorbs radiation at a particular wavelength, the intensity of radiation being transmitted decreases. This results in a *decrease* in %T and appears in the spectrum as a dip, called an **absorption peak**, or **absorption band**. The portion of the spectrum where %T measures 100 (or near 100) is called the **base line**, which is recorded at the top of an infrared spectrum.

Visible and ultraviolet spectra (Chapter 21) are usually presented as graphs of A versus wavelength. In these cases, the base line (zero absorbance)

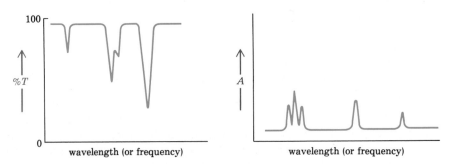

Figure 9.4 Spectra are graphs of percent transmission (%T) or absorbance (A) by a sample versus wavelength or frequency of radiation.

runs along the bottom of the spectrum, and absorption is recorded as an *increase* of the signal. The general appearance of spectra using $\%T$ and A is shown in Figure 9.4.

Although the physical appearance of a nuclear magnetic resonance spectrum is similar to that of a visible or ultraviolet spectrum (base line at the bottom), the physical principles of nuclear magnetic resonance spectroscopy are different from those of other types of spectroscopy. Therefore, we will defer a discussion of these spectra until Section 9.6.

Section 9.3
Absorption of Infrared Radiation

Nuclei of atoms bonded by covalent bonds undergo vibrations, or oscillations, in a manner similar to two balls attached by a spring. When molecules absorb infrared radiation, the absorbed energy causes an increase in the amplitude of the vibrations of the bonded atoms. The molecule is then in an **excited vibrational state**. The absorbed energy is subsequently dissipated as heat when the molecule returns to the ground state.

The vibrational states of bonds occur at fixed, or quantized, energy levels. The particular wavelength of absorption by a given type of bond depends on the energy difference between the ground state and the excited state. Therefore, different types of bonds (C—H, C—C, O—H, and so forth) absorb infrared radiation at different characteristic wavelengths.

A bond within a molecule may undergo different types of oscillations; therefore, a particular bond may absorb energy at more than one wavelength. For example, an O—H bond absorbs energy at about 3330 cm^{-1} (3.0 μm). Energy of this wavelength causes increased **stretching vibrations** of the O—H bond. An O—H bond also absorbs at about 1250 cm^{-1} (8.0 μm); energy of this wavelength causes increased **bending vibrations**. These different types of vibrations are called different **fundamental modes of vibration**.

$$\underset{stretching}{\overset{O}{\underset{CH_3\quad H}{\diagdown}}} \qquad \underset{bending}{\overset{O}{\underset{CH_3\quad H}{\diagup}}}$$

The amount of energy that a bond absorbs depends on the change in the bond moment as the bonded atoms vibrate—a greater change in bond moment results in the absorption of a larger amount of energy. Nonpolar bonds do not absorb infrared radiation because there is no change in bond moment when the atoms oscillate. Relatively nonpolar bonds (most C—C and C—H bonds in organic molecules) give rise to weak absorption. By contrast, polar bonds (such as C=O) exhibit strong absorption.

Section 9.4
The Infrared Spectrum

The instrument used to measure absorption of infrared radiation is called an **infrared spectrophotometer**. A diagram of a typical instrument is shown in Figure 9.5. At one end of the instrument is the light source, which emits all wavelengths of infrared radiation. The light from this source is split by mirrors (not shown) into two beams, the reference beam and the sample beam. After passing through the reference cell (which contains solvent, if used in the

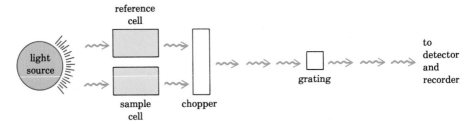

Figure 9.5 The infrared spectrophotometer.

sample, or nothing if the sample is pure) and the sample cell, the two beams are combined in the chopper (another mirror system) into one beam that alternates from reference beam to sample beam. This alternating beam is diffracted by a grating that separates the beam into its different wavelengths. The detector measures the difference in intensities of the two segments of the beam at each wavelength and passes this information on to the recorder, which produces the spectrum.

Figure 9.6 shows two infrared spectra of 1-hexanol. The top spectrum is a typical example of the type of spectrum obtained directly from a spec-

Figure 9.6 Infrared spectra of 1-hexanol. The upper spectrum is a reproduction of an actual spectrum, while the lower one is an artist's rendition. Upper spectrum © Sadtler Research Laboratories, Division of Bio-Rad Laboratories, Inc., 1986.

trophotometer. The lower spectrum is an artist's rendition of the real spectrum. In the artist-rendered spectrum, many small extraneous peaks arising from impurities and electronic "noise" have been deleted. For the sake of clarity, we will use this second type of spectrum in this text.

The scales at the bottoms of these spectra are in wavenumbers, decreasing from 4000 cm^{-1} to about 670 cm^{-1} or lower. The corresponding wavelengths in μm (or μ) are shown at the top. The wavelength or frequency of the *minimum point* (maximum absorption) of an absorption band is used to identify each band. This point is more reproducible than the range of a wide band, which may vary with concentration or with the sensitivity of the instrument.

The infrared bands in a spectrum can be classified by intensity: *strong* (*s*), *medium* (*m*), and *weak* (*w*). A weaker band overlapping a stronger band is called a *shoulder* (*sh*). These terms are relative, and the assignment of any given band as *s*, *m*, *w*, or *sh* is qualitative at best. In Figure 9.6, a few bands in the lower spectrum are labeled according to this classification method.

The number of identical groups in a molecule alters the relative strengths of the absorption bands in a spectrum. For example, a single OH group in a molecule produces a relatively strong absorption, while a single CH absorption is relatively weak. However, if a compound has many CH bonds, the collective effect of the CH absorption gives a peak that is medium or even strong.

Section 9.5
Interpretation of Infrared Spectra

Chemists have studied thousands of infrared spectra and have determined typical ranges of the wavelengths of absorption for each of the functional groups. **Correlation charts** provide summaries of this information.

A typical correlation chart for the stretching and bending frequencies of various groups is shown inside the back cover of this book. From the chart we see that OH and NH stretching bands are found between 3000 and 3700 cm^{-1} (2.7–3.3 μm). (Figure 9.6, the infrared spectrum of 1-hexanol, shows absorption in this region.) If the infrared spectrum of a compound of unknown structure shows absorption in this region, then we suspect that the compound contains either an OH or an NH group in its structure. If this region does not contain an absorption band, we conclude that the structure probably does not have an OH group or an NH group.

The region from 1400 to 4000 cm^{-1} (2.5 μm to about 7.1 μm), to the left in the infrared spectrum, is especially useful for identification of the various functional groups. This region shows absorption arising from stretching modes. The region to the right of 1400 cm^{-1} is often quite complex because both stretching and bending modes give rise to absorption here. In this region, correlation of an individual band with a specific functional group usually cannot be made with accuracy. However, each organic compound has its own unique absorption here. This part of the spectrum is therefore called the **fingerprint region**. Although the left-hand portion of a spectrum may appear the same for similar compounds, the fingerprint region must also match for two spectra to represent the same compound.

Figure 9.7 shows the infrared spectra of two alkanes of the formula C_8H_{18}, *n*-octane and 2-methylheptane. Note that the two spectra are practically identical from 1400 to 4000 cm^{-1}, but that the fingerprint regions are slightly different.

Figure 9.7 Infrared spectra of *n*-octane and 2-methylheptane, showing slight differences in the fingerprint regions. Also note the CH absorption peaks.

In the following sections, we will discuss the characteristic infrared absorption of compounds containing aliphatic C—C and C—H bonds and a few functional groups. Our intent is to develop familiarity with the features of typical infrared spectra. As we discuss the various functional groups in future chapters, we will also include discussions of their infrared spectral characteristics.

A. Carbon–Carbon and Carbon–Hydrogen

Bonds between sp^3-hybridized carbon atoms (C—C single bonds) give rise to weak absorption bands in the infrared spectrum. In general, these absorption bands are not very useful for structure identification. Bonds between sp^2-hybridized carbons (C=C) often exhibit characteristic absorption (variable in strength) around 1600–1700 cm^{-1} (5.8–6.2 μm). Aryl carbon–carbon bonds show absorption at slightly lower frequencies (to the right in the spectrum). Bonds between sp-hybridized carbons (C≡C) show weak, but extremely characteristic, absorption at 2100–2250 cm^{-1} (4.4–4.8 μm), a region of the spectrum where most other groups show no absorption.

sp^2 C=C: 1600–1700 cm^{-1} (5.8–6.2 μm)

sp^2 C—C (aryl):   1450–1600 cm^{-1} (6.25–6.9 μm)

sp C≡C: 2100–2250 cm^{-1} (4.4–4.8 μm)

Almost all organic compounds contain CH bonds. Absorption arising from CH stretching is seen at about 2800–3300 cm^{-1} (3.1–3.75 μm). The CH stretching peaks are often useful in determining the hybridization of the carbon atom. In Figure 9.7, the CH absorption is indicated. Spectra showing C—C and CH absorption by alkenes, alkynes, and benzene compounds will be presented in Chapters 10 and 11.

sp^3 C—H (alkanes or alkyl groups):   2800–3000 cm^{-1} (3.3–3.6 μm)

sp^2 C—H (=CH—):   3000–3300 cm^{-1} (3.0–3.3 μm)

sp C—H (≡CH):   ∼ 3300 cm^{-1} (3.0 μm)

The **geminal-** or **gem-dimethyl** grouping (two methyl groups on the same carbon) often exhibits a double CH bending peak in the 1360–1385 cm^{-1} (7.22–7.35 μm) region (see Figure 9.7). Unfortunately, the two peaks are not visible in all spectra; sometimes only a single peak is observed.

⟩C(CH$_3$)$_2$:   1360–1385 cm^{-1} (7.22–7.35 μm) (two peaks)

B. Haloalkanes

The stretching absorption of the CX bond of a haloalkane falls in the fingerprint region of the infrared spectrum, from 500 to 1430 cm^{-1} (7–20 μm). (See Figure 9.8.) Without additional information, the presence or absence of a band in this region cannot be used for verifying the presence of a halogen in an organic compound.

Figure 9.8 Infrared spectrum of chloroform, a trihaloalkane.

C. Alcohols and Amines

Alcohols and amines exhibit conspicuous OH or NH stretching absorption at $3000-3700$ cm^{-1} ($2.7-3.3$ μm), to the *left* of CH absorption. If there are two hydrogens on an amine nitrogen ($-NH_2$), the NH absorption appears as a double peak. If there is only one H on the N, only one peak is observed. Of course, if there is no NH (as in the case of a tertiary amine, R_3N), there is no absorption in this region. Figure 9.9 shows the spectrum of an alcohol, while Figure 9.10 shows spectra of the three types of amines.

Alcohols and amines also exhibit C—O and C—N absorption in the fingerprint region. These bands are not always easy to identify because this region of the spectrum often contains a large number of peaks.

OH or NH: $3000-3700$ cm^{-1} ($2.7-3.3$ μm)

C—O or C—N: $900-1300$ cm^{-1} ($8-11$ μm)

Hydrogen bonding changes the position and appearance of an infrared absorption band. The spectra in Figures 9.9 and 9.10 are those of pure liquids in which hydrogen bonding is extensive. The OH absorption in Figure 9.9 is a wide band at about 3330 cm^{-1} (3.0 μm). When hydrogen bonding is less extensive, a sharper, less intense OH peak is observed. Figure 9.11 (page 334) shows two partial spectra of an alcohol. One spectrum is of a pure liquid (hydrogen bonded); the other spectrum is of the alcohol in the vapor phase (not hydrogen bonded). The differences in OH absorption are apparent in this figure.

Absorption by NH bonds is less intense than OH absorption, partly because of weaker hydrogen bonds in amines and partly because NH bonds are less polar.

Study Problem

9.3 Figure 9.12 (page 334) gives the infrared spectra of two compounds, I and II, both of which appear in the following list. Which is I and which is II?

(a) $CH_3(CH_2)_6CH_3$ (b) $CH_3(CH_2)_6CH_2OH$ (c) $CH_3(CH_2)_5N(CH_3)_2$
(d) $CH_3CH_2CH_2CH_2NH_2$ (e) $CH_3(CH_2)_6CH_2I$ (f) $CH_3CH_2CH_2NHCH_3$

Figure 9.9 Infrared spectrum of cyclopentanol, an alcohol.

Figure 9.10 Infrared spectra of a primary amine, *n*-propylamine (top); a secondary amine, dipropylamine (center); and a tertiary amine, tripropylamine (bottom).

Figure 9.11 Partial infrared spectra of an alcohol in the vapor and liquid phases, showing hydrogen-bonded and nonhydrogen-bonded OH absorption.

Figure 9.12 Infrared spectra for Problem 9.3.

Figure 9.13 Infrared spectrum of di-*n*-propyl ether, an ether.

D. Ethers

Ethers have a C—O stretching band that falls in the fingerprint region at $1050\text{–}1260 \text{ cm}^{-1}$ (7.9–9.5 μm). Because oxygen is electronegative, the stretching causes a large change in bond moment. Therefore, the C—O absorption is usually strong (see Figure 9.13). Alcohols, esters, and other compounds containing C—O single bonds also show absorption here.

Study Problem

9.4 A compound yielded only iodoethane as a product when heated with HI. The infrared spectrum of the original compound is shown in Figure 9.14. What is its structure? Make assignments to the prominent spectral bands.

Figure 9.14 Infrared spectrum of unknown compound for Problem 9.4.

Table 9.2 Stretching vibrations for some carbonyl compounds[a]

Type of Compound	Position of Absorption	
	cm^{-1}	μm
aldehyde, $RCHO$	1720–1740	5.75–5.81
ketone, $RCOR$	1705–1750	5.71–5.87
carboxylic acid, $RCOOH$	1700–1725	5.80–5.88
ester, $RCOOR$	1735–1750	5.71–5.76

[a] In each case, R is saturated and aliphatic. Compounds with ring strain, conjugation, and other substituents may fall outside the listed absorption range.

E. Carbonyl Compounds

One of the most distinctive bands in an infrared spectrum is the one arising from the carbonyl stretching mode. This is a strong peak observed somewhere between 1640 and 1820 cm^{-1} (5.5–6.1 μm).

The carbonyl group is part of a number of functional groups. The exact position of the carbonyl absorption, the positions of other absorption bands in the infrared spectrum, and other spectral techniques (particularly nmr) may be needed to identify the functional group. The positions of C=O absorption for aldehydes, ketones, carboxylic acids, and esters are listed in Table 9.2.

Ketones give the simplest spectra of the carbonyl compounds. If a compound is an aliphatic ketone, all strong stretching infrared absorption arises from either C=O or CH. Other functionality may increase the complexity of the spectrum, of course. The infrared spectrum of a ketone is shown in Figure 9.15.

Figure 9.15 Infrared spectrum of butanone, a ketone.

Figure 9.16 Infrared spectrum of butanal, an aldehyde.

Aldehydes give infrared spectra very similar to those of ketones. The important difference between an aldehyde and a ketone is that an aldehyde has an H bonded to the carbonyl carbon. This particular CH bond shows two characteristic stretching bands (just to the right of the aliphatic CH band) at 2820–2900 cm^{-1} (3.45–3.55 μm) and 2700–2780 cm^{-1} (3.60–3.70 μm). Both these CH peaks are sharp, but weak, and the peak at 2900 cm^{-1} (3.45 μm) may be obscured by overlapping absorption of other CH bonds (see Figure 9.16). The aldehyde CH also has a very characteristic nmr absorption. (See Section 9.7.) If the infrared spectrum of a compound suggests that the structure is an aldehyde, the nmr spectrum should be checked.

Carboxylic acids exhibit typical C=O absorption and also show a very distinctive O—H band, which begins at about 3330 cm^{-1} (3.0 μm) and slopes into the aliphatic CH absorption band (Figure 9.17). The reason that a carboxyl OH gives a different-looking spectrum from that of an alcohol OH is that carboxylic

Figure 9.17 Infrared spectrum of propanoic acid, a carboxylic acid.

Figure 9.18 Infrared spectrum of ethyl pentanoate, an ester.

acids form hydrogen-bonded *dimers*:

Esters exhibit both a typical carbonyl band and a C—O band. The C—O band, like that in ethers, is observed in the fingerprint region, 1100–1300 cm^{-1} (7.7–9.1 μm) and is sometimes difficult to assign. However, this C—O band is strong and, in some cases, may be used to distinguish between esters and ketones. See Figure 9.18 for the infrared spectrum of a typical ester.

Study Problem

9.5 One of the two spectra in Figure 9.19 is that of a ketone; the other is of an ester. Which is which? Explain your answer.

Figure 9.19 Infrared spectra of unknown compounds for Problem 9.5.

Figure 9.19 *(continued)* Infrared spectra for Problem 9.5.

▼ ■ Section 9.6
Nuclear Magnetic Resonance Spectroscopy

The infrared spectrum of a compound gives a picture of the different functional groups in an organic molecule, but gives only meager clues about the hydrocarbon portion of the molecule. **Nuclear magnetic resonance (nmr) spectroscopy** fills this gap by providing information about the carbon and hydrogen atoms in the molecule.

Nmr spectroscopy is based upon the absorption of radio waves by certain nuclei in organic molecules when they are in a strong magnetic field. Before proceeding with a discussion of nmr spectra and their use in organic chemistry, let us consider the physical principles that give rise to the nmr phenomenon.

A. Origin of the Nmr Phenomenon

The nuclei of atoms of all elements can be classified as either *having spin* or *not having spin*. A nucleus with spin gives rise to a small magnetic field, which is described by a **nuclear magnetic moment**, a vector.

To the organic chemist, the most important isotopes that have nuclear spin are ^1_1H and $^{13}_6C$. Equally important is the fact that the common isotopes of carbon ($^{12}_6C$) and oxygen ($^{16}_8O$) do not have nuclear spin. Different isotopes (such as ^1H and ^2H) or different elements (such as ^1H and ^{13}C) with nuclear spin do not absorb energy at the same combination of magnetic field strength and radio-frequency. Thus we must run separate types of nmr spectra for each isotope. We will first describe nmr spectroscopy using ^1H, or proton, nmr spectra and then discuss ^{13}C spectroscopy in Section 9.12.

In nmr spectroscopy, an **external magnetic field** is generated by a permanent magnet or an electromagnet. The strength of this external field is symbolized by H_0, and its direction is represented by an arrow.

Symbol representing the external magnetic field: $\underset{H_0}{\uparrow}$

A spinning proton with its nuclear magnetic moment is similar, in many respects, to a tiny bar magnet. When molecules containing hydrogen atoms are placed in an external magnetic field, the magnetic moment of each hydrogen nucleus, or proton, aligns itself in one of two different orientations with respect to the external magnetic field. (Keep in mind that it is only the magnetic moments of *hydrogen nuclei*, not molecules, that become aligned.) The two orientations that the nuclear magnetic moment may assume are **parallel** and **antiparallel** to the external field. In the parallel state, the magnetic moment of the proton points in the *same direction* as that of the external field. In the antiparallel state, the magnetic moment of the proton *opposes* the external field. At any given time, approximately half the protons in a sample are in the parallel state and half are in the antiparallel state.

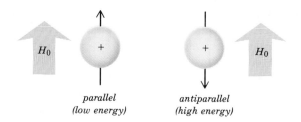

parallel
(low energy)   antiparallel
(high energy)

The parallel state of a proton is slightly more stable (lower energy) than the antiparallel state. When exposed to the proper frequency of radio waves, the magnetic moments of a small fraction of the parallel protons absorb energy and turn around, or *flip*, to the higher-energy antiparallel state. After being converted to the higher-energy antiparallel state, a nucleus can lose the absorbed energy to its surroundings as heat and thereby return to the lower-energy parallel state. When the sample is scanned with a properly adjusted instrument, the absorption and loss of energy by the nuclei are sufficiently rapid that a slight excess of nuclei in the parallel state can be maintained. Therefore, the sample continues to absorb energy, and the signal in the spectrum is observed.

The amount of energy required to flip the magnetic moment of a proton from parallel to antiparallel depends, in part, upon the strength of H_0. If H_0 is increased, the energy difference between the parallel and antiparallel states increases. Thus, if H_0 is increased, the nucleus is more resistant to being flipped and higher-energy, higher-frequency radiation is required.

When the particular combination of external magnetic field strength and radiofrequency causes a proton to flip from parallel to antiparallel, the proton is said to be in **resonance** (a different type of resonance from that of "resonance" structures of benzene). The term **nuclear magnetic resonance** means "nuclei in resonance in a magnetic field."

It would seem that all protons should come into resonance at the same combination of H_0 and radiofrequency. However, this is not the case. The magnetic field actually observed by a proton in a particular molecule is a combination of two fields: (1) the applied external magnetic field (H_0) and (2) **induced molecular magnetic fields**, small magnetic fields induced in the bonds of a molecule by H_0. Thus, protons within different types of molecules and within different parts of the same molecule may be exposed to magnetic fields of different intensities. The magnetic field observed by a proton is also modified by the spin states of nearby protons, a topic we will discuss later in this chapter. The protons of a molecule flip at different combinations of H_0 and radiofrequency because they are in different molecular (and magnetic) environments. Because of the differences in the energy absorption by protons, we are able to obtain a spectrum of the different types of protons.

B. The Nmr Spectrum

A diagram of an nmr spectrometer is shown in Figure 9.20. The sample is placed between the poles of a magnet and is irradiated with radio waves. When the protons flip from the parallel to the antiparallel state, the absorption of energy is detected by a power indicator.

In one type of nmr spectrometer, the radiofrequency is held constant at 60 MHz (60 megaHertz, or 60×10^6 Hz), H_0 is varied over a small range, and the frequency of energy absorption is recorded at the various values for H_0. Thus, the nmr spectrum is a graph of the amount of energy absorbed (I, or intensity) versus magnetic field strength.

Figure 9.21 shows two nmr spectra of methanol: a reproduction of an actual spectrum and an artist's copy. On the lower spectrum, we have indicated the direction of field sweep from weak H_0 to strong H_0. You can see two principal peaks in these spectra. The left-hand peak arises from the OH proton; the right-hand peak, from the CH_3 protons. The protons that flip more easily

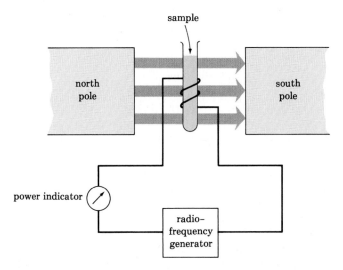

Figure 9.20 Schematic diagram of an nmr spectrometer.

Figure 9.21 ^{1}H nmr spectra of methanol (solvent, CCl_4). The upper spectrum is a reproduction of an actual spectrum, while the lower one is an artist's rendition. The lower spectrum shows the field sweep and chemical shifts. (The fluctuations to the right of each signal, called "ringing," arise from distortion of the magnetic field when the sample is scanned rapidly. Ringing is an indication that the instrument is properly adjusted.) Upper spectrum © Sadtler Research Laboratories, Division of Bio-Rad Laboratories, Inc., 1986.

absorb energy at a lower H_0. They give rise to an absorption peak **downfield** (farther to the left). The protons that flip with greater difficulty absorb energy at a higher H_0 and give a peak that is **upfield** (to the right).

In an nmr spectrum, the position of absorption by a proton depends on the net strength of the localized magnetic field surrounding it. This localized field is a result of the applied field H_0 and the induced molecular field surrounding the proton and opposing the applied field. If the induced field around a particular

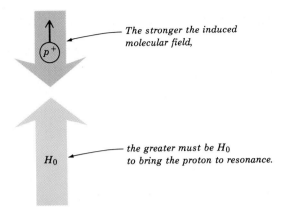

Figure 9.22 Induced molecular fields oppose H_0, the external applied magnetic field. The strength of H_0 needed to flip the spin states of protons depends partly on how strong the induced molecular magnetic field is.

proton is relatively strong, it opposes H_0 more strongly and a greater applied field is required to bring the proton into resonance. This particular proton is said to be **shielded** and we observe its signal *upfield* in the spectrum. On the other hand, if the induced field around a proton is relatively weak, then a weaker applied field will bring this proton into resonance. This proton is said to be **deshielded** and its signal appears *downfield* (Figure 9.22).

Shielding and deshielding are relative terms. In order to obtain quantitative measurements, we need a reference point. The compound that has been chosen for this reference point is tetramethylsilane (TMS), $(CH_3)_4Si$, the protons of which absorb to the far right in the nmr spectrum. The absorption for most other protons is observed downfield from that of TMS. In practice, a small amount of TMS is added directly to the sample, and the peak for TMS is observed on the spectrum along with any absorption peaks from the sample compound. The difference between the position of absorption of TMS and that of a particular proton is called the **chemical shift**. You can see the TMS peak in the spectra in Figure 9.21, where we have also indicated the chemical shifts.

Chemical shifts are reported in **δ values**, which are expressed as *parts per million (ppm) of the applied radiofrequency*. At 60 MHz, 1.0 ppm is 60 Hz. Therefore, a δ value of 1.0 ppm is 60 Hz downfield from the position of absorption of TMS, which is set at 0 ppm, or $\delta 0$. As may be seen in Figure 9.21, the two types of protons in CH_3OH have δ values of 3.4 ppm and 4.15 ppm, respectively. (Note that the lower scale on an nmr spectrum shows δ values in ppm. The use of the Hz scale at the top of an nmr spectrum will be discussed in Section 9.10.)

Section 9.7
Types of Induced Molecular Magnetic Fields

A. Fields Induced by Sigma Electrons

Any hydrogen atom in an organic compound is bonded to carbon, oxygen, or some other atom by a sigma bond. The external magnetic field causes these sigma-bond electrons to circulate. The result is a small molecular magnetic field that opposes H_0 (Figure 9.23). It takes a slightly higher external field strength to

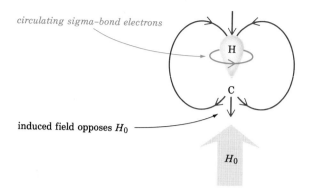

circulating sigma–bond electrons

induced field opposes H_0

H_0

Figure 9.23 The induced field from circulating sigma-bond electrons opposes H_0 in the vicinity of the proton. (For simplicity, the figure shows only one CH bond of an organic molecule.)

overcome the effect of the induced field to bring the proton into resonance. Therefore, the proton absorbs upfield compared to a hypothetical naked proton. The strength of the induced field depends upon the electron density near the hydrogen atom in the sigma bond. The higher the electron density, the greater the induced field (and the farther upfield the observed absorption).

The electron density of a carbon–hydrogen covalent bond is affected by the electronegativity of the other atoms bonded to the carbon. Let us consider a specific example. The CF bond in CH_3F is polar: the fluorine atom carries a partial negative charge and the carbon atom carries a partial positive charge. Because the carbon has a partial positive charge, the electrons in each C—H sigma bond are drawn toward the carbon and away from the hydrogen atom. Recall from our discussion of carbocation stability that the polarization of bonds by positive or negative centers is called the **inductive effect**. This shift in electron density toward an electronegative element (F) is another example of the inductive effect. In this case, the electron-withdrawing effect of F results in a greater electron density around F and a lesser electron density around each hydrogen. The protons of CH_3F are deshielded and absorb downfield compared to the protons of CH_4.

$$H \rightarrow \overset{\displaystyle H}{\underset{\displaystyle H}{C}} \rightarrow F$$

F causes a decrease in e^- density around each H

The following series of methyl halides show increasing shielding of the hydrogen nuclei with decreasing electronegativity of the atom attached to —CH_3.

H_3C—F	H_3C—Cl	H_3C—Br	H_3C—I
$\delta 4.3$	$\delta 3.0$	$\delta 2.7$	$\delta 2.1$

increased shielding of H

Sample Problem

Which of the circled protons is the most shielded? The most deshielded? What are the relative positions of nmr absorption?

(a) $CH_3C\,\boxed{H}_2Cl$ (b) $CH_3C\,\boxed{H}_3$ (c) $CH_3C\,\boxed{H}_2I$

Solution: Chlorine is more electronegative than iodine, which in turn is more electronegative than hydrogen. Therefore, (a) is the most deshielded and absorbs downfield; (b) is the most shielded and absorbs upfield, closer to TMS; and (c) is intermediate.

In a single molecule, a proton that is bonded to the same carbon as an electronegative atom is more deshielded than protons on other carbons. Figure 9.31 (page 352) shows the spectrum of chloroethane. The peaks at $\delta1.5$ arise from the CH_3 protons, while the deshielded CH_2Cl protons absorb farther downfield, at $\delta3.55$. (The complexity of these peaks will be discussed later in this chapter.)

The inductive effect of an electronegative atom falls off rapidly when it is passed through a number of sigma bonds. In nmr spectra, the inductive effect is negligible three carbons away from the electronegative atom.

effect of X is   H   H   H   *effect of X is important*
of little importance | | | *in decreasing*
in decreasing   H—C—C—C—X   *e⁻ density around*
e⁻ density | | | *this proton*
around this proton   H   H   H

Most elements encountered in organic compounds are more electronegative than carbon. Their inductive effect is one of electron-withdrawal, and protons affected by them are deshielded. However, silicon is *less* electronegative than carbon. The silicon–carbon bond is polarized such that carbon carries a partial negative charge. The electrons in the CH bond of an $SiCH_3$ group are repelled from the negative carbon and pushed toward the hydrogen atoms. In this case, the inductive effect is *electron-releasing*. The protons of an $SiCH_3$ group are highly shielded because of the increased electron density around them. This is the reason that the protons of TMS absorb upfield and that TMS provides a good reference peak in an nmr spectrum.

greater e⁻ density on H; highly shielded

$$\text{H}_3\text{C}\leftarrow\underset{\underset{\text{CH}_3}{\downarrow}}{\overset{\overset{\text{CH}_3}{\uparrow}}{\text{Si}}}\rightarrow\text{CH}_3$$

B. Fields Induced by Pi Electrons

The magnetic fields induced by *pi electrons* are *directional* (that is, unsymmetrical). A measurement that varies depending on the direction in which the measurement is taken is said to be **anisotropic**. Because the effects of molecular fields induced by pi electrons are direction-dependent, they are called

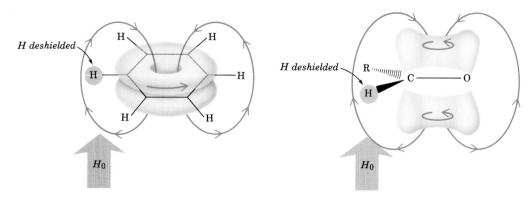

Figure 9.24 Circulating pi electrons in benzene or aldehydes induce a magnetic field that deshields the adjacent protons. (All six benzene protons are deshielded.)

anisotropic effects. (These effects are contrasted to inductive effects, which are symmetrical around the proton.) Anisotropic effects occur in addition to the ever-present molecular fields induced by sigma-bond electrons.

In benzene, the pi electrons are delocalized around the ring. Under the influence of an external magnetic field, these pi electrons *circulate around the ring*. This circulation, called **ring current**, induces a molecular magnetic field with the geometry shown in Figure 9.24. (All the benzene rings, of course, do not line up as we have represented the ring in this figure. This figure shows the net effect of the vector forces.)

Referring to Figure 9.24, you can see that the field induced by the ring current actually augments the applied field H_0 outside the ring in the vicinity of the benzene protons. Relatively less applied field is required to bring

Figure 9.25 ^{1}H nmr spectrum of 2-phenylpropene, showing absorption by aryl, vinylic, and methyl protons.

Figure 9.26 ¹H nmr spectrum of ethanal (acetaldehyde), showing the offset scan for the aldehyde proton.

aryl protons into resonance than is necessary for alkyl protons. Therefore, aryl protons are deshielded and absorb farther downfield than alkyl protons.

    An analogous situation is observed with a vinylic hydrogen or an aldehyde hydrogen. In either case, the pi electrons are set in motion and induce a field that adds to the applied field in the vicinity of the =CH proton (see Figure 9.24). A proton bonded to an sp^2-hybridized carbon of C=O or C=C absorbs downfield from an alkyl proton. Figure 9.25 shows the nmr spectrum of a compound with aryl protons, vinylic protons, and a CH_3 group. In this spectrum, it is evident that the vinylic protons absorb downfield from CH_3 protons and that aryl protons absorb even farther downfield.

    Figure 9.26 shows the spectrum of an aldehyde. The aldehyde proton is shifted downfield both by anisotropic effects and by electron-withdrawal by the carbonyl oxygen. The combination of effects results in absorption that is far downfield (δ9–10), outside the normal scan for some nmr spectrometers. Instrument design allows scanning at field strengths that are lower than normal; the scan of this region (δ8–20), traced above the standard nmr scan, is called an **offset scan**.

C. Summary of Induced-Field Effects

The presence of an electronegative atom causes a decrease in electron density around a proton by the **inductive effect**. Such a proton is deshielded and absorbs downfield. In aromatic compounds, alkenes, and aldehydes, a proton bonded to the sp^2-hybridized carbon is deshielded by **anisotropic effects** and absorbs even farther downfield. The absorption positions are summarized in Figure 9.27. A chart inside the back cover lists the δ values for a number of types of protons.

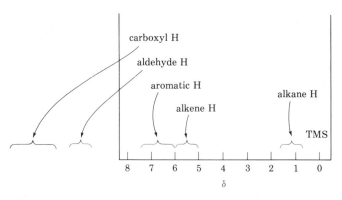

Figure 9.27 Relative positions of proton absorption in the ^{1}H nmr spectrum. (See also the chart inside the back cover of this book.)

▼ Section 9.8
Counting the Protons

A. Equivalent and Nonequivalent Protons

Protons that are in the same magnetic environment in a molecule have the same chemical shift in an nmr spectrum. Such protons are said to be **magnetically equivalent protons**. Protons that are in different magnetic environments have different chemical shifts and are said to be **nonequivalent protons**.

Magnetically equivalent protons in nmr spectroscopy are generally the same as *chemically equivalent protons*. In chloroethane, the three methyl protons are magnetically equivalent and are also chemically equivalent. To see that they are chemically equivalent, imagine a chemical reaction in which one of these three protons is replaced by another atom, such as bromine. If only one product can result, regardless of which proton is replaced, then the protons are chemically equivalent. In the following example, note that replacement of any of the methyl protons by Br leads to the same compound, 1-bromo-2-chloroethane.

equivalent

$$H-\overset{\overset{\displaystyle H}{|}}{\underset{\underset{\displaystyle H}{|}}{C}}-CH_2Cl \longrightarrow Br-\overset{\overset{\displaystyle H}{|}}{\underset{\underset{\displaystyle H}{|}}{C}}-CH_2Cl$$

If any H is replaced by Br, only one bromo compound is obtained.

The two protons of the CH_2Cl group are also magnetically and chemically equivalent to each other. However, the three protons of the CH_3 group are not equivalent to the two CH_2Cl protons.

three equivalent protons (but nonequivalent to CH_2)   *two equivalent protons (but nonequivalent to CH_3)*

$$CH_3CH_2Cl$$

The three CH_3 protons have the same chemical shift and absorb at the same position in the nmr spectrum. The two CH_2 protons are deshielded

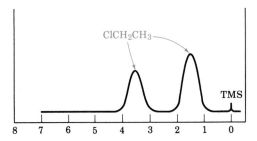

Figure 9.28 Stylized low-resolution ^{1}H nmr spectrum of CH_3CH_2Cl.

compared to the methyl protons and have a greater chemical shift. A low-resolution spectrum of chloroethane would look like the stylized spectrum shown in Figure 9.28.

Equivalent protons need not necessarily be on the same carbon atom. For example, diethyl ether contains only two types of equivalent protons: the CH_3 protons and the CH_2 protons.

six equivalent protons

$$CH_3CH_2OCH_2CH_3$$

four equivalent protons

Restricted rotation and the geometry of a molecule can affect the equivalence of protons. For example, rotation about a carbon–carbon double bond is restricted. Therefore, in chloroethene, the proton *cis* to the Cl atom is in a different magnetic environment from that of the *trans* proton. Both these protons are in different magnetic environments from that of the proton on the C—Cl carbon. In chloroethene, all three protons are nonequivalent.

Cl   H — *cis to Cl*

    C=C

H   H — *trans to Cl*

chloroethene
(vinyl chloride)

three nonequivalent protons

Sample Problem

Which protons in *p*-chlorotoluene (1-chloro-4-methylbenzene) are chemically equivalent, and which are nonequivalent?

Solution:

H_b H_a

H_3C———Cl

H_b H_a

the three methyl protons are equivalent;
the two H_a's are equivalent;
the two H_b's are equivalent;
the methyl protons, H_a and H_b, are nonequivalent

Replacement of either proton labeled H_a by Br would result in the same compound. Similarly, replacement of either H_b would yield the same compound. However, replacement of H_a by Br would result in a different compound from that obtained by replacing H_b by Br.

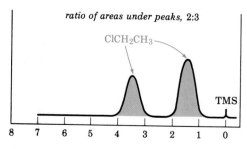

B. Areas Under the Peaks

If we measure the areas under the peaks in an nmr spectrum, we find that *the areas are in the same ratio as the number of protons that give rise to each signal.* In the case of chloroethane (Figure 9.29), the ratio is 2:3. (Note that the *height* of the peak is not the important feature, but rather the *area* under the peak.)

Sample Problem

How many types of equivalent protons are there in each of the following structures? What would be the relative areas under the nmr absorption bands?

(a) $(CH_3)_2CHCl$ (b) $CH_3CH_2OCH_2CH_3$ (c) $Cl\!-\!\!\bigcirc\!\!-\!OCH_3$

Solution: (a) two, 6:1   (b) two, 3:2,   (c) three, 3:2:2 (3 for CH_3, 2 for aryl protons adjacent to the oxygen, 2 for aryl protons adjacent to Cl).

Most nmr spectrometers are equipped with **integrators**, which give a signal that shows the relative areas under the peaks in a spectrum. The integration appears as a series of steps superimposed upon the nmr spectrum. The height of the step over each absorption peak is proportional to the area under that peak.

ratio of areas under peaks, 2:3

ClCH₂CH₃

TMS

8 7 6 5 4 3 2 1 0

Figure 9.29 Stylized low-resolution ^1H nmr spectrum of CH_3CH_2Cl, showing areas under the peaks.

Figure 9.30 ¹H nmr spectrum of 1-bromo-2,4,6-trimethylbenzene, showing an integration curve.

From the relative heights of the steps on the integration curve, the relative areas under the peaks may be determined. In Figure 9.30, the heights of the steps of the integration curve were measured with a ruler and found to be 33 mm, 100 mm, and 50 mm. For determining the relative numbers of equivalent protons, these values are converted to ratios of small whole numbers, 2:6:3. (In the laboratory the numbers rarely turn out to be exactly whole numbers; therefore, some rounding of numbers must be done.) We can compare these values with the numbers of protons in the known structure of 1-bromo-2,4,6-trimethylbenzene and, indeed, the values agree.

In this text, we will not show the integration lines in the spectra, but will indicate relative areas as numbers directly above the proton absorption peaks, as shown in Figure 9.31.

Sample Problem

The heights of four steps of an integration curve were measured to be 90 mm, 36 mm, 37 mm, and 54 mm. Calculate the ratio of the different types of protons in the sample compound.

Solution:

1. Divide each height by the smallest one (36 mm in this case):

$$\frac{90 \text{ mm}}{36 \text{ mm}} = 2.5 \qquad \frac{36 \text{ mm}}{36 \text{ mm}} = 1.0 \qquad \frac{37 \text{ mm}}{36 \text{ mm}} = 1.0 \qquad \frac{54 \text{ mm}}{36 \text{ mm}} = 1.5$$

2. Multiply the quotients by the integer that will convert them to the smallest whole numbers possible (rounding off, if necessary). In this case, multiplying by 2 gives the ratio of 5:2:2:3. Therefore, the ratio of the numbers of nonequivalent protons is 5:2:2:3.

Study Problem

9.6 The integration curve of an nmr spectrum shows two steps with heights of 62.5 mm and 20.8 mm. Which of the following compounds is compatible with this spectrum?

    (a) CH_3CH_3 **(b)** $CH_3CH_2CH_3$ **(c)** $CH_3CH_2CH_2CH_3$

▼ **Section 9.9**
Spin–Spin Coupling

The spectrum of chloroethane in Figures 9.28 and 9.29 is a stylized low-resolution spectrum. If we increase the resolution (that is, the sensitivity), the peaks are resolved, or split, into groups of peaks. (See Figure 9.31.) This type of splitting is called **spin–spin splitting** and is caused by the presence of *vicinal, or neighboring, protons (protons on an adjacent carbon) that are nonequivalent to the proton in question.* Protons that split each other's signals are said to have undergone **spin–spin coupling**.

*these protons split
the signal for* ——
the CH_2 *protons*

*these protons split
— the signal for
the* CH_3 *protons*

$$CH_3CH_2Cl$$

    Why do protons undergo spin–spin coupling? The splitting of the signal arises from the two spin states (parallel and antiparallel) of the neighboring protons.

Figure 9.31 ^1H nmr spectrum of chloroethane, CH_3CH_2Cl.

Signal for H_a split by the two spin states of H_b:

$$H_a \quad H_b \qquad\qquad H_a \quad H_b$$

$$-\overset{|}{\underset{|}{C}}-\overset{|}{\underset{|}{C}}- \qquad H_0 \qquad -\overset{|}{\underset{|}{C}}-\overset{|}{\underset{|}{C}}-$$

parallel   *antiparallel*

The spin of a proton generates a magnetic moment. If the spin of the neighboring proton (H_b in the preceding formula) is parallel, its magnetic moment *adds* to the applied magnetic field. Consequently, the first proton (H_a in the preceding formula) sees a slightly stronger field and comes into resonance at a slightly lower applied field strength. If the neighboring proton is in the antiparallel state, its magnetic moment *decreases* the magnetic field around the first proton. In this case, it takes slightly more applied H_0 for that proton to come into resonance. Because approximately half the H_b nuclei are parallel and half are antiparallel at any given time, there are two types of H_a in the sample: those with a neighboring H_b in a parallel spin state and those with a neighboring H_b in an antiparallel spin state. Consequently, we observe two peaks for H_a instead of just one.

For many compounds, we can predict the number of spin–spin splitting peaks in the nmr absorption of a particular proton (or a group of equivalent protons) by counting *the number (n) of neighboring protons nonequivalent to the proton in question and adding 1*. This is called the **n + 1 rule**.

These three equivalent protons see *two* neighboring, nonequivalent protons. Their nmr band is split into 2 + 1, or 3, peaks.

These two equivalent protons see *three* neighboring, nonequivalent protons. Their nmr band is split into 3 + 1, or 4, peaks.

$$CH_3CH_2Cl$$

Protons that have the same chemical shift do not split each other's signals. Only neighboring protons that have different chemical shifts cause splitting. Some examples follow:

no neighboring H:
not split

one neighboring, nonequivalent H:
H_b split into two

one neighboring, nonequivalent H:
H_a split into two

$$CH_3O-\langle\text{ring}\rangle-NO_2$$

with Cl, Cl at top and H_a, H_b at bottom

$$H_3C-\underset{\underset{CH_3}{|}}{\overset{\overset{CH_3}{|}}{Si}}-CH_3$$

twelve equivalent H's:
no splitting

benzene ring with H at each position

six equivalent H's:
no splitting

In some cases, the magnetic environments of chemically nonequivalent protons are so similar that the protons exhibit identical chemical shifts. In this case, no splitting is observed. For example, toluene has four groups of chemically nonequivalent protons, yet the nmr spectrum shows only two absorption peaks (one for the CH_3 protons and one for the ring protons).

In summary, the **chemical shift** for a particular proton is determined by the molecular magnetic field surrounding it. The **area** under an absorption band is determined by the number of equivalent protons giving rise to the particular signal. **Spin–spin splitting** of a signal is dependent on the number of neighboring protons nonequivalent to the proton giving the signal.

Sample Problem

Predict the spin–spin splitting patterns for the protons in 2-chloropropane.

Solution:

Study Problem

9.7 Using the $n + 1$ rule, predict the number of nmr peaks for each of the indicated protons:

(a) $CH_3C\underline{H}_2CH_3$ (b) $(CH_3)_3C\underline{H}$ (c) $(C\underline{H}_3)_3CH$ (d)

▼ ## Section 9.10
Splitting Patterns

A. The Singlet

A proton with no neighboring protons magnetically nonequivalent to it shows a single peak, called a **singlet**, in the nmr spectrum. Figure 9.32, the nmr spectrum of *p*-methoxybenzaldehyde, shows two singlets: one for the $CH_3O—$ protons and one for the —CHO proton.

Figure 9.32 ^{1}H nmr spectrum of *p*-methoxybenzaldehyde. (Note the offset signal for the aldehyde proton; see Section 9.7B.)

Study Problems

9.8 Which of the following compounds would show at least one singlet in the nmr spectrum?

  (**a**) CH_3CH_3 (**b**) $(CH_3)_2CHCH(CH_3)_2$ (**c**) $(CH_3)_3CCl$
  (**d**) Cl_3CCHCl_2 (**e**) $ClSi(CH_3)_3$

9.9 How many peaks would you expect to observe in the nmr spectra of (**a**) cyclohexane, and (**b**) benzene?

B. The Doublet

A proton with one neighboring, nonequivalent proton gives a signal that is split into a double peak, or **doublet**. In the following example, a pair of doublets is produced, one for each proton.

H_a *gives a doublet*   — H_b *also gives a doublet*

$$\underset{\overset{|}{\text{C}}}{\text{H}_a}\ \underset{\overset{|}{\text{C}}}{\text{H}_b}$$

In the nmr spectrum of the preceding hypothetical structure, the δ value for each proton is the value at the *center* of the doublet. (See Figure 9.33.) The relative areas under the entire doublets in this case are 1:1, reflecting the fact that each doublet arises from the absorption by one proton. (The two peaks *within* any doublet also have an area ratio that is ideally 1:1, but may be slightly different, as may be seen for the aryl protons in Figure 9.32.)

The separation between the two peaks of a doublet is called the **coupling constant J**, and varies with the environment of the protons and their geometric

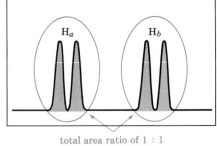

Figure 9.33 Spin–spin splitting pattern for
$$-\overset{\displaystyle H_a}{\underset{\displaystyle |}{\overset{|}{C}}}-\overset{\displaystyle H_b}{\underset{\displaystyle |}{\overset{|}{C}}}-.$$

relationship to each other. (Because splitting patterns are caused by internal forces, coupling constants are independent of the strength of H_0.) The symbol J_{ab} means the coupling constant for H_a split by H_b or for H_b split by H_a. For any pair of coupled protons, the J value is the same in each of the two doublets. By convention, J values are reported in Hz. Therefore, the Hz scale at the top of an nmr spectrum is used to determine coupling constants. For a pair of neighboring, nonequivalent protons attached to freely rotating carbons, the J value is about 7 Hz.

Figure 9.32, the nmr spectrum of *p*-methoxybenzaldehyde, shows a pair of doublets in the aromatic region. The two protons labeled H_a are equivalent, as are the two H_b protons. H_a and H_b are nonequivalent to each other and are neighboring. Their signals are split into a pair of doublets. Note that the peaks within the doublets are not perfectly symmetrical. The inside peaks of a doublet are taller. This phenomenon is called **leaning**.

Study Problem

9.10 Assuming that the nmr spectrum of each of the following partial structures would exhibit two groups of peaks, tell how many peaks would be in each group. (Use the $n + 1$ rule.)

(a) $-\overset{|}{\underset{|}{CH}}-\overset{|}{\underset{|}{CH}}-$ (b) $-CH_2-\overset{|}{\underset{|}{CH}}-$ (c) $CH_3-\overset{|}{\underset{|}{CH}}-$

Sample Problems

Which of the following compounds would show a doublet (as well as other signals) in its nmr spectrum?

(a) CH_3CHCl_2 (b) $CH_3CH_2CH_2Cl$ (c) $CH_3CHClCH_3$

Solution: Underlined protons would appear as doublets: (a) $C\underline{H}_3CHCl_2$; (b) no doublet;   (c) $C\underline{H}_3CHClC\underline{H}_3$ as one doublet and not as a pair of doublets. (Why?)

Give the total relative areas under all of the absorption bands for the compounds in the preceding problem.

Solution: (a) 3:1   (b) 3:2:2   (c) 6:1

Tell how many singlets and how many doublets would appear in the nmr spectrum of each of the following substituted benzenes:

(a) CH_3—⟨⟩—OCH_3 (b) Cl—⟨⟩—NH_2

Solution: (a) two singlets ($C\underline{H}_3$ and $OC\underline{H}_3$) and two doublets for the aryl protons; (b) one singlet ($N\underline{H}_2$) and two doublets. (A pair of doublets in the aryl region is typical of 1,4-disubstituted benzenes.)

Study Problem

$$\overset{\displaystyle O}{\overset{\displaystyle \|}{}}$$

9.11 The nmr spectrum of an aryl ketone ($Ar\overset{O}{\overset{\|}{C}}R$), C_8H_7ClO, is shown in Figure 9.34. What is the structure of this compound?

Figure 9.34 ^1H nmr spectrum for unknown aryl ketone in Problem 9.11.

C. The Triplet

If a proton (H_a) sees two neighboring protons equivalent to each other, but not equivalent to itself, the nmr signal of H_a is a **triplet** ($2 + 1 = 3$). If the two protons labeled H_b are equivalent, they give a signal that is split into a doublet by H_a.

H_a sees two neighboring H's,
signal is a triplet

H_b sees one neighboring H,
signal is a doublet

$$\begin{array}{cc} H_a & H_b \\ | & | \\ -C-C-H_b \\ | & | \end{array}$$

Figure 9.35 **Spin–spin splitting pattern for** $-\underset{|}{\overset{H_a}{C}}-\underset{|}{\overset{H_b}{C}}-H_b$.

Figure 9.36 **^{1}H nmr spectrum of 1,1,2-trichloroethane.**

The nmr absorption pattern for all three protons in the preceding partial structure consists of a doublet and a triplet. The peaks in the triplet each are separated by the *same J value as that for the doublet*. The total width of the triplet (from side peak to side peak) is therefore $2J$ (see Figure 9.35). The areas under the entire triplet and the entire doublet in our example are in the ratio of 1 for H_a to 2 for H_b. (The relative areas *within* the triplet, however, are in a ratio of 1:2:1. The reasons for this will be discussed in Section 9.11.)

The nmr spectrum of 1,1,2-trichloroethane, which has a doublet and a triplet, is shown in Figure 9.36.

$$Cl-\underset{\underset{\underline{H}}{|}}{\overset{\overset{Cl}{|}}{C}}-C\underline{H}_2-Cl$$

$\begin{cases} \textit{more shielded: upfield} \\ \textit{sees one H: doublet} \\ \textit{relative total area: 2} \end{cases}$

$\begin{cases} \textit{less shielded: downfield} \\ \textit{sees two H's: triplet} \\ \textit{relative total area: 1} \end{cases}$

Study Problem

9.12 Suggest a reason for the fact that the proton in the Cl_2CH group in 1,1,2-trichloroethane is less shielded than the protons in the CH_2Cl group.

Sample Problem

Which of the following compounds shows a triplet (among other signals) in the nmr? How many triplets will each compound show?

(a) $\overset{Cl}{\underset{|}{CH_2}}\overset{OCH_3}{\underset{|}{CH_2}}$ (b) CH_3CH_2—⬡   (c) $Cl_2CHCH_2CHCl_2$

Solution: (a) two triplets, one for each CH_2; (b) one triplet for CH_3; (c) two triplets, H_a split by two H_b's and H_b split by two H_a's.

$$Cl_2C\overset{H_a}{\underset{|}{-}}\overset{H_b}{\underset{\underset{H_b}{|}}{C}}\overset{H_a}{\underset{|}{-}}CCl_2$$

Study Problem

9.13 Figure 9.37 shows the nmr spectrum of 2-phenylethyl acetate. Assign each signal to the proper protons.

Figure 9.37 ^1H nmr spectrum of 2-phenylethyl acetate.

D. The Quartet

Consider a compound with a methyl group and one nonequivalent proton on the adjacent carbon.

The three equivalent methyl protons (H_b) see one neighboring proton and appear in the spectrum as a doublet, the total relative area of which is 3 (for three protons).

The signal arising from H_a is observed as a **quartet** (3 + 1) because it sees three neighboring protons. The J values between each pair of peaks in the quartet is the same as the J value between the peaks in the doublet. In our example, the total area under the quartet for H_a is 1. (The area ratio *within* a quartet is 1:3:3:1; see Figure 9.38.)

The ethyl group (CH_3CH_2—), which is very common in organic compounds, exhibits a characteristic nmr pattern—a triplet and a quartet.

Figure 9.38 Spin–spin splitting pattern for the compound.

Figure 9.39 ^1H nmr spectrum showing a typical ethyl pattern.

CH$_3$ sees two neighboring H's and is split into a triplet

$$C\underline{H}_3C\underline{H}_2-$$

CH$_2$ sees three neighboring H's and is split into a quartet

The chemical shifts of an ethyl group are also characteristic. The CH$_2$ is often bonded to an electronegative atom, such as oxygen, which deshields the CH$_2$ protons. The quartet is thus observed downfield, while the triplet for the more shielded CH$_3$ group is observed upfield. The nmr spectrum of chloroethane (Figure 9.31, page 352) shows typical ethyl absorption. Figure 9.39 contains another example of an nmr spectrum that shows an ethyl pattern: an upfield triplet and a downfield quartet.

Study Problem

9.14 Match each nmr spectrum in Figure 9.40 (page 362) with a compound in the following list:

(a) CH$_3$CO$_2$CH$_2$CH$_3$ (b) CH$_3$CH$_2$—⟨O⟩—I   (c) CH$_3$CO$_2$CH(CH$_3$)$_2$

(d) CH$_3$CH$_2$CH$_2$NO$_2$ (e) CH$_3$CH$_2$I   (f) (CH$_3$)$_2$CHNO$_2$

E. Chemical Exchange and Hydrogen Bonding

On the basis of the preceding discussions, one would expect the nmr spectrum of methanol (Figure 9.21, page 342) to show a doublet (for the CH$_3$ protons) and a quartet (for the OH proton). If the spectrum of methanol is run at a very low temperature ($-40°$) or with specially prepared solvents such as CCl$_4$ stored over Na$_2$CO$_3$, this is exactly what is observed. However, if the spectrum is run at room temperature or with ordinary spectroscopic solvents, only two singlets are observed.

The reason for this behavior of methanol is that alcohol molecules undergo rapid reaction with one another at room temperature in the presence of a trace of acid, exchanging OH protons in a process called **chemical exchange**. This exchange is so rapid that neighboring protons cannot distinguish any differences in spin states and thus see only the average value of zero. Amines (RNH$_2$ and R$_2$NH) also undergo chemical exchange.

$$CH_3\overset{..}{\underset{|}{O}}: + H'^+ \rightleftharpoons CH_3\overset{+}{\underset{|}{O}}H' \rightleftharpoons CH_3\overset{..}{\underset{..}{O}}H' + H^+$$
$$HH$$

Of more practical importance to most organic chemists is the fact that the chemical shifts of OH and NH protons are *solvent-*, *temperature-*, and *concentration-dependent* because of hydrogen bonding. In a nonhydrogen-bonding solvent (such as CCl$_4$) and at low concentrations (1% or less), OH proton absorption is observed at a δ value of about 0.5 ppm. At the more usual,

Figure 9.40 Nmr spectra for Problem 9.14.

higher concentrations, the absorption is observed in the 4–5.5 ppm region because of hydrogen bonding between CH_3OH molecules. In hydrogen-bonding solvents, the OH proton absorption can be shifted even farther downfield.

F. Other Factors That Affect Splitting Patterns

The spin–spin splitting patterns we have been discussing in this chapter are idealized cases. Most nmr spectra are more complex than those we have been using as examples. The complexity arises from a number of factors, of which we will mention only two. One factor that adds complexity to an nmr spectrum is the *nonequivalence of neighboring protons*:

$$\underset{\overset{|}{CHCl_2}}{ClCH_2}\overset{\overset{O}{\parallel}}{C}\ H\ CH \qquad \text{\textit{signal is split by three types of}}$$
$$\text{\textit{nonequivalent, neighboring H's}}$$

Another factor that adds complexity is the *magnitude of the chemical shift*. The $n + 1$ splitting patterns are truly apparent in an nmr spectrum only if the signals for coupled protons are separated from each other by a fairly large chemical shift. When the chemical shifts are close, the inside peaks of a multiplet increase in size, while the outside peaks decrease in size. This is the phenomenon of **leaning**, mentioned previously. When chemical shifts become very close, the peaks can coalesce into a singlet.

Figure 9.41 is the nmr spectrum of a typical alkane, *n*-octane. The chemical shifts of the CH_2 groups are all about the same ($\delta 1.27$); the CH_2 signal is a single, rather broad peak. The two methyl groups, which are equivalent to each other, are shielded compared with the methylene protons and absorb at a slightly higher field strength ($\delta 0.83$). According to spin–spin splitting rules, the methyl absorption should be a triplet. With a little imagination, you can see the triplet leaning toward the methylene absorption peak.

Figure 9.41 ^{1}H nmr spectrum of *n*-octane.

Section 9.11
Spin–Spin Splitting Diagrams

A **spin–spin splitting diagram**, also called a **tree diagram**, is a convenient technique for the analysis of splitting patterns. Let us consider H_a in the simple partial structure $\diagup CH_a$—$CH_b\diagdown$. The splitting of the signal for H_a into a doublet by H_b can be symbolized by the following tree diagram, which we "read" starting at the top and moving toward the bottom.

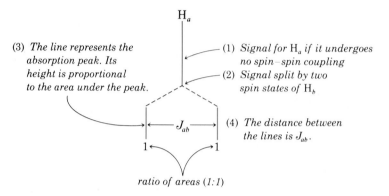

(3) *The line represents the absorption peak. Its height is proportional to the area under the peak.*

(1) *Signal for H_a if it undergoes no spin–spin coupling*
(2) *Signal split by two spin states of H_b*

(4) *The distance between the lines is J_{ab}.*

ratio of areas (1:1)

The splitting of the signal for H_b by the spin states of H_a may be represented by a similar tree diagram. These two diagrams can then be superimposed on an nmr spectrum (see Figure 9.42).

The tree diagram describing a triplet is a direct extension of that for a doublet. Consider the absorption pattern for H_a in the following grouping:

$$-\underset{|}{\overset{|}{C}}-\underset{|}{\overset{H_a \quad H_b}{C}}-H_b$$

In this case, H_a sees two neighboring protons and is split into a triplet. The coupling constant is J_{ab}. The triplet comes about because the peak for H_a is split

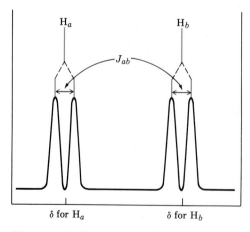

Figure 9.42 Spin–spin splitting pattern of two nonequivalent neighboring H's.

twice, once for each H_b. Using a tree diagram, we can see the result of each of the two splits. H_a is first split into a pair of doublets, and then the resulting two peaks are split again. We observe a triplet because the center two peaks absorb at the same position in the nmr spectrum. Consequently, the area of the center peak of the triplet is twice that of the two outside peaks.

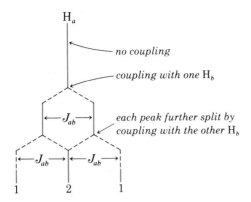

Sample Problems

Draw a spin–spin splitting diagram for H_m of the following system, where $J_{am} = 10$ Hz and $J_{mx} = 5$ Hz:

$$-\overset{|}{C}-\overset{|}{C}-\overset{|}{C}-$$
$$\quad\ H_a\ H_m\ H_x$$

Solution:

Here, the absorption pattern is not a triplet, as would be predicted by the $n + 1$ rule. Instead, four distinct lines of equal heights are observed because the coupling constants J_{am} and J_{mx} are not equal.

Draw a splitting diagram for the absorption of H_a in the partial structure

$$-\overset{|}{\underset{|}{C}}CH_3$$
$$\quad\ H_a$$

Solution:

$$H_a$$

all J values $= J_{ax}$

1   3   3   1

The original signal is split three times. The 1:3:3:1 ratio of the areas arises from the fact that all the protons have the same coupling constant and, consequently, the protons have superimposed absorption positions.

Study Problems

9.15 Construct a tree diagram for each of the indicated protons:

(a) $ClCH_2C\underline{H}_2CH_2Cl$ (b) $Cl_2CHC\underline{H}_3$ (c) $CH_3C\underline{H}_2OCH_3$

9.16 Draw a tree diagram for H_x in the following partial structure, where $J_{ax} = 10$ Hz and $J_{bx} = 5$ Hz. (Be careful to relate the height of each line to the relative area of each peak.)

$$-\overset{|}{\underset{H_a}{C}}-\overset{|}{\underset{H_x}{C}}-\overset{|}{\underset{H_b}{C}}-H_b$$

▼ Section 9.12
Carbon-13 Nmr Spectroscopy

Proton, or ^1H, nmr spectroscopy provides structural information about the hydrogen atoms in an organic molecule. Carbon-13, or ^{13}C, nmr spectroscopy produces structural information about the *carbons* in an organic molecule.

In ^1H nmr spectroscopy, we deal with the common natural isotope of hydrogen; 99.985% of natural hydrogen atoms are ^1_1H. However, 98.9% of the carbon atoms in nature are $^{12}_6C$, an isotope with nuclei that have no spin. Carbon-13 constitutes only 1.1% of naturally occurring carbon atoms. Also, the transition of a ^{13}C nucleus from parallel to antiparallel is a low-energy transition. Consequently, ^{13}C nmr spectra can be obtained only on very sensitive spectrometers. These spectrometers have become widely available in recent years, and ^{13}C spectroscopy is becoming increasingly important in organic laboratories.

The common technique for enhancing a weak ^{13}C nmr signal and eliminating random electronic noise is a relatively new technique called **Fourier transform nmr spectroscopy**. A Fourier transform spectrometer can generate a ^1H or ^{13}C nmr spectrum in a few seconds and store a large number of repetitive scans in computer memory. The stored scans are computer-averaged, the process that averages the noise to zero and increases the intensities of the real signals. Then, the data are converted by a mathematical technique called **Fourier transform analysis** to provide the spectrum that we see.

One feature that makes the new spectrometers practical is their speed compared to that of the conventional continuous-wave spectrometers. In a Fourier transform nmr spectrometer, the sample is irradiated with a brief intense pulse of a range of radiofrequencies that brings all the nuclei under study into resonance simultaneously. As the nuclei return to their normal distribution, the frequencies of energy losses are measured simultaneously and stored in computer memory. Fourier transform analysis makes interpretation of the stored signals possible.

While the low abundance of ^{13}C nuclei complicates instrument design, it also reduces the complexity of ^{13}C spectra compared to ^1H spectra. Although adjacent ^{13}C nuclei in a molecule will split each other's signals, the chances of finding adjacent ^{13}C nuclei are very small. For this reason, no $^{13}C-^{13}C$ splitting patterns are observed in ^{13}C spectra.

In a typical ^{13}C spectrum, the areas under the peaks are not proportional to the numbers of equivalent carbon atoms. Consequently, ^{13}C nmr spectra are not integrated.

There are two principal types of ^{13}C spectra: those that show $^{13}C-^1H$ spin–spin splitting patterns and those that do not. These two types of spectra are frequently used in conjunction with each other. In both types of spectra, TMS is used as an internal standard, and chemical shifts are measured downfield from the TMS peak. The chemical shifts in ^{13}C nmr are far greater than those observed in ^1H nmr. Most protons in ^1H nmr spectra show absorption between δ values of 0–10 ppm downfield from TMS. Only a few, such as aldehyde or carboxyl protons, show peaks outside this range. Carbon-13 absorption is observed over a range of 0–200 ppm downfield from TMS. This large range of chemical shifts is another factor that simplifies ^{13}C spectra compared with ^1H spectra. In ^{13}C spectra we are less likely to observe overlapping absorption.

The relative shifts in ^{13}C nmr spectroscopy are roughly parallel to those in ^1H nmr spectroscopy. TMS absorbs upfield, while aldehyde and carboxyl carbons absorb far downfield. Figure 9.43 shows the general locations of absorption by different types of carbon atoms.

A. Proton Decoupling

A ^{13}C **proton-decoupled spectrum** is, as the name implies, a spectrum in which ^{13}C is not coupled with ^1H and thus shows no spin–spin splitting. Decoupling is accomplished electronically by the application of a second radiofrequency to the sample. This extra energy causes rapid interconversions between the parallel and antiparallel spin states of the protons. As a result, a ^{13}C nucleus sees only an average of the two spin states of the protons, and its signal is not split.

Figure 9.43 Relative positions of ^{13}C nmr absorption.

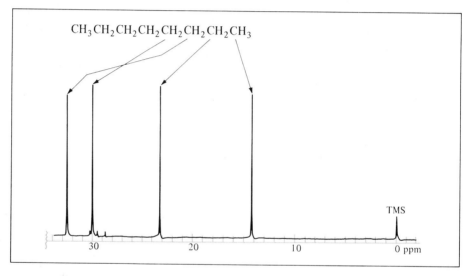

Figure 9.44 The proton-decoupled ^{13}C nmr spectrum of n-octane.

Because there is no splitting in a proton-decoupled spectrum, the signal for each group of magnetically equivalent carbon atoms appears as a *singlet*. By simply counting the number of peaks in the spectrum, we can determine the number of different types of carbon atoms in a molecule of the sample.

Figure 9.44 shows the proton-decoupled ^{13}C nmr spectrum of n-octane, a compound containing four types of nonequivalent carbons (carbons 1 and 8, carbons 2 and 7, carbons 3 and 6, carbons 4 and 5). The spectrum shows four singlets. Compare this spectrum with the ^{1}H spectrum of n-octane in Figure 9.41 (page 363). In the ^{1}H spectrum, the proton signals are crowded together in the $\delta 0$–1.5 area, and analysis of the peaks is impossible. In the ^{13}C spectrum, the signals are farther apart, in the $\delta 0$–35 range, and are easily distinguished.

B. Off-Resonance Decoupling

The second common type of ^{13}C nmr spectra, called **off-resonance-decoupled spectra**, is the type in which the ^{13}C–^{1}H splitting is not suppressed. In these spectra, the signal for each carbon is split by *protons bonded directly to it*. As in ${}^1H$ nmr spectroscopy, the $n + 1$ rule is followed. In ${}^{13}C$ nmr, n is the number of hydrogen atoms bonded to the carbon in question.

An impressive amount of information can be deduced about the structure of a compound when the off-resonance-decoupled and the proton-decoupled spectra of the compound are compared. Consider the two spectra of vinyl acetate shown in Figure 9.45. Inspection of the spectra reveals that the carbonyl carbon at $\delta 169$ is a singlet in both spectra; therefore, this carbon is not bonded to any hydrogens. The signal for the $CH_3CO_2\underline{C}H{=}CH_2$ carbon is split into a

Figure 9.45 The off-resonance-decoupled (top) and the proton-decoupled (bottom) ${}^{13}C$ nmr spectra of vinyl acetate.

doublet in the coupled spectrum, indicating one bonded H. Similarly, the C̲H$_2$ signal at $\delta 98$ is a triplet, and the C̲H$_3$ signal at $\delta 20$ is a quartet.

Study Problem

9.17 Figure 9.46 shows the ^{13}C nmr spectra of an ester with the molecular formula $C_5H_7O_2Br$. What is the structure of this compound?

Figure 9.46 ^{13}C nmr spectra for Problem 9.17.

Section 9.13
Using Infrared and Nmr Spectra for the Identification of Organic Structures

From an infrared spectrum, we can deduce the identities of functional groups. From a proton nmr spectrum, we can often deduce the structure of the hydrocarbon portion of a molecule. Sometimes it is possible to deduce the complete structure of a compound from just the infrared and nmr spectra. More commonly, additional information is needed (such as chemical reactivity, elemental analysis, or other spectra). In this text, we provide additional information generally in the form of a molecular formula. From the molecular formula, we can calculate the number of rings or double bonds. (*Example*: C_6H_{12}

is C_nH_{2n}; the structure contains either one double bond or a ring.) We can determine fragments of the structure from the spectra. Then, we try to match the fragments with the molecular formula.

Figure 9.47 Spectra for C_3H_6O, Sample Problem.

$$CH_3-\overset{\overset{\textstyle O}{\|}}{C}-CH_3$$

Sample Problems

A compound has a molecular formula of C_3H_6O. Its infrared and ^1H nmr spectra are given in Figure 9.47. What is the structure of this compound?

Solution: From the molecular formula, we know that the compound has one oxygen; therefore, the compound must be an alcohol, an ether, an aldehyde, or a ketone. Because the molecular formula is of the type $C_nH_{2n}O$, we know that the structure contains either one double bond (C=C or C=O) or one ring. To distinguish between the possible functional groups, we use the infrared spectrum. We see strong absorption in the C=O region of 1750 cm^{-1} (5.8 μm). We conclude that the oxygen is not in a hydroxyl group or ether group, but rather is in a carbonyl group (aldehyde or ketone). This means, then, that no ring is present.

In the ^{1}H nmr spectrum, we see no downfield, offset absorption; we conclude that the compound is a ketone instead of an aldehyde. The rest of the nmr spectrum shows only a singlet; therefore, all six hydrogens are equivalent and have no neighboring, nonequivalent hydrogens. The compound must be propanone (acetone): $(CH_3)_2C\!=\!O$.

A compound has a molecular formula of $C_8H_8O_2$. Its infrared spectrum, ^{1}H nmr spectrum, and proton-decoupled ^{13}C nmr spectrum are shown in Figure 9.48. What is the structure of the compound?

Solution: Using the infrared spectrum, we can identify the compound as a carboxylic acid. The —OH absorption near 3200 cm^{-1} (3.1 μm) slopes into the C—H region of the spectrum at 3000 cm^{-1} (3.3 μm), and carbonyl absorption is observed near 1700 cm^{-1} (5.9 μm). The offset singlet at $\delta10.9$ in the ^{1}H nmr spectrum and the absorption at $\delta178$ in the ^{13}C nmr spectrum are consistent with the assignment.

The total proton count in the ^{1}H nmr spectrum is eight. This value is the same as the number of hydrogens in the molecular formula. Thus, the ratio of proton signals in the ^{1}H nmr spectrum reflects the actual number of protons in the structure.

The ^{1}H nmr signal at $\delta7.2$ (area 5) suggests a phenyl group (C_6H_5—). This information is not corroborated by the infrared spectrum because the regions where aryl absorption is expected (3100 cm^{-1} and about 1600 cm^{-1}) are masked by wide absorption bands. However, corroborating evidence is found in the ^{13}C nmr spectrum at about $\delta130$. The appearance of four peaks in this region is consistent with phenyl absorption. There are only four absorption peaks for the six carbons of the phenyl group because of symmetry. The four bands are assigned as follows:

The phenyl group and the carbonyl group account for seven of the eight carbons in the unknown. The remaining carbon must be bonded to both the phenyl group and the carboxyl group. This is evidenced by (1) the benzene ring being monosubstituted (phenyl group) and (2) a two-proton absorption in the ^{1}H nmr spectrum (not three as it would be for —CH_3).

Therefore, the structure of the unknown is as follows.

phenylacetic acid

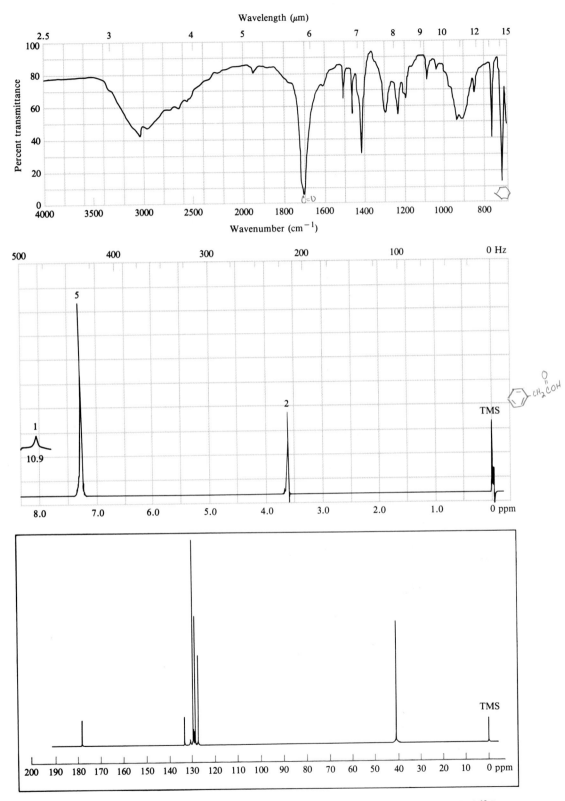

Figure 9.48 Infrared spectrum, ¹H nmr spectrum, and proton-decoupled ¹³C nmr spectrum for $C_8H_8O_2$ in Sample Problem, page 372.

Study Problems

9.18 A compound has the formula C_3H_8O. Its infrared and ^1H nmr spectra are shown in Figure 9.49. What is the structure of this compound?

Figure 9.49 Spectra for C_3H_8O, Problem 9.18.

9.19 Figure 9.50 shows the infrared and ^1H nmr spectra for a compound with the molecular formula C_7H_8O. What is the structure of this compound?

Figure 9.50 Spectra for C_7H_8O, Problem 9.19.

9.20 Figure 9.51 shows ¹H nmr and infrared spectra for a compound with the molecular formula $C_4H_7ClO_2$. What is the structure of the compound?

Figure 9.51 Spectra for $C_4H_7ClO_2$, Problem 9.20.

9.21 Figure 9.52 shows the infrared spectrum, ¹H nmr spectrum, and proton-decoupled ¹³C nmr spectrum for a compound with the molecular formula $C_{10}H_{12}O$. What is the structure of the compound?

$$CH_3CH_2CH{-}\overset{\overset{O}{\|}}{C}OH$$
$$\underset{Cl}{}$$

Figure 9.52 Infrared spectrum, ¹H nmr spectrum, and proton-decoupled ¹³C nmr spectrum for $C_{10}H_{12}O$, Problem 9.21.

▼ Summary

Organic compounds can absorb **electromagnetic radiation** of various wavelengths. Absorption in the **infrared region** results in **vibrational excitations** of bonds. Different types of bonds require differing amounts of energy for vibrational excitations. In an infrared spectrum, the region of $1400-4000$ cm^{-1} ($2.5-7.1$ μm) is useful for determination of functional groups, while the region beyond is the **fingerprint region**.

   Nuclear magnetic resonance is the result of certain nuclei in a magnetic field (H_0) absorbing electromagnetic radiation in the radiofrequency region and flipping from the *parallel* to the *antiparallel spin state*. An **induced molecular magnetic field** can *shield* protons (oppose H_0) or *deshield* protons ("augment" H_0) and cause a *chemical shift* (δ) of the absorption band. The induced field is a result of **anisotropic effects** and **inductive effects**. In ^{1}H nmr, a shielded proton absorbs *upfield*, close to the reference TMS, while a deshielded proton absorbs *downfield*.

$$\underset{\text{RCO\underline{H}}}{\overset{O}{\overset{\|}{C}}} \quad \underset{\text{RC\underline{H}}}{\overset{O}{\overset{\|}{C}}} \quad \text{\Large\bigcirc}\!\!-\text{\underline{H}} \quad \text{R}_2\text{C}=\text{C\underline{H}R} \quad \underset{\text{R}_2\text{C\underline{H}}}{\overset{Cl}{\overset{|}{C}}} \quad \text{R\underline{H}} \quad \text{TMS}$$

increasing shielding →

   Spin-spin splitting of an absorption band in ^{1}H nmr results from the spin states of neighboring nonequivalent protons. The signal of a particular proton (or group of equivalent protons) is split into $n+1$ peaks, where n is the number of neighboring protons equivalent to each other, but nonequivalent to the proton in question.

   The distance (in Hz) between any two peaks in a split band is the **coupling constant J**. For protons that are *coupled* (splitting each other's signals), the J values are the same.

   The *area* under an entire absorption band is proportional to the relative number of protons giving rise to that signal.

   Carbon-13 nmr spectra show the absorption frequencies of ^{13}C atoms instead of protons. Spectra may be **proton-decoupled**, in which case no splitting is observed, or they may be **off-resonance-decoupled**, in which case the ^{13}C signals are split by protons bonded to them.

Study Problems*

9.22 Make the following conversions: **(a)** 3000 cm^{-1} to μm; **(b)** 1760 cm^{-1} to μm; **(c)** 5.60 μm to cm^{-1}; **(d)** 8.20 μm to cm^{-1}.

9.23 In otherwise similar compounds, which one of each of the following pairs of partial structures would give stronger infrared absorption, and why?
    (a) C=O or C=C   **(b)** C=C—Cl or C=C—H   **(c)** O—H or N—H

---

*For information concerning the organization of the *Study Problems* and *Additional Problems*, see the *Note to student* on page 39.

9.24 Tell how you could distinguish each of the following pairs of compounds by their infrared spectra:

(a) $CH_3CH_2CH_2N(CH_3)_2$ and $CH_3CH_2CH_2NH_2$

(b) $CH_3CH_2CH_2CO_2H$ and $CH_3CH_2CO_2CH_3$

(c) $CH_3CH_2\overset{\overset{\displaystyle O}{\|}}{C}CH_3$ and $CH_3CH_2CO_2CH_3$

9.25 A chemist is oxidizing cyclohexanol to cyclohexanone. How can infrared spectroscopy be used to determine when the reaction is completed?

9.26 A compound with the molecular formula $C_5H_{10}O$ contains no double bonds and undergoes reaction when heated with HI to yield $C_5H_{10}I_2$. The infrared spectrum of the original compound shows absorption at about 2850 cm^{-1} (3.5 μm) and 1110 cm^{-1} (9 μm), but none near 3330 cm^{-1} (3 μm) nor 1720 cm^{-1} (5.8 μm). What are the possible structures of the compound and the diiodide?

9.27 At 60 MHz, how many Hz downfield from TMS is a chemical shift of $\delta 7.5$?

9.28 Indicate which underlined proton in each of the following groups of compounds will absorb farther upfield:

(a) $CH_3CH_2C\underline{H}_2Cl$, $CH_3CHClC\underline{H}_3$, $C\underline{H}_2ClCH_2CH_3$

(b) $CH_3C\underline{H}_2Cl$, $CH_3C\underline{H}Cl_2$ (c) ⬡—$\underline{H}$,   ⌬—$\underline{H}$

(d) ⬡=$C\underline{H}_2$,   ⬡—$\overset{\overset{\displaystyle O}{\|}}{C}\underline{H}$

9.29 How many different groups of chemically equivalent protons are present in each of the following structures? If more than one, indicate each group.

(a) $CH_3CH_2CH_2CH_3$ (b) $CH_3CH_2OCH_2CH_3$
(c) $CH_3CH{=}CH_2$ (d) *trans*-CHCl=CHCl
(e) *cis*-CHCl=CHCl   (f) $CH_3CH_2CO_2CH_2CH_3$
(g) (*R*)-2-chlorobutane   (h) (*S*)-2-chlorobutane
(i) $(CH_3)_2CHN(CH_3)CH(CH_3)_2$ (j) $BrCH_2CH_2CH(CH_3)_2$

(k) I—⬡—CH_2CH_3 (l) CH_3O—⬡—OCH_3

(m) CH_3O—⬡—CH_3 (n) 18-crown-6 ether (page 314)

9.30 In Problem 9.29, tell how many principal signals would probably be observed in the proton nmr spectrum of each compound, and predict the relative areas under the signals.

9.31 An nmr spectrum shows two principal signals with an area ratio of 3:1. Based on this information only, which of the following structures are possibilities?

(a) $CH_2{=}CHCH_3$ (b) $CH_2{=}CHCO_2CH_3$
(c) $CH_3CH_2CH_2CH_2CH_2CH_3$ (d) $CH_3CH_2CH_3$
(e) $CH_2{=}C(CH_3)_2$ (f) $CH_3CO_2CH_3$

9.32 Calculate the ratios of the different types of hydrogen atoms in a sample when the steps of the integration curve measure 81.5, 28, 55, and 80 mm.

9.33 A ^{1}H nmr spectrum shows four signals with integration curve heights of 4.0, 3.5, 5.4, and 5.5 cm. What is the ratio of the types of protons in this sample?

9.34 For each of the following compounds, predict the *multiplicity* (number of peaks arising from spin–spin coupling) and the *total relative area* under the signal of each set of equivalent protons in the ^1H nmr spectrum.

(a) $CH_3CH_2CO_2CH_3$ (b) $CH_3OCH_2CH_2OCH_3$ (c) $CH_3\overset{\overset{\displaystyle O}{\|}}{C}CH_2CH_3$

(d) CH_3O—⟨◯⟩—Cl (e) Cl_2CHCH_2Br

9.35 How would you distinguish between each of the following pairs of compounds by ^1H nmr spectroscopy?

(a) $CH_3CH_2CH_2CH_2CH=CH_2$ and $(CH_3)_2C=C(CH_3)_2$

(b) CH_3CH_2CHO and $CH_3\overset{\overset{\displaystyle O}{\|}}{C}CH_3$

(c) $CH_3\overset{\overset{\displaystyle O}{\|}}{C}CH_3$ and $CH_3\overset{\overset{\displaystyle O}{\|}}{C}OCH_3$

(d) CH_3CH_2—⟨◯⟩   and CH_3—⟨◯⟩—CH_3

9.36 How would you use either infrared or ^1H nmr spectroscopy to distinguish between:

(a) 1-propanol and propylene oxide?   (b) diisopropyl ether and di-*n*-propyl ether?

(c) ethanol and 1,2-ethanediol?   (d) ⟨◯⟩NCH_3 and ⟨◯⟩NH?

(e) ethanol and chloroethane?   (f) acetic acid and acetone?

9.37 Draw a tree diagram for H_m in the following partial structure, where $J_{am} = 5$ Hz and $J_{mx} = 11$ Hz.

$$H_a-\overset{\overset{\displaystyle H_a}{|}}{\underset{\underset{\displaystyle H_a}{|}}{C}}-\overset{|}{\underset{\underset{\displaystyle H_m}{|}}{C}}-\overset{|}{\underset{\underset{\displaystyle H_x}{|}}{C}}-$$

9.38 Predict **(a)** the number of principal peaks in the ^1H nmr spectrum of the local anesthetic *xylocaine* (following); **(b)** the splitting patterns of these peaks; **(c)** the characteristic absorption peaks in the infrared spectrum.

⟨◯⟩ with CH_3 above and CH_3 below —$NH\overset{\overset{\displaystyle O}{\|}}{C}CH_2N(CH_2CH_3)_2$

xylocaine

9.39 Sketch the expected ^1H nmr spectrum for 1,1-dichloroethane. Be sure to include (*qualitatively*) anticipated chemical shifts, splitting patterns, and appropriate areas.

9.40 Sketch the expected ^1H nmr spectrum for each of the following compounds:

(a) $CH_3CHClCHClCH_3$ (b) CH_3CH_2O—⟨◯⟩—$CH_2O\overset{\overset{\displaystyle O}{\|}}{C}CH_3$

9.41 A compound (molecular weight 72) containing only C, H, and O shows four signals in its proton-decoupled ^{13}C nmr spectrum: $\delta 8$, 30, 37, and 208. The off-resonance-decoupled spectrum shows these signals as a quartet, a quartet, a triplet, and a singlet. What is the structure of the compound?

Additional Problems

9.42 Under the influence of H_0, [18]annulene has an induced ring current not too different from that of benzene. Predict the shielding and deshielding of the protons of this ring system.

[18]annulene

9.43 The methyl protons of 15,16-dimethylpyrene do not have the same chemical shift as those of $C_6H_5C\underline{H}_3$. Instead, their absorption is *upfield* from TMS ($\delta -4.2$). Why?

15,16-dimethylpyrene

9.44 Predict the multiplicity of each absorption band in the off-resonance-decoupled ^{13}C nmr spectra of the following compounds:

(a) $(CH_3)_2\overset{\displaystyle OH}{\underset{\displaystyle |}{C}}C\equiv CH$ (b)

9.45 Determine the structures of the following compounds:

(a) The ^{1}H nmr spectrum of an alcohol ($C_5H_{12}O$) shows the following absorption: one singlet (relative area 1); two doublets (areas 3 and 6); and two multiplets (areas both 1). When treated with HBr, the alcohol yields an alkyl bromide ($C_5H_{11}Br$). Its nmr spectrum shows only a singlet (area 6); a triplet (area 3); and a quartet (area 2). What are the structures of the alcohol and the alkyl bromide?

(b) A compound (C_4H_7NO) shows moderately strong infrared absorption at 3000 cm^{-1} (3.33 μm) and weak absorption at 2240 cm^{-1} (4.46 μm). The only other absorption is in the fingerprint region. The ^{1}H nmr spectrum shows a triplet ($\delta 2.6$), a singlet (3.3), and a triplet (3.5). The area ratio of these signals is 2 (triplet): 5 (singlet and second triplet combined).

(c) A compound (molecular weight 122) containing only C, H, and O shows the following ^{1}H nmr absorption: singlet ($\delta 2.15$), singlet (3.45), and a group of four peaks (centered around 6.7); area ratio of 3:3:4.

(d) A compound with the molecular formula $C_5H_{10}O_2$ shows strong infrared absorption at 1750 cm^{-1} (5.71 μm). The nmr spectrum shows a doublet ($\delta 1.2$), a singlet (2.0), and a septet (4.95); area ratio of 6:3:1.

(e) A compound ($C_8H_{12}O_2$) shows its principal infrared absorption at 2900–3050 cm^{-1} (3.28–3.45 μm), 1760 cm^{-1} (5.68 μm), 1460 cm^{-1} (6.85 μm), and 1380 cm^{-1} (7.25 μm). Its ^{1}H nmr spectrum shows only a singlet at $\delta 1.3$.

9.46 Nitrogen-15 is nmr active and can split the signal of a proton just as a neighboring proton can. For example, the signal for the proton in the partial structure ^{15}N—H would be a doublet. The ^{1}H nmr spectrum, run on a 220 MHz instrument, of ^{15}N-enriched α-aminomethyleneglutaconic anhydride shows signals with the following δ values and coupling constants: $\delta 9.20$ (H_a), 9.44 (H_b), 7.98 (H_x); $J_{Na} = 92$ Hz, $J_{Nb} = 95$ Hz, $J_{ab} = 4$ Hz, $J_{ax} = 17$ Hz, $J_{bx} = 8$ Hz, $J_{Nx} < 1$ Hz. Sketch a tree diagram for this compound, showing the splitting patterns for H_a, H_b, and H_x.

α-aminomethyleneglutaconic anhydride

9.47 Figures 9.53 through 9.59 each give a molecular formula, an infrared spectrum, and a ^{1}H nmr spectrum for an unknown compound. What is the structure of each compound?

Figure 9.53 Spectra for Problem 9.47(a): $C_8H_{10}O_2$.

OH   C=O   C-O

Figure 9.54 Spectra for Problem 9.47(b): $C_{10}H_{12}O_2$.

$C_4H_8O_2$

$- CH_3$

$C_3H_5O_2$

$- CH_2$

CH_3O_2

$- CH_2$

OH

HO——O-$CH_2CH_2CH_3$

Figure 9.55 Spectra for Problem 9.47(c): $C_6H_{11}O_2Cl$.

Figure 9.56 Spectra for Problem 9.47(d): $C_6H_{10}O_2$.

Figure 9.56 *(continued)* Spectra for Problem 9.47(d): $C_6H_{10}O_2$.

$$CH_3 \overset{O}{\underset{\|}{C}} CH_2 \ CH_2 \overset{O}{\underset{\|}{C}} CH_3$$

Figure 9.57 Spectra for Problem 9.47(e): $C_4H_8O_3$.

$$CH_3-\underset{\underset{CH_3}{|}}{\overset{\overset{OH}{|}}{CH}}-\overset{O}{\underset{\|}{C}}-O-CH_3$$

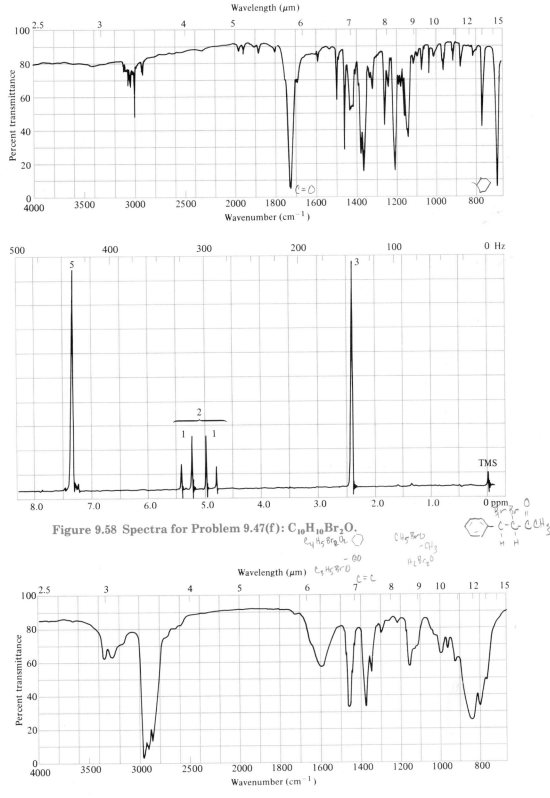

Figure 9.58 Spectra for Problem 9.47(f): $C_{10}H_{10}Br_2O$.

Figure 9.59 Spectra for Problem 9.47(g): $C_4H_{11}N$.

Figure 9.59 (*continued*) Spectra for Problem 9.47(g): C$_4$H$_{11}$N.

$-NH_2$

$C_4 H_9$

9.48   Figures 9.60–9.62 show ^{13}C nmr spectra, along with additional information concerning compounds of unknown structure. What are the structures of these compounds?

Figure 9.60 ^{13}C nmr spectra for Problem 9.48(a): C$_8$H$_{12}$.

Figure 9.61 ^{13}C nmr spectra for Problem 9.48(b): an ester with the molecular formula $C_5H_8O_2$.

Figure 9.62 ^{13}C nmr spectra for Problem 9.48(c): a compound with the molecular formula $C_7H_{16}O_4$. The proton nmr spectrum of this compound shows only a triplet ($\delta 1.95$), a singlet (3.3), and a triplet (4.5), with relative areas of 1:6:1.

9.49 Figure 9.63 shows the infrared spectrum, ^1H nmr spectrum, and proton-decoupled ^{13}C nmr spectrum for a compound with the molecular formula $C_7H_{14}O$. What is the structure of the compound?

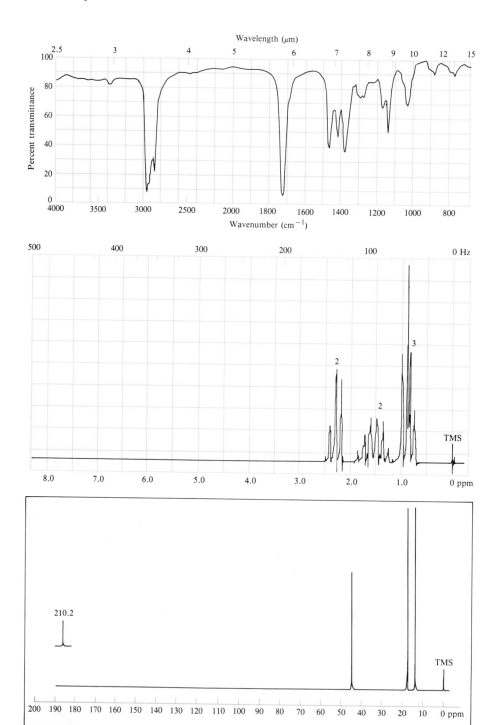

Figure 9.63 Infrared spectrum, ^1H nmr spectrum, and proton-decoupled ^{13}C nmr spectrum for $C_7H_{14}O$, Problem 9.49.

10 ▼

Alkenes and Alkynes

An **alkene** is a hydrocarbon with one double bond. Alkenes are sometimes called **olefins**, from *olefiant gas* ("oil-forming gas"), an old name for ethylene (CH_2=CH_2). An **alkyne** is a hydrocarbon with one triple bond. Acetylene (CH≡CH) is the simplest alkyne.

A carbon–carbon double bond is a common functional group in natural products. Most frequently, the double bond occurs in conjunction with other functional groups. However, alkenes with no other functionality are not at all rare and are often found as plant products and in petroleum. Two interesting examples of naturally occurring compounds containing carbon–carbon double bonds follow. Both these compounds are examples of *insect pheromones*, compounds emitted by one insect to transmit information to another insect of the same species (see Section 19.2).

neocembrene

a termite trail marker

$$(CH_3)_2C=CHCH_2CH_2\overset{\overset{\displaystyle CH_3}{|}}{C}=CHCH_2CH_2\overset{\overset{\displaystyle CH_2}{\|}}{C}CH=CH_2$$

3-methylene-7,11-dimethyl-1,6,10-dodecatriene

an aphid alarm pheromone

Section 10.1
Bonding in Alkenes and Alkynes; Acidity of Alkynes

The bonding in ethylene and acetylene was discussed in detail in Chapter 2. Recall that the two carbon atoms in ethylene are in the sp^2-hybrid state. The three sp^2 bonds from each carbon atom lie in the same plane, with bond angles of approximately 120°. The pi bond joining the two sp^2-hybridized carbons lies above and below the plane of the sigma bonds (see Figure 2.14, page 57).

side view

All the atoms in the ethylene molecule lie in the same plane. However, in a molecule that also has sp^3-hybridized carbons, only those atoms bonded to the double-bond carbons lie in the same plane.

— these atoms lie in a plane

The two carbon atoms in acetylene are in the sp-hybrid state. The sigma bonds formed by the sp-hybridized carbon atoms are linear. In acetylene, *two* pi bonds join the sp-hybridized carbon atoms (Figure 2.18, page 61).

acetylene

A triple-bond carbon is in the sp-hybrid state. The sp orbital is one-half s, while an sp^2 orbital is one-third s, and an sp^3 orbital is only one-fourth s. Because an sp orbital has more s character, the electrons in this orbital are closer to the carbon nucleus than are electrons in an sp^2 or sp^3 orbital. (See

Section 2.4F). In an alkyne, the *sp*-hybridized carbon is therefore *more electronegative* than most other carbon atoms. Thus, an alkynyl CH bond is *more polar* than an alkane CH bond or an alkene CH bond.

$$\overset{\delta+}{R}-\overset{\delta-}{C}\equiv\overset{\delta-}{C}-\overset{\delta+}{H}$$

sp-hybridized carbons electron-withdrawing

One of the most important results of the polarity of the alkynyl carbon–hydrogen bond is that $RC\equiv CH$ can lose a hydrogen ion to a strong base. The resulting anion ($RC\equiv C:^-$) is called an **acetylide ion**. With a pK_a of 25, alkynes are not strong acids. They are weaker acids than water (pK_a 15.7), but stronger acids than ammonia (p$K_a \sim 35$). Alkynes undergo reaction with a strong base like sodamide ($NaNH_2$) or a Grignard reagent. Alkanes and alkenes are not sufficiently acidic to react under these conditions.

$$CH_3C\equiv C{-}H + :NH_2^- \xrightarrow{\text{liq. NH}_3} CH_3C\equiv C:^- + :NH_3$$

propyne   *an acetylide ion*

$$CH_3C\equiv CH + CH_3MgI \longrightarrow CH_3C\equiv CMgI + CH_4$$

▼ ■ Section 10.2
Nomenclature of Alkenes and Alkynes

In the IUPAC system, the continuous-chain alkenes are named after their alkane parents, but with the **-ane** ending changed to **-ene**. For example, CH_3CH_3 is ethane and $CH_2{=}CH_2$ is ethene (trivial name, ethylene).

$$CH_2{=}CH_2 \qquad CH_3CH{=}CH_2$$

IUPAC: ethene   propene   cyclohexene

A hydrocarbon with two double bonds is called a **diene**, while one with three double bonds is called a **triene**. The following examples illustrate diene and triene nomenclature:

$$CH_2{=}CHCH{=}CH_2 \qquad \overset{\displaystyle CH_3}{\underset{}{CH_2{=}CCH{=}CH_2}} \qquad CH_2{=}CHCH{=}CHCH{=}CH_2$$

1,3-butadiene   2-methyl-1,3-butadiene   1,3,5-hexatriene

a diene   *a diene*   *a triene*

In the names of most alkenes, we need a prefix number to show the position of the double bond. Unless there is functionality of higher nomenclature priority, we number the chain from the end that gives the lower number to the double bond. The prefix number specifies the carbon atom in the chain where the double bond begins.

$$CH_3CH_3C\!=\!CHCH_2CH_3 \qquad CH_2\!=\!CHCH_2CH_2OH \qquad CH_3CH\!=\!CHCOH$$

2-methyl-2-pentene   3-buten-1-ol   2-butenoic acid

Sample Problem

Write the IUPAC name for $CH_3CH_2CH_2CCH_2CH_2CH_3$.
$\qquad\qquad\qquad\qquad\qquad\quad \| \atop CH_2$

Solution: While indeed the molecule does have a seven-carbon chain, the longest continuous chain that *contains the double bond* has only five carbons. Consequently, the structure is named as a pentene. The IUPAC name is 2-propyl-1-pentene.

Some alkenes and alkenyl groups have trivial names that are in common use. A few of these are shown in Tables 10.1 and 10.2.

Table 10.1 Trivial names of some alkenyl groups

Structure	Name	Example
$CH_2\!=$	methylene[a]	$\bigcirc\!=\!CH_2$
		methylenecyclohexane
$CH_2\!=\!CH-$	vinyl	$CH_2\!=\!CHCl$
		vinyl chloride
$CH_2\!=\!CHCH_2-$	allyl	$CH_2\!=\!CHCH_2Br$
		allyl bromide

[a] The term *methylene* is also used to refer to a disubstituted sp^3-hybridized carbon ($-CH_2-$); for example, CH_2Cl_2 is called methylene chloride.

Table 10.2 Boiling points of some alkenes and alkynes

Name	Structure	Bp, °C
Alkenes:		
ethene (ethylene)	$CH_2\!=\!CH_2$	-102
propene (propylene)	$CH_3CH\!=\!CH_2$	-48
methylpropene (isobutylene)	$(CH_3)_2C\!=\!CH_2$	-7
1-butene	$CH_3CH_2CH\!=\!CH_2$	-6
1-pentene	$CH_3CH_2CH_2CH\!=\!CH_2$	30
allene	$CH_2\!=\!C\!=\!CH_2$	-34.5
isoprene	$CH_2\!=\!C(CH_3)CH\!=\!CH_2$	34
Alkynes:		
ethyne (acetylene)	$CH\!\equiv\!CH$	-75
propyne	$CH_3C\!\equiv\!CH$	-23
1-butyne	$CH_3CH_2C\!\equiv\!CH$	8.1
2-butyne	$CH_3C\!\equiv\!CCH_3$	27

A pi bond prevents free rotation of groups around a double bond. Consequently, some alkenes exhibit geometric isomerism. This topic and the nomenclature of geometric isomers were covered in Section 4.1. (Because alkynes are linear molecules, they do not exhibit geometric isomerism.)

$$H_3C \quad CH_3 \qquad H_3C \quad H$$
$$\diagdown C{=}C \diagup \qquad\qquad \diagdown C{=}C \diagup$$
$$\diagup \qquad \diagdown \qquad \diagup \qquad \diagdown$$
$$H \qquad H \qquad H \qquad CH_3$$

cis-2-butene   *trans*-2-butene
or (*Z*)-2-butene   or (*E*)-2-butene

The IUPAC nomenclature of alkynes is directly analogous to that of the alkenes. The suffix for an alkyne is **-yne**, and a position number is used to signify the position of the triple bond in the parent hydrocarbon chain. Unless there is functionality of higher nomenclature priority in the molecule, the chain is numbered to give the triple bond the lower number.

In an older, trivial system of nomenclature for the simple alkynes, acetylene (CH≡CH) is considered the parent. Groups attached to the *sp*-hybridized carbons are named as substituents on acetylene. In this text we will use the IUPAC nomenclature system for the alkynes except for acetylene itself.

$$\langle\bigcirc\rangle{-}C{\equiv}CH \qquad CH_3C{\equiv}CCH_2CH_3$$

IUPAC: phenylethyne   2-pentyne
trivial: phenylacetylene   ethylmethylacetylene

Study Problems

10.1 Give the structures of

(a) (*E*)-1-chloro-3,4-dimethyl-3-hexene;   (b) *cis*-1,3-pentadiene;
(c) cyclohexylethyne;   (d) diphenylethyne.

10.2 Name the following compounds:

$$\overset{\displaystyle CH_3}{\underset{\displaystyle |}{}}$$
(a) $CH_2{=}CCH_2CH{=}CH_2$ (b) $CH{\equiv}CCH_2OH$ (c) $CH{\equiv}CC{\equiv}CH$

Section 10.3
Physical Properties of Alkenes and Alkynes

The physical properties of alkenes (but not the chemical properties) are practically identical to those of their alkane parents. Table 10.2 lists the boiling points of a few alkenes and alkynes. The boiling points of a homologous series of alkenes increase about 30° per CH_2 group. This is the same increase observed for a homologous series of alkanes. As with alkanes, branching in an alkene lowers the boiling point slightly.

Although alkenes are considered to be nonpolar, they are slightly more soluble in water than the corresponding alkanes because the pi electrons, which are somewhat exposed, are attracted to the partially positive hydrogen of water.

▼ ## Section 10.4
Spectra of Alkenes and Alkynes

A. Infrared Spectra

Alkenes. Ethylene, tetrachloroethylene, and other alkenes containing a nonpolar C=C double bond do not absorb infrared radiation in the C=C range. Alkenes such as RR'C=CHR and other unsymmetrical alkenes contain a polar C=C double bond and thus do absorb infrared radiation. Stretching of the C=C double bond gives rise to absorption at 1600–1700 cm^{-1} (5.8–6.2 μm). Because the stretching usually results in only a small change in bond moment, the absorption is weak, 10 to 100 times less intense than that of a carbonyl group. The absorption due to the stretching of the alkenyl, or vinylic, carbon–hydrogen bond (=C—H) at about 3000–3100 cm^{-1} (3.2–3.6 μm) is also weak. Alkenyl carbon–hydrogen bonds exhibit bending absorption in the fingerprint region of the infrared spectrum. (See Table 10.3.) Figure 10.1 shows the spectra of heptane and 1-heptene. The differences between an alkane and an alkene are evident in these spectra.

Alkynes. The C≡C stretching frequency of alkynes is at 2100–2250 cm^{-1} (4.4–4.8 μm). This absorption is quite weak and can easily be lost in the background noise of the spectrum. However, with the exception of C≡N and Si—H, no other groups show absorption in this region. The ≡C—H stretching frequency is about 3300 cm^{-1} (3.0 μm) as a sharp peak (see Figure 10.2).

B. Proton Nmr Spectra

Alkenes. The chemical shift for a vinylic proton is at an approximate δ value of 5.0 ppm. The exact position of the absorption depends on the location of the double bond in the hydrocarbon chain. In general, protons on terminal alkenyl carbons absorb near δ4.7, while the protons on nonterminal carbons absorb slightly farther downfield, at δ values of about 5.3 ppm.

$$CH_3CH_2\underline{CH}=\underline{CH}_2$$
$$\quad\quad\quad\;\; \delta5.3 \quad\quad \delta4.7$$

Table 10.3 Infrared absorption characteristics of alkenes and alkynes

Type of Vibration	Position of Absorption	
	cm^{-1}	*μm*
Alkenes:		
=C—H stretching	3000–3100	3.2–3.3
=C—H bending	800–1000	10.0–12.5
=CH$_2$ bending	855–995	10.0–11.7
C=C stretching	1600–1700	5.8–6.2
Alkynes:		
≡C—H stretching	~3300	~3.0
C≡C stretching	2100–2250	4.4–4.8

Figure 10.1 Infrared spectra of heptane and 1-heptene.

Figure 10.2 Infrared spectrum of 1-decyne.

The splitting patterns for vinylic protons are more complex than those for alkyl protons. The complexity arises from the lack of rotation around the double bond. Let us look at a general example:

$$
\begin{array}{ll}
J_{ab} = & 0\text{–}3.5 \text{ Hz} \\
J_{ax} = & 6\text{–}14 \text{ Hz} \\
J_{bx} = & 11\text{–}18 \text{ Hz}
\end{array}
$$

In this example, all three vinylic protons (H_a, H_b, and H_x) are nonequivalent. Therefore, they exhibit different chemical shifts and give rise to three separate signals. In addition, the coupling constants between any two of the protons (J_{ax}, J_{bx}, and J_{ab}) are different. Each of the three signals is therefore split into four peaks. (For example, the signal for H_x is split into two by H_b and again into two by H_a.) A total of twelve peaks may be observed in the nmr spectrum for these three protons, depending on the resolution in the spectrum.

This pattern of twelve peaks can be seen in the ^1H nmr spectrum of *p*-chlorostyrene in Figure 10.3. Figure 10.4 shows tree diagrams for the splitting patterns of the double-bond protons in *p*-chlorostyrene. With the aid of the tree diagram, we can see that none of the absorption peaks are superimposed because of the differences in *J* values.

In the spectrum of *p*-chlorostyrene (Figure 10.3), the chemical shift for H_a is at $\delta 5.3$, while the chemical shift for H_b, which is *cis* (and closer) to the benzene ring, is at $\delta 5.7$. The signal for H_b is downfield because H_b is somewhat deshielded by the induced field of the benzene ring. The signal for H_x is even farther downfield because H_x is more deshielded by the induced field of the ring.

Although any terminal vinylic group of the type $RCH{=}CH_2$ should give a spectrum with twelve vinyl-proton peaks, the twelve peaks are not always evident. For example, see Problem 10.3.

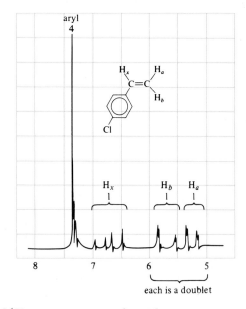

Figure 10.3 Partial ^1H nmr spectrum of *p*-chlorostyrene.

$$J_{ax} = 11 \text{ Hz}, \quad J_{ab} = 2 \text{ Hz}, \quad J_{bx} = 18 \text{ Hz}$$

Figure 10.4 Tree diagrams for the splitting patterns of the three alkenyl protons in *p*-chlorostyrene.

Study Problem

10.3 Figure 10.5 is the nmr spectrum of methyl vinyl sulfone, CH_2=$CHSO_2CH_3$. The spectrum contains eight peaks (not 12) in the alkene region. Construct a tree diagram that would explain the "missing" peaks.

Alkynes.   An alkyne of the type RC≡CR has no acetylenic protons. Therefore, a disubstituted alkyne has no characteristic ^1H nmr absorption. (The rest of the molecule may give rise to absorption, however.) A monosubstituted alkyne, RC≡CH, shows absorption for the alkynyl proton at a δ value of about

Figure 10.5 ^1H nmr spectrum of CH_2=$CHSO_2CH_3$ for Problem 10.3.

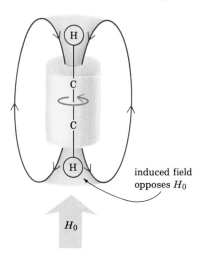

Figure 10.6 An alkynyl proton is shielded by the induced magnetic field of the triple bond.

3 ppm. This absorption is not nearly as far downfield as that for a vinylic or aryl proton because the alkynyl proton is *shielded* by the induced field of the triple bond. Figure 10.6 shows how the circulation of the pi electrons results in this field. Note the difference between this anisotropic effect and the effect for a vinylic proton, $=CHR$ (Section 9.7B). In the case of an alkyne, the induced field opposes, rather than augments, H_0.

▼ Section 10.5
Preparation of Alkenes and Alkynes

Alkenes can be prepared by the elimination reactions of alcohols (in strong acid) or alkyl halides (in base). Table 10.4 (page 403) lists the sections where these reactions are discussed in detail.

Primary alcohols undergo elimination reactions slowly. In hot, concentrated H_2SO_4, the product alkene may also undergo isomerization and other reactions. Therefore, primary alcohols are not usually useful in alkene preparation. *Primary alkyl halides* also undergo elimination reactions slowly by an E2 path. However, if a bulky base such as the *t*-butoxide ion is used to minimize reaction by an S_N2 path, an alkene may be obtained in good yield.

1° RX (E2 and S_N2):

$$CH_3CH_2CH_2CH_2CH_2Br \xrightarrow[\text{warm}]{K^+ \ ^-OC(CH_3)_3}$$

1-bromopentane

$$\begin{cases} CH_3CH_2CH_2CH=CH_2 \\ \text{1-pentene (85\%)} \\ + \\ CH_3CH_2CH_2CH_2CH_2OC(CH_3)_3 \\ \text{\textit{t}-butyl pentyl ether (12\%)} \end{cases}$$

Secondary alcohols undergo elimination by an E1 path when heated with a strong acid, and rearrangement of the intermediate carbocation may occur.

Except in simple cases, secondary alcohols are therefore not useful intermediates for the preparation of alkenes. *Secondary alkyl halides* can undergo E2
reactions. Although product mixtures might be expected, the predominant
product is usually the more highly substituted *trans*-alkene.

2° RX (E2):

$$CH_3CH_2CH_2CHCH_3 \xrightarrow[\text{warm}]{Na^+ \ ^-OCH_2CH_3}$$

Br on CH of 2-bromopentane

2-bromopentane

$$\begin{matrix} CH_3CH_2 & H \\ & C=C \\ H & CH_3 \end{matrix}$$

trans-2-pentene (51%)

+

$$\begin{matrix} CH_3CH_2 & CH_3 \\ & C=C \\ H & H \end{matrix}$$

cis-2-pentene (18%)

+

$$CH_3CH_2CH_2CH=CH_2$$

1-pentene (31%)

Tertiary alcohols undergo elimination readily through carbocations (E1)
when treated with a strong acid. *Tertiary alkyl halides* undergo elimination
with base principally by the E2 reaction. In both cases, excellent yields can be
obtained if all three R groups of R_3CX or R_3COH are the same. Otherwise,
mixtures might be obtained.

Study Problem

10.4 Each of the following reactions yields one alkene in a yield of over 80 percent. Give the
structure of each alkene product.

(a) cyclohexyl—OH $\xrightarrow[\text{heat}]{\text{conc. } H_2SO_4}$

(b) $(CH_3)_3CBr \xrightarrow[CH_3CH_2OH]{Na^+ \ ^-OCH_2CH_3}$

(c) $CH_3CH_2\overset{\underset{OH}{|}}{C}(CH_3)_2 \xrightarrow[\text{heat}]{50\% \ H_2SO_4}$

Alkynes can also be prepared by elimination reactions, but stronger conditions are required to remove the second halide ion. Under these conditions,
the triple bond may migrate from one position to another.

General:

$$\underset{\underset{Br}{|}}{R}CH-\underset{\underset{Br}{|}}{C}HR \xrightarrow[-HBr]{(1)} RCH=\underset{\underset{Br}{|}}{C}R \xrightarrow[-HBr]{(2)} RC≡CR$$

(1) OH^- or OR^- in CH_3CH_2OH
(2) $NaNH_2$, molten KOH, or very concentrated KOH + heat

$$\underset{\text{1,2-dibromo-1,2-diphenylethane}}{\text{C}_6\text{H}_5-\underset{\underset{\text{Br}}{|}}{\text{CH}}-\underset{\underset{\text{Br}}{|}}{\text{CH}}-\text{C}_6\text{H}_5} \xrightarrow[-2\,\text{HBr}]{2\,\text{KOH, CH}_3\text{CH}_2\text{OH}} \underset{\text{diphenylethyne (85\%)}}{\text{C}_6\text{H}_5-\text{C}\equiv\text{C}-\text{C}_6\text{H}_5}$$

$$\underset{\text{1,2-dibromohexadecane}}{\text{CH}_3(\text{CH}_2)_{13}\underset{\underset{\text{Br}}{|}}{\text{CH}}-\underset{\underset{\text{Br}}{|}}{\text{CH}_2}} \xrightarrow[-2\,\text{HBr}]{2\,\text{NaNH}_2} [\text{CH}_3(\text{CH}_2)_{13}\text{C}\equiv\text{CH}] \xrightarrow{\text{NaNH}_2}$$

$$\text{CH}_3(\text{CH}_2)_{13}\text{C}\equiv\text{C:}^- \ \text{Na}^+ \xrightarrow{\text{H}_2\text{O}} \underset{\text{1-hexadecyne (65\%)}}{\text{CH}_3(\text{CH}_2)_{13}\text{C}\equiv\text{CH}}$$

In Section 10.1, we mentioned that treatment of an alkyne with a strong base yields an acetylide. An acetylide ion may be used as a nucleophile in S_N2 reactions with primary alkyl halides. (Secondary and tertiary alkyl halides are more likely to give elimination products.) This reaction provides a synthetic route to substituted or more complex alkynes from simpler ones.

Preparation of acetylide:

$$\text{CH}_3\text{C}\equiv\text{CH} + \text{NaNH}_2 \xrightarrow[-30°]{\text{liq. NH}_3} \text{CH}_3\text{C}\equiv\text{C:}^- \ \text{Na}^+ + \text{NH}_3$$

Reaction with an alkyl halide:

$$\text{CH}_3\text{C}\equiv\text{C:}^- + \underset{\text{1-chloropropane}}{\text{CH}_3\text{CH}_2\text{CH}_2-\text{Cl}} \xrightarrow{S_N2} \underset{\text{2-hexyne}}{\text{CH}_3\text{CH}_2\text{CH}_2\text{C}\equiv\text{CCH}_3} + \text{Cl}^-$$

As we also mentioned in Section 10.1, alkynyl Grignard reagents can be prepared by the reaction of a Grignard reagent and a 1-alkyne. In this reaction, the Grignard reagent acts as a base while the alkyne acts as an acid.

Preparation of $RC\equiv CMgX$:

$$\underset{\text{propyne}}{\text{CH}_3\text{C}\equiv\text{C}\overset{\delta+}{\text{H}}} + \overset{\delta-}{\text{CH}}_3\text{MgI} \longrightarrow \text{CH}_3\text{C}\equiv\text{CMgI} + \text{CH}_4$$

As with other Grignard reagents, the nucleophilic carbon of an alkynyl Grignard reagent attacks partially positive centers, such as the carbon of a carbonyl group. The advantage of this type of Grignard synthesis is that more-complex alkynes can be prepared more easily this way than by S_N2 reactions.

Reaction with a ketone:

$$\text{CH}_3\text{C}\equiv\overset{\delta-}{\text{C}}-\overset{\delta+}{\text{MgI}} + \underset{\delta+}{\text{CH}_3\overset{\overset{\text{O:}^{\delta-}}{\|}}{\text{C}}\text{CH}_3} \longrightarrow \text{CH}_3\underset{\underset{\text{C}\equiv\text{CCH}_3}{|}}{\overset{\overset{\text{:OMgI}}{|}}{\text{C}}}\text{CH}_3 \xrightarrow[\text{H}_2\text{O}]{\text{H}^+} \underset{\text{2-methyl-3-pentyn-2-ol}}{\text{CH}_3\underset{\underset{\text{C}\equiv\text{CCH}_3}{|}}{\overset{\overset{\text{:OH}}{|}}{\text{C}}}\text{CH}_3}$$

Table 10.4 summarizes the methods of synthesizing alkenes and alkynes.

Table 10.4 Summary of laboratory syntheses of alkenes and alkynes

Reaction	*Section Reference*
Alkenes:	
$\overset{\text{X}}{\underset{\vert}{R_2CCHR_2}} + OH^- \xrightarrow{\text{heat}} R_2C{=}CR_2$	5.8 5.9
$\overset{\text{OH}}{\underset{\vert}{R_2CCHR_2}} + H_2SO_4 \xrightarrow{\text{heat}} R_2C{=}CR_2$	7.7
$R_2C{=}O + (C_6H_5)_3P{=}CR'_2 \longrightarrow R_2C{=}CR'_2$	13.6D
$RC{\equiv}CR + H_2 \xrightarrow{\text{catalyst}} RCH{=}CHR$ (*cis*)	10.13B
$RC{\equiv}CR \xrightarrow[\text{liq. NH}_3]{\text{Na or Li}} RCH{=}CHR$ (*trans*)	10.13B
Alkynes:	
$\overset{\text{X X}}{\underset{\vert\ \ \vert}{RCHCH_2}} \xrightarrow[\text{(2) H}_2\text{O}]{\text{(1) 3 NaNH}_2\text{, liq. NH}_3} RC{\equiv}CH$	10.5
$RC{\equiv}CH \xrightarrow[\text{liq. NH}_3]{\text{NaNH}_2} RC{\equiv}C{:}^-\ Na^+ \xrightarrow{R'X} RC{\equiv}CR'$	10.5
$RC{\equiv}CH \xrightarrow{\text{CH}_3\text{MgI}} RC{\equiv}CMgI \xrightarrow[\text{(2) H}_2\text{O, H}^+]{\text{(1) R}'_2\text{C}=\text{O}} RC{\equiv}C\overset{\text{OH}}{\underset{\vert}{C}}R'_2$	10.5

Study Problems

10.5 Suggest synthetic routes, starting with propyne, to: **(a)** 2-pentyne, and **(b)** 5-phenyl-2-pentyne.

10.6 Show how you would synthesize 1-pentyn-3-ol from acetylene, propanal (CH_3CH_2CHO), and ethylmagnesium bromide.

▼ ## Section 10.6
Preview of Addition Reactions

Three typical reactions of alkenes are the reactions with hydrogen, with chlorine, and with a hydrogen halide:

Each of these reactions is an **addition reaction**. In each case, a reagent adds to the alkene without the loss of any other atoms. We will find that the characteristic reaction of unsaturated compounds is the *addition of reagents to pi bonds*.

In an addition reaction of an alkene, the pi bond is broken and its pair of electrons is used in the formation of two new sigma bonds. In each case, the sp^2-hybridized carbon atoms are rehybridized to sp^3. Compounds containing pi bonds are usually of higher energy than comparable compounds containing only sigma bonds. Consequently, an addition reaction is usually exothermic.

In general, carbon–carbon double bonds are not attacked by nucleophiles because there is no partially positive carbon atom to attract a nucleophile. However, the exposed pi electrons in the carbon–carbon double bond are attractive to *electrophiles* (E^+), such as H^+. Many reactions of alkenes and alkynes are therefore initiated by an *electrophilic attack*, a reaction step that results in a carbocation. The carbocation can then be attacked by a nucleophile to yield the product. We will discuss this type of addition reaction first and then proceed to other types of alkene reactions.

two electrons in the pi bond

an alkene

a carbocation

Section 10.7
Addition of Hydrogen Halides to Alkenes and Alkynes

Hydrogen halides add to the pi bonds of alkenes to yield alkyl halides. Alkynes react in an analogous manner and yield either vinylic halides or 1,1-dihalo-alkanes, depending on the amount of HX used. We will not stress alkynes in our discussions, however, because alkenes are more important both in the laboratory and in nature.

$$CH_2{=}CH_2 + HX \longrightarrow CH_3CH_2X$$

ethylene   *an ethyl halide*

$$CH{\equiv}CH \xrightarrow{\text{HX}} CH_2{=}CHX \xrightarrow{\text{HX}} CH_3CHX_2$$

acetylene   *a vinyl halide*   *a 1,1-dihaloethane*

The addition of hydrogen halides to alkenes is often used as a synthetic reaction. Usually, the gaseous HX is bubbled through a solution of the alkene. (Concentrated aqueous solutions of hydrogen halides give mixtures of products because water can also add to double bonds.) The relative reactivity of HX in this reaction is HI > HBr > HCl > HF. The strongest acid (HI) is the most reactive toward alkenes, while the weakest acid (HF) is the least reactive.

A hydrogen halide contains a highly polar H—X bond and can easily lose H^+ to the pi bond of an alkene. The result of the attack of H^+ is an intermediate carbocation, which quickly undergoes reaction with a negative halide ion to yield an alkyl halide. Because the initial attack is by an electrophile, the addition of HX to an alkene is called an **electrophilic addition reaction**.

Step 1 (slow):

$$H{-}\overset{\cdot\cdot}{\underset{\cdot\cdot}{Cl}}:$$

$$CH_3CH{=}CHCH_3 \longrightarrow \left[CH_3\overset{+}{C}HCHCH_3 \right] + :\overset{\cdot\cdot}{\underset{\cdot\cdot}{Cl}}:^-$$

2-butene   *intermediate carbocation*

Step 2 (fast):

$$:\overset{\cdot\cdot}{\underset{\cdot\cdot}{Cl}}:$$

$$[CH_3\overset{+}{C}HCH_2CH_3] + :\overset{\cdot\cdot}{\underset{\cdot\cdot}{Cl}}:^- \longrightarrow CH_3CHCH_2CH_3$$

2-chlorobutane

A. Markovnikov's Rule

If an alkene is *unsymmetrical* (that is, if the groups attached to the two sp^2-hybridized carbons differ), there is the possibility of two different products from HX addition.

$$CH_3CH{=}CHCH_3 \qquad \qquad CH_3CH{=}CH_2$$

symmetrical alkenes   *unsymmetrical alkenes*

$$CH_3CH{=}CHCH_3 \xrightarrow{\;HCl\;} CH_3\overset{H}{\underset{}{C}H}{-}\overset{Cl}{\underset{}{C}H}CH_3$$

2-butene   2-chlorobutane

symmetrical   *only one possible product*

$$CH_3CH{=}CH_2 \xrightarrow{\;HCl\;}$$

propene

unsymmetrical

$CH_3CH_2{-}CH_2Cl$ 1-chloropropane

$CH_3\overset{Cl}{\underset{}{C}H}{-}CH_3$ 2-chloropropane

two possible products

In an electrophilic addition that can lead to two products, one product usually predominates over the other. In 1869, the Russian chemist Vladimir

Markovnikov formulated the following empirical rule. *In additions of HX to unsymmetrical alkenes, the H+ of HX goes to the double-bonded carbon that already has the greatest number of hydrogens.*

By Markovnikov's rule, we would predict that the reaction of HCl with propene yields 2-chloropropane (and not the 1-chloro isomer). Examples of reactions that obey Markovnikov's rule follow:

$$CH_3CH{=}CH_2 \xrightarrow{\text{HCl}} CH_3\underset{|}{\overset{Cl}{CH}}{-}CH_3$$

H goes here

propene   2-chloropropane

$$(CH_3)_2C{=}CHCH_3 \xrightarrow{\text{HBr}} (CH_3)_2\underset{|}{\overset{Br}{C}}{-}CH_2CH_3$$

H goes here

2-methyl-2-butene   2-bromo-2-methylbutane

H goes here

$$\xrightarrow{\text{HI}}$$

1-methylcyclohexene   1-iodo-1-methylcyclohexane

The addition of HX to an alkene is referred to as a **regioselective reaction** (from the Latin *regio*, "direction"), a reaction in which one direction of addition to an unsymmetrical alkene predominates over the other direction. Therefore, a regioselective addition reaction is a reaction in which two isomeric products could be obtained, but one predominates. The addition of HX to an alkene is regioselective because the H+ of HX becomes bonded to the alkenyl carbon that already has the greater number of hydrogens. The reason for this selectivity is that this path of addition yields the more stable of the two possible intermediate carbocations, as explained in the next section.

Study Problem

10.7 The electrophilic addition of HX to alkynes also follows Markovnikov's rule. Predict the product of the reaction of 1.0 mol of 1-hexyne with   **(a)** 1.0 mol of HBr, and   **(b)** 2.0 mol of HBr.

B. The Reason for Markovnikov's Rule

Markovnikov formulated his rule because of experimental observations. Why is this empirical rule followed? To answer this question, let us return to the mechanism of HX addition. *Step 1* is the formation of a carbocation. For propene, two possible carbocations could be formed:

$$CH_3CH{=}CH_2 \longrightarrow \left[CH_3\overset{\delta+}{CH}{=\!=}\overset{\delta+}{CH_2} \right] \longrightarrow CH_3CH_2\overset{+}{CH_2}$$

H^+ $H^{\delta+}$

transition state   *1°, less stable*

$$CH_3CH\!\!=\!\!\overset{H^+}{\overset{\curvearrowright}{CH_2}} \longrightarrow \left[CH_3\overset{\delta+}{CH}\!\!=\!\!\overset{H^{\delta+}}{\overset{|}{CH_2}} \right] \longrightarrow CH_3\overset{+}{CH}CH_3$$

$$\text{\textit{transition state}} \qquad \text{\textit{2°, more stable}}$$

The order of stability of carbocations is 3° > 2° > 1°. For propene, the two positions of H⁺ addition lead to (1) a high-energy, unstable, primary carbocation, or (2) a lower-energy, more-stable, secondary carbocation. The transition states leading to these intermediates have carbocation character. Therefore, the secondary carbocation has a lower-energy transition state and a faster rate of formation. (See Figure 10.7.)

Addition of a reagent to an unsymmetrical alkene proceeds by way of the more stable carbocation. This is the reason that Markovnikov's rule is followed.

$$CH_3CH\!\!=\!\!CH_2 \xrightarrow{H^+} [CH_3\overset{+}{CH}CH_3] \xrightarrow{Cl^-} CH_3\overset{\overset{\displaystyle Cl}{|}}{CH}CH_3$$

$$(CH_3)_2C\!\!=\!\!CHCH_3 \xrightarrow{H^+} [(CH_3)_2\overset{+}{C}CH_2CH_3] \xrightarrow{Br^-} (CH_3)_2\overset{\overset{\displaystyle Br}{|}}{C}CH_2CH_3$$

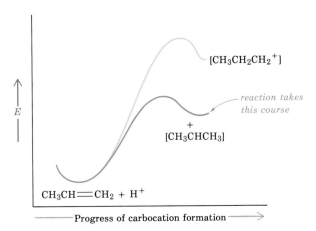

Sample Problem

Predict the relative rates of reaction of the following alkenes toward HBr (lowest rate first):

(a) $CH_3CH_2CH\!\!=\!\!CH_2$ (b) $CH_2\!\!=\!\!CH_2$ (c) $(CH_3)_2C\!\!=\!\!CHCH_3$

Solution: The alkene that can form the most stable carbocation has the lowest E_{act} and the fastest reaction.

$$CH_2\!\!=\!\!CH_2 \qquad CH_3CH_2CH\!\!=\!\!CH_2 \qquad (CH_3)_2C\!\!=\!\!CHCH_3$$

increasing rate of reaction ⟶

Figure 10.7 Energy diagram for the protonation of propene.

Study Problems

10.8 For the alkenes in (a) and (c) in the preceding sample problem, give the structures of the carbocation intermediate and the major product of the reaction with HBr.

10.9 When propene is treated with HCl in ethanol, one of the products is ethyl isopropyl ether. Suggest a mechanism for its formation.

10.10 Because the addition of HX to an alkene proceeds by way of a carbocation, rearrangements can occur (see Section 5.6F). For each of the following reactions, predict the products, including the rearranged product (if any).

(a) 3,3-dimethyl-1-butene + HI $\longrightarrow$
(b) styrene ($C_6H_5CH{=}CH_2$) + HCl $\longrightarrow$
(c) 3-methyl-1-butene + HCl $\longrightarrow$
(d) 4,4-dimethyl-2-pentene + HCl $\longrightarrow$

C. Anti-Markovnikov Addition of HBr

The addition of HBr to alkenes sometimes proceeds by Markovnikov's rule, but sometimes it does not. (This effect is not observed with HCl or HI.)

$$CH_3CH{=}CH_2 \xrightarrow{HBr} \underset{\text{2-bromopropane}}{CH_3\overset{\overset{\displaystyle Br}{|}}{C}HCH_3} \quad \text{but sometimes} \quad \underset{\substack{\text{1-bromopropane} \\ \textit{anti-Markovnikov product}}}{CH_3CH_2CH_2Br}$$

It has been observed that the primary alkyl bromide is obtained only when a peroxide or O_2 is present in the reaction mixture. Oxygen is a stable diradical (see Section 6.6B), and peroxides (ROOR) are easily cleaved into radicals. When O_2 or peroxides are present, HBr addition proceeds through a *free-radical* mechanism instead of an ionic one.

Formation of Br·:

$$ROOR \longrightarrow 2\,RO·$$

$$RO· + HBr \longrightarrow ROH + Br·$$

Addition of Br· to alkene:

$$CH_3CH{=}CH_2 + Br· \longrightarrow \underset{\substack{Br \\ \textit{2°, more stable}}}{CH_3\dot{C}HCH_2} \quad \text{and not} \quad \underset{\substack{Br \\ \textit{1°, less stable}}}{CH_3CH\dot{C}H_2}$$

Formation of product:

$$CH_3\dot{C}HCH_2Br + H{-}Br \longrightarrow CH_3CH_2CH_2Br + Br·$$

When Br· attacks the alkene, the more stable radical is formed. (The stability of radicals, like that of carbocations, is in the order $3° > 2° > 1°$.) The result of the free-radical addition in our example is 1-bromopropane.

Hydrogen chloride does not undergo free-radical addition to alkenes because of the relative slowness of the homolytic cleavage of HCl into radicals. Hydrogen iodide does not undergo this reaction because the addition of I· to alkenes is endothermic and too slow to sustain a chain reaction.

Study Problem

10.11 Predict the products:

(a) $(CH_3)_2C=CH_2$ + HBr $\xrightarrow{ROOR}$

(b) $(CH_3)_2C=CH_2$ + HBr $\xrightarrow{\text{no peroxides}}$

▼ ## Section 10.8
Addition of H$_2$SO$_4$ and H$_2$O to Alkenes and Alkynes

Sulfuric acid undergoes addition to an alkene just as a hydrogen halide does. The product is an alkyl hydrogen sulfate, which can be used to synthesize alcohols or ethers. (See Section 8.3A.)

$$CH_3CH=CH_2 + H-OSO_3H \longrightarrow \underset{\text{2-propyl hydrogen sulfate}}{CH_3\overset{\overset{\displaystyle OSO_3H}{|}}{CH}-CH_3}$$

propene

In strongly acidic solution (such as aqueous sulfuric acid), water adds to a double bond to yield an alcohol. This reaction is called the **hydration of an alkene**.

$$CH_3CH=CH_2 + H_2O \xrightarrow{H^+} \underset{\text{2-propanol (60\%)}}{CH_3\overset{\overset{\displaystyle OH}{|}}{CH}-CH_3}$$

propene

Both reactions occur in two steps, similar to the addition of a hydrogen halide. The first step is the protonation of the alkene to yield a carbocation. The second step is the addition of a nucleophile to the carbocation. Because initially a carbocation is formed, both reactions follow Markovnikov's rule. Rearrangements are to be expected if the carbocation can undergo a 1,2-shift of H or R to yield a more stable carbocation.

Step 1:

$$R_2C=CHR + H^+ \rightleftharpoons [R_2\overset{+}{C}-CH_2R]$$

Step 2:

$$[R_2\overset{+}{C}-CH_2R] + H_2\ddot{O} \rightleftharpoons \underset{\text{a protonated alcohol}}{R_2\overset{\overset{\displaystyle H-\overset{+}{\ddot{O}}H}{|}}{C}-CH_2R} \rightleftharpoons \underset{\text{an alcohol}}{R_2\overset{\overset{\displaystyle :\ddot{O}H}{|}}{C}-CH_2R} + H^+$$

Hydration reactions of alkenes are the reverse of dehydration reactions of alcohols. The products depend on the conditions used. (See Sections 7.7 and 8.3A.)

Alkynes also undergo hydration, but the initial product is a vinylic alcohol, or **enol**. An enol is in equilibrium with an aldehyde or ketone. (See Section 13.10.) The equilibrium almost always favors the carbonyl compound. Therefore, hydration of an alkyne actually results in an aldehyde or ketone. (The hydration of alkynes proceeds more smoothly when a mercuric salt is added to catalyze the reaction.)

$$\text{CH}_3(\text{CH}_2)_3\text{C}\equiv\text{CH} + \text{H}_2\text{O} \xrightarrow[\text{H}_2\text{SO}_4]{\text{HgSO}_4} \left[\text{CH}_3(\text{CH}_2)_3\overset{\overset{\displaystyle \text{OH}}{|}}{\text{C}}=\text{CH}_2 \right] \rightleftharpoons \text{CH}_3(\text{CH}_2)_3\overset{\overset{\displaystyle \text{O}}{||}}{\text{C}}\text{CH}_3$$

an enol   2-hexanone (80%)

Acid-catalyzed hydration of an alkene or alkyne is seldom used in the laboratory because of the relatively low yields and possibilities of rearrangements and polymerization. However, many simple alcohols, such as ethanol and 2-propanol, are synthesized industrially by this technique.

Study Problems

10.12 Show the mechanism for the addition of H_2SO_4 to 1-butene to yield a butyl hydrogen sulfate.

10.13 Predict the major products:

(a) $\text{CH}_3\text{CH}_2\text{CH}=\text{CH}_2 + \text{H}_2\text{O} \xrightarrow{\text{H}^+}$

(b) $(\text{CH}_3)_3\text{CCH}=\text{CH}_2 + \text{H}_2\text{O} \xrightarrow{\text{H}^+}$

(c) 1 + 1 $\text{H}_2\text{O} \xrightarrow{\text{H}^+}$

10.14 Alcohols can add to carbon–carbon double bonds in much the same way water does. The following unsaturated ether is often used to react with alcohols to protect the —OH group while reactions are carried out on other parts of the molecule. Predict the product of the reaction of methanol with dihydropyran in the presence of an acidic catalyst. Write a mechanism for the formation of this product and show resonance structures for any intermediates, if appropriate.

dihydropyran

Section 10.9
Hydration Using Mercuric Acetate

Mercuric acetate, $\text{Hg}(\text{O}_2\text{CCH}_3)_2$, and water add to alkenes in a reaction called **oxymercuration**. Unlike the addition reactions we have discussed so far, oxymercuration proceeds *without rearrangement*. The product of oxymercuration is usually reduced with sodium borohydride ($NaBH_4$) in a subsequent re-action called **demercuration** to yield an alcohol, the same alcohol that would be formed if water had been added across the double bond. Oxymercuration–demercuration reactions usually give better yields of alcohols than the addition of water with H_2SO_4.

Oxymercuration:

$$CH_3CH_2CH_2CH{=}CH_2 \xrightarrow[H_2O]{Hg\left(\overset{\overset{O}{\|}}{O\ddot{C}CH_3}\right)_2} \overset{\overset{\displaystyle OH}{|}}{CH_3CH_2CH_2\underset{\underset{\displaystyle HgO_2CCH_3}{|}}{CH}CH_2}$$

1-pentene

Demercuration:

$$\overset{\overset{\displaystyle OH}{|}}{CH_3CH_2CH_2CH}CH_2{-}HgO_2CCH_3 \xrightarrow{NaBH_4} \overset{\overset{\displaystyle OH}{|}}{CH_3CH_2CH_2CH}CH_3 + Hg$$

2-pentanol
(90% overall)

Like the addition of other reagents to alkenes, oxymercuration is a two-step process. The addition proceeds by electrophilic attack of $^+HgO_2CCH_3$ followed by nucleophilic attack of H_2O. Because rearrangements do not occur, the intermediate formed by electrophilic attack cannot be a true carbocation. On the other hand, since Markovnikov's rule is followed, the intermediate must have some carbocation character. Both these facts are explained by postulating a **bridged ion**, or **cyclic ion**, as the intermediate.

Dissociation of mercuric acetate:

$$Hg(O_2CCH_3)_2 \rightleftharpoons {}^+HgO_2CCH_3 + {}^-O_2CCH_3$$

Electrophilic attack:

$$R_2C{=}CHR \xrightarrow{} \left[\begin{array}{c} \overset{\delta+}{R_2C}{-}CHR \\ {}_{\diagdown}{}_{\diagup} \\ HgO_2CCH_3 \\ {}_{\delta+} \end{array}\right]$$
$$\underset{{}^+HgO_2CCH_3}{\diagup}$$

a bridged intermediate

Attack of H_2O and proton loss:

$$\left[\begin{array}{c} \overset{\delta+}{R_2C}{-}CHR \\ {}_{\diagdown}{}_{\diagup} \\ HgO_2CCH_3 \\ {}_{\delta+} \end{array}\right] \xrightarrow{H_2\ddot{O}:} \left[\begin{array}{c} \overset{\overset{\displaystyle +:OH_2}{|}}{R_2C}{-}CHR \\ {}_{|} \\ HgO_2CCH_3 \end{array}\right] \xrightarrow{-H^+} \overset{\overset{\displaystyle :\ddot{O}H}{|}}{R_2C}{-}\underset{\underset{\displaystyle HgO_2CCH_3}{|}}{CHR}$$

The formation of a bridged intermediate is not very different from the formation of a carbocation. The reaction of this intermediate is also very similar to that of a carbocation. The difference between this bridged intermediate and a true carbocation is that Hg is partially bonded to *each* double-bond carbon and rearrangements cannot occur. The more positive carbon in the bridged intermediate (the carbon attacked by H_2O) can be predicted from carbocation stabilities ($3° > 2° > 1°$). We can compare the reaction of this bridged ion to the reaction of the protonated intermediate formed during the acid-catalyzed substitution reaction of an epoxide (Section 8.5B).

more positive carbon;
H₂O attacks here

$$R_2\overset{\delta+}{C}\text{---}CHR \quad \overset{|}{\underset{\delta+}{HgO_2CCH_3}} \qquad \text{similar to} \qquad R_2\overset{+}{C}\text{---}CHR \quad \overset{|}{HgO_2CCH_3}$$

bridged intermediate   *true carbocation*

The reducing agent in the demercuration reaction, sodium borohydride, is an important reducing agent in organic chemistry. It forms stable solutions in aqueous base, but decomposes and releases H_2 in acidic solution. We will encounter this reagent again as a reducing agent for aldehydes and ketones (Section 13.7B).

$$Na^+ \quad H\text{---}\overset{\overset{\displaystyle H}{|}}{\underset{\underset{\displaystyle H}{|}}{B}}\text{---}H$$

sodium borohydride

Study Problems

10.15 **(a)** Write equations for the steps in the oxymercuration–demercuration of 3,3-dimethyl-1-butene.

(b) Compare the product of this reaction sequence with the product from the reaction of 3,3-dimethyl-1-butene and dilute aqueous HCl.

10.16 Oxymercuration–demercuration results in alcohols as products. If the reaction is carried out in an alcohol instead of in water (a sequence called **solvomercuration–demercuration**), an *ether* is obtained as a product. Write equations to show how you would prepare 2-methoxy-2-methylbutane from 2-methyl-1-butene by this technique.

Section 10.10
Addition of Borane to Alkenes

Diborane (B_2H_6) is a toxic gas prepared by the reaction of sodium borohydride and boron trifluoride ($3\,NaBH_4 + 4\,BF_3 \rightarrow 2\,B_2H_6 + 3\,NaBF_4$). In diethyl ether solution, diborane dissociates into borane (BH_3) solvated by an ether molecule: $(CH_3CH_2)_2\ddot{O}{:}\text{----}BH_3$. Borane undergoes rapid and quantitative reaction with most alkenes to form **organoboranes** (R_3B). The overall reaction is the result of three separate reaction steps. In each step, one alkyl group is added to borane until all three hydrogen atoms have been replaced by alkyl groups. This sequence of reactions is called **hydroboration**.

Step 1:

$$CH_2{=}CH_2 + \underset{H}{\overset{H}{\underset{|}{\overset{|}{B}}}}{-}H \longrightarrow CH_3{-}\underset{}{\overset{BH_2}{\underset{|}{CH_2}}}$$

Step 2:

$$CH_2{=}CH_2 + CH_3CH_2BH_2 \longrightarrow (CH_3CH_2)_2BH$$

Step 3:

$$CH_2{=}CH_2 + (CH_3CH_2)_2BH \longrightarrow (CH_3CH_2)_3B$$

triethylborane

an organoborane

Organoboranes were discovered in the 1950s by Herbert C. Brown at Purdue University, who was awarded a Nobel Prize in 1979 for his work with organoboron compounds. The value of these compounds arises from the variety of other compounds that can be synthesized from them. Let us first consider the addition of BH_3 to alkenes and then look at some of the products that can be obtained from the organoboranes.

Borane is different from the other addition reagents we have mentioned, because H is the *electronegative* portion of the molecule instead of the electropositive portion as it is in HCl or H_2O. When borane adds to a double bond, the hydrogen (as a hydride ion, H^-) becomes bonded to the *more substituted carbon*. The result is what appears to be anti-Markovnikov addition.

$$CH_3CH{=}CH_2 \longrightarrow CH_3\underset{H}{\overset{}{\underset{|}{CH}}}{-}\underset{BH_2}{\overset{}{\underset{|}{CH_2}}}$$

$$\underset{\delta-}{H}{-}\underset{\delta+}{BH_2}$$

H on more substituted carbon

Steric hindrance also plays a role in the course of this reaction. Best yields of the anti-Markovnikov organoborane are obtained when one carbon of the double bond is substantially more hindered than the other.

less hindered

more hindered

$$CH_3CH_2CH{=}CH_2 \qquad CH_3CH_2\underset{}{\overset{CH_3}{\underset{|}{C}}}{=}CH_2$$

94% yield of anti-Markovnikov product

99% yield of anti-Markovnikov product

Organoboranes are easily oxidized to alcohols by alkaline hydrogen peroxide. The result of borane addition, followed by H_2O_2 oxidation, appears as if water had been added to the double bond in an anti-Markovnikov manner.

$$3 \ CH_3CH_2CH_2CH_2CH{=}CH_2 \xrightarrow{BH_3} (CH_3CH_2CH_2CH_2CH_2CH_2)_3B$$

1-hexene

$$\xrightarrow{H_2O_2,\ OH^-} 3 \ CH_3(CH_2)_4CH_2OH$$

1-hexanol (90%)

1-methylcyclopentene   (1) BH$_3$ (2) H$_2$O$_2$, OH$^-$ *trans*-2-methylcyclopentanol (87%)

Now, we have discussed three routes to alcohols from alkenes. These are summarized in the following flow diagram:

Besides being oxidized to alcohols, organoboranes can be converted to alkanes, alkyl halides, or other products. In each case, the newly introduced atom or group becomes bonded to the less-substituted carbon of the double bond.

$$3 \ CH_3CH{=}CH_2 \xrightarrow{BH_3} (CH_3CH_2CH_2)_3B$$

$$\xrightarrow{3\ CH_3CO_2D} 3 \ CH_3CH_2CH_2D$$
1-deuteriopropane
a deuteriated alkane

$$\xrightarrow{3\ Br_2,\ OH^-} 3 \ CH_3CH_2CH_2Br$$
1-bromopropane
an alkyl halide

A. Stereochemistry of Hydroboration

When borane adds to a double bond, the boron atom and the hydride ion add to the two carbon atoms of the double bond *simultaneously*. The result is that the B and H must add to the same side of the double bond. An addition reaction in which two species add to the same side is called a **cis-addition**, or **syn-addition**. (*Syn*, like *cis*, means "on the same side, or face.")

A syn-addition:

transition state

If the addition product is capable of geometric isomerism, such as the addition product of 1-methylcyclohexene, the B and the H are thus *cis* to each other in the product.

1-methylcyclohexene

BH$_2$ and H are cis

When an organoborane is subsequently oxidized to an alcohol, the hydroxyl group ends up in the *same position* as the boron atom that it replaced, that is, with retention of configuration at that carbon.

trans-2-methylcyclohexanol

OH replaces BR$_2$ with retention of configuration

The reason that the configuration is retained is that the oxidation proceeds by a 1,2-shift (similar in some respects to a carbocation rearrangement), followed by hydrolysis of the BO bond to yield the alcohol. The RO bond is not affected in this hydrolysis.

from H$_2$O$_2$ + OH$^-$

1,2-shift of R

OR

$$\xrightarrow[-2\ OH^-]{2\ ^-OOH} RO-B-OR \xrightarrow[\text{hydrolysis}]{3\ OH^-} 3\ ROH + BO_3{}^{3-}$$

Study Problems

10.17 Use a transition-state structure to show why *trans*-2-methylcyclohexylborane cannot yield *cis*-2-methylcyclohexanol when treated with alkaline hydrogen peroxide.

10.18 Predict the major organic products (including stereochemistry) of the hydroboration and oxidation of   **(a)** 1-ethylcyclopentene;   **(b)** (*Z*)-3-methyl-2-pentene.

▼ ## Section 10.11
Addition of Halogens to Alkenes and Alkynes

Like acids, chlorine and bromine add to carbon–carbon double bonds and triple bonds. A common laboratory test for the presence of a double or triple bond in a compound of unknown structure is the treatment of the compound with a dilute solution of bromine in CCl_4. The test reagent has the reddish-brown color of Br_2; disappearance of this color is a positive test. The decolorization of a Br_2/CCl_4 solution by an unknown is suggestive, but not definitive proof, that a double or triple bond is present. A few other types of compounds, such as aldehydes, ketones, and phenols, also decolorize Br_2/CCl_4 solution.

$$CH_3CH=CHCH_3 + Br_2 \longrightarrow CH_3\overset{\overset{\displaystyle Br}{|}}{CH}-\overset{\overset{\displaystyle Br}{|}}{CH}CH_3$$

2-butene   *red*   2,3-dibromobutane

colorless

$$CH_3C{\equiv}CCH_3 + 2\,Br_2 \longrightarrow CH_3\overset{\overset{\displaystyle Br}{|}}{\underset{\underset{\displaystyle Br}{|}}{C}}-\overset{\overset{\displaystyle Br}{|}}{\underset{\underset{\displaystyle Br}{|}}{C}}CH_3$$

2-butyne   *red*

2,2,3,3-tetrabromobutane

colorless

Neither F_2 nor I_2 is a generally useful reagent for alkene addition reactions. Fluorine undergoes explosive reaction with organic compounds and its use requires special techniques. Iodine adds to a double bond, but the 1,2-diiodo product is unstable and loses I_2 to re-form the alkene.

$$R_2CI-CIR_2 \rightleftharpoons R_2C=CR_2 + I_2$$

Therefore, we can say this addition reaction is general only for *chlorine* and *bromine*.

A more substituted alkene is more reactive toward X_2 than a less substituted alkene (see Table 10.5). This is the same general order of reactivity as that toward HX.

$CH_2=CH_2$ $RCH=CH_2$ $R_2C=CH_2$ $R_2C=CHR$ $R_2C=CR_2$

increasing reactivity toward X_2 or HX addition ⟶

Table 10.5 Relative reactivities of some alkenes toward Br_2 in methanol

Compound	Relative Rate
$CH_2=CH_2$	1.0
$CH_3CH_2CH=CH_2$	97
cis-$CH_3CH_2CH=CHCH_3$	4,300
$(CH_3)_2C=C(CH_3)_2$	930,000

A. Electrophilic Attack of X_2

Halogens undergo addition reactions with alkenes. But, what is the source of the electrophile in X_2? When X_2 approaches the pi-bond electrons, polarity is induced in the X_2 molecule by repulsion between the pi electrons and the electrons in the X_2 molecule.

As the X—X bond becomes more polarized, it becomes progressively weaker until it finally breaks. The result is a halide ion and a positively charged organohalogen ion, called a **halonium ion**. The halonium ion from bromine, the **bromonium ion**, is bridged, similar to the intermediate in oxymercuration. (Chloronium ions are also sometimes bridged.) In the addition of Br_2 to ethylene or some other symmetrical alkene, the bridged bromonium ion is symmetrical, with Br bonded equally to each carbon.

a symmetrical bridged bromonium ion

·If the alkene is unsymmetrical, most of the positive charge is carried on the more substituted carbon. Carbocation stability is followed.

propene

an unsymmetrical bridged bromonium ion

B. *anti*-Attack of X^-

The bridged intermediate ion is positively charged and of high energy. Like a carbocation, it exists only momentarily in solution. Reaction is completed by attack by a nucleophile (in this case, Br^-). Br^- cannot attack a carbon of the bridged intermediate from the top (as we have shown the structure); that path is blocked by the Br bridge. Therefore, Br^- attacks from the opposite side of the intermediate. The result is **anti-addition** of Br_2 to the double bond. (Contrast this mode of addition to the *syn*-addition of BH_3, Section 10.10A.)

1,2-dibromopropane

General mechanism:

Step 1 (slow):

$$R_2C{=}CHR + X_2 \longrightarrow \left[R_2\overset{\delta+}{C}{-}CHR \atop \overset{\delta+}{X} \right] + X^-$$

Step 2 (fast):

$$\left[R_2\overset{\delta+}{C}{-}CHR \atop \overset{\delta+}{X} \right] + X^- \longrightarrow R_2C{-}CHR \atop \underset{X}{\overset{X}{|}}$$

Study Problem

10.19 Predict the products of Br_2 addition to:

    (a) $(CH_3)_2C{=}CH_2$ **(b)** $(CH_3)_2C{=}CHCH_3$

C. Evidence for *anti*-Addition

Considerable experimental evidence supports a bromination mechanism involving a bridged-ion intermediate (*anti*-addition) instead of a mechanism involving a simple carbocation intermediate. Bromination reactions of alkenes and alkynes are **stereoselective**—they produce predominantly one stereoisomeric product in reactions where two or more products might be expected. For example, when cyclohexene is brominated, the product is *trans*-1,2-dibromocyclohexane. If the intermediate were a simple carbocation, both *cis*- and *trans*-1,2-dibromocyclohexane should be formed.

cyclohexene

bridge on top

Br^-

bridged intermediate

trans-1,2-dibromocyclohexane (95%)

and not a carbocation intermediate

Attack of Br^- here would lead to the cis isomer.

Attack here would lead to the trans isomer.

Study Problems

10.20 2,3-Dibromobutane exists as three stereoisomeric forms: a pair of enantiomers, $(2R,3R)$ and $(2S,3S)$, and a *meso* form, $(2R,3S)$. Use an equation to show which isomers are formed when *cis*-2-butene is brominated.

10.21 Write equations to show the stereochemistry of the reaction of bromine with **(a)** (E)-1,2-dideuterioethene;   **(b)** (Z)-1,2-dideuterioethene. Name the products.

D. Mixed Addition

Bromination reactions of alkenes proceed by way of a bromonium-ion intermediate, followed by attack by a bromide ion to yield the dibromide. Is the second step limited to attack only by bromide ion? Can other nucleophiles compete with bromide ion in the second step to yield other products? Consider the case of a bromination reaction carried out with Br_2 in a solution containing Cl^- (from, say, NaCl). In this reaction, two nucleophiles (Br^- and Cl^-) are present. In such a case, *mixed dihalide products* are observed—along with the dibromoalkane, we find some bromochloroalkane.

$$CH_2{=}CH_2 + Br_2 \longrightarrow \left[\overset{+}{\underset{CH_2-CH_2}{Br}} \right] \xrightarrow[Cl^-]{Br^-} \begin{array}{c} \overset{Br}{\underset{CH_2-CH_2Br}{|}} \\ \text{1,2-dibromoethane} \\ \\ \overset{Br}{\underset{CH_2-CH_2Cl}{|}} \\ \text{1-bromo-2-chloroethane} \end{array}$$

Study Problem

10.22 Would you expect to find 1,2-dichloroethane as a product in the preceding example? Explain.

Sample Problem

When propene is treated with $Br_2 + Cl^-$, only one bromochloropropane is isolated as a product. What is its structure? Show, by equations, its formation.

Solution:

Step 1:

$$CH_3CH{=}CH_2 + Br_2 \xrightarrow{-Br^-} \left[CH_3\overset{\delta+}{\underset{\delta+}{CH}}\overset{Br}{\underset{}{|}}CH_2 \right] , \quad \text{not} \quad \left[CH_3CH\overset{\delta+}{\underset{\delta+}{—}}CH_2 \right]$$

Step 2:

$$\left[CH_3\underset{\delta+}{CH}\overset{\overset{\delta+}{Br}}{—}CH_2 \right] + Cl^- \longrightarrow CH_3\underset{Cl}{\overset{Br}{CHCH_2}}, \quad \text{not} \quad CH_3\overset{Br}{\underset{}{CHCH_2Cl}}$$

1-bromo-2-chloropropane

E. Addition of Halogens and Water

When an alkene is treated with a mixture of Cl_2 or Br_2 in water, a **1,2-halohydrin** (a compound with X and OH on adjacent carbon atoms) is formed. Recall from Section 8.3C that 1,2-halohydrins can be used to synthesize epoxides.

General:

$$R_2C{=}CH_2 \xrightarrow{\ X_2,\, H_2O\ } \underset{\underset{\displaystyle OH}{|}}{R_2C}\overset{\overset{\displaystyle X}{|}}{-}CH_2$$

an alkene ⟶ *a 1,2-halohydrin*

The path is similar to that for mixed halogen addition:

Step 1:

$$(CH_3)_2C{=}CH_2 + Br_2 \longrightarrow \left[(CH_3)_2\underset{\delta+}{C}\overset{\overset{\displaystyle \overset{\delta+}{Br}}{\diagup\,|}}{-}CH_2 \right] + Br^-$$

Step 2:

$$\left[(CH_3)_2\underset{\delta+}{C}\overset{\overset{\displaystyle \overset{\delta+}{Br}}{|}}{-}CH_2 \right] + H_2\ddot{O}\!: \longrightarrow$$

$$\underset{\underset{\displaystyle \overset{+}{:}OH_2}{|}}{(CH_3)_2C}\overset{\overset{\displaystyle Br}{|}}{-}CH_2 \underset{\xleftarrow{\hspace{1.2em}}}{\overset{-H^+}{\rightleftharpoons}} \underset{\underset{\displaystyle OH}{|}}{(CH_3)_2C}\overset{\overset{\displaystyle Br}{|}}{-}CH_2$$

1-bromo-2-methyl-2-propanol (73%)

Study Problem

10.23 When cyclohexene is treated with aqueous Cl_2, one product is obtained in 70% yield. Write equations for the steps in this reaction (complete with stereochemistry).

Section 10.12
Addition of Carbenes to Alkenes

If a student were asked whether a compound with the structure CH_2 exists, the answer might be, "No, because the carbon only has two bonds." However, in 1959 spectroscopic evidence showed that such a species does have a fleeting existence. The structure $:CH_2$ is called *methylene* and belongs to a class of highly reactive intermediates called **carbenes** ($R_2C:$). There are two types of methylene: *singlet methylene*, with an sp^2-hybridized carbon, and *triplet methylene*, with an sp-hybridized carbon (see Figure 10.8). Singlet methylene is

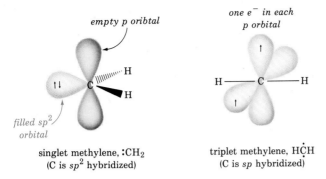

Figure 10.8 The bonding in singlet methylene and triplet methylene.

more useful in organic reactions and thus we will confine our discussion to this form of methylene.

Singlet methylene can be prepared by the photolysis (cleavage by light) of diazomethane (CH_2N_2), an unusual and reactive compound itself.

$$[:\bar{C}H_2-\overset{..}{N}\overset{+}{=}N: \longleftrightarrow :\bar{C}H_2-\overset{+}{N}\overset{+}{=}N: \longleftrightarrow CH_2\overset{+}{=}N=\overset{..}{N}:^- \longleftrightarrow {}^+CH_2-\overset{..}{N}=\overset{..}{N}:^-]$$

resonance structures for diazomethane

$$:\bar{C}H_2-\overset{+}{N}\equiv N: \xrightarrow{\ h\nu\ } :CH_2 + :N\equiv N:$$

The carbon in $:CH_2$, with only six valence electrons, is electrophilic and adds to the double bond of an alkene to yield a substituted cyclopropane. This reaction is a stereospecific *syn*-addition. Thus, *cis* alkenes yield *cis* cyclopropanes, and *trans* alkenes yield *trans* cyclopropanes.

$$
\begin{array}{c}
\underset{H}{\overset{R}{\diagdown}}C=C\underset{H}{\overset{R}{\diagup}} \quad + \quad :CH_2
\end{array}
$$

cis   singlet
          methylene

a cis-dialkylcyclopropane
(*meso*)

$$
\begin{array}{c}
\underset{H}{\overset{R}{\diagdown}}C=C\underset{R}{\overset{H}{\diagup}} \quad + \quad :CH_2
\end{array}
$$

trans   singlet
          methylene

a trans-dialkylcyclopropane
(*racemic*)

Dichlorocyclopropanes can be prepared with dichlorocarbene ($Cl_2C:$), which is formed by the reaction of a strong base and chloroform. The conversion of $CHCl_3$ to $Cl_2C:$ is related to elimination reactions of alkyl halides in that the base removes HCl from the molecule. The elimination reaction of $CHCl_3$ is an *α* **elimination** instead of a *β* elimination. (A molecule that has a *β* hydrogen will lose the *β* hydrogen in preference. However, $CHCl_3$ has no *β* hydrogen.)

Step 1:

$(CH_3)_3C\ddot{O}:^- $ *t*-butoxide ion $ + H-\overset{\displaystyle Cl}{\underset{\displaystyle Cl}{C}}-Cl$ chloroform $\rightleftharpoons (CH_3)_3C\ddot{O}H + \left[:\overset{\displaystyle Cl}{\underset{\displaystyle Cl}{C}}-Cl \right]^-$

Step 2:

$\left[:\overset{\displaystyle Cl}{\underset{\displaystyle Cl}{C}}-\ddot{Cl}: \right]^- \longrightarrow :CCl_2$ dichlorocarbene $+ :\ddot{\underset{..}{Cl}}:^-$

Dichlorocarbene adds to double bonds to yield *gem*-dihalocyclopropanes. (*Geminal*, or *gem*, means "on the same carbon.")

$+ :CCl_2 \longrightarrow$

(60%)

Study Problem

10.24 The following compound can be prepared from a monocyclic compound in 63% yield. Write one or more equations that show this synthesis.

Another way to prepare cyclopropanes is with a zinc–copper alloy and diiodomethane, which react to yield a **carbenoid** ("like a carbene"). The carbenoid, in turn, reacts with alkenes just as a carbene would. This reaction sequence, called the **Simmons–Smith reaction**, is fairly general.

$CH_2I_2 + Zn(Cu) \xrightarrow{\text{diethyl ether}} I-CH_2-ZnI$

a carbenoid,
which reacts like a carbene

$CH_3-\bigcirc-CH=CH_2 \xrightarrow[\text{diethyl ether}]{\text{Zn(Cu), } CH_2I_2} CH_3-\bigcirc-\triangle$

(67%)

The electrophilic addition reactions of alkenes are outlined in Figure 10.9.

Markovnikov additions:

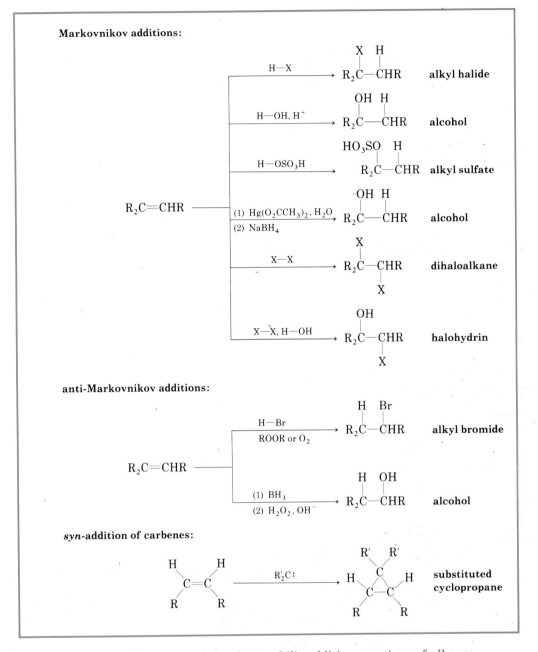

Figure 10.9 Summary of the electrophilic addition reactions of alkenes.

▼
Section 10.13
Catalytic Hydrogenation

The catalytic addition of hydrogen gas to an alkene or alkyne is a reduction of the pi-bonded compound. The reaction is general for alkenes, alkynes, and other compounds with pi bonds.

Alkenes and alkynes:

$$CH_3CH=CH_2 + H_2 \xrightarrow{Pt} CH_3CH_2CH_3$$

propene   propane

$$CH_3C≡CH + 2 H_2 \xrightarrow{Pt} CH_3CH_2CH_3$$

propyne   propane

Other pi systems:

$$\underset{\text{acetone}}{CH_3\overset{O}{\overset{||}{C}}CH_3} + H_2 \xrightarrow{Pt} \underset{\text{2-propanol}}{CH_3\overset{OH}{\overset{|}{C}H}CH_3}$$

$$\underset{\substack{\text{ethanenitrile}\\\text{(acetonitrile)}}}{CH_3C≡N} + 2 H_2 \xrightarrow{Pt} \underset{\text{ethylamine}}{CH_3CH_2NH_2}$$

$$\underset{\text{benzene}}{\text{⬡}} + 3 H_2 \xrightarrow[\text{heat, pressure}]{Pt} \underset{\text{cyclohexane}}{\text{⬡}}$$

A. Action of the Catalyst

Hydrogenation reactions are exothermic, but they do not proceed sponta-neously because the energies of activation are extremely high. Heating cannot supply the energy needed to get the molecules to the transition state. However, reaction proceeds smoothly when a catalyst is added.

A finely divided metal or a metal adsorbed onto an insoluble, inert carrier (such as elemental carbon or barium carbonate) is often used as a hydrogena-tion catalyst. The metal chosen depends on the compound to be reduced and the conditions of the hydrogenation. For example, platinum, palladium, nickel, rhenium, and copper are all suitable for the reduction of alkenes. For esters, which are far more difficult to reduce, a copper–chromium catalyst (plus heat and pressure) is usually used.

A **poisoned catalyst** (one that is partly deactivated) is used for hydro-genation of an alkyne to an alkene instead of to an alkane. Palladium that has been treated with quinoline (page 788) is a typical poisoned catalyst.

$$\underset{\text{propyne}}{CH_3C≡CH} + H_2 \xrightarrow{\text{deactivated Pd}} \underset{\text{propene}}{CH_3CH=CH_2}$$

How does a hydrogenation catalyst ease the course of a hydrogenation reaction? Experimental evidence supports the theory that first the hydrogen molecules are adsorbed onto the metallic surface; then the H_2 sigma bonds are broken and metal–H bonds are formed. The alkene is also adsorbed onto the metallic surface, with its pi bond interacting with the empty orbitals of the metal. The alkene molecule moves around on the surface until it collides with a metal-bonded hydrogen atom, undergoes reaction, and then leaves as the hy-drogenated product (see Figure 10.10).

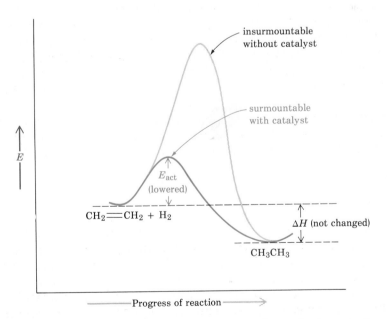

Figure 10.10 Hydrogenation of an alkene.

The overall effect of the catalyst is to provide a surface on which the reaction can occur and to weaken the bonds of both H_2 and the alkene. The result is a *lowering of the energy of activation for the reaction*. Figure 10.11 shows energy diagrams for a hydrogenation reaction. Note that *the catalyst does not affect the energies of reactants or products*. The ΔH for the reaction is not changed by catalytic action; only the E_{act} is changed.

In recent years, soluble catalysts have been developed so that hydrogenation can occur in a homogeneous solution, rather than on a surface. These catalysts, organic metal complexes such as $[(C_6H_5)_3P]_3RhCl$, selectively catalyze the reduction of nonhindered double bonds. For example, if two double

Figure 10.11 Idealized energy diagrams for a hydrogenation reaction. (The reaction actually occurs stepwise.)

bonds of a diene differ in steric hindrance, the less hindered double bond can be selectively hydrogenated.

(90%)

If the soluble catalyst is complexed with an optically active chelating agent, asymmetry can be introduced by the hydrogenation reaction.

configuration depends on configuration of catalyst

$$CH_2{=}C{-}NHCCH_3 + H_2 \xrightarrow{\text{chiral complex Rh salt}}$$

achiral

one enantiomer

B. Stereochemistry of Hydrogenation

Evidence shows that the two hydrogen atoms add to the same side of, or *syn* to, the double bond when a solid catalyst is used. The *syn*-addition arises from the reaction occurring on a surface. Access by H_2 to only one side of the pi bond is more favorable than access to both sides. If the hydrogenation products are capable of geometric isomerism, the *cis*-product is usually observed as the predominant product. (In some cases, however, isomerization to the more stable *trans*-product occurs.)

$$CH_3{-}C{\equiv}C{-}CH_3 + H_2 \xrightarrow[\text{syn-addition}]{\text{deactivated Pd}}$$

2-butyne

cis-2-butene

1,2-dimethylcyclohexene

cis-1,2-dimethylcyclohexane (82%)

Sample Problem

The *cis*-isomers of alkenes can be prepared by the catalytic hydrogenation of alkynes. The *trans*-isomers of alkenes can be obtained in excellent yields by a **dissolving-metal reduction** of the alkynes: a reduction using a solution of sodium or lithium metal in liquid ammonia at $-33°$. In a dissolving-metal reduction, the alkali metal is oxidized to a cation, while protons for the reduction are obtained from NH_3. Write balanced equations showing the reduction of 3-octyne by   (a) hydrogenation and   (b) Na in liquid NH_3. Name the products.

Solution:

(a) $CH_3CH_2C{\equiv}C(CH_2)_3CH_3 + H_2 \xrightarrow{\text{catalyst}}$

$$\underset{H \qquad\quad H}{\overset{CH_3CH_2 \qquad (CH_2)_3CH_3}{C=C}}$$

cis-3-octene

(b) $CH_3CH_2C{\equiv}C(CH_2)_3CH_3 + 2\,Na\cdot + 2:NH_3 \xrightarrow[-33°]{\text{liq. } NH_3}$

$$\underset{H \qquad (CH_2)_3CH_3}{\overset{CH_3CH_2 \qquad H}{C=C}} \qquad + 2\,Na^+ + 2:\overset{\cdot\cdot}{N}H_2{}^-$$

trans-3-octene

C. How Heats of Hydrogenation Show Alkene Stability

The **heat of hydrogenation** of an alkene is the energy difference between the starting alkene and the product alkane. It is calculated from the amount of heat released in a hydrogenation reaction. Table 10.6 lists the heats of hydrogenation of a few alkenes.

     Let us consider the three alkenes that can be reduced to butane:

$$\begin{array}{l} CH_3CH_2CH{=}CH_2 \searrow{\scriptstyle H_2,\,Pt} \\ \textit{cis-}CH_3CH{=}CHCH_3 \xrightarrow{H_2,\,Pt} \quad CH_3CH_2CH_2CH_3 \\ \textit{trans-}CH_3CH{=}CHCH_3 \nearrow{\scriptstyle H_2,\,Pt} \qquad \text{butane} \end{array}$$

The product butane has the same energy regardless of the starting alkene. Any differences in ΔH for the three reactions reflect *differences in the energies of the starting alkenes*. The greater the value of the ΔH of hydrogenation, the higher is the energy of the starting alkene (see Figure 10.12).

Table 10.6 Heats of hydrogenation for some alkenes and dienes

Name	*Structure*	$-\Delta H$, *kcal/mol*
ethene (ethylene)	$CH_2{=}CH_2$	32.8
propene (propylene)	$CH_3CH{=}CH_2$	30.1
1-butene	$CH_3CH_2CH{=}CH_2$	30.3
cis-2-butene	*cis*-$CH_3CH{=}CHCH_3$	28.6
trans-2-butene	*trans*-$CH_3CH{=}CHCH_3$	27.6
2-methyl-2-butene	$\underset{\vert}{CH_3}$ $CH_3C{=}CHCH_3$	26.9
3-methyl-1-butene	$\underset{\vert}{CH_3}$ $CH_3CHCH{=}CH_2$	30.3
1,3-pentadiene	$CH_2{=}CHCH{=}CHCH_3$	54.1
1,4-pentadiene	$CH_2{=}CHCH_2CH{=}CH_2$	60.8

Figure 10.12 Comparison of the heats of hydrogenation of the three butenes that yield butane upon reduction.

From the differences in ΔH, we can see that 1-butene contains 1.7 kcal/mol more energy than does *cis*-2-butene. This means that 1-butene is 1.7 kcal/mol less stable than *cis*-2-butene. On the other hand, *cis*-2-butene contains 1.0 kcal/mol more energy than *trans*-2-butene. The relative heats of hydrogenation thus show that *trans*-2-butene is the most stable butene and that 1-butene is the least stable.

$$CH_2{=}CHCH_2CH_3 \qquad cis\text{-}CH_3CH{=}CHCH_3 \qquad trans\text{-}CH_3CH{=}CHCH_3$$

decreasing energy content; increasing stability →

From just such ΔH comparisons, the relative stabilities of a large number of alkenes have been determined. The following statements summarize what we have learned about alkene stabilities:

1. Increasing the alkyl substitution on the carbons of a double bond stabilizes an alkene. The reasons for this trend are not well-understood. Hyperconjugation (Section 5.6E), bond dissociation energies, and other effects have been proposed to explain the stability differences.

$$CH_2{=}CH_2 \quad RCH{=}CH_2 \quad RCH{=}CHR \quad R_2C{=}CH_2 \quad R_2C{=}CHR \quad R_2C{=}CR_2$$

increasing stability →

2. Conjugated dienes are more stable than dienes with isolated double bonds (because of delocalization of the pi-electron density in the conjugated system).
3. *trans*-Alkenes are more stable than *cis*-alkenes (because there are fewer steric repulsions in *trans*-isomers).

repulsion: ⟍

H H H H	H H
C C	C H
H C=C H	H C=C H
H H	H C
	H H

less stable   *more stable*

Study Problem

10.25 Which of each of the following pairs of alkenes would you expect to show the greater difference in energy between the *cis-* and *trans-*isomers? Why?

(a) $(CH_3)_3CCH{=}CHCH_2CH_3$ and $CH_3CH{=}CHCH_3$

(b) $ClCH{=}CHCl$ and $CH_3CH{=}CHCl$

D. Hydrogenation of Fats and Oils

The molecules of animal fats and vegetable oils contain long hydrocarbon chains. In vegetable oils, these chains are **polyunsaturated** (have several double bonds). Solid fats, on the other hand, usually contain few, if any, double bonds.

A vegetable oil can be converted to a substance of more solid consistency by partial hydrogenation of the carbon–carbon double bonds. The process of converting liquid oils to solid fats by this technique is called **hardening**. Although the polyunsaturates may be more healthful, the hydrogenated products are generally more palatable. Partially hydrogenated peanut oil is used to make peanut butter, and partially hydrogenated corn oil or safflower oil is used in margarine. Note that the carbonyl groups in the vegetable oil do not undergo hydrogenation under these conditions because they are more difficult to hydrogenate. Also, any or all of the carbon–carbon double bonds could be hydrogenated. Therefore, a mixture of partially hydrogenated products would result.

$$
\begin{array}{l}
\overset{O}{\overset{\|}{CH_2OC(CH_2)_{14}CH_3}} \\
\overset{O}{\overset{\|}{CHOC(CH_2)_7CH{=}CH(CH_2)_7CH_3}} \\
\overset{O}{\overset{\|}{CH_2OC(CH_2)_7CH{=}CHCH_2CH{=}CH(CH_2)_4CH_3}}
\end{array}
\quad\xrightarrow{\ 2\,H_2,\ Pt\ }\quad
\begin{array}{l}
\overset{O}{\overset{\|}{CH_2OC(CH_2)_{14}CH_3}} \\
\overset{O}{\overset{\|}{CHOC(CH_2)_7CH{=}CH(CH_2)_7CH_3}} \\
\overset{O}{\overset{\|}{CH_2OC(CH_2)_{16}CH_3}}
\end{array}
$$

a typical vegetable oil   *a typical fat*

Section 10.14
Oxidation of Alkenes

Alkenes can be oxidized to a variety of products depending on the alkene structure and on the reagent used. Reactions involving oxidation of a carbon–carbon bond can be classified into two general groups: (1) oxidation of the pi bond *without cleavage of the sigma bond*, and (2) oxidation of the pi bond *with cleavage of the sigma bond*.

The products of oxidation without cleavage are generally either 1,2-diols or epoxides.

Without cleavage:

an epoxide   a 1,2-diol, or glycol

When both the sigma bond and the pi bond of an alkene are cleaved in an oxidation, the products are ketones, aldehydes, or carboxylic acids.

With cleavage:

$$\underset{}{\overset{}{C}}=\underset{}{\overset{}{C}} \xrightarrow{\text{[O]}} \quad \overset{O}{\underset{}{\overset{\parallel}{-C-}}} \quad \text{or} \quad \overset{O}{\underset{}{\overset{\parallel}{-CH}}} \quad \text{or} \quad \overset{O}{\underset{}{\overset{\parallel}{-COH}}}$$

$$\text{ketones} \qquad \text{aldehydes} \qquad \text{carboxylic acids}$$

A variety of reagents are used to oxidize alkenes. Some of the more common ones are listed in Table 10.7.

A. Oxidation Without Cleavage

Diol formation. The most popular reagent used to convert an alkene to a 1,2-diol is a cold, alkaline, aqueous solution of potassium permanganate (even though this reagent usually gives low yields). Osmium tetroxide (OsO_4), followed by reduction with a reagent such as Na_2SO_3 or $NaHSO_3$, gives better yields of diols, but the use of this reagent is limited because it is both expensive and toxic. (However, osmium tetroxide need be present in only catalytic amounts if used in conjunction with hydrogen peroxide.) Both the permanganate and the OsO_4 oxidations proceed by way of cyclic inorganic esters, which yield *cis*-diols if the product is capable of geometric isomerism.

$$CH_2=CH_2 \xrightarrow[25^\circ]{MnO_4^-} CH_2-CH_2 \xrightarrow{OH^-} CH_2-CH_2 + MnO_2$$

ethylene

1,2-ethanediol
(ethylene glycol)

cyclohexene $\xrightarrow{OsO_4}$ $\xrightarrow[H_2O]{Na_2SO_3}$ + Os

cis-1,2-cyclohexanediol (62%)

Table 10.7 Common reagents for oxidation of alkenes

Reagent	Products
Oxidation Without Cleavage:	
$KMnO_4$ with OH^- (cold)	1,2-diols
OsO_4 followed by Na_2SO_3	1,2-diols
$C_6H_5CO_3H$	epoxides
O_2 (catalyst)	methyl ketones
Oxidation With Cleavage:	
$KMnO_4$ (hot)	carboxylic acids and ketones
O_3 followed by oxidation	carboxylic acids and ketones
O_3 followed by reduction	aldehydes and ketones

General:

$$R_2C{=}CR_2 \xrightarrow[\text{\textit{syn}-addition}]{\text{cold MnO}_4^- \text{ or OsO}_4} R_2C{-}CR_2$$
$$\overset{OH}{|} \quad \overset{OH}{|}$$

an alkene ———————————————— *a 1,2-diol*

The reaction with cold permanganate solution constitutes the **Baeyer test** for unsaturation in compounds of unknown structure. The test solution ($KMnO_4$) is purple. As the reaction proceeds, the purple color disappears and a brown precipitate of MnO_2 forms. The Baeyer test for double bonds, while widely used, has a serious limitation—any easily oxidized group (aldehyde, alkene, alkyne) gives a positive test result.

Study Problem

10.26 The following compounds are treated with OsO_4, followed by Na_2SO_3. What products would you expect? (Indicate any stereoisomerism.)

(a) $CH_3CH{=}C(CH_3)_2$ (b) ⬠—CH_3

Epoxide formation.   In Section 8.3C, we described how the oxidation of an alkene by a peroxycarboxylic acid yields the epoxide.

cycloheptene   cycloheptene oxide (78%)

In Chapter 8, we discussed the S_N2 cleavage of epoxides of cycloalkanes, which results in *trans*-1,2-diols. By contrast, oxidation of an alkene with OsO_4 or cold $KMnO_4$ yields *cis*-1,2-diols. Thus, either type of diol may be prepared from an alkene, depending on the choice of reagents.

cis-1,2-cyclopentanediol

cyclopentene

trans-1,2-cyclopentanediol

Methyl ketone formation.   When treated with oxygen and a catalyst, alkenes with a —$CH{=}CH_2$ group can be converted to methyl ketones.

$$n\text{-}C_8H_{17}CH{=}CH_2 + O_2 \xrightarrow[\text{CuCl}_2]{\text{PdCl}_2} C_8H_{17}\overset{\displaystyle O}{\overset{\|}{C}}CH_3 \ .$$

This general synthetic method yields the same products as the hydration of the corresponding terminal alkyne (Section 10.8).

B. Oxidation with Cleavage

The products of oxidation with cleavage depend both upon the oxidizing conditions and upon the structure of the alkene. Let us consider first the structure of the alkene.

The structural feature of the alkene that determines the products of oxidative cleavage is the *presence or absence of hydrogen atoms on the sp²* *carbons*. If neither alkene carbon is bonded to a hydrogen atom (that is, each alkene carbon is disubstituted), then oxidative cleavage results in a pair of *ketone* molecules.

If, on the other hand, each alkene carbon has a hydrogen bonded to it, then the products of oxidative cleavage are either *aldehydes* or *carboxylic acids*, depending upon the reaction conditions.

Cleavage with KMnO₄. A hot solution of $KMnO_4$ is a vigorous oxidizing agent that leads to only ketones and carboxylic acids. (Aldehydes cannot be isolated from $KMnO_4$ solutions. They would be oxidized promptly to carboxylic acids.)

Under these vigorous oxidizing conditions, the carbon of a terminal double bond is oxidized to CO_2.

methylene-
cyclopentane   cyclopentanone

The reason for CO_2 formation is that the methylene group is first oxidized to formic acid, which is further oxidized to carbonic acid. The latter undergoes spontaneous decomposition to CO_2 and H_2O.

formic acid   carbonic acid

Ozonolysis. Ozonolysis (cleavage by ozone) has been used for determining the structures of unsaturated compounds because this reaction results in the degradation of large molecules into smaller, identifiable fragments. Today, spectroscopy has largely replaced ozonolysis as an analytical technique.

An ozone molecule consists of three oxygen atoms bonded in a chain. The two O—O bonds are of equal length (1.29 Å) with a bond angle of 116°. The structure is best represented as a resonance hybrid.

resonance structures of ozone

Ozonolysis consists of two separate reactions: (1) oxidation of the alkene by ozone to an **ozonide**, and (2) either oxidation or reduction of the ozonide to the final products. The initial oxidation is usually carried out by bubbling ozone through a solution of the alkene in an inert solvent such as CCl_4. Ozone attacks the pi bond to yield an unstable intermediate called a 1,2,3-trioxolane. This intermediate then goes through a series of transformations in which the carbon–carbon sigma bond is cleaved. The product is an ozonide (a 1,2,4-trioxolane), which is rarely isolated, but is carried on to the second step.

2-methyl-2-butene   *a 1,2,3-trioxolane*

an ozonide
(a 1,2,4-trioxolane)

The second reaction in ozonolysis is either the oxidation or the reduction of the ozonide. If the ozonide is subjected to a *reductive work-up*, the part of the original alkene with a monosubstituted carbon yields an aldehyde. If an *oxidative work-up* is used, the part of the alkene with a monosubstituted carbon yields a carboxylic acid. In either case, the portion of an alkene containing a disubstituted carbon yields a ketone. Common reducing agents are zinc metal in aqueous acid and dimethyl sulfide. Hydrogen peroxide is a common oxidizing agent.

Reductive work-up to aldehydes and ketones:

Oxidative work-up to carboxylic acids and ketones:

Sample Problem

Predict the products of reductive ozonolysis of γ-terpinene, a compound found in coriander oil:

Solution:

Study Problem

10.27 What would be the products of both reductive and oxidative ozonolyses of each of the following alkenes or dienes?

(a)   (b)   (c)   (d)

OH   OH
| |
aqueous KMnO$_4$, OH$^-$, cold
or (1) OsO$_4$, (2) Na$_2$SO$_3$ → R$_2$C——CHR'   *cis*-1,2-diol

ArCO$_3$H → R$_2$C—CHR'   epoxide
(O across the two carbons — epoxide)

aqueous KMnO$_4$, heat
or (1) O$_3$, (2) H$_2$O$_2$, H$^+$ → R$_2$C=O + HO$_2$CR'   ketone and/or
                                                              carboxylic
                                                              acid

R$_2$C=CHR'

(1) O$_3$, (2) Zn, H$^+$ → R$_2$C=O + HCR'   ketone and/or
                                              aldehyde
(HCR' with O double bond)

Figure 10.13 Summary of the oxidation of alkenes. (Alkenes with terminal double bonds can also be oxidized to methyl ketones.)

Figure 10.13 summarizes the oxidation reactions of alkenes.

Section 10.15
1,2-Addition and 1,4-Addition to Conjugated Dienes

Many of the reactions of conjugated dienes are the same as those of compounds with isolated double bonds. Acidic reagents and the halogens can add across one or both of the pi bonds. In conjugated-diene systems, these simple addition reactions are called **1,2-additions**, a term that refers to addition to adjacent carbons of a conjugated-diene system and not necessarily to nomenclature numbers.

1,2-Addition reactions:

$$CH_2=CHCH=CH_2 \xrightarrow{HBr} \overset{H}{\underset{|}{C}}H_2-\overset{Br}{\underset{|}{C}}HCH=CH_2$$

1,3-butadiene   3-bromo-1-butene

$$CH_3CH=CHCH=CHCH_3 \xrightarrow{Cl_2} \overset{Cl}{\underset{|}{C}}H_3CH-\overset{Cl}{\underset{|}{C}}HCH=CHCH_3$$

2,4-hexadiene   4,5-dichloro-2-hexene

Study Problem

10.28 Predict the product of 1,2-addition to the second double bond in each of the two preceding examples.

Along with 1,2-addition, conjugated dienes can also undergo **1,4-addition**, also called **conjugate addition**. In these reactions an equimolar quantity of the reagent adds to the two end carbons (carbons 1 and 4) of the diene system. The remaining double bond ends up in the center of the original diene system.

1,4-Addition reactions:

$$CH_2{=}CHCH{=}CH_2 \xrightarrow{\ HBr\ } \overset{\overset{H}{|}}{C}H_2{-}CH{=}CH{-}\overset{\overset{Br}{|}}{C}H_2$$

1,3-butadiene   1-bromo-2-butene

$$CH_3CH{=}CHCH{=}CHCH_3 \xrightarrow{\ Cl_2\ } CH_3\overset{\overset{Cl}{|}}{C}H{-}CH{=}CH{-}\overset{\overset{Cl}{|}}{C}HCH_3$$

2,4-hexadiene   2,5-dichloro-3-hexene

A. Mechanisms of 1,2- and 1,4-Addition

Let us look at the mechanism of each type of addition. The mechanism for 1,2-addition of HX is the same as that for addition to an isolated double bond. (The reaction of 1,3-butadiene goes through the more stable allylic carbocation and not through the less stable $^+CH_2CH_2CH{=}CH_2$.)

1,2-Addition:

$$CH_2{=}CHCH{=}CH_2 \xrightarrow{\ H^+\ } \left[\overset{\overset{H}{|}}{C}H_2{-}\overset{+}{C}HCH{=}CH_2\right] \xrightarrow{:\ddot{B}r:^-} \overset{\overset{H}{|}}{C}H_2{-}\overset{\overset{:\ddot{B}r:}{|}}{C}HCH{=}CH_2$$

a 2° allylic carbocation   3-bromo-1-butene

The mechanism for 1,4-addition is a direct extension of that for 1,2-addition. The carbocation in the preceding example is an *allylic cation* (Section 5.7A) and is resonance stabilized. Because of the resonance stabilization of the allylic cation, there is a partial positive charge on carbon 4 of the diene system as well as on carbon 2. Attack at carbon 4 leads to the 1,4-addition product.

1,4-Addition:

$$CH_2{=}CHCH{=}CH_2 \xrightarrow{\ H^+\ } \left[\overset{\overset{H}{|}}{C}H_2{-}\overset{+}{C}H{-}CH{=}CH_2 \longleftrightarrow \overset{\overset{H}{|}}{C}H_2{-}CH{=}CH{-}\overset{+}{C}H_2\right]$$

$$\xrightarrow{:\ddot{B}r:^-} \overset{\overset{H}{|}}{C}H_2{-}CH{=}CH{-}\overset{\overset{:\ddot{B}r:}{|}}{C}H_2$$

1-bromo-2-butene

If an equimolar quantity of reagent is added to 1,3-butadiene, a mixture of two products results: 3-bromo-1-butene from 1,2-addition, and 1-bromo-2-butene from 1,4-addition.

Sample Problem

(a) Write the structures of *all possible* carbocation intermediates in the addition of 1.0 mol of HI to 1.0 mol of 2,4-hexadiene.

(b) Which carbocation would you expect to be formed at the faster rate?

Solution:

(a) $CH_3CH{=}CH{-}CH{=}CHCH_3 \xrightarrow{\ H^+\ }$

$\quad\quad [CH_3CH_2{-}\overset{+}{C}H{-}CH{=}CHCH_3]$ or $[CH_3\overset{+}{C}H{-}CH_2{-}CH{=}CHCH_3]$

(Addition of H^+ to the other double bond gives identical intermediates.)

(b) The first carbocation shown would be formed at the faster rate because it is a resonance-stabilized allylic carbocation.

Study Problems

10.29 Write the *resonance structures* for the principal intermediate in the preceding sample problem, and give the structures of the principal products.

10.30 Predict the products of the addition of an equimolar quantity of bromine to the following dienes. Write resonance structures for any intermediates formed.

(a)   (b)

In the reaction of 1,3-butadiene with an equimolar quantity of HBr, the ratio of 1,2-addition to 1,4-addition varies with the temperature at which the reaction is carried out. At $-80°$, the 1,2-addition product predominates. At $40°$, the 1,4-addition product predominates.

$$CH_2{=}CHCH{=}CH_2 + HBr$$

1,3-butadiene

$-80°$ path:

$$\overset{\displaystyle Br}{\underset{|}{CH_3CHCH}}{=}CH_2 + CH_3CH{=}CHCH_2Br$$

1,2-product (80%)   1,4-product (20%)

$40°$ path:

$$\overset{\displaystyle Br}{\underset{|}{CH_3CHCH}}{=}CH_2 + CH_3CH{=}CHCH_2Br$$

(20%)   (80%)

It has also been observed that warming 3-bromo-1-butene (the 1,2-addition product) to $40°$ with a trace of acid results in an equilibrium mixture that is predominantly 1-bromo-2-butene (the 1,4-addition product).

$$\overset{\displaystyle Br}{\underset{|}{CH_3CHCH}}{=}CH_2 \underset{\xrightarrow{\quad}}{\overset{H^+,\,40°}{\xleftarrow{\quad}}} CH_3CH{=}CHCH_2Br$$

(20%)   (80%)

How can we explain these observations? At low temperature, the reaction yields predominantly the 1,2-addition product because the 1,2-addition has the lower E_{act} (the 2° carbon carries more + charge than the 1° carbon) and thus

Figure 10.14 Partial energy diagram for the reaction of 1,3-butadiene with HBr. At −80°, the 1,2-addition product predominates.

the faster rate. The relative rates of the reactions control the product ratio. The reaction is under *kinetic control* at low temperatures. Figure 10.14 shows the energy diagram for the competing reactions.

At higher temperatures, a greater percentage of molecules can reach the higher-energy transition state, and the two products are in *equilibrium*. The more stable 1,4-product (which is the more substituted alkene) predominates. At the higher temperatures, the relative stabilities of the products control the product ratios, and the reaction is under *thermodynamic control*. Figure 10.15 shows the energy diagram for the equilibrium.

Study Problems

10.31 Which *geometric isomer* of 1-bromo-2-butene would predominate at 40°?

10.32 1,3-Butadiene is treated with an equimolar quantity of Br_2 at −15°. Two structural isomers are obtained: 46% A and 54% B. When the reaction is carried out at 60°, the product mixture contains 90% A. What are the structures of A and B?

Figure 10.15 Energy diagram for the equilibrium between the 1,2- and 1,4-addition products. At 40°, the 1,4-addition product predominates.

B. 1,4-Addition Polymers

Conjugated dienes can be polymerized by 1,4-addition. The product still contains unsaturation; therefore, the polymer could contain all *cis* units, all *trans* units, or a mixture of *cis* and *trans* units. The following equation shows the 1,4-polymerization of isoprene.

$$x\ CH_2{=}\overset{\overset{\displaystyle CH_3}{|}}{C}{-}CH{=}CH_2 \quad \xrightarrow[\text{1,4-addition}]{\text{catalyst}} \quad {+}CH_2\overset{\overset{\displaystyle CH_3}{|}}{C}{=}CHCH_2{-}CH_2\overset{\overset{\displaystyle CH_3}{|}}{C}{=}CHCH_2{+}_x$$

2-methyl-1,3-butadiene   polyisoprene
(isoprene)

Natural rubber is polyisoprene with *cis* double bonds. The *trans* polymer, called **gutta-percha**, is a hard polymer used as a covering for golf balls and in temporary dental fillings.

natural rubber (all *cis*)

gutta-percha (all *trans*)

The *cis* double bonds in natural rubber force the molecules into a distorted conformation that can be stretched and yet return to its original shape. The *trans* double bonds in gutta-percha allow the polymer molecules to pack more closely to one another and form a more crystalline, nonelastic substance.

Neither natural rubber nor gutta-percha is formed in nature by the direct polymerization of isoprene, as was once believed. Instead, they are formed by a biosynthetic path starting from acetylcoenzyme A. (See Section 23.7C.)

Study Problem

10.33 *Neoprene,* developed in 1932, was the first synthetic rubber. It is used in washers, gaskets, and the like. Unlike natural rubber, neoprene does not swell when in contact with petroleum products. Therefore, it is also used extensively in the fabrication of petroleum containers, such as gasoline tanks. Neoprene is the all *trans,* head-to-tail polymer formed by a 1,4-addition of 2-chloro-1,3-butadiene. What is the structure of this polymer?

▼ ## Section 10.16
The Diels–Alder Reaction

One very important type of 1,4-addition is represented by the **Diels–Alder reaction**, which is a route to cyclohexenes. The reaction is named after the German chemists Otto Diels and Kurt Alder, who jointly received the 1950

Nobel prize for their work in this area. In a Diels–Alder reaction, a conjugated diene is heated with a second unsaturated compound, called the **dienophile** ("diene lover"), to yield a product containing a six-membered ring.

1,3-butadiene	propenal	3-cyclohexene-1-carbaldehyde
the diene	*the dienophile*	(100%)

The Diels–Alder reaction is but one example of a larger class of reactions called **pericyclic reactions**. In this section, we will limit our discussion to the Diels–Alder reaction. In Chapter 20, we will discuss the theory of pericyclic reactions.

A. Line Formulas and Conformations

Because the Diels–Alder reaction converts open-chain compounds to cyclic compounds, the use of **line formulas** is very convenient for representing the open-chain compounds in this reaction. These line formulas are directly analogous to polygon formulas that represent rings.

Another convention is the use of the terms **s-*cis*** and **s-*trans*** to describe the conformations of conjugated dienes. (The letter "s" is used because it is the geometry around the central *single bond* that determines the conformation.) The following compounds illustrate the use of the terms. For open-chain compounds, these formulas represent interconvertible conformers because sigma-bond rotation (requiring about 4 kcal/mol for 1,3-butadiene) is all that is needed for conversion from one to another.

s-*cis*   s-*trans*

H$_3$C―/=\―CH$_3$	
s-cis	H$_3$C CH$_3$
even though the	s-cis
stereochemistry at	and (Z), or
each double bond is	cis, at each
(E), or trans	double bond

When the diene function is part of a cyclic system, the s-*cis* and s-*trans* structures represent different compounds. Interconversion cannot occur without bonds being broken.

"s-cis"   "s-trans"

Study Problem

10.34 Classify each of the following dienes as s-*cis* or s-*trans*. Indicate which is interconvertible with the other form.

(a)   (b)   (c)

In a Diels–Alder reaction, the diene must have the s-*cis*, not the s-*trans*, conformation. In some structures (such as 1,3-butadiene), the s-*cis* and s-*trans* conformers are readily interconvertible. In other diene systems (such as in ring systems), the s-*trans* isomer does not undergo reaction.

Some s-cis dienes that can be used in a Diels–Alder reaction:

B. Examples of Diels–Alder Reactions

Using the line-formula symbolism, we can show the Diels–Alder reaction that was presented on page 440 as:

the diene   the dienophile   the product

The dienophile usually contains other unsaturation (an aldehyde group in the preceding example) that does not participate directly in the addition reaction. This group does, however, enhance the reactivity of the dienophile's carbon–carbon double bond (the site of reaction) by electron-withdrawal. (Remember, the carbon in C—O carries a partial positive charge.)

You can see from the following examples the versatility of the Diels–Alder reaction for the synthesis of cyclic compounds containing cyclohexene rings. For practice, find the cyclohexene rings in the following examples.

(39%)

(100%)

Study Problems

10.35 Predict the Diels–Alder products:

10.36 Suggest a synthesis for the following compound. (*Hint*: An alkyne may be used as a dienophile.)

C. Stereochemistry of the Diels–Alder Reaction

A Diels–Alder reaction is a *cis*, or *syn*-addition, and is thus stereospecific.

a *cis*-dienophile   a *cis*-product

There are two ways a diene can approach a dienophile. In some cases, two possible products can be formed. The two modes of addition are called **endo** (inside) and **exo** (outside), as the following examples show. The *endo* addition is usually favored, probably because of favorable interactions of the pi orbitals of the developing double bond and the pi orbitals in the unsaturated group. In the *endo* product of our example, the carbonyl group of the dienophile is *trans* to the bridge. In the *exo* product, the dienophile C=O is *cis* to the bridge.

Study Problems

10.37 Predict the product (including stereochemistry):

10.38 The following Diels–Alder reaction can yield two isomeric cyclohexenes. What are their structures?

$$CH_2=CHC=CH_2 + CH_2=CHCHO \xrightarrow{\text{heat}}$$
with CH_3 substituent on the central carbon

▼ ## Section 10.17
Use of Alkenes and Alkynes in Synthesis

From Table 10.8, you can see that alkenes are valuable starting materials for synthesizing other organic compounds. Alkynes are not as widely used in synthesis (nor are they as readily available).

The addition reactions that use H^+ as a catalyst proceed through carbocations and yield Markovnikov products (and possible rearrangement products). These reactions include those with HX and with H_2O (H^+).

Markovnikov products:

Table 10.8 Types of compounds that can be obtained from alkenes and dienes

Reaction	Product	Section Reference

Markovnikov Addition:

$$R_2C{=}CHR + HX \longrightarrow R_2\overset{\displaystyle X}{\underset{|}{C}}CH_2R$$

alkyl halide 10.7A–C

$$+ H_2O \xrightarrow{H^+} R_2\overset{\displaystyle OH}{\underset{|}{C}}CH_2R$$

alcohol 10.8

$$\xrightarrow[\text{(2) NaBH}_4]{\text{(1) Hg(O}_2\text{CCH}_3)_2, \text{ H}_2\text{O}} R_2\overset{\displaystyle OH}{\underset{|}{C}}CH_2R$$

alcohol 10.9

$$\xrightarrow[\text{(2) NaBH}_4]{\text{(1) Hg(O}_2\text{CCH}_3)_2, \text{ R'OH}} R_2\overset{\displaystyle OR'}{\underset{|}{C}}CH_2R$$

ether 10.9

$$+ X_2 \longrightarrow R_2\overset{\displaystyle X}{\underset{\underset{\displaystyle X}{|}}{C}}{-}CHR$$

dihaloalkane 10.11A–C

$$+ X_2 \xrightarrow{H_2O} R_2\overset{\displaystyle OH}{\underset{\underset{\displaystyle X}{|}}{C}}CHR$$

1,2-halohydrin 10.11D

"Anti-Markovnikov" Addition:

$$R_2C{=}CHR \xrightarrow[\text{(2) H}_2\text{O}_2, \text{ OH}^-]{\text{(1) BH}_3} R_2CH\overset{\displaystyle OH}{\underset{|}{C}}HR$$

alcohol 10.10

$$\xrightarrow[\text{(2) RCO}_2\text{H}]{\text{(1) BH}_3} R_2CHCH_2R$$

alkane 10.10

$$\xrightarrow[\text{(2) Br}_2, \text{ OH}^-]{\text{(1) BH}_3} R_2CH\overset{\displaystyle Br}{\underset{|}{C}}HR$$

alkyl bromide 10.10

$$+ HBr \xrightarrow[\text{peroxide}]{\text{O}_2 \text{ or}} R_2CH\overset{\displaystyle Br}{\underset{|}{C}}HR$$

alkyl bromide 10.7D

1,4-Addition:[a]

$$R_2C{=}CHCH{=}CR_2 + HX \longrightarrow R_2CHCH{=}CH\overset{\displaystyle X}{\underset{|}{C}}R_2$$

allylic halide 10.15

$$R_2C{=}CHCH{=}CR_2 + X_2 \longrightarrow R_2\overset{\displaystyle X}{\underset{|}{C}}CH{=}CH\overset{\displaystyle X}{\underset{|}{C}}R_2$$

dihaloalkene 10.15

Addition Leading to Cyclic Products:

$$R_2C{=}CHR + :CH_2 \longrightarrow R_2\overset{\displaystyle CH_2}{\overset{\displaystyle \diagup \diagdown}{C{-}CHR}}$$

a cyclopropane 10.12

a cyclohexene 10.16

where Y is usually C=O or other unsaturated group

(continued)

Table 10.8 (*continued*) **Types of compounds that can be obtained from alkenes and dienes**

Reaction	Product	Section Reference
Reduction:		
$R_2C{=}CHR + H_2 \xrightarrow{\text{Pt}} R_2CHCH_2R$	**alkane**	10.13
Oxidation:		
$R_2C{=}CHR + MnO_4^- \xrightarrow{25°}$ R$_2$C(OH)CHR(OH)	**1,2-diol**[b]	10.14A
$+ C_6H_5CO_3H \longrightarrow$ R$_2$C—CHR (epoxide)	**epoxide**[b]	10.14A
$+ MnO_4^- \xrightarrow{\text{heat}} R_2C{=}O + HOCR(O)$	**ketone, carboxylic acid**	10.14B
$\xrightarrow[\text{(2) Zn, H}^+\text{, H}_2\text{O}]{\text{(1) O}_3} R_2C{=}O + HCR(O)$	**ketone, aldehyde**[c]	10.14B
$RCH{=}CH_2 \xrightarrow[\text{catalyst}]{O_2} RCCH_3(O)$	**methyl ketone**	10.14A

[a] 1,2-Addition may also occur.
[b] A *cis*-1,2-diol can be prepared from an alkene and cold MnO_4^-; the *trans*-1,2-diol can be prepared from hydrolysis of the epoxide.
[c] Ozonolysis with an oxidative work-up yields ketones and carboxylic acids.

Alcohols (or ethers) can be prepared *without rearrangement*, in excellent yield, by oxymercuration–demercuration reactions, a sequence that also leads to Markovnikov products.

$$R_2C{=}CHR \xrightarrow[-CH_3CO_2H]{\substack{Hg(O_2CCH_3)_2 \\ H_2O}} R_2C(OH){-}CHR(HgO_2CCH_3) \xrightarrow{NaBH_4} R_2C(OH)CH_2R$$

"Anti-Markovnikov" products, with the functional group added to the less substituted carbon of the alkene, can be prepared with HBr (and O_2 or a peroxide catalyst) or by hydroboration.

"anti-Markovnikov" products:

$$3\,R_2C{=}CHR \xrightarrow{BH_3} (R_2CH{-}CHR)_3B \begin{cases} \xrightarrow{H_2O_2,\,OH^-} 3\,R_2CHCHR(OH) \\ \xrightarrow{Br_2,\,OH^-} 3\,R_2CHCHR(Br) \end{cases}$$

Difunctional compounds arise from the addition of X_2 (or $X_2 + H_2O$) to a double bond; by 1,2- or 1,4-addition to a diene; or by oxidation of an alkene to a 1,2-diol. For example,

$$R_2C=CHR \xrightarrow{X^+} R_2\overset{\delta+}{C} \cdots \overset{\delta+}{\overset{X}{C}} \text{---} CHR$$

$$\xrightarrow{X^-} \overset{\overset{X}{|}}{R_2CCHR} \overset{|}{X}$$

$$\xrightarrow[-H^+]{H_2O} R_2C\overset{\overset{X}{|}}{\text{---}}\underset{OH}{\overset{|}{CHR}}$$

Oxidation of alkenes is a way to introduce other functionality, such as an epoxide ring. Vigorous oxidizing conditions result in cleavage of the double bond and yield carbonyl compounds, as shown in Table 10.7.

A tremendous variety of *cyclic compounds* containing six-membered rings can be made by Diels–Alder reactions of conjugated dienes with dienophiles. Cyclopropane rings can be made by the reactions of carbenes or carbenoids with alkenes.

Sample Problem

Suggest a synthesis for 3-methyl-2-pentanone from 3-bromo-3-methylpentane.

Solution:

1. Write the structures:

$$\underset{\underset{CH_3}{|}}{CH_3CH_2\overset{\overset{Br}{|}}{C}CH_2CH_3} \longrightarrow \underset{\underset{CH_3}{|}}{CH_3\overset{\overset{O}{||}}{C}CHCH_2CH_3}$$

2. There is no one-step reaction from RX to a ketone. Therefore, working backwards, ask yourself what reactions lead to ketones (without cleavage of the carbon skeleton). Oxidation of a 2° alcohol is a standard reaction leading to a ketone.

$$\underset{\underset{CH_3}{|}}{CH_3\overset{\overset{OH}{|}}{C}HCHCH_2CH_3} \xrightarrow{H_2CrO_4} \underset{\underset{CH_3}{|}}{CH_3\overset{\overset{O}{||}}{C}CHCH_2CH_3}$$

3. The hydroxyl group in the alcohol is not in the same position as the Br in the starting material. Therefore, a simple substitution reaction (RX + OH⁻) would not be a route to the alcohol; however, an alcohol can be obtained from an alkene.

$$\underset{\underset{CH_3}{|}}{CH_3CH=CCH_2CH_3} \longrightarrow \underset{\underset{CH_3}{|}}{CH_3\overset{\overset{OH}{|}}{C}H\text{---}CHCH_2CH_3}$$

Because the alcohol is the *anti-Markovnikov product*, a reaction with H_2O + H^+ or oxymercuration–demercuration will not yield the proper alcohol, but hydroboration–oxidation will. The reagents for the preceding conversion are therefore (1) BH_3 (or, more correctly, B_2H_6); (2) H_2O_2, ^-OH.

4. Can the preceding alkene be prepared from the starting alkyl bromide? Yes, this is an elimination reaction that proceeds by an E2 path.

$$\underset{\underset{CH_3}{|}}{\overset{\overset{Br}{|}}{CH_3CH_2CCH_2CH_3}} \xrightarrow[\text{E2}]{Na^+\ ^-OCH_3} \underset{\underset{CH_3}{|}}{CH_3CH=CCH_2CH_3}$$

5. Now, the entire sequence can be written.

$$\underset{\underset{CH_3}{|}}{\overset{\overset{Br}{|}}{CH_3CH_2CCH_2CH_3}} \xrightarrow{Na^+\ ^-OCH_3} \underset{\underset{CH_3}{|}}{CH_3CH=CCH_2CH_3} \xrightarrow[\text{(2) } H_2O_2,\, OH^-]{\text{(1) } BH_3}$$

$$\underset{\underset{CH_3}{|}}{\overset{\overset{OH}{|}}{CH_3CHCHCH_2CH_3}} \xrightarrow{H_2CrO_4} \underset{\underset{CH_3}{|}}{\overset{\overset{O}{\|}}{CH_3CCHCH_2CH_3}}$$

Study Problem

10.39 Suggest syntheses for the following compounds:

(a) $\underset{\underset{CH_3}{|}}{(CH_3)_2C=CCH_2CH_3}$ from organic compounds containing four or fewer carbon atoms

(b) $CH_3CH_2C\equiv CH$ from an alkene

(c) from organic compounds containing six or fewer carbon atoms

▼ Summary

The reactivities of alkenes and alkynes arise from the weakness and exposure of the pi bond, as well as from the exposure of the trigonal linear carbon. An alkyne with a $-C\equiv CH$ group is more *acidic* than other hydrocarbons.

Alkenes are susceptible to **electrophilic attack**. If the alkene and reagent are unsymmetrical, reaction goes through the most stable carbocation. (**Markovnikov's rule**: H^+ adds to the carbon that already has the most H's.) As reagents, hydrogen halides, H_2SO_4, or $H_2O + H^+$ can be used. Reactions with halogens, halogens plus water, and $Hg(O_2CCH_3)_2$ proceed through bridged ions and thus are *stereoselective*. (These reactions are summarized in Table 10.8, page 444.) The addition of BH_3 yields products that appear to be anti-Markovnikov.

$$R_2C{=}CHR \xrightarrow{H^+} \left[\underset{+}{R_2C}\text{—CHR} \overset{H}{|} \right] \xrightarrow{Nu:^-} \underset{|}{R_2C}\overset{Nu}{|}\text{—CH}_2R$$

$$R_2C{=}CHR \xrightarrow{E^+} \left[\underset{\delta+}{R_2C}\overset{\overset{\delta+}{E}}{\diagdown}\text{CHR} \right] \xrightarrow[\substack{anti\text{-addition,} \\ \text{stereoselective}}]{Nu:^-} \underset{\underset{Nu}{|}}{R_2C}\overset{E}{|}\text{—CHR}$$

Some addition reactions proceed by **syn-additions** instead of by way of carbocations or bridged ions. Hydrogenation, hydroboration, carbene-addition, and Diels–Alder reactions are in this category. *syn*-Additions are usually stereospecific and often lead to *cis* products. (Note that hydroboration also leads to "anti-Markovnikov" products.)

$$\underset{R_2C{=}CHR}{\overset{H{-}BH_2}{}} \xrightarrow[\text{stereospecific}]{syn\text{-addition}} \underset{R_2C\text{—CHR}}{\overset{H\;\;BH_2}{|\;\;\;|}}$$

Hydrogenation reactions can be used to determine relative *stabilities of alkenes*. The most highly substituted alkenes are the most stable. Acyclic *trans*-alkenes are usually more stable than *cis*-alkenes. Conjugated dienes are more stable than nonconjugated dienes.

Oxidation of alkenes can lead to *cis*-1,2-diols, epoxides, ketones, aldehydes, or carboxylic acids. These reactions are summarized in Figure 10.13 and Table 10.8. Conjugated dienes can undergo **1,2-addition** or **1,4-addition**, the latter by way of allylic carbocations.

$$CH_2{=}CHCH{=}CH_2 \begin{cases} \xrightarrow[1,2]{X_2} XCH_2CHXCH{=}CH_2 \\ \\ \xrightarrow[1,4]{X_2} XCH_2CH{=}CHCH_2X \end{cases}$$

Study Problems*

10.40 Write the IUPAC name for each of the following compounds:

(a) $BrCH_2CH{=}CH_2$ (b) $(CH_3)_2CHC{\equiv}CH$

3-bromopropenes

$$CH_3CHCH_2C{\equiv}CH$$
$$CH_3$$

* For information concerning the organization of the *Study Problems* and *Additional Problems*, see the *Note to student* on page 39.

(c)

Cyclohexadiene with two Cl substituents

(d) CH_2=$CHCO_2H$

(e) $(CH_3)_2CHCH_2$—C(H)(OH)—CH=CH—CH₃ structure with H and OH on carbon, and C=C with H and CH₃

10.41 Give the structure for each of the following compounds:

- **(a)** 2-methyl-1-butene
- **(b)** 2-pentyne
- **(c)** 1,3-hexadiene
- **(d)** *cis*-2-hexene
- **(e)** *trans*-1,2-diphenylethene
- **(f)** *cis*-1,2-dibromoethene
- **(g)** (2*Z*,4*Z*)-hexadiene
- **(h)** (*E*)-3-phenyl-2-butenoic acid

10.42 Draw a tree diagram for the alkene protons H_x and H_y in the following structure (J_{xy} = 10 Hz, J_{ay} = 2 Hz, J_{ax} = 6 Hz).

$$
\begin{array}{c}
H_x \qquad H_y \\
C=C \\
(CH_3)_3C \qquad H_a
\end{array}
$$

10.43 A compound isolated from a reaction mixture is thought to be either 1-hexene or 1-hexyne. An infrared spectrum is obtained. What would you look for in the spectrum to determine if the compound is the alkene or the alkyne? Sketch the expected spectrum of each.

10.44 Write equations for the syntheses of the following compounds from alcohols or alkyl halides.

(a) cyclohexene with —CH_3

(b) cyclohexane with =CH_2

(c)
$$
\begin{array}{c}
C_6H_5 \qquad C_6H_5 \\
C=C \\
H \qquad CH_3
\end{array}
$$

(d)
$$
\begin{array}{c}
C_6H_5 \qquad CH_3 \\
C=C \\
H \qquad C_6H_5
\end{array}
$$

10.45 Starting with propyne, write flow equations, including all steps, that show how you would prepare the following compounds.

(a) 2-butyne

(b) cyclohexane with OH and C≡CCH₃ substituents

(c) CH_3C≡CD

10.46 Predict the products:

(a) $(CH_3)_2C$=CH_2 + HCl $\xrightarrow{\text{diethyl ether}}$

(b) indene structure + excess gaseous HCl $\xrightarrow{5-10°}$

(c) 1 $CH_3CH_2CH_2CH_2C$≡CH + 1 HBr $\xrightarrow[\text{no peroxides}]{15°, \text{ dark}}$

(d) cyclopentane ring =$CHCH_3$ + HBr $\xrightarrow{\text{no peroxides}}$

(e) $\text{[cyclopentane ring]}=\text{CHCH}_3 + \text{HBr} \xrightarrow{\text{C}_6\text{H}_5\overset{\overset{\displaystyle O}{\|}}{\text{C}}\text{OO}\overset{\overset{\displaystyle O}{\|}}{\text{C}}\text{C}_6\text{H}_5}$

10.47 (1) Arrange the following compounds in order of increasing reactivity toward H_2SO_4 (least reactive first). (2) Write formulas for the intermediate and the major product in each case.

(a) propene   **(b)** 2-methylpropene   **(c)** 2-butene

10.48 Predict the major organic products of the following reactions:

(a) 3-methyl-2-pentene with aqueous H_2SO_4
(b) 2-methylpropene with H_2SO_4 in ethanol
(c) 2,2-dimethyl-3-hexene with aqueous H_2SO_4
(d) 1-butene with $0.1M$ aqueous HI
(e) methylenecyclohexane (page 394) with the strong acid trifluoroacetic acid (CF_3CO_2H).

10.49 Predict the major organic products:

(a) $(CH_3)_3CCH{=}CHCH_3 \xrightarrow[\text{(2) NaBH}_4]{\text{(1) Hg(O}_2\text{CCH}_3)_2, \text{ H}_2\text{O}}$

(b) $\text{[cyclohexane ring]}{-}CH{=}CH_2 \xrightarrow[\text{(2) NaBH}_4]{\text{(1) Hg(O}_2\text{CCH}_3)_2, \text{ CH}_3\text{CH}_2\text{OH}}$

(c) H$_3$C   CH$_3$

$\xrightarrow[\text{(2) H}_2\text{O}_2, \text{ OH}^-]{\text{(1) BH}_3}$

CH$_3$

α-pinene

found in turpentine

(d) $\text{[cyclopentene ring]}{-}CH_3 \xrightarrow[\text{(2) Br}_2, \text{ OH}^-]{\text{(1) BH}_3}$

10.50 Write equations for the mechanism to predict the stereochemistry of the oxymercuration product of 1-methylcyclohexene.

10.51 Predict the products. Show stereochemistry where applicable.

(a) $(E)\text{-}C_6H_5CH{=}CH\overset{\overset{\displaystyle O}{\|}}{C}OCH_2CH_3 + Br_2 \longrightarrow$

(b) $(Z)\text{-}CH_3(CH_2)_7CH{=}CH(CH_2)_7\overset{\overset{\displaystyle O}{\|}}{C}OCH_3 + Br_2 \longrightarrow$

(c) $BrCH_2CH{=}CH_2 + Br_2 \longrightarrow$

(d) $trans\text{-}C_6H_5CH{=}CHC_6H_5 + Br_2 \longrightarrow$

10.52 Predict the likely organic products of the following reactions: **(a)** *cis*-2-pentene with Cl_2;   **(b)** 1-methylcyclohexene with Br_2;   **(c)** 1-methylcyclohexene with aqueous Br_2.

10.53 Fumaric acid (*trans*-$HO_2CCH{=}CHCO_2H$), which is essential for cell respiration, is found in all plant and animal tissues. The *Boletaceae* family of mushrooms is especially rich in this acid. (One trivial name of fumaric acid is *boletic acid*.) When fumaric acid is treated with bromine, an 84% yield of a dibrominated product is obtained. Predict the stereochemistry of the product.

10.54 Styrene, $C_6H_5CH=CH_2$, was mixed with bromine and anhydrous acetic acid. Two products were obtained. Write equations to show the identities of these products, and write a mechanism that shows how they are formed.

10.55 Suggest syntheses for the following compounds from compounds that do not contain a cyclopropane ring.

(a)

(b) $CH_3(CH_2)_4CH$—CH_2 with CH_2 cyclopropane

(c)

(d) $(CH_3)_2C$—$C(CH_3)_2$ with $CHCl$ cyclopropane

10.56 Predict the major organic products (showing stereochemistry):

(a) $=O + H_2 \xrightarrow{Pt}$

(b) $CH_3CH_2C\equiv CCH(CH_3)_2 + H_2 \xrightarrow{\text{deactivated Pd}}$

(c) α-pinene (Problem 10.49) $+ H_2 \xrightarrow{Pt}$

(d) $CH_3C\equiv C\overset{\overset{\displaystyle O}{\|}}{C}OCH_2CH_3 + 1\,H_2 \xrightarrow{\text{deactivated Pd}}$

(e) $+ 1\,H_2 \xrightarrow{[(C_6H_5)_3P]_3RhCl}$

10.57 Of each of the following pairs of compounds, tell which one would be the more stable. Explain.

(a) $=CH_2$ or $=CHCH_3$

(b) *cis-* or *trans*-$CH_3CH_2CH=CHCH_3$

(c) (*E*)- or (*Z*)-$CH_3\overset{\overset{\displaystyle O}{\|}}{C}OCH=CH\overset{\overset{\displaystyle O}{\|}}{C}OCH_3$

(d) $CH_2=CHCH_2\overset{\overset{\displaystyle O}{\|}}{C}OCH_3$ or $CH_2=CH\overset{\overset{\displaystyle O}{\|}}{C}OCH_2CH_3$

10.58 Predict the products when (*R*)(*S*)-limonene (1-methyl-4-isopropenylcyclohexene) is treated with an excess of the following reagents: (a) $C_6H_5CO_3H$; (b) hot $KMnO_4$ solution; (c) cold $KMnO_4$ solution.

CH$_3$

CH$_3$C=CH$_2$

limonene

found in citrus oils

10.59 During an investigation of naturally occurring antitumor agents, the following compound was subjected to OsO$_4$ oxidation, followed by treatment with NaHSO$_3$. A single stereoisomeric product was obtained in 93% yield. What was this product?

CH$_3$

H$_3$C—

H

O

—CO$_2$CH$_3$

10.60 Predict the ozonolysis products when each of the following alkenes is subjected to (1) reaction with O$_3$, followed by (2) reaction with H$_2$O$_2$ and H$^+$:

(a)

H$_3$C   CH$_3$

CH$_3$

α-pinene

(b)

H$_3$C   CH$_3$

CH$_2$

β-pinene

10.61 Each of a series of unknown alkenes is treated with (1) O$_3$, and (2) Zn, H$^+$, H$_2$O. The following products are obtained. Give the structure (or structures, if there is more than one possibility) of each unknown alkene.

(a) HĊCH$_2$CH$_2$CH$_2$ĊH only

(b) CH$_3$CH$_2$ĊH + CH$_3$ĊCH$_3$

(c) CH$_3$ĊH only

(d) HĊH +

O

(e) CH$_3$Ċ—ĊH + 2 CH$_3$ĊCH$_3$

(f) CH$_3$ĊH + HĊH + HĊCH$_2$ĊH

10.62 An equimolar quantity of HCl is added to each of the following alkenes. Give the structure of the likely products in each case.   (a) 1,3-pentadiene;   (b) 2,2-dimethyl-3-heptene;   (c) 1,3-cyclopentadiene.

10.63 The reaction of 2-methyl-1,3-butadiene (isoprene), used in the manufacture of rubber and other polymers, with HCl at 25° yields compounds A and B in a ratio of 75:25. Write mechanisms for the formation of these compounds.

Cl
|
(CH$_3$)$_2$CCH=CH$_2$ (CH$_3$)$_2$C=CHCH$_2$Cl

A   B

10.64 *Styrene-butadiene rubber* (SBR) is a synthetic rubber used in tires. It is a copolymer of three parts of 1,3-butadiene and one part styrene. About 80% of the polymer is formed by 1,4-addition, while the remaining 20% is formed by 1,2-addition. Most of the double bonds in SBR are *trans*. (These features of SBR are randomly incorporated in the polymer.) Write a formula for a hypothetical segment of this polymer that illustrates the above characteristics.

10.65 Some polymers are formed by **cationic addition**, a reaction that proceeds through a carbocation intermediate. A Lewis acid, such as BF_3 with a trace of water, is an effective catalyst for cationic polymerization of some alkenes:

$$x\ CH_2{=}\underset{\underset{CH_3}{|}}{\overset{\overset{CH_3}{|}}{C}} \quad \xrightarrow{BF_3,\ H_2O} \quad \left(CH_2{-}\underset{\underset{CH_3}{|}}{\overset{\overset{CH_3}{|}}{C}}\right)_x$$

methylpropene   polyisobutylene
(isobutylene)

(a) What is the function of the trace of water in the catalyst system?
(b) Write an equation for the mechanism showing how polyisobutylene is formed.
(c) Use the mechanism to explain why the structure of the polymer is *not*

$$\left(CH_2\underset{\underset{CH_3}{|}}{\overset{\overset{CH_3}{|}}{C}}{-}\underset{\underset{CH_3}{|}}{\overset{\overset{CH_3}{|}}{C}}CH_2\right)_x$$

(d) Would cationic polymerization be the method of choice to prepare polyethylene? Explain.

10.66 Each of the following dienes is shown in the s-*trans* form. Which of these compounds are interconvertible with their s-*cis* forms? Show the s-*cis* conformations in these cases.

(a)   (b)   (c)

10.67 Predict the products of the following Diels–Alder reactions, showing stereochemistry where appropriate:

(a) [diene] $+\ CH_2{=}CHCCH_3$ (with C=O) $\xrightarrow{heat}$

(b) [diene] $+$ [maleic anhydride] $\xrightarrow{heat}$

(c) CH_3O [diene] $+$ [indene with CO_2H] $\xrightarrow{heat}$

(d) [furan] $+$ [norbornadiene] $\xrightarrow{heat}$

10.68 Suggest a Diels–Alder reaction that would lead to each of the following compounds.

(a)

(b)

(c)

(d)

Additional Problems

10.69 Predict the organic product of the reaction of *trans*-2-butene with each of the following reagents. (Show the stereochemistry where appropriate.) (a) HI; (b) OsO_4, followed by treatment with Na_2SO_3; (c) $C_6H_5CO_3H$, followed by H_2O + HCl; (d) H_2SO_4; (e) O_3, followed by Zn + HCl; (f) $CHCl_3$ and K^+ $^-OC(CH_3)_3$.

10.70 Give the structure of the product formed when 1-methylcyclopentene is treated with each of the following reagents: (a) H_2 (Pt catalyst); (b) Cl_2 in H_2O; (c) Cl_2 in CCl_4; (d) cold alkaline $KMnO_4$; (e) hot $KMnO_4$ solution; (f) dilute solution of Br_2 in CH_3OH; (g) HCl; (h) HBr with H_2O_2; (i) mercuric acetate in water followed by $NaBH_4$; (j) OsO_4, followed by $NaHSO_3$.

10.71 Suggest a synthesis from an alkene for each of the following compounds:

(a)

(b)

10.72 Alkynes undergo reaction with potassium permanganate; the products depend on the reaction conditions. With aqueous $KMnO_4$, oxidative cleavage is the main reaction. However, with anhydrous dichloromethane solvent and a phase-transfer agent (see Sections 8.6 and 17.7), 1,2-diketones can be obtained. Predict the products when 4-octyne is subjected to each set of these reaction conditions.

10.73 Write an equation or equations for the preparation of each of the following compounds from methylenecyclohexane (page 394):

(a)

(b)

(c)

10.74 Show how you could synthesize the following compounds from propyne:

(a) $CH_3\overset{O}{\overset{\|}{C}}CH_3$

(b) $CH_3\overset{O}{\overset{\|}{C}}CH_2CH_3$

(c) $CH_3\overset{OCH_3}{\overset{|}{C}H}C\equiv CCH_3$

(d) $\overset{Br\ Br}{\overset{|\ \ |}{CH_2CHCH_2Br}}$

10.75 Suggest synthetic paths for the following conversions:

(a) 1,2-dideuteriopropane from propane

(**b**) acetic acid from 1-bromopropane
(**c**) *trans*-1,2-cyclohexanediol from cyclohexane
(**d**) *trans*-2-chloro-1-cyclopentanol from chlorocyclopentane
(**e**) diphenylethyne from 1,2-diphenylethene

10.76 The toxic components of poison ivy and poison oak are a group of 1,2-hydroquinones (dihydroxybenzenes) called *urushiols*. One of these urushiols (A), $C_{21}H_{34}O_2$, is treated with dilute NaOH, followed by iodomethane, and yields a dimethylated compound (B), $C_{23}H_{38}O_2$. When B is treated with ozone, followed by treatment with Zn dust and water, the products are:

and $HC(CH_2)_5CH_3$ with O double bonded

What are the structures of A and B?

10.77 The hydration of the following compound in an aqueous solution of H_2SO_4 and HgO gives a 65% yield of one product. What is the structure of this product?

10.78 Starting with only acetylene, ethylene, iodomethane, and any appropriate inorganic reagents, show how you would prepare the following compounds:

(**a**) *cis*-4-octene;   (**b**) 1-butene;   (**c**) 1-chloro-2-butanol;
(**d**) 2-methyl-2-butanol;   (**e**) *meso*-2,3-butanediol; (**f**) racemic 2,3-butanediol.

10.79 Suggest a mechanism showing all steps for each of the following reactions:

(**a**) $(CH_3)_2C=CH(CH_2)_2CH=C(CH_3)_2$ $\xrightarrow{H^+}$

(**b**)

(**c**) $(CH_3)_2C=CH(CH_2)_2\overset{CH_3}{\underset{}{C}}=CHCH_3$ $\xrightarrow{H^+}$

10.80 How would you make the following conversions?

(**a**)

$\longrightarrow$ $CH_3\overset{O}{\overset{||}{C}}(CH_2)_8\overset{O}{\overset{||}{C}}H$

(**b**)

$\longrightarrow$ $CH_2=CH\overset{OH}{\underset{}{C}}H(CH_2)_4\overset{OH}{\underset{CH_3}{C}}CH=CH_2$

10.81 Suggest syntheses for the following compounds from alkenes or dienes containing six or fewer carbons and other appropriate reagents:

(a) 1-cyclohexyl-1-propanone   (b) cyclohexylmethanol
(c) *trans*-2-methyl-1-cyclohexanol   (d) *cis*-1,2-dibromocyclohexane

(e) racemic HO$_2$C┉┉⟨   ⟩┉┉CO$_2$H

10.82 *Muscalure*, the sex attractant of the common housefly (*Musca domestica*), is a hydrocarbon with the formula C$_{23}$H$_{46}$. Catalytic hydrogenation of this compound yields C$_{23}$H$_{48}$, while oxidation with a hot, alkaline solution of KMnO$_4$ followed by acidification yields CH$_3$(CH$_2$)$_{12}$CO$_2$H and CH$_3$(CH$_2$)$_7$CO$_2$H. Addition of bromine to the hydrocarbon yields one pair of enantiomeric dibromides (C$_{23}$H$_{46}$Br$_2$). What is the structure of the housefly sex attractant?

10.83 *Cembrene A* is a compound isolated from the frontal gland secretion of a termite soldier and from other sources. From the following experimental results, suggest a structure for cembrene A. (Do not attempt to determine the stereochemistry.)

A combination of gas chromatographic and mass spectral data provided the molecular formula C$_{20}$H$_{32}$. The proton nmr spectrum indicated: four =C—CH$_3$ groups,

three =C—H groups, and one =CH$_2$ group. Other evidence showed that cembrene A

does not contain any tetrasubstituted double bonds; that is, each double bond contains at least one =CH or =CH$_2$ grouping.

Upon catalytic hydrogenation, 1.0 mol of cembrene A took up 4.0 mol of H$_2$. Reductive ozonolysis gave two products in addition to formaldehyde, HCHO. One product was compound A, the same diketo aldehyde obtained from the reductive ozonolysis of limonene (see Problem 10.58). The second was compound B. The molar ratio of B to A was 2:1.

$$
\begin{array}{cc}
O & O \\
\| & \| \\
CH_3CCH_2CH_2CH \\
\end{array}
$$
B

10.84 The following compound, 1*H*-indene, when heated with 85% H$_3$PO$_4$, yields an interesting solid product that emits flashes of blue light when scratched (a phenomenon called *triboluminescence*).

1*H*-indene

The molecular formula of this product was determined to be C$_{18}$H$_{16}$. The proton nmr spectrum showed a multiplet at δ2.22, a multiplet at 2.90, a singlet at 3.21, a triplet at 4.75, a singlet at 6.53, and a multiplet at 7.12. The area ratio for these signals was 2:2:2:1:1:8. What is the structure of the product?

10.85 The ^{13}C nmr spectrum in Figure 10.16 is of a hydrocarbon containing a methylcyclohexane skeleton and one double bond. Locate the position of the double bond within the structure.

10.86 A six-carbon compound, A, can be converted to B by hydrogenation. The proton nmr spectrum of A and the infrared spectrum of B are shown in Figure 10.17. What are the structures of A and B?

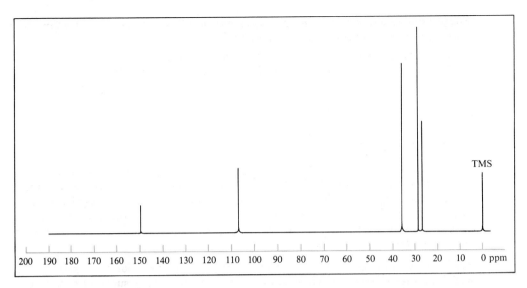

Figure 10.16 ^{13}C nmr spectrum for Problem 10.85.

Figure 10.17 Spectra for unknown compounds in Problem 10.86.

10.87 Suggest a structure for each spectrum given in Figure 10.18.

Compound A, C_8H_{16}

Compound B, C_5H_{10}

Figure 10.18 ¹H nmr spectra for unknown compounds in Problem 10.87.

Compound C, C_5H_8

Compound D, C_8H_{14}

Figure 10.18 (*continued*) ¹H nmr spectra for unknown compounds in Problem 10.87.

11 ▽

Aromaticity and Benzene; Electrophilic Aromatic Substitution

Benzene is the simplest of the aromatic compounds and one that we have already encountered many times. In this chapter and in Chapter 12, we will formalize the definition of aromaticity, and we will discuss the properties and reactions of benzene and substituted benzenes. In Chapter 18, we will discuss the chemistry of polycyclic and heterocyclic aromatic compounds. (*Heterocyclic* aromatic compounds have at least two different kinds of ring atoms, such as carbon and nitrogen.)

Benzene was first isolated in 1825 by Michael Faraday from oily residues that had accumulated in London gas mains. Today the main source of benzene, substituted benzenes, and other aromatic compounds is petroleum. Until the 1940s, coal tar was the principal source. The types of aromatic compounds obtained from these sources are hydrocarbons, phenols, and aromatic heterocycles.

Aromatic hydrocarbons:

toluene

p-xylene

substituted benzenes

naphthalene   phenanthrene

polycyclic compounds

Aromatic nitrogen heterocycles:

pyridine   quinoline

Compounds containing benzene rings and aromatic heterocyclic rings are also exceedingly common in biological systems.

nicotine

in tobacco

estrone

an estrogen, or female hormone

uric acid

associated with gout

▼ Section 11.1
Nomenclature of Substituted Benzenes

The naming of monosubstituted benzenes was introduced in Section 3.3K. Many common benzene compounds have their own names, names that are not necessarily systematic. Some of the more commonly used names are listed in Table 11.1.

Disubstituted benzenes are named either with position numbers or with the prefixes **ortho**, **meta**, and **para**. The prefix *ortho* signifies that two substituents are 1,2 to each other on a benzene ring. *Meta* signifies a 1,3-relationship, and *para* means a 1,4-relationship. The use of *ortho*, *meta*, and *para* in lieu of position numbers is reserved exclusively for disubstituted benzenes. The system is never used with cyclohexanes or other ring systems.

ortho-, or o-   meta-, or m-   para, or p-

Table 11.1 Structures and names of some common substituted benzenes

Structure	Name	Structure	Name
⬡—CH₃	toluene	⬡—OH	phenol
CH₃—⬡—CH₃	*p*-xylene	⬡—CO₂H	benzoic acid
⬡—CH=CH₂	styrene	⬡—CH₂OH	benzyl alcohol
⬡—NH₂	aniline	CH₃—⬡—SO₂Cl	*p*-toluenesulfonyl chloride (tosyl chloride)
⬡—NHCCH₃ (with O double bond)	acetanilide	CH₃C—⬡ (with O double bond)	acetophenone
		⬡—C—⬡ (with O double bond)	benzophenone

The use of these prefixes in the naming of some disubstituted benzenes follows:

o-dibromobenzene   *m*-chloroaniline   *p*-chlorophenol

In reactions of benzene compounds, we will speak of *ortho-*, *meta-*, or *para-* substitution. A monosubstituted benzene has two *ortho* and two *meta* positions, but only one *para* position.

ortho   meta   para

If there are three or more substituents on a benzene ring, the *o*-, *m*-, and *p*- system is no longer applicable. In this case, numbers must be used. As in numbering any compound, we number the benzene ring in such a way as to keep the prefix numbers as low as possible and give preference to the group of highest nomenclature priority. If a substituted benzene, such as aniline or toluene, is used as the parent, that substituent is understood to be at position 1 on the ring.

1,2,4-tribromobenzene   2-chloro-4-nitroaniline   2,4,6-trinitrotoluene
(TNT)

Benzene as a substituent is called a **phenyl group**. How a *toluene* substituent is named depends on the point of attachment.

phenyl   benzyl   *p*-tolyl   *o*-tolyl

Sample Problem

Name the following substituted benzenes:

(a)   (b) Br—⟨⟩—CH$_2$Cl   (c)

Solution: (a) 2,6-dimethylaniline;   (b) *p*-bromobenzyl chloride;   (c) *m*-tolylcyclo-hexane (or *m*-cyclohexyltoluene).

Study Problem

11.1 Write formulas for: **(a)** ethylbenzene;   **(b)** 2,4,6-tribromoaniline;   **(c)** *p*-ethylphenol;   **(d)** 2-phenylethanol;   **(e)** benzyl bromide.

Section 11.2
Physical Properties of Aromatic Hydrocarbons

Like aliphatic and alicyclic hydrocarbons, benzene and other aromatic hydro-carbons are nonpolar. They are insoluble in water, but are soluble in organic solvents such as diethyl ether, carbon tetrachloride, or hexane. Benzene itself has been widely used as a solvent. It has the useful property of forming an azeotrope with water. (The azeotrope, a mixture that distills with a constant composition, consists of 91% benzene–9% H_2O and boils at 69.4°C.) Compounds dissolved in benzene are easily dried by distillation of the azeotrope.

Although the boiling points and melting points of aromatic hydrocarbons (listed in Table 11.2) are typical of those of nonpolar organic compounds, note that *p*-xylene has a higher melting point than *o*- or *m*-xylene. A higher melting

Table 11.2 Melting points and boiling points of some aromatic hydrocarbons

Name	Structure	Mp, °C	Bp, °C
benzene		5.5	80
toluene	—CH₃	−95	111
o-xylene	CH₃ / —CH₃	−25	144
m-xylene	CH₃ / —CH₃	−48	139
p-xylene	CH₃— —CH₃	13	138

point is typical of *p*-disubstituted benzenes. A *p*-isomer is more symmetrical and can form a more orderly and stronger crystal lattice in the solid state than the less symmetrical *o*- or *m*-isomers.

Many of the compounds found in coal tar and cigarette tar that contain four or more fused benzene rings are carcinogenic (cancer-causing). Benzene itself is toxic and somewhat carcinogenic; therefore, it should be used in the laboratory only when necessary. (In many cases, toluene, which is less toxic, may be used as a substitute.)

Two highly carcinogenic hydrocarbons:

benzo[*a*]pyrene   benzanthracene

▼ Section 11.3
Spectra of Substituted Benzenes

Both infrared and nmr spectra provide data useful in the structure determination of substituted benzenes. The proton nmr spectrum provides the most clear-cut answer to the absence or presence of aryl protons (and thus an aromatic ring).

A. Infrared Spectra

The infrared absorption bands of substituted benzenes are summarized in Table 11.3. The presence of a benzene ring in a compound of unknown structure

Table 11.3 Infrared absorption characteristics of benzene compounds

Type of Vibration	Position of Absorption	
	cm^{-1}	μm
aryl C—H	3000–3300	3.0–3.3
aryl C—C (four peaks)	1450–1600	6.25–6.9

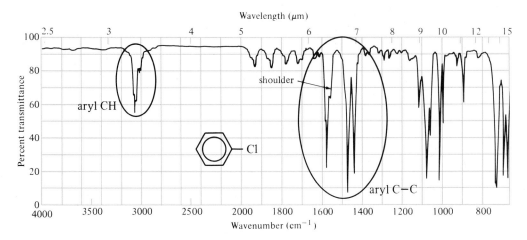

Figure 11.1 Infrared spectrum of chlorobenzene.

can often be determined by inspection of two regions of the infrared spectrum. The absorption for aryl CH stretching, which is generally weak, falls near 3030 cm^{-1} (3.3 μm), just to the left of aliphatic CH absorption. Absorption for aryl C—C vibrations gives a series of four peaks, generally between 1450 and 1600 cm^{-1} (about 6–7 μm). However, all four peaks are not always apparent. In the spectrum of chlorobenzene (Figure 11.1), the first peak is visible, but the second peak is a barely visible shoulder. In this spectrum, the third and fourth

Table 11.4 The C—H bending absorption of substituted benzenes

Substitution		Appearance	Position of Absorption	
			cm^{-1}	μm
monosubstituted,		two peaks	730–770	12.9–13.7
			690–710	14.0–14.5
o-disubstituted,		one peak^a	735–770	12.9–13.6
m-disubstituted,		three peaks	860–900	11.1–11.6
			750–810	12.3–13.3
			680–725	13.7–14.7
p-disubstituted,		one peak	800–860	11.6–12.5

^{a} An additional band is often observed at about 680 cm^{-1} (14.7 μm).

peaks are quite evident. Yet, in some spectra, the fourth peak at 1450 cm^{-1} (6.90 μm) is obscured by aliphatic CH_2 bending absorption.

We can sometimes determine the positions of substitution on a benzene ring from the infrared spectrum. Differently substituted benzene rings often give characteristic absorption at about 680–900 cm^{-1} (11–15 μm). The patterns observed are summarized in Table 11.4. Figure 11.2 shows infrared spectra of the

Figure 11.2 Infrared spectra for *o*-, *m*-, and *p*-chlorotoluene.

three isomeric chlorotoluenes. By comparison of these spectra with Table 11.4, you can see how the absorption in the fingerprint region may be used. Also compare these spectra with that of chlorobenzene in Figure 11.1. Unfortunately, the absorption in this region is not always so clear-cut.

B. Proton Nmr Spectra

The proton (^{1}H) nmr spectra of aromatic compounds are distinctive. Protons on an aromatic ring absorb downfield, with δ values between 6.5 ppm and 8 ppm. This downfield absorption is due to the ring current, which gives rise to a molecular magnetic field that deshields protons bonded to the ring. The chemical shift for the protons in benzene itself is at $\delta7.27$. Electronegative substituents on the ring shift the absorption of adjacent protons farther *downfield*, while electron-releasing groups shift absorption *upfield* from that of unsubstituted benzene. Simple splitting patterns of aryl protons are sometimes observed. In many cases, however, the splitting patterns are very complex.

Benzylic protons are not as affected by the aromatic ring current as are ring protons; their absorption is observed farther upfield, in the region of $\delta2.3$. (See the nmr spectrum of toluene in Figure 11.3.)

Figure 11.3 ^{1}H nmr spectrum of toluene, showing aryl and benzylic CH absorption.

Study Problem

11.2 Figure 11.4 shows two proton nmr spectra (without the relative areas under the peaks). One spectrum is that of *p*-chloroaniline, while the other is that of *p*-iodoanisole (p-$IC_6H_4OCH_3$). On the basis of the chemical shifts (compared to the chemical shift of benzene protons), assign structures to the two spectra.

Figure 11.4 ^{1}H nmr spectra of *p*-chloroaniline and *p*-iodoanisole for Problem 11.2.

Section 11.4
Stability of the Benzene Ring

The heat of hydrogenation of cyclohexene is 28.6 kcal/mol. If benzene contained three alternate single and double bonds without any pi-electron delocalization, we would expect its heat of hydrogenation to be 3 × 28.6, or 85.8, kcal/mol. However, benzene liberates only 49.8 kcal/mol of energy when it is hydrogenated.

The hydrogenation of benzene liberates 36 kcal/mol less energy than would be liberated by the hydrogenation of the hypothetical cyclohexatriene. Therefore, benzene, with delocalized pi electrons, contains 36 kcal/mol less energy than it would contain if the pi electrons were localized in three isolated double bonds. This difference in energy between benzene and the imaginary cyclohexatriene is called the **resonance energy** of benzene. The resonance energy is the energy lost (stability gained) by the complete delocalization of electrons in the pi system. It is a measure of the added stability of the aromatic system compared to that of the localized system. Figure 11.5 shows the energy diagrams for these hydrogenations.

What does the resonance energy of benzene mean in terms of chemical reactivity? It means that more energy is required for a reaction in which the aromatic character of the ring is lost. Hydrogenation is one example of such a reaction. Whereas an alkene can be hydrogenated at room temperature under

Figure 11.5 Energy diagrams for the hydrogenation of cyclohexatriene (hypothetical) and benzene.

atmospheric pressure, benzene requires high temperature and high pressure. Also, benzene does not undergo most reactions that are typical of alkenes. For example, benzene does not undergo addition of HX or X_2, nor does it undergo oxidation with $KMnO_4$ solution.

Section 11.5
The Bonding in Benzene

Although the molecular formula of benzene (C_6H_6) was determined shortly after its discovery in 1825, 40 years elapsed before Kekulé's brilliant proposal for the structure of benzene—a hexagon with three double bonds. To explain the existence of only three (not four) isomeric disubstituted benzenes, Kekulé later suggested that the benzene ring is in rapid equilibrium with the structure in which the double bonds are in the alternative positions. This idea survived for more than 50 years before it was replaced by the theories of resonance and molecular orbitals.

benzene in 1865   benzene in 1872   benzene in 1940

The Kekulé formulas for benzene, showing three double bonds instead of a circle in the ring, do not explain the unique stability of the benzene ring. However, these formulas have one advantage—they allow us to count the number of pi electrons at a glance. In conjunction with resonance theory, these formulas are quite useful. Therefore, we will use Kekulé formulas when we discuss the reactions of benzene.

Let us now consider the π molecular orbitals of benzene. Benzene has six sp^2-hybridized carbons in a ring. The ring is planar, and each carbon atom has a p orbital perpendicular to this plane. Figure 2.23 (page 72) shows a representation of the p orbitals of benzene and how they overlap in the lowest-energy bonding molecular orbital.

Overlap of six atomic p orbitals leads to the formation of six π molecular orbitals. When we look at all six possible π molecular orbitals of benzene (Figure 11.6), we see that our representation of the aromatic pi cloud as a "double donut" represents only one of the six molecular orbitals. In the π_1 orbital, all six p orbitals of benzene are in phase and overlap equally. This orbital is of lowest energy because it has no nodes between carbon nuclei.

The π_2 orbital and the π_3 orbital each has one nodal plane between the carbon nuclei. These two bonding orbitals are degenerate (equal in energy) and of higher energy than the π_1 molecular orbital. Benzene, with six p electrons,

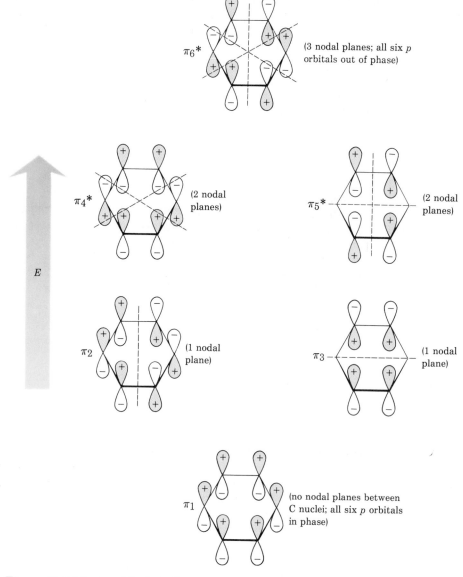

Figure 11.6 The π orbitals in benzene. Nodal planes are represented by dashed lines; the "missing *p* orbitals" in π_3 and $\pi_5{}^*$ are a result of nodes at these positions. (The + and − signs are mathematical signs of phase, not electrical charges.)

has the π_1, π_2, and π_3 orbitals each filled with a pair of electrons. These are the bonding orbitals of benzene.

Along with the three bonding orbitals in benzene are three antibonding orbitals. Two of these antibonding orbitals (π_4* and π_5*) have two nodal planes each and the highest-energy orbital (π_6*) has three nodal planes between carbon nuclei. Recall that a node or a nodal plane is a region of very low electron density. A molecular orbital with a nodal plane between nuclei is of higher energy than a molecular orbital without a nodal plane between nuclei. As we progress from π_1 to π_6*, the number of nodal planes increases; this is the reason that the energy associated with these molecular orbitals increases.

Section 11.6
What Is an Aromatic Compound?

Benzene is one member of the large class of aromatic compounds—compounds that are *substantially stabilized by pi-electron delocalization*. The resonance energy of an aromatic compound is a measure of its gain in stability. (The structural features that give rise to aromaticity will be discussed shortly.)

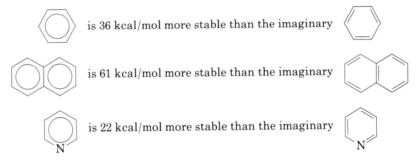

is 36 kcal/mol more stable than the imaginary

is 61 kcal/mol more stable than the imaginary

is 22 kcal/mol more stable than the imaginary

The most convenient way to determine if a compound is aromatic is by observing the chemical shifts of hydrogens bonded to the ring atoms. Hydrogens on the outer periphery of an aromatic ring are highly deshielded and absorb far downfield from most other hydrogens, usually beyond $\delta 7$. (See Section 9.7B.)

	benzene	pyridine	naphthalene
δ of aromatic CH, ppm:	7.3	7.1–8.5	7.3–7.8

$$CH_3\underline{CH}=\underline{CH}CH_3$$

δ of nonaromatic sp^2 CH, ppm: 5.3

Study Problem

11.3 The ^{1}H nmr spectrum of cyclooctatetraene shows only a singlet at $\delta 5.7$. On the basis of this chemical shift, would you say that cyclooctatetraene is aromatic or not?

cyclooctatetraene

▼ Section 11.7
Requirements for Aromaticity

What structural features are necessary for a molecule to be aromatic? The molecule must be *cyclic* and *planar*, and each atom of the ring or rings must have a *p orbital perpendicular to the plane of the ring.* If a system does not fit these criteria, there cannot be complete delocalization of the pi electrons. Whether these three criteria are met can often be deduced from inspection of the formula of an organic compound. The line-bond formula of an aromatic compound usually shows a ring with alternate single and double bonds. There are cases, however, of cyclic organic compounds with alternate single and double bonds that are *not* aromatic. Cyclooctatetraene is such a compound. Cyclooctatetraene undergoes addition reactions with the hydrogen halides and with the halogens. These reactions are typical of alkenes, but are not typical of benzene and other aromatic compounds. Cyclooctatetraene is not planar, but has been shown to be shaped like a tub.

cyclooctatetraene

a cyclic tetraene; not aromatic

Why is cyclooctatetraene not aromatic? To answer this question, we must proceed to a fourth criterion for aromaticity, a criterion usually called **the Hückel rule**.

A. The Hückel Rule

In 1931, the German chemist Erich Hückel proposed that, to be aromatic, a monocyclic (one ring), planar compound must have **(4n + 2) pi electrons**, where **n** is an integer. According to the Hückel rule, a ring with 2, 6, 10, or 14 pi electrons may be aromatic, but a ring with 8 or 12 pi electrons may not. Thus, cyclooctatetraene (with 8 pi electrons) would not be aromatic even if it were planar.

six π e⁻   *eight π e⁻*

$4n + 2 = 8$

$4n = 6$

2

(4n + 2)   4n
n = 1   n = 2

aromatic   *not aromatic*

Why can a monocyclic compound with six to ten pi electrons be aromatic, but not a compound with eight pi electrons? The answer is found in the number of pi electrons versus the number of pi orbitals available. To be aromatic, a molecule must have *all pi electrons paired and all bonding orbitals filled*. This system provides the maximum and complete overlap required for aromatic stabilization. If some pi orbitals are *not* filled (that is, there are unpaired pi electrons), overlap is not maximized, and the compound is not aromatic. Benzene has six pi electrons and three bonding pi orbitals. The three bonding pi orbitals are filled to capacity, all pi electrons are paired, and benzene is aromatic.

Let us look at the π molecular orbitals for cyclooctatetraene. This compound has eight p orbitals on the ring. Overlap of eight p orbitals would result in eight π molecular orbitals.

If cyclooctatetraene were planar and had a pi system similar to that in benzene, the π_1, π_2, and π_3 orbitals would be filled with six of the pi electrons. The remaining two electrons would be found, one each, in the degenerate π_4 and π_5 orbitals (Hund's rule; see Section 1.1B). The pi electrons of cyclooctatetraene would not all be paired, and overlap would not be maximal. Thus, cyclooctatetraene cannot be aromatic.

B. The Ions of Cyclopentadiene

Cyclopentadiene is a conjugated diene and is not aromatic. The principal reason that cyclopentadiene is not aromatic is that one of the carbon atoms is sp^3 hybridized. This sp^3-hybridized carbon has no p orbital available for bonding. However, removal of a hydrogen ion from this carbon changes the hybridization to sp^2. This carbon now has a p orbital containing a pair of electrons.

cyclopentadiene   cyclopentadienyl
                                anion

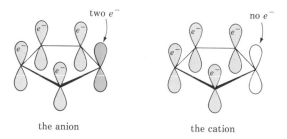

Figure 11.7 Orbital pictures of the anion and the cation of cyclopentadiene. (Each carbon is sp^2 hybridized and is bonded to one H.)

The cyclopentadiene cation also has all its carbon atoms in the sp^2 state. (The p orbital pictures for the ions of cyclopentadiene are shown in Figure 11.7.)

cyclopentadienyl
cation

Would either or both of these ions be aromatic? Either ion would have five π molecular orbitals (from five p orbitals, one per carbon). The cyclopentadienyl anion, with six pi electrons ($4n + 2$), has three pi orbitals filled and all pi electrons paired. *The anion is aromatic.* The cation, however, would have only four p electrons ($4n$) in the three orbitals. The pi electrons would *not* all be paired, and there would *not* be maximum overlap. *The cation is not aromatic.*

Cyclopentadiene is a stronger acid than most other hydrocarbons because its anion is aromatic and thus is resonance stabilized. Although cyclopentadiene is not nearly as strong an acid as a carboxylic acid, it is a proton donor in the presence of a strong base. (The pK_a of cyclopentadiene is 16, similar to that of an alcohol. By contrast, the pK_a of cyclopentane is about 50.)

Study Problem

11.4 Analyze cycloheptatriene as we have just done for cyclopentadiene. **(a)** Is cyclo-heptatriene aromatic? **(b)** Is its conjugated base aromatic? **(c)** Is the cycloheptatri-enyl cation aromatic? Explain your answers.

(a)

cycloheptatriene

(b)

its conjugate base

(c)

cycloheptatrienyl cation

Section 11.8
Electrophilic Aromatic Substitution

The aromaticity of benzene confers a unique stability to the pi system, and benzene does not undergo most of the reactions typical of alkenes. However, benzene is far from inert. Under the proper conditions, benzene readily undergoes **electrophilic aromatic substitution reactions**—reactions in which an electrophile is substituted for one of the hydrogen atoms on the aromatic ring. Two examples of this type of substitution reaction follow. Note that the aromaticity of the ring is retained in each product.

Halogenation:

chlorobenzene (90%)

Nitration:

nitrobenzene (85%)

The preceding examples show **monosubstitution** of the benzene ring. Further substitution is possible:

chlorobenzene

o-chloronitrobenzene (30%)

p-chloronitrobenzene (70%)

We will first consider the mechanism of monosubstitution (that is, the first substitution) and then the mechanism of further substitution leading to disubstituted benzenes.

▼ ## Section 11.9
The First Substitution

In each of the two monosubstitution reactions shown, a Lewis acid is used as a catalyst. The Lewis acid undergoes reaction with the reagent (such as X_2 or HNO_3) to generate an electrophile, which is the actual agent of substitution. For example, H_2SO_4 (a very strong acid) can remove a hydroxyl group from nitric acid to yield the nitronium ion, $^+NO_2$.

Formation of an electrophile by a Lewis acid:

$$H\overset{..}{\underset{..}{O}}{-}NO_2 + H_2SO_4 \underset{\overset{-HSO_4^-}{\rightleftharpoons}}{} H_2\overset{+}{\underset{..}{O}}{-}NO_2 \rightleftharpoons H_2\overset{..}{O}{:} + \quad ^+NO_2$$

an electrophile

An electrophile can attack the pi electrons of a benzene ring to yield a resonance-stabilized carbocation intermediate called a *benzenonium ion*, or an *arenium ion*. Like other carbocations, this carbocation reacts further. A hydrogen ion is removed from the intermediate (by HSO_4^-, for example) to yield the substitution product. In showing these structures, we use Kekulé formulas, which allow us to keep track of the number of pi electrons. In the following equation, we also show the hydrogen atoms bonded to the ring carbons so you can see how they are affected (or not) in the reaction.

benzene   intermediate   product
                   carbocation

As we discuss the various types of electrophilic aromatic substitution reactions, you will see that the mechanisms are all simply variations of this general one.

A. Halogenation

Aromatic halogenation is typified by the bromination of benzene. The catalyst in aromatic bromination is $FeBr_3$ (often generated *in situ* from Fe and Br_2). The function of the catalyst is to generate the electrophile Br^+. This may occur by direct reaction and fission of the Br—Br bond. More likely, Br_2 is not completely cleaved upon reaction with the $FeBr_3$ catalyst, but is polarized. For the sake of simplicity, we will show Br^+ as the electrophile.

$$:\overset{..}{\underset{..}{Br}}{-}\overset{..}{\underset{..}{Br}}{:} + FeBr_3 \rightleftharpoons \overset{\delta+}{:\overset{..}{Br}}{-}\overset{\delta-}{\overset{..}{\underset{..}{Br}}{:}}{-}{-}{-}FeBr_3 \rightleftharpoons :\overset{..}{\underset{..}{Br}}{^+} + FeBr_4^-$$

electrophilic

polarized   *cleaved*

When an electrophile such as Br^+ collides with the electrons of the aromatic pi cloud, a pair of the pi electrons forms a sigma bond with the electrophile. This step is the slow step, and therefore the rate-determining step, in the reaction.

Step 1 (slow):

The intermediate carbocation loses a proton to a base in the reaction mixture. The product is bromobenzene, a product in which the aromatic character of the ring has been restored.

Step 2 (fast):

bromobenzene

The third step in the reaction mechanism (which is concurrent with the loss of H^+) is the regeneration of the Lewis acid catalyst. The proton released in Step 2 undergoes reaction with the $FeBr_4^-$ ion to yield HBr and $FeBr_3$.

Step 3 (fast):

$$H^+ + FeBr_4^- \rightleftharpoons FeBr_3 + HBr$$

Ignoring the function of the catalyst, we can write an equation for the overall reaction of the aromatic bromination of benzene:

benzene   intermediate   bromobenzene

Sample Problem

Write resonance structures for the intermediate carbocation in the aromatic bromination of benzene.

Solution:

The intermediate is often represented as

Note the similarity between an electrophilic aromatic substitution reaction and an E1 reaction. In an E1 reaction, an intermediate alkyl carbocation eliminates a proton to form an alkene.

$$\left[-\overset{+}{\underset{|}{C}} \overset{H}{\underset{|}{C}} - \right] \xrightarrow[-H^+]{E1} \quad \overset{}{\underset{}{C}} = \overset{}{\underset{}{C}}$$

An alkyl carbocation also can undergo reaction with a nucleophile in an S_N1 reaction. However, the intermediate carbocation in aromatic substitution reaction does *not* undergo this reaction with a nucleophile. The addition of the nucleophile would destroy the aromatic stabilization of the benzene ring, whereas loss of a proton restores the aromatic stabilization.

does not occur

B. Isotope Effect in Aromatic Substitution

Recall from Section 5.9A that a CD bond is stronger than a CH bond. If the breaking of a CH bond is part of the rate-determining step of a reaction, the rate of reaction for a CD compound is slower than the rate for the corresponding CH compound.

If, as we have said, the rate-determining step of electrophilic aromatic substitution is the formation of the intermediate carbocation, then reaction of deuteriated benzene would proceed at the *same rate* as the reaction of normal benzene. Experiments have shown this to be true. Benzene and perdeuterio-benzene (C_6D_6) undergo electrophilic bromination at the same rate; no kinetic isotope effect is observed.

Step 2 in the reaction mechanism, loss of H^+ or D^+, does involve breaking of the CH or CD bond. Undoubtedly, the rate of elimination of D^+ is slower than that of H^+, but in either case the second step is so fast compared to Step 1 that no change in overall rate of reaction is observed.

C. Nitration

Benzene undergoes nitration when treated with concentrated HNO_3. The Lewis acid catalyst in this reaction is concentrated H_2SO_4. Like halogenation, aromatic nitration is a two-step reaction. The first step (the slow step) is electrophilic attack. In nitration, the electrophile is $^+NO_2$; the equation for its formation is shown on page 478. The result of the attack is a carbocation, which undergoes rapid loss of H^+ in the second step. This H^+ combines with HSO_4^- to regenerate the catalyst, H_2SO_4.

benzene   nitrobenzene

D. Alkylation

The alkylation of benzene is the substitution of an alkyl group for a hydrogen on the ring. Alkylation with an alkyl halide and a trace of $AlCl_3$ as catalyst is often referred to as a **Friedel–Crafts alkylation**, after Charles Friedel, a French chemist, and James Crafts, an American chemist, who developed this reaction in 1877. The reaction of 2-chloropropane with benzene in the presence of $AlCl_3$ is a typical Friedel–Crafts alkylation reaction.

2-chloropropane   isopropylbenzene
(isopropyl chloride)   (cumene)

The first step in alkylation is the generation of the electrophile: an alkyl cation.

$$R\!-\!\ddot{\underset{..}{Cl}}\!: + \ AlCl_3 \ \rightleftharpoons \ R^+ + AlCl_4^-$$

The second step is electrophilic attack on benzene, while the third step is elimination of a hydrogen ion. The product is an alkylbenzene.

an alkylbenzene

One problem in Friedel–Crafts alkylations is that the substitution of an alkyl group on the benzene ring activates the ring so that a *second* substitution may also occur. (We will discuss ring activation and second substitutions later in this chapter.) To suppress this second reaction, an excess of the aromatic compound is commonly used.

$$\text{\Large⬡} + (CH_3)_2CHCl \xrightarrow{\ AlCl_3\ } \text{\Large⬡}-CH(CH_3)_2 + \text{some } (CH_3)_2CH-\text{\Large⬡}-CH(CH_3)_2$$

excess

Another problem in Friedel–Crafts alkylations is that the attacking electrophile may undergo rearrangement by 1,2-shifts of H or R.

$$\text{\Large⬡} + CH_3CH_2CH_2Cl \xrightarrow[25°]{AlCl_3} \text{\Large⬡}-CH(CH_3)_2 + \text{\Large⬡}-CH_2CH_2CH_3$$

1-chloropropane   isopropylbenzene (70%)   *n*-propylbenzene (30%)

$$\text{\Large⬡} + \overset{\displaystyle CH_3}{\underset{\displaystyle}{CH_3\overset{|}{C}HCH_2Cl}} \xrightarrow[25°]{AlCl_3} \text{\Large⬡}-C(CH_3)_3 \text{ and } no \text{ \Large⬡}-CH_2CH(CH_3)_2$$

1-chloro-2-methylpropane   *t*-butylbenzene (100%)   isobutylbenzene

The rearrangements shown are of primary alkyl halides, which do not readily form carbocations. In these cases, reaction probably proceeds through an RX—AlCl$_3$ complex.

$$CH_3CH_2CH_2-\overset{..}{\underset{..}{Cl}}: + AlCl_3 \rightleftharpoons CH_3CH_2\overset{\delta+}{CH_2}-\overset{\delta-}{\underset{..}{\overset{..}{Cl}}} ------ AlCl_3$$

a complex

This complex may (1) undergo reaction with benzene to yield the nonrearranged product, or (2) undergo rearrangement to a secondary or tertiary carbocation, which leads to the rearranged product.

No rearrangement:

$$\text{\Large⬡}-H + \overset{+}{C}H_2\frown\overset{..}{\underset{..}{Cl}} ------ AlCl_3 \xrightarrow{-AlCl_4^-}$$

$$\underset{CH_3}{\overset{CH_2}{|}}$$

$$\left[\text{\Large⬡}\overset{\displaystyle +CH_2CH_2CH_3}{\underset{\displaystyle H}{}}\right] \xrightarrow{-H^+} \text{\Large⬡}-CH_2CH_2CH_3$$

n-propylbenzene

Rearrangement:

$$\overset{H}{\underset{}{CH_3\overset{|}{C}HCH_2}}\frown\overset{..}{\underset{..}{Cl}} ------ AlCl_3 \xrightarrow{-AlCl_4^-} [CH_3\overset{+}{C}HCH_3] \xrightarrow[-H^+]{\text{\Large⬡}} \text{\Large⬡}-CH(CH_3)_2$$

a 2° carbocation   isopropylbenzene

Alkylations can also be accomplished with alkenes in the presence of HCl and AlCl$_3$. The mechanism is similar to alkylation with an alkyl halide and proceeds by way of the more stable alkyl carbocation.

$$CH_3CH=CH_2 + HCl + AlCl_3 \longrightarrow [CH_3\overset{+}{C}HCH_3] + AlCl_4^-$$

propene

more stable
2° carbocation

isopropylbenzene (85%)

Study Problem

11.5 Predict the *major* organic product of the reaction of benzene in the presence of AlCl₃ (and HCl in the case of an alkene) with: **(a)** 1-chlorobutane; **(b)** methylpropene; **(c)** neopentyl chloride (1-chloro-2,2-dimethylpropane); **(d)** dichloromethane (using an excess of benzene).

E. Acylation

The $\underset{\substack{\|\\O}}{RC}$— group or the $\underset{\substack{\|\\O}}{ArC}$— group is called an **acyl group**. Substitution of an acyl group on an aromatic ring by reaction with an acid halide is called an **aromatic acylation reaction**, or a **Friedel–Crafts acylation**.

acetyl chloride

an acid halide

acetophenone
(97%)

This reaction is often the method of choice for preparing an aryl ketone. The carbonyl group of the aryl ketone can be reduced to a CH_2 group (Section 13.7C). By the combination of a Friedel–Crafts acylation and reduction, an alkylbenzene may be prepared without rearrangement of the alkyl group.

propanoyl chloride

phenyl ethyl ketone
(90%)

n-propylbenzene
(80%)

The mechanism of the Friedel–Crafts acylation reaction is similar to that of the other electrophilic aromatic substitution reactions. The attacking electrophile is an **acylium ion** ($R\overset{+}{C}=O$). Rearrangements do not occur because the acylium ion is resonance stabilized.

$$\underset{\substack{\text{polarized}}}{RCCl: + AlCl_3 \longrightarrow RC\text{-}Cl\text{---}AlCl_3} \xrightarrow{-AlCl_4^-} \left[R\text{---}C\overset{+}{=}\overset{..}{O}: \longleftrightarrow R\text{---}C\equiv\overset{+}{O}: \right]$$

the electrophilic carbon

resonance structures for
an acylium ion

Because $AlCl_3$ forms complexes with carbonyl groups ($C=\overset{..}{O}:\text{---}AlCl_3$), a slight excess of $AlCl_3$, such as 1.1 mol of $AlCl_3$ to 1.0 mol of RCOCl, is used in a Friedel–Crafts acylation. The product ketone is also complexed with $AlCl_3$, but treatment with water during the work-up of the reaction mixture liberates the ketone.

Study Problems

11.6 Write a mechanism for the reaction of an acylium ion with benzene to yield a ketone.

11.7 Predict the major organic products of the reaction of benzene with each of the following compounds and $AlCl_3$ as the catalyst:

(a) $CH_3CH_2\underset{\underset{\text{CH}_3}{|}}{C}HCH_2Cl$ (b) $CH_3CH_2\underset{\underset{\text{CH}_3}{|}}{C}H\overset{\overset{\text{O}}{\|}}{C}Cl$, followed by Zn/Hg, HCl

F. Sulfonation

The sulfonation of benzene with fuming sulfuric acid ($H_2SO_4 + SO_3$) yields benzenesulfonic acid.

$$\bigcirc + SO_3 \xrightleftharpoons[40°]{H_2SO_4} \left[\bigcirc\overset{+}{\underset{H}{\diagdown}}{}^{SO_3^-} \right] \Longrightarrow \bigcirc\text{---}SO_3H$$

benzenesulfonic acid (50%)

Unlike the other electrophilic substitution reactions of benzene, sulfonation is a readily reversible reaction if water is present, and it shows a moderate kinetic isotope effect. Perdeuteriobenzene undergoes sulfonation at about half the rate of ordinary benzene. From these data, we conclude that the intermediate benzenonium ion in sulfonation can revert to benzene or go on to benzenesulfonic acid with almost equal ease (or the reaction would not be readily reversible). Also, the rates of reaction of Steps 1 and 2 must be more nearly equal than for other electrophilic aromatic substitution reactions (or the reaction would not show a kinetic isotope effect). Figure 11.8 shows an energy diagram for the sulfonation of benzene. Note that the energies of the transition states of Steps 1 and 2 are roughly the same.

The sulfonic acid group is easily displaced by a variety of other groups; therefore, arylsulfonic acids are useful synthetic intermediates. We will introduce a few of these reactions of arylsulfonic acids in Section 12.5.

Figure 11.9 summarizes the monosubstitution reactions of benzene.

transition states 1 and 2 of nearly equal energies

When perdeuteriobenzene is the reactant, this transition state is higher energy relative to the energies of reactants and products.

E

$+ SO_3^-$

H

$C_6H_6 + SO_3$

$C_6H_5SO_3H$

Progress of reaction

Figure 11.8 Energy diagram for the sulfonation of benzene.

benzene

$\dfrac{X_2}{FeX_3}$ → —X   **bromo- or chlorobenzene**

$\dfrac{HNO_3}{H_2SO_4}$ → —NO_2 **nitrobenzene**

$\dfrac{RX + AlX_3}{\text{or alkene} + HX + AlX_3}$ → —R   **an alkylbenzene** (possible rearrangement of R)

$\dfrac{\overset{O}{\overset{\|}{RCX}}}{AlX_3}$ → —$\overset{O}{\overset{\|}{C}}R$ **a phenyl ketone** (no rearrangement of R)

$\dfrac{SO_3}{H_2SO_4}$ → —SO_3H **benzenesulfonic acid** (reversible reaction)

Figure 11.9 Summary of the monosubstitution reactions of benzene.

▼ Section 11.10
The Second Substitution

A substituted benzene may undergo substitution of a second group. Some substituted benzenes undergo reaction *more easily* than benzene itself, while other substituted benzenes undergo reaction *less easily*. For example, aniline undergoes electrophilic substitution a million times faster than benzene. Nitrobenzene, on the other hand, undergoes reaction at approximately one-millionth the rate of benzene! (Rather than allowing a reaction to proceed over a much longer period of time, a chemist commonly uses stronger reagents and higher temperatures for a less reactive compound.)

aniline

2,4,6-tribromoaniline (100%)

*no catalyst needed
as it would be for benzene*

nitrobenzene

m-dinitrobenzene (93%)

*much slower to nitrate than benzene,
unless stronger reagents and
higher temperatures are used*

In these examples, we would say that NH_2 is an **activating group**. Its presence causes the ring to be *more* susceptible to further electrophilic substitution. On the other hand, the NO_2 group is a **deactivating group**. Its presence causes the ring to be *less* susceptible than benzene to substitution.

Besides the differences in reaction rates of substituted benzenes, the *position* of the second attack varies:

chlorobenzene   *ortho* (30%)   *para* (70%)

nitrobenzene   *meta* (93%)

Study Problem

11.8 If chlorobenzene were nitrated at equal rates at each possible position of substitution, what would be the ratio of *o*-, *m*-, and *p*-products?

Chlorobenzene is nitrated in the *ortho*- and *para*-positions, but not in the *meta*-position. However, nitrobenzene undergoes a second nitration in the *meta*-position; very little substitution at the *ortho*- or *para*-positions occurs. These examples show that the nature of the incoming group has little effect on its own positioning on the ring. The position of second substitution is determined by the group that is already on the ring.

Table 11.5 Orientation of the nitro group in aromatic nitration

| Reactant | Approximate Percents of Products | | |
	o	*p*	*m*
C_6H_5OH	50	50	—
$C_6H_5CH_3$	60	40	—
C_6H_5Cl	30	70	—
C_6H_5Br	40	60	—
$C_6H_5NO_2$	7	—	93
$C_6H_5CO_2H$	20	—	80

Table 11.6 Effect of the first substituent on the second substitution

	o,p-Directors	*m*-Directors (*All Deactivating*)	
increasing activation	$-\ddot{N}H_2, -\ddot{N}HR, -\ddot{N}R_2$	$-\overset{\overset{\displaystyle O}{\|}}{C}R$	
	$-\ddot{O}H$	$-CO_2R$	
	$-\ddot{O}R$	$-SO_3H$	
	$-\overset{\overset{\displaystyle O}{\|}}{\ddot{N}H}CR$	$-CHO$	
		$-CO_2H$	**increasing deactivation**
	$-C_6H_5$ (aryl)	$-CN$	
	$-R$ (alkyl)	$-NO_2$	
	$-\ddot{\underset{\cdot\cdot}{X}}:$ (*deactivating*)	$-NR_3^+$	

To differentiate between these two types of substituent, Cl is called an **ortho**, **para**-**director**, while NO$_2$ is called a **meta-director**. Any substituent on a benzene ring is either an *o,p*-director or an *m*-director, although to varying degrees, as is shown in Table 11.5.

Table 11.6 contains a summary of commonly encountered benzene substituents classified as *activating* or *deactivating* and as *o,p-directors* or *m-directors*. Note that all *o,p*-directors except the halogens are also activating groups. All *m*-directors are deactivating. Also note that all *o,p*-directors except aryl and alkyl groups have an unshared pair of electrons on the atom bonded to the ring. None of the *m*-directors has an unshared pair of electrons on the atom bonded to the ring.

o,p-director,
activating

m-director,
deactivating

Study Problem

11.9 What would be the major organic products of second substitution in each of the following reactions?

(a) $C_6H_5NO_2$ $\xrightarrow{Br_2, \text{ FeBr}_3}$ (b) $C_6H_5CH_3$ $\xrightarrow[\text{AlCl}_3, \text{ HCl}]{CH_2=CH_2}$

A. Mechanism of the Second Substitution with an *o,p*-Director

Why are most *o,p*-directors activating groups? Why do they direct incoming groups to the *o*- and *p*-positions? To answer these questions, let us consider aniline, a compound with the *o,p*-directing NH_2 group on the ring.

The amino group in aniline activates the benzene ring toward substitution to such an extent that (1) no Lewis acid catalyst is needed, and (2) it is very difficult to obtain a monobromoaniline. Aniline quickly undergoes reaction to form the 2,4,6-tribromoaniline (both *o*-positions and the *p*-position brominated).

The mechanism of the bromination of aniline is similar to the mechanism of the bromination of benzene itself.

aniline   *p*-bromoaniline

The difference between the mechanism of the bromination of aniline and that of benzene lies in the stabilization of the intermediate carbocation. The carbocation from aniline is resonance stabilized just as is the carbocation from benzene but, in the case of aniline, the amino group can increase the stabilization. Increased stabilization of the intermediate means a transition state of lower energy in Step 1 (see Figure 11.10) and therefore a faster reaction.

Resonance structures for the p-intermediate:

added stabilization

In the intermediate for *m*-substitution, the nitrogen of the amino group cannot help share the positive charge. (Draw the Kekulé structures for the *m*-substituted intermediate and verify this statement for yourself.) Therefore, the intermediate for *m*-substitution is of higher energy than are the intermediates

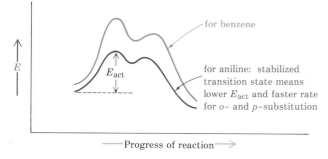

Figure 11.10 Energy diagrams for the bromination of aniline and benzene.

leading to either *o*- or *p*-substituted products. Because this intermediate is of higher energy, its transition state also has a higher energy, and the rate of reaction at the *m*-position is lower.

The amino group, along with —OH, —OR, —NHCOR, and phenyl groups, activates the benzene ring toward electrophilic substitution by donating a pair of electrons to the ring through resonance. Substitution occurs at the *o*- and *p*-positions because the group helps share the positive charge in these intermediates.

Although the amino group is an *o,p*-director and a ring activator, its character changes in a reaction mixture containing a Lewis acid such as H_2SO_4, HNO_3, or $AlCl_3$. The reason for this is that the amino group reacts with a Lewis acid to yield a *meta*-directing, deactivating ammonium ion group.

$$\langle\bigcirc\rangle\!-\!\overset{..}{N}H_2 + H_2SO_4 \longrightarrow \langle\bigcirc\rangle\!-\!\overset{+}{N}H_3 + HSO_4^-$$

o,p-director,
activating

m-director,
deactivating

Study Problems

11.10 Draw the resonance structures for the intermediate in the nitration of phenol to yield *o*-nitrophenol.

11.11 The benzene ring in acetanilide (Table 11.1, page 463) is *less* reactive toward electrophilic substitution than the ring in aniline. Suggest a reason for this lesser activation by the —NHCCH$_3$ group.
‖
O

Halogens are different from the other *o,p*-directors. They direct an incoming group *ortho* or *para*, but they *deactivate* the ring to electrophilic substitution. A halogen substituent on the benzene ring directs an incoming group to the *o*- or *p*-position for the same reason the amino or hydroxyl group does. The halogen can donate electrons and help share the positive charge in the intermediate.

But why does a halogen deactivate the ring? A halogen, oxygen, or nitrogen withdraws electronic charge from the ring by the inductive effect. We would expect that any electronegative group would decrease the electron density of the ring and would make the ring less attractive to an incoming electrophile.

e⁻ withdrawal should deactivate the ring (and does deactivate the ring when X is Cl, Br, I)

In phenol or aniline, the effect of ring deactivation by electron withdrawal is counterbalanced by the release of electrons by resonance. Why is the same effect not observed in halobenzenes? In phenol or aniline, the resonance structures of the intermediate that confer added stability arise from overlap of $2p$ orbitals of carbon and $2p$ orbitals of N or O. These $2p$ orbitals are about the same size, and overlap is maximal.

In chlorobenzene, bromobenzene, or iodobenzene, the overlap in the intermediate is $2p$–$3p$, $2p$–$4p$, or $2p$–$5p$, respectively. The overlap is between orbitals of different size and is not so effective. The intermediate is less stabilized, the transition-state energy is higher, and the rate of reaction is lower. (Fluorobenzene, with $2p$–$2p$ overlap in the intermediate, is more reactive than the other halobenzenes toward electrophilic substitution in spite of its greater electronegativity.

In summary, OH, NH₂, or a halogen determines the orientation of an incoming group by *donating electrons* by resonance and by adding resonance-stabilization to *o*- and *p*-intermediates. For OH and NH₂, electron release by resonance activates the ring toward electrophilic substitution. Electron release by resonance is less effective for Cl, Br, or I than it is for OH or NH₂. Chlorobenzene, bromobenzene, and iodobenzene contain deactivated rings because the inductive electron-withdrawal by these substituents is relatively more effective.

An *alkyl group* does not have unshared electrons to donate for resonance-stabilization. However, an alkyl group is electron-releasing by the inductive effect, a topic we discussed in Section 5.6E. Because an alkyl group releases electrons to the benzene ring, the ring gains electron density and becomes more attractive toward an incoming electrophile.

R groups activate because they are e⁻-releasing

To see why alkyl groups direct incoming electrophiles to *o*- and *p*-positions, we must again look at the resonance structures of the intermediates.

ortho:

added stabilizations; + next to e⁻-releasing R

para:

meta:

The intermediates for *o*- or *p*-substitution both have resonance structures in which the positive charge is adjacent to the R group. These structures are especially important contributors to resonance-stabilization because the R group can help delocalize the positive charge by electron release and lower the energy of the transition state leading to these intermediates. The situation is directly analogous to that of an R group stabilizing a carbocation. The resonance structures for the intermediate in *m*-substitution have no such contributor. The *m*-intermediate is of higher energy. Attack on an alkylbenzene occurs *ortho* and *para* at a rate that is much faster than attack at a *meta* position.

Sample Problem

Which compound would you expect to undergo aromatic nitration more readily, $C_6H_5CH_3$ or $C_6H_5CCl_3$?

Solution: While the CH_3 group is electron-releasing and activates the ring, the CCl_3 group is strongly *electron-withdrawing* because of the influence of the electronegative chlorines. $C_6H_5CH_3$ has an activated ring; $C_6H_5CCl_3$ has a *deactivated* ring and undergoes substitution more slowly.

Study Problem

11.12 Predict the major organic products of the reaction of each of the following compounds with 2-chloropropane and $AlCl_3$. Indicate the relative order of the reaction rates.

(a) bromobenzene   (b) phenol   (c) toluene   (d) benzene

B. Mechanism of the Second Substitution with a *meta*-Director

In benzene substituted with a *meta*-director (such as NO_2 or CO_2H), the atom bonded to the benzene ring has no unshared pair of electrons and carries a positive or a partial positive charge. It is easy to see why the *m*-directors are deactivating. Each one is *electron-withdrawing* and cannot donate electrons by resonance. Each one decreases the electron density of the ring and makes it less attractive to an incoming electrophile. The energy of the Step-1 transition state is higher than it would be for unsubstituted benzene.

A *meta*-director does not activate the *m*-position toward electrophilic substitution. A *meta*-director *deactivates all positions in the ring*, but it deactivates the *m*-position less than the other positions. The resonance structures of the intermediates resulting from attack at the various positions show that the *o*- and *p*-intermediates are destabilized by the nearness of two positive charges. The *m*-intermediate has no such destabilizing resonance structure. The following resonance structures are for the intermediate in the bromination of nitrobenzene:

ortho:

high energy
(adjacent + charges)

para:

meta:

favored: no adjacent + charges

Figure 11.11 summarizes how the three main types of substituents affect the course of a second substitution.

Section 11.11
The Third Substitution

What if a benzene ring has two substituents already? Where does a third substituent go? A few general rules cover the majority of cases.

1. If two substituents direct an incoming group to the same position, that will be the principal position of the third substitution.

o to CH₃ and m to NO₂

p-nitrotoluene   2-bromo-4-nitrotoluene (90%)

Electron-releasing substituent: an *o,p*-director; activates the ring toward E$^+$.

Halogen substituent: an *o,p*-director; deactivates the ring toward E$^+$.

where X = Cl, Br, or I

Electron-withdrawing substituent: a *meta*-director, deactivates the ring toward E$^+$

Figure 11.11 Summary of substituent effects in the second substitution. (Table 11.6, page 487, lists some common substituents.)

2. If two groups conflict in their directive effects, the more powerful activator (Table 11.6) will exert the predominant directive effect.

p-chlorophenol   2,4-dichlorophenol (94%)

3. If two deactivating groups are on the ring, regardless of their positions, it may be difficult to effect a third substitution.

4. If two groups on a ring are *meta* to each other, the ring does not usually undergo substitution at the position between them, even if the ring is activated. The lack of reactivity at this position is probably due to steric hindrance.

m-chloroanisole   3,4-dichloroanisole (64%)

2,5-dichloroanisole (18%)

Study Problem

11.13 Predict the products of the next substitution:

(a) $\xrightarrow{\text{Cl}_2}$ (b) $\xrightarrow{\text{CH}_3\text{Cl, AlCl}_3}$

(c) $\xrightarrow{\text{HNO}_3,\ \text{H}_2\text{SO}_4}$

▼ Summary

An **aromatic compound** is a type of compound that gains substantial stabilization by pi-electron delocalization. To be aromatic, a compound must be cyclic and planar. Each ring atom must have a *p* orbital perpendicular to the plane of the ring, and the *p* orbitals must contain $(4n + 2)$ pi electrons **(Hückel rule)**.

Benzene and other aromatics undergo **electrophilic aromatic substitution reactions**. These reactions are summarized in Table 11.7.

Table 11.7 Summary of electrophilic substitution reactions of benzene and its derivatives

Reaction		Product	Section Reference
$C_6H_6 + X_2$	$\xrightarrow{\text{FeX}_3} C_6H_5X$	**halobenzene**	11.9A
$+ \text{HNO}_3$	$\xrightarrow{\text{H}_2\text{SO}_4} C_6H_5NO_2$	**nitrobenzene**[a]	11.9C
$+ \text{RX}$	$\xrightarrow{\text{AlX}_3} C_6H_5R$	**alkylbenzene**	11.9D
$+ R_2C{=}CHR$	$\xrightarrow[\text{HX}]{\text{AlX}_3} C_6H_5CR_2CH_2R$	**alkylbenzene**	11.9D
$\overset{\displaystyle O}{\underset{\displaystyle \|}{+ \text{RCX}}}$	$\xrightarrow{\text{AlX}_3} C_6H_5\overset{\displaystyle O}{\underset{\displaystyle \|}{\text{CR}}}$	**aryl ketone**[b]	11.9E
$+ \text{SO}_3$	$\xrightarrow{\text{H}_2\text{SO}_4} C_6H_5SO_3H$	**benzenesulfonic acid**[c]	11.9F
$C_6H_5{-}\ddot{Y} + E^+ \longrightarrow$ where Y is OH, NH$_2$, X, R, etc.		***o*- and *p*-products**	11.10
$C_6H_5{-}\overset{\delta+}{Z} + E^+ \longrightarrow$ where Z is NO$_2$, CO$_2$H, etc.		***m*-products**	11.10

[a] Can be reduced to aniline (Section 12.3).
[b] Can be reduced to an alkylbenzene (Section 13.7C).
[c] Can be converted to C_6H_6 or C_6H_5OH (Section 12.5).

A second substitution results in *o*- and *p*-isomers or else *m*-isomers, depending on the first substituent (see Table 11.6, page 487). The *o,p*-directors (except R) have electrons that can be donated to the ring by resonance.

Electron release by resonance:

Electron release by inductive effect:

All *o,p*-directors except X activate the entire ring toward electrophilic substitution. The *o*- and *p*-positions are the preferred positions of substitution because of added resonance-stabilization of their intermediates.

All *m*-directors and X *deactivate* the ring toward further electrophilic substitution by electron withdrawal.

deactivated toward E⁺

Study Problems*

11.14 Draw formulas for the following compounds:

(a) "penta" (pentachlorophenol), a toxic insecticide and wood preservative;

(b) PABA (*p*-aminobenzoic acid), used in sunscreen lotions; also a part of the structure of the folic acids, which are nutritional factors;

(c) 2,4-dinitroaniline, a toxic compound used in the dye industry;

(d) Dowicide 9 (2-cyclopentyl-4-chlorophenol), a germicide;

(e) anthranilic acid (*o*-aminobenzoic acid), the cadmium salt of which has been used to kill intestinal roundworms in hogs;

(f) *m*-toluidine (*m*-aminotoluene), a toxic compound used in the dye industry.

11.15 Name each of the following compounds:

(a) $C_6H_5CH=CH_2$ (b)   (c)

* For information concerning the organization of the *Study Problems* and *Additional Problems*, see the *Note to student* on page 39.

(d) CH$_3$—⬡(NO$_2$)   (e) ⬡ with I, I —CH$_3$ (f) C$_6$H$_5$CH=CHCH=CHC$_6$H$_5$

11.16 Give the structures and the names of all the isomeric: **(a)** monobromoanilines; **(b)** dinitrotoluenes.

11.17 Which of the following compounds could be aromatic? (*Hint:* In many cases, the Hückel rule may be extrapolated to structures with more than one ring.)

(a)   Cl$^-$ **(b)**

azulene

(c) $-$: ⬡ :$-$

cyclooctatetraene
dianion

11.18 The heat of hydrogenation of benzene is -49.8 kcal/mol, while that of 1,4-cyclo-hexadiene is -59.3 kcal/mol. Calculate ΔH for the conversion of benzene to 1,4-cyclohexadiene.

11.19 From the following heats of hydrogenation, calculate the energy of the added stabilization by conjugation of the double bond in styrene (C$_6$H$_5$CH=CH$_2$): styrene, -76.9 kcal/mol; benzene, -49.8 kcal/mol; propene, -28.6 kcal/mol.

11.20 Naphthalene undergoes electrophilic aromatic substitution at the α-position:

(1) Predict the organic products of the monosubstitution reactions of naphthalene with each of the following sets of reagents. (2) Write equations to show how the electrophile is generated in each case.

(a) Br$_2$, FeBr$_3$ (b) HNO$_3$, H$_2$SO$_4$
(c) ClCH$_2$CH$_3$, AlCl$_3$ (d) CH$_2$=CH$_2$, AlCl$_3$, HCl

(e) CH$_3\overset{\overset{\text{O}}{\|}}{\text{C}}$Cl, AlCl$_3$

11.21 Write equations showing the preparation of the following compounds from benzene:
   (a) nitrobenzene   **(b)** chlorobenzene
   (c) 1-phenyl-1-pentanone   **(d)** benzenesulfonic acid
   (e) 1-butylbenzene   **(f)** bromobenzene
   (g) isopropylbenzene (two routes)

11.22 Write a flow equation for the mechanism of the nitration of benzene.

11.23 Write a flow equation for the mechanism and probable products of the reaction of benzene with 1-chlorobutane in the presence of aluminum chloride.

11.24 Predict the product (including stereochemistry) of the following reaction:

11.25 Friedel–Crafts acylation reactions can be accomplished with acid anhydrides as well as with acid halides:

an acid anhydride

Predict the product of the AlCl$_3$-catalyzed reaction of benzene with:

(a)   **(b)**

11.26 Which one of each of the following pairs is more reactive toward aromatic bromination?

(a) acetanilide (C$_6$H$_5$NHCCH$_3$) or benzene
(b) bromobenzene or toluene
(c) *p*-xylene (*p*-dimethylbenzene) or *p*-toluic acid (*p*-methylbenzoic acid)
(d) *m*-dinitrobenzene or *m*-nitrotoluene
(e) chlorobenzene or *m*-dichlorobenzene

11.27 Label each of the following groups as an *o,p*-director or as an *m*-director:

(a) —N⟨⟩   **(b)** ⟨⟩NH—   **(c)** —OC$_6$H$_5$

(d) —CNHCH$_3$ **(e)** —CC$_6$H$_5$ **(f)**

(g) —OCCH$_3$ **(h)** —COCH$_3$ **(i)**

11.28 Draw resonance structures for the following carbocation intermediates:

11.29 In acidic solution, nitrous acid (HONO) reacts with some aromatic compounds in a manner similar to that of nitric acid.

(a) Write an equation showing the formation of the electrophile.

(b) Write the equation showing the electrophilic substitution of phenol by this species. (Include the structure of the intermediate.)

11.30 Predict the major organic products:

(a) $C_6H_5CH_2CH_2CH_2\overset{\overset{\displaystyle O}{\|}}{C}Cl \xrightarrow{AlCl_3}$

(b) toluene + 1-chloro-2-methylpropane $\xrightarrow{AlCl_3}$

(c) benzene + 1-chloro-2-butene $\xrightarrow{AlCl_3}$

11.31 Explain the following observations: (a) Treatment of benzene with an excess of fuming nitric acid and sulfuric acid yields *m*-dinitrobenzene, but not the trinitrated product. (b) Under the same conditions, toluene yields 2,4,6-trinitrotoluene (TNT).

11.32 Ring monobromination of a diethylbenzene yields three isomeric bromodiethylbenzenes (two formed in minor amounts). Give the structures of the diethylbenzene and the brominated products.

11.33 Predict the major products of aromatic monochlorination of: (a) chlorobenzene; (b) *o*-dichlorobenzene; (c) *m*-bromochlorobenzene; (d) *m*-xylene; (e) acetophenone (see Table 11.1); (f) (trichloromethyl)benzene ($C_6H_5CCl_3$).

11.34 Write equations for the mononitration of the following compounds:

(a) *p*-bromophenol (b) *p*-nitrotoluene

(c) *m*-methoxybenzoic acid (d) 3-bromo-1-methoxybenzene

Additional Problems

11.35 Predict the major organic product of the reaction of benzene with ICl (with iron filings added as catalyst).

11.36 When bromobenzene is chlorinated, two isomeric compounds (A and B, C_6H_4BrCl) can be isolated. Bromination of A gives a number of isomeric products of the composition $C_6H_3Br_2Cl$, while bromination of B yields two isomers (C and D) of the composition $C_6H_3Br_2Cl$. Compound C is identical with one of the compounds obtained from the bromination of A; however, D is different from any of the isomeric compounds obtained from the bromination of A. Give the structures of A, B, C, and D.

11.37 Suggest mechanisms for the following reactions:

(a) benzene + $F\overset{\overset{\displaystyle O}{\|}}{C}H \xrightarrow{BF_3}$ benzaldehyde $\left(C_6H_5\overset{\overset{\displaystyle O}{\|}}{C}H\right)$

(b) *t*-butylbenzene + $Br_2 \xrightarrow{AlBr_3}$ bromobenzene + 2-methylpropene

(c) $\xrightarrow[HCl]{AlCl_3}$

11.38 2-Naphthol can be hydrogenated to the decahydro derivative using rhodium-on-alumina as catalyst under rather mild conditions (55 psi; room temperature; 12 hr). A 94–97% yield of a mixture of geometric isomers is obtained.

2-naphthol

(a) Write an equation for the reaction. (b) Draw the structures of the geometric isomers. (c) Analysis of the mixture indicates that it consists of 50–53% of one geometric isomer. Which is it?

11.39 When 2,3-di-*t*-butylcyclopropenone is treated with *t*-butyllithium and the resulting intermediate subjected to dehydrating conditions, a stable cation is formed. Why is this cation stable?

2,3-di-*t*-butylcyclopropenone

tri-*t*-butylcyclopropenyl tetrafluoroborate

11.40 In the proton nmr spectrum of 1,6-methano[10]annulene, the ring protons absorb at low field ($\delta 6.8$–7.5), while the —CH_2— protons absorb far upfield ($\delta - 0.5$, to the *right* of TMS). Explain.

1,6-methano[10]annulene

11.41 The chemical shift of the proton bonded to carbon 5 of 4-ethynylphenanthrene is 1.71 ppm downfield from the chemical shift of the absorption of the same proton on phenanthrene itself. Explain why this is true.

4-ethynylphenanthrene   phenanthrene

11.42 A compound with the formula $C_{15}H_{14}O$ gives the proton nmr spectrum shown in Figure 11.12. Its infrared spectrum shows strong absorption about 1750 cm^{-1} (5.75 μm). What is the structure of the compound?

Figure 11.12 ^{1}H nmr spectrum for Problem 11.42.

12 ▽

Substituted Benzenes

In Chapter 11, we described electrophilic aromatic substitution reactions of substituted and unsubstituted benzene rings. We discussed how a substituent on the ring can change the reactivity of the ring. However, the reverse is also true. A substituent on an aromatic ring often exhibits different types of reactivity than the same substituent in other compounds. For example, the methyl group in toluene reacts differently from the methyl group in methylpropane. A phenol hydroxyl group reacts differently from an alcohol hydroxyl group.

In this chapter we will discuss the reactions of the substituents on aromatic rings. Then, we will summarize the different types of syntheses that can be carried out both with electrophilic aromatic substitution reactions and with substituent reactions.

Section 12.1
Alkylbenzenes

Although alkyl groups in general are relatively nonreactive, alkyl groups bonded to an aromatic ring show a unique reactivity. The carbon adjacent to an aromatic ring is a benzylic carbon.

A benzyl cation, a benzyl radical, and a benzyl carbanion are all resonance-stabilized by the benzene ring. Consequently, the benzylic position is a site of attack in many reactions.

Benzylic carbons:

Study Problem

12.1 Draw the resonance structures of the carbocation intermediate in the S_N1 reaction of benzyl bromide with water.

Although an alkene is readily oxidized by such reagents as hot $KMnO_4$ solution, a benzene ring is not oxidized under these laboratory conditions. However, the *alkyl group* of an alkylbenzene can be oxidized. Because of the reactivity of the benzylic position, alkylbenzenes with a benzylic H atom all yield the same product, benzoic acid, upon vigorous oxidation.

$$\text{toluene} \xrightarrow{\text{hot MnO}_4^-}$$
$$\text{ethylbenzene} \xrightarrow{\text{hot MnO}_4^-} \text{benzoic acid}$$
$$\text{isopropylbenzene} \xrightarrow{\text{hot MnO}_4^-}$$

toluene — CH_3

ethylbenzene — CH_2CH_3

isopropylbenzene — $CH(CH_3)_2$

benzoic acid — CO_2H

Study Problem

12.2 What would be the oxidation product of tetralin with hot $KMnO_4$ solution?

tetralin

Free-radical halogenation is another reaction that takes place preferentially at the benzylic position (Section 6.5B). (Note that the conditions of radical halogenation are different from those for electrophilic aromatic halogenation, which is an ionic reaction, not a radical reaction.)

$$\text{C}_6\text{H}_5{-}\text{CH}_2\text{CH}_3 \xrightarrow[hv]{\text{Br}_2} \text{C}_6\text{H}_5{-}\overset{\displaystyle \text{Br}}{\underset{\displaystyle}{\text{CHCH}_3}}$$

ethylbenzene   1-bromo-1-phenylethane

▼ ## Section 12.2
Phenols

A phenol (ArOH) is a compound with a hydroxyl group bonded to an aromatic ring. As we mentioned in Section 11.10, the OH group is a powerful activator in electrophilic aromatic substitution reactions.

Because a bond from an sp^2-hybridized carbon is stronger than a bond from an sp^3-hybridized carbon (Section 2.4F) and because of geometric restraints, the C—O bond of a phenol is not easily broken. Phenols do not undergo S_N1 or S_N2 reactions or elimination reactions as alcohols do.

$$\text{R—OH} + \text{HBr} \xrightarrow{S_N1 \text{ or } S_N2} \text{RBr} + \text{H}_2\text{O}$$

an alcohol

$$\text{Ar—OH} + \text{HBr} \longrightarrow \text{no reaction}$$

a phenol

A. Acidity of Phenols

Although the C—O bond of a phenol is not easily broken, the OH bond is readily broken. Phenol, with a pK_a of 10.00, is a stronger acid than an alcohol or water. Because a phenoxide ion (ArO⁻) is a weaker base than OH⁻, a phenoxide can be prepared by treatment of a phenol with aqueous NaOH. This reactivity is in direct contrast to that of alcohols.

$$\text{CH}_3\text{CH}_2\text{OH} + \text{NaOH} \xrightleftharpoons{\quad} \overset{\textit{not favored}}{\text{CH}_3\text{CH}_2\text{O}^- \text{Na}^+} + \text{H}_2\text{O}$$

ethanol, $pK_a = 15.9$ sodium ethoxide $pK_a = 15.74$

$$\text{C}_6\text{H}_5{-}\text{OH} + \text{NaOH} \xrightleftharpoons{\quad} \overset{\textit{favored}}{\text{C}_6\text{H}_5{-}\text{O}^- \text{Na}^+} + \text{H}_2\text{O}$$

phenol, $pK_a = 10.00$ sodium phenoxide $pK_a = 15.74$

Recall that phenoxides are useful in the preparation of aryl alkyl ethers.

$$\text{ArO}^- + \text{RX} \xrightarrow{S_N2} \text{ArOR} + \text{X}^-$$

a phenoxide   *a 1° alkyl*   *an ether*
ion   *halide*

The degree of ionization of a weak acid is determined by the relative stabilities of the un-ionized compound and the anion:

$$HA \rightleftharpoons H^+ + A^-$$

If A$^-$ is stabilized relative to HA, acidity is increased.

The reason for the greater acidity of phenols compared to that of alcohols is that the product anion is *resonance-stabilized*, with the negative charge delocalized by the aromatic ring.

Resonance structures for the phenoxide ion:

The negative charge in an alkoxide ion (RO$^-$) cannot be delocalized. Therefore, an alkoxide ion is of higher energy relative to the alcohol, and alcohols are not as strong acids as phenols.

Because of its acidity, phenol was originally called *carbolic acid*. In the 1800s, the British surgeon Joseph Lister urged that phenol be used as a hospital antiseptic. Prior to that time, no antiseptics were used because it was thought that odors, not microorganisms, were the cause of infection. As an antiseptic, phenol itself has been replaced by less irritating compounds. Interestingly, many modern antiseptics still contain phenolic groups.

hexachlorophene

banned for most uses because it can be absorbed through the skin

n-hexylresorcinol

Study Problems

12.3 What are the principal ions in solution when the following reagents are mixed?
(a) sodium ethoxide and phenol   (b) sodium phenoxide and ethanol

12.4 The pK_a of *p*-nitrophenol is 7.15. Use resonance structures to explain why this compound is more acidic than phenol.

p-nitrophenol

B. Esterification of Phenols

Esterification of phenol does not involve cleavage of the strong C—O bond of the phenol, but depends on the cleavage of the OH bond. (We will discuss the mechanism of esterification reactions in Chapters 14 and 15.) Therefore, esters of phenols can be synthesized by the same reactions that lead to alkyl esters.

Although a carboxylic acid could be used in the esterification of a phenol, yields are usually low. Better results are obtained if a more reactive derivative of a carboxylic acid is used instead of the acid itself. Acetic anhydride is a reactive derivative that leads to acetate esters (Section 15.4C).

acetic anhydride   phenol   phenyl acetate

an ester

Study Problem

12.5 Suggest syntheses for the following compounds:

(a)   (b)

C. The Kolbe Reaction

Because the OH group is such a powerful ring activator in electrophilic substitution reactions, phenols and phenoxides react with weak electrophiles. The **Kolbe reaction** is the reaction of sodium phenoxide with CO_2 to yield sodium salicylate. This salt yields salicylic acid when acidified.

salicylate ion

salicylic acid

Salicylic acid is used to synthesize acetylsalicylic acid, commonly called *aspirin.*

$$\text{(structure: salicylic acid)} -CO_2H + (CH_3\overset{O}{\overset{\|}{C}})_2O \longrightarrow \text{(structure: aspirin)} -CO_2H + CH_3\overset{O}{\overset{\|}{C}}OH$$

acetic anhydride

acetylsalicylic acid
(aspirin)

D. The Reimer–Tiemann Reaction

Another interesting reaction of phenol is its reaction with chloroform in aqueous base, followed by treatment with aqueous acid. The product is salicylaldehyde. This reaction is called the **Reimer–Tiemann reaction**.

$$\text{(phenol)} -OH \quad \xrightarrow[\text{(2) } H_2O, H^+]{\text{(1) } CHCl_3, OH^-, 70°} \quad \text{(salicylaldehyde)}$$

salicylaldehyde

The first step in this reaction is the formation of the very reactive intermediate dichlorocarbene from the action of the base on chloroform. (This type of reaction has been discussed in Section 10.12.)

$$HÖ:^- + H\overset{\frown}{-}CCl_3 \underset{}{\overset{-H_2O}{\rightleftharpoons}} \;^-:CCl_3 \xrightarrow{-Cl^-} \quad :CCl_2$$

dichlorocarbene

The carbon in $:CCl_2$ contains only six valence electrons; thus, $:CCl_2$ can act as an electrophile in aromatic substitution.

$$\text{(phenoxide)} \overset{..}{O}:^- + :CCl_2 \longrightarrow \left[\text{(intermediate)} \right] \longrightarrow \text{(o-(dichloromethyl)phenoxide ion)}$$

o-(dichloromethyl)phenoxide ion

The (dichloromethyl)phenoxide ion contains two chlorides in a benzylic position. Recall that benzylic halides undergo rapid S_N1 and S_N2 reactions with OH^-. The dichloro compound undergoes just such a substitution to yield an unstable 1,1-chlorohydrin that subsequently loses H^+ (to the base) and Cl^-.

$$\text{(CHCl}_2\text{-phenoxide)} \xrightarrow[-Cl^-]{^-OH} \text{(1,1-chlorohydrin)} \xrightarrow[\substack{-H_2O \\ -Cl^-}]{^-OH} \text{(salicylaldehyde anion)}$$

a 1,1-chlorohydrin

Acidification of the alkaline solution yields salicylaldehyde.

salicylaldehyde

E. Oxidation of Phenols

Phenol itself resists oxidation because formation of a carbonyl group at a ring carbon would lead to loss of aromatic stabilization. However, 1,2- and 1,4-dihydroxybenzenes, called **hydroquinones**, can be oxidized to **quinones**. The oxidation proceeds with very mild oxidizing agents, such as Ag^+ or Fe^{3+}, and is readily reversible. Although simple hydroquinones are colorless, quinones are colored.

$+ 2H^+ + 2e^-$

hydroquinone   1,4-benzoquinone
(quinone)

$+ 2H^+ + 2e^-$

catechol   1,2-benzoquinone

The ability of hydroquinone to reduce silver ions to silver metal is the chemical basis of photography. Silver ions in a silver halide crystal that has been exposed to light are more easily reduced than the silver ions of an unexposed crystal. Hydroquinone in the developer fluid reduces these light-activated silver ions at a faster rate than the nonexposed silver ions. In the fixing process, unreacted silver halide is converted to a water-soluble silver complex by sodium thiosulfate, $Na_2S_2O_3$ (called *hypo*), and washed from the film. The result is the familiar photographic negative.

Bombardier beetles rely on the easy oxidation of hydroquinones as a defensive mechanism against spiders, mice, and frogs. These beetles contain glands that store H_2O_2 and other glands that store hydroquinones. When the beetle is threatened, secretions from the two types of glands are mixed together, along with enzymes that catalyze the oxidation of the hydroquinones. The products are irritating quinones and O_2 (which acts as a propellant) that are squirted at the predator.

Section 12.3
Benzenediazonium Salts

A. Preparation and Reactivity of Benzenediazonium Chloride

Arylamines such as aniline can be prepared in the laboratory by the nitration of aromatic compounds followed by reduction of the nitro group. A mixture of iron filings and concentrated HCl is a common reducing agent for this reaction. Because the reaction is run in acid, a *protonated* amine is the product. Subsequent treatment with base generates the amine itself.

$$\text{(benzene)} \xrightarrow[\text{H}_2\text{SO}_4]{\text{HNO}_3} \text{(C}_6\text{H}_5)-\text{NO}_2 \xrightarrow[\text{HCl}]{\text{Fe}} \text{(C}_6\text{H}_5)-\overset{+}{\text{N}}\text{H}_3 \ \text{Cl}^- \xrightarrow{\text{OH}^-} \text{(C}_6\text{H}_5)-\text{NH}_2$$

Aniline and other arylamines undergo reaction with cold nitrous acid (HONO) in HCl solution to yield **aryldiazonium chlorides** ($\text{ArN}_2^+ \ \text{Cl}^-$). The nitrous acid is usually generated *in situ* by the reaction of sodium nitrite ($\text{Na}^+ \ ^-\text{ONO}$) with HCl. Diazonium salts are very reactive. Therefore, the reaction mixture must be chilled. (Alkyldiazonium salts, $\text{RN}_2^+ \ \text{X}^-$, are not stable even in the cold; see Section 17.9.)

$$\text{(C}_6\text{H}_5)-\text{NH}_2 + \text{NaNO}_2 + \text{HCl} \xrightarrow{0°} \text{(C}_6\text{H}_5)-\overset{+}{\text{N}}\equiv\text{N} \ \text{Cl}^- + \text{H}_2\text{O}$$

aniline   benzenediazonium
chloride

The high reactivity of diazonium salts arises from the excellent leaving ability of nitrogen gas, N_2. Because of this leaving ability, the diazonium group may be displaced by a variety of nucleophiles, such as I^-. Some of the substitution reactions that we will present are thought to proceed by a radical mechanism. In other cases, the substitution reaction proceeds through an aryl cation, by a mechanism similar to that of an $\text{S}_\text{N}1$ reaction.

$$\text{(C}_6\text{H}_5)-\overset{+}{\text{N}}\equiv\text{N:} \xrightarrow{-\text{N}_2} \left[\text{(C}_6\text{H}_5)}^+ \right] \xrightarrow{\text{Nu:}^-} \text{(C}_6\text{H}_5)-\text{Nu}$$

In these displacement reactions, the diazonium salt is generally prepared (but not isolated), the nucleophilic reagent is added, and the mixture is allowed to warm or is heated. Yields of substitution products are generally good to excellent: 70–95% from the starting arylamine.

B. Reactions of Benzenediazonium Salts

Displacement by halide ions and cyanide ions. The diazonium group is readily displaced as N_2 by halide ions (F^-, Cl^-, Br^-, or I^-) or the cyanide ion (^-CN). These reactions provide synthetic routes to aryl fluorides, iodides, and nitriles (ArCN), none of which can be obtained by direct electrophilic substitution. Although aryl bromides and chlorides can be synthesized by electrophilic substitution reactions, the products are often contaminated with disubstituted by-products. Diazonium displacement yields pure monochloro or monobromo compounds, uncontaminated by disubstituted products.

For the substitution of —Cl, —Br, or —CN, a copper(I) salt is used as the source of the nucleophile, and the reaction mixture is warmed to 50–100°. (The Cu^+ ion acts as a catalyst in these reactions.) This reaction, using the copper(I) salt, is called the **Sandmeyer reaction** after the Swiss chemist Traugott Sandmeyer, who first reported it in 1884.

$$C_6H_5{-}NH_2 \xrightarrow[0°]{\text{NaNO}_2 \text{ HCl}} C_6H_5{-}N_2^+ \text{ Cl}^-$$

$$\xrightarrow{\text{CuCl}} C_6H_5{-}Cl + N_2 \quad \text{chlorobenzene}$$

$$\xrightarrow{\text{CuBr}} C_6H_5{-}Br + N_2 \quad \text{bromobenzene}$$

$$\xrightarrow{\text{CuCN + KCN}} C_6H_5{-}CN + N_2 \quad \text{benzonitrile}$$

Aryl iodides and fluorides are prepared with KI (no catalyst needed) or fluoroboric acid (HBF_4), respectively.

$$C_6H_5{-}N_2^+ \text{ Cl}^- \xrightarrow{\text{KI}} C_6H_5{-}I \quad \text{iodobenzene (76\%)}$$

$$\xrightarrow{\text{HBF}_4} C_6H_5{-}N_2^+ \text{ BF}_4^- \xrightarrow{\text{heat}} C_6H_5{-}F \quad \text{fluorobenzene (56\%)}$$

isolated and dried

One use of these diazonium reactions is the preparation of compounds that cannot be readily synthesized by other routes. For example, consider the synthesis of *p*-toluic acid (*p*-methylbenzoic acid). The usual route to an aryl carboxylic acid is the oxidation of an alkylarene (R—Ar). Oxidation cannot be used for the preparation of *p*-toluic acid from *p*-xylene (*p*-dimethylbenzene) because *both* methyl groups would be oxidized to carboxyl groups. However, *p*-toluic acid can be obtained by way of a diazonium intermediate, followed by treatment with CuCN–KCN and subsequent hydrolysis of the —CN group to —CO_2H, a reaction that will be discussed in Section 15.11.

$$CH_3{-}C_6H_4{-}NH_2 \xrightarrow[0°]{\text{NaNO}_2 \text{ HCl}} CH_3{-}C_6H_4{-}N_2^+ \text{ Cl}^- \xrightarrow{\text{CuCN + KCN}}$$

p-toluidine

$$CH_3{-}C_6H_4{-}CN \xrightarrow[\text{(2) HCl}]{\text{(1) NaOH, H}_2\text{O, heat}} CH_3{-}C_6H_4{-}CO_2H$$

p-tolunitrile

p-toluic acid
(59% overall)

Displacement by —OH. Phenols can be prepared from diazonium salts by reaction with hot aqueous acid. This reaction provides one of the few laboratory routes to phenols.

m-nitrophenol (86%)

Displacement by —H. The —N_2^+ group may be replaced with —H by treatment of the diazonium salt with hypophosphorous acid, H_3PO_2. This sequence provides a method of *removal of an NH₂ group from an aromatic ring.*

benzene

Sample Problem

Show how you would synthesize 1,3,5-tribromobenzene from aniline. (Remember that bromine is an *o,p*-director in electrophilic substitution reactions.)

Solution: Recall from page 488 that aniline undergoes tribromination readily. Therefore, the following reaction sequence will yield the desired product:

1,3,5-tribromobenzene

Coupling reactions. Coupling reactions of aryldiazonium salts are used to prepare dyes from aniline and from substituted anilines. In these reactions, the diazonium ion acts as an *electrophile*. Resonance structures for the diazonium ion show that each nitrogen carries a partial positive charge.

Figure 12.1 Reactions of an aryldiazonium chloride.

The terminal nitrogen attacks predominantly the *para*-position of an activated benzene ring (one substituted with an electron-releasing group like NH_2 or OH). The coupling product contains an **azo group** (—N=N—) and is generally referred to as an **azo compound**. Many azo compounds are used as dyes.

p-hydroxyazobenzene

orange

methyl red (63%)

The reactions of aryldiazonium salts are summarized in Figure 12.1.

Study Problem

12.6 How would you make the following conversions? In (c), use no aromatic reagent other than benzene.

(a)

(b) CH_3—⟨benzene ring⟩—NO_2 ⟶ CH_3—⟨benzene ring⟩—Br

(c) ⟨benzene ring⟩ ⟶ ⟨benzene ring⟩—$N{=}N$—⟨benzene ring⟩—OCH_3

Section 12.4
Halobenzenes and Nucleophilic Aromatic Substitution

In Chapter 5, we mentioned that aryl halides do not undergo the substitution and elimination reactions characteristic of alkyl halides.

⟨benzene ring⟩—X + Nu:⁻ ⟶ no S_N1 or S_N2 reaction

an aryl halide

Under certain circumstances, however, an aryl halide can undergo a **nucleophilic aromatic substitution reaction**.

⟨benzene ring⟩—X + Nu:⁻ ⟶ ⟨benzene ring⟩—Nu + X⁻

Although this reaction appears to be similar to an S_N1 or S_N2 reaction, it is quite different. It is also different from *electrophilic* substitution, which is initiated by E^+, not Nu:⁻.

The reactivity of aryl halides toward nucleophiles is *enhanced by the presence of electron-withdrawing substituents on the ring*. This is the opposite of a substituent's effect in electrophilic aromatic substitution. Here, an electron-withdrawing substituent makes the ring less rich in electrons and more attractive to an attacking nucleophile.

⟨benzene ring⟩—Cl $\xrightarrow[\substack{350° \\ 300\ atm}]{10\%\ NaOH}$ ⟨benzene ring⟩—OH

chlorobenzene   phenol

increasing reactivity toward Nu:⁻

O_2N—⟨benzene ring⟩—Cl $\xrightarrow[160°]{15\%\ NaOH}$ O_2N—⟨benzene ring⟩—OH

p-chloronitrobenzene   *p*-nitrophenol

O_2N—⟨benzene ring with NO_2 top and NO_2 bottom⟩—Cl $\xrightarrow[warm]{H_2O}$ O_2N—⟨benzene ring with NO_2 top and NO_2 bottom⟩—OH

1-chloro-2,4,6-trinitrobenzene

2,4,6-trinitrophenol (picric acid)

an explosive

Other nucleophiles besides OH^- and H_2O can also undergo reaction with activated aryl chlorides.

1-chloro-2,4-dinitrobenzene   2,4-dinitroaniline (76%)

The mechanism of nucleophilic aromatic substitution is thought to proceed by one of two routes. The first is through a carbanion intermediate. If the ring is activated toward nucleophilic substitution by an electron-withdrawing group, the reaction proceeds by a two-step mechanism: (1) addition of the nucleophile to form a carbanion (stabilized by the electron-withdrawing group), and (2) subsequent loss of the halide ion.

carbanion intermediate

Although the carbanion intermediate is unstable and reactive, it is stabilized to some extent by resonance and by dispersal of the negative charge by the electron-withdrawing group. Resonance structures of this intermediate show that an electron-withdrawing *o*- or *p*-substituent lends more stability to the carbanion intermediate than does a *meta*-substituent. When the substituent is *o*- or *p*-, the negative charge is adjacent to the electron-withdrawing group in one resonance structure. In the case of the nitro group, added stability is gained because the nitro group helps disperse the negative charge by resonance.

Resonance structures for p-substituted intermediate:

added stabilization

For m-substituted intermediate (no added stabilization):

Study Problem

12.7 Draw resonance structures for the carbanion intermediate in the reaction of sodium hydroxide and *o*-chloronitrobenzene.

If there is no electron-withdrawing substituent on the ring, nucleophilic aromatic substitution is very difficult and proceeds by the second mechanistic sequence. In these cases, the reaction is thought to proceed by way of a **benzyne** intermediate.

The structure of benzyne is not like that of an alkyne. The two ring carbons that have a triple bond between them are joined by an sp^2–sp^2 sigma bond and p–p overlap (the aromatic pi cloud). These two bonds are the same as in benzene. The third bond is the side-to-side overlap of two sp^2 orbitals, the ones that originally were used in the bonds to H and X. Figure 12.2 shows this overlap. Because of the rigid geometry of the ring and the unfavorable angles of normal sp^2 orbitals, this overlap cannot be very good. The new bond is very weak, and benzyne is a highly reactive intermediate.

The formation of the benzyne intermediate in the reaction of chlorobenzene with $NaNH_2$ in liquid NH_3 is thought to occur by the following route:

Benzyne undergoes rapid addition of a nucleophile to yield a carbanion that can abstract a proton from NH_3 to yield the product (aniline, here) and another NH_2^- ion.

Figure 12.2 The bonding in the benzyne intermediate.

Figure 12.3 Summary of aromatic nucleophilic substitution reactions. (A bond to the center of a benzene ring denotes an unspecified position of substitution.)

One interesting reaction of a benzyne intermediate is its Diels–Alder reaction with anthracene (its structure is over the arrow in the following equation) to yield an unusual compound called triptycene.

Figure 12.3 summarizes the aromatic nucleophilic substitution reactions of halobenzenes.

Study Problem

12.8 In an experiment designed to test the existence of benzyne as an intermediate in nucleophilic aromatic substitution, chlorobenzene was treated with sodamide ($NaNH_2$) in liquid ammonia. The chlorobenzene was labeled with carbon-14 at the C—Cl position.

(a) If this reaction proceeds in accordance with the benzyne mechanism, what would be the distribution of carbon-14 in the product aniline—that is, at what positions and in what percentages would carbon-14 be found?

(b) What would be the distribution of carbon-14 if the reaction proceeds by simple nucleophilic displacement?

▼ ## Section 12.5
Syntheses Using Substituted Benzene Compounds

Table 11.7 (page 494) and Table 12.1 show some possible applications of benzene and its derivatives in synthesis. In the laboratory, electrophilic aromatic

Table 12.1 Summary of substitution and substituent reactions of substituted benzenes

Reaction	Product	Section References
Electrophilic Substitution:		
$C_6H_5Y + E^+ \longrightarrow$ o- and p-products		11.10
$C_6H_5Z + E^+ \longrightarrow$ m-products		11.10
where Y is e^--releasing and Z is e^--withdrawing		
Nucleophilic Substitution:		
$C_6H_5X + OH^- \longrightarrow C_6H_5OH$	**phenol**[a]	12.4
$C_6H_5X + NH_3 \longrightarrow C_6H_5NH_2$	**aniline**[a]	12.4
Substituent Reactions:		
$C_6H_5R + MnO_4^- \xrightarrow[\text{catalyst}]{\text{heat}} C_6H_5CO_2H$	**carboxylic acid**	12.1
$C_6H_5CH_2R + Br_2$ or NBS $\longrightarrow C_6H_5CHBrR$	**benzylic bromide**	6.5B, 12.1
$C_6H_5OH + OH^- \longrightarrow C_6H_5O^-$	**phenoxide**	12.2A
$C_6H_5OH + (R\overset{O}{\overset{\|}{C}})_2O \longrightarrow C_6H_5O\overset{O}{\overset{\|}{C}}R$	**ester**	12.2B
$C_6H_5OH \xrightarrow[\text{(2) } H_2O, H^+]{\text{(1) } CHCl_3, OH^-}$ (CHO, OH aryl ring)	**o-hydroxy aryl aldehyde**	12.2D
HO—(ring)—OH + [O] $\longrightarrow$ O=(ring)=O	**quinone**	12.2E
Reactions of Aryldiazonium Salts:[b]		
$C_6H_5N_2^+ Cl^- + Nu:^- \longrightarrow C_6H_5Nu$	**substituted (or unsubstituted) benzene**	12.3B
$C_6H_5N_2^+ Cl^- + C_6H_5Y \longrightarrow C_6H_5N{=}NC_6H_4Y$	**azo compound**	12.3B
where Y is OH, NH$_2$, etc.		

[a] Practical in the laboratory only if one or more other electron-withdrawing groups is present on the ring.
[b] The preparation of aryldiazonium salts is discussed in Section 12.3A. Some specific reactions of these salts are summarized in Figure 12.1.

substitutions are more widely used as synthetic reactions than are nucleophilic aromatic substitution reactions because there are fewer limitations on the starting material. (For nucleophilic aromatic substitution, a chemist must either use a ring substituted with electron-withdrawing groups, or use "forcing" conditions, as shown in the equations on page 512.)

When using electrophilic aromatic substitution reactions to prepare substituted benzene compounds, a chemist may need to use ingenuity. For example, in the synthesis of m-chloronitrobenzene, chlorination would be a poor choice as the first step because this reaction places an o, p-director on the ring. Subsequent nitration would give o- and p-chloronitrobenzene, but not the desired m-chloronitrobenzene. The first substitution should be nitration because the nitro group is a *meta*-director. Therefore, in synthetic work with substituted benzenes, the *order of substitution reactions* is important.

o-chloro-
nitrobenzene

p-chloro-
nitrobenzene

m-chloronitrobenzene

Conversion of one group to another may also be necessary. For instance, reduction of the nitro group to an amino group gives a route to *m*-substituted anilines. Benzene may first be nitrated, then subjected to *m*-substitution, and finally the nitro group may be reduced.

an amine salt

m-bromoaniline

Arylamines can also be converted to aryldiazonium salts, making possible a diversity of substitution products, as was shown in Figure 12.1, page 511. In viewing aromatic compounds from a synthetic standpoint, consider every nitro group as a potential diazonium group.

Knowledge of individual chemical characteristics of aromatic compounds is often required for successful solution of synthetic problems. For example, aniline does not undergo Friedel–Crafts reactions because an amino group (a basic group) undergoes reaction with Lewis acids to yield a strongly deactivating group.

a Lewis acid

aniline

a base

An amino group is an *o*, *p*-director, but an ammonium group ($-NR_3^+$) is an *m*-director, and deactivating.

Like aniline, nitrobenzene also does not undergo Friedel–Crafts reactions. In this case, lack of reaction is due to deactivation of the ring by the electron-withdrawing nitro group.

nitrobenzene

Like the diazonium group, the sulfonic acid group is easily removed from an aromatic ring. and therefore can be displaced by a variety of reagents. Phenols, for example, can be prepared from arylsulfonic acids as well as from diazonium salts.

benzenesulfonic   *as steam*   benzene
acid

phenol

p-hydroxybenzene-   2,4,6-tribromophenol
sulfonic acid

Study Problem

12.9 Show how you would synthesize the following compounds from benzene: **(a)** *m*-chloroaniline; **(b)** *n*-pentylbenzene; **(c)** 1,2-diphenylethane; **(d)** triphenylmethane.

▼ Summary

Alkylbenzenes contain a benzylic position that is active toward many reagents, such as oxidizing agents and free-radical halogenating agents.

Phenols are more acidic than alcohols and react with aqueous sodium hydroxide to yield sodium phenoxides.

$$ArOH + Na^+ \ ^-OH \longrightarrow ArO^- \ Na^+ + H_2O$$

Phenols can be *esterified* with a reactive esterification agent, such as a carboxylic acid anhydride (see page 505).

A phenoxide can undergo reaction with CO_2 to yield a phenolic carboxylic acid (**Kolbe reaction**). Treatment of a phenol with chloroform ($CHCl_3$) and base, followed by acidification, yields a phenol-aldehyde (**Reimer-Tiemann reaction**).

Hydroquinones (1,2- and 1,4-dihydroxybenzenes) undergo reversible oxidation to **quinones**.

Arylamines are converted to **aryldiazonium salts** by reaction with HONO. These salts are stable at 0°, but are highly reactive toward a variety of nucleophiles (see Figure 12.1, page 511).

$$ArNH_2 \xrightarrow[\substack{HCl \\ 0°}]{NaNO_2} ArN_2^+ \ Cl^- \xrightarrow{Nu:^-} ArNu + N_2 + Cl^-$$

Halobenzenes do not undergo S_N1 or S_N2 reactions, but X^- can be displaced in **nucleophilic aromatic substitution reactions**, especially if the ring is activated by electron-withdrawing groups such as NO_2.

Study Problems*

12.10 Predict the products.

(a)

(b) $CH_3-\!\!\!\bigcirc\!\!\!-NO_2 + Br_2 \xrightarrow{light}$

(c) $CH_3-\!\!\!\bigcirc\!\!\!-NO_2 + Na_2Cr_2O_7 \xrightarrow[heat]{H_2O, H^+}$

(d)

12.11 TNT (2,4,6-trinitrotoluene) is treated with an excess of concentrated sulfuric acid and sodium dichromate. One product is obtained in 69% yield. What is the structure of this product?

12.12 Write equations for the expected reactions of 3,5-dimethylphenol with the following reagents. If little or no reaction is expected, write "no reaction."

(a) Na	**(b)** NaOH	**(c)** HI
(d) H_2SO_4, heat	**(e)** NaH	**(f)** $SOCl_2$
(g) NaCN	**(h)** PCl_3	**(i)** $(CH_3O)_2SO_2$, NaOH

---

*For information concerning the organization of the *Study Problems* and *Additional Problems*, see the *Note to student* on page 39.

12.13 Write formulas for the principal product(s) formed by the reaction of *p*-cresol (*p*-methylphenol) with each of the following reagents:

(a) CO_2, NaOH   (b) $CHCl_3$, OH^- (c) $KMnO_4$, heat
(d) HNO_3, cold   (e) Br_2, H_2O

12.14 *Anisic acid*, the oxidation product of a licorice flavor ingredient, can be synthesized from *p*-nitrotoluene by the following sequence of reactions. Fill in the blanks.

$$1.0 \text{ mol } p\text{-nitrotoluene} \xrightarrow{H_2CrO_4} \underline{\text{(a)}} \xrightarrow[\text{(2) NaOH}]{\text{(1) Fe, HCl}} \underline{\text{(b)}} \xrightarrow[\text{(2) } H_2O, \text{ heat}]{\text{(1) NaNO}_2, \text{ HCl}}$$

$$\underline{\text{(c)}} \xrightarrow[\substack{\text{(2) 1.0 mol CH}_3\text{I} \\ \text{(3) H}^+}]{\text{(1) NaOH}} \underline{\text{(d)}}$$

12.15 β-Resorcylic acid can be prepared in 60% yield by the following reaction sequence. What is the structure of β-resorcylic acid?

$$\xrightarrow[\text{(2) conc. HCl}]{\text{(1) } CO_2, \text{ KHCO}_3, H_2O, \text{ heat}} \beta\text{-resorcylic acid}$$

m-dihydroxybenzene
(resorcinol)

12.16 Suggest a synthesis for the following compound from benzene and any other reagents needed.

12.17 Complete the following equations. If no reaction occurs write "no reaction."

(a) H_3C—⟨○⟩—$\overset{+}{N}H_3$ Cl^- + Fe + HCl $\longrightarrow$

(b) H_3C—⟨○⟩—$\overset{+}{N}_2$ Cl^- + CuCl $\longrightarrow$

(c) H_3C—⟨○⟩—$\overset{+}{N}_2$ Cl^- + H_3PO_2 $\longrightarrow$

(d) H_3C—⟨○⟩—$\overset{+}{N}_2$ Cl^- $\xrightarrow[\text{(2) heat}]{\text{(1) HBF}_4}$

12.18 Predict the major organic products:

(a) *p*-ethylaniline + $NaNO_2$ + HCl $\xrightarrow{0°}$
(b) β-naphthylamine (page 743) + $NaNO_2$ + HCl $\xrightarrow{0°}$
(c) [the product from (a)] + CuCN $\xrightarrow[\text{KCN}]{100°}$
(d) [the product from (b)] + phenol $\xrightarrow{100°}$

12.19 How would you make the following conversions?

(a) [structure: benzene ring with Cl and NH₂] ⟶ [structure: benzene ring with Cl and Br]

(b) HO—[benzene ring]—NH₂ ⟶ HO—[benzene ring]—I

(c) CH₃—[benzene ring with NO₂ and NO₂] ⟶ CH₃—[benzene ring with NH₂ and NH₂]

(d) [purine structure with H₂N and CH₃] ⟶ [purine structure with F and CH₃]

(e) CH₃—[benzene ring with Br and NH₂] ⟶ CH₃—[benzene ring with Br and OH]

12.20 Complete the following equations:

(a) O₂N—[benzene ring with NO₂]—Cl + ⁻OCH₃ $\xrightarrow{heat}$

(b) O₂N—[benzene ring with NO₂]—Cl + H₂NNH₂ $\xrightarrow{heat}$

(c) O₂N—[benzene ring with NO₂]—Cl + NaI ⟶

(d) O₂N—[benzene ring]—Cl + [benzene ring]—O⁻ K⁺ ⟶

12.21 2,4-Dinitrofluorobenzene is used in structure determinations of polypeptides (small proteins). This compound undergoes reaction with the free amino group at one end of a peptide chain. Write a mechanism for its reaction with glycylglycine.

$$H_2NCH_2\overset{\overset{\displaystyle O}{\|}}{C}NHCH_2CO_2H$$

glycylglycine

12.22 Using benzene or toluene as a starting material, show how you could prepare the following compounds:

(a) ethylbenzene   (b) *m-n*-butylnitrobenzene
(c) styrene   (d) *p*-bromotoluene

12.23 How would you make the following conversions?

(a) O₂N—[benzene ring with NO₂]—Cl ⟶ O₂N—[benzene ring with NO₂]—SCH₂—[benzene ring]

(b)

H$_3$C — (ring with NH$_2$) — H$_3$C $\longrightarrow$ H$_3$C — (ring with OH) — H$_3$C

(c) H$_3$C —(ring)— NH$_2$ $\longrightarrow$ H$_3$C —(ring)— N=N —(ring)— NH$_2$

(d) H$_3$C —(ring)— NH$_2$ $\longrightarrow$ H$_3$C —(ring)— I

(e) benzene $\longrightarrow$ C$_6$H$_5$CN

(f) CH$_3$ —(ring with Br)— CH$_3$ $\longrightarrow$ CH$_3$ —(ring with CH$_2$CH$_2$CH$_3$)— CH$_3$

12.24 Suggest a procedure by which each of the following conversions could be carried out:

(a) phenol to cyclohexanol
(b) phenol to *p*-aminophenol
(c) phenol to phenyl propanoate
(d) nitrobenzene to salicylaldehyde

Additional Problems

12.25 Predict the major organic products:

(a) ethylbenzene + Cl$_2$ $\xrightarrow{\text{FeCl}_3}$
(b) ethylbenzene + Br$_2$ $\xrightarrow{hv}$
(c) styrene + MnO$_4^-$ $\xrightarrow{\text{heat}}$
(d) toluene + 1-chloropropane $\xrightarrow{\text{AlCl}_3}$
(e) toluene + fuming H$_2$SO$_4$ $\longrightarrow$
(f) styrene + excess H$_2$ $\xrightarrow[200°, 100 \text{ atm}]{\text{Ni}}$

12.26 Suggest syntheses for the three isomeric mononitrobenzoic acids from toluene.

12.27 How would you synthesize each of the following compounds from benzene (assuming that you can separate *o*-, *m*-, and *p*-isomers)?

(a) phenylcyclohexane   (b) 1-chloro-2-methyl-2-phenylpropane
(c) anisole (C$_6$H$_5$OCH$_3$)   (d) terephthalic acid (*p*-HO$_2$C—C$_6$H$_4$—CO$_2$H)
(e) *m*-dibromobenzene   (f) *m*-nitrophenol
(g) 2-phenyl-2-propanol   (h) 1,1-diphenylethane
(i) benzyl methyl ether

12.28 What ions are formed when triphenylmethanol is treated with concentrated H$_2$SO$_4$? Suggest a reason for the formation of these ions. Write a mechanism, showing resonance structures if necessary.

12.29 When 1-phenylpropene is treated with HBr in the absence of peroxides, the product is 1-bromo-1-phenylpropane. However, when a peroxide is present, the reaction yields 2-bromo-1-phenylpropane. Explain.

12.30 (a) Suggest a reason for the fact that aniline (with a pK_b of 9.37) is much less basic than cyclohexylamine, C$_6$H$_{11}$NH$_2$ (pK_b = 3.34).
(b) Would you expect *p*-nitroaniline to be more or less basic than aniline? Why?

12.31 Suggest syntheses for the following compounds, starting with benzene, toluene, or phenol and other appropriate reagents. (Assume that you can separate o-, m-, and p-isomers.)

(a) 2,4,6-trichlorophenyl acetate   (b) bromodiphenylmethane
(c) p-nitroacetophenone   (d) 1-phenylcyclohexene

(e)

12.32 Suggest a synthetic pathway for the following conversion:

12.33 Upon reaction with potassium nitrate and sulfuric acid, m-dichlorobenzene yields a dichlorodinitrobenzene in 70% yield. This latter compound, when heated to 140° with ammonia, yields a new product in 88% yield. Write equations for these reactions.

12.34 In the industrial synthesis of cresols (methylphenols), heating p-chlorotoluene with NaOH yields the m- and p-cresols in nearly equal amounts. Using an equation, explain this result.

12.35 Using resonance structures, explain why the bond dissociation energy of the carbon-chlorine bond of chlorobenzene is about 15 kcal/mol greater than that of an alkyl chloride, such as 2-chloropropane.

12.36 A nitrogen atom in an aromatic ring is analogous to a nitro group in its ability to aid nucleophilic aromatic substitution. 2-Chloropyrimidine, for example, reacts rapidly with dimethylamine to yield 2-(dimethylamino)pyrimidine.

Write resonance structures for the intermediate(s) to explain this reactivity.

12.37 (a) The herbicide (2,4,5-trichlorophenoxy)acetic acid (called 2,4,5-T), a component of the defoliant Agent Orange, is synthesized by heating a tetrachlorinated benzene with methanolic NaOH, followed by reaction with $ClCH_2CO_2Na$ and acidification. Write equations for these reactions.

(b) A by-product of this synthesis is 2,3,7,8-tetrachlorodibenzodioxin (often called 2,3,7,8-TCDD, or simply *dioxin*), which is highly toxic and not readily biodegradable. Suggest a mechanism for the formation of 2,3,7,8-TCDD.

2,4,5-T   2,3,7,8-TCDD

12.38 A compound C_6H_5Cl was treated with HNO_3 and H_2SO_4. The products of this reaction were then treated with Fe and HCl, followed by neutralization. The proton nmr spectrum of one of the products of this reaction is given in Figure 12.4. What is the structure of this component? What would be the structures of the other components of the product mixture?

Figure 12.4 ^1H nmr spectrum for Problem 12.38.

12.39 One of the isomeric cresols (methylphenols) was treated with NaOH and iodomethane. The proton nmr spectrum of the product is shown in Figure 12.5. What are the structures of the cresol and the product?

Figure 12.5 ^1H nmr spectrum for Problem 12.39.

12.40 Upon treatment with HNO_3, a monosubstituted aromatic compound yielded two isomeric products, A and B. Treatment of A with NaOH followed by CH_3I gave C. In an
identical manner, compound B yielded compound D. The infrared spectra of A, B, C, and
D are given in Figure 12.6. Reaction of C with Fe and HCl, followed by treatment with
base, yielded E. Following this same procedure, compound D yielded compound F. The
proton nmr spectra of E and F are given in Figure 12.7. What are the structures of the
original compound, A, B, C, D, E, and F?

Figure 12.6 Infrared spectra for Problem 12.40.

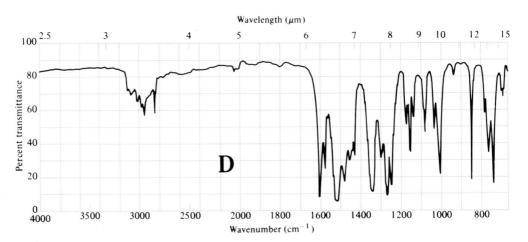

Figure 12.6 (*continued*) Infrared spectra for Problem 12.40.

Figure 12.7 ^{1}H nmr spectra for Problem 12.40.

13 ▼

Aldehydes and Ketones

Aldehydes and ketones are but two of many classes of organic compounds that contain carbonyl groups. A **ketone** has *two alkyl (or aryl) groups* attached to the carbonyl carbon, while an **aldehyde** has at least one hydrogen atom attached to the carbonyl carbon. The other group in an aldehyde (R in the following formulas) can be alkyl, aryl, or H.

$$\underset{\text{\textit{an aldehyde}}}{R-\overset{\overset{\textstyle O}{\|}}{C}-H \;\; \text{or} \;\; RCHO} \qquad \underset{\text{\textit{a ketone}}}{R-\overset{\overset{\textstyle O}{\|}}{C}-R \;\; \text{or} \;\; RCOR}$$

Aldehydes and ketones are common in living systems. The sugar ribose and the female hormone progesterone are two examples of biologically important aldehydes and ketones.

ribose

a carbohydrate

progesterone

a female hormone

Many aldehydes and ketones have distinctive odors. Aldehydes are generally pungent-smelling and ketones are sweet-smelling. For example, *trans*-cinnamaldehyde is the principal component of cinnamon oil, and the enantiomers of carvone are responsible for the odors of caraway seeds and spearmint.

(*E*)-3-phenylpropenal
(*trans*-cinnamaldehyde)

(+)-carvone: spearmint oil
(−)-carvone: caraway-seed oil

the chiral carbon

▼ ▬▬ ## Section 13.1
Nomenclature of Aldehydes and Ketones

In the IUPAC system, the name of an aldehyde is derived from the name of the parent alkane by changing the final **-e** to **-al**. No number is needed; the —CHO group always contains carbon 1.

$$CH_3CH \qquad CH_3CHCH \qquad CH_3CH=CHCH$$
$$\overset{O}{\underset{}{\|}} \qquad \overset{O}{\underset{Cl}{\|}} \qquad \overset{O}{\underset{}{\|}}$$

IUPAC: ethanal   2-chloropropanal   2-butenal

Ketones are named by changing the **-e** of the alkane name to **-one**. The -e is retained in dione names. A number is used where necessary.

cyclohexanone

$$CH_3CCH_2CH_2CH_3 \qquad CH_3CCH_2CCH_3$$

IUPAC: cyclohexanone   2-pentanone   2,4-pentanedione

Trivial names for the common aldehydes and ketones are widely used. Aldehydes are named after the parent carboxylic acids with the *-oic acid* or *-ic acid* ending changed to *-aldehyde*. Table 13.1 lists a few examples.

Table 13.1 Trivial names for some carboxylic acids and aldehydes

Carboxylic Acid		Aldehyde	
$\overset{O}{\overset{\|}{HC}}OH$	formic acid	$\overset{O}{\overset{\|}{HC}}H$	formaldehyde
$CH_3\overset{O}{\overset{\|}{C}}OH$	acetic acid	$CH_3\overset{O}{\overset{\|}{C}}H$	acetaldehyde
$CH_3CH_2\overset{O}{\overset{\|}{C}}OH$	propionic acid	$CH_3CH_2\overset{O}{\overset{\|}{C}}H$	propionaldehyde
$CH_3CH_2CH_2\overset{O}{\overset{\|}{C}}OH$	butyric acid	$CH_3CH_2CH_2\overset{O}{\overset{\|}{C}}H$	butyraldehyde
benzene ring $-\overset{O}{\overset{\|}{C}}OH$	benzoic acid	benzene ring $-\overset{O}{\overset{\|}{C}}H$	benzaldehyde

Propanone is usually called *acetone*, while the other simple ketones are sometimes named by a functional-group name. The alkyl or aryl groups attached to the carbonyl group are named, then the word *ketone* is added.

$$CH_3\overset{O}{\overset{\|}{C}}CH_3 \qquad CH_3\overset{O}{\overset{\|}{C}}CH_2CH_3 \qquad (CH_3)_2CH\overset{O}{\overset{\|}{C}}C(CH_3)_3$$

IUPAC: propanone   butanone   2,2,4-trimethyl-3-pentanone
trivial: acetone   methyl ethyl ketone   isopropyl *t*-butyl ketone

Other positions in a molecule in relation to the carbonyl group may be referred to by Greek letters. The carbon adjacent to the C=O group (or any other functional group for that matter) is called the **alpha** (α) carbon. The next carbon is **beta** (β), then **gamma** (γ), **delta** (δ), and so forth, according to the Greek alphabet. Occasionally, **omega** (ω), the last letter in the Greek alphabet, is used to designate the terminal carbon of a long chain, regardless of the actual length of the chain. The groups (or atoms) attached to an α carbon are called α groups; those attached to the β carbon are called β groups.

$$CH_3CH_2CH_2\overset{O}{\overset{\|}{C}}H \qquad CH_3\overset{O}{\overset{\|}{C}}CH_2\overset{O}{\overset{\|}{C}}CH_3 \qquad CH_3\overset{Br}{\overset{\|}{C}}HCHO$$

      α carbon   *a β-diketone*   *an α-bromoaldehyde*
   β carbon

$$BrCH_2CH_2CH_2CH_2\overset{O}{\overset{\|}{C}}C_6H_5 \qquad Br(CH_2)_{10}\overset{O}{\overset{\|}{C}}CH_3 \qquad Br(CH_2)_{25}\overset{O}{\overset{\|}{C}}CH_2CH_3$$

all are ω-bromoketones

The Greek-letter designations may be used in the trivial names of carbonyl compounds, but *not in the IUPAC names* because this latter practice would be mixing two systems of nomenclature.

$$\text{C}_6\text{H}_5\text{—CH}_2\text{CH}_2\overset{\displaystyle O}{\overset{\|}{\text{CH}}} \qquad \text{CH}_3\overset{\displaystyle O}{\underset{\underset{\displaystyle Br}{|}}{\overset{\|}{\text{CH}}\text{CH}}}$$

| **IUPAC:** | 3-phenylpropanal | 2-bromopropanal |
| **trivial:** | *β*-phenylpropionaldehyde | *α*-bromopropionaldehyde |

▼ ■ Section 13.2

Preparation of Aldehydes and Ketones

In the laboratory, the most common way of synthesizing a simple aldehyde or ketone is by the *oxidation of an alcohol*. Aryl ketones can be prepared by *Friedel–Crafts acylation reactions*. General equations for these reactions are shown in Table 13.2.

$$\text{CH}_3(\text{CH}_2)_5\text{CH}_2\text{OH} + \text{CrO}_3\cdot 2\,\text{N}\bigcirc \longrightarrow \text{CH}_3(\text{CH}_2)_5\overset{\displaystyle O}{\overset{\|}{\text{CH}}}$$

1-heptanol   heptanal (84%)

+ $\text{H}_2\text{CrO}_4 \xrightarrow{60°}$

menthol   menthone (84%)

$(\text{CH}_3)_2\text{CH}\!-\!\bigcirc\!-\!\text{CH}_3 + \text{CH}_3\overset{\displaystyle O}{\overset{\|}{\text{C}}}\text{Cl} \xrightarrow{\text{AlCl}_3}$

p-isopropyltoluene   acetyl chloride   5-isopropyl-2-methylacetophenone (55%)

One of the more important aldehydes, *formaldehyde*, which is used as a reagent and as a preservative for biological specimens, is a gas. However, it is conveniently shipped or stored in water solution (formalin = 37% formaldehyde and 7–15% methanol in H_2O) or as a solid polymer or trimer. Heating any one of these preparations yields the gaseous formaldehyde.

formalin (HCHO + H_2O) ⟍ heat

ɛ—$\text{CH}_2\text{OCH}_2\text{OCH}_2\text{OCH}_2\text{O}$—ɛ $\xrightarrow{\text{heat}}$ $\overset{\displaystyle O}{\overset{\|}{\text{HCH}}}$

paraformaldehyde   heat ↗   methanal
(formaldehyde)

trioxane
mp 62°C

Table 13.2 Summary of laboratory syntheses of aldehydes and ketones[a]

Reaction	Section Reference
Aldehydes:	
$RCH_2OH \xrightarrow[\text{pyridine}]{CrO_3} RCHO$ *1° alcohol*	7.9C
$\underset{\text{acid chloride}}{RCCl} \xrightarrow[\text{(2) } H_2O, H^+]{\text{(1) LiAlH[OC(CH}_3)_3]_3} RCHO$	15.3C
Ketones:	
$R_2CHOH \xrightarrow[H^+]{H_2CrO_4} R_2C{=}O$ *2° alcohol*	7.9C
$RC{\equiv}CR \xrightarrow[Hg^{2+}]{H_2O, H^+} RCCH_2R$ *alkyne*	10.8
$\underset{\text{acid chloride}}{RCCl} \xrightarrow{R'_2CuLi} RCR'$	15.3C
Aryl Ketones:	
$\text{benzene} + RCCl \xrightarrow{AlCl_3} \text{benzene–CR}$	11.9E

[a] Polyfunctional aldehydes and ketones can be prepared by condensation and alkylation reactions, which will be discussed in Chapter 16.

Acetaldehyde, with a boiling point near room temperature (20°), is also conveniently stored or shipped in a cyclic trimer or tetramer form. Acetaldehyde is used as an intermediate in industrial syntheses of acetic acid, acetic anhydride (Section 15.4A), and other compounds.

paraldehyde

a sedative and hypnotic
bp 125°C

metaldehyde

used as snail poison
mp 246°C

$\xrightarrow{\text{heat}}$

$CH_3CH{=}O$

ethanal
(acetaldehyde)

▼
Section 13.3
Physical Properties of Aldehydes and Ketones

The carbonyl group consists of an sp^2-hybridized carbon atom joined to an oxygen atom by a sigma bond and a pi bond. (See Figure 2.21, page 68.) The sigma bonds of the carbonyl carbon lie in a *plane*, with bond angles of approximately 120°. The pi bond joining the C and the O lies above and below the plane of these sigma bonds. The carbonyl group is *polar*, the electrons in the sigma bond, and especially those in the pi bond, being drawn toward the electronegative oxygen. The oxygen of the carbonyl group has *two pairs of unshared valence electrons*. All these structural features—the flatness, the pi bond, the polarity, and the unshared electrons—contribute to the properties and the reactivity of the carbonyl group.

$$\mu = 2.85 \text{ D}$$

Because they are polar, and therefore undergo intermolecular dipole–dipole attractions, aldehydes and ketones have higher boiling points than nonpolar compounds of similar molecular weight (see Table 13.3). To a limited extent, aldehydes and ketones can solvate ions. (For example, NaI is soluble in acetone.)

CH_3	O	OH
CH_3CHCH_3	CH_3CCH_3	CH_3CHCH_3
bp $-12°$	bp $56°$	bp $82.5°$

Because of the unshared electrons on the oxygen, a carbonyl compound can form hydrogen bonds (but not with another carbonyl compound unless it has an acidic hydrogen available for hydrogen bonding).

Table 13.3 Physical properties of some aldehydes and ketones

Trivial Name	Structure	Bp, °C	Solubility in H_2O
Aldehydes:			
formaldehyde	$HCHO$	-21	∞
acetaldehyde	CH_3CHO	20	∞
propionaldehyde	CH_3CH_2CHO	49	16 g/100 mL
butyraldehyde	$CH_3CH_2CH_2CHO$	76	7 g/100 mL
benzaldehyde	C_6H_5CHO	178	slightly
Ketones:			
acetone	CH_3CCH_3	56	∞
methyl ethyl ketone	$CH_3CCH_2CH_3$	80	26 g/100 mL
acetophenone	$C_6H_5CCH_3$	202	insoluble
benzophenone	$C_6H_5CC_6H_5$	306	insoluble

$$:\overset{..}{O}:\text{---}H\text{---}O$$

$$\underset{R\text{---}C\text{---}R}{\overset{\|}{\ }}\quad \underset{H}{\overset{|}{\ }}$$

The result of this ability to form hydrogen bonds is that the aldehydes and ketones of low molecular weight, like alcohols, are soluble in water (Table 13.3). However, since they cannot undergo hydrogen bonding with molecules of their own kind, the boiling points are substantially lower than those of corresponding alcohols.

Section 13.4
Spectral Properties of Aldehydes and Ketones

A. Infrared Spectra

The infrared spectrum is useful in the detection of a carbonyl group in a ketone or an aldehyde. (Characteristic absorption bands are listed in Table 13.4.) However, carbonyl groups are also found in other compounds (carboxylic acids, esters, and so forth). For this reason, the fact that a carbonyl group is present does not mean that an unknown is necessarily an aldehyde or a ketone.

For aldehydes, corroborating evidence can be found in both the infrared and nmr spectra because of the unique absorption by the aldehyde hydrogen. Ketones, unfortunately, cannot be positively identified by spectral methods. The usual procedure is to eliminate the other carbonyl compounds as possibilities. If a carbonyl compound is not an aldehyde, carboxylic acid, ester, amide, etc., it is probably a ketone.

The C=O absorption of both aldehydes and ketones appears somewhere near 1700 cm^{-1} (about 5.8 μm). If the carbonyl group is in conjugation with a double bond or benzene ring, the position of absorption is shifted to a slightly lower frequency (about 1675 cm^{-1}, or 6 μm, for ketones). Figure 13.1 shows the infrared spectra of cyclohexanone (nonconjugated) and 2-cyclohexenone (conjugated).

The CH stretching of the aldehyde group, which shows absorption just to the right of aliphatic CH absorption, is characteristic of an aldehyde. Usually two peaks appear in this region. The two CH peaks of the aldehyde are clearly evident in the spectrum of butanal in Figure 13.2, but the peak closer to the aliphatic CH absorption is often obscured by that absorption.

Table 13.4 Characteristic infrared absorption of aldehydes and ketones

Type of Vibration	Position of Absorption[a]	
	cm^{-1}	*μm*
Aldehydes:		
C—H stretching of —CHO	2700–2900	3.45–3.7
C=O stretching	1700–1740	5.7–5.9
Ketones:		
C=O stretching	1660–1750	5.7–6.0

[a] Other substituents or ring strain cause the carbonyl absorption to fall outside the listed ranges.

Figure 13.1 Infrared spectra of cyclohexanone and 2-cyclohexenone, illustrating the slight shift of C=O absorption to lower frequency (longer wavelength) by conjugation.

Figure 13.2 Infrared spectrum of butanal.

Study Problem

13.1 Match the infrared spectrum in Figure 13.3 to one of the following structures:

Figure 13.3 Infrared spectrum for Problem 13.1.

B. Proton Nmr Spectra

The molecular magnetic field induced by the electrons in a carbonyl group has a profound effect on the nmr absorption of the aldehyde proton. You will recall from Section 9.7B that nmr absorption for an aldehyde proton is shifted far

Figure 13.4 ¹H nmr spectrum of butanal, showing the relative chemical shifts of α, β, and γ protons and the aldehyde proton (at δ9.74).

Figure 13.5 ^{1}H nmr spectrum of 1-phenyl-2-propanone.

downfield ($\delta 9$–10). This large shift arises from the additive effects of both anisotropic deshielding by the pi electrons and inductive deshielding by the electropositive carbon of the carbonyl group.

The α hydrogens of either aldehydes or ketones are not affected to such a large extent by the carbonyl group. The nmr absorption for the α protons ($\delta 2.1$–2.6) appears slightly downfield from that of other CH absorption (about $\delta 1.5$) because of electron withdrawal by the electronegative oxygen atom. The effects of this deshielding by the inductive effect are evident in the nmr spectra of butanal (Figure 13.4) and 1-phenyl-2-propanone (Figure 13.5). In an aldehyde, the splitting of the aldehyde proton can sometimes be used to determine the number of α hydrogens. The spectrum of butanal shows a triplet for the —CHO proton, an indication of two α hydrogens.

Figure 13.6 ^{1}H nmr spectrum for Problem 13.2. (The offset absorption is at $\delta 10.0$.)

Study Problem

13.2 A compound with the molecular formula C_7H_6O has the nmr spectrum shown in Figure 13.6. What is the structure of this compound?

Section 13.5
Addition Reactions of Aldehydes and Ketones

Isolated carbon–carbon double bonds are nonpolar. For reaction, an electrophile is generally needed to attack the pi-bond electrons. However, the carbon–oxygen double bond is polar even without electrophilic attack. A carbonyl compound may be attacked either by a nucleophile or by an electrophile.

Many reactions of carbonyl groups involve an initial protonation of the oxygen. This protonation enhances the positive charge of the carbonyl carbon so that this carbon is more easily attacked by weaker nucleophiles.

major contributor   *minor contributor*

*resonance structures for
a protonated carbonyl group*

Like alkenes, aldehydes and ketones undergo addition of reagents to the pi bond. Many of these addition reactions, especially those with weak nucleophiles, are acid-catalyzed for the foregoing reason.

The relative reactivities of aldehydes and ketones in addition reactions may be attributed partly to the *amount of positive charge on the carbonyl carbon*. A greater positive charge means a higher reactivity. If this partial positive charge is dispersed throughout the molecule, then the carbonyl compound is more stable and less reactive.

The carbonyl group is stabilized by adjacent alkyl groups, which are electron-releasing. A ketone, with two R groups, is about 7 kcal/mol more stable than an aldehyde, with only one R group. Formaldehyde, with no alkyl groups, is the most reactive of the aldehydes and ketones.

$$\underset{\text{a ketone}}{R-\overset{\overset{\displaystyle O}{\|}}{C}-R} \qquad \underset{\text{an aldehyde}}{R-\overset{\overset{\displaystyle O}{\|}}{C}-H} \qquad \underset{\text{formaldehyde}}{H-\overset{\overset{\displaystyle O}{\|}}{C}-H}$$

increasing reactivity →

Sample Problem

List the following aldehydes in terms of increasing reactivity: CH_3CHO, $ClCH_2CHO$, Cl_2CHCHO, Cl_3CCHO.

Solution: Cl is *electron-withdrawing*. The carbonyl carbon becomes *increasingly positive* and *increasingly reactive* as more Cl atoms are added to the α carbon. Therefore, the order of reactivity is as already shown, with CH_3CHO being the least reactive and Cl_3CCHO being the most reactive.

$$\underset{\text{stabilized}}{CH_3 \rightarrow \overset{\delta+}{C}HO} \qquad \underset{\text{destabilized}}{Cl_3C \leftarrow \overset{\delta+}{C}HO}$$

Steric factors also play a role in the relative reactivities of aldehydes and ketones. An addition reaction of the carbonyl group leads to an increase in steric hindrance around the carbonyl carbon.

$$\overset{sp^2,\ less\ hindered}{\underset{R}{\overset{R}{\diagup}}\overset{\overset{\displaystyle O}{\|}}{C}} + CH_3CH_2OH \underset{}{\overset{H^+}{\rightleftharpoons}} \overset{sp^3,\ more\ hindered}{R-\overset{\overset{\displaystyle OH}{|}}{\underset{\underset{\displaystyle OCH_2CH_3}{|}}{C}}-R}$$

Bulky groups around the carbonyl group lead to more steric hindrance in the product (and in the transition state). The product is of higher energy because of steric repulsions. A more hindered ketone is therefore less reactive than an aldehyde or a less hindered ketone. The lack of steric hindrance is another reason that formaldehyde is more reactive than other aldehydes.

$$\underset{\text{an ethyl ketone}}{CH_3CH_2CH_2\overset{\overset{\displaystyle O}{\|}}{C}CH_2CH_3} \qquad \underset{\text{a methyl ketone}}{CH_3CH_2CH_2\overset{\overset{\displaystyle O}{\|}}{C}CH_3} \qquad \underset{\text{an aldehyde}}{CH_3CH_2CH_2\overset{\overset{\displaystyle O}{\|}}{C}H}$$

increasing reactivity →

A. Reaction with Water

Water can add to a carbonyl group to form a 1,1-diol, called a **gem-diol**, or **hydrate**. The reaction is reversible, and the equilibrium generally lies on the carbonyl side.

General:

$$
\underset{\substack{\text{an aldehyde} \\ \text{or ketone}}}{\text{R}-\overset{\overset{\displaystyle O}{\|}}{\text{C}}-\text{R}} + \text{H}_2\text{O} \overset{\text{H}^+}{\rightleftarrows} \underset{\substack{\text{a hydrate} \\ \text{(two OH's on C)}}}{\text{R}-\overset{\overset{\displaystyle \text{OH}}{|}}{\underset{\underset{\displaystyle \text{OH}}{|}}{\text{C}}}-\text{R}}
$$

$$
\underset{\text{acetone}}{\text{CH}_3-\overset{\overset{\displaystyle O}{\|}}{\text{C}}-\text{CH}_3} + \text{H}_2\text{O} \overset{\text{H}^+}{\rightleftarrows} \text{CH}_3-\overset{\overset{\displaystyle \text{OH}}{|}}{\underset{\underset{\displaystyle \text{OH}}{|}}{\text{C}}}-\text{CH}_3
$$

2,2-propanediol
(acetone hydrate)

Stable hydrates are known, but they are the exception rather than the rule. *Chloral hydrate* (a hypnotic and chief ingredient of the "Mickey Finn," or knock-out drops) is an example of a stable solid hydrate. *Formalin* contains a stable aqueous hydrate of formaldehyde.

$$
\underset{\text{chloral}}{\text{Cl}_3\text{C}\overset{\overset{\displaystyle O}{\|}}{\text{C}}\text{H}} + \text{H}_2\text{O} \overset{\text{H}^+}{\rightleftarrows} \underset{\text{chloral hydrate}}{\text{Cl}_3\text{CCH(OH)}_2}
$$

$$
\underset{\text{formaldehyde}}{\text{H}\overset{\overset{\displaystyle O}{\|}}{\text{C}}\text{H}} + \text{H}_2\text{O} \overset{\text{H}^+}{\rightleftarrows} \underset{\text{in formalin}}{\text{HCH(OH)}_2}
$$

Both formaldehyde and chloral are more reactive than most other aldehydes or ketones because the carbonyl carbon in each has a fairly large amount of positive charge. In formaldehyde, there are no alkyl groups to help disperse the positive charge. In chloral, the strongly electron-withdrawing $\text{Cl}_3\text{C}-$ group enhances the positive charge by removing electron density.

Let us compare the equilibrium constants for the hydration reactions of chloral (with a more positive carbonyl carbon) and acetone (with the positive charge more dispersed). The equilibrium constants differ by a factor of 10^7!

$$
\underset{}{\text{Cl}_3\text{C}\overset{\overset{\displaystyle O}{\|}}{\text{C}}\text{H}} + \text{H}_2\text{O} \overset{\text{H}^+}{\rightleftarrows} \text{Cl}_3\text{C}\overset{\overset{\displaystyle \text{OH}}{|}}{\underset{\underset{\displaystyle \text{OH}}{|}}{\text{C}}}\text{H} \qquad K = \frac{[\text{Cl}_3\text{CCH(OH)}_2]}{[\text{Cl}_3\text{CCHO}][\text{H}_2\text{O}]} = 30{,}000
$$

$$
\text{CH}_3\overset{\overset{\displaystyle O}{\|}}{\text{C}}\text{CH}_3 + \text{H}_2\text{O} \overset{\text{H}^+}{\rightleftarrows} \text{CH}_3\overset{\overset{\displaystyle \text{OH}}{|}}{\underset{\underset{\displaystyle \text{OH}}{|}}{\text{C}}}\text{CH}_3 \qquad K = \frac{[\text{CH}_3\text{C(OH)}_2\text{CH}_3]}{[\text{CH}_3\text{COCH}_3][\text{H}_2\text{O}]} = 0.002
$$

Hydrate formation is the reason that aldehydes are more readily oxidized to carboxylic acids in aqueous media than in nonaqueous media.

$$\underset{\text{RCH}}{\overset{\text{O}}{\|}} \underset{\text{H}_2\text{O}}{\overset{\longleftrightarrow}{}} \underset{\underset{\text{H}}{|}}{\overset{\text{OH}}{\underset{|}{\text{RC}-\text{OH}}}} \overset{[\text{O}]}{\longrightarrow} \underset{\text{RCOH}}{\overset{\text{O}}{\|}}$$

easily oxidized

Study Problem

13.3 Which of the following compounds would you predict to form stable hydrates? Explain.

(a) $\text{Cl}_3\text{CCCCl}_3$ with O double-bonded   (b) cyclohexane ring $=\text{O}$ (c) $\text{C}_6\text{H}_5\text{CHO}$

B. Reaction with Alcohols

Like water, an alcohol can add to a carbonyl group. In most cases, the equilibrium lies on the aldehyde or ketone side of the equation, just as in the reaction with water.

The product of addition of *one* molecule of an alcohol to an aldehyde is called a **hemiacetal**, while the product of addition of *two* molecules of alcohol (with the loss of H_2O) is called an **acetal**. (**Hemiketal** and **ketal** are the corresponding terms used for ketone products.) All these reactions are catalyzed by a trace of strong acid.

General:

$$\underset{\text{an aldehyde}}{\underset{\text{RCH}}{\overset{\text{O}}{\|}}} \underset{\text{H}^+}{\overset{\text{R'OH,}}{\underset{\longleftrightarrow}{}}} \underset{\overset{|}{\text{OH}}}{\overset{\text{OR'}}{\overset{|}{\text{RCH}}}} \underset{\text{H}^+}{\overset{\text{R'OH,}}{\underset{\longleftrightarrow}{}}} \underset{\overset{|}{\text{OR'}}}{\overset{\text{OR'}}{\overset{|}{\text{RCH}}}} + \text{H}_2\text{O}$$

a hemiacetal
(OH and OR on C)

an acetal
(two OR's on C)

$$\underset{\text{acetaldehyde}}{\underset{\text{CH}_3\text{CH}}{\overset{\text{O}}{\|}}} \underset{\text{H}^+}{\overset{\text{CH}_3\text{CH}_2\text{OH,}}{\underset{\longleftrightarrow}{}}} \underset{\overset{|}{\text{OH}}}{\overset{\text{OCH}_2\text{CH}_3}{\overset{|}{\text{CH}_3\text{CH}}}} \underset{\text{H}^+}{\overset{\text{CH}_3\text{CH}_2\text{OH,}}{\underset{\longleftrightarrow}{}}} \underset{\overset{|}{\text{OCH}_2\text{CH}_3}}{\overset{\text{OCH}_2\text{CH}_3}{\overset{|}{\text{CH}_3\text{CH}}}} + \text{H}_2\text{O}$$

a hemiacetal

an acetal

Sample Problem

Give the structures of the organic compounds present in a methanol solution of cyclohexanone that contains a trace of HCl.

Solution:

cyclohexanone $=\text{O}$ $\underset{\text{H}^+}{\overset{\text{CH}_3\text{OH,}}{\underset{\longleftrightarrow}{}}}$ cyclohexane with OCH_3 and OH $\underset{\text{H}^+}{\overset{\text{CH}_3\text{OH,}}{\underset{\longleftrightarrow}{}}}$ cyclohexane with OCH_3 and OCH_3

hemiketal

ketal

Study Problems

13.4 Which of the following structures contains a hemiacetal or hemiketal group, and which contains an acetal or ketal group? Circle and identify each group.

(a) (cyclohexane ring with OH and OCH₂CH₂OH substituents)

(b) (spiro dioxolane structure with two O atoms)

(c) (cyclohexane with $-$OCHCH₃ and OH substituents)

13.5 Give the structures of the alcohol and the aldehyde or ketone that are needed to prepare each of the compounds shown in the preceding problem.

The mechanism for the reversible reaction of an aldehyde or a ketone with an alcohol is typical of the mechanisms for many acid-catalyzed addition reactions of carbonyl compounds: a series of protonations and deprotonations of oxygen-containing groups.

Protonation:

$$\underset{\text{an aldehyde}}{\overset{\overset{\displaystyle\ddot{O}{:}}{\|}}{RCH}} \;+\; H^+ \;\rightleftharpoons\; \left[\underset{\text{a protonated aldehyde}}{\overset{\overset{\displaystyle\overset{+}{O}H}{\|}}{RCH}}\right]$$

Attack of R'OH:

$$\left[\underset{}{R{-}\overset{\overset{+}{O}H}{\underset{}{C}}{-}H}\right] + R'\ddot{O}H \rightleftharpoons \underset{\substack{\text{a protonated}\\ \text{hemiacetal}}}{\left[R{-}\overset{:\ddot{O}H}{\underset{\overset{+}{H\underset{\cdot\cdot}{O}R'}}{C}}{-}H\right]} \overset{-H^+}{\rightleftharpoons} \underset{\text{a hemiacetal}}{R{-}\overset{:\ddot{O}H}{\underset{:\ddot{O}R'}{C}}{-}H}$$

In the mechanism for acetal formation from the hemiacetal, again protonation and deprotonation, along with loss of water, are the major reaction steps. Acetal formation from a hemiacetal is therefore a two-step *substitution* of an OR group for an OH group.

Protonation and loss of water:

$$\underset{\text{a hemiacetal}}{R{-}\overset{:\ddot{O}H}{\underset{OR'}{C}}{-}H} \overset{H^+}{\rightleftharpoons} \left[R{-}\overset{\overset{+}{O}H_2}{\underset{:OR'}{C}}{-}H\right] \overset{-H_2O}{\rightleftharpoons} \left[R{-}\overset{}{\underset{\overset{+}{O}R'}{C}}{-}H\right]$$

Attack of R'OH:

$$\left[R{-}\overset{}{\underset{\overset{+}{O}R'}{C}}{-}H\right] \overset{R'\ddot{O}H}{\rightleftharpoons} \left[R{-}\overset{\overset{+}{H\underset{}{O}R'}}{\underset{:\ddot{O}R'}{C}}{-}H\right] \overset{-H^+}{\rightleftharpoons} \underset{\text{an acetal}}{R{-}\overset{:\ddot{O}R'}{\underset{:\ddot{O}R'}{C}}{-}H}$$

In the equilibrium between an aldehyde, a hemiacetal, and an acetal, the aldehyde is generally favored. In an equilibrium mixture, we would usually find a large amount of aldehyde and only small amounts of hemiacetal and acetal. There is one important exception to this generality. A molecule that has an OH group γ or δ (1,4 or 1,5) to an aldehyde or ketone carbonyl group undergoes an intramolecular reaction to form a five- or six-membered hemiacetal ring. *These cyclic hemiacetals are favored over the open-chain aldehyde forms.*

the hemiacetal carbon has OH and OR

94% at equilibrium

The reason that cyclic hemiacetals are important is that glucose and other sugars contain hydroxyl groups γ and δ to carbonyl groups. Sugars, therefore, form cyclic hemiacetals in water solution. This topic will be discussed in Section 22.4.

the hemiacetal carbon

glucose

a sugar

favored

Study Problem

13.6 Predict the cyclic hemiacetal or hemiketal products in water solutions of (a) 5-hydroxy-2-hexanone, and (b) 1,3,4,5,6-pentahydroxy-2-hexanone. [One stereoisomer of (b) is *fructose*, or *fruit sugar*.]

In most cases, a hemiacetal cannot be isolated. Acetals, however, are stable in nonacidic solution and can be isolated. (In acidic solution, of course, they are in equilibrium with their aldehydes.) If an acetal is the desired product from reaction of an aldehyde and an alcohol, an *excess of alcohol* is used to drive the series of reaction steps to that product. *Removing water* as it is formed also helps drive the reversible reactions to the acetal. Best results in this type of reaction are obtained when the acetal is *cyclic*:

benzaldehyde   2,3-butanediol   *a cyclic acetal (96%)*

If a desired reaction can be carried out under alkaline conditions, acetal and ketal groups are effective blocking groups for aldehydes and ketones. For example, after changing an aldehyde group to an acetal, we can oxidize a double bond in the same molecule without oxidizing the aldehyde to a carboxylic acid.

Blocking:

$$CH_2=CHCH + \begin{array}{c} HOCH_2 \\ | \\ HOCH_2 \end{array} \xrightarrow{H^+} CH_2=CHCH \begin{array}{c} O-CH_2 \\ | \\ O-CH_2 \end{array} + H_2O$$

propenal
(acrolein)

an acetal

Oxidation of double bond and regeneration of aldehyde:

$$CH_2=CHCH \begin{array}{c} O \\ \diagdown \\ O \end{array} \xrightarrow[\text{cold}]{MnO_4^-,\, OH^-} \begin{array}{c} OH\ OH \\ |\ \ | \\ CH_2CH-CH \end{array} \begin{array}{c} O \\ \diagdown \\ O \end{array} \xrightarrow{H_2O,\, H^+} \begin{array}{c} OH\ OH\ \ O \\ |\ \ |\ \ \| \\ CH_2CH-CH \end{array}$$

2,3-dihydroxypropanal

C. Reaction with Hydrogen Cyanide

Hydrogen cyanide (bp 26°) can be considered to be either a gas or a liquid with a low boiling point. In typical laboratory operations, it is used as a gas, but with special apparatus it can be used as a liquid (and, in some cases, even as a solvent). Often, HCN is generated directly in a reaction mixture from KCN or NaCN and a strong acid. Hydrogen cyanide is toxic and is particularly insidious because some people can detect its odor only at levels that may be lethal.

Like water and alcohols, hydrogen cyanide can add to the carbonyl group of an aldehyde or a ketone. The product in either case is referred to as a **cyanohydrin**.

$$\begin{array}{c} O \\ \| \\ CH_3CH \end{array} + H\,CN \xrightarrow{CN^-} \begin{array}{c} OH \\ | \\ CH_3CH-CN \end{array}$$

acetaldehyde

acetaldehyde cyanohydrin (75%)

General:

$$\begin{array}{c} O \\ \| \\ R-C-R \end{array} + HCN \xrightarrow{CN^-} \begin{array}{c} OH \\ | \\ R-C-R \\ | \\ CN \end{array}$$

an aldehyde
or ketone

a cyanohydrin

Hydrogen cyanide does not add directly to a carbonyl group. Successful addition requires slightly alkaline reaction conditions such as those found in a NaCN–HCN buffer solution. In alkaline solution, the concentration of cyanide ion is increased, and addition proceeds by nucleophilic attack of CN^- on the carbonyl group. Although weak nucleophiles (such as H_2O and ROH) require acid catalysis for addition to the carbonyl group, the strongly nucleophilic CN^- does not require a catalyst.

$$HCN\colon + \ ^-\colon\!\ddot{O}H \rightleftharpoons H_2\ddot{O}\colon + \ ^-\colon CN\colon$$

$$\begin{array}{c} \ddot{O}\colon \\ \| \\ R-C-H \\ | \\ \colon CN\colon \end{array} \rightleftharpoons \begin{array}{c} \colon\ddot{O}\colon^- \\ | \\ R-C-H \\ | \\ CN\colon \end{array} \xrightarrow{HCN} \begin{array}{c} \colon\ddot{O}H \\ | \\ R-C-H \\ | \\ CN\colon \end{array} + \ ^-\colon CN\colon$$

Cyanohydrins are useful synthetic intermediates. For example, the CN group can be hydrolyzed to a carboxyl group or converted to an ester group. Also, the —OH group of a cyanohydrin is far more reactive than that of an ordinary alcohol and can be replaced by ammonia to yield an amino group. This latter property finds use in the synthesis of amino acids.

$$CH_3\overset{\overset{O}{\|}}{C}CH_3 \xrightarrow[CN^-]{HCN} CH_3-\overset{\overset{OH}{|}}{\underset{\underset{CH_3}{|}}{C}}-CN \xrightarrow[H_2SO_4]{CH_3OH} CH_2{=}\overset{\underset{\underset{CH_3}{|}}{}}{C}-\overset{\overset{O}{\|}}{C}OCH_3$$

methyl methacrylate

used to prepare methacrylate polymers

$$CH_3\overset{\overset{O}{\|}}{C}H \xrightarrow[NH_4Cl]{NaCN} CH_3\underset{\underset{OH}{|}}{C}HCN \xrightarrow{NH_3} CH_3\underset{\underset{NH_2}{|}}{C}HCN \xrightarrow[H_2O]{HCl} CH_3\underset{\underset{NH_2}{|}}{C}H\overset{\overset{O}{\|}}{C}OH$$

alanine (60%)

an amino acid

The millipede *Apheloria corrugata* carries its own poison-gas generator in the form of *mandelonitrile*, a cyanohydrin stored in its defensive glands. When the millipede is attacked, the cyanohydrin is mixed with an enzyme that causes a rapid dissociation to a mixture of benzaldehyde and HCN, which is squirted on the predator to ward off the attack. A single millipede can emit enough HCN to kill a mouse! It is interesting to note that mandelonitrile, benzaldehyde, and HCN all have the odor of bitter almonds, despite their disparity in structure.

$$C_6H_5\underset{\underset{OH}{|}}{C}HCN \xrightarrow{enzyme} C_6H_5\overset{\overset{O}{\|}}{C}H + HCN$$

mandelonitrile   benzaldehyde

In plants of the genus *Prunus* (which includes plums, apricots, cherries, and peaches), cyanohydrins are biosynthesized and stored as sugar derivatives in the kernels of the pits. Amygdalin and laetrile are the best known of these cyanohydrins. (These two compounds are closely related structurally; indeed, amygdalin is often sold as laetrile.) Because these cyanohydrins can be hydrolyzed enzymatically to HCN, the pits of cherries and other *Prunus* species should not be eaten in quantity.

amygdalin   laetrile

("*lae*vorotatory glycosidic ni*trile*")

Study Problems

13.7 Suggest a synthesis for lactic acid, $CH_3CH(OH)CO_2H$, from ethanol and HCN.

13.8 Aldehydes and nonhindered ketones (primarily methyl ketones) yield water-soluble **bisulfite addition products** when treated with concentrated, aqueous sodium bisulfite.

$$\underset{O}{\overset{\displaystyle O}{\underset{\|}{RCH}}} + Na^+\ {}^-SO_3H \longrightarrow \underset{OH}{\overset{\displaystyle OH}{\underset{|}{RCH}}}{-}SO_3^-\ Na^+$$

bisulfite addition product

This reaction is sometimes used to separate an aldehyde or ketone from water-insoluble organic compounds. (The aldehyde or ketone can be regenerated by treatment of the aqueous bisulfite solution with acid or base.)

(a) Suggest a mechanism for the formation of the bisulfite addition product of acetaldehyde.

(b) Explain how the following compounds (in a diethyl ether solution) could be separated from one another by a series of extractions: heptanoic acid, 4-heptanone, and heptanal.

D. Reaction with Grignard Reagents

The reaction of a Grignard reagent with a carbonyl compound is another example of a nucleophilic addition to the positive carbon of a carbonyl group. However, the addition of a Grignard reagent is *not* a reversible reaction. (Why not?)

$$\overset{\displaystyle \ddot{O}:}{\underset{\displaystyle -\overset{\|}{C}-}{}} + \overset{\delta-}{C}H_3\overset{\delta+}{-}MgI \longrightarrow \underset{\displaystyle CH_3}{-\overset{\displaystyle :\ddot{O}:^-\ \overset{+}{M}gI}{\underset{|}{\overset{|}{C}}}-}$$

Reaction of a Grignard reagent with an aldehyde or a ketone provides an excellent method for the synthesis of alcohols and was discussed in that context earlier (Section 7.4D). The reaction sequence consists of two separate steps: (1) the reaction of the Grignard reagent with the carbonyl compound, and (2) hydrolysis of the resulting magnesium alkoxide to yield the alcohol. Recall that the Grignard reaction of formaldehyde yields a *primary alcohol*. Other aldehydes yield *secondary alcohols*. Ketones yield *tertiary alcohols*.

$$RMgX
\begin{cases}
\xrightarrow{HCHO} \underset{R}{\overset{OMgX}{\underset{|}{\overset{|}{HCH}}}} \xrightarrow{H_2O,\ H^+} RCH_2OH \quad \textit{1° alcohol} \\[2em]
\xrightarrow{R'CHO} \underset{R}{\overset{OMgX}{\underset{|}{\overset{|}{R'CH}}}} \xrightarrow{H_2O,\ H^+} \underset{}{\overset{OH}{\underset{|}{R'CHR}}} \quad \textit{2° alcohol} \\[2em]
\xrightarrow{\overset{O}{\overset{\|}{R'CR''}}} \underset{R}{\overset{OMgX}{\underset{|}{\overset{|}{R'CR''}}}} \xrightarrow{H_2O,\ H^+} \underset{R}{\overset{OH}{\underset{|}{\overset{|}{R'CR''}}}} \quad \textit{3° alcohol}
\end{cases}$$

Figure 13.7 Summary of the addition reactions of aldehydes and ketones.

Figure 13.7 summarizes the addition reactions of aldehydes and ketones.

Sample Problems

Suggest *two* synthetic routes to 2-butanol from an aldehyde or ketone and a Grignard reagent.

Solution: In the synthesis of 2-butanol, either (1) CH_3MgX and CH_3CH_2CHO or (2) CH_3CH_2MgX and CH_3CHO can be used. In the laboratory, the choice would depend on a number of factors, including availability and cost of the appropriate alkyl halides and aldehydes.

The two sequences to 2-butanol:

(1) CH_3CH_2CHO $\xrightarrow[\text{(2) } H_2O, H^+]{\text{(1) } CH_3MgI}$

OH
|
$CH_3CHCH_2CH_3$

(2) CH_3CHO $\xrightarrow[\text{(2) } H_2O, H^+]{\text{(1) } CH_3CH_2MgBr}$

Suggest three different Grignard reactions leading to 2-phenyl-2-butanol.

Solution:

$$
\begin{array}{lll}
\text{(1)} \quad C_6H_5MgBr & \dfrac{\text{(1) } CH_3\overset{\displaystyle O}{\overset{\|}{C}}CH_2CH_3}{\text{(2) } H_2O,\ H^+} & \\[3mm]
\text{(2)} \quad CH_3MgI & \dfrac{\text{(1) } C_6H_5\overset{\displaystyle O}{\overset{\|}{C}}CH_2CH_3}{\text{(2) } H_2O,\ H^+} & \longrightarrow\ C_6H_5-\overset{\displaystyle OH}{\underset{\displaystyle CH_3}{C}}-CH_2CH_3 \\[3mm]
\text{(3)} \quad CH_3CH_2MgBr & \dfrac{\text{(1) } C_6H_5\overset{\displaystyle O}{\overset{\|}{C}}CH_3}{\text{(2) } H_2O,\ H^+} &
\end{array}
$$

Study Problems

13.9 What Grignard reagents would you use to effect the following conversions?
  (a) formaldehyde to benzyl alcohol;
  (b) cyclohexanone to 1-propylcyclohexanol.

13.10 Which of the following compounds could *not* be used as a carbonyl starting material in a Grignard synthesis? (*Hint*: See Section 7.4D.)

(a) $CH_3\overset{\displaystyle OH}{\overset{|}{C}}HCH_2\overset{\displaystyle O}{\overset{\|}{C}}H$ **(b)** $H_2N-\!\!\bigcirc\!\!-\overset{\displaystyle O}{\overset{\|}{C}}H$ **(c)** $H\overset{\displaystyle O}{\overset{\|}{C}}CH_2CH_2CH_2\overset{\displaystyle O}{\overset{\|}{C}}H$

▼   # Section 13.6
Addition–Elimination Reactions of Aldehydes and Ketones

In the preceding section, we discussed the addition reactions of aldehydes and ketones. Some reagents undergo addition to aldehydes and ketones followed by elimination of water or other small molecule to yield a product containing a double bond.

$$
R-\overset{\displaystyle O}{\overset{\|}{C}}-R + H-NuH \xrightarrow{\text{addition}} \left[R-\overset{\displaystyle OH}{\underset{\displaystyle NuH}{C}}-R \right] \xrightarrow[-H_2O]{\text{elimination}} R-\overset{\displaystyle \ \ }{\underset{\displaystyle Nu}{\underset{\|}{C}}}-R
$$

an unstable addition product

A. Reaction with Ammonia and Primary Amines

Ammonia is a nucleophile that can attack a carbonyl group of an aldehyde or a ketone in an addition–elimination reaction. The reaction is catalyzed by a trace of acid. The product is an **imine**, a compound that contains the C=N grouping.

$$\underset{\text{RCH}}{\overset{\text{O}}{\|}} + \text{H-NH}_2 \underset{\text{H}^+}{\rightleftharpoons} \left[\underset{\text{RCH-NH}_2}{\overset{\text{OH}}{|}} \right] \overset{-\text{H}_2\text{O}}{\rightleftharpoons} \text{RCH=NH}$$

an imine

Unsubstituted imines formed from NH_3 are unstable and polymerize on standing. However, if a *primary amine* (RNH_2) is used instead of ammonia, a more stable, substituted imine (sometimes called a **Schiff base**) is formed. Aromatic aldehydes (such as benzaldehyde) or arylamines (such as aniline) give the most stable imines, but other aldehydes, ketones, or primary amines can be used.

benzaldehyde   methylamine   *an imine* (95%)

benzaldehyde   aniline   *an imine* (87%)

The mechanism for imine formation is essentially a two-step process. The first step is the *addition* of the nucleophilic amine to the partially positive carbonyl carbon, followed by the loss of a proton from the nitrogen and the gain of a proton by the oxygen.

Step 1, addition:

Step 2 is the protonation of the OH group, which then can be lost as water in an *elimination* reaction.

Step 2, elimination:

the imine

Imine formation is a reaction that is *pH-dependent.* Why? Consider the two steps in the mechanism. The first step is the addition of the free nonprotonated amine to the carbonyl group. If the solution is too acidic, the concentration of the amine becomes negligible. If this happens, the usually fast addition step becomes slow and actually becomes the rate-determining step in the sequence.

In acid:

not nucleophilic

$$R\ddot{N}H_2 + H^+ \;\rightleftharpoons\; R\overset{+}{N}H_3$$

The second step in the reaction is the elimination of the protonated OH group as water. Unlike the first step (amine addition), the rate of the second step increases with increasing acid concentration. (Remember, OH^- is a strong base and a poor leaving group, while $-OH_2^+$ can leave as the weak base and good leaving group H_2O.)

An increase in acidity causes Step 2 to go faster, but Step 1 to go slower, while decreasing acidity causes Step 1 to go faster, but Step 2 to go slower. Between these two extremes is the optimum pH (about pH 3–4), at which the rate of the overall reaction is greatest. At this pH, some of the amine is protonated, but some is free to initiate the nucleophilic addition. At this pH, too, enough acid is present so that elimination can proceed at a reasonable rate.

Study Problems

13.11 Why does an arylamine yield a more stable imine than isopropylamine?

13.12 What is the geometry of $CH_3CH=NCH_3$? Would you expect this compound to have any stereoisomeric forms?

13.13 How would you prepare each of the following imines from a carbonyl compound?

(a) $\langle \rangle$—CH=NCH₂CH₃   **(b)** CH₃—$\langle \rangle$—CH=N—$\langle \rangle$

(c) $(CH_3)_2C=N$—$\langle \rangle$—CH₃   **(d)**

$$NH$$

Biological transamination.   Imines are important intermediates in the biosynthesis of α-amino acids, $RCH(NH_2)CO_2H$, which are used by an organism in the synthesis of proteins. If a diet does not contain the required proportions of necessary amino acids, an organism can, in some cases, convert an unneeded amino acid to a desired amino acid in a **transamination reaction**. The process involves the transfer of an amino group from the unneeded amino acid to a keto acid.

Transamination:

$$\underset{\substack{old \\ amino\ acid}}{\overset{\displaystyle CO_2H}{\underset{\displaystyle R}{H_2N-C-H}}} + \underset{\substack{old \\ keto\ acid}}{\overset{\displaystyle CO_2H}{\underset{\displaystyle R'}{C=O}}} \xrightarrow[\text{enzymes}]{\text{transaminase}} \underset{\substack{new \\ keto\ acid}}{\overset{\displaystyle CO_2H}{\underset{\displaystyle R}{C=O}}} + \underset{\substack{new \\ amino\ acid}}{\overset{\displaystyle CO_2H}{\underset{\displaystyle R'}{H_2N-C-H}}}$$

The reaction is thought to proceed through a series of imine intermediates:

$$
\underset{\substack{\text{enzyme-bound}\\\text{aldehyde}}}{\overset{R''}{\underset{|}{CHO}}} + \underset{\substack{\text{old}\\\text{amino acid}}}{\overset{R''}{\underset{R}{\overset{|}{\underset{|}{H_2NCH}}}}\overset{CO_2H}{\underset{}{}}} \underset{-H_2O}{\rightleftharpoons} \overset{R''}{\underset{R}{\overset{|}{\underset{|}{CH=NCH}}}}\overset{CO_2H}{} \rightleftharpoons \overset{R''}{\underset{R}{\overset{|}{\underset{|}{CH_2N=C}}}}\overset{CO_2H}{} \underset{H_2O}{\rightleftharpoons} \overset{R''}{\underset{}{\overset{|}{\underset{}{CH_2NH_2}}}} + \underset{\substack{\text{new}\\\text{keto acid}}}{\overset{|}{\underset{R}{\overset{|}{C=O}}}}\overset{CO_2H}{}
$$

$$
\overset{R''}{\underset{}{\overset{|}{\underset{}{CH_2NH_2}}}} + \underset{\substack{\text{old}\\\text{keto acid}}}{\overset{|}{\underset{R'}{\overset{|}{C=O}}}}\overset{CO_2H}{} \underset{-H_2O}{\rightleftharpoons} \overset{R''}{\underset{R'}{\overset{|}{\underset{|}{CH_2N=C}}}}\overset{CO_2H}{} \rightleftharpoons \overset{R''}{\underset{R'}{\overset{|}{\underset{|}{CH=NCH}}}}\overset{CO_2H}{} \underset{H_2O}{\rightleftharpoons} \underset{\substack{\text{new}\\\text{amino acid}}}{\overset{R''}{\underset{R'}{\overset{|}{\underset{|}{CHO + H_2NCH}}}}\overset{CO_2H}{}}
$$

B. Reaction with Secondary Amines

With primary amines, aldehydes and ketones yield imines. With *secondary amines* (R_2NH), aldehydes and ketones yield **iminium ions**, which undergo further reaction to **enamines** (vinylamines). The enamine is formed by loss of a proton from a carbon atom β to the nitrogen, which results in a double bond between the α and β carbon atoms. Enamines are useful synthetic intermediates. You will encounter them again in Chapter 16.

$$
\underset{\substack{\text{a } 2° \text{ amine}}}{\overset{O}{\underset{\substack{\text{dimethylamine}}}{CH_3\overset{\|}{C}H + (CH_3)_2NH}}} \underset{-H_2O}{\overset{H^+}{\rightleftharpoons}} \underset{\substack{\text{an iminium ion}}}{\left[\overset{H}{\underset{}{\overset{|}{CH_2\text{-}CH=\overset{+}{N}(CH_3)_2}}}\right]} \underset{-H^+}{\rightleftharpoons} \underset{\substack{\text{an enamine}}}{CH_2=CH\overset{..}{N}(CH_3)_2}
$$

double bond between carbons α and β to the nitrogen

Study Problems

13.14 Predict the product of the reaction of cyclohexanone with:

(a) CH_3NH_2 (b) $(CH_3)_2NH$ (c) ⬡NH

13.15 Predict the enamine product of the following reaction:

+ HN⬠ $\xrightarrow[-H_2O]{H^+}$

C. Reaction with Hydrazine and Related Compounds

Imines are easily hydrolyzed (cleaved by water). The initial step of hydrolysis is protonation of the imine nitrogen. If an *electronegative group* is attached to the

Table 13.5 Some nitrogen compounds that form stable substitution products with aldehydes and ketones

Name	Structure	Product with RCHO
hydroxylamine	$HONH_2$	$RCH{=}NOH$ *an oxime*
hydrazine	H_2NNH_2	$RCH{=}NNH_2$ *a hydrazone*
phenylhydrazine	⬡—$NHNH_2$	$RCH{=}NNHC_6H_5$ *a phenylhydrazone*
2,4-dinitro- phenylhydrazine	O_2N—⬡(NO_2)—$NHNH_2$	$RCH{=}NNH$—⬡(NO_2)—NO_2 *a 2,4-dinitrophenylhydrazone*
semicarbazide	$H_2NNHCNH_2$ (O above C)	$RCH{=}NNHCNH_2$ (O above C) *a semicarbazone*

imine nitrogen, the basicity of the nitrogen is reduced and the hydrolysis is suppressed.

an imine

electron-withdrawing by the inductive effect

not favored

Imine-type products formed from aldehydes or ketones and a nitrogen compound of the type $H_2N{-}NH_2$ or $H_2N{-}OH$ (reagents with an electronegative group attached to the N) are quite stable. Table 13.5 lists the variety of nitrogen compounds that react with aldehydes and most ketones to form stable imine-type products.

acetone   hydrazine   acetone hydrazone

cyclopentanone   phenylhydrazine   cyclopentanone
                                                        phenylhydrazone

The hydrazones and other products listed in Table 13.5, especially the high-molecular-weight 2,4-dinitrophenylhydrazones, or DNP's, are generally solids. Before the wide use of spectrometers, these derivatives were used extensively for identification purposes. A liquid ketone of unknown structure could be converted to the solid DNP, purified by crystallization, and its melting point compared to those of DNP's of known structure.

Study Problem

13.16 Predict the product of the reaction of: (a) butanone with semicarbazide;   (b) cyclohexanone with 2,4-dinitrophenylhydrazine; and   (c) acetophenone ($C_6H_5COCH_3$) with hydrazine.

D. Reaction with Phosphonium Ylides (the Wittig Reaction)

In 1954, Georg Wittig reported a general synthesis of alkenes from carbonyl compounds using *phosphonium ylides*. This synthesis is called the **Wittig reaction**.

an aldehyde   *a phosphonium*   *an alkene*   triphenylphosphine
or ketone   *ylide*   *(where at least one*   oxide
                                      R or R' = H)

An **ylide** is a molecule with *adjacent + and − charges* (see the following resonance structures). An ylide is formed by removal of a proton from the carbon adjacent to a positively charged heteroatom (such as P^+, S^+, or N^+). The phosphonium ylide for a Wittig reaction is prepared by (1) nucleophilic substitution (S_N2) of an alkyl halide with a tertiary phosphine, such as triphenylphosphine (a good nucleophile, a weak base), and (2) treatment with a strong base, such as *n*-butyllithium ($CH_3CH_2CH_2CH_2Li$), a reaction in which the intermediate phosphonium ion eliminates a proton to form the ylide.

triphenylphosphine   *a phosphonium ion*

$$CH_3CH_2CH_2CH_3 + (C_6H_5)_3\overset{+}{P}\!-\!\overset{..}{C}R'_2 \longleftrightarrow (C_6H_5)_3P{=}CR'_2$$

resonance structures for the ylide

The Wittig reaction is versatile. The alkyl halide used to prepare the ylide may be methyl, primary, or secondary, but not tertiary (why not?). The halide may also contain other functionality, such as double bonds or alkoxyl groups. The product of the Wittig reaction is an alkene with the double bond in the desired position, even if it is not the most stable alkene. Yields are generally good (about 70%) for mono-, di-, and trisubstituted alkenes. (The reaction fails for tetrasubstituted alkenes, probably because of steric hindrance.) Unfortunately, it is sometimes difficult to predict whether the *cis* or the *trans* product will predominate in a particular reaction.

$$\text{cyclohexanone} \quad \text{O} + (C_6H_5)_3P=CH_2 \xrightarrow{\text{THF}} \text{=CH}_2 + (C_6H_5)_3P=O$$

cyclohexanone   methylenecyclohexane (84%)

$$\overset{\overset{\displaystyle O}{\displaystyle \|}}{\text{CH}} + (C_6H_5)_3P=CHCH=CH- \xrightarrow[\text{heat}]{\text{xylene}}$$

benzaldehyde

$$\text{—CH=CHCH=CH—} + (C_6H_5)_3P=O$$

1,4-diphenyl-1,3-butadiene (67%)

The mechanism of the Wittig reaction involves nucleophilic attack of the carbonyl group by the negative carbon of the ylide.

nucleophilic carbon

$$(C_6H_5)_3P=CR'_2 \longleftrightarrow (C_6H_5)_3\overset{+}{P}-\overset{..}{C}R'_2$$

resonance structures

Addition to carbonyl:

$$\begin{array}{c} R \\ | \\ R-C + :CR'_2 \\ \overset{\|}{O}: \quad {}^+P(C_6H_5)_3 \end{array} \longrightarrow \begin{array}{c} R \\ | \\ R-C—CR'_2 \\ | \quad | \\ {}^-:\overset{..}{O}: \quad {}^+P(C_6H_5)_3 \end{array}$$

a betaine

The addition product of the ylide and an aldehyde or ketone is a **betaine** (a molecule having *nonadjacent* opposite charges). The betaine undergoes cyclization and elimination of triphenylphosphine oxide to form the alkene.

Elimination to form alkene:

$$\begin{array}{c} R_2C—CR'_2 \\ | \quad | \\ :\overset{..}{O}: \quad P(C_6H_5)_3 \\ {}^- \quad {}^+ \end{array} \longrightarrow \begin{array}{c} R_2C—CR'_2 \\ | \quad | \\ :\overset{..}{O}—P(C_6H_5)_3 \end{array} \longrightarrow R_2C=CR'_2 + :\overset{..}{O}=P(C_6H_5)_3$$

Figure 13.8 Summary of the addition–elimination reactions of aldehydes and ketones.

Study Problem

13.17 Write equations to show how you would prepare the following compounds by Wittig reactions. (Begin with organic halides and carbonyl compounds.)

(a) $C_6H_5CH=CHCH_2CH_3$ (b) [structure] =CH–[structure]   (c) [structure]

Figure 13.8 summarizes the addition–elimination reactions of aldehydes and ketones.

Section 13.7
Reduction of Aldehydes and Ketones

An aldehyde or a ketone can be reduced to an alcohol, a hydrocarbon, or an amine. The product of the reduction depends on the reducing agent and on the structure of the carbonyl compound.

$$\underset{\substack{\text{an aldehyde} \\ \text{or ketone}}}{\overset{\text{O}}{\underset{\|}{R\overset{}{C}R}}} \xrightarrow{[\text{H}]} \underset{\substack{\text{an alcohol}}}{\overset{\text{OH}}{\underset{|}{RCHR}}} \quad \text{or} \quad \underset{\substack{\text{a hydrocarbon}}}{RCH_2R} \quad \text{or} \quad \underset{\substack{\text{an amine}}}{\overset{\text{NR}'_2}{\underset{|}{RCHR}}}$$

A. Hydrogenation

The pi bond of a carbonyl group can undergo hydrogenation, just as can the pi bond of an alkene. The reaction conditions depend on the compound being reduced and on the catalyst. Unhindered ketones, such as cyclohexanone, can be hydrogenated at room temperature and four atmospheres of pressure with a platinum or ruthenium catalyst. Other carbonyl compounds and other catalyst systems, such as copper chromite or Raney nickel, may require harsher reaction conditions.

cyclohexanone

a ketone

cyclohexanol (91%)

a 2° alcohol

$$CH_3CH{=}O \ + \ H_2 \ \xrightarrow{Pt} \ CH_3CH_2OH$$

acetaldehyde

an aldehyde

ethanol

a 1° alcohol

Study Problem

13.18 2-Heptanol can be obtained by two Grignard reactions as well as by a hydrogenation reaction. Write the equations for the three sets of reactions that would lead to this alcohol.

If both a double bond and a carbonyl group are present in a structure, the double bond may be hydrogenated, leaving the carbonyl intact, or both may be hydrogenated. However, the carbonyl group cannot be catalytically hydrogenated independently of the double bond. To reduce a carbonyl group while leaving a carbon–carbon double bond intact, a metal hydride reduction is the method of choice.

C=C reduced (but not C=O):

$$CH_3CH{=}CHCH_2\overset{\overset{\displaystyle O}{\|}}{C}H \ + \ H_2 \ \xrightarrow[25°]{Ni} \ CH_3CH_2CH_2CH_2\overset{\overset{\displaystyle O}{\|}}{C}H$$

3-pentenal

pentanal

C=C and C=O reduced:

$$CH_3CH{=}CHCH_2\overset{\overset{\displaystyle O}{\|}}{C}H \ + \ 2\,H_2 \ \xrightarrow[\text{heat, pressure}]{Ni} \ CH_3CH_2CH_2CH_2CH_2OH$$

3-pentenal

1-pentanol

If the hydrogenation of a keto group yields a chiral alcohol, a racemic mixture is formed unless the hydrogenation catalyst is chelated with one enantiomer of a chiral compound. Under these conditions, asymmetry is induced into the product, and only one enantiomer of the product alcohol is isolated. This type of reaction is called **asymmetric hydrogenation**.

Asymmetric hydrogenation of a ketone:

$$\text{an achiral} \atop \text{keto lactone} \qquad\qquad \text{a chiral hydroxy lactone (98\%)} \atop [\alpha]_D^{20} = -49.87°$$

B. Metal Hydrides

Hydrogen gas is inexpensive on a molar basis; however, a hydrogenation reaction is rather inconvenient. The apparatus usually consists of gas tanks and a metal pressure vessel. An alternative reduction procedure involves the use of a metal hydride. Two valuable reducing agents are *lithium aluminum hydride* (often abbreviated LAH) and *sodium borohydride*. Both reduce aldehydes and ketones to alcohols.

lithium aluminum hydride (LAH)   sodium borohydride

These two metal hydrides are quite different in their reactivities. $LiAlH_4$ is a powerful reducing agent that reduces not only aldehydes and ketones, but also carboxylic acids, esters, amides, and nitriles. $LiAlH_4$ undergoes violent reaction with water; reductions are usually carried out in a solvent such as anhydrous ether.

Sodium borohydride is a milder reducing agent than $LiAlH_4$. Its reactions can be carried out in water or aqueous alcohol as the solvent. For the reduction of an aldehyde or ketone, $NaBH_4$ is the preferred reagent; it is certainly more convenient to use because of its lower reactivity toward water. While $NaBH_4$ reduces aldehydes and ketones rapidly, it reduces esters very slowly. Therefore, an aldehyde or ketone carbonyl group can be reduced without the simultaneous reduction of an ester group in the same molecule. This selectivity is not possible with $LiAlH_4$.

Neither $NaBH_4$ nor $LiAlH_4$ reduces isolated carbon–carbon double bonds, although C=C in conjugation with a carbonyl group is sometimes attacked. Consequently, a structure that contains both a double bond and a carbonyl group can often be reduced selectively at the carbonyl position. In this respect, the metal hydrides are complementary to hydrogen gas as reducing agents.

C=O reduced (but not C=C):

$$CH_3\overset{O}{\underset{\|}{C}}CH_2CH=CHCH_2\overset{O}{\underset{\|}{C}}CH_3 \xrightarrow[(2)\ H_2O,\ H^+]{(1)\ LiAlH_4} CH_3\overset{OH}{\underset{|}{C}}HCH_2CH=CHCH_2\overset{OH}{\underset{|}{C}}HCH_3$$

4-octene-2,7-dione   4-octene-2,7-diol (70%)

Diisobutylaluminum hydride (DBAH), $[(CH_3)_2CHCH_2]_2AlH$, is a newer, but popular, metal hydride reducing agent that is similar to $LiAlH_4$ in reducing power. Besides reducing aldehydes or ketones to alcohols, DBAH reduces carboxylic acids and esters to aldehydes or alcohols (depending on the reaction conditions). Other hydrides that have specialized reducing activity are also available; for example, see Section 15.3C.

Study Problems

13.19 Show how each of the following alcohols could be prepared by the $NaBH_4$ reduction of an aldehyde or ketone:

(a) $CH_3CH_2CH_2OH$ **(b)** (structure)—OH   **(c)** (structure)—$\overset{OH}{\underset{|}{C}}HCH_3$

13.20 When glucose (page 892) is treated with $NaBH_4$, then with aqueous acid, an artificial sweetener called *sorbitol* results. What is the structure of sorbitol?

Metal hydrides react by transferring a negative hydride ion to the positive carbon of a carbonyl group, just as a Grignard reagent transfers R to the carbonyl group.

$$R\overset{\ddot{O}:}{\underset{|}{\overset{\|}{C}}}R + H\overset{H}{\underset{|}{\underset{H}{B}}}H \longrightarrow R\overset{:\ddot{O}:^-}{\underset{|}{\underset{H}{C}}}R + BH_3 \qquad \textit{three more H's}$$

Each hydride ion can reduce one carbonyl group. Therefore, one mole of $NaBH_4$ can reduce *four* moles of aldehyde or ketone, theoretically. After the reaction is completed, treatment with water or aqueous acid liberates the alcohol from its salt. (Of course, if water, methanol, or ethanol is used as a solvent for a borohydride reduction, this step occurs spontaneously.) In the hydrolysis, the boron portion of the organoborate is converted to boric acid, H_3BO_3.

Step 1:

$$4\ R\overset{O}{\underset{\|}{C}}R + NaBH_4 \longrightarrow Na^+\ ^-B\left(\underset{\underset{\displaystyle OCHR}{\underset{|}{R}}}{}\right)_4$$

Step 2:

$$\underset{\text{R}\overset{\displaystyle |}{\text{C}}\text{HR}}{\overset{\displaystyle\text{O}^-}{}} + \text{H}^+ \longrightarrow \underset{\text{R}\overset{\displaystyle |}{\text{C}}\text{HR}}{\overset{\displaystyle\text{OH}}{}}$$

Camphor is a bridged, cyclic compound with a ketone group. Reduction of camphor with LiAlH$_4$ leads to 90% of the isomer in which the OH group is *cis* to the bridge. Why is this so? Let us look at the structure of camphor. Note that the bridge provides substantial steric hindrance on one side of the carbonyl group—on top, as it is shown here.

camphor

— more hindered (*cis to bridge*)

— less hindered (*trans to bridge*)

When a ketone is reduced by LiAlH$_4$, it is not just a tiny hydride ion attacking; it is the relatively bulky AlH$_4^-$ ion or an alkoxyaluminum hydride ion, such as $^-$AlH$_2$(OR)$_2$. There is evidence that the lithium ion forms a complex with the carbonyl oxygen while the hydride ion is transferred from the AlH$_4^-$ ion to the carbonyl carbon. A look at the likely structure of the transition state shows how AlH$_4^-$ attacks the *less hindered* side of the camphor structure; that is, AlH$_4^-$ attacks the carbonyl group from the side that is *trans* to the bridge. The resultant OH group is then formed *cis* to the bridge.

transition state

cis to bridge (*after hydrolysis*)

Sample Problem

When a *cis*-3-alkyl-4-*t*-butylcyclohexanone is treated with LiAlH$_4$, the predominant product is that in which the OH is *cis* to the *t*-butyl group and the alkyl group. Suggest a reason.

Solution: The equatorial *t*-butyl group forces the R group into the axial position (see Section 4.5). The transition state leading to an equatorial OH shows more steric hindrance (1,3-diaxial interactions) than that leading to the axial OH. Therefore, the transition state leading to an axial OH is of lower energy and is favored, and the *cis*-product results.

To equatorial:

more hindered, not favored

To axial:

less hindered, favored

C. Wolff–Kishner and Clemmensen Reductions

The Clemmensen reduction and the Wolff–Kishner reduction are primarily used to reduce aryl ketones obtained from Friedel–Crafts reactions (Section 11.9E), but may sometimes be used to reduce other aldehydes and ketones. Both these methods of reduction result in the conversion of a C=O group to a CH_2 group.

In the Wolff–Kishner reduction, the aldehyde or ketone is first converted to a hydrazone by reaction with hydrazine. The hydrazone is then treated with a strong base, such as potassium hydroxide or potassium *t*-butoxide in dimethyl sulfoxide (DMSO) as solvent. The reaction is therefore limited to carbonyl compounds that are stable in base.

Wolff–Kishner reduction:

acetophenone   ethylbenzene (73%)

In the Clemmensen reduction, on the other hand, a zinc amalgam (an alloy of zinc and mercury) and concentrated HCl are used. These reagents would be the reagents of choice for a compound unstable in base but stable in acid.

Clemmensen reduction:

(45%)

D. Reductive Amination

If an amine is the desired reduction product, the carbonyl compound is treated with ammonia or a primary amine to form an imine in the presence of hydrogen and a catalyst. The imine C=N group then undergoes catalytic hydrogenation in the same way that a C=C or a C=O group does.

$$C_6H_5-\overset{\overset{O}{\|}}{CH} \xrightarrow[-H_2O]{NH_3} \left[C_6H_5-CH=NH \right] \xrightarrow{H_2,\ Ni} C_6H_5-CH_2NH_2$$

benzaldehyde   an imine   benzylamine (89%)

$$CH_3CH_2\overset{\overset{O}{\|}}{C}CH_3 \xrightarrow[-H_2O]{CH_3NH_2} \left[CH_3CH_2\overset{\overset{NCH_3}{\|}}{C}CH_3 \right] \xrightarrow{H_2,\ Pt} CH_3CH_2\overset{\overset{NHCH_3}{|}}{C}HCH_3$$

butanone   an imine   N-methyl-2-butylamine (69%)

Reductive amination is a good method for the preparation of an amine with a secondary alkyl group: R_2CHNH_2. (Treating the secondary alkyl halide R_2CHX with NH_3 in an S_N2 reaction may result in elimination or in dialkyl-amines, a reaction we will discuss in Chapter 17.)

$$C_6H_{11}-Br \xrightarrow{NH_3} \text{a mixture of products}$$

bromocyclohexane

$$C_6H_{10}=O \xrightarrow[H_2,\ Ni]{NH_3} C_6H_{11}-NH_2$$

cyclohexanone   cyclohexylamine (80%)

Figure 13.9 summarizes the reduction reactions of aldehydes and ketones.

$$\overset{\overset{O}{\|}}{RCR} \begin{cases} \xrightarrow[\text{or (1) NaBH}_4\ (2)\ H_2O,\ H^+]{H_2,\ \text{catalyst}} & \overset{\overset{OH}{|}}{RCHR} \quad \textbf{an alcohol} \\\\ \xrightarrow[\text{or Zn/Hg, HCl}]{(1)\ NH_2NH_2,\ H^+\ (2)\ KOH} & RCH_2R \quad \textbf{a hydrocarbon} \\\\ \xrightarrow[H_2,\ \text{catalyst}]{NH_3\ \text{or}\ R'NH_2} & \overset{\overset{NH_2}{|}}{RCHR} \ \text{or}\ \overset{\overset{NHR'}{|}}{RCHR} \quad \textbf{an amine} \end{cases}$$

an aldehyde or ketone

Figure 13.9 Summary of reductions of aldehydes and ketones.

▼ Section 13.8
Oxidation of Aldehydes and Ketones

Ketones are not easily oxidized (see Sections 13.10 and 13.11 for exceptions), but aldehydes are very easily oxidized to carboxylic acids. Almost any reagent that oxidizes an alcohol also oxidizes an aldehyde (see Section 7.9C). Permanganate or dichromate salts are the most popular oxidizing agents, but are by no means the only reagents that can be used.

$$CH_3(CH_2)_5\overset{O}{\overset{\|}{C}}H \xrightarrow[\text{H}_2\text{O}]{\text{KMnO}_4, \text{H}^+} CH_3(CH_2)_5\overset{O}{\overset{\|}{C}}OH$$

heptanal   heptanoic acid (78%)

an aldehyde   *a carboxylic acid*

$$CH_3\overset{O}{\overset{\|}{C}}CH_3 \xrightarrow{\text{KMnO}_4} \text{no reaction}$$

acetone

a ketone

In addition to oxidation by permanganate or dichromate, aldehydes are oxidized by very mild oxidizing agents, such as Ag^+ or Cu^{2+}. **Tollens reagent** (an alkaline solution of the silver–ammonia complex ion) is used as a test for aldehydes. The aldehyde is oxidized to a carboxylate anion; the Ag^+ in the Tollens reagent is reduced to Ag metal. A positive test is indicated by the formation of a silver mirror on the wall of the test tube. With the widespread use of spectroscopy, the Tollens test is no longer the test of choice for an aldehyde, but mirrors are sometimes still made this way.

$$R\overset{O}{\overset{\|}{C}}H + \quad Ag(NH_3)_2^+ \xrightarrow{\text{OH}^-} R\overset{O}{\overset{\|}{C}}O^- + Ag$$

from Tollens reagent   *mirror*

▼ Section 13.9
Reactivity of the Alpha Hydrogens

A carbon–hydrogen bond is usually stable, nonpolar, and certainly not acidic. But the presence of a carbonyl group results in an *acidic alpha hydrogen*. If a hydrogen is alpha to *two* carbonyl groups, it is acidic enough that salts can be formed by treatment with an alkoxide. The pK_a of ethyl acetoacetate ($CH_3COCH_2CO_2CH_2CH_3$) is 11; it is more acidic than ethanol ($pK_a = 16$) or water ($pK_a = 15.7$). Treatment of this β-dicarbonyl compound with sodium ethoxide (or any other strong base, such as NaH or NaNH$_2$) yields the sodium salt. (Sodium hydroxide is usually not used with keto esters because the ester groups undergo hydrolysis in NaOH and water. This reaction will be discussed in Section 15.5C.)

$$\underset{\substack{\text{acetone}\\ \text{p}K_a = 20}}{\overset{\displaystyle \overset{O}{\|}}{CH_3CCH_3}} \qquad\qquad \underset{\substack{\text{ethyl acetoacetate}\\ \text{p}K_a = 11}}{CH_3CCH_2COCH_2CH_3}$$

α to one C=O — *α to two C=O groups*

$$\underset{\substack{|\\H}}{CH_2CCH_3} + Na^+ \; {^-}:\ddot{O}CH_2CH_3 \; \rightleftharpoons \; Na^+ \; {^-}:CH_2\overset{\displaystyle \overset{O}{\|}}{C}CH_3 + CH_3CH_2\ddot{O}H$$

not favored

$$\underset{\substack{|\\H}}{CH_3CCHCOCH_2CH_3} + Na^+ \; {^-}:\ddot{O}CH_2CH_3 \; \rightleftharpoons \; CH_3C\underset{}{\overset{}{C}}HCOCH_2CH_3 + CH_3CH_2OH$$

$$Na^+$$

favored

Why is a hydrogen alpha to a carbonyl group acidic? The answer is two-fold. First, the alpha carbon is adjacent to one or more partially positive carbon atoms. The alpha carbon, too, partakes of some of this positive charge (inductive effect by electron-withdrawal), and C—H bonds are consequently weakened.

$$-\overset{\displaystyle \overset{O}{\|}}{C}\leftarrow\overset{\delta+}{CH_2}\rightarrow\overset{\displaystyle \overset{O}{\|}}{C}-$$

Second, and more important, is the resonance stabilization of the **enolate ion**, the anion formed when the proton is lost. From the resonance structures, we can see that the negative charge is carried by the carbonyl oxygens as well as by the α carbon. This delocalization of the charge stabilizes the enolate ion and favors its formation.

Adjacent to one carbonyl group:

$$\left[-\ddot{C}H-\overset{\displaystyle :\ddot{O}}{\|}C- \;\longleftrightarrow\; -CH=\overset{\displaystyle :\ddot{O}:^-}{C}- \right] \text{ or } -\overset{\delta-}{C}H{=\!=}\overset{\displaystyle O^{\delta-}}{C}-$$

Adjacent to two carbonyl groups:

$$\left[-\overset{\displaystyle :\ddot{O}}{\|}C-CH=\overset{\displaystyle :\ddot{O}:^-}{C}- \;\longleftrightarrow\; -\overset{\displaystyle :\ddot{O}}{\|}C-\ddot{C}H-\overset{\displaystyle \ddot{O}:}{\|}C- \;\longleftrightarrow\; -\overset{\displaystyle :\ddot{O}:^-}{C}=CH-\overset{\displaystyle \ddot{O}:}{\|}C- \right] \text{ or } -\overset{\displaystyle \overset{\delta-}{O}}{C}{=\!=}\underset{\delta-}{CH}{=\!=}\overset{\displaystyle \overset{\delta-}{O}}{C}-$$

Study Problem

13.21 Give the resonance structures of the enolate ions formed when the following diones are treated with sodium ethoxide:

(a) $\text{C}_6\text{H}_5-\overset{\displaystyle \overset{O}{\|}}{C}CH_2\overset{\displaystyle \overset{O}{\|}}{C}CH_3$ (b) $CH_3\overset{\displaystyle \overset{O}{\|}}{C}CH_2\overset{\displaystyle \overset{O}{\|}}{C}CH_3$

Section 13.10
Tautomerism

Even if a strong base is not present, the acidity of an alpha hydrogen may be evident. A carbonyl compound with an acidic alpha hydrogen may exist in two forms called **tautomers**: a special type of interconvertible structural isomers that differ from each other only in the location of a double bond and a hydrogen atom relative to the oxygen. The two tautomers of a simple ketone are called the *keto tautomer* and the *enol tautomer*. The keto tautomer of a carbonyl compound has the expected carbonyl structure. The enol tautomer (from *-ene* + *-ol*), which is a vinylic alcohol, is formed by transfer of an acidic hydrogen from the α carbon to the carbonyl oxygen. Because a hydrogen atom is in different positions, the two tautomeric forms are not resonance structures, but are two different structures in equilibrium. (Remember that resonance structures vary only in the positions of *electrons*.)

keto form   enol form

keto form   enol form
of acetone   of acetone

The relative quantities of enol versus keto in a pure liquid can be estimated by infrared or nmr spectroscopy. Acetone exists primarily in the keto form (99.99%, as determined by a specialized titration procedure). Most other simple aldehydes and ketones also exist primarily in their keto forms. However, 2,4-pentanedione exists as 80% enol! How can this tremendous difference be explained? Let us consider the structures of the 2,4-pentanedione tautomers:

keto form (20%)   enol form (80%)

The enol form not only has conjugated double bonds, which add a small amount of stability, but it is also structurally arranged for internal hydrogen bonding, which helps stabilize this tautomer.

Sample Problem

Suggest reasons why 1,2-cyclohexanedione exists 100% in an enol form.

Solution:

dipole–dipole
repulsions

repulsions relieved;
stabilized by hydrogen
bonding

Study Problem

13.22 A trace of either acid or base will catalyze a keto-enol tautomerization. Consequently, it is difficult to isolate a pure tautomer free of the other form. In 1911, the German chemist Ludwig Knorr isolated the keto tautomer of ethyl acetoacetate by crystallization at $-78°$, a temperature at which the rate of equilibration is slow. The formula for the keto tautomer follows. Write the formula for the enol tautomer that illustrates its hydrogen bonding.

$$CH_3\overset{O}{\overset{\|}{C}}CH_2\overset{O}{\overset{\|}{C}}OCH_2CH_3$$

ethyl acetoacetate

Tautomerism can affect the reactivity of a compound. An exception to the generality that ketones are not easily oxidized is the oxidation of a ketone with at least one α hydrogen. A ketone that can undergo tautomerization can be oxidized by a strong oxidizing agent at the carbon–carbon double bond of the enol tautomer. Yields in this reaction are poor because, under these conditions, other C—C bonds may be cleaved. This reaction is not used in synthetic work, but is used often in structure determinations.

A. Tautomerism in Carbohydrate Metabolism

The first step in the metabolism of carbohydrates (starches and sugars) is their breakdown to glucose in the digestive tract. This breakdown is the hydrolysis of the acetal bonds. In the cells of an organism, glucose is eventually converted to CO_2 and H_2O. The first step in this cellular sequence of reactions is the formation of glucose 6-phosphate, followed by isomerism to fructose 6-phosphate. The isomerism reaction is simply an enzyme-catalyzed tautomerization that proceeds by way of an *enediol intermediate*, an intermediate that can lead to the two carbonyl products.

$$
\underset{\text{an aldehyde}}{\overset{\displaystyle HC=O}{\underset{\displaystyle H-C-OH}{|}}}
\; \rightleftharpoons \;
\underset{\text{an enediol}}{\left[\overset{\displaystyle HCOH}{\underset{\displaystyle COH}{\|}}\right]}
\; \rightleftharpoons \;
\underset{\text{a ketone}}{\overset{\displaystyle H_2COH}{\underset{\displaystyle C=O}{|}}}
$$

glucose 6-phosphate   *an enediol*   fructose 6-phosphate

(Structures: glucose 6-phosphate $\rightleftharpoons$ an enediol $\rightleftharpoons$ fructose 6-phosphate, each with $CH_2OPO_3H^-$ terminal group.)

Study Problem

13.23 After fructose 6-phosphate is formed, it is converted enzymatically to the 1,6-diphosphate, which is turn is cleaved into two three-carbon compounds:

$$
\underset{\substack{\text{dihydroxyacetone}\\\text{phosphate}}}{\overset{\displaystyle CH_2OPO_3H^-}{\underset{\displaystyle CH_2OH}{\underset{\displaystyle |}{\overset{\displaystyle |}{C=O}}}}}
\qquad\qquad
\underset{\substack{\text{glyceraldehyde}\\\text{3-phosphate}}}{\overset{\displaystyle CHO}{\underset{\displaystyle CH_2OPO_3H^-}{H-C-OH}}}
$$

Before being carried on to the next step, the dihydroxyacetone phosphate is also converted to glyceraldehyde 3-phosphate. Suggest an intermediate in this conversion.

Section 13.11
Alpha Halogenation

Ketones are readily halogenated at the α carbon. The reaction requires either alkaline conditions or an acidic catalyst. (Note that base is a *reactant*, whereas acid is a *catalyst*.)

In base:

$$
\underset{\text{acetone}}{CH_3\overset{\displaystyle O}{\overset{\|}{C}}CH_3} + Br_2 + OH^- \longrightarrow \underset{\text{bromoacetone}}{BrCH_2\overset{\displaystyle O}{\overset{\|}{C}}CH_3} + Br^- + H_2O
$$

$$
\underset{\text{cyclohexanone}}{\bigcirc\!\!=\!\!O} + Br_2 + OH^- \longrightarrow \underset{\substack{\text{2-bromo-}\\\text{cyclohexanone}}}{\bigcirc\!\!=\!\!O} + Br^- + H_2O
$$

$$\textit{In acid:} \left\{ \begin{array}{l} CH_3\overset{\displaystyle O}{\overset{\|}{C}}CH_3 + Br_2 \xrightarrow{\ H^+\ } BrCH_2\overset{\displaystyle O}{\overset{\|}{C}}CH_3 + HBr \\[2em] \bigcirc\!\!-O + Br_2 \xrightarrow{\ H^+\ } \bigcirc\!\!=O + HBr \\ \qquad\qquad\qquad\qquad\qquad\quad Br \end{array} \right.$$

The first step (the slow step) in the reaction under alkaline conditions is the formation of the enolate ion. The anion of a ketone with only one carbonyl group is a much stronger base than the hydroxide ion. Therefore, the acid–base equilibrium favors the hydroxide ion rather than the enolate ion. Nonetheless, a *few* enolate ions exist in alkaline solution. As these few anions are used up, more are generated to go on to Step 2. In Step 2, the enolate ion quickly undergoes reaction with halogen to yield the α-halogenated ketone and a halide ion.

In base:

Step 1 (slow):

$$CH_3\overset{\displaystyle O}{\overset{\|}{C}}CH_3 + OH^- \underset{-H_2O}{\overset{\longrightarrow}{\longleftarrow}} \left[CH_3\overset{\displaystyle \overset{..}{\overset{..}{O}}:}{\overset{\|}{C}}\!\!\!\overset{\curvearrowleft}{}\overset{..}{C}H_2 \longleftrightarrow CH_3\overset{\displaystyle :\overset{..}{O}:^-}{\overset{|}{C}}=CH_2 \right]$$

resonance structures for the enolate ion

Step 2 (fast):

$$CH_3\overset{\displaystyle O}{\overset{\|}{C}}CH_2 + :\overset{..}{\underset{..}{Br}}\!\!-\!\!\overset{..}{\underset{..}{Br}}: \longrightarrow CH_3\overset{\displaystyle O}{\overset{\|}{C}}CH_2\overset{..}{\underset{..}{Br}}: + :\overset{..}{\underset{..}{Br}}:^-$$

Study Problems

13.24 Upon which species does the rate of base-promoted α halogenation depend? Would the reaction follow first-order or second-order kinetics?

13.25 One of the disadvantages of base-promoted α halogenation of a ketone is that a second halogen atom is introduced more easily than the first.

$$CH_3\overset{\displaystyle O}{\overset{\|}{C}}CH_2Br + Br_2 + OH^- \longrightarrow CH_3\overset{\displaystyle O}{\overset{\|}{C}}CHBr_2 + Br^- + H_2O$$

Write the steps in the mechanism of the second halogenation and suggest a reason why this reaction is faster than the first halogenation.

Alpha halogenation in acid usually gives higher yields than the reaction in base. The acid-catalyzed reaction proceeds by way of the enol, the formation of which is the rate-determining step. The carbon–carbon double bond of the enol undergoes electrophilic addition, just like any carbon–carbon double bond, to form the more stable carbocation. In this case, the more stable carbocation is the one in which the positive charge is on the carbon of the carbonyl group (because this intermediate is resonance-stabilized). This carbocation intermediate quickly loses a proton and forms the ketone, which is now halogenated in the alpha position.

In acid:

Step 1 (fast):

$$CH_3\overset{\overset{\displaystyle :O:}{\|}}{C}CH_3 + H^+ \rightleftharpoons \left[CH_3\overset{\overset{\displaystyle ^+\ddot{O}H}{\|}}{C}CH_3 \right]$$

Step 2 (slow):

$$\left[CH_3\overset{\overset{\displaystyle ^+\ddot{O}H}{\|}}{C}\underset{\underset{\displaystyle H}{|}}{\curvearrowleft}CH_2 \right] \rightleftharpoons \left[CH_3\overset{\overset{\displaystyle :\ddot{O}H}{|}}{C}=CH_2 \right] + H^+$$

the enol

Step 3 (fast):

$$\left[CH_3\overset{\overset{\displaystyle :\ddot{O}H}{|}}{C}=CH_2 \right] + Br\!\!-\!\!Br \longrightarrow \left[CH_3\overset{\overset{\displaystyle ^+\ddot{O}H}{\|}}{C}\!-\!CH_2Br \right] + Br^-$$

Step 4 (fast):

$$\left[CH_3\overset{\overset{\displaystyle ^+\ddot{O}H}{\|}}{C}CH_2Br \right] \rightleftharpoons CH_3\overset{\overset{\displaystyle :\ddot{O}:}{\|}}{C}CH_2Br + H^+$$

Study Problem

13.26 **(a)** What species are involved in the rate-determining step of acid-catalyzed α halogenation of acetone?
(b) What would be the relative rates of α bromination and iodination?

A. Haloform Reaction

Alpha halogenation is the basis of the **iodoform test** for methyl ketones. The methyl group of a methyl ketone is iodinated stepwise until the yellow solid iodoform (CHI_3) is formed.

Iodoform test:

cyclohexyl
methyl ketone

*intermediate
triiodomethyl ketone*

cyclohexyl-
carboxylate ion

iodoform

yellow solid

The intermediate is unstable in base because a trihalomethyl anion is a good leaving group.

The test is not uniquely specific for methyl ketones. Iodine is a mild oxidizing agent, and any compound that can be oxidized to a methyl carbonyl compound also gives a positive test.

$$\underset{\text{2-propanol}}{\underset{\text{OH}}{\overset{\displaystyle |}{\text{CH}_3\text{CHCH}_3}}} \xrightarrow[\text{OH}^-]{\text{I}_2} \underset{\text{acetone}}{\overset{\displaystyle \text{O}}{\overset{||}{\text{CH}_3\text{CCH}_3}}} \xrightarrow[\text{OH}^-]{\text{I}_2} \underset{\text{acetate ion}}{\overset{\displaystyle \text{O}}{\overset{||}{\text{CH}_3\text{CO}^-}}} + \text{CHI}_3$$

Bromine and chlorine also react with methyl ketones to yield bromoform (CHBr_3) and chloroform (CHCl_3), respectively. "Haloform" is the general term used to describe CHX_3; hence this reaction is often referred to as the **haloform reaction**. Because bromoform and chloroform are nondistinctive liquids, their formations are not useful for test purposes. However, the reaction of a methyl ketone with any of these halogens provides a method for the conversion of these compounds to carboxylic acids.

$$\underset{\text{3,3-dimethyl-2-butanone}}{\overset{\displaystyle \text{O}}{\overset{||}{(\text{CH}_3)_3\text{CCCH}_3}}} \xrightarrow[\text{(2) H}^+]{\text{(1) 3 Br}_2,\text{ OH}^-} \underset{\text{2,2-dimethylpropanoic acid (74\%)}}{\overset{\displaystyle \text{O}}{\overset{||}{(\text{CH}_3)_3\text{CCOH}}}}$$

Study Problems

13.27 Which of the following compounds gives a positive iodoform test?

$$\textbf{(a)}\ \underset{}{\overset{\displaystyle \text{O}}{\overset{||}{\text{ICH}_2\text{CH}}}} \qquad \textbf{(b)}\ \underset{}{\overset{\displaystyle \text{O}}{\overset{||}{\text{CH}_3\text{CH}_2\text{CH}}}} \qquad \textbf{(c)}\ \underset{}{\overset{\displaystyle \text{OH}}{\overset{|}{\text{CH}_3\text{CH}_2\text{CHCH}_3}}}$$

13.28 What methyl ketones can be used to prepare the following carboxylic acids by haloform reactions?

(a) (structure: decalin with CO$_2$H)   **(b)** $(\text{CH}_3)_2\text{CHCO}_2\text{H}$ **(c)** HO_2C—(benzene ring)—CO_2H

▼ ## Section 13.12
1,4-Addition to α,β-Unsaturated Carbonyl Compounds

A. Electrophilic 1,4-Addition

When an alkene reacts with HCl, the reaction proceeds by electrophilic attack of H^+ to yield the more stable carbocation, followed by attack of Cl^-.

$$\underset{\text{propene}}{\text{CH}_3\text{CH}{=}\text{CH}_2} \xrightarrow{\text{H}^+} [\text{CH}_3\overset{+}{\text{CH}}\text{CH}_3] \xrightarrow{\text{Cl}^-} \underset{\text{2-chloropropane}}{\overset{\displaystyle \text{Cl}}{\overset{|}{\text{CH}_3\text{CHCH}_3}}}$$

An α,β-*unsaturated* aldehyde or ketone has a carbon–carbon double bond in conjugation with a carbonyl group. The carbon–carbon double bond in an alkene is nonpolar. However, a carbon–carbon double bond in conjugation with a carbonyl group is *polar*, as the following resonance structures indicate.

$$CH_2{=}CH{-}\overset{\overset{\displaystyle \ddot{O}:}{\|}}{CH} \longleftrightarrow CH_2{=}CH{-}\overset{\overset{\displaystyle :\ddot{O}:^-}{|}}{\underset{+}{CH}} \longleftrightarrow {}^+CH_2{-}CH{=}\overset{\overset{\displaystyle :\ddot{O}:^-}{|}}{CH}$$

The resonance structures show that the β carbon, as well as the carbonyl carbon, carries a partial positive charge, while the carbonyl oxygen carries a partial negative charge.

$$CH_2{=}CH{-}\overset{\overset{\displaystyle O^{\delta-}}{\|}}{CH}$$

β carbon is δ+ ⟋   ⟍ *carbonyl carbon is δ+*

Because the C=C grouping in an α,β-unsaturated carbonyl compound is polarized, the mechanism for electrophilic addition is somewhat different from that for electrophilic addition to an isolated, nonpolar, alkene double bond. Let us consider two examples of electrophilic addition reactions of α,β-unsaturated carbonyl compounds. Then we will discuss the mechanism.

$$CH_2{=}CH\overset{\overset{\displaystyle O}{\|}}{C}H + HCl \longrightarrow \overset{\overset{\displaystyle Cl}{|}}{CH_2}CH_2\overset{\overset{\displaystyle O}{\|}}{C}H$$

propenal   3-chloropropanal

$$CH_3CH{=}CH\overset{\overset{\displaystyle O}{\|}}{C}CH_3 + H_2O \xrightarrow{\;H^+\;} CH_3\overset{\overset{\displaystyle OH}{|}}{C}HCH_2\overset{\overset{\displaystyle O}{\|}}{C}CH_3$$

3-penten-2-one   4-hydroxy-2-pentanone

In each reaction, the nucleophilic part of the reagent (not H⁺) becomes bonded to the β carbon. The reason for this is that the β carbon has a partial positive charge. The initial attack by H⁺ occurs not at this positive carbon, but at the partially negative oxygen of the carbonyl group.

$$\overset{\delta+}{CH_2}{=}CH\overset{\overset{\displaystyle \delta-\ddot{O}:}{\|}}{C}H \quad \curvearrowright H^+$$

The protonated intermediate is resonance stabilized. In this intermediate, the β carbon still carries a partial positive charge and can be attacked by a nucleophile.

Protonation:

$$CH_2{=}CH\overset{\overset{\displaystyle \ddot{O}:}{\|}}{C}H \; \underset{\displaystyle H^+}{\rightleftarrows} \; \left[CH_2{=}CH\overset{\overset{\displaystyle {}^+\ddot{O}H}{\|}}{C}H \longleftrightarrow CH_2{=}CH{-}\overset{\overset{\displaystyle :\ddot{O}H}{|}}{CH} \longleftrightarrow {}^+CH_2CH{=}\overset{\overset{\displaystyle :\ddot{O}H}{|}}{CH} \right]$$

resonance stabilized

Attack of Nu:⁻:

$$\left[\begin{array}{c} \text{OH} \\ | \\ {}^{+}\text{CH}_2\text{CH}{=}\text{CH} \end{array} \right] + :\ddot{\text{Cl}}:^{-} \longrightarrow \left[\begin{array}{c} \text{OH} \\ | \\ :\ddot{\text{Cl}}\text{CH}_2\text{CH}{=}\text{CH} \end{array} \right] \rightleftarrows \text{ClCH}_2\text{CH}_2\overset{\text{O}}{\overset{\|}{\text{CH}}}$$

an enol of an aldehyde   *an aldehyde*

Note that this reaction is a **1,4-addition**, or **conjugate addition**, the same type of addition observed with conjugated dienes. The difference is that the initial addition product is an enol, which undergoes tautomerization to the final keto form of the aldehyde.

You might wonder why the nucleophile cannot attack the carbonyl carbon (which also carries a partial positive charge in the intermediate). This attack on the carbonyl carbon can occur, but the product is unstable and reverts to starting material. This is a concurrent, but nonproductive, side reaction.

$$\left[\begin{array}{c} :\ddot{\text{O}}\text{H} \\ | \\ \text{CH}_2{=}\text{CH}\overset{+}{\text{CH}} \end{array} \right] + :\ddot{\text{Cl}}:^{-} \rightleftarrows \left[\begin{array}{c} :\ddot{\text{O}}{-}\text{H} \\ | \\ \text{CH}_2{=}\text{CHCH}{-}\ddot{\text{Cl}}: \end{array} \right] \rightleftarrows \text{CH}_2{=}\text{CH}\overset{\cdot\ddot{\text{O}}\cdot}{\overset{\|}{\text{CH}}} + \text{H}\ddot{\text{Cl}}:$$

B. Nucleophilic 1,4-Addition

The pi bond of an alkene is not normally attacked by a nucleophile unless there has been prior attack by an electrophile. However, a double bond in conjugation with a carbonyl group is polarized. In this case, nucleophilic addition can occur *either at the C=C double bond or at the C=O double bond* (at either of the two partially positive carbons).

$$\text{CH}_2{=}\text{CH}\overset{\text{O}}{\overset{\|}{\text{C}}}\text{CH}_3 + {}^{-}\text{CN} \xrightarrow{\text{HCN}} \text{CH}_2{=}\text{CH}\overset{\text{OH}}{\underset{\text{CN}}{\overset{|}{\underset{|}{\text{C}}}}}\text{CH}_3 \quad \text{or} \quad \text{CH}_2{-}\text{CH}_2\overset{\text{O}}{\overset{\|}{\text{C}}}\text{CH}_3$$

Let us look at the mechanism of each reaction. First, we will consider the attack of the cyanide ion (from HCN and dilute base) on the carbonyl group. In this case, the nucleophilic CN⁻ attacks the partially positive carbon of the carbonyl group. The reaction is no different from cyanohydrin formation by an ordinary ketone.

Attack of CN⁻ on the carbonyl carbon:

$$\text{CH}_2{=}\text{CH}\overset{\ddot{\text{O}}:}{\overset{\|}{\text{C}}}\text{CH}_3 + {}^{-}:\text{CN}: \longrightarrow \left[\begin{array}{c} :\ddot{\text{O}}:^{-} \\ | \\ \text{CH}_2{=}\text{CH}\overset{|}{\text{C}}\text{CH}_3 \\ | \\ \text{CN}: \end{array} \right] \xrightleftharpoons[]{\text{HCN}} \text{CH}_2{=}\text{CH}\overset{:\ddot{\text{O}}\text{H}}{\underset{\text{CN}}{\overset{|}{\underset{|}{\text{C}}}}}\text{CH}_3 + \text{CN}^{-}$$

Now we will consider attack of the nucleophilic CN⁻ on the β carbon. This reaction is a 1,4-addition of CN⁻ and H⁺ to the conjugated system. The product of the 1,4-addition is an enol, which forms the product ketone.

Attack of CN⁻ on the β carbon:

$$CH_2\!\!=\!\!CH\!-\!\overset{\displaystyle \cdot\cdot\!O\!\cdot\cdot}{\overset{\|}{C}}CH_3 \;\rightleftharpoons\; \left[CH_2\!-\!\overset{}{\underset{|}{C}H}\!-\!\overset{\displaystyle :\!\ddot{O}}{\overset{\|}{C}}CH_3 \;\longleftrightarrow\; CH_2\!-\!CH\!\!=\!\!\overset{\displaystyle :\!\ddot{O}\!:^-}{C}CH_3 \right] \xrightarrow{\;HCN\;}$$

with $\overset{}{\underset{|}{CN}}$ groups below the left and middle structures.

$$CN^- + \left[CH_2\!-\!CH\!\!=\!\!\overset{\displaystyle :\!\ddot{O}H}{\underset{\underset{\displaystyle CN}{|}}{C}}CH_3 \right] \rightleftharpoons \overset{\displaystyle O}{\overset{\|}{CH_2CH_2CCH_3}} \; \underset{\displaystyle CN}{|}$$

an enol

Which of the two addition reactions occurs? Sometimes both do, and a mixture of products results. In most cases, however, one product or the other predominates. Steric hindrance around the double bond or the carbonyl group may lead to preferred attack at the nonhindered position. Aldehydes, less hindered than ketones, usually undergo carbonyl attack.

$$(CH_3CH_2)_2C\!\!=\!\!CH\overset{\displaystyle O}{\overset{\|}{C}}H \qquad CH_2\!\!=\!\!\overset{\displaystyle O}{\overset{\|}{C}}\underset{\underset{\displaystyle CH_3}{|}}{C}CH_2CH_3$$

Nu:⁻ attacks here   *Nu:⁻ attacks here*

A highly basic nucleophile (such as RMgX or LiAlH₄) attacks preferentially at the carbonyl group, while a weaker base (such as CN⁻ or R₂NH) usually attacks at the carbon–carbon double bond.

Stronger bases attack 1,2 (at C=O):

(98%)

$$CH_3CH\!\!=\!\!CH\overset{\displaystyle O}{\overset{\|}{C}}CH_3 + CH_3MgI \longrightarrow \xrightarrow{H_2O,\,H^-} CH_3CH\!\!=\!\!CH\overset{\displaystyle OH}{\underset{\underset{\displaystyle CH_3}{|}}{C}}CH_3$$

(72%)

Weaker bases attack 1,4 (at C=C):

$$(CH_3)_2C\!\!=\!\!CH\overset{\displaystyle O}{\overset{\|}{C}}CH_3 + :NH_3 \longrightarrow (CH_3)_2\underset{\underset{\displaystyle NH_2}{|}}{C}CH_2\overset{\displaystyle O}{\overset{\|}{C}}CH_3$$

(70%)

(96%)

Lithium dialkylcopper reagents, also called **cuprates**, are synthesized from an alkyllithium and a copper(I) halide, such as CuI.

$$2 \text{ CH}_3\text{Li} + \text{CuI} \longrightarrow (\text{CH}_3)_2\text{CuLi} + \text{LiI}$$

a cuprate

These compounds attack an α,β-unsaturated carbonyl compound almost exclusively at the carbon–carbon double bond by 1,4-addition.

(97%)

Study Problem

13.29 Predict the major organic product of the reaction of each of the following reagents with 2-cyclohexenone: **(a)** CH_3MgI (followed by H^+, H_2O); **(b)** 1 equivalent of H_2 with Ni catalyst (25°); **(c)** NaBH_4 (followed by H^+, H_2O); **(d)** NH_3; **(e)** $(\text{CH}_2{=}\text{CH})_2\text{CuLi}$ (followed by H^+, H_2O).

▼ ## Section 13.13
Use of Aldehydes and Ketones in Synthesis

Aldehydes and ketones are readily available by the oxidation of alcohols and can be converted to a variety of other types of compounds, as may be seen in Table 13.6. When viewed from a synthesis standpoint, the reactions in this table can be grouped into three major categories:

1. reactions in which the carbonyl group is retained (α halogenation and 1,4-addition);
2. reactions in which the carbonyl group is converted to another functional group (for example, reduction or conversion to a hemiacetal);
3. reactions in which extension of the carbon skeleton occurs at the carbonyl group (Grignard and Wittig reactions).

In designing synthetic sequences, you should keep all three classes of reactions in mind.

Table 13.6 Types of compounds that can be obtained from aldehydes and ketones[a]

Reaction	Product	Section Reference
Addition:[b]		
RCHO + 2 R′OH $\xrightarrow{H^+}$ $\overset{\overset{\displaystyle OR'}{\displaystyle |}}{RCHOR'}$	acetal	13.5B
RCHO + HCN $\xrightarrow{CN^-}$ $\overset{\overset{\displaystyle OH}{\displaystyle |}}{RCHCN}$	cyanohydrin	13.5C
$\overset{\overset{\displaystyle O}{\displaystyle \|}}{RCR}$ $\xrightarrow[\text{(2) H}_2\text{O, H}^+]{\text{(1) R′MgX}}$ $\overset{\overset{\displaystyle OH}{\displaystyle |}}{\underset{\underset{\displaystyle R'}{\displaystyle |}}{RCR}}$	alcohol	7.4D, 13.5D
Addition–Elimination:[b]		
RCHO + R′NH$_2$ $\longrightarrow$ RCH=NR′	imine	13.6A
RCH$_2$CHO + R′$_2$NH $\longrightarrow$ RCH=CHNR′$_2$	enamine	13.6B, 16.5
RCHO + R′NHNH$_2$ $\longrightarrow$ RCH=NNHR′	hydrazone	13.6C
R$_2$C=O + (C$_6$H$_5$)$_3$P=CR′$_2$ $\longrightarrow$ R$_2$C=CR′$_2$	alkene	13.6D
Reduction:		
R$_2$C=O $\xrightarrow[\text{or metal hydride}]{\text{H}_2,\text{ catalyst}}$ R$_2$CHOH	alcohol	13.7A, B
R$_2$C=O $\xrightarrow[\text{or Zn/Hg, HCl}]{\text{(1) NH}_2\text{NH}_2,\text{ (2) OH}^-}$ R$_2$CH$_2$	alkane or alkylbenzene	13.7C
R$_2$C=O $\xrightarrow[\text{H}_2,\text{ Ni}]{\text{R}'_2\text{NH}}$ R$_2$CHNR′$_2$	amine	13.7D, 17.5B
α Halogenation:		
$\overset{\overset{\displaystyle O}{\displaystyle \|}}{RCH_2CR}$ + X$_2$ $\xrightarrow{H^+}$ $\overset{\overset{\displaystyle O}{\displaystyle \|}}{\underset{\underset{\displaystyle X}{\displaystyle |}}{RCHCR}}$	α-halo carbonyl	13.11
$\overset{\overset{\displaystyle O}{\displaystyle \|}}{RCCH_3}$ + 3 X$_2$ $\xrightarrow[\text{(2) H}^+]{\text{(1) OH}^-}$ RCO$_2$H	carboxylic acid	13.11A
1,4-Addition:		
$\overset{\overset{\displaystyle O}{\displaystyle \|}}{RCH=CHCR}$ + Nu:⁻ or HNu $\longrightarrow$ $\overset{\overset{\displaystyle Nu\quad O}{\displaystyle |\quad\ \|}}{RCHCH_2CR}$	β-substituted carbonyl	13.12

[a] The condensation and alkylation reactions of aldehydes and ketones will be discussed in Chapter 16.
[b] Nonhindered ketones, such as methyl ketones, can also be used.

Sample Problems

How would you make the following conversion?

$$\overset{\overset{\displaystyle O}{\displaystyle \|}}{CH_3CCH_3} \longrightarrow \overset{\overset{\displaystyle O}{\displaystyle \|}}{CH_3CCH_2OH}$$

Solution: In this conversion, the carbonyl group is retained, but a functional group is inserted α to the C=O group. Alpha halogenation is a way to insert an α functional group.

$$CH_3\overset{\overset{\displaystyle O}{\|}}{C}CH_3 \quad \xrightarrow{Cl_2, H^+} \quad CH_3\overset{\overset{\displaystyle O}{\|}}{C}CH_2Cl$$

The alcohol can then be obtained by treatment of the α-chloroketone with aqueous NaOH (S_N2 reaction).

$$CH_3\overset{\overset{\displaystyle O}{\|}}{C}CH_2Cl \quad \xrightarrow{^-OH} \quad CH_3\overset{\overset{\displaystyle O}{\|}}{C}CH_2OH$$

How would you make the following conversion?

$$\text{C}_6\text{H}_5\text{—}\overset{\overset{\displaystyle O}{\|}}{C}H \longrightarrow \text{C}_6\text{H}_5\text{—}CH\text{=}CHCH_2CH_3$$

Solution: This conversion involves an extension of the carbon skeleton at the carbonyl group. The Wittig reaction could be used for this conversion.

$$C_6H_5\overset{\overset{\displaystyle O}{\|}}{C}H \quad \xrightarrow{(C_6H_5)_3P\text{=}CHCH_2CH_3} \quad \text{product}$$

The product could also be obtained from the alcohol $C_6H_5CH(OH)CH_2CH_2CH_3$ by dehydration (spontaneous in this case because the product C=C is in conjugation with a benzene ring). The alcohol, in turn, can be obtained by a Grignard reaction, another reaction that allows us to build up a carbon skeleton.

$$C_6H_5\overset{\overset{\displaystyle O}{\|}}{C}H \quad \xrightarrow[(2)\ H_2O,\ H^+]{(1)\ CH_3CH_2CH_2MgBr} \quad C_6H_5\overset{\overset{\displaystyle OH}{|}}{C}HCH_2CH_2CH_3 \quad \xrightarrow{-H_2O} \quad \text{product}$$

Suggest a synthesis for *N*-ethyl-3-hexylamine from organic compounds containing three or fewer carbon atoms.

Solution:

1. Write the structure: $CH_3CH_2\overset{\overset{\displaystyle NHCH_2CH_3}{|}}{C}HCH_2CH_2CH_3$
2. It is apparent that more than one step will be needed to synthesize this compound. We must both build up a carbon skeleton and also insert the ethylamino group. Let us consider the amino group first. It could be placed in this structure by reductive amination of a ketone.

$$CH_3CH_2\overset{\overset{\displaystyle O}{\|}}{C}CH_2CH_2CH_3 \quad \xrightarrow[H_2,\ Ni]{CH_3CH_2NH_2} \quad \text{product}$$

3. Is there a one-step reaction to 3-hexanone from starting materials containing three or fewer carbons? No, but a Grignard reaction can provide us with an alcohol, which is readily oxidized to a ketone. Thus, we can write the initial series of reactions.

$$CH_3CH_2\overset{\overset{\displaystyle O}{\|}}{C}H \quad \xrightarrow[(2)\ H_2O,\ H^+]{(1)\ CH_3CH_2CH_2MgBr} \quad CH_3CH_2\overset{\overset{\displaystyle OH}{|}}{C}HCH_2CH_2CH_3 \quad \xrightarrow{H_2CrO_4}$$

$$CH_3CH_2\overset{\overset{\displaystyle O}{\|}}{C}CH_2CH_2CH_3$$

The Grignard reagent can be prepared by the reaction of RX with Mg.

$$CH_3CH_2CH_2Br \xrightarrow[\text{diethyl ether}]{\text{Mg}} CH_3CH_2CH_2MgBr$$

The proposed synthesis is complete, and the series of reactions may be written forward.

Study Problem

13.30 Suggest synthetic routes to the following compounds from aldehydes or ketones of six or fewer carbon atoms and other appropriate reagents:

(a) 2-phenylpropene   (b) —CO$_2$H   (c) 2-cyclopentyl-1-heptene

▼ Summary

The **carbonyl group** is *planar* and *polar*, and the oxygen has *two pairs of unshared valence electrons*. The carbonyl group can undergo *electrophilic* or *nucleophilic* attack.

Because of (1) *inductive stabilization* of the partial positive charge on the carbonyl carbon, and (2) *steric hindrance*, ketones are less reactive than aldehydes:

$$\overset{O}{\overset{\|}{RCR}} < \overset{O}{\overset{\|}{RCH}} < \overset{O}{\overset{\|}{HCH}}$$

Many reactions of aldehydes and ketones are simple **addition reactions** at the C=O pi bond. These reactions can lead to hydrates, hemiacetals (or hemiketals) and cyanohydrins.

$$R-\overset{O}{\overset{\|}{C}}-H + H-Nu \rightleftharpoons R-\overset{\overset{OH}{|}}{\underset{\underset{Nu}{|}}{C}}-H$$

where HNu = H$_2$O, ROH, or HCN

Other addition reactions are **reduction** and **Grignard reactions**.

$$
\underset{\text{RCR'}}{\overset{O}{\|}} + H—H \xrightarrow{\text{catalyst}} \underset{\text{RCHR'}}{\overset{OH}{|}}
$$

$$
\underset{\text{RCR}}{\overset{O}{\|}} + R'MgX \xrightarrow{\text{diethyl ether}} \underset{\underset{R'}{|}}{\overset{OMgX}{\underset{|}{RCR}}}
$$

Substitution reactions of aldehydes and ketones are the result of initial addition reactions followed by elimination, or **addition–elimination reactions**. The formation of imines, enamines, hydrazones, and alkenes (by the Wittig reaction) all fall into this category. These reactions are summarized in Table 13.6.

$$
\underset{\text{RCR}}{\overset{O}{\|}} \xrightarrow[\text{addition}]{R'NH_2} \left[\underset{\underset{NHR'}{|}}{\overset{OH}{\underset{|}{RCR}}} \right] \xrightarrow[\text{elimination}]{-H_2O} \underset{\underset{NR'}{\|}}{RCR}
$$

Because the carbonyl group is polar and because its pi electrons can participate in resonance-stabilization, an α *hydrogen is acidic*, especially if it is α to two carbonyl groups. This acidity can give rise to **tautomerism**. Because of tautomerism, ketones can undergo **α halogenation**, as shown in Table 13.6, or **oxidative cleavage** between the carbonyl carbon and the α carbon.

$$
\underset{\text{CH}_3\overset{O}{\overset{\|}{C}}H}{} + OH^- \rightleftharpoons {}^-{:}CH_2\overset{O}{\overset{\|}{C}}H + H_2O
$$

acidic

$$
CH_3\overset{O}{\overset{\|}{C}}CH_2\overset{O}{\overset{\|}{C}}CH_3 + OH^- \rightleftharpoons CH_3\overset{O}{\overset{\|}{C}}\overset{..}{\overset{-}{C}}HCCH_3 + H_2O
$$

$$
\underset{\text{keto tautomer}}{CH_3\overset{O}{\overset{\|}{C}}—\overset{H}{\underset{|}{C}}H—\overset{O}{\overset{\|}{C}}CH_3} \rightleftharpoons \underset{\text{enol tautomer}}{CH_3\overset{OH}{\overset{|}{C}}=CH—\overset{O}{\overset{\|}{C}}CH_3}
$$

When a C=C double bond is in conjugation with C=O, addition reactions may occur 1,2 or 1,4.

Electrophilic:

$$
R_2C=CH—\overset{O}{\overset{\|}{C}}R \xrightarrow[1,4]{HCl} \left[\underset{\underset{Cl}{|}}{R_2C}—CH=\overset{OH}{\overset{|}{C}}R \right] \rightleftharpoons \underset{\underset{Cl}{|}}{R_2C}—CH_2—\overset{O}{\overset{\|}{C}}R
$$

Nucleophilic:

$$
R_2C=CH\overset{O}{\overset{\|}{C}}H + R'MgX \xrightarrow{1,2} R_2C=CH\underset{\underset{R'}{|}}{\overset{OMgX}{\overset{|}{C}}H}
$$

nonhindered C=O   strong base

$$CH_2=CHCR + NH_3 \xrightarrow{1,4} \begin{bmatrix} OH \\ | \\ CH_2-CH=CR \\ | \\ NH_2 \end{bmatrix} \rightleftharpoons CH_2CH_2CR$$

hindered C=O   weak base

Study Problems*

13.31 Write formulas for:

(a) bromoacetone, a lachrymator that has been used as a war gas;

(b) biacetyl (2,3-butanedione), used to give margarine a butter flavor;

(c) 4-(4-hydroxyphenyl)-2-butanone, responsible for the flavor of raspberries;

(d) geranial, (E)-3,7-dimethyl-2,6-octadienal, which is found in lemongrass oil and is used in perfumes and flavorings.

13.32 Write an acceptable name for each of the following structures:

(a) [structure]   (b) [structure]

(c) $(CH_3)_2CHCCH_2CH_2CH_3$ (d) $HOCH_2CCH_3$

(e) $C_6H_5CH_2CHO$ (f) $BrCH_2CH_2CH_2CHO$

13.33 Give formulas for: (a) a β-ketoaldehyde; (b) an α,β-unsaturated ketone; (c) an α-bromoaldehyde; (d) a β-hydroxyketone.

13.34 Write the equation for the formation of the hydrate of each of the following aldehydes. Which hydrate would be more stable?

(a) Br_2CHCHO (b) Br_2CHCH_2CHO

13.35 Write equations for the reactions leading to the hemiacetal and acetal (or hemiketal and ketal) of: (a) propanal and methanol; (b) 5-hydroxy-2-hexanone and methanol; (c) acetone and glycerol (1,2,3-propanetriol).

13.36 Each of the following compounds is dissolved in water to which a trace of HCl has been added. Give the structures of any other compounds (besides HCl, water, and the compound in question) that would be found in each solution.

(a) [structure]   (b) CH_3—[structure]   (c) [structure]

13.37 Write a flow equation showing the mechanism of preparation of the following compound from an aldehyde:

[structure] CH_2OH / CH_2OH

* For information concerning the organization of the *Study Problems* and *Additional Problems*, see the *Note to student* on page 39.

13.38 Write equations for the reactions of (a) cyclohexanone with 1,2-ethanethiol ($HSCH_2CH_2SH$) and a trace of acid; (b) formaldehyde (HCHO) with aqueous potassium cyanide and sulfuric acid; (c) *cis-* or *trans-*tetrahydropyran-2,3-diol with methanol and a trace of acid.

tetrahydropyran-2,3-diol

13.39 Predict the major organic products:

(a) $\xrightarrow[\text{(2) } H^+,\, H_2O]{\text{(1) } CH_3MgI}$

(b) $\xrightarrow[\text{(2) } H^+,\, H_2O]{\text{(1) } C_6H_5MgBr}$

(c) HCHO $\xrightarrow[\text{(2) } H^+,\, H_2O]{\text{(1)}}$

13.40 Suggest a synthesis for each of the following compounds, starting with iodomethane and other appropriate reagents: (a) 3-methyl-3-pentanol; (b) ethanol; (c) 2-pentanol.

13.41 Write equations for Grignard syntheses of the following alcohols from aldehydes or ketones. Give all possible correct answers.

(a) $-CH_2CH_2CH_2OH$ (b)

13.42 The following reactions proceed in yields of 85–95%. Predict the major organic products.

(a) A mixture of *o*-nitrobenzaldehyde and aniline is heated with a buffered acid catalyst.
(b) A mixture of 2-methylpropanal and pyrrolidine is heated with a buffered acid catalyst.

pyrrolidine

(c) A benzene solution of benzaldehyde and *t*-butylamine is heated with a buffered acid catalyst.
(d) Heptanal is treated with an aqueous solution of hydroxylamine hydrochloride ($NH_2OH \cdot HCl$) and sodium carbonate.

13.43 Show how you would prepare the following compounds from compounds containing six or fewer carbon atoms:

(a) $CH_3CH_2CH=CHN(CH_2CH_3)_2$ (b)

(c) $CH_3CH_2CH=NC_6H_5$ (d) $CH_3CH_2CH=NNHC_6H_5$

13.44 List the following compounds in terms of reactivity toward 2,4-dinitrophenylhydrazine (least reactive first): **(a)** 2-pentanone; **(b)** 3-pentanone; **(c)** pentanal.

13.45 Predict the organic products:

(a)
$$\xrightarrow[150°, 35\ atm]{H_2, \text{ copper chromite catalyst}}$$

(b)
$$\xrightarrow[(2)\ H_2O, H^+]{(1)\ NaBH_4, CH_3CH_2OH}$$

(c)
$$\xrightarrow[(HOCH_2CH_2)_2O, 240°]{NH_2NH_2, KOH, H_2O}$$

(d)
$$\xrightarrow[heat]{Zn(Hg), conc.\ HCl}$$

vanillin

(e) $CH_3\overset{O}{\underset{\|}{C}}CH_3 + NH_3 \xrightarrow{H_2, Pt}$

(f) $CH_3\overset{O}{\underset{\|}{C}}CH_3 + HOCH_2CH_2NH_2 \xrightarrow{H_2, Pt}$

(g)
$$\xrightarrow{H_2, Pt}$$

13.46 Predict the product and its stereochemistry:

$$\xrightarrow[(2)\ H_2O, H^-]{(1)\ NaBH_4, CH_3OH}$$

13.47 How would you effect the following conversions by Wittig reactions? (Start with organic halides.)

(a) $=O \longrightarrow$ $=CHCH=CH_2$

(b) $\longrightarrow$

(c) CH_3O——$\overset{O}{\underset{\|}{C}}CH_3 \longrightarrow CH_3O$——$\underset{\underset{CH_3}{|}}{C}=CHOCH_3$

(d) $(CH_3)_2C=O \longrightarrow (CH_3)_2C=CF_2$

13.48 Complete the following equations:

(a) $\langle\ \rangle{=}O + MnO_4^-\ \xrightarrow[25°]{CH_3CO_2H}$

(b) $\langle\ \rangle{-}CHO + MnO_4^-\ \xrightarrow[25°]{OH^-}$

13.49 Which of the following compounds would give a positive Tollens test?

(a) CH_3CHO (b) $CH_3\overset{O}{\overset{\|}{C}}CH_3$ (c) $\langle\ \rangle{-}OH$ (d) $\langle\ \rangle{-}OCH_3$

13.50 Suggest a synthetic scheme to convert 3-hydroxypropanal to HO_2CCH_2CHO.

13.51 Rank the following compounds in terms of increasing acidity (least acidic first):

(a) 2,4-pentanedione   (b) butanal   (c) water

13.52 Complete the following equations for acid–base reactions:

(a) $CH_3\overset{O}{\overset{\|}{C}}H + OH^- \rightleftharpoons$

(b) $\overset{O}{\diagdown}\diagup\overset{O}{\diagup} + CO_3^{2-} \rightleftharpoons$

(c) $(C_6H_5)_2CH\overset{O}{\overset{\|}{C}}OCH_2CH_3 + Na^+\ ^-OCH_2CH_3 \rightleftharpoons$

(d) $CH_3CH_2\overset{O}{\overset{\|}{C}}CH_3 + \text{excess } D_2O \underset{}{\overset{^-OD}{\rightleftharpoons}}$

(e) $(S)\text{-}CH_3CH_2\overset{CH_3}{\overset{|}{C}H}CH_2CHO + OH^- \rightleftharpoons$

(f) $(S)\text{-}CH_3CH_2CH_2\overset{CH_3}{\overset{|}{C}H}CHO + OH^- \rightleftharpoons$

13.53 Each of the following compounds is about as acidic as a β-diketone. Show the resonance structures for the anions of each that would account for this acidity.

(a) CH_3NO_2 (b) $CH_3\overset{O}{\overset{\|}{C}}CH_2CN$ (c) $C_6H_5CH_2\overset{O}{\overset{\|}{C}}H$

13.54 Write equations that illustrate tautomerism of the following compounds:

(a) [structure]   (b) [structure]   (c) [structure]

(d) [structure]   (e) [structure]

13.55 Predict the organic products:

(a) $+ Cl_2$ $\xrightarrow{H^+}$ **(b)** $+ Br_2$ $\xrightarrow{H^+}$

13.56 What would be observed if each of the following compounds were placed in a test tube and a solution of I_2 in dilute aqueous NaOH were added? Write equations where appropriate.

(a)   **(b)**   **(c)**

13.57 Complete the following equations:

(a) $CH_2{=}CHCCH_3 + HBr \longrightarrow$

(b) $CH_2{=}CHCH_2CH_2CCH_3 + HCl \longrightarrow$

(c) $+ CH_3OH$ $\xrightarrow{H^+}$

(d) $(CH_3)_2C{=}C(CO_2CH_2CH_3)_2 + (CH_3)_2CuLi \longrightarrow$

13.58 When 2-cyclohexenone is treated with 2-propenylmagnesium bromide, the following compound is obtained in 68% yield. Suggest a mechanism for this reaction.

Additional Problems

13.59 Predict the major organic products:

(a) $\xrightarrow[\text{warm}]{H_2SO_4}$

(b) $+ (C_6H_5)_3P{=}CH_2 \longrightarrow$

(c) $\xrightarrow[\text{(2) } H_2O, H^+]{\text{(1) } NaC{\equiv}CCH_3}$

(d) CH_3O_2C—⟨benzene⟩—$SCH_2\overset{O}{\underset{||}{C}}CH=CH_2$ + ⟨phthalimide⟩ $:\overset{..}{N}:^-$ ⟶

(e) ⟨bicyclic structure with CH_3O and OCH_3⟩ $\xrightarrow[H_2O]{H_2SO_4}$

(f) $CH_3CH_2\overset{O}{\underset{||}{C}}H$ $\xrightarrow[\text{(2) } H_2O, H^+]{\text{(1) } C_6H_5\overset{OLi}{\underset{|}{C}}HCH_2Li, THF}$

(g) $trans\text{-}CH_3CH=CHCO_2CH_3$ + $BrCl$ $\xrightarrow{CH_2Cl_2}$

(h) ⟨dihydropyran ring with CH_3, O, and OCH_3⟩ $\xrightarrow[H_2O]{HCl}$

13.60 Predict the product of the reaction of cyclopentanone with:

(a) Br_2 in acetic acid
(b) $NaBH_4$, followed by H^+, H_2O
(c) phenylhydrazine with H^+
(d) $(CH_3)_2CHMgBr$, followed by H^+, H_2O
(e) $(C_6H_5)_3P=CHCH_3$
(f) hydrazine, followed by heating in K^+ $^-OC(CH_3)_3$ solution

13.61 Show how the following compounds could be prepared from four-carbon aldehydes or ketones: (a) 3-chloropropanoic acid; (b) 4-octene; (c) 3-methyl-3-heptanol; (d) di-n-butyl ether; (e) n-butylamine; (f) 3-butenoic acid; (g) 3-buten-1-ol.

13.62 Show how you could synthesize the following compounds from organic compounds containing only C, H, and O and any inorganic reagents you wish.

(a) $CH_3CH_2NHCH_2CH_2\overset{O}{\underset{||}{C}}CH_2CH_3$

(b) ⟨cyclohexane ring with CN and $-\overset{O}{\underset{||}{C}}CH_3$⟩

(c) ⟨cyclohexene ring with C_6H_5, C_6H_5, and CN⟩

(d) $ICH_2CH_2\overset{O}{\underset{O}{C}}H$ ⟨cyclic acetal⟩

13.63 In the reductive amination of RCHO with ammonia, a secondary amine is a common by-product. Suggest a general structure for this secondary amine and explain how it is formed.

13.64 When quinone (page 507) is treated with HCl, a chlorohydroquinone is formed. Write the equation for this reaction, showing the intermediates.

13.65 Treatment of one mole of $LiAlH_4$ with three moles of CH_3CH_2OH yields lithium triethoxyaluminum hydride, $LiAlH(OCH_2CH_3)_3$. When 3,3,5-trimethylcyclohexanone is treated with this reducing agent, followed by hydrolysis, an 83% yield of one diastereomer of an alcohol is obtained. Write the formula for this diastereomer, and explain why it is the principal product.

13.66 The following compound has a pair of geometric isomers: (*E*) and (*Z*).

$$\underset{\text{CH}_3\text{CH}_2\text{O}\overset{\displaystyle\text{O}}{\overset{\|}{\text{C}}}\text{CH}=\overset{\displaystyle\overset{\text{CH}_3}{|}}{\text{C}}-\text{NHCH}_2\text{C}_6\text{H}_5}{}$$

(a) One isomer melts at 23°, while the other isomer melts at 75–80°. Which isomer melts at 23°? Explain.

(b) In solution, the two isomers of the preceding enamine–ester are in equilibrium. Write a mechanism for this equilibration.

13.67 The LiAlH$_4$ reduction of a stereoisomeric 3-phenyl-2-pentanone yielded 75% of the (2*R*,3*S*) alcohol and 25% of the (2*S*,3*S*) alcohol.

(a) What was the stereochemistry of the starting ketone?

(b) Why was the (2*R*,3*S*) alcohol the predominant product? (*Hint:* Use models.)

13.68 The formation of a hemiacetal from an aldehyde and an alcohol is catalyzed by base, as well as by acid. The mechanism for the acid-catalyzed reaction was presented in Section 13.5B. Suggest a mechanism for the base-catalyzed reaction of acetaldehyde and methanol.

13.69 Write flow equations for syntheses of the following compounds:

(a) 4-methoxy-2-phenyl-2-butene from compounds containing six or fewer carbons

(b) 4-methyl-1,3-pentadiene from acetone

(c) 3,5-heptanediol from compounds containing four or fewer carbons

(d) 1,3-cyclohexanedione from cyclohexanone

(e) from cyclopentanol

(f) from

(g) from

(h) from monocyclic compounds

(i) $\text{CH}_2=$$=\text{CHCH}=\text{CH}_2$ from $\text{CH}_2=$$=\text{O}$

(j) $\text{CH}_3\text{CHClCCl}_2\text{CHO}$ from a compound containing only C, H, and O.

(k) 1,2,3-decanetriol from *n*-octanal

(l) from

13.70 The final step in the preparation of squaric acid ($C_4H_2O_4$) is an acid hydrolysis of the following tetramethoxy compound:

$$\xrightarrow[100°]{18\% \text{ aqueous HCl}} \text{squaric acid (89\%)} + 4\ CH_3OH$$

(a) Write the formulas for the tautomers of squaric acid.
(b) Describe an experimental technique for determining which structure best represents this compound.

13.71 When alkynes are subjected to oxidative cleavage with aqueous $KMnO_4$ solution, the expected products are carboxylic acids. For example,

$$CH_3CH_2CH_2C{\equiv}CCH_2CH_2CH_3 \xrightarrow[\text{heat}]{KMnO_4.\ H_2O.\ H^+} 2\ CH_3CH_2CH_2CO_2H$$

However, when the above alkyne is subjected to this oxidation, another isolated product is propanoic acid. Using equations, suggest a reason for this.

13.72 In part of the synthesis of a fungitoxin found in brown seaweed, the following conversion was carried out. Suggest a reaction sequence for this conversion.

13.73 *Glutaraldehyde* forms a cyclic hydrate that contains one equivalent of water. What is its structure, and how is it formed? (*Hint*: If *one* aldehyde group is hydrated, what reaction might the molecule undergo?)

$$\overset{O}{\overset{\|}{H C}}(CH_2)_3\overset{O}{\overset{\|}{C}}H$$

glutaraldehyde

used as an antiseptic

13.74 Benzene reacts with formaldehyde and HCl in the presence of zinc chloride to yield benzyl chloride. Write a mechanism for this reaction.

13.75 Incubation of an α-keto ester with fermenting baker's yeast results in an enzymatic asymmetric reduction.

Two diastereomers are formed because the β carbon of the starting diketo ester (and thus of the products) is (R) and (S), but only one configuration of the α carbon is generated. For determination of the configuration of the α carbon, the product **I** was treated with Zn and HCl to yield another ester **II**, which was then converted to carboxylic acid **III** by reaction with NaOH followed by acidification. The carboxylic acid **III** was shown to be the same as that obtained by the hydrogenation of the benzene ring of (R)-mandelic acid.

$$CO_2H$$
$$H \blacktriangleright C \blacktriangleleft OH$$
$$C_6H_5$$

(R)-mandelic acid

(1) How many chiral carbons does compound **III** have?
(2) What is the absolute configuration of the α carbon resulting from the reduction by baker's yeast?
(3) Write equations that relate the configurations of **I** and (R)-mandelic acid.

13.76 *Civetone* is an active ingredient of *civet*, a mixture isolated from the scent glands of the African civet cat and used in perfumes. Civetone, which has the formula $C_{17}H_{30}O$, shows strong absorption in the infrared spectrum at 1700 cm^{-1} (5.8 μm) and shows no offset downfield absorption in the proton nmr spectrum. Treatment of civetone with Br$_2$ in CCl$_4$ yields a single dibromide A, $C_{17}H_{30}Br_2O$. Oxidation of civetone with KMnO$_4$ solution yields a diacid B, $C_{17}H_{30}O_5$. Oxidation of civetone with hot, concentrated HNO$_3$ yields principally $HO_2C(CH_2)_7CO_2H$ and $HO_2C(CH_2)_6CO_2H$. Hydrogenation of civetone (Pd catalyst, no heat or pressure), followed by oxidation with hot HNO$_3$, yields a diacid $HO_2C(CH_2)_{15}CO_2H$. What are the structures of civetone, A, and B?

13.77 *Muscone* $(C_{16}H_{30}O)$ is the active ingredient of *musk*, which is obtained from the scent glands of the male musk deer and is also used in perfumes. Muscone is oxidized by hot HNO$_3$ to a mixture of diacids, two of which are

$$\underset{\displaystyle HO_2C(CH_2)_{11}CHCH_2CO_2H}{\overset{\displaystyle CH_3}{|}} \quad \text{and} \quad \underset{\displaystyle HO_2C(CH_2)_{12}CHCO_2H.}{\overset{\displaystyle CH_3}{|}}$$

Reduction of muscone with a zinc-mercury amalgam and HCl yields methylcyclopentadecane:

What is the structure of muscone?

13.78 Queen bees secrete a compound A, $C_{10}H_{16}O_3$, that attracts worker bees. Compound A is insoluble in water, but soluble in alkaline solution. It gives a positive iodoform test. When A is subjected to ozonolysis (oxidative work-up), Compound B, $C_8H_{14}O_3$, is obtained. Compound B is also soluble in aqueous base and also gives a positive iodoform test. Oxidation of A with hot HNO_3 yields, among other products, hexanedioic acid and heptanedioic acid. What are the structures of A and B?

13.79 **(a)** Sulcatol, the aggregation pheromone of the ambrosia beetle (*Gnathotricus sulcatus*), can be obtained by the following sequence of reactions: Treatment of 5-methyl-1,4-hexadiene with 9-BBN (a sterically hindered borohydride reagent used to convert nonhindered alkenes to boranes), followed by decomposition of the intermediate with H_2O_2 + NaOH, gave an 80% yield of compound A. Treatment of A with the pyridine–chromium trioxide complex gave compound B in 75% yield. Compound B, in turn, when treated with methylmagnesium iodide, followed by an ammonium chloride work-up, gave sulcatol in 72% yield. What are the structures of A, B, and sulcatol?

(b) When the alarm pheromone of the ant *Iridomyrmex detectus* is treated with $NaBH_4$ in ethanol, sulcatol is obtained in 95% yield. What is the structure of this alarm pheromone?

13.80 One mole of 1,4-butanediol was added to a mixture of 1.0 mole of 1,4-cyclohexanedione and 0.5 g of *p*-toluenesulfonic acid over a 3.5-hr period. Analysis showed the resulting mixture contained 6% starting ketone and two products: 80% A and 14% B. Compound A was isolated in 59% yield as a clear liquid. The infrared spectrum (neat) showed absorption at 1710, 1250, 1125, and 850 cm^{-1} (5.85, 8.00, 8.90, and 11.72 μm). The proton nmr spectrum (CDCl$_3$) showed peaks at δ3.80 (multiplet), 2.50 (triplet), 2.00 (triplet), and 1.72 (multiplet) ppm. The area ratios are 1:1:1:1. What is the structure of Compound A? What is the most probable structure for Compound B?

13.81 When the following iodo ketone was treated with KOH in CH$_3$OH, Compound A (C$_{10}$H$_{16}$O$_2$) was obtained in 93% yield. The infrared spectrum of A showed absorption at 2940, 2860, 1200, 1150, 1090, 1010, and 995 cm^{-1} (3.40, 3.50, 8.33, 8.70, 9.17, 9.90, and 10.05 μm). The ^{13}C nmr spectrum (off-resonance decoupled) showed the following peaks: δ95.9 (singlet), 71.6 (doublet), 48.2 (quartet), 38.9 (triplet), 35.1 (triplet), 35.0 (triplet), 29.2 (doublet). What is the structure of compound A?

the iodo ketone

13.82 Match each of the spectra in Figure 13.10 with one of the following structures:

(a) $(CH_3)_3CCHO$

(b)

(c) $NCCH_2CH_2\overset{\displaystyle CH_3}{\underset{\displaystyle CH_3}{C}}CHO$

(d) $(CH_3)_2N$——CHO

(e) $(CH_3)_2N$——CH_2CHO

Figure 13.10 ¹H nmr spectra for Problem 13.82.

13.83 Compound A has a molecular weight of 132. Treatment of A with $NaBH_4$ in aqueous methanol yields Compound B. The proton nmr spectrum of A and the infrared spectrum of B are shown in Figure 13.11. What are the structures of A and B?

Figure 13.11 Spectra for Problem 13.83.

14 ▼

Carboxylic Acids

A carboxylic acid is an organic compound containing the **carboxyl group**, —CO_2H. The carboxyl group contains a carbonyl group and a hydroxyl group. The interactions of these two groups lead to a chemical reactivity that is unique to carboxylic acids.

planar *polar* *unshared electrons*

Carboxylic acids are important biologically and commercially. Aspirin is a carboxylic acid, as are oleic acid (Section 23.1) and the prostaglandins (Section 23.3).

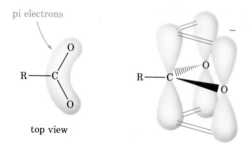

acetylsalicylic acid
(aspirin)

$cis\text{-}CH_3(CH_2)_7CH{=}CH(CH_2)_7CO_2H$

oleic acid

a fatty acid: a component of fats

prostaglandin E_1 (PGE$_1$)

a moderator of hormone activity

Because the carboxyl group is polar and generally nonhindered, its reactions are not usually affected to a great extent by the rest of the molecule. The carboxyl groups in aspirin, oleic acid, and other carboxylic acids undergo similar reactions.

The most notable chemical property of carboxylic acids is their acidity. Ranked against mineral acids such as HCl or HNO$_3$ (pK_a values about 1 or smaller), carboxylic acids are weak acids (pK_a values typically about 5). However, carboxylic acids are more acidic than alcohols or phenols, primarily because of resonance-stabilization of the carboxylate anion, RCO$_2^-$. A p-orbital picture of the carboxylate ion is shown in Figure 14.1.

$$CH_3\overset{O}{\overset{\|}{C}}OH + H_2O \rightleftharpoons \left[CH_3\overset{\overset{..}{\overset{\ddot{O}:}{}}}{\overset{\|}{C}}{-}\overset{..}{\overset{..}{O}}:^- \longleftrightarrow CH_3\overset{:\overset{..}{O}:^-}{\overset{|}{C}}{=}\overset{..}{\overset{..}{O}}: \right] + H_3O^+$$

resonance-stabilized

pi electrons

R—C

O

O

top view

R—C

O

O

side view showing *p*–orbital overlap
(unshared valence electrons not shown)

Figure 14.1 **Bonding in the carboxylate ion, RCO$_2^-$.**

▼ ## Section 14.1
Nomenclature of Carboxylic Acids

The IUPAC name of an aliphatic carboxylic acid is that of the alkane parent with the **-e** changed to **-oic acid**. The carboxyl carbon is carbon 1, just as the aldehyde carbon is. (Note that the final -e of the alkane name is retained in the name of a dioic acid.)

$$HCO_2H \qquad CH_3CH_2\overset{Cl}{\overset{|}{C}}HCO_2H \qquad HO_2CCH_2CO_2H$$

IUPAC: methanoic acid   2-chlorobutanoic acid   propanedioic acid

Table 14.1 Trivial names of the first ten carboxylic acids

Number of Carbons	Structure	Trivial Name	Occurrence and Derivation of Name
1	HCO_2H	formic	ants (L. *formica*)
2	CH_3CO_2H	acetic	vinegar (L. *acetum*)
3	$CH_3CH_2CO_2H$	propionic	milk, butter, and cheese (Gr. *protos*, first; *pion*, fat)
4	$CH_3(CH_2)_2CO_2H$	butyric	butter (L. *butyrum*)
5	$CH_3(CH_2)_3CO_2H$	valeric	valerian root (L. *valere*, to be strong)
6	$CH_3(CH_2)_4CO_2H$	caproic	goat (L. *caper*)
7	$CH_3(CH_2)_5CO_2H$	enanthic	(Gr. *oenanthe*, vine blossom)
8	$CH_3(CH_2)_6CO_2H$	caprylic	goat
9	$CH_3(CH_2)_7CO_2H$	pelargonic	Its ester is found in *Pelargonium roseum*, a geranium.
10	$CH_3(CH_2)_8CO_2H$	capric	goat

For the first four carboxylic acids, the trivial names are used more often than the IUPAC names. (See Table 14.1.) The name **formic acid** comes from *formica* (Latin for "ants"). In medieval times, alchemists obtained formic acid by distilling red ants! **Acetic acid** is from the Latin *acetum*, "vinegar." In its pure form, it is called *glacial* acetic acid. The term "glacial" arises from the fact that pure acetic acid is a viscous liquid that solidifies into an icy-looking solid. The name for **propionic acid** literally means "first fat." Propionic acid is the first carboxylic acid (the one of lowest molecular weight' to exhibit some of the properties of **fatty acids**, which are carboxylic acids obtained from the hydrolysis of fats (discussed in Section 23.1). **Butyric acid** is found in rancid butter (Latin, *butyrum*).

Some other commonly encountered carboxylic acids and names follow:

p-toluic acid   o-phthalic acid   cyclohexanecarboxylic acid   acrylic acid

As with aldehydes and ketones, Greek letters may be used in the trivial names of carboxylic acids to refer to a position in the molecule relative to the carboxyl group.

$$CH_3CH_2CO_2H$$

β carbon   α carbon

$$CH_3CH_2\overset{\underset{\textstyle |}{Br}}{C}HCO_2H$$

α-bromobutyric acid
or 2-bromobutanoic acid

It is sometimes convenient to refer to the RCO— group as an **acyl group** and to RCO_2^- as an **acyloxy group**. For example, the *acylation* of benzene is the substitution of an RCO—group on the aromatic ring.

$$\underset{\text{an acyl group}}{\overset{\displaystyle O}{\overset{\|}{RC{-}}}} \qquad \underset{\text{an acyloxy group}}{\overset{\displaystyle O}{\overset{\|}{RCO{-}}}}$$

$$\underset{\substack{\text{acetyl group} \\ (Ac{-})}}{\overset{\displaystyle O}{\overset{\|}{CH_3C{-}}}} \qquad \underset{\substack{\text{benzoyl group} \\ (Bz{-})}}{\overset{\displaystyle O}{\overset{\|}{\bigcirc{-}C{-}}}} \qquad \underset{\substack{\text{acetoxy group} \\ (AcO{-})}}{\overset{\displaystyle O}{\overset{\|}{CH_3CO{-}}}}$$

The nomenclature of carboxylic acids is also discussed in the appendix.

Study Problems

14.1 Write IUPAC names for the following carboxylic acids:

(a) $CH_2{=}CHCO_2H$ (b) $HO_2CCH_2CH_2CH_2CO_2H$ (c) $BrCHCO_2H$
$\qquad\qquad\qquad\qquad\qquad\qquad\qquad\qquad\qquad\qquad\qquad\qquad\quad\,\, |$
$\qquad\qquad\qquad\qquad\qquad\qquad\qquad\qquad\qquad\qquad\qquad\qquad\,\, CH_3$

14.2 Write trivial names for

(a) $CH_3CHBrCO_2H$ and   (b) $HOCH_2CH_2CO_2H$.

▼ ## Section 14.2
Physical Properties of Carboxylic Acids

The carboxyl group is ideally structured for forming two hydrogen bonds between a pair of molecules. A pair of hydrogen-bonded carboxylic acid molecules is often referred to as a **carboxylic acid dimer**. Because of the strength
of these hydrogen bonds (a total of about 10 kcal/mol for the two hydrogen
bonds), carboxylic acids are found as dimers to a limited extent even in the
vapor phase.

a carboxylic acid dimer   hydrogen bonds with H_2O

Study Problem

14.3 Show the principal hydrogen bonds between molecules of the following compounds:

(a) $\bigcirc{-}CO_2H$ (b) $\overset{CO_2H}{\underset{OH}{\bigcirc}}$

(c) $\bigcirc{-}CO_2H$ and CH_3OH

Table 14.2 Physical properties of some carboxylic acids

Name	Structure	Mp, °C	Bp, °C	Solubility in H_2O at 20°C
formic	HCO_2H	8	100.5	∞
acetic	CH_3CO_2H	16.6	118	∞
propionic	$CH_3CH_2CO_2H$	−22	141	∞
butyric	$CH_3(CH_2)_2CO_2H$	−4.5	165.5	∞
valeric	$CH_3(CH_2)_3CO_2H$	−34	187	3.7 g/100 mL
caproic	$CH_3(CH_2)_4CO_2H$	−3	205	1.0 g/100 mL
cyclohexane- carboxylic	$C_6H_{11}CO_2H$	31	233	0.2 g/100 mL
benzoic	$C_6H_5CO_2H$	122	250	0.3 g/100 mL

The physical properties of carboxylic acids reflect the strong hydrogen bonding between carboxylic acid molecules. The melting points and boiling points are relatively high. The infrared spectra of carboxylic acids also show the effects of hydrogen bonding. (See Section 14.3A.) The lower-molecular-weight acids are water soluble, as well as soluble in organic solvents. The melting points, boiling points, and water solubilities of some carboxylic acids are listed in Table 14.2.

A notable property (not a physical property, but a physiological one) of the lower-molecular-weight carboxylic acids is their odor. Formic and acetic acids have pungent odors. Propionic acid has a pungent odor reminiscent of rancid fat. The odor of rancid butter arises in part from butyric acid. Caproic acid smells like a goat. (Goat sweat, incidentally, contains caproic acid.) Valeric acid (from the Latin *valere*, "to be strong") is not a strong acid, but it does have a strong odor somewhere between that of rancid butter and goat sweat. (Interestingly, valeric acid is the sex attractant of the sugar beet wireworm.) Dogs can differentiate the odors of individual humans because of differing proportions of carboxylic acids in human sweat. The odors of the aliphatic carboxylic acids of ten and more carbons diminish, probably because of their lack of volatility.

Section 14.3
Spectral Properties of Carboxylic Acids

A. Infrared Spectra

Carboxylic acids, either as pure liquids or in solution at concentrations in excess of about 0.01 *M*, exist primarily as hydrogen-bonded dimers rather than as discrete monomers. The infrared spectrum of a carboxylic acid is therefore the spectrum of the dimer. Because of the hydrogen bonding, the OH stretching absorption of carboxylic acids is very broad and very intense. This OH absorption starts around 3300 cm^{-1} (3.0 μm) and slopes into the region of aliphatic carbon–hydrogen absorption. (See Figure 14.2.) The broadness of the carboxylic acid OH band can often obscure both aliphatic and aromatic CH absorption, as well as any other OH or NH absorption in the spectrum.

The carbonyl absorption is observed at about 1700–1725 cm^{-1} (5.8–5.88 μm) and is moderately strong. Conjugation shifts this absorption to lower frequencies: 1680–1700 cm^{-1} (5.9–5.95 μm).

Table 14.3 Characteristic infrared absorption for carboxylic acids

| | Position of Absorption | |
Type of Vibration	cm^{-1}	μm
O—H stretching	2860–3300	3.0–3.5
C=O stretching	1700–1725	5.8–5.88
C—O stretching	1210–1330	7.5–8.26
O—H bending	1300–1440	6.94–7.71
O—H bending (dimer)	~925	~10.8

Figure 14.2 Infrared spectrum of 2-methylbutanoic acid.

Figure 14.3 ^{1}H nmr spectrum of 2-methylpropanoic acid.

The fingerprint region in an infrared spectrum of a carboxylic acid often shows C—O stretching and OH bending. (See Table 14.3.) Another OH bending vibration of the dimer results in a broad absorption near 925 cm^{-1} (10.8 μm).

B. Proton Nmr Spectra

In the nmr spectrum, the absorption of the acidic proton of a carboxylic acid is seen as a singlet far downfield (δ10–13).

The alpha protons are only slightly affected by the C=O group. Their absorption is slightly downfield (about δ2.2) because of the inductive effect of the partially positive carbonyl carbon. There is no unique splitting pattern associated with the carboxylic acid group because the carboxyl proton has no neighboring protons. (See Figure 14.3.)

▼ ## Section 14.4
Preparation of Carboxylic Acids

The numerous synthetic paths that lead to carboxylic acids can be grouped into three types of reactions: (1) *hydrolysis of the derivatives of the carboxylic acids*; (2) *oxidation reactions*; and (3) *Grignard reactions*. These reactions are summarized in Table 14.4.

Hydrolysis of carboxylic acid derivatives results from the attack of water or OH$^-$ on the carbonyl carbon (or the —CN carbon of a nitrile) of the derivatives. The hydrolysis of an ester to yield a carboxylic acid and an alcohol is typical of this class of reactions. We will discuss these reactions in detail in Chapter 15.

$$\underset{\substack{\text{ethyl acetate}\\ \textit{an ester}}}{CH_3\overset{\overset{O}{\|}}{C}OCH_2CH_3} + H_2O \underset{}{\overset{H^+}{\rightleftharpoons}} \underset{\substack{\text{acetic acid}\\ \textit{a carboxylic acid}}}{CH_3\overset{\overset{O}{\|}}{C}OH} + \underset{\substack{\text{ethanol}\\ \textit{an alcohol}}}{HOCH_2CH_3}$$

Oxidation of primary alcohols and aldehydes to yield carboxylic acids was discussed in Sections 7.9C and 13.8. The principal limitation of alcohol oxidation is that the necessary strength of the oxidizing agent precludes the presence of another oxidizable functional group in the molecule (unless it is protected by a blocking group, such as the dibromide of an alkene or an acetal of an aldehyde). Even with this limitation, the oxidation of primary alcohols is the most common oxidative procedure leading to carboxylic acids because alcohols are often available.

$$RCH_2OH \xrightarrow[\text{heat}]{H_2CrO_4} RCO_2H$$

no unblocked double bond, aldehyde, benzyl group, or other —OH group

Table 14.4 Summary of laboratory syntheses of carboxylic acids

Reaction	Section Reference

Hydrolysis:[a]

ester:

$$\underset{O}{\overset{\|}{RC}}-OR' \ + H_2O \ \xrightarrow{H^+ \text{ or } OH^-} \ RCO_2H + HOR'$$ 15.5C

amide:

$$\underset{O}{\overset{\|}{RC}}-NR'_2 \ + H_2O \ \xrightarrow{H^+ \text{ or } OH^-} \ RCO_2H + HNR'_2$$ 15.9C

anhydride:

$$\underset{O}{\overset{\|}{RC}}-\underset{O}{\overset{\|}{OCR'}} \ + H_2O \ \xrightarrow{H^+ \text{ or } OH^-} \ RCO_2H + HO_2CR'$$ 15.4C

acid halide:

$$\underset{O}{\overset{\|}{RC}}-X \ + H_2O \ \xrightarrow{H^+ \text{ or } OH^-} \ RCO_2H + X^-$$ 15.3C

nitrile: $RC{\equiv}N \ + H_2O \ \xrightarrow{H^+ \text{ or } OH^-} \ RCO_2H + NH_3$ 15.11D

Oxidation:

1° alcohol: $RCH_2OH \ + [O]^b \ \longrightarrow \ RCO_2H$ 7.9C

aldehyde: $\underset{O}{\overset{\|}{RCH}} \ + [O] \ \longrightarrow \ RCO_2H$ 13.8

alkene: $RCH{=}CR_2 \ + [O] \ \longrightarrow \ RCO_2H + R_2C{=}O$ 10.14

alkylarene: $Ar-R \ + [O] \ \longrightarrow \ ArCO_2H$ 12.1

methyl ketone:[a] $\underset{O}{\overset{\|}{RCCH_3}} \ + X_2 \ \xrightarrow{OH^-} \ RCO_2H + CHX_3$ 13.11A

Grignard reaction:

$RX \ \xrightarrow[\substack{(2)\ CO_2 \\ (3)\ H_2O,\ H^+}]{(1)\ Mg,\ ether} \ RCO_2H$ 14.4

[a] In alkaline solution, the carboxylate is obtained. The acid can be generated by acidification: $RCO_2^- + H^+ \rightarrow RCO_2H$.

[b] Typical oxidizing agents are $KMnO_4$ or H_2CrO_4 solutions.

The oxidation of aldehydes proceeds with mild oxidizing agents (such as Ag^+) that do not oxidize other groups. However, aldehydes are not as readily available as primary alcohols.

Oxidation of alkenes is primarily an analytical technique, but also can be used in synthesis of carboxylic acids. Like alcohols, alkenes require vigorous oxidizing agents.

cyclohexene   → $\xrightarrow[\text{heat}]{MnO_4^-}$ $HO_2CCH_2CH_2CH_2CH_2CO_2H$

hexanedioic acid
(adipic acid)

Oxidation of substituted alkylbenzenes is an excellent route to substituted benzoic acids. A carboxyl group is a *meta*-director, but an alkyl group is an *o,p*-director. Electrophilic substitution of an alkylbenzene, followed by oxidation, yields *o*- and *p*-substituted benzoic acids.

2,4,6-trinitrotoluene   2,4,6-trinitrobenzoic acid (69%)

*A **Grignard reaction*** between a Grignard reagent (1°, 2°, 3°, vinylic, or aryl) and carbon dioxide (as a gas or as dry ice) is often the method of choice for preparing a carboxylic acid.

This Grignard reaction is directly analogous to the Grignard reactions of ketones or other carbonyl compounds, However, the reaction of a Grignard reagent with carbon dioxide does not yield an alcohol but a *magnesium carboxylate salt*. The magnesium salt is insoluble in the ether solvent used in a Grignard reaction. Therefore, only one of the two pi bonds of CO_2 reacts. Treatment of the insoluble magnesium salt with aqueous acid liberates the *carboxylic acid*.

Step 1:

a carboxylate
(insoluble in ethers)

Step 2:

a carboxylic acid

$$(CH_3)_3CMgCl \xrightarrow[\text{(2) H}_2\text{O, H}^+]{\text{(1) CO}_2} (CH_3)_3CCO_2H$$

(63%)

Let us summarize the common synthetic routes to carboxylic acids from a different viewpoint, that of what happens to the molecule as a whole. Synthesis from an alkyl halide by way of the nitrile or Grignard reagent leads to a carboxylic acid with *one more carbon* than in the alkyl halide.

Chain-lengthening:

$$1° \; RX \xrightarrow[\text{S}_N2]{\text{CN}^-} RCN \xrightarrow{\text{H}_2\text{O, H}^+} RCO_2H$$

$$RX \xrightarrow[\text{ether}]{\text{Mg}} RMgX \xrightarrow[\text{(2) H}_2\text{O, H}^+]{\text{(1) CO}_2} RCO_2H$$

Oxidation of a primary alcohol (or aldehyde) does not affect the length of the carbon chain, nor does hydrolysis of a carboxylic acid derivative.

Same number of carbons:

$$RCH_2OH \xrightarrow[\text{(2) H}^+]{\text{(1) MnO}_4^-, \text{ heat}} RCO_2H$$

Oxidation of an alkene (unless the alkene is cyclic) causes fragmentation of the parent chain.

Fragmentation:

$$RCH=CR_2 \xrightarrow[\text{(2) H}^+]{\text{(1) MnO}_4^-, \text{ heat}} RCO_2H + O=CR_2$$

Study Problem

14.4 Considering only nitrile formation and Grignard reactions, propose a feasible route for each of the following conversions:

(a) $HOCH_2CH_2Cl \longrightarrow HOCH_2CH_2CO_2H$

(b) $(CH_3)_3CCH_2Cl \longrightarrow (CH_3)_3CCH_2CO_2H$

(c)

(d)

▼ ## Section 14.5
How Structure Affects Acid Strength

Carboxylic acids, sulfonic acids (RSO_3H), and alkyl hydrogen sulfates ($ROSO_3H$) are the only classes of organic compounds more acidic than carbonic acid (H_2CO_3). Of these three classes, carboxylic acids are by far the most common.

	RCH_3	RNH_2	$RC\equiv CH$	ROH	H_2O	$ArOH$	H_2CO_3	RCO_2H
approx. pK_a:	~45–50	~35	~25	15–19	15.7	10	6.4	5

increasing acid strength

Acid strength is a term describing the extent of ionization of a Brønsted acid in water: the greater the amount of ionization, the stronger is the acid. The strength of an acid is expressed as its K_a or its pK_a. Table 14.5 lists some carboxylic acids and their pK_a values. (You may want to review the discussion of K_a and pK_a in Section 1.10 before proceeding.) In this section, we will discuss the general structural features that affect the acid strength of an organic compound. Our emphasis will be on carboxylic acids, but we will not limit the discussion to just these compounds.

The reaction of a weak acid with water is reversible. The equilibrium favors the lower-energy side of the equation. Any structural feature that *stabilizes the anion* with respect to its conjugate acid *increases the acid strength* by driving the equilibrium to the H_3O^+ and anion (A^-) side.

$$HA + H_2O \rightleftharpoons H_3O^+ + A^-$$

lower energy means HA is stronger acid

Table 14.5 pK_a values for some carboxylic acids

Trivial Name	Structure	pK_a
formic	HCO_2H	3.75
acetic	CH_3CO_2H	4.75
propionic	$CH_3CH_2CO_2H$	4.87
butyric	$CH_3(CH_2)_2CO_2H$	4.81
trimethylacetic	$(CH_3)_3CCO_2H$	5.02
fluoroacetic	FCH_2CO_2H	2.66
chloroacetic	$ClCH_2CO_2H$	2.81
bromoacetic	$BrCH_2CO_2H$	2.87
iodoacetic	ICH_2CO_2H	3.13
dichloroacetic	Cl_2CHCO_2H	1.29
trichloroacetic	Cl_3CCO_2H	0.7
α-chloropropionic	$CH_3CHClCO_2H$	2.8
β-chloropropionic	$ClCH_2CH_2CO_2H$	4.1
lactic	$\overset{\displaystyle OH}{\overset{\displaystyle |}{CH_3CHCO_2H}}$	3.87
vinylacetic	$CH_2{=}CHCH_2CO_2H$	4.35

The principal factors that affect the stability of A^-, and thus the acid strength of HA, are: (1) electronegativity of the atom carrying the negative charge in A^-; (2) size of A^-; (3) hybridization of the atom carrying the negative charge in A^-; (4) inductive effect of other atoms or groups attached to the negative atom in A^-; (5) resonance-stabilization of A^-; and (6) solvation of A^-. We will discuss each of these features in turn, but keep in mind that they do not work independently of one another.

A. Electronegativity

As we described in Section 1.10, a more electronegative atom holds its bonding electrons more tightly than does a less electronegative atom. In comparisons of anions, the anion with a more electronegative atom carrying the negative ionic charge is generally the more stable anion. Therefore, as we proceed from left to right in the periodic table, we find that the elements form progressively more stable anions and that the conjugate acids are progressively stronger acids.

C   N   O   F

increasing electronegativity of elements

$R_3C{-}H$ $R_2N{-}H$ $RO{-}H$ $F{-}H$

increasing acid strength

Just the reverse is true when we consider the base strengths of the conjugate bases. The anion of a very weak acid is a very strong base, while the anion of a stronger acid is a weaker base.

R_3C^- R_2N^- RO^- F^-

increasing base strength

Consider, for example, the ionization reactions of ethanol and HF in water. As an element, F is more electronegative than O; thus, a fluoride ion is better able to carry a negative charge than is the alkoxide ion. Although HF is a weak acid, it is a much stronger acid than ethanol. Conversely, the fluoride ion is a weaker base than the ethoxide ion.

$$CH_3CH_2OH + H_2O \rightleftharpoons CH_3CH_2O^- + H_3O^+$$

weaker acid   *stronger base*
$pK_a = 15.9$

$$HF + H_2O \rightleftharpoons F^- + H_3O^+$$

stronger acid   *weaker base*
$pK_a = 3.45$

B. Size

A larger atom is better able to disperse a negative charge than is a smaller atom. Dispersal of a charge results in stabilization. Thus, as the size of an atom attached to H increases through a series of compounds in any group of the periodic table, the stability of the anion increases and so does the acid strength. The effect of size can be more important than the effect of electronegativity. Because of the small size of the fluorine atom, HF is a weaker acid than the other hydrogen halides, even though fluorine is more electronegative than the other halogens.

F^- Cl^- Br^- I^-

increasing ionic radii

	HF	HCl	HBr	HI
pK_a:	3.45	-7	-9	-9.5

increasing acid strength

C. Hybridization

In Section 10.1, we discussed why an alkyne with a $\equiv$CH group is weakly acidic. The increasing s character of the hybrid orbitals of carbon in the series $sp^3 \rightarrow sp^2 \rightarrow sp$ means increasing electronegativity of the carbon, and thus increasing polarity of the CH bond and increasing acid strength. A greater electronegativity of the atom bonded to H also enhances anion stability and thus the acidity of the compound. For these reasons, an alkyne C$\equiv$CH loses a proton more readily than does an alkene C=CH. An alkene, in turn, can lose a proton more readily than an alkane.

	CH_3CH_3	$CH_2{=}CH_2$	$CH{\equiv}CH$
approx. pK_a:	~ 50	~ 45	~ 26

increasing acid strength

Again, the strongest acid of the series yields the anion that is the weakest base.

$$CH_3CH_2^- \qquad CH_2{=}CH^- \qquad CH{\equiv}C^-$$

increasing base strength

D. Inductive Effect

So far, we have discussed how the atom bonded directly to a hydrogen affects acid strength. However, other parts of a molecule can also affect acid strength. Compare the pK_a values for acetic acid and chloroacetic acid:

$$CH_3CO_2H \qquad\qquad ClCH_2CO_2H$$

acetic acid   chloroacetic acid
$pK_a = 4.75$ $pK_a = 2.81$

Chloroacetic acid is a much stronger acid than acetic acid. This enhanced acidity arises from the inductive effect of the electronegative chlorine. In the un-ionized carboxylic acid, the electron-withdrawing Cl decreases the electron density at the α carbon. The result is a relatively high-energy structure with adjacent positive charges.

$$\overset{\delta-}{Cl}{\leftarrow}\underset{\delta+}{CH_2}{-}\overset{\displaystyle O}{\overset{\|}{\underset{\delta+}{C}OH}}$$ *adjacent $\delta+$ charges destabilize acid*

The presence of chlorine, however, *reduces* the energy of the anion. The negative charge of the carboxylate group is partially dispersed by the nearby $\delta+$ charge.

$$\underset{\delta-}{Cl}{\leftarrow}\underset{\delta+}{CH_2}{\leftarrow}\overset{\displaystyle O}{\overset{\|}{C}}{-}O^-$$ *nearby $\delta+$ and $-$ charges stabilize anion*

The effect of an electronegative group near the carboxyl group is to strengthen the acid by destabilizing the acid and stabilizing the anion relative to each other.

$$ClCH_2\overset{\displaystyle O}{\overset{\|}{C}}OH + H_2O \rightleftharpoons ClCH_2\overset{\displaystyle O}{\overset{\|}{C}}O^- + H_3O^+$$

less stable   *more stable*
than CH_3CO_2H   *than $CH_3CO_2^-$*

A list of groups in order of their electron-withdrawal power follows:

$$CH_3-\quad H-\quad CH_2=CH-\quad C_6H_5-\quad HO-\quad CH_3O-\quad I-\quad Br-\quad Cl-$$

increasing power of electron-withdrawal →

The pK_a values of the following carboxylic acids reflect the differences in electron-withdrawal by groups attached to $-CH_2CO_2H$:

$CH_3CH_2CO_2H$	CH_3CO_2H	$CH_2=CHCH_2CO_2H$	$C_6H_5CH_2CO_2H$	$HOCH_2CO_2H$	$ClCH_2CO_2H$
pK_a: 4.87	4.75	4.35	4.31	3.87	2.81

increasing acid strength →

Additional electron-withdrawing groups amplify the inductive effect. Dichloroacetic acid is a stronger acid than is chloroacetic acid, and trichloroacetic acid is the strongest of the three.

$ClCH_2CO_2H$	Cl_2CHCO_2H	Cl_3CCO_2H
chloroacetic acid	dichloroacetic acid	trichloroacetic acid
$pK_a = 2.81$	$pK_a = 1.29$	$pK_a = 0.7$

The influence of the inductive effect upon acid strength diminishes with an increasing number of atoms between the carboxyl group and the electronegative group. 2-Chlorobutanoic acid is a substantially stronger acid than butanoic acid itself. However, 4-chlorobutanoic acid has a pK_a value very close to that of the unsubstituted acid.

$CH_3CH_2CH_2CO_2H$	$\overset{Cl}{\underset{\mid}{C}}H_2CH_2CH_2CO_2H$	$CH_3\overset{Cl}{\underset{\mid}{C}}HCH_2CO_2H$	$CH_3CH_2\overset{Cl}{\underset{\mid}{C}}HCO_2H$
pK_a: 4.8	4.5	4.0	2.9

increasing acid strength as distance between —Cl and —CO₂H decreases →

Study Problem

14.5 Which is the stronger acid: (a) phenylacetic acid or bromoacetic acid?   (b) dibromoacetic acid or bromoacetic acid?   (c) 2-iodopropanoic acid or 3-iodopropanoic acid? Explain your answers.

E. Resonance Stabilization

Alcohols, phenols, and carboxylic acids all contain OH groups. Yet these classes of compounds vary dramatically in acid strength. The differences may be attributed directly to the resonance stabilization (or lack of it) of the anion with respect to its conjugate acid.

	ROH	ArOH	RCO_2H
approx. pK_a:	15–19	10	5

In the case of *alcohols*, the anion is not resonance-stabilized. The negative charge of an alkoxide ion resides entirely on the oxygen and is not delocalized. At the opposite end of the scale are the *carboxylic acids*. The negative charge of the carboxylate ion is equally shared by two electronegative oxygen atoms. *Phenols* are intermediate between carboxylic acids and alcohols in acidity. The oxygen of a phenoxide ion is adjacent to the aromatic ring and the negative charge is partially delocalized by the ring, as shown in Section 12.2A.

F. Solvation

The solvation of the anion can play a major role in the acidity of a compound. Solvent molecules can stabilize the anion by helping disperse the negative charge through dipole–dipole interactions. Any factor that increases the degree of solvation of the anion increases the acidity of that compound in solution. For example, water has a greater ability to solvate ions than does ethanol. A water solution of a carboxylic acid is more acidic than an ethanol solution by a factor of about 10^5!

G. Acid Strengths of Substituted Benzoic Acids

You might expect that resonance stabilization by the aromatic ring would play a large role in the relative acid strengths of benzoic acid and substituted benzoic acids; however, this is not the case. The negative charge of the carboxylate ion is shared by the two carboxylate oxygen atoms but cannot be effectively delocalized by the aromatic ring. (The oxygens of the carboxylate anion are not directly adjacent to the aromatic ring; resonance structures in which the negative charge is delocalized by the ring cannot be drawn.)

Even though the negative charge of the benzoate ion is not delocalized by the benzene ring, benzoic acid is a stronger acid than phenol. In the benzoate ion, the negative charge is equally shared by two electronegative oxygen atoms. In the phenoxide ion, however, most of the negative charge resides on the single oxygen atom.

Because the benzene ring does not participate in resonance-stabilization of the carboxylate group, substituents on a benzene ring influence acidity primarily by the inductive effect. Thus, an *electron-withdrawing* group usually enhances the acidity of a benzoic acid. (See Table 14.6.)

all are stronger acids than benzoic acid

Table 14.6 pK$_a$ values for some substituted benzoic acids

| Acid[a] | Position of Substitution versus pK$_a$ | | |
	Ortho	Meta	Para
⬡—CO$_2$H	4.2	4.2	4.2
CH$_3$⬡—CO$_2$H	3.9	4.3	4.4
HO⬡—CO$_2$H	3.0	4.1	4.5
CH$_3$O⬡—CO$_2$H	4.1	4.1	4.5
Br⬡—CO$_2$H	2.9	3.8	4.0
Cl⬡—CO$_2$H	2.9	3.8	4.0
O$_2$N⬡—CO$_2$H	2.2	3.5	3.4

[a] A bond to the center of a benzene ring denotes an unspecified position of substitution.

The reasons that electronegative substituents increase the acid strength are, again, destabilization of the acid and stabilization of the anion.

$$O_2N \leftarrow \bigcirc \overset{\delta+}{\underset{\delta+}{}} \overset{\delta+}{C}O_2H \xrightarrow[\longleftarrow]{-H^+} O_2N \leftarrow \bigcirc \leftarrow CO_2^-$$

*destabilized by
electron-withdrawal*   *stabilized by
electron-withdrawal*

An *electron-releasing* alkyl substituent that is *m-* or *p-* to the carboxyl group decreases the acid strength of a benzoic acid. By releasing electrons, the substituent stabilizes the un-ionized acid and destabilizes the anion.

$$\overset{R}{\underset{}{\bigcirc}} -CO_2H \qquad R \rightarrow \bigcirc -CO_2H$$

*m- or p-alkyl group decreases
acid strength*

Almost all *ortho*-substituents (whether electron-releasing or electron-withdrawing) increase the acid strength of a benzoic acid. The reasons for this

***ortho*-effect**, as it is called, are probably a combination of both steric and electronic factors.

o-substituents increase acid strength

Study Problems

14.6 **(a)** Would you expect *m*-isopropylbenzoic acid to be a stronger or weaker acid than benzoic acid?
(b) Which would be the stronger acid, *m*-nitrobenzoic acid or 3,5-dinitrobenzoic acid?

14.7 The pK_a of *o*-methoxybenzoic acid is 4.1. It is a stronger acid than benzoic acid because of the *ortho*-effect. However, *o*-hydroxybenzoic acid (salicylic acid) is a much stronger acid than the methoxy acid—its pK_a is 3.0. Write the structures of the two anions and suggest why the hydroxyl group has the greater stabilizing effect.

14.8 Rank the following hydrocarbons in order of increasing acidity of the methyl hydrogen (weakest acid first): **(a)** diphenylmethane; **(b)** triphenylmethane; **(c)** toluene; **(d)** methane.

Section 14.6
Reaction of Carboxylic Acids with Bases

The reaction of a carboxylic acid with a base results in a *salt*. An organic salt has many of the physical properties of its inorganic counterparts. Like NaCl or KNO_3, an organic salt is high melting, water soluble, and odorless.

$$HCO_2H \; + \; Na^+ \, OH^- \longrightarrow \; HCO_2^- \, Na^+ \; + \; H_2O$$

formic acid   sodium formate

cyclohexane-carboxylic acid   sodium cyclohexane-carboxylate

The carboxylate anion is named by changing the **-ic acid** ending of the carboxylic acid name to **-ate**. In the name of the salt, the name of the cation precedes the name of the anion as a separate word.

$$CH_3CO_2H \quad CH_3CO_2^- \, Na^+$$

acetic acid   sodium acetate

benzoic acid   ammonium benzoate

glutamic acid

an amino acid found in proteins

monosodium glutamate

a flavor enhancer (MSG)

The carboxylate ion is a weak base and can act as a nucleophile. Esters, for example, are occasionally prepared by the reaction of especially reactive alkyl halides and carboxylates (see Section 5.11).

$$
\underset{\text{acetate ion}}{CH_3\overset{\displaystyle O}{\overset{\|}{C}}O^-} + \underset{\text{benzyl bromide}}{\overset{\displaystyle C_6H_5}{\underset{}{C}H_2-Br}} \longrightarrow \underset{\text{benzyl acetate}}{C_6H_5CH_2O\overset{\displaystyle O}{\overset{\|}{C}}CH_3} + Br^-
$$

an ester

Because it is more acidic than carbonic acid, a carboxylic acid undergoes an acid–base reaction with sodium bicarbonate as well as with stronger bases such as NaOH.

$$
CH_3\overset{\displaystyle O}{\overset{\|}{C}}OH + HCO_3^- \longrightarrow CH_3\overset{\displaystyle O}{\overset{\|}{C}}O^- + [H_2CO_3] \longrightarrow H_2O + CO_2
$$

Although carboxylic acids react with sodium bicarbonate, phenols require the stronger base NaOH, and alcohols require a yet stronger base, such as NaNH$_2$.

With NaHCO$_3$:

$$
RCO_2H \xrightarrow{\ HCO_3^-\ } RCO_2^- + CO_2 + H_2O
$$

$$
ArOH \quad \text{and} \quad ROH \xrightarrow{\ HCO_3^-\ } \text{no appreciable reaction}
$$

With NaOH:

$$
RCO_2H \xrightarrow{\ OH^-\ } RCO_2^- + H_2O
$$

$$
ArOH \xrightarrow{\ OH^-\ } ArO^- + H_2O
$$

$$
ROH \xrightarrow{\ OH^-\ } \text{no appreciable reaction}
$$

This difference between phenols and carboxylic acids in reactivity toward NaOH and NaHCO$_3$ is the basis of a simple classification and separation procedure. If a water-insoluble compound dissolves in NaOH solution, but not in NaHCO$_3$ solution, it is likely to be a phenol. On the other hand, if the compound dissolves in *both* NaOH and NaHCO$_3$ solution, it is probably a carboxylic acid.

A carboxylic acid can be extracted from a mixture of water-insoluble organic compounds with sodium bicarbonate solution. The acid forms a sodium salt and becomes water-soluble, while the other organic compounds remain insoluble. The free carboxylic acid is obtained from the water solution by acidification.

When used either as a test or as a separation procedure, the reaction with sodium bicarbonate has its limitations. If the hydrocarbon portion of the carboxylic acid is long, the compound may fail to dissolve in NaHCO$_3$ solution and may even fail to dissolve in NaOH solution. In addition, some phenols, such as the nitrophenols, have acid strengths comparable to those of carboxylic acids; these phenols dissolve in NaHCO$_3$ as well as in NaOH solutions.

Study Problems

14.9 An aldehyde obtained from a storeroom is contaminated with some carboxylic acid (from air oxidation of the aldehyde). Describe in words and equations how you would remove the unwanted acid.

14.10 You are confronted with an ether solution containing a mixture of heptanoic acid, 1-naphthol, and 1-octanol. How would you separate these three components?

1-naphthol

Section 14.7
Esterification of Carboxylic Acids

An **ester of a carboxylic acid** is a compound that contains the —CO$_2$R group, where R may be alkyl or aryl. An ester may be formed by the direct reaction of a carboxylic acid with an alcohol, a reaction called an **esterification reaction**. Esterification is acid-catalyzed and reversible.

General:

$$\underset{\text{a carboxylic acid}}{\text{RCOH}} + \underset{\text{an alcohol}}{\text{R'OH}} \underset{}{\overset{\text{H}^+,\ \text{heat}}{\rightleftarrows}} \underset{\text{an ester}}{\text{RCOR'}} + \text{H}_2\text{O}$$

$$\underset{\text{acetic acid}}{\text{CH}_3\text{COH}} + \underset{\text{ethanol}}{\text{CH}_3\text{CH}_2\text{OH}} \overset{\text{H}^+,\ \text{heat}}{\rightleftarrows} \underset{\text{ethyl acetate}}{\text{CH}_3\text{COCH}_2\text{CH}_3} + \text{H}_2\text{O}$$

benzoic acid   cyclohexanol   cyclohexyl benzoate

The rate at which a carboxylic acid is esterified depends primarily upon the steric hindrance in the alcohol and the carboxylic acid. The acid strength of the carboxylic acid plays only a minor role in the rate at which the ester is formed.

Reactivity of alcohols toward esterification:

3° ROH   2° ROH   1° ROH   CH$_3$OH

increasing reactivity →

Reactivity of carboxylic acids toward esterification:

R$_3$CCO$_2$H   R$_2$CHCO$_2$H   RCH$_2$CO$_2$H   CH$_3$CO$_2$H   HCO$_2$H

increasing reactivity →

Like many reactions of aldehydes and ketones, esterification of a carboxylic acid proceeds through a series of protonation and deprotonation steps. The carbonyl oxygen is protonated, the nucleophilic alcohol attacks the positive carbon, and elimination of water yields the ester.

We can abbreviate this mechanism in the following way:

a carboxylic acid   an ester

In an esterification reaction, it is the C—O bond of the carboxylic acid that is broken and not the O—H bond of the acid nor the C—O bond of the alcohol. Evidence for the mechanism is the reaction of a labeled alcohol such as $CH_3^{18}OH$ with a carboxylic acid. In this case, the ^{18}O stays with the methyl group.

Study Problem

14.11 Write a flow equation for the mechanism of the esterification of acetic acid with $CH_3^{18}OH$.

The esterification reaction is reversible. To obtain a high yield of an ester, we must shift the equilibrium to the ester side. One technique for accomplishing this is to use an excess of one of the reactants (the cheaper one). Another technique is to remove one of the products from the reaction mixture (for example, by the azeotropic distillation of water).

As the amount of steric hindrance in the intermediate increases, the yield of ester decreases. The reason is that esterification is a reversible reaction and the less hindered species (the reactants) are favored. If bulky esters are to be prepared, it is better to use another synthetic route, such as the reaction of an alcohol with an acid anhydride or an acid chloride, which are more reactive than the carboxylic acid (see Chapter 15) and undergo irreversible reactions with alcohols.

Phenyl esters ($RCO_2C_6H_5$) are not generally prepared directly from phenols and carboxylic acids because the equilibrium favors the acid–phenol side rather than the ester side. Phenyl esters, like bulky esters, can be obtained from phenol and the more reactive acid derivatives.

Study Problems

14.12 Predict the esterification products of: **(a)** *p*-toluic acid (page 593) and 2-propanol;   **(b)** terephthalic acid (p-HO$_2$C—C$_6$H$_4$—CO$_2$H) and excess ethanol;   **(c)** acetic acid and (*R*)-2-butanol.

14.13 4-Hydroxybutanoic acid spontaneously forms a cyclic ester, or *lactone*. What is the structure of this lactone?

Section 14.8
Reduction of Carboxylic Acids

The carbonyl carbon of a carboxylic acid is at the highest oxidation state it can attain and still be part of an organic molecule. (The next higher oxidation state is in CO_2.) Other than combustion or oxidation by very strong reagents, such as hot H_2SO_4–CrO_3 (cleaning solution), the carboxylic acid group is inert toward further oxidative reaction.

Surprisingly, the carboxylic acid group is also inert toward most common reducing agents (such as hydrogen plus catalyst). This inertness made necessary the development of alternative reduction methods, such as conversion of the carboxylic acid to an ester and then reduction of the ester. However, the introduction of lithium aluminum hydride in the late 1940s simplified the reduction because LiAlH$_4$ reduces a carboxyl group directly to a —CH$_2$OH group. (Other carbonyl functionality in the molecule is, of course, reduced as well. See Section 13.7.)

$$CH_3CO_2H \xrightarrow[\text{(2) H}_2\text{O, H}^+]{\text{(1) LiAlH}_4} CH_3CH_2OH$$

acetic acid   ethanol

$$\underset{\text{benzoic acid}}{\text{C}_6\text{H}_5\text{—CO}_2\text{H}} \xrightarrow[\text{(2) H}_2\text{O, H}^+]{\text{(1) LiAlH}_4} \underset{\text{benzyl alcohol}}{\text{C}_6\text{H}_5\text{—CH}_2\text{OH}}$$

Study Problem

14.14 Give the structures for the LiAlH$_4$ reduction products of:

(a) (cyclohexenyl)—CO$_2$H   **(b)** O=(cyclohexadienyl)—CO$_2$H

Figure 14.4 Summary of carboxylic acid reactions presented in this chapter. Table 14.8 (page 616) shows some other reactions, along with their section references.

Figure 14.4 summarizes the reactions of carboxylic acids presented in Sections 14.6–14.8.

Section 14.9
Polyfunctional Carboxylic Acids

Dicarboxylic acids and carboxylic acids containing other functional groups often show unique chemical properties. In this section, we will consider a few of the more important polyfunctional carboxylic acids. Hydroxy acids will be mentioned in Chapter 15 and amino acids will be discussed in Chapter 24.

A. Acidity of Dibasic Acids

A **dibasic acid** is one that reacts with *two equivalents of base* (1.0 mol acid reacts with 2.0 mol base). The term **diprotic acid** (two acidic protons) would perhaps be a better term to describe such a compound. In general, dibasic carboxylic acids have a chemistry similar to that of monocarboxylic acids, but let us examine a few of the differences.

With any dibasic acid (inorganic or organic), the first hydrogen ion is removed more easily than the second. Thus, K_1 (the acidity constant for the ionization of the first H^+) is larger than K_2 (that for the ionization of the second H^+), and the value for pK_1 is smaller than that for pK_2. The difference between pK_1 and pK_2 decreases with increasing distance between the carboxyl groups. (Why?) The first and second pK_a values for a few diacids are listed in Table 14.7.

$$HO_2CCH_2CO_2H + H_2O \rightleftharpoons HO_2CCH_2CO_2^- + H_3O^+$$

malonic acid
$pK_1 = 2.83$

$$HO_2CCH_2CO_2^- + H_2O \rightleftharpoons {}^-O_2CCH_2CO_2^- + H_3O^+$$

$pK_2 = 5.69$

Table 14.7 pK_a values for some diacids

Trivial Name[a]	Structure	pK_1	pK_2
oxalic	$HO_2C—CO_2H$	1.2	4.2
malonic	$HO_2CCH_2CO_2H$	2.8	5.7
succinic	$HO_2C(CH_2)_2CO_2H$	4.2	5.6
glutaric	$HO_2C(CH_2)_3CO_2H$	4.3	5.4
adipic	$HO_2C(CH_2)_4CO_2H$	4.4	5.4
pimelic	$HO_2C(CH_2)_5CO_2H$	4.5	5.4
o-phthalic		2.9	5.5

[a] To memorize the trivial names of the diacids with two to seven carbons, note that the first letters (o, m, s, g, a, p) fit the phrase: "Oh my, such good apple pie." (o-Phthalic acid is, of course, not part of the homologous series.)

B. Anhydride Formation by Dibasic Acids

An **anhydride of a carboxylic acid** has the structure of two carboxylic acid molecules joined together with the loss of water.

Although it would seem reasonable at first glance, heating most carboxylic acids to drive off water does *not* result in the anhydride. Exceptions are dicarboxylic acids that can form five- or six-membered cyclic anhydrides. These diacids yield anhydrides when heated to 200–300°, but milder conditions can be used if acetic anhydride is added. Acetic anhydride reacts with the diacid to form the cyclic anhydride and acetic acid. Distillation of the acetic acid drives the reaction to completion. Anhydrides will be discussed further in Section 15.4.

Study Problem

14.15 What would be the product when each of the following diacids is heated with a dehydrating agent such as acetic anhydride?

(a) o-phthalic acid (Table 14.7)   (b)

C. Decarboxylation of β-Keto Acids and β-Diacids

Simply heating most carboxylic acids does not result in any chemical reaction. However, a carboxylic acid with a β carbonyl group undergoes **decarboxylation** (loss of CO_2) when heated. The temperature necessary depends on the individual compound.

General:

$$
\underset{\text{a }\beta\text{-keto acid}}{\text{RCCH}_2\text{COH}} \xrightarrow{\text{heat}} \underset{\text{a ketone}}{\text{RCCH}_3} + CO_2
$$

$$
\text{CH}_3\text{CCH}_2\text{COH} \xrightarrow{\text{heat}} \text{CH}_3\text{CCH}_3 + CO_2
$$

$$
\text{C}_6\text{H}_5\text{-CCH}_2\text{COH} \xrightarrow{\text{heat}} \text{C}_6\text{H}_5\text{-CCH}_3 + CO_2
$$

Decarboxylation takes place through a cyclic transition state:

transition state   *an enol*

The cyclic transition state requires only a carbonyl group beta to the carboxyl group. This carbonyl group need not necessarily be a keto group. A β-diacid also undergoes decarboxylation when heated. Decarboxylation of substituted malonic acids is especially important in organic synthesis, and you will encounter this reaction again in Chapter 16.

$$
\underset{\text{ethylmalonic acid}}{\text{H-OC-CHCOH}} \xrightarrow{\text{heat}} \underset{\text{butanoic acid}}{\text{CH}_3\text{CH}_2\text{CH}_2\text{COH}} + CO_2
$$

Study Problem

14.16 Write an equation for the mechanism of the decarboxylation of ethylmalonic acid.

A few α-carbonyl acids, such as oxalic acid, can also undergo decarboxylation. The decarboxylation of α-keto acids is common in biological systems in which enzymes catalyze the reaction.

$$\underset{\text{oxalic acid}}{\text{HO}\overset{\overset{\text{O}}{\|}}{\text{C}}-\overset{\overset{\text{O}}{\|}}{\text{C}}\text{O H}} \xrightarrow{150°} \underset{\text{formic acid}}{\text{HCO}_2\text{H}} + \text{CO}_2$$

$$\underset{\substack{\text{pyruvic acid} \\ \textit{formed in glucose} \\ \textit{metabolism}}}{\text{CH}_3\overset{\overset{\text{O}}{\|}}{\text{C}}-\overset{\overset{\text{O}}{\|}}{\text{C}}\text{O H}} \xrightarrow{\text{enzymes}} \underset{\text{acetaldehyde}}{\text{CH}_3\overset{\overset{\text{O}}{\|}}{\text{C}}\text{H}} + \text{CO}_2$$

Study Problem

14.17 Predict the products (if any) when each of the following acids is heated:

(a) $CH_2(CO_2H)_2$ (b) $CH_3\overset{\overset{\text{O}}{\|}}{C}\underset{\underset{CH_3}{|}}{CH}CO_2H$ (c) $CH_3\overset{\overset{\text{O}}{\|}}{C}CH_2CH_2CH_2CO_2H$

D. α,β-Unsaturated Carboxylic Acids

An isolated double bond in an unsaturated carboxylic acid behaves independently of the carboxyl group. A *conjugated* double bond and carboxyl group, however, can undergo typical 1,4-addition reactions. These reactions take place in the same fashion as those of α,β-unsaturated aldehydes or ketones (Section 13.12).

$$^-\!:\!C\equiv N\!: + CH_2\!=\!CH-\overset{\overset{:\ddot{O}}{\|}}{C}-OH \longrightarrow \left[NCCH_2CH\!=\!\overset{\overset{:\ddot{O}:^-}{|}}{C}OH \right] \underset{\textit{an enol}}{\rightleftharpoons} NCCH_2CH_2\overset{\overset{\cdot\dot{O}\cdot}{\|}}{C}\ddot{O}:^-$$

$$\ddot{N}H_3 + \underset{\underset{HO_2C}{}}{\overset{H\quad CO_2H}{C\!=\!C}}_{\quad H} \xrightarrow[180°]{NH_4Cl} \underset{\substack{\text{aspartic acid} \\ \textit{an amino acid found} \\ \textit{in proteins}}}{^-O_2C\overset{\overset{^+NH_3}{|}}{C}HCH_2CO_2H}$$

fumaric acid

Section 14.10
Use of Carboxylic Acids in Synthesis

The types of products that can be obtained from carboxylic acids are shown in Table 14.8. In many of these reactions, an acid halide or an acid anhydride gives better yields. Esterification is such a reaction. However, the acids themselves are usually more readily available.

Table 14.8 Compounds that can be obtained from carboxylic acids

Reaction	Product	Section Reference
Neutralization:[a]		
$RCO_2H + Na^+ OH^- \xrightarrow{H_2O} RCO_2^- Na^+$	carboxylate salt	14.5, 14.6
Esterification:		
$RCO_2H + R'OH \xrightarrow{H^+} RCO_2R'$	ester	14.7
Reduction:		
$RCO_2H \xrightarrow[\text{(2) } H_2O, H^+]{\text{(1) LiAlH}_4} RCH_2OH$	1° alcohol	14.8
Anhydride Formation:		
$HO_2C(CH_2)_nCO_2H \xrightarrow[\text{heat}]{(CH_3C)_2O}$ cyclic anhydride structure where $n = 2$ or 3	cyclic anhydride	14.9B 15.4B
Decarboxylation:		
$RCCR'_2CO_2H \xrightarrow{\text{heat}} RCCHR'_2$	ketone	14.9C
$HO_2CCR_2CO_2H \xrightarrow{\text{heat}} R_2CHCO_2H$	carboxylic acid	14.9C
1,4-Addition:		
$R_2C{=}CHCO_2H + HNu: \longrightarrow R_2CCH_2CO_2H$ (with Nu substituent)	β-substituted carboxylic acid	14.9D
Acid Halide Formation:[b]		
$RCO_2H + SOCl_2 \text{ or } PCl_3 \longrightarrow RCCl$	acid halide	15.3B
α Halogenation:		
$RCH_2CO_2H + Cl_2 \xrightarrow{PCl_3} RCHClCO_2H$	α-chloro acid	15.3C

[a] Carboxylate salts can be used to make esters by S_N2 reactions with reactive halides:

$$RCO_2^- + R'X \longrightarrow RCO_2R' \quad \text{(See Section 5.11.)}$$

[b] Acid halides may be used to make a variety of compounds, including anhydrides and esters. These reactions are discussed in Chapter 15.

Study Problems

14.18 Show two routes to methyl butanoate from butanoic acid.

14.19 Show by flow equations how you would make the following conversions:

(a) propanoic acid $\longrightarrow$ butanoic acid

(b) 3-chloropropanoic acid $\longrightarrow$ butanedioic acid

(c) $\longrightarrow$ (phenyl)$-CH_2CO_2H$

▼ Summary

Carboxylic acids (RCO_2H) undergo hydrogen bonding to form **dimers**, a structural feature that has an effect on their physical and spectral properties.

Carboxylic acids can be synthesized by: (1) the *hydrolysis of their derivatives* (esters, amides, anhydrides, acid halides, or nitriles); (2) *oxidation of primary alcohols, aldehydes, alkenes, or alkylbenzenes*; and (3) *Grignard reactions* of RMgX and CO_2.

The strength of an acid is determined by the relative stabilities of the acid and its anion. Acid strength is affected by *electronegativity* (HF > ROH > R_2NH > RH), by *size* (HI > HBr > HCl > HF), and by *hybridization* ($\equiv CH > {=}CH_2 > -CH_3$). The *inductive effect* of electron-withdrawing groups causes an increase in acid strength ($ClCH_2CO_2H > CH_3CO_2H$). *Resonance stabilization* of the anion also strengthens an acid ($RCO_2H > ArOH > ROH$). The anion of an acid can be partially stabilized by *solvation*—increased solvation of the anion strengthens an acid.

The strength of a benzoic acid is determined largely by inductive effects because the $-CO_2^-$ group does not enter into resonance with the aromatic ring. Electron-withdrawing substituents strengthen the acid, while electron-releasing groups weaken the acid. Substitution at the *o*-position almost always increases acid strength.

$$O_2N \leftarrow\!\!\!\bigcirc\!\!\!-CO_2H > \bigcirc\!\!\!-CO_2H > CH_3 \rightarrow\!\!\!\bigcirc\!\!\!-CO_2H$$

Carboxylic acids are one of the few general classes of organic compounds that are more acidic than H_2CO_3; carboxylic acids react with HCO_3^-. **Carboxylate salts** result from the reaction of a carboxylic acid with base.

$$\underset{\substack{a\ carboxylic \\ acid}}{RCO_2H} + \underset{a\ base}{NaOH} \longrightarrow \underset{\substack{a\ sodium \\ carboxylate}}{RCO_2^-\ Na^+} + H_2O$$

The reactions of carboxylic acids are summarized in Table 14.8, page 616.

Study Problems*

14.20 Write an IUPAC name for each of the following carboxylic acids or salts.

(a) $(CH_3)_3CCO_2H$

(b) $CH_3CH_2CHBrCHBrCO_2H$

(c)
$$\underset{HO}{\overset{OH}{\bigcirc}}\!\!-CO_2H$$
an analgesic

(d) $Cl-\!\!\!\bigcirc\!\!\!-OCH_2CO_2H$
a plant growth regulator and fruit thinner

** For information concerning the organization of the *Study Problems* and *Additional Problems*, see the *Note to student* on page 39.*

(e) $\begin{array}{c}CO_2^-\\|\\CO_2^-\end{array}$ Ca^{2+} (f) $Cl-\langle\bigcirc\rangle-CO_2^-\ Na^+$

   used in ceramic glazes   *a food preservative*

14.21 Give the structures for: **(a)** 4-iodobutanoic acid; **(b)** potassium formate; **(c)** disodium *o*-phthalate; **(d)** sodium benzoate; **(e)** *m*-methylbenzoic acid.

14.22 Give the structure of each of the following groups: **(a)** propionyl group; **(b)** butyryl group; **(c)** *m*-nitrobenzoyl group.

14.23 Show the principal types of hydrogen bonding (there may be more than one type) that occur in each of the following systems: **(a)** an aqueous solution of propanoic acid; **(b)** an aqueous solution of lactic acid, $CH_3CH(OH)CO_2H$.

14.24 Show how you could synthesize butanoic acid from each of the following compounds:

(a) 1-bromopropane   **(b)** 1-butanol
(c) butanal   **(d)** 4-octene
(e) $CH_3CH_2CH_2CO_2CH_2CH_3$ **(f)** $CH_3CH_2CH_2CO_2-\langle\bigcirc\rangle$

14.25 Write equations showing how benzoic acid could be synthesized from benzene by way of **(a)** a Grignard reaction and **(b)** a nitrile hydrolysis.

14.26 Calculate the pK_a for each of the following acids (see Section 1.10):
(a) *cis*-$C_6H_5CH=CHCO_2H$, $K_a = 1.3 \times 10^{-4}$
(b) *trans*-$C_6H_5CH=CHCO_2H$, $K_a = 3.65 \times 10^{-5}$

14.27 The pH of a $0.0100M$ solution of an acid was found to be 2.50. Calculate the K_a of this acid.

14.28 How do you account for the following acidities?

   FCH_2CO_2H $ClCH_2CO_2H$ $BrCH_2CO_2H$

   p$K_a = 2.66$ p$K_a = 2.81$ p$K_a = 2.87$

14.29 List the following compounds in order of increasing acid strength (least acidic first):
(a) $CH_3CH_2CHBrCO_2H$ **(b)** $CH_3CHBrCH_2CO_2H$
(c) $CH_3CH_2CH_2CO_2H$ **(d)** $CH_3CH_2CH_2CH_2OH$
(e) C_6H_5OH **(f)** H_2CO_3
(g) Br_3CCO_2H **(h)** H_2O

14.30 Which one of each of the following pairs of carboxylic acids is the stronger acid? **(a)** benzoic acid or *p*-bromobenzoic acid; **(b)** benzoic acid or *m*-bromobenzoic acid; **(c)** *m*-bromobenzoic acid or 3,5-dibromobenzoic acid. Explain.

14.31 Arrange the following compounds in order of increasing acidity (weakest acid first). Give a reason for your answer.

   p-nitrophenol *p*-methylphenol phenol

14.32 Which of the following acids is most acidic? least acidic?

14.33 In each pair, which is the stronger base? Explain.

  (a) $CH_3CH=CH^-$ or $CH_3C\equiv C^-$
  (b) Cl^- or $CH_3CO_2^-$
  (c) $ClCH_2CO_2^-$ or $Cl_2CHCO_2^-$
  (d) $(CH_3)_3CO^-$ or $(CH_3)_3CCO_2^-$
  (e) $CH_3CHClCO_2^-$ or $ClCH_2CH_2CO_2^-$

14.34 Predict the major organic products:

  (a)
+ excess NaOH $\longrightarrow$

  (b) $C_6H_5CH_2Cl + CH_3CH_2\overset{\overset{\textstyle O}{\|}}{C}O^-\ Na^+ \longrightarrow$
  (c) $CH_3CO_2H + CH_3O^- \longrightarrow$
  (d) $CH_3CO_2H + CH_3NH_2 \longrightarrow$
  (e) $CH_3CO_2^- + ClCH_2CO_2H \longrightarrow$
  (f) $HO_2CCH_2CO_2H + 1\ NaHCO_3 \longrightarrow$

  (g)
+ 2 $NaHCO_3 \longrightarrow$

  (h) $C_6H_5CO_2H + CH_3I + CO_3^{2-} \longrightarrow$
  (i) $C_6H_5CO_2H + LiOH \longrightarrow$
  (j) $C_6H_5OH + LiOH \longrightarrow$

14.35 A mixture contains *p*-ethylphenol, benzoic acid, and benzaldehyde. The mixture is dissolved in ether and washed with an aqueous solution of $NaHCO_3$. The bicarbonate solution is Solution A. The mixture is then washed with an aqueous solution of NaOH (Solution B) and finally with water (Solution C). The remaining ether solution is Solution D. Identify the main organic components in Solutions A, B, C, and D.

14.36 How would you separate each of the following pairs of compounds?

  (a) octanoic acid   and   ethyl octanoate
  (b) *p*-butylphenol   and   *p*-butylphenyl acetate
  (c) *p*-butylphenol   and   4-butylcyclohexanol

14.37 Complete the following equations, giving the organic products:

  (a) $HCO_2H + $ $\xrightarrow[\text{heat}]{H^+}$

  (b) excess $-CO_2H + HOCH_2CH_2OH \xrightarrow[\text{heat}]{H^+}$

  (c) $\xrightarrow[\text{heat}]{H^+}$

14.38 List the following acids and alcohols in order of increasing rates of reaction leading to the ester (slowest first). Explain your answer.

  (a) acetic acid with methanol and HCl
  (b) cyclohexanecarboxylic acid with *t*-butyl alcohol and HCl
  (c) cyclohexanecarboxylic acid with ethanol and HCl

14.39 Racemic 2-bromopropanoic acid reacts with (*R*)-2-butanol to produce an ester mixture.

(a) What are the structures (stereochemistry included) of the esters?   (b) Are these esters enantiomers or diastereomers?

14.40 Write equations for the preparation of the following compounds from carboxylic acids:

(a) ethyl formate, $HCO_2CH_2CH_3$, used as a flavoring agent in lemonade
(b) diethyl oxalate, $CH_3CH_2O_2C-CO_2CH_2CH_3$, an industrial intermediate and solvent
(c) ethyl p-aminobenzoate, a topical anesthetic

14.41 Predict the products when the following compounds are treated with $LiAlH_4$ followed by treatment with aqueous acid:

(a) 2,2-dimethylpropanoic acid
(b) succinic acid, $HO_2CCH_2CH_2CO_2H$

(c)

14.42 Suggest reasons why o-phthalic acid has a pK_1 of 2.89 and a pK_2 of 5.51, while terephthalic acid has a pK_1 of 3.51 and a pK_2 of 4.82.

o-phthalic acid   terephthalic acid

14.43 What are the major organic products when the following compounds are heated with acetic anhydride?

(a)

(b) $HO_2CCH_2CH_2CH_2CO_2H$

14.44 Predict the products of heating the following compounds:

(a)

(b)

14.45 Write flow equations that would explain each of the following reactions.

(a) $(HOCH_2)_2C(CO_2CH_2CH_3)_2 + HBr \xrightarrow{heat} CH_2=CCO_2H$ (43%)
with CH_2Br substituent

(b)

$\xrightarrow[\text{heat}]{H_2O, H^+}$

(89%)

14.46 One of the steps in the biological oxidation of glucose to CO_2 and H_2O is the β decarboxylation of oxalosuccinic acid. What is the product of the decarboxylation?

$$
\begin{array}{c}
CO_2H \\
| \\
HO_2CCH_2CHCCO_2H \\
\underset{O}{\parallel}
\end{array}
$$

oxalosuccinic acid

14.47 Predict the products:

(a) ⬡—CO_2H + HCl ⟶

(b) ⬡=CO_2H + H_2O $\xrightarrow{H^+}$

(c) $CH_3CH{=}CHCO_2H$ + 2 CH_3NH_2 ⟶

Additional Problems

14.48 Predict the major organic products when $CH_3CH_2CH_2CH(CO_2H)_2$ is treated with an excess of each of the following reagents: **(a)** aqueous $NaHCO_3$; **(b)** $LiAlH_4$ followed by aqueous acid; **(c)** methanol + HCl; **(d)** methanol + NaOH; **(e)** a mixture of xylenes (dimethylbenzenes) and heat; **(f)** methylmagnesium iodide in diethyl ether; **(g)** aqueous NH_3.

14.49 Suggest reagents for the following conversions:
(a) $H_2NCH_2CH_2CH_2Br$ ⟶ $^+H_3NCH_2CH_2CH_2CO_2^-$
(b) 2-bromobutane ⟶ 2-methylbutanoic acid
(c) CH_3CH_2OH ⟶ $CH_3\underset{\underset{OH}{|}}{CH}CO_2H$

(d) $BrCH_2CH_2CH_2Br$ ⟶ $HO_2CCH_2CH_2CH_2CO_2H$

(e) ⬠=CH_2 ⟶ ⬠—CH_2CO_2H

(f) 1-bromo-2,4,6-trimethylbenzene ⟶ 2,4,6-trimethylbenzoic acid

(g) ⬡ with $CO_2CH_2CH_3$ and $\underset{\underset{O}{\parallel}}{C}CH_3$ ⟶ ⬡—$\underset{\underset{O}{\parallel}}{C}CH_3$

(h) $CH_3CH_2CH_2Br$ ⟶ $CH_3(CH_2)_3CO_2H$

14.50 Predict the major organic products:

(a) a concentrated solution (inert solvent) of $HOCH_2(CH_2)_8CO_2H$ $\xrightarrow[\text{heat}]{H^+}$

(b) a dilute solution (inert solvent) of $HOCH_2(CH_2)_8CO_2H$ $\xrightarrow[\text{heat}]{H^+}$

(c) [bicyclic structure with H, CO_2H] $\xrightarrow{\text{excess } H_2, \text{ Pd}}$

(d) $CH_3C{\equiv}CH$ $\xrightarrow[\text{(3) H}_2\text{O, HCl}]{\substack{\text{(1) NaNH}_2\text{, liq. NH}_3 \\ \text{(2) CO}_2\text{, THF, (CH}_3\text{CH}_2)_2\text{O}}}$

14.51 *Nicotinic acid* (also called *niacin*), a B vitamin, can be obtained by the vigorous oxidation of nicotine (page 462). What is the structure of nicotine acid?

14.52 *Dacron* is a synthetic polymer that can be made by the reaction of terephthalic acid (Problem 14.42) and 1,2-ethanediol. What is the structure of Dacron?

14.53 With $^{14}CH_3I$ as the only source of ^{14}C, show how you would prepare the following labeled compounds:

(a) $CH_3CH_2CO_2{}^{14}CH_3$ (b) $^{14}CH_3CO_2H$ (c) $CH_3{}^{14}CO_2H$

14.54 Write equations to show how benzoic acid can be converted to the following compounds:

14.55 Predict the relative rates for the following reaction when Y is (a) —CH_3; (b) —NO_2; (c) —H; (d) —Br. Explain your answer.

14.56 Predict the product(s), including stereochemistry, when the following compound is heated under reflux with an excess of 1 *M* HCl. (Use equations in your answer.)

14.57 Propose mechanisms for the following reactions:

(a) $CH_3\overset{\text{O}}{\overset{\|}{C}}OH + H_2{}^{18}O \rightleftharpoons CH_3\overset{\text{O}}{\overset{\|}{C}}-{}^{18}OH + CH_3\overset{{}^{18}\text{O}}{\overset{\|}{C}}OH + H_2O$

(b)

(c)

cubanecarboxylic acid

(d) $BrCH_2(CH_2)_4CH_2CO_2H$ $\xrightarrow[\text{(2) H}^+\text{, heat}]{\text{(1) OH}^-}$

(e) $C_6H_5CCl_3$ $\xrightarrow[\text{(2) H}_2\text{O, H}^+]{\text{(1) OH}^-\text{, heat}}$ $C_6H_5CO_2H$

14.58 When tartaric acid (page 153) is heated, it is converted to pyruvic acid. Write a flow equation to show the steps in this reaction.

$$\overset{O}{\overset{\|}{CH_3CCO_2H}}$$

pyruvic acid

14.59 The salt of a carboxylic acid does not show carbonyl absorption at 1660–2000 cm^{-1} (5–6 μm) in the infrared spectrum. Explain.

14.60 Compound A, $C_4H_6O_4$, yielded Compound B, $C_4H_4O_3$, when heated. When A was treated with an excess of methanol and a trace of sulfuric acid, Compound C, $C_6H_{10}O_4$, was obtained. Upon treatment with LiAlH$_4$ followed by hydrolysis, Compound A yielded Compound D, $C_4H_{10}O_2$. What are the structures of Compounds A, B, C, and D?

14.61 Suggest syntheses for the following compounds from compounds containing six or fewer carbons:

(a) $CH_3(CH_2)_7CO_2H$ (b) $CH_3(CH_2)_2\overset{CO_2CH_3}{\overset{|}{CH}}CH_2CO_2CH_3$

(c)

14.62 Suggest synthetic schemes for the following conversions:

(a) ethyl propanoate from bromoethane as the only organic starting material

(b)   from cyclopentanol

(c)   from

14.63 Compound A ($C_4H_8O_2$) was treated with LiAlH$_4$. After hydrolysis, Compound B was isolated. The infrared spectrum of A and the proton nmr spectrum of B are shown in Figure 14.5. What are the structures of A and B?

Figure 14.5 Spectra for Problem 14.63.

14.64 A chemist treated Compound A ($C_4H_7O_2Br$) with potassium *t*-butoxide in *t*-butyl alcohol. After acidification of the product mixture, the chemist isolated two isomeric products: B and C. The infrared spectrum of B and the proton nmr spectrum of C are shown in Figure 14.6. What are the possible structures of A, B, and C?

Figure 14.6 Spectra for Problem 14.64. (The offset absorption for *C* is at $\delta 12.18$.)

15 ▼

Derivatives of Carboxylic Acids

A derivative of a carboxylic acid is a compound that yields a carboxylic acid upon reaction with water.

$$\underset{\text{ethyl acetate}}{CH_3\overset{O}{\overset{\|}{C}}OCH_2CH_3} + H_2O \underset{}{\overset{H^+,\ heat}{\rightleftharpoons}} \underset{\text{acetic acid}}{CH_3\overset{O}{\overset{\|}{C}}OH} + HOCH_2CH_3$$

$$\underset{\text{acetamide}}{CH_3\overset{O}{\overset{\|}{C}}NH_2} + H_2O + H^+ \overset{heat}{\longrightarrow} CH_3\overset{O}{\overset{\|}{C}}OH + NH_4^+$$

In this chapter, we will discuss carboxylic acid halides, anhydrides, esters, amides, and nitriles. Table 15.1 shows some representative examples of these compounds. Note that all the derivatives except nitriles contain **acyl groups**,

Table 15.1 Some derivatives of carboxylic acids

Class	Examples	
	Structure	Trivial Name
acid halide	CH_3COCl	acetyl chloride
	◯—COCl	benzoyl chloride
acid anhydride	$(CH_3CO)_2O$	acetic anhydride
	(◯—CO)$_2$O	benzoic anhydride
ester	$CH_3CO_2CH_2CH_3$	ethyl acetate
	◯—CO_2CH_3	methyl benzoate
amide	CH_3CONH_2	acetamide
	◯—$CONH_2$	benzamide
nitrile	CH_3CN	acetonitrile
	◯—CN	benzonitrile

RCO—. In each case, an electronegative atom is attached to the carbonyl carbon of the acyl group. For this reason, the chemistry of each of these compound classes is similar.

electronegative groups

an ester   an acid halide   an amide   an anhydride

Carboxylic acids and some of their derivatives are found in nature. Fats are triesters, waxes are monoesters, and proteins are polyamides, to name just a few. Acid halides are never found in nature, and anhydrides are found only rarely. One example of a naturally occurring anhydride is *cantharidin*, a cyclic anhydride found in "Spanish fly." Cantharidin is a toxic irritant of the urinary tract. The dried insects were used by the ancient Greeks and Romans as an aphrodisiac. Cantharidin is also reputed to remove warts.

cantharidin

Section 15.1
Reactivity of Carboxylic Acid Derivatives

Why are carboxylic acids, esters, and amides commonly found in nature, while acid halides and anhydrides are not? Why are the carboxylic acid derivatives different from aldehydes or ketones? We can answer these questions by considering the relative reactivities of carboxylic acid derivatives and how they undergo reaction.

The derivatives of carboxylic acids contain *leaving groups* bonded to the acyl carbons, whereas aldehydes and ketones do not. Reagents generally *add* to the carbonyl group of ketones and aldehydes, but *substitute* for the leaving groups of acid derivatives:

Addition:

$$\underset{\text{CH}_3\overset{\overset{\displaystyle O}{\|}}{\text{C}}\text{CH}_3}{} \underset{\longleftarrow}{\overset{\text{HCN}}{\longrightarrow}} \text{CH}_3-\overset{\overset{\displaystyle \text{OH}}{|}}{\underset{\underset{\displaystyle \text{CN}}{|}}{\text{C}}}-\text{CH}_3$$

Substitution:

$$\text{CH}_3\overset{\overset{\displaystyle O}{\|}}{\text{C}}-\text{Cl} \xrightarrow{\text{H}_2\text{O}} \text{CH}_3\overset{\overset{\displaystyle O}{\|}}{\text{C}}-\text{OH} + \text{HCl}$$

good leaving group

In Chapter 5, we mentioned that a good leaving group is a weak base. Therefore, Cl^- is a good leaving group, but ^-OH and ^-OR are poor leaving groups. The reactivity of carbonyl compounds toward substitution at the carbonyl carbon may be directly attributed to the basicity of the leaving group:

$$^-\text{CH}_3 \quad ^-\text{NH}_2 \quad ^-\text{OR} \quad ^-\text{O}\overset{\overset{\displaystyle O}{\|}}{\text{C}}\text{R} \quad \text{X}^-$$

decreasing basicity
(increasing ease of displacement)

$$\underset{\text{R}-\text{C}-\text{R}}{\overset{\overset{\displaystyle O}{\|}}{}} \qquad \underset{\text{R}\text{C}-\text{NH}_2}{\overset{\overset{\displaystyle O}{\|}}{}} \qquad \underset{\text{R}\text{C}-\text{OR}'}{\overset{\overset{\displaystyle O}{\|}}{}} \qquad \underset{\text{R}\text{C}-\text{O}\text{C}\text{R}}{\overset{\overset{\displaystyle O \quad O}{\| \quad \|}}{}} \qquad \underset{\text{R}\text{C}-\text{Cl}}{\overset{\overset{\displaystyle O}{\|}}{}}$$

increasing reactivity toward water

Acid chlorides and acid anhydrides, with good leaving groups, are readily attacked by water. Therefore, we would not expect to find these compounds in the cells of plants or animals. Because of their high reactivity, however, these acid derivatives are invaluable in the synthesis of other organic compounds. A relatively nonreactive carboxylic acid may be converted to one of these more reactive derivatives and then converted to a ketone, an ester, or an amide (Figure 15.1).

Esters and amides are relatively stable toward water. In the laboratory, these compounds require an acid or a base and, usually, heating to undergo

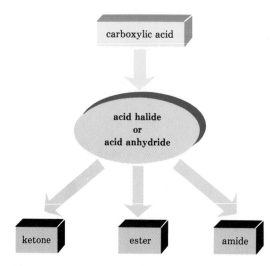

Figure 15.1 The synthetic relationship of acid halides and anhydrides to the other carboxylic acid derivatives and ketones.

hydrolysis. In nature, enzymes can perform the functions of acid or base and heat.

Study Problem

15.1 Both acetyl chloride and vinyl chloride ($CH_2{=}CHCl$) have a chlorine atom bonded to an sp^2-hybridized carbon. Suggest a reason why acetyl chloride is quite reactive while vinyl chloride is almost inert to substitution reactions.

Section 15.2
Spectral Properties of Carboxylic Acid Derivatives

The nmr spectra of carboxylic acid derivatives provide little information about the functionality in these compounds. The signals for α hydrogens of these carbonyl compounds are shifted slightly downfield from the signals for ordinary aliphatic hydrogens because of deshielding by the partially positive carbonyl carbon atom. The α hydrogens of an acid chloride exhibit a greater chemical shift than those of the other acid derivatives. This large chemical shift arises from the greater ability of the Cl (compared to O or N) to withdraw electron density from nearby bonds.

	CH_3CN	CH_3COCH_3	CH_3CNH_2	CH_3COH	CH_3CCl
δ:	2.00	2.03	2.08	2.10	2.67

increasing chemical shift of α hydrogens

The infrared spectra of acid derivatives provide more information about the type of functional group than do nmr spectra. With the exception of the nitriles, the principal distinguishing feature of the infrared spectra of all the

Table 15.2 Infrared carbonyl absorption for carboxylic acid derivatives

		Position of Absorption	
Class	*Structure*	*cm^{-1}*	*μm*
acid chloride	$\overset{\displaystyle O}{\overset{\displaystyle \|}{RCCl}}$	1785–1815	5.51–5.60
acid anhydride	$\overset{\displaystyle O\ \ \ O}{\overset{\displaystyle \|\ \ \ \|}{RCOCR}}$	1740–1840 (usually two peaks)	5.45–5.75
ester	$\overset{\displaystyle O}{\overset{\displaystyle \|}{RCOR}}$	1740	5.75
amide	$\overset{\displaystyle O}{\overset{\displaystyle \|}{RCNH_2}}$	1630–1700	5.9–6.0

carboxylic acid derivatives is the carbonyl absorption found at about 1630–1840 cm^{-1} (5.4–6 μm). The positions of carbonyl absorption for the various acid derivatives are summarized in Table 15.2. Anhydrides and esters also show C—O absorption in the region of 1050–1250 cm^{-1} (8–9.5 μm).

$$\overset{O}{\overset{\|}{RCCl}} \quad \overset{O}{\overset{\|}{RCNH_2}} \quad \overset{O}{\overset{\|}{RC{-}OR}} \quad \overset{O}{\overset{\|}{RC{-}O{-}CR}}$$

— show C=O   — also show C—O

A. Acid Chlorides

The carbonyl infrared absorption of acid chlorides is observed at slightly higher frequencies than that of other acid derivatives. There is no other distinguishing feature in the infrared spectrum that signifies "This is an acid chloride." See Figure 15.2 for the infrared spectrum of a typical acid chloride.

Figure 15.2 Infrared spectrum of decanoyl chloride.

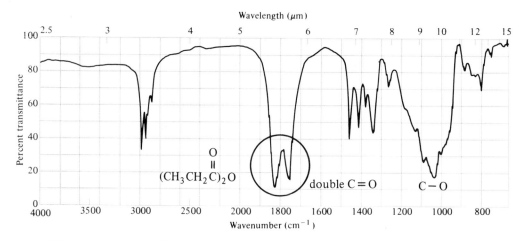

Figure 15.3 **Infrared spectrum of propanoic anhydride.**

B. Anhydrides

A carboxylic acid anhydride, which has two C=O groups, generally exhibits a *double carbonyl peak* in the infrared spectrum. Anhydrides also exhibit a C—O stretching band around 1100 cm^{-1} (9 μm). Figure 15.3 shows the infrared spectrum of a typical aliphatic anhydride.

C. Esters

The carbonyl infrared absorption of aliphatic esters is observed at about 1740 cm^{-1} (5.75 μm). However, conjugated esters (either α,β-unsaturated esters or α-aryl esters) absorb at slightly lower frequencies, about 1725 cm^{-1} (5.8 μm). Esters also exhibit C—O stretching absorption in the fingerprint region. See Figure 15.4 for the spectrum of a typical ester.

Figure 15.4 **Infrared spectrum of *sec*-butyl propanoate.**

Figure 15.5 Infrared spectra of a 1° amide (top), a 2° amide (center), and a 3° amide (bottom).

D. Amides

The position of absorption of the carbonyl group of an amide is variable and depends upon the extent of hydrogen bonding between molecules. The infrared spectrum of a pure liquid amide (maximum hydrogen bonding) shows a carbonyl peak called the **amide I band** around 1650 cm^{-1} (6.0 μm). (The spectra in Figure 15.5 show this C=O peak.) As the sample is diluted with a nonhydrogen-bonding solvent, the extent of the hydrogen bonding diminishes, and the C=O absorption is shifted to a higher frequency (1700 cm^{-1}; 5.88 μm).

The **amide II band** appears between 1515–1670 cm^{-1} (6.0–6.6 μm), just to the right of the C=O absorption. This absorption arises from NH bending. Therefore, a disubstituted, or tertiary, amide does not show an amide II band.

The NH stretching vibrations give rise to absorption to the left of aliphatic CH absorption at 3125–3570 cm^{-1} (2.8–3.2 μm). (This is about the same region where the NH of amines and OH absorb.) *Primary amides* (RCONH$_2$) show a double peak in this region. *Secondary amides* (RCONHR), with only one NH bond, show a single peak. *Tertiary amides* (RCONR$_2$), with no NH, show no absorption in this region. Figure 15.5 shows the infrared spectra of the three types of amides. Compare the NH stretching and bending absorptions of these three compounds.

E. Nitriles

The C≡N absorption is found in the triple-bond region of the infrared spectrum (2200–2300 cm^{-1}; 4.3–4.5 μm) and is medium to weak in intensity. (See Figure 15.6.)

Figure 15.6 Infrared spectrum of pentanenitrile.

Study Problem

15.2 An unknown has the molecular formula $C_5H_7O_2N$. Its infrared and ^1H nmr spectra are given in Figure 15.7. What is the structure of the unknown?

Figure 15.7 Spectra for Problem 15.2.

Section 15.3
Acid Halides

Acid fluorides, chlorides, bromides, and iodides all undergo similar reactions. Because acid chlorides are the most common of the acid halides, our discussions will emphasize these compounds.

A. Nomenclature of Acid Chlorides

Acid chlorides are named after the parent carboxylic acid with the **-ic acid** ending changed to **-yl chloride**.

$$\underset{\text{O}}{\overset{\text{O}}{\|}}$$

	$CH_3\overset{\text{O}}{\overset{\|}{C}}Cl$	$CH_3CH_2\overset{\text{O}}{\overset{\|}{C}}Cl$	$CH_3CH_2CH_2\overset{\text{O}}{\overset{\|}{C}}Cl$
IUPAC:	ethanoyl chloride	propanoyl chloride	butanoyl chloride
trivial:	acetyl chloride	propionyl chloride	butyryl chloride

B. Preparation of Acid Chlorides

Acid chlorides can be obtained directly from their parent carboxylic acids by reaction with thionyl chloride ($SOCl_2$) or some other inorganic acid chloride such as phosphorus trichloride (PCl_3):

General:

$$\underset{\substack{a\ carboxylic \\ acid}}{R\overset{\text{O}}{\overset{\|}{C}}OH} + SOCl_2 \longrightarrow \underset{\substack{an\ acid \\ chloride}}{R\overset{\text{O}}{\overset{\|}{C}}Cl} + SO_2 + HCl$$

$$3\ R\overset{\text{O}}{\overset{\|}{C}}OH + PCl_3 \longrightarrow R\overset{\text{O}}{\overset{\|}{C}}Cl + H_3PO_3$$

$$\underset{\text{butanoic acid}}{CH_3CH_2CH_2\overset{\text{O}}{\overset{\|}{C}}OH} + SOCl_2 \xrightarrow{\text{heat}} \underset{\text{butanoyl chloride (85\%)}}{CH_3CH_2CH_2\overset{\text{O}}{\overset{\|}{C}}Cl} + SO_2 + HCl$$

$$\underset{\text{acetic acid}}{3\ CH_3\overset{\text{O}}{\overset{\|}{C}}OH} + PCl_3 \xrightarrow{\text{heat}} \underset{\text{acetyl chloride (70\%)}}{3\ CH_3\overset{\text{O}}{\overset{\|}{C}}Cl} + H_3PO_3$$

Note the similarity between these reactions and the corresponding reactions of alcohols (Section 7.6).

$$ROH + SOCl_2 \longrightarrow RCl + SO_2 + HCl$$

Study Problem

15.3 Write an equation for the preparation of each of the following acid chlorides:

(a) $CH_3CH_2\overset{\text{O}}{\overset{\|}{C}}Cl$ **(b)** $Cl\overset{\text{O}}{\overset{\|}{C}}{-}\overset{\text{O}}{\overset{\|}{C}}Cl$ **(c)** $C_6H_5CH_2\overset{\text{O}}{\overset{\|}{C}}Cl$

C. Reactions of Acid Chlorides

The acid halides are the most reactive of all the derivatives of carboxylic acids. The halide ion is a good leaving group. Bonded to the positive carbon of a carbonyl group, it is displaced even more easily than when it is bonded to an alkyl carbon. In the following general mechanism for the reaction of an acid chloride with a nucleophile, note that displacement of Cl^- is not a simple displacement like an S_N2 reaction. Rather, the reaction consists of two steps: (1) *addition of the nucleophile to the carbonyl group*, followed by (2) *elimination of the chloride ion*. The result of this reaction is a **nucleophilic acyl substitution**, which means "nucleophilic substitution on an acyl (RCO—) carbon."

$$R-\overset{\overset{\displaystyle \ddot{O}:}{\|}}{\underset{\underset{\displaystyle :Nu}{|}}{C}}-Cl \xrightarrow{\text{(1) addition}} \left[R-\overset{\overset{\displaystyle :\ddot{O}:^-}{|}}{\underset{\underset{\displaystyle Nu}{|}}{C}}-\ddot{\underset{\cdot\cdot}{C}}l: \right] \xrightarrow{\text{(2) elimination}} R-\overset{\overset{\displaystyle \ddot{O}:}{\|}}{\underset{\underset{\displaystyle Nu}{|}}{C}} + :\ddot{\underset{\cdot\cdot}{C}}l:^-$$

intermediate

Hydrolysis. Cleavage by water, called **hydrolysis**, is a typical reaction of an acid chloride with a nucleophile. (Because of hydrogen bonding and solvation effects, the actual mechanism is more complex than we show here.)

Nucleophilic attack and elimination of Cl^-:

$$CH_3\overset{\overset{\displaystyle \ddot{O}:}{\|}}{C}Cl + H_2\ddot{O}: \xrightarrow{\text{addition}} \left[CH_3\overset{\overset{\displaystyle :\ddot{O}:^-}{|}}{\underset{\underset{\displaystyle ^+\ddot{O}H_2}{|}}{C}}-\ddot{\underset{\cdot\cdot}{C}}l: \right] \xrightarrow{\text{elimination}} CH_3\overset{\overset{\displaystyle \ddot{O}:}{\|}}{\underset{\underset{\displaystyle ^+\ddot{O}H_2}{|}}{C}} + :\ddot{\underset{\cdot\cdot}{C}}l:^-$$

Loss of a proton:

$$CH_3\overset{\overset{\displaystyle O}{\|}}{\underset{\underset{\displaystyle ^+\ddot{O}H_2}{|}}{C}} \rightleftharpoons CH_3\overset{\overset{\displaystyle O}{\|}}{\underset{\underset{\displaystyle :\ddot{O}H}{|}}{C}} + H^+$$

Overall reaction:

$$CH_3\overset{\overset{\displaystyle O}{\|}}{C}Cl + H_2O \longrightarrow CH_3\overset{\overset{\displaystyle O}{\|}}{C}OH + HCl$$

acetyl chloride   acetic acid

Although all acid chlorides undergo acidic hydrolysis to yield carboxylic acids and alkaline hydrolysis to yield carboxylic acid salts, their rates of reaction vary. An acid chloride with a bulky alkyl group bonded to the carbonyl group undergoes reaction more slowly than an acid chloride with a small alkyl group. For example, acetyl chloride reacts almost explosively with water, but butanoyl chloride requires gentle reflux.

The effect of the size of the alkyl group on the reaction rate is one of *solubility in water* instead of steric hindrance. An acid chloride with a small alkyl group is more soluble and undergoes reaction faster. An increase in the size of the alkyl portion of the molecule renders the acid chloride less water-soluble; the reaction is slower. If hydrolysis of a variety of acid chlorides is carried out in an inert solvent that dissolves both the acid chloride and water, the rates of hydrolysis are similar.

$$
\underset{\text{CH}_3\text{CH}_2\text{CH}_2\overset{\displaystyle\text{O}}{\overset{\|}{\text{C}}}\text{Cl}}{}\qquad
\underset{\text{CH}_3\text{CH}_2\overset{\displaystyle\text{O}}{\overset{\|}{\text{C}}}\text{Cl}}{}\qquad
\underset{\text{CH}_3\overset{\displaystyle\text{O}}{\overset{\|}{\text{C}}}\text{Cl}}{}
$$

increasing rate of hydrolysis in pure H$_2$O

Study Problems

15.4 Suggest a mechanism for the hydrolysis of butanoyl chloride in dilute, aqueous NaOH.

15.5 Complete the following equation and predict the position(s) of ^{18}O in the product. Explain your answer.

$$
\text{C}_6\text{H}_{11}\!-\!\overset{\displaystyle\text{O}}{\overset{\|}{\text{C}}}\text{Cl} + \text{H}_2{}^{18}\text{O} \longrightarrow
$$

Reaction with alcohols.   Acid chlorides react with alcohols to yield esters and HCl in a reaction that is directly analogous to hydrolysis. The cleavage of an organic compound with an alcohol is referred to as **alcoholysis**. Alcoholysis of acid chlorides is valuable for the synthesis of hindered esters or phenyl esters.

General:

$$
\underset{}{\text{R}\overset{\displaystyle\text{O}}{\overset{\|}{\text{C}}}\text{Cl}} + \underset{\textit{an alcohol}}{\text{R}'\text{OH}} \longrightarrow \underset{\textit{an ester}}{\text{R}\overset{\displaystyle\text{O}}{\overset{\|}{\text{C}}}\text{OR}'} + \text{HCl}
$$

It is usually wise to remove HCl from the reaction mixture as it is formed because HCl can undergo reaction with alcohols. A tertiary amine or pyridine is usually added as a scavenger for HCl.

$$
\underset{\textit{acetyl chloride}}{\text{CH}_3\overset{\displaystyle\text{O}}{\overset{\|}{\text{C}}}\text{Cl}} + \underset{\textit{t-butyl alcohol}}{(\text{CH}_3)_3\text{COH}} + \underset{\textit{N,N-dimethylaniline}}{\text{C}_6\text{H}_5\!-\!\ddot{\text{N}}(\text{CH}_3)_2} \longrightarrow
$$

$$
\underset{\textit{t-butyl acetate (68\%)}}{\text{CH}_3\overset{\displaystyle\text{O}}{\overset{\|}{\text{C}}}\text{OC}(\text{CH}_3)_3} + \underset{\substack{\textit{N,N-dimethylanilinium}\\\textit{chloride}}}{\text{C}_6\text{H}_5\!-\!\overset{+}{\text{N}}\text{H}(\text{CH}_3)_2 \ \text{Cl}^-}
$$

$$\underset{\text{benzoyl chloride}}{\overset{\overset{\text{O}}{\|}}{\bigcirc\!\!\!-\!\!\overset{}{C}Cl}} + \underset{\text{ethanol}}{CH_3CH_2OH} + \underset{\text{pyridine}}{\bigcirc\!\!\!N:} \longrightarrow$$

$$\underset{\text{ethyl benzoate (80\%)}}{\overset{\overset{\text{O}}{\|}}{\bigcirc\!\!\!-\!\!\overset{}{C}OCH_2CH_3}} + \underset{\text{pyridinium chloride}}{\bigcirc\!\!\!\overset{+}{N}H\ Cl^-}$$

Study Problem

15.6 Write the mechanism for the reaction of butanoyl chloride with phenol in the presence of pyridine.

Reaction with ammonia and amines. Ammonia and amines are good nucleophiles. Like other nucleophiles, they react with acid chlorides. The organic product of the reaction is an *amide*. As protons are lost in the deprotonation step, they react with the basic NH_3 or amine. For this reason, at least two equivalents of NH_3 or amine must be used.

$$\underset{\text{ammonia}}{\overset{\overset{\text{O}}{\|}}{CH_3CCl}} + NH_3 \longrightarrow \underset{\text{a 1° amide}}{\overset{\overset{\text{O}}{\|}}{CH_3CNH_2}} + HCl \xrightarrow{\;NH_3\;} NH_4^+\ Cl^-$$

$$\underset{\substack{\text{methylamine} \\ \text{a 1° amine}}}{\overset{\overset{\text{O}}{\|}}{CH_3CCl}} + 2\,CH_3NH_2 \longrightarrow \underset{\text{a 2° amide}}{\overset{\overset{\text{O}}{\|}}{CH_3CNHCH_3}} + CH_3NH_3^+\ Cl^-$$

$$\underset{\substack{\text{dimethylamine} \\ \text{a 2° amine}}}{\overset{\overset{\text{O}}{\|}}{CH_3CCl}} + 2\,(CH_3)_2NH \longrightarrow \underset{\text{a 3° amide}}{\overset{\overset{\text{O}}{\|}}{CH_3CN(CH_3)_2}} + (CH_3)_2NH_2^+\ Cl^-$$

If an amine is expensive, a chemist may not want to use an excess in the reaction with acid chloride. Only one mole of amine is needed to undergo reaction with acid chloride. A second mole is wasted when it is acting merely as a scavenger for HCl. In this case, a less expensive base can be used to remove the HCl. For example, an inexpensive tertiary amine might be used. (A tertiary amine is reactive toward HCl, but cannot form an amide with an acid chloride. Why not?)

Sample Problem

A tertiary amine does not form an amide with an acid chloride, but it does react to form a reactive intermediate called an *acylammonium salt*. (This unstable intermediate does not interfere with amide formation between the acid chloride and NH_3, RNH_2, or R_2NH, however.) Write an equation for the formation of the acylammonium salt of benzoyl chloride and R_3N.

Solution:

an acylammonium salt,

which reacts with $R_2'NH$ to yield $ArC{-}NR_2'$

If the acid halide (such as benzoyl chloride) is not very reactive toward water, aqueous NaOH can be added to remove HX. The reactants and the aqueous NaOH form two layers. As HX is formed, it reacts with NaOH in the water layer. This reaction of an acid chloride and an amine in the presence of NaOH solution is called the **Schotten–Baumann reaction**.

| benzoyl chloride | piperidine | | an amide (80%) | |

Study Problem

15.7 The Schotten–Baumann reaction is not applicable if the acid chloride and amine are water-soluble. What products would you expect from a mixture of acetyl chloride, methylamine, NaOH, and H_2O?

Conversion to anhydrides. Carboxylate ions are nucleophiles, and carboxylate salts (RCO_2Na) can be used for displacement of the chloride of acid chlorides. The product of the reaction is an acid anhydride.

$$CH_3(CH_2)_5CCl + CH_3(CH_2)_5CO^- \longrightarrow CH_3(CH_2)_5COC(CH_2)_5CH_3 + Cl^-$$

heptanoyl chloride   heptanoate ion   heptanoic anhydride (60%)

Study Problem

15.8 Write the mechanism for the reaction of sodium acetate and benzoyl chloride.

Conversion to aryl ketones. Acid halides are usually the reagents of choice for Friedel–Crafts acylation reactions (Section 11.9E). This reaction is a route to aryl alkyl ketones without rearrangement of the alkyl side chain. (Recall that similar *alkylations* go through alkyl carbocations and that rearrangements are common.)

$$\text{benzene} + CH_3CH_2\overset{\displaystyle O}{\overset{\displaystyle \|}{C}}Cl \xrightarrow{\text{AlCl}_3} \text{1-phenyl-1-propanone} + HCl$$

benzene   propanoyl chloride   1-phenyl-1-propanone (90%)

Study Problem

15.9 Without referring to Chapter 11, tell how you would prepare isobutylbenzene from benzene.

Reaction with organometallic compounds.   An acid chloride undergoes re-action with a variety of nucleophiles, including organometallic compounds. Reaction of an acid chloride with a Grignard reagent first yields a ketone, which then undergoes further reaction with the Grignard reagent to yield a *tertiary alcohol* after hydrolysis. (If an excess of the acid halide is used and temperatures of about $-25°$ are maintained, the intermediate ketone can be isolated.)

General:

$$R-\overset{\displaystyle O}{\overset{\displaystyle \|}{C}}-Cl \xrightarrow{R'MgX} R-\overset{\displaystyle O}{\overset{\displaystyle \|}{C}}-R' \xrightarrow{R'MgX} R-\overset{\displaystyle OMgX}{\underset{\displaystyle R'}{C}}-R' \xrightarrow{H_2O, H^+} R-\overset{\displaystyle OH}{\underset{\displaystyle R'}{C}}-R'$$

an acid chloride   *a ketone*   *a 3° alcohol*

$$\text{benzoyl chloride} \xrightarrow[\text{(2) H}_2\text{O, H}^+]{\text{(1) 2 CH}_3\text{MgI}} \text{2-phenyl-2-propanol}$$

benzoyl chloride   2-phenyl-2-propanol

a 3° alcohol

Lithium dialkylcuprates (R_2CuLi) (Section 13.12B) are commonly used to convert acid halides to ketones, often in very high yields. These reactions are usually carried out in an ether solvent at low temperatures (0 to $-78°$). The other derivatives of carboxylic acids do not react with cuprates.

General:

$$R\overset{\displaystyle O}{\overset{\displaystyle \|}{C}}Cl + R'_2CuLi \longrightarrow R\overset{\displaystyle O}{\overset{\displaystyle \|}{C}}R'$$

    a cuprate   *a ketone*

$$(CH_3)_3C\overset{\displaystyle O}{\overset{\displaystyle \|}{C}}Cl + (CH_3)_2CuLi \xrightarrow[-78°]{\text{ether}} (CH_3)_3C\overset{\displaystyle O}{\overset{\displaystyle \|}{C}}CH_3$$

2,2-dimethylpropanoyl chloride   *from CuI and CH₃Li*   3,3-dimethyl-2-butanone (60%)

Study Problem

15.10 Show by equations how you would prepare the following ketones from alkyl halides and acid chlorides:

(a) [structure: benzene ring—C(=O)—CH(CH$_3$)$_2$]   (b) [structure: cyclohexane ring—C(=O)—CH$_3$].

Reduction. Reduction of acid chlorides with lithium aluminum hydride yields primary alcohols. Because primary alcohols also can be obtained by LiAlH$_4$ reduction of the parent acids, this reaction finds little synthetic use.

[reaction scheme: benzoic acid (benzene—COH with C=O) →(SOCl$_2$)→ benzene—CCl (C=O) →((1) LiAlH$_4$ (2) H$_2$O)→ benzene—CH$_2$OH benzyl alcohol; curved arrow below: (1) LiAlH$_4$ (2) H$_2$O, H$^+$]

A much more valuable reaction is the partial reduction of an acid chloride to an aldehyde. (A carboxylic acid itself is not readily reduced to an aldehyde.) A milder reducing agent than LiAlH$_4$ is required to reduce RCOCl to RCHO instead of to RCH$_2$OH. A suitable reagent is **lithium tri-*t*-butoxyaluminum hydride**, which is obtained from *t*-butyl alcohol and LiAlH$_4$. This reducing reagent is less reactive than LiAlH$_4$ because of both steric hindrance and electron-withdrawal by the oxygen atoms.

Preparation of reducing agent:

$$3\,(CH_3)_3COH + LiAlH_4 \longrightarrow Li^+\ H{-}\overset{\displaystyle OC(CH_3)_3}{\underset{\displaystyle OC(CH_3)_3}{Al^-{-}OC(CH_3)_3}} + 3\,H_2$$

t-butyl alcohol

lithium tri-*t*-butoxyaluminum hydride

Reaction with RCOCl:

[structure: benzene ring with O$_2$N at two positions (3,5), and —CCl (C=O)] →((1) LiAlH[OC(CH$_3$)$_3$]$_3$, −78° (2) H$_2$O, H$^+$)→ [structure: benzene ring with O$_2$N at two positions, and —CH (C=O)]

3,5-dinitrobenzoyl chloride   3,5-dinitrobenzaldehyde (63%)

Alpha halogenation. Ketones can be halogenated in the α position by treatment with X$_2$ and H$^+$ or OH$^-$. This reaction, which was discussed in Section 13.11, proceeds by way of the enol. Acid halides also undergo tautomerization and therefore undergo α halogenation.

$$
\underset{\substack{\text{"keto" form}}}{CH_3\overset{\displaystyle O}{\overset{\|}{C}}Cl} \rightleftharpoons \underset{\substack{\text{"enol" form}}}{CH_2{=}\overset{\displaystyle OH}{\overset{|}{C}}Cl} \xrightarrow{Cl_2} \left[CH_2{-}\underset{Cl}{\overset{OH}{\overset{|}{\underset{|}{C}}}}Cl \right] \xrightarrow{-\,HCl} \underset{Cl}{CH_2{-}\overset{\displaystyle O}{\overset{\|}{C}}Cl}
$$

chloroacetyl chloride

an α-halo acid halide

Carboxylic acids do not tautomerize readily and thus do not undergo α halogenation. However, the halogenation of acid halides provides a technique by which we can obtain α-halocarboxylic acids. If a catalytic amount of PCl_3 or PBr_3 is added to a carboxylic acid along with the halogenating agent, the PCl_3 or PBr_3 converts a small proportion of the acid to the acid halide, which undergoes α halogenation.

$$
R_2CH\overset{\displaystyle O}{\overset{\|}{C}}OH \xrightarrow{PBr_3} R_2CH\overset{\displaystyle O}{\overset{\|}{C}}Br \xrightarrow{Br_2} R_2\underset{Br}{\overset{\displaystyle O}{\overset{\|}{C}}}\overset{\displaystyle O}{\overset{\|}{C}}Br
$$

the halogenated
acid halide

In the reaction mixture, the acid halide is in equilibrium with the carboxylic acid itself. Because the acid is present in excess, the α-halo acid halide is converted to the α-halo acid.

can undergo
further reaction

$$
\underset{\substack{\text{the halogenated}\\\text{acid halide}}}{R_2\underset{Br}{\overset{\displaystyle O}{\overset{\|}{C}}}\overset{\displaystyle O}{\overset{\|}{C}}Br} + \underset{\substack{\text{the acid}\\\text{(excess)}}}{R_2CH\overset{\displaystyle O}{\overset{\|}{C}}OH} \rightleftharpoons \underset{\substack{\text{the α-halo acid}}}{R_2\underset{Br}{\overset{\displaystyle O}{\overset{\|}{C}}}\overset{\displaystyle O}{\overset{\|}{C}}OH} + \underset{\substack{\text{the nonhalogenated}\\\text{acid halide}}}{R_2CH\overset{\displaystyle O}{\overset{\|}{C}}Br}
$$

In the process, more nonhalogenated acid halide is formed. This acid halide undergoes α halogenation, and the reaction sequence is repeated. The overall result of this sequence is, in effect, the α halogenation of a carboxylic acid. This reaction sequence is called the **Hell–Volhard–Zelinsky reaction** after the chemists who developed the technique.

Overall reaction:

$$
R_2CHCO_2H + X_2 \xrightarrow{PX_3 \text{ catalyst}} R_2\overset{X}{\overset{|}{C}}CO_2H + HX
$$

Some specific examples of α halogenation of carboxylic acids (by way of acid halides) follow:

$$
\underset{\substack{\text{hexanoic acid}}}{CH_3(CH_2)_4CO_2H} + Br_2 \xrightarrow[\text{heat}]{PBr_3 \text{ catalyst}} \underset{\substack{\text{2-bromohexanoic acid (89\%)}}}{CH_3(CH_2)_3\overset{Br}{\overset{|}{C}}HCO_2H} + HBr
$$

$$(CH_3)_2CHCO_2H \ + \ Br_2 \ \xrightarrow[\text{heat}]{\text{PBr}_3} \ (CH_3)_2\overset{\overset{\displaystyle Br}{|}}{C}-\overset{\overset{\displaystyle O}{\|}}{C}Br \ \xrightarrow{\text{H}_2\text{O}} \ (CH_3)_2\overset{\overset{\displaystyle Br}{|}}{C}-CO_2H$$

2-methylpropanoic acid   2-bromo-2-methylpropanoyl   2-bromo-2-methylpropanoic
              bromide (83%)   acid

Study Problem

15.11 Predict the organic products of each of the following reactions:

(a) $\langle \bigcirc \rangle$—CH_2CO_2H + Br_2 $\xrightarrow{\text{PBr}_3 \text{ catalyst}}$

(b) $ClCH_2CH_2CO_2H$ + Cl_2 $\xrightarrow{\text{PCl}_3 \text{ catalyst}}$

Figure 15.8 summarizes the reactions of acid halides.

Figure 15.8 Summary of reactions of acid halides.

▼ Section 15.4
Anhydrides of Carboxylic Acids

An **anhydride of a carboxylic acid** has the structure of two carboxylic acid molecules with a molecule of water removed. (*Anhydride* means "without water.")

$$CH_3\overset{\overset{\displaystyle O}{\|}}{C}O\,H \quad HO\,\overset{\overset{\displaystyle O}{\|}}{C}CH_3 \xrightarrow{\,-H_2O\,} CH_3\overset{\overset{\displaystyle O}{\|}}{C}O\overset{\overset{\displaystyle O}{\|}}{C}CH_3$$

an anhydride

A. Nomenclature of Anhydrides

Symmetrical anhydrides are those in which the two acyl groups are the same. They are named after the parent carboxylic acid followed by the word **anhydride**.

$$CH_3\overset{\overset{\displaystyle O}{\|}}{C}O\overset{\overset{\displaystyle O}{\|}}{C}CH_3 \qquad CH_3CH_2\overset{\overset{\displaystyle O}{\|}}{C}O\overset{\overset{\displaystyle O}{\|}}{C}CH_2CH_3$$

IUPAC: ethanoic anhydride   propanoic anhydride
trivial: acetic anhydride   propionic anhydride

Unsymmetrical anhydrides are named by the word **anhydride** preceded by the names of the *two* parent acids.

$$CH_3\overset{\overset{\displaystyle O}{\|}}{C}O\overset{\overset{\displaystyle O}{\|}}{C}CH_2CH_3$$

IUPAC: ethanoic propanoic anhydride
trivial: acetic propionic anhydride

Study Problem

15.12 Write a suitable name for each of the following anhydrides:

(a) $CH_3CH_2CH_2CH_2\overset{\overset{\displaystyle O}{\|}}{C}O\overset{\overset{\displaystyle O}{\|}}{C}CH_2CH_2CH_2CH_3$ (b) $C_6H_5\overset{\overset{\displaystyle O}{\|}}{C}O\overset{\overset{\displaystyle O}{\|}}{C}CH_3$

B. Preparation of Anhydrides

With few exceptions, acid anhydrides cannot be formed directly from their parent carboxylic acids, but must be prepared from the more reactive derivatives of carboxylic acids. One route to an anhydride is from an acid chloride with a carboxylate, a reaction mentioned in Section 15.3C.

$$R\overset{\overset{\displaystyle O}{\|}}{C}\!\!-\!\!Cl \quad + \quad {}^-O\overset{\overset{\displaystyle O}{\|}}{C}R' \quad \longrightarrow \quad R\overset{\overset{\displaystyle O}{\|}}{C}\!\!-\!\!O\overset{\overset{\displaystyle O}{\|}}{C}R' \; + \; Cl^-$$

an acid chloride   a carboxylate ion   an anhydride

Another route to an anhydride is by treatment of the carboxylic acid with acetic anhydride. A reversible reaction occurs between a carboxylic acid and an anhydride. The equilibrium can be shifted in the desired direction by distilling the acetic acid as it is formed. This reaction has been mentioned previously (Section 14.9B).

$$2 \;\bigcirc\!\!-CO_2H \;+\; CH_3\overset{O}{\overset{\|}{C}}O\overset{O}{\overset{\|}{C}}CH_3 \quad\underset{}{\overset{heat}{\rightleftharpoons}}\quad \bigcirc\!\!-\overset{O}{\overset{\|}{C}}O\overset{O}{\overset{\|}{C}}\!\!-\!\bigcirc \;+\; 2\,CH_3CO_2H\uparrow$$

benzoic acid   acetic anhydride   benzoic anhydride (74%)   acetic acid

A five- or six-membered cyclic anhydride can be obtained by simply warming the appropriate diacid with acetic anhydride for a few minutes. See Section 14.9B.

C. Reactions of Anhydrides

Like acid halides, acid anhydrides are more reactive than carboxylic acids and can be used to synthesize ketones, esters, or amides. Acid anhydrides undergo reactions with the same nucleophiles that the acid chlorides react with. However, the rates of reaction are slower. (A carboxylate ion is not quite as good a leaving group as a halide ion.) The other product in these reactions is a carboxylic acid or, when the reaction mixture is alkaline, its anion.

General:

$$\underset{RC-OCR}{\overset{\overset{\cdot\cdot}{O}:\;\;O}{\|\;\;\;\;\|}} + :Nu^- \quad\xrightarrow{addition}\quad \left[\underset{\underset{Nu}{|}}{\overset{:\overset{\cdot\cdot}{O}:^-\;O}{RC-OCR}}\right] \quad\xrightarrow{elimination}\quad \underset{\underset{Nu}{|}}{\overset{\cdot O\cdot}{\overset{\|}{RC}}} + :\overset{O}{\overset{\|}{OCR}}$$

intermediate

Hydrolysis. Anhydrides react with water to yield carboxylic acids. The rate of reaction, like the rate of hydrolysis of an acid chloride, depends on the solubility of the anhydride in water.

General:

$$\underset{an\ anhydride}{\overset{O\ \ O}{\overset{\|\ \ \|}{RCOCR'}}} + H_2O \;\longrightarrow\; \underset{carboxylic\ acids}{RCO_2H + R'CO_2H}$$

$$\underset{acetic\ anhydride}{\overset{O\ \ O}{\overset{\|\ \ \|}{CH_3COCCH_3}}} + H_2O \;\longrightarrow\; 2\,\underset{acetic\ acid}{\overset{O}{\overset{\|}{CH_3COH}}}$$

Reaction with alcohols and phenols. The acid-catalyzed reaction of an anhydride with an alcohol or a phenol yields an ester. The reaction is particularly useful with commercially available acetic anhydride.

$$2 \ CH_3\overset{\displaystyle O}{\overset{\|}{C}}O\overset{\displaystyle O}{\overset{\|}{C}}CH_3 \ + \ HO{-}\langle\bigcirc\rangle{-}OH \ \xrightarrow{H_2SO_4} \ CH_3\overset{\displaystyle O}{\overset{\|}{C}}O{-}\langle\bigcirc\rangle{-}O\overset{\displaystyle O}{\overset{\|}{C}}CH_3 \ + \ 2 \ CH_3\overset{\displaystyle O}{\overset{\|}{C}}OH$$

hydroquinone   hydroquinone diacetate (93%)

Phenyl esters can be prepared under either acidic or alkaline conditions. Under alkaline conditions, the sodium salt of the phenol is first prepared and then treated with the anhydride.

$$\langle\bigcirc\rangle{-}OH \ \xrightarrow{NaOH} \ \langle\bigcirc\rangle{-}O^- \ Na^+ \ \xrightarrow{(CH_3C)_2O} \ \langle\bigcirc\rangle{-}O\overset{\displaystyle O}{\overset{\|}{C}}CH_3 \ + \ CH_3\overset{\displaystyle O}{\overset{\|}{C}}O^- \ Na^+$$

phenol   phenyl acetate

Study Problem

15.13 Complete the following equations:

(a) [2-methoxyphenol with OH] $+ \ CH_3\overset{\displaystyle O}{\overset{\|}{C}}O\overset{\displaystyle O}{\overset{\|}{C}}CH_3 \ \xrightarrow{H^+}$

(b) [1-naphthol] $\xrightarrow[\text{(2) } (CH_3CO)_2O]{\text{(1) NaOH}}$

(c) [cyclohexyl]$-OH \ + \ CH_3\overset{\displaystyle O}{\overset{\|}{C}}O\overset{\displaystyle O}{\overset{\|}{C}}CH_3 \ \xrightarrow{H^+}$

(d) [phthalic anhydride] $+ \ CH_3CH_2OH \ \longrightarrow$

Reaction with ammonia and amines. Ammonia, primary amines, and secondary amines react with anhydrides to yield *amides.* For example, ammonia and acetic anhydride yield acetamide, while amines and acetic anhydride give substituted acetamides. One mole of amine is consumed in neutralizing the acetic acid formed in the reaction.

$$CH_3\overset{\displaystyle O}{\overset{\|}{C}}O\overset{\displaystyle O}{\overset{\|}{C}}CH_3 \ + \ 2 \ NH_3 \ \longrightarrow \ CH_3\overset{\displaystyle O}{\overset{\|}{C}}NH_2 \ + \ CH_3CO_2^- \ NH_4^+$$

ammonia   acetamide

Figure 15.9 Summary of reactions of anhydrides of carboxylic acids.

Figure 15.9 summarizes the reactions of anhydrides of carboxylic acids.

Sample Problem

Predict the product of reaction of one mole of succinic anhydride with two moles of ammonia.

Solution:

Study Problem

15.14 The acetyl group is often used as a blocking group to protect an amine function. For example, *o*-aminotoluene can be nitrated at carbon 3 (the carbon *ortho* to the nitrogen) in 55% overall yield by converting the amino group to an amide group, nitrating the amide, and hydrolyzing the amide back to an amine.

(a) Write equations for these reactions.

(b) What product probably constitutes the remaining 45% of the yield? Write equations for its formation.

(c) Predict what would happen if *o*-aminotoluene were treated with nitric acid directly.

Section 15.5
Esters of Carboxylic Acids

Esters, one of the most useful classes of organic compounds, can be converted to a variety of other compounds (Figure 15.10). Esters are common in nature. Fats and waxes are esters (Section 23.1). Esters are also used for synthetic polymers. For example, Dacron is a polyester (Section 15.7). Table 15.3 lists some representative esters.

Volatile esters lend pleasant aromas to many fruits and perfumes. (You might find it interesting to compare the odors of some of these esters to the odors of carboxylic acids, Section 14.2.) Natural fruit flavors are complex blends of many esters with other organic compounds. Synthetic fruit flavors are usually simple blends of just a few esters with a few other substances. Therefore, these synthetic flavors seldom completely duplicate the fuller-bodied natural flavors.

A. Nomenclature of Esters

The name of an ester consists of two words. The first word is the name of the **alkyl group** attached to the ester oxygen. The second word is derived from the

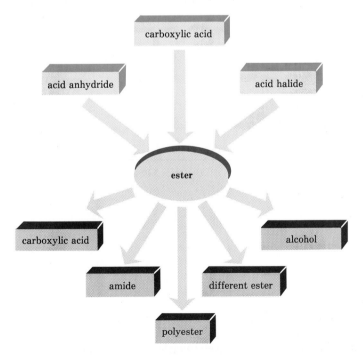

Figure 15.10 The synthetic relationship of esters to other compounds.

Table 15.3 Names, odors, and boiling points of selected esters

Trivial Name	Structure	Odor	Bp, °C
methyl acetate	$CH_3CO_2CH_3$	pleasant	57.5
ethyl acetate	$CH_3CO_2CH_2CH_3$	pleasant	77
propyl acetate	$CH_3CO_2CH_2CH_2CH_3$	like pears	102
ethyl butyrate	$CH_3(CH_2)_2CO_2CH_2CH_3$	like pineapple	121
isoamyl acetate	$CH_3CO_2(CH_2)_2CH(CH_3)_2$	like bananas	142
isobutyl propionate	$CH_3CH_2CO_2CH_2CH(CH_3)_2$	like rum	137
methyl salicylate		like wintergreen	220

carboxylic acid name with **-ic acid** changed to **-ate**. Note the similarity between the name of an ester and that of a carboxylate salt.

IUPAC:	propanoic acid	sodium propanoate	methyl propanoate
trivial:	propionic acid	sodium propionate	methyl propionate

B. Preparation of Esters

Most methods for ester synthesis have been covered elsewhere in this text. In this section, we will provide a summary of these methods.

From carboxylic acids and alcohols (Section 14.7):

4-phenylbutanoic acid   ethanol   ethyl 4-phenylbutanoate (85%)

From acid halides and alcohols (for hindered systems and for phenols) (Section 15.3C):

malonyl chloride   *t*-butyl alcohol   di-*t*-butyl malonate (80%)

From an anhydride and an alcohol or phenol (Section 15.4C):

phthalic anhydride   2-butyl hydrogen phthalate (97%)

From a carboxylate and a reactive alkyl halide (Sections 5.11 and 14.6):

$$CH_3\overset{\displaystyle O}{\overset{\displaystyle \|}{C}}O^- + \langle \bigcirc \rangle\!-\!CH_2Cl \longrightarrow \langle \bigcirc \rangle\!-\!CH_2O\overset{\displaystyle O}{\overset{\displaystyle \|}{C}}CH_3 + Cl^-$$

acetate ion   benzyl chloride   benzyl acetate (93%)

Study Problem

15.15 Write equations to show two different preparations of: **(a)** methyl *n*-butanoate (odor of apples); and   **(b)** 3-methyl-2-butenyl acetate ("Juicy fruit" odor).

C. Reactions of Esters

In *acidic solution*, the carbonyl oxygen of an ester can be protonated. The partially positive carbon can then be attacked by a weak nucleophile such as water.

Protonation:

$$R-\overset{\displaystyle \overset{\cdot\cdot}{O}\cdot}{\underset{}{\overset{\|}{C}}}-OR \;\rightleftharpoons\; \left[R-\overset{\displaystyle \overset{+}{O}H}{\underset{}{\overset{\|}{C}}}-OR \right] \qquad :Nu^-$$

In *alkaline solution*, the carbonyl carbon of an ester can be attacked by a good nucleophile without prior protonation. This is the same addition–elimination path as for nucleophilic attack on acid chlorides or anhydrides.

$$R-\overset{\displaystyle \overset{\cdot\cdot}{O}:^{\delta-}}{\underset{\delta+}{\overset{\|}{C}}}-OR + :Nu^- \xrightarrow{\text{addition}} \left[R-\overset{\displaystyle :\overset{\cdot\cdot}{O}:^-}{\underset{Nu}{\overset{|}{C}}}-\overset{\cdot\cdot}{O}R \right] \xrightarrow{\text{elimination}} R-\overset{\displaystyle \overset{\cdot\cdot}{O}\cdot}{\underset{Nu}{\overset{\|}{C}}} + \;^-:\overset{\cdot\cdot}{O}R$$

intermediate

Acid hydrolysis.   The esterification of a carboxylic acid with an alcohol (Section 14.7) is a reversible reaction. When a carboxylic acid is esterified, an excess of the alcohol is used. To cause the reverse reaction—that is, *acid-catalyzed hydrolysis of an ester to a carboxylic acid*—an excess of water is used. The excess of water shifts the equilibrium to the carboxylic acid side of the equation.

Esterification:

$$\langle \bigcirc \rangle\!-\!\overset{\displaystyle O}{\overset{\displaystyle \|}{C}}OH + \text{excess } CH_3OH \;\underset{}{\overset{H^+,\ heat}{\rightleftharpoons}}\; \langle \bigcirc \rangle\!-\!\overset{\displaystyle O}{\overset{\displaystyle \|}{C}}OCH_3 + H_2O$$

benzoic acid   methanol   methyl benzoate

Hydrolysis:

$$\text{methyl benzoate} \quad C_6H_5-\overset{\overset{\displaystyle O}{\|}}{C}OCH_3 + H_2O \underset{}{\overset{H^+, \text{ heat}}{\rightleftharpoons}} C_6H_5-\overset{\overset{\displaystyle O}{\|}}{C}OH + CH_3OH$$

methyl benzoate   excess   benzoic acid   methanol

If water labeled with oxygen-18 is used in the hydrolysis, the labeled oxygen ends up in the carboxylic acid.

$$R\overset{\overset{\displaystyle O}{\|}}{C}OR + H_2{}^{18}O \overset{H^+, \text{ heat}}{\rightleftharpoons} R\overset{\overset{\displaystyle O}{\|}}{C}{}^{18}OH + ROH \quad \overset{no\ ^{18}O}{\frown}$$

The reason for this is that the water attacks the *carbonyl group*. The RO bond is not broken in the hydrolysis.

$$R\overset{\overset{\displaystyle O}{\|}}{C}-O-R$$

— this bond is broken (between C and O)

— this bond is not broken (between O and R)

The following mechanism accounts for this observation. The first step is *protonation*, followed by *addition of H$_2$O*. A deprotonation, another protonation, and then *elimination of R'OH*, followed by a final deprotonation complete the mechanism.

$$\ddot{O}: \quad R\overset{\overset{\displaystyle O}{\|}}{C}-OR' \overset{H^+}{\rightleftharpoons} \left[R\overset{\overset{\displaystyle +\ddot{O}H}{\|}}{C}-OR' \right] \overset{H_2{}^{18}\ddot{O}:}{\rightleftharpoons} \left[R\overset{\overset{\displaystyle :\ddot{O}H}{|}}{\underset{\underset{\displaystyle {}^{18}\overset{+}{O}H_2}{|}}{C}}-OR' \right] \overset{-H^+}{\rightleftharpoons} \left[R\overset{\overset{\displaystyle :\ddot{O}H}{|}}{\underset{\underset{\displaystyle {}^{18}\ddot{O}H}{|}}{C}}-\ddot{O}R' \right] \overset{H^+}{\rightleftharpoons}$$

$$\left[R\overset{\overset{\displaystyle :\ddot{O}H}{|}}{\underset{\underset{\displaystyle {}^{18}\ddot{O}H}{|}}{C}}-\overset{+}{\underset{H}{\ddot{O}}}R' \right] \overset{-R'OH}{\rightleftharpoons} \left[R\overset{\overset{\displaystyle +\ddot{O}H}{\|}}{\underset{\underset{\displaystyle {}^{18}\ddot{O}H}{|}}{C}} \longleftrightarrow R\overset{\overset{\displaystyle :\ddot{O}H}{|}}{\underset{\underset{\displaystyle {}^{18}\overset{+}{O}H}{\|}}{C}} \right] \overset{-H^+}{\rightleftharpoons} R\overset{\overset{\displaystyle :\ddot{O}}{\|}}{\underset{\underset{\displaystyle {}^{18}\ddot{O}H}{|}}{C}} \quad \text{and} \quad R\overset{\overset{\displaystyle :\ddot{O}H}{|}}{\underset{\underset{\displaystyle {}^{18}\ddot{O}:}{\|}}{C}}$$

resonance structures
for protonated acid

An abbreviated mechanism for ester hydrolysis can be written:

$$R\overset{\overset{\displaystyle O}{\|}}{C}OR' + H_2O \overset{H^+}{\rightleftharpoons} \left[R-\overset{\overset{\displaystyle OH}{|}}{\underset{\underset{\displaystyle OH}{|}}{C}}-OR' \right] \rightleftharpoons R\overset{\overset{\displaystyle O}{\|}}{C}OH + HOR'$$

an ester   *a carboxylic*   *an alcohol*
                                   acid

Study Problem

15.16 What would you expect as acid-hydrolysis products from the following labeled ester?

$$CH_3C(=O)-{}^{18}OCH_2CH_3$$

Alkaline hydrolysis (saponification).   Hydrolysis of an ester in base, or **saponification**, is an *irreversible reaction*. Because it is irreversible, saponification often gives better yields of carboxylic acid and alcohol than does acidic hydrolysis. Because the reaction occurs in base, the product of saponification is the carboxylate salt. The free acid is generated when the solution is acidified. Note that OH⁻ is a reactant, not a catalyst, in this reaction.

Saponification:

methyl benzoate   benzoate ion

Acidification:

benzoic acid

A large body of evidence supports the following mechanistic scheme, which is typical of nucleophilic attack on a carboxylic acid derivative.

Step 1 (addition of OH⁻) (slow):

Step 2 (elimination of ⁻OR′ and proton transfer) (fast):

What is the evidence supporting this mechanism? First, the reaction follows *second-order kinetics*—that is, both the ester and OH⁻ are involved in the rate-determining step. Second, if the alcohol portion of the ester contains a chiral carbon, saponification proceeds with *retention of configuration* in the alcohol. This evidence supports the cleavage of the carbonyl–oxygen bond, not cleavage of the alkyl–oxygen bond.

(R)-2-butyl benzoate   benzoate ion   *(R)*-2-butanol

no racemization or inversion observed;
this bond is cleaved

Study Problem

15.17 Predict the products of saponification of $CH_3\overset{\text{O}}{\underset{\|}{C}}{}^{18}OCH_2CH_3$.

The word *saponification* means "the making of soap." Soaps, which are synthesized by the saponification of fats, will be discussed at greater length in Chapter 23.

glyceryl tripalmitate

a fat, or triglyceride

Sample Problem

Which would have a faster rate of saponification: (a) ethyl benzoate, or   (b) ethyl *p*-nitrobenzoate? Why?

Solution: (b) has the faster rate because it has an electron-withdrawing nitro group. The transition state leading to the intermediate in Step 1 is more stabilized by dispersal of the negative charge:

$$O_2N \leftarrow \overset{\overset{\displaystyle O^-}{|}}{\underset{\underset{\displaystyle OH}{|}}{\bigcirc}} \leftarrow C - OCH_2CH_3$$

Study Problem

15.18 Write equations for the saponification of the following esters with aqueous NaOH:

(a) $\bigcirc - CO_2CH_2CH_3$ (b)   (c)

Transesterification.   Exchange of the alcohol portion of an ester can be accomplished in acidic or basic solution by a reversible reaction between the ester and an alcohol. These **transesterification reactions** are directly analogous to hydrolysis in acid or base. Because the reactions are reversible, an excess of the initial alcohol is generally used.

$$CH_2 {=} CHC\overset{\overset{\displaystyle O}{\|}}{O}CH_3 + CH_3CH_2OH \underset{}{\overset{H^+}{\rightleftarrows}} CH_2 {=} CHC\overset{\overset{\displaystyle O}{\|}}{O}CH_2CH_3 + CH_3OH$$

methyl propenoate   *excess*   ethyl propenoate (99%)
(methyl acrylate)

Study Problem

15.19 Suggest mechanisms for the transesterification reactions of ethyl acetate with
(a) methanol and HCl, and   (b) methanol and sodium methoxide.

Reaction with ammonia.   Esters undergo reaction with aqueous ammonia or amines to yield amides. The reaction is slow compared to the reactions of acid halides or anhydrides with ammonia. This slowness of the ester reaction can be an advantage because the reaction of an acid chloride with an amine can sometimes be violent.

$$ClCH_2C\overset{\overset{\displaystyle O}{\|}}{O}CH_2CH_3 + NH_3 \overset{0°}{\underset{1\ hr}{\longrightarrow}} ClCH_2C\overset{\overset{\displaystyle O}{\|}}{N}H_2 \quad + CH_3CH_2OH$$

ethyl chloroacetate   chloroacetamide (80%)

The ester route to amides is also the reaction of choice for the synthesis of an amide with another functional group that would not be stable toward an acid chloride.

Study Problem

15.20 Why is it not possible to convert 2-hydroxypropanoic acid to 2-hydroxypropanamide by way of an acid chloride? How could this carboxylic acid be converted to its amide?

Reduction. Esters can be reduced under pressure by catalytic hydrogenation, a reaction sometimes called **hydrogenolysis of esters**, or by lithium aluminum hydride. An older technique is the reaction of the ester with sodium metal in ethanol. Regardless of the reducing agent, a pair of alcohols (with at least one being primary) results from the reduction of an ester.

General:

$$\underset{\text{to a primary}}{\overset{\overset{\displaystyle O}{\parallel}}{RC}}\!\!-\!OR' \xrightarrow{\text{[H]}} RCH_2OH + HOR'$$

to a primary alcohol   *to the other alcohol*

$$CH_3(CH_2)_8\overset{\overset{\displaystyle O}{\parallel}}{C}OCH_2CH_3 \xrightarrow[CH_3CH_2OH]{Na} CH_3(CH_2)_8CH_2OH + HOCH_2CH_3$$

ethyl decanoate   1-decanol (70%)   ethanol

$$CH_3CH_2O\overset{\overset{\displaystyle O}{\parallel}}{C}(CH_2)_4\overset{\overset{\displaystyle O}{\parallel}}{C}OCH_2CH_3 \xrightarrow[255°,\ 200\ atm]{\underset{CuCr_2O_4\ catalyst}{H_2}} HOCH_2(CH_2)_4CH_2OH + 2\ HOCH_2CH_3$$

diethyl adipate   1,6-hexanediol (90%)   ethanol

1° hydroxyl from the carbonyl carbon

$$\underset{\substack{\text{a cyclic ester,}\\\text{or lactone}}}{\text{(structure)}} \xrightarrow[\text{(2) H}_2\text{O, H}^+]{\text{(1) LiAlH}_4} CH_3\overset{\overset{\displaystyle OH}{|}}{C}HCH_2CH_2CH_2OH$$

1,4-pentanediol (86%)

Study Problem

15.21 Show by equations how you would prepare $CH_3(CH_2)_{14}CH_2OH$ from glyceryl tripalmitate (page 654).

Reaction with Grignard reagents. The reaction of esters with Grignard reagents is an excellent technique for the preparation of *tertiary alcohols with two identical R groups.*

General:

$$\underset{\text{an ester}}{\overset{\overset{\displaystyle O}{\parallel}}{RC}OR'} \xrightarrow[\text{(2) H}_2\text{O, H}^+]{\text{(1) 2 R''MgX}} R\!-\!\overset{\overset{\displaystyle OH}{|}}{\underset{\underset{\displaystyle R''}{|}}{C}}\!-\!R'' \quad \begin{array}{l}\text{two R'' groups}\\\text{the same}\end{array}$$

a 3° alcohol

$$CH_3\overset{\overset{\displaystyle O}{\parallel}}{C}OCH_2CH_3 \xrightarrow[\text{(2) H}_2\text{O, NH}_4\text{Cl}]{\text{(1) 2 C}_6\text{H}_5\text{MgBr}} CH_3\overset{\overset{\displaystyle OH}{|}}{C}(C_6H_5)_2 \xrightarrow{-H_2O} CH_2\!=\!C(C_6H_5)_2$$

ethyl acetate   1,1-diphenylethanol   1,1-diphenylethene (70%)

If a formate ester is subjected to a Grignard reaction, a *secondary alcohol with two identical R groups* is obtained. Formates are a special case because the carbonyl carbon is attached to an H atom, not an alkyl or aryl group.

$$
\underset{\text{a formate ester}}{HCOR} \quad \xrightarrow[\text{(2) } H_2O, H^+]{\text{(1) } 2\ R'MgX} \quad \underset{\text{a 2° alcohol}}{H-\overset{\overset{\displaystyle OH}{|}}{\underset{\underset{\displaystyle R'}{|}}{C}}-R'} \quad \begin{array}{l}\textit{two R' groups}\\\textit{the same}\end{array}
$$

The mechanism of the reaction of a Grignard reagent with an ester is similar to that of the reaction of a Grignard reagent with an aldehyde or ketone; that is, the nucleophilic carbon of the Grignard reagent attacks the positive carbon of the carbonyl group. With esters, as with acid halides (Section 15.3C), *two* equivalents of Grignard reagent attack the carbonyl carbon atom. To see why this is so, let us consider the reaction stepwise. First, the negative carbon of the Grignard reagent attacks the carbon of the carbonyl group. The product of this step has a hemiketal-like structure that loses an alkoxyl group to yield a ketone.

Initial attack:

$$
RC{-}OR' + \overset{\delta-}{C}\overset{\delta+}{H_3}MgX \longrightarrow \left[RC\overset{CH_3}{\underset{}{\cdots OR'}} \right] \longrightarrow \left[\underset{\text{a ketone}}{RC\overset{O}{\underset{CH_3}{}}} \right] + R'\overset{\cdot\cdot}{\underset{\cdot\cdot}{O}}{:}^-\ {}^+MgX
$$

a type of hemiketal

The ketone then reacts with a *second* molecule of Grignard reagent. The second reaction is faster than the first; hence the ketone cannot be isolated.

Second attack and hydrolysis:

$$
\left[RCCH_3 \right] + CH_3MgX \longrightarrow \underset{CH_3}{RCCH_3} \xrightarrow{H_2O, H^+} \underset{\underset{\text{a 3° alcohol}}{CH_3}}{RCCH_3} + Mg^{2+} + X^- + H_2O
$$

Study Problems

15.22 Show by equations how you would prepare the following alcohols from methyl propanoate: (**a**) 2-methyl-2-butanol;   (**b**) 3-ethyl-3-pentanol.

15.23 Predict the product of the reaction of ethyl formate with ethylmagnesium bromide, followed by work-up with aqueous acid.

Figure 15.11 summarizes the reactions of esters.

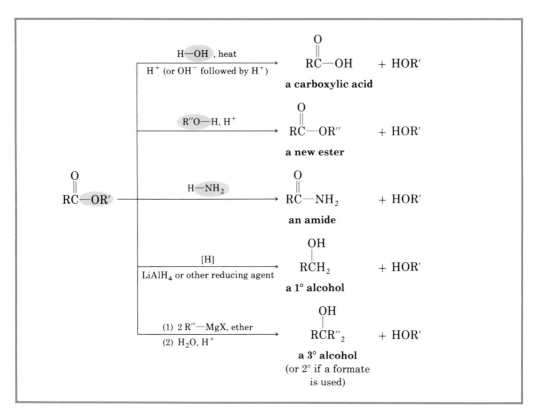

Figure 15.11 Summary of reactions of esters.

Section 15.6
Lactones

A **lactone** is a cyclic ester. Lactones are fairly common in natural sources. For example, vitamin C and nepetalactone, the cat attractant in catnip, are both lactones. It is interesting to note the close structural relationship of nepetalactone to iridomyrmecin, an odorous compound found in the *iridomyrmex* species of ants.

vitamin C
(ascorbic acid)

nepetalactone

cat attractant in catnip

iridomyrmecin

in ants

Lactones are formed from molecules that contain a carboxyl group and a hydroxyl group. These molecules can undergo an intramolecular esterification.

$$\underset{\substack{\text{4-hydroxybutanoic}\\ \text{acid}}}{\underset{\text{a } \gamma\text{-hydroxy acid}}{\begin{array}{c}\text{CH}_2\text{OH}\\ |\quad\diagup\text{CO}_2\text{H}\\ \text{CH}_2\text{CH}_2\end{array}}} \xrightleftharpoons{\text{H}^+} \underset{\substack{\text{4-hydroxybutanoic}\\ \text{acid lactone}\\ (\gamma\text{-butyrolactone})}}{\boxed{\text{O}}} + \text{H}_2\text{O}$$

$$\underset{\substack{\text{5-hydroxypentanoic}\\ \text{acid}}}{\underset{\text{a } \delta\text{-hydroxy acid}}{\begin{array}{c}\diagup\text{CH}_2\text{OH}\\ \text{CH}_2\quad\diagdown\text{CO}_2\text{H}\\ \diagdown\text{CH}_2\text{CH}_2\end{array}}} \xrightleftharpoons{\text{H}^+} \underset{\substack{\text{5-hydroxypentanoic}\\ \text{acid lactone}\\ (\delta\text{-valerolactone})}}{\boxed{\text{O}}} + \text{H}_2\text{O}$$

With γ- or δ-hydroxy acids, which form lactones that are five- or six-membered rings, the cyclization is so facile that the hydroxy acids often cannot be isolated. Although the reaction is catalyzed by acid or base, even a trace of acid from the glassware is sufficient to catalyze lactone formation if a five- or six-membered ring is the product.

Study Problem

15.24 Lactic acid (2-hydroxypropanoic acid) does not form a lactone. However, when it is heated, it yields a dimeric cyclic ester called a **lactide**. What is the structure of this lactide?

Carboxylic acids with hydroxyl groups in the α or β position do not form ordinary cyclic lactones readily because of the small, strained rings that would result. Carboxylic acids with hydroxyl groups farther away than the γ or δ position do not form lactones spontaneously. However, the lactones of these hydroxy acids may be synthesized under the usual conditions for esterification. In these cases, a *dilute solution* of hydroxy acid in an inert solvent is used. An intramolecular reaction is favored by dilute solution because collisions between molecules are less apt to occur. If the solution is *concentrated*, the hydroxy acid molecules undergo reaction with one another to yield a **polyester**. In either case, a solvent such as benzene allows the product water to be distilled as an azeotrope and drives the reaction toward the lactone (or polyester).

$$\xrightarrow[\text{dilute solution}]{\text{H}^+, \text{ heat}} \underset{\textit{the lactone}}{\boxed{\text{O}}} + \text{H}_2\text{O}$$

$$\underset{\text{7-hydroxyheptanoic acid}}{\text{HOCH}_2(\text{CH}_2)_5\overset{\displaystyle\overset{\text{O}}{\|}}{\text{C}}\text{OH}}$$

$$\xrightarrow[\text{concentrated solution}]{\text{H}^+, \text{ heat}} \underset{\textit{the polyester}}{\left[\text{OCH}_2(\text{CH}_2)_5\overset{\displaystyle\overset{\text{O}}{\|}}{\text{C}}\right]_x} + \text{H}_2\text{O}$$

Section 15.7
Polyesters

The synthetic fiber **Dacron** (Figure 15.12) is a polyester made by the transesterification reaction of dimethyl terephthalate and ethylene glycol. The reason that polymer formation can occur is that the reactants are *bifunctional*. Thus, each reactant can undergo reaction with two other molecules.

When the monomers are bifunctional, such as dimethyl terephthalate and ethylene glycol, polymer growth must occur in a *linear* fashion. Linear polymers often make excellent textile fibers. If more than two reactive sites are present in one of the monomers, then the polymer can grow into a cross-linked network. **Glyptal** (a polymer of glycerol and phthalic anhydride) is an example of a cross-linked polyester.

Figure 15.12 Synthesis of the polyester Dacron by a transesterification reaction.

Section 15.8
Amides

A. Nomenclature of Amides

A **amide** is a compound that has a trivalent nitrogen bonded to a carbonyl group. An amide is named from the parent carboxylic acid with the **-oic** (or **-ic**) **acid** ending changed to **-amide**.

$$\underset{\text{O}}{\overset{\text{O}}{\underset{\|}{\text{CH}_3\text{CNH}_2}}} \qquad \text{CH}_3\text{CH}_2\text{CH}_2\text{CNH}_2$$

IUPAC: ethanamide   butanamide
trivial: acetamide   butyramide

Amides with alkyl substituents on the nitrogen have their names preceded by *N*-alkyl, where *N* refers to the nitrogen atom.

$$\text{C}_6\text{H}_5\text{CNHCH}_3 \qquad \text{HCN(CH}_3)_2$$

N-methylbenzamide   *N,N*-dimethylformamide

A few amides of interest follow; the amide groups are highlighted in color:

nicotinamide,
or niacinamide

a B vitamin

caffeine

a lactam, or cyclic amide

lysergic acid diethylamide
(LSD)

B. Preparation of Amides

Amides are synthesized from derivatives of carboxylic acids and ammonia or the appropriate amine. These reactions have been discussed previously in this chapter.

$$\underset{\text{RCCl}}{\overset{\text{O}}{\|}} \xrightarrow{\text{R}'_2\text{NH}}$$

$$\underset{\text{RCOCR}}{\overset{\text{O}\;\;\text{O}}{\|\;\;\|}} \xrightarrow{\text{R}'_2\text{NH}} \underset{\text{RCNR}'_2}{\overset{\text{O}}{\|}}$$

$$\underset{\text{RCOR}'}{\overset{\text{O}\quad\text{R}'_2\text{NH}}{\|}}$$

C. Reactions of Amides

An amide contains a nitrogen with a pair of unshared valence electrons. It would be reasonable to expect amides to undergo reaction with acids, as do amines; however, they do not. Amides are *very* weak bases with pK_b values of 15–16. (By contrast, methylamine has a pK_b of 3.34.) The resonance structures for an amide show why the nitrogen of an amide is neither particularly basic nor nucleophilic.

$$CH_3\ddot{N}H_2 \ + \ \text{dilute HCl} \ \longrightarrow \ CH_3NH_3^+ \ Cl^-$$

methylamine   methylammonium
                                    chloride

$$\overset{\displaystyle O}{\overset{\|}{CH_3C}}\ddot{N}H_2 + \text{dilute HCl} \ \longrightarrow \ \text{no appreciable salt formation}$$

acetamide

Resonance structures for an amide:

$$R{-}\overset{\overset{\displaystyle \ddot{O}:}{\|}}{C}{-}\ddot{N}H_2 \ \longleftrightarrow \ R{-}\overset{\overset{\displaystyle :\ddot{O}:^-}{|}}{C}{=}\overset{+}{N}H_2 \qquad \begin{array}{l}\textit{less basic than} \\ \textit{an amine nitrogen}\end{array}$$

The effect of the partial double-bond character of the bond between the carbonyl carbon and the nitrogen of an amide is evident in the proton nmr spectrum of *N,N*-dimethylformamide (Figure 15.13). The spectrum shows one peak for *each* methyl group. If the $-N(CH_3)_2$ group underwent free rotation around its bond to the carbonyl group, the methyl groups would be magnetically equivalent and would give rise to one singlet. Because of the restricted rotation around the carbonyl carbon–nitrogen bond, the two methyl groups are in

Figure 15.13 The ^{1}H nmr spectrum of *N,N*-dimethylformamide, showing a pair of peaks for the *N*-methyl groups.

different magnetic environments. (The energy barrier for rotations around the carbonyl carbon–nitrogen bond in an amide has been found to be about 18 kcal/mol at room temperature.)

Hydrolysis.   Like esters, amides can be hydrolyzed in either acidic or alkaline solution. In either case, the acid or base is a reactant, not a catalyst, and must be used in a 1:1 molar ratio or in excess. Neither type of hydrolysis reaction is reversible.

In acid:

$$\underset{\textit{N,N-dimethylpropanamide}}{CH_3CH_2\overset{\overset{\displaystyle O}{\|}}{C}N(CH_3)_2} + H_2O + H^+ \longrightarrow \underset{\substack{\text{propanoic}\\\text{acid}}}{CH_3CH_2CO_2H} + \underset{\substack{\text{dimethyl-}\\\text{ammonium ion}}}{H_2\overset{+}{N}(CH_3)_2}$$

In base:

$$CH_3CH_2\overset{\overset{\displaystyle O}{\|}}{C}N(CH_3)_2 + OH^- \longrightarrow \underset{\text{propanoate ion}}{CH_3CH_2\overset{\overset{\displaystyle O}{\|}}{C}O^-} + \underset{\text{dimethylamine}}{HN(CH_3)_2}$$

Hydrolysis of an amide in acidic solution proceeds in a fashion similar to hydrolysis of an ester. The carbonyl oxygen is protonated, the carbonyl carbon is attacked by H_2O, protons are transferred, and an amine is expelled. This amine then reacts with H^+ to yield the amine salt. The formation of the amine salt explains (1) why H^+ is a reactant, not a catalyst, and (2) why the reverse reaction does not proceed. (Although R_2NH is a nucleophile, $R_2NH_2^+$ is not; this ion cannot attack the carbonyl group.)

In acid:

Alkaline hydrolysis of an amide is similar to saponification of an ester. The products are the carboxylate salt of the acid and a free amine or ammonia.

In base:

$$CH_3CH_2\overset{\ddot{O}}{\overset{\|}{C}}NH_2 + :\ddot{O}H^- \xrightleftharpoons[]{\text{addition}} \left[CH_3CH_2\overset{:\ddot{O}:^-}{\underset{:\underset{\cdot\cdot}{O}\text{—}H}{C}\text{—}\ddot{N}H_2} \right] \xrightarrow[\text{and proton transfer}]{\text{elimination}} CH_3CH_2\overset{\dot{O}\cdot}{\overset{\|}{C}} + NH_3$$
$$\overset{}{\underset{:\ddot{O}:^-}{}}$$

Study Problems

15.25 *Aspartame* is a low-calorie synthetic sweetener introduced in 1983. This compound is a *dipeptide*, a small protein that yields two amino acid molecules upon hydrolysis. Predict the products of the hydrolysis of aspartame in (**a**) aqueous HCl, and (**b**) aqueous NaOH.

$$\underset{\underset{\text{aspartame}}{}}{HO\overset{O}{\overset{\|}{C}}CH_2\underset{\underset{NH_2}{|}}{C}H\overset{O}{\overset{\|}{C}}NH\underset{\underset{CH_2C_6H_5}{|}}{C}H\overset{O}{\overset{\|}{C}}OCH_3}$$

15.26 What would be the hydrolysis products of (**a**) nicotinamide (page 661) in acid? (**b**) LSD (page 661) in base? (**c**) the following lactam (cyclic amide) in base?

Reduction.   Reduction of amides with lithium aluminum hydride results in conversion of the carbonyl group to —CH$_2$—. The product is an amine. Sodium borohydride does not reduce amides.

General:

$$\underset{\text{an amide}}{R\overset{O}{\overset{\|}{C}}NR'_2} \xrightarrow[\text{(2) H}_2\text{O, H}^+]{\text{(1) LiAlH}_4} \underset{\text{an amine}}{RCH_2NR'_2}$$

$$\underset{N\text{-methyldodecanamide}}{CH_3(CH_2)_{10}\overset{O}{\overset{\|}{C}}NHCH_3} \xrightarrow[\text{(2) H}_2\text{O, H}^+]{\text{(1) LiAlH}_4} \underset{N\text{-methyldodecylamine (95\%)}}{CH_3(CH_2)_{10}CH_2NHCH_3}$$

5,5-dimethyl-2-pyrrolidone   2,2-dimethylpyrrolidine (79%)

The first step in the reduction is the addition of H:⁻ to the carbonyl carbon. Oxygen is then abstracted by the aluminum hydride to yield an imine or iminium ion, which in turn is reduced by another hydride transfer.

Section 15.9
Polyamides

There can be no question that the most important polyamides are the *proteins*. Chapter 24 is devoted to this subject. The most notable example of a man-made polyamide is the synthetic polyamide **nylon 6,6**, which is prepared from adipic acid (a diacid) and hexamethylenediamine (a diamine). As in the synthesis of the polyester Dacron, the result of the reaction of two types of bifunctional molecules is a linear polymer.

Nylon 6,6 is but one member of the family of synthetic nylons. Nylon 6,6 is made from a *six-carbon* diacid and *six-carbon* diamine. **Nylon 6**, on the other hand, is prepared from caprolactam, a monomer that contains the acid and amine in the same molecule (with *six carbons*). In this reaction, caprolactam undergoes ring opening with water; then, in the polymerization, water is eliminated.

A newer fiber, "para-aramid," is so strong and indestructible that it is widely used in military protective (bulletproof) vests and helmets.

"para-aramid"

▼ ## Section 15.10
Compounds Related to Amides

Some types of compounds related to amides are shown in Table 15.4. **Urea** is one of the most important amide relatives. Excess nitrogen from the metabolism of proteins is excreted by the higher animals as urea. Some lower animals excrete ammonia, while reptiles and birds excrete **guanidine**. Both guanidine and urea, as well as ammonia, are widely used as nitrogen fertilizers and as starting materials for synthetic polymers and drugs.

$$H_2N-\overset{\overset{\displaystyle O}{\|}}{C}-NH_2 \qquad R_2N-\overset{\overset{\displaystyle O}{\|}}{C}-NR_2 \qquad H_2N-\overset{\overset{\displaystyle NH}{\|}}{C}-NH_2$$

urea   *a substituted urea*   guanidine

Urea is used for the synthesis of barbiturates (used as sedatives) by reaction with α-substituted diethyl malonates. This reaction is similar to the reaction of an ester with an amine to yield an amide.

urea   diethyl malonate   barbituric acid

not substituted

Study Problem

15.27 What two reactants would you use to synthesize each of the following barbiturates?

(a) phenobarbital   (b) pentobarbital (Nembutal)

An **imide**, a compound with the $-\overset{\overset{\displaystyle O}{\|}}{C}NH\overset{\overset{\displaystyle O}{\|}}{C}-$ group, is the nitrogen analog of an acid anhydride. Like amides, an imide can be made from ammonia and an acid anhydride.

phthalic anhydride   phthalimide *a cyclic imide*

Table 15.4 Some types of compounds related to amides

Partial Structure	Class of Compound	Example
$-\overset{\overset{\displaystyle O}{\|\|}}{C}N\diagdown$	amide	$CH_3\overset{\overset{\displaystyle O}{\|\|}}{C}NH_2$
$-\overset{\overset{\displaystyle O}{\|\|}}{C}N\diagdown$ in ring	lactam	(ring structure with C=O and NH)
$-\overset{\overset{\displaystyle O}{\|\|}}{C}N\overset{\overset{\displaystyle O}{\|\|}}{C}-$	imide	(ring structure with two C=O and NH)
$\diagup N\overset{\overset{\displaystyle O}{\|\|}}{C}N\diagdown$	urea	$H_2N\overset{\overset{\displaystyle O}{\|\|}}{C}NH_2$
$\diagup N\overset{\overset{\displaystyle O}{\|\|}}{C}O-$	carbamate, or urethane	$H_2N\overset{\overset{\displaystyle O}{\|\|}}{C}OCH_3$
$-\overset{\overset{\displaystyle O}{\|\|}}{\underset{\underset{\displaystyle O}{\|\|}}{S}}N\diagdown$	sulfonamide	(benzene ring)$-\overset{\overset{\displaystyle O}{\|\|}}{\underset{\underset{\displaystyle O}{\|\|}}{S}}NH_2$

A **carbamate**, or **urethane**, is a compound in which the $-NH_2$, $-NHR$, or $-NR_2$ group is bonded to an ester carbonyl group. A carbamate is related to a carbonate structure, with one O replaced by N.

$$RO-\overset{\overset{\displaystyle O}{\|\|}}{C}-OR \qquad H_2N-\overset{\overset{\displaystyle O}{\|\|}}{C}-OR$$

a carbonate   *a carbamate*

$$H_2N\overset{\overset{\displaystyle O}{\|\|}}{C}OCH_2\overset{\overset{\displaystyle CH_3}{\underset{\underset{\displaystyle CH_2CH_2CH_3}{|}}{C}}}CH_2O\overset{\overset{\displaystyle O}{\|\|}}{C}NH_2$$

meprobamate

a dicarbamate used as
a tranquilizer (Miltown, Equanil)

(naphthalene ring with) $O\overset{\overset{\displaystyle O}{\|\|}}{C}NHCH_3$

1-naphthyl-*N*-methylcarbamate
(Sevin)

a biodegradable insecticide

One way in which a carbamate may be prepared is by the action of an alcohol or phenol on an **isocyanate**, a compound containing the $-N=C=O$ group.

phenyl isocyanate   phenol   phenyl *N*-phenylcarbamate

An analogous reaction is used to make **polyurethanes**, such as that used for polyurethane foam insulation. The foaming effect in polyurethane foam is achieved by the addition of a low-boiling liquid, such as dichloromethane, which vaporizes during polymerization.

a polyurethane

Sulfa drugs are **sulfonamides**, compounds in which the nitrogen is attached to a sulfonyl group rather than to an acyl group. A sulfonamide is prepared by the action of an arylsulfonyl chloride on ammonia or on a primary or secondary amine.

benzenesulfonyl   *a sulfonamide*
chloride

Many *p*-aminosulfonamides are effective bacteriostatic agents. Although the bacteriological properties of one of the sulfonamides were observed in 1909, these compounds were not widely used against infections in humans until around 1940.

essential part of structure
for drug activity

H_2N—◯—SO_2NH_2

sulfanilamide

H_2N—◯—SO_2NH—[thiazole]

sulfathiazole

H_2N—◯—SO_2NH—[pyridazine]—OCH_3

sulfamethoxypyridazine

H_2N—◯—SO_2NH—[pyrimidine]—CH_3

sulfamerazine

Sulfa drugs inhibit the growth and multiplication of some types of bacteria that require *p*-aminobenzoic acid (PABA) for the biosynthesis of folic acids. The sulfa drugs are hydrolyzed *in vivo* to sulfanilamide, which is mistaken for PABA by certain bacterial enzymes. When the enzyme is associated with sulfanilamide, it is inhibited from catalyzing the incorporation of PABA into a folic acid molecule. This process is called **competitive inhibition of an enzyme-catalyzed reaction**. Sulfa drugs are not effective against all bacteria because some strains do not require PABA and some strains can synthesize their own PABA, which successfully competes against the sulfonamide for a position on the enzyme.

$$H_2N-\!\!\bigcirc\!\!-CO_2H$$

p-aminobenzoic acid
(PABA)

folic acid

Study Problem

15.28 Sodium cyclohexylsulfamate (a **cyclamate**) is an artificial sweetener that is thirty times sweeter than cane sugar. This compound can be prepared by the reaction of cyclohexylamine and chlorosulfonic acid, followed by treatment with sodium hydroxide. What is the structure of this cyclamate?

cyclohexylamine   chlorosulfonic
acid

▼ Section 15.11
Nitriles

A. Nomenclature of Nitriles

Nitriles are organic compounds containing the C≡N group. They are also sometimes called *cyano compounds* or *cyanides*. In the IUPAC system, the number of carbon atoms, including that in the CN group, determines the alkane parent. The alkane name is suffixed with **-nitrile**. Some nitriles are named after the trivial names of their carboxylic acid parent with the **-ic acid** changed to **-nitrile**, or to **-onitrile** if the parent name lacks an *o*-.

$$CH_3C\equiv N \qquad \text{◯—CN}$$

IUPAC:	ethanenitrile	benzenecarbonitrile
trivial:	acetonitrile	benzonitrile

Study Problem

15.29 Write formulas for **(a)** propanenitrile; **(b)** butyronitrile.

B. Bonding in Nitriles

The cyano group contains a triple bond—one sigma bond and two pi bonds (Figure 15.14). Although the nitrogen has a pair of unshared electrons, a nitrile is a very weak base. The pK_b of a nitrile is about 24, while the pK_b of NH_3 is about 4.5 (about 20 powers of ten difference). The lack of basicity of a —CN: group results from the unshared electrons being in an sp orbital. The greater amount of s character in an sp orbital (compared with that in an sp^2 or sp^3 orbital) means that these sp electrons are more tightly held and less available for bonding to a proton.

C. Preparation of Nitriles

The CN^- ion (from NaCN, for example) is a good nucleophile for S_N2 displacement of a halide ion from an alkyl halide. This reaction is the principal route to aliphatic nitriles. However, because of elimination reactions, high yields are obtained only with *primary alkyl halides* and, to a lesser extent, *secondary alkyl halides*.

$$CH_3CH_2CH_2CH_2Br + CN^- \xrightarrow{\;S_N2\;} CH_3CH_2CH_2CH_2CN + Br^-$$

1-bromobutane   pentanenitrile

a 1° alkyl halide   (90%)

Aryl nitriles are best obtained through the **diazonium salts**, compounds that were discussed in Section 12.3B.

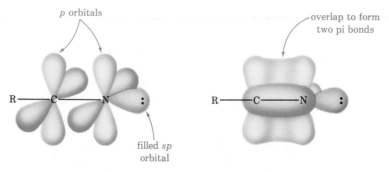

p orbitals

overlap to form
two pi bonds

R——C——N :   R——C——N :

filled *sp*
orbital

Figure 15.14 The bonding in a nitrile, $RC\equiv N$:.

aniline   benzenediazonium   benzonitrile
                          chloride

Another route to nitriles is the dehydration of amides with such dehydrating agents as $SOCl_2$, P_2O_5, or acetic anhydride.

$$CH_3(CH_2)_3\overset{\overset{O}{\|}}{\underset{\underset{CH_3CH_2}{|}}{C}}HCNH_2 + SOCl_2 \longrightarrow CH_3(CH_2)_3\underset{\underset{CH_3CH_2}{|}}{C}HCN$$

2-ethylhexanamide   2-ethylhexanenitrile (94%)

D. Reactions of Nitriles

Hydrolysis. Nitriles are included as carboxylic acid derivatives because their hydrolysis yields carboxylic acids. The hydrolysis of a nitrile can be carried out by heating with either aqueous acid or base.

phenylacetonitrile   phenylacetic acid (78%)

In *acidic hydrolysis*, the weakly basic nitrogen is protonated and then water attacks the electropositive carbon atom. The reaction goes through an amide, which is further hydrolyzed to the carboxylic acid and ammonium ion. Because of the formation of the NH_4^+ ion (from NH_3 and H^+), an excess of acid must be used in nitrile hydrolysis.

In acid:

intermediate
amide

If comparatively mild conditions are used, the amide may be isolated as the product of the reaction.

phenylacetonitrile   2-phenylacetamide (86%)

Alkaline hydrolysis occurs by nucleophilic attack on the partially positive carbon of the nitrile group. The reaction again results in an amide, which is further hydrolyzed to the carboxylate and ammonia. The free acid is obtained when the solution is acidified.

In base:

$$RC{\equiv}N{:} + {:}\ddot{O}H^- \rightleftharpoons \left[RC{=}\ddot{N}{:}^- \atop {:}\underset{\cdot\cdot}{O}H \right] \xrightarrow[-\text{OH}^-]{H{-}\ddot{O}H} \left[RC{=}\ddot{N}H \atop {:}\underset{\cdot\cdot}{O}{-}H \right] \xrightarrow[-H_2O]{H{-}\ddot{O}H}$$

$$RC{-}\ddot{N}H_2 \atop {:}\underset{\cdot\cdot}{O} \xrightarrow{\text{OH}^-} \underset{O}{RC} + NH_3 \atop \overset{O^-}{\underset{}{}}$$

intermediate
amide

$$\xrightarrow{H^+} RCO_2H$$

Reduction. Nitriles can be reduced to *primary amines* of the type RCH_2NH_2 either by catalytic hydrogenation or by lithium aluminum hydride.

General:

$$RC{\equiv}N \xrightarrow{[H]} RCH_2NH_2$$

$$\bigcirc{-}CH_2C{\equiv}N \xrightarrow[\text{heat} \atop \text{pressure}]{2 H_2, Ni} \left[\bigcirc{-}CH_2CH{=}NH \right] \longrightarrow \bigcirc{-}CH_2CH_2NH_2$$

phenylacetonitrile   (2-phenylethyl)amine (87%)

$$CH_3CH_2CH_2C{\equiv}N \xrightarrow[(2)\ H_2O]{(1)\ LiAlH_4} CH_3CH_2CH_2CH_2NH_2$$

butanenitrile   *n*-butylamine (85%)

Figure 15.15 summarizes the reactions of amides and nitriles.

$$\underset{RC{-}NR'_2}{\overset{O}{\|}} \begin{cases} \xrightarrow[\text{(or OH}^-\text{ followed by H}^+)]{HO{-}H + H^+} & \underset{\text{a carboxylic acid}}{\overset{O}{\underset{}{RC{-}OH}}} + H_2\overset{+}{N}R'_2 \\ \xrightarrow[(2)\ H_2O, H^+]{(1)\ LiAlH_4} & RCH_2{-}NR'_2 \quad \textbf{an amine} \end{cases}$$

$$RC{\equiv}N \begin{cases} \xrightarrow[\text{(or OH}^-\text{ followed by H}^+)]{HO{-}H + H^+} \underset{\text{a 1° amide}}{\overset{O}{RC{-}NH_2}} \xrightarrow{HO{-}H} \underset{\text{a carboxylic acid}}{\overset{O}{RC{-}OH}} + NH_4^+ \\ \xrightarrow[\text{or (1) LiAlH}_4, (2)\ H_2O, H^+]{H_2, \text{ catalyst, heat, pressure}} RCH_2NH_2 \quad \textbf{a 1° amine} \end{cases}$$

Figure 15.15 **Summary of reactions of amides and nitriles.**

Section 15.12
Use of Carboxylic Acid Derivatives in Synthesis

Carboxylic acids and their derivatives are all synthetically interconvertible. However, of the carboxylic acid derivatives, the acid halides and anhydrides are probably the most versatile because they are more reactive than other carbonyl compounds. Either of these two reactants can be used to synthesize hindered esters or phenyl esters, which cannot be prepared in good yield by heating RCO_2H and $R'OH$ with an acidic catalyst because of an unfavorable equilibrium. (Suggest a reason why the reactions of acid halides or anhydrides with $R'OH$ are essentially *irreversible*.) These two derivatives are also the most useful reagents for making *N*-substituted amides.

$$
\left. \begin{array}{c} \overset{O}{\overset{\|}{RC}}-Cl \\ \text{or} \\ \overset{O}{\overset{\|}{RC}}-\overset{O}{\overset{\|}{OCR}} \end{array} \right\} + R'OH \longrightarrow \overset{O}{\overset{\|}{RC}}OR' \quad \text{where R and R' can be hindered or aryl}
$$

$$
\left. \begin{array}{c} \overset{O}{\overset{\|}{RC}}-Cl \\ \text{or} \\ \overset{O}{\overset{\|}{RC}}-\overset{O}{\overset{\|}{OCR}} \end{array} \right\} + R'_2NH \longrightarrow \overset{O}{\overset{\|}{RC}}NR'_2
$$

The reduction of an acid chloride with $LiAlH(OR)_3$ yields an aldehyde. (Can you think of another route to aldehydes?)

Although esters are not as reactive as acid chlorides or anhydrides, they are useful in the synthesis of alcohols (by reduction or by Grignard reactions) and are valuable starting materials in the synthesis of complex molecules. We will discuss these reactions of esters in Chapter 16.

The synthesis of nitriles affords one of the most convenient techniques for extending an aliphatic carbon chain by one or for introducing a carboxyl group or an NH_2 group.

$$
RX \xrightarrow{CN^-} RCN \begin{array}{c} \nearrow \xrightarrow{H_2O,\ H^+} RCO_2H \\ \\ \searrow \xrightarrow{[H]} RCH_2NH_2 \end{array}
$$

As we have mentioned, the reaction of RX and CN^- gives best yields with *primary* alkyl halides. Secondary alkyl halides can also be used, but give lower yields. (What other products can be expected?)

The preparations and reactions of the carboxylic acid derivatives are summarized in Tables 15.5 and 15.6.

Table 15.5 Summary of laboratory syntheses of carboxylic acid derivatives

Reaction	*Section Reference*

Acid Chlorides:

$$RCO_2H + SOCl_2 \ \text{ or } \ PCl_3 \longrightarrow \overset{\displaystyle O}{\overset{\|}{R}}CCl$$

15.3B

Acid Anhydrides:

$$\overset{\displaystyle O}{\overset{\|}{R}}CCl + {}^-OCR' \longrightarrow R\overset{\displaystyle O\ O}{\overset{\| \ \|}{C}}OCR'$$

15.4B

$$RCO_2H + \text{excess } (CH_3C)_2O \longrightarrow RCOCR$$

15.4B

$$HO_2C(CH_2)_nCO_2H \xrightarrow{\text{heat}} \overset{O}{\underset{(CH_2)_n}{\diagdown C \diagup}} \overset{O}{\diagup} \overset{O}{\diagdown}$$
$$n = 2 \text{ or } 3$$

14.9B

Esters:[a]

$$RCO_2H + R'OH \xrightarrow{H^+} RCO_2R'$$ 14.7

$$\overset{\displaystyle O}{\overset{\|}{R}}CCl + R'OH \longrightarrow RCO_2R'$$ 15.3C

$$(\overset{\displaystyle O}{\overset{\|}{R}}C)_2O + R'OH \longrightarrow RCO_2R'$$ 15.4C

$$RCO_2^- + R'X \longrightarrow RCO_2R'^b$$ 5.11, 14.6

$$RCO_2R' + R''OH \xrightarrow{H^+ \text{ or } {}^-OR''} RCO_2R''$$ 15.5C

Amides:

$$\overset{\displaystyle O}{\overset{\|}{R}}CCl + HNR'_2 \longrightarrow \overset{\displaystyle O}{\overset{\|}{R}}CNR'_2$$ 15.3C

$$(\overset{\displaystyle O}{\overset{\|}{R}}C)_2O + HNR'_2 \longrightarrow \overset{\displaystyle O}{\overset{\|}{R}}CNR'_2$$ 15.4C

$$RCO_2R' + NH_3 \longrightarrow \overset{\displaystyle O}{\overset{\|}{R}}CNH_2$$ 15.5C

Nitriles:

$$RX + CN^- \longrightarrow RCN$$ 15.11C

$$ArNH_2 \xrightarrow[0°]{\overset{NaNO_2}{HCl}} ArN_2^+ \, Cl^- \xrightarrow[\text{heat}]{CuCN + KCN} ArCN$$ 12.3B, 15.11C

$$\overset{\displaystyle O}{\overset{\|}{R}}CNH_2 \xrightarrow[-H_2O]{\text{dehydrating agent}} RCN$$ 15.11C

[a] The syntheses of some complex esters will be discussed in Chapter 16.
[b] For this reaction to be successful, a reactive halide must be used.

Table 15.6 Types of compounds that can be obtained from carboxylic acid derivatives

Reaction	Product	Section Reference
Acid Chlorides:[a]		
$\overset{O}{\overset{\|}{R C}Cl} + H_2O \longrightarrow RCO_2H$	carboxylic acid	15.3C
$\overset{O}{\overset{\|}{R C}Cl} + R'OH \longrightarrow RCO_2R'$	ester	15.3C
$\overset{O}{\overset{\|}{R C}Cl} + R'_2NH \longrightarrow \overset{O}{\overset{\|}{R C}NR'_2}$	amide	15.3C
$\overset{O}{\overset{\|}{R C}Cl} + R'CO_2^- \longrightarrow \overset{O}{\overset{\|}{R C}}O\overset{O}{\overset{\|}{C}R'}$	anhydride	15.3C
$\overset{O}{\overset{\|}{R C}Cl} + C_6H_6 \xrightarrow{AlCl_3} \overset{O}{\overset{\|}{R C}C_6H_5}$	aryl ketone	11.9E, 15.3C
$\overset{O}{\overset{\|}{R C}Cl} \xrightarrow[(2)\ H_2O,\ H^+]{(1)\ 2\ R'MgX} \overset{OH}{\overset{\|}{R C}R'_2}$	3° alcohol	15.3C
$\overset{O}{\overset{\|}{R C}Cl} + LiCuR'_2 \longrightarrow \overset{O}{\overset{\|}{R C}R'}$	ketone	15.3C
$\overset{O}{\overset{\|}{R C}Cl} \xrightarrow[(2)\ H_2O]{(1)\ LiAlH(OR')_3} \overset{O}{\overset{\|}{R C}H}$	aldehyde	15.3C
$RCH_2\overset{O}{\overset{\|}{C}}Cl + Cl_2 \longrightarrow RCHCl\overset{O}{\overset{\|}{C}}Cl$	α-chloro acid chloride	15.3C
Acid Anhydrides:		
$R\overset{O}{\overset{\|}{C}}O\overset{O}{\overset{\|}{C}}R + H_2O \longrightarrow 2\ RCO_2H$	carboxylic acid	15.4C
$(RCO)_2O + R'OH \longrightarrow RCO_2R'$	ester	15.4C
$(RCO)_2O + R'_2NH \longrightarrow \overset{O}{\overset{\|}{R C}NR'_2}$	amide	15.4C
Esters:[a]		
$RCO_2R' + H_2O \xrightarrow{H^+\ or\ OH^-} RCO_2H$	carboxylic acid	15.5C
$RCO_2R' + R''OH \xrightarrow{H^+\ or\ ^-OR''} RCO_2R''$	ester	15.5C
$RCO_2R' + NH_3 \longrightarrow \overset{O}{\overset{\|}{R C}NH_2}$	amide	15.5C
$RCO_2R' + [H] \longrightarrow RCH_2OH + HOR'$	alcohols	15.5C
$RCO_2R' \xrightarrow[(2)\ H_2O,\ H^+]{(1)\ 2\ R''MgX} \overset{OH}{\overset{\|}{R C}R''_2}$	3° alcohol	15.5C

(continued)

Table 15.6 (*continued*) Types of compounds that can be obtained from carboxylic acid derivatives

Reaction	Product	Section Reference
Amides:		
$\overset{\displaystyle O}{\overset{\|}{R\text{C}}}NR'_2 + H_2O \xrightarrow{\text{H}^+ \text{ or OH}^-} RCO_2H + HNR'_2$	**carboxylic acid and amine**	15.8C
$\overset{\displaystyle O}{\overset{\|}{R\text{C}}}NR'_2 + [H] \longrightarrow RCH_2NR'_2$	**amine**	15.8C, 17.5B
$\overset{\displaystyle O}{\overset{\|}{R\text{C}}}NH_2 + Br_2 + \ ^-OH \longrightarrow RNH_2$	**amine**	17.5C
Nitriles:[a]		
$RCN + H_2O \xrightarrow{\text{H}^+ \text{ or OH}^-} RCO_2H$	**carboxylic acid**	15.11D
$RCN + H_2O \xrightarrow{\text{H}^+ \text{ or OH}^-} \overset{\displaystyle O}{\overset{\|}{R\text{C}}}NH_2$	**amide**	15.11D
$RCN + [H] \longrightarrow RCH_2NH_2$	**amine**	15.11D

[a] Acid chlorides, esters, and nitriles can be used to synthesize more-complex compounds. Some of these reactions will be discussed in Chapter 16.

Study Problem

15.30 Suggest synthetic routes to the following compounds:

(a) ⬡—$\overset{\displaystyle O}{\overset{\|}{\text{C}}}N(CH_3)_2$ from a carboxylic acid

(b) 2-methyl-4-propyl-4-heptanol from compounds containing six or fewer carbon atoms

(c) $CH_3CH_2\overset{\displaystyle O}{\overset{\|}{\text{C}}}NH(CH_2)_3CH_3$ from 1-propanol and *no other organic reagents*

▼ Summary

The derivatives of carboxylic acids are usually prepared from the carboxylic acids themselves or from other more reactive derivatives, as shown in Table 15.5.

The reactions of the various carboxylic acid derivatives with nucleophiles are similar to one another. Differences arise from differences in reactivity of the various derivatives.

$$\text{increasing reactivity} \uparrow \begin{cases} \underset{\overset{\|}{O}}{RC}-Cl \\[2pt] \underset{\overset{\|}{O}}{RC}-\underset{\overset{\|}{O}}{OCR} \\[2pt] \underset{\overset{\|}{O}}{RC}-OR \\[2pt] \underset{\overset{\|}{O}}{RC}-NH_2 \\[2pt] RCN \end{cases} \xrightarrow[H^+ \text{ or } OH^-]{H_2O} \underset{\overset{\|}{O}}{RC}-OH \quad \textit{a carboxylic acid}$$

$$\text{increasing reactivity} \uparrow \begin{cases} RCOCl \\ (RCO)_2O \\ RCO_2R \end{cases} \xrightarrow{NH_3} \underset{\overset{\|}{O}}{RC}-NH_2 \quad \textit{an amide}$$

$$\text{increasing reactivity} \uparrow \begin{cases} RCOCl \\ (RCO)_2O \\ RCO_2R \end{cases} \xrightarrow{R'OH} \underset{\overset{\|}{O}}{RC}-OR' \quad \textit{an ester}$$

In addition to these reactions, the more reactive acid halides undergo Friedel–Crafts reactions with aromatic compounds and also undergo reaction with lithium dialkylcuprates to yield ketones. (See Table 15.6.)

Esters react with Grignard reagents to yield tertiary alcohols.

$$RCO_2R \xrightarrow[(2)\ H_2O,\ H^+]{(1)\ 2R'MgX} R\overset{\overset{\displaystyle OH}{|}}{C}R'_2 \quad \textit{a 3° alcohol}$$

All the derivatives can be reduced. For example:

$$RCO_2R' \xrightarrow[(2)\ H_2O,\ H^+]{(1)\ LiAlH_4} RCH_2OH + HOR' \quad \textit{alcohols}$$

$$\underset{\overset{\|}{O}}{RC}NR'_2 \xrightarrow[(2)\ H_2O,\ H^+]{(1)\ LiAlH_4} RCH_2NR'_2 \quad \textit{an amine}$$

$$RCN \xrightarrow[(2)\ H_2O,\ H^+]{(1)\ LiAlH_4} RCH_2NH_2 \quad \textit{a 1° amine}$$

Study Problems*

15.31 List the following compounds in order of increasing bond moment of the indicated bond:

(a) $CH_3\overset{\displaystyle O}{\overset{\|}{C}}\!-\!OCH_3$ (b) $CH_3\overset{\displaystyle O}{\overset{\|}{C}}\!-\!NH_2$ (c) $CH_3\overset{\displaystyle O}{\overset{\|}{C}}\!-\!Cl$

15.32 List the compounds in the preceding problem in order of their reactivity toward water (least reactive first).

15.33 Name the following compounds:

(a) $(CH_3)_2CHCH_2\overset{\displaystyle O}{\overset{\|}{C}}Cl$ (b) $Br\!-\!\bigcirc\!-\!\overset{\displaystyle O}{\overset{\|}{C}}Br$ (c) $CH_2\!=\!CHCH_2\overset{\displaystyle O}{\overset{\|}{C}}Cl$

15.34 Oleic acid, cis-$CH_3(CH_2)_7CH\!=\!CH(CH_2)_7CO_2H$, is a carboxylic acid that can be obtained from the hydrolysis of animal fats and vegetable oils. Oleoyl chloride can be synthesized from this acid in 99% yield. Write equations that illustrate two ways in which this synthesis could be carried out.

15.35 What would be the products of the reactions of propanoyl chloride with the following reagents?

(a) water   (b) cyclohexanol
(c) *p*-bromophenol   (d) piperidine (page 740)
(e) sodium ethoxide in ethanol
(f) excess aqueous NaOH
(g) sodium formate in tetrahydrofuran
(h) LiAlH$_4$ in ether, followed by aqueous acid
(i) excess CH_3CH_2MgBr in ether, followed by acidification
(j) bromine, followed by aqueous acid

15.36 *Phosgene*, $Cl\overset{\displaystyle O}{\overset{\|}{C}}Cl$, is a toxic gas used as a war gas in World War I. Phosgene undergoes the usual reactions of an acid chloride, but can undergo these reactions twice. Predict the product of the reaction of 1.0 mol of phosgene with each of the following reagents: (a) excess H_2O; (b) excess ethanol; (c) 1.0 mol ethanol; (d) 1.0 mol ethanol, followed by 2.0 mol NH$_3$.

15.37 Name the following compounds:

(a) $CH_3CH_2\overset{\displaystyle O}{\overset{\|}{C}}O\overset{\displaystyle O}{\overset{\|}{C}}CH_2CH_2CH_2CH_3$ (b) $Cl\!-\!\bigcirc\!-\!\overset{\displaystyle O}{\overset{\|}{C}}O\overset{\displaystyle O}{\overset{\|}{C}}CH_2\!-\!\bigcirc$

(c) $H\overset{\displaystyle O}{\overset{\|}{C}}O\overset{\displaystyle O}{\overset{\|}{C}}\!-\!\bigcirc$ (d) $H\overset{\displaystyle O}{\overset{\|}{C}}O\overset{\displaystyle O}{\overset{\|}{C}}CH_3$

15.38 Write equations showing how the anhydrides in Problem 15.37 could be synthesized.

* For information concerning the organization of the *Study Problems* and *Additional Problems*, see the *Note to student* on page 39.

15.39 Predict the products when the following cyclic anhydride is treated with the following reagents:

(a) aqueous NaOH   (b) aqueous HCl   (c) ethanol
(d) sodium ethoxide in ethanol   (e) methylamine

15.40 The pain reliever *phenacetin* is commercially prepared by the reaction of an anhydride with an amine. Write an equation for this reaction.

$$CH_3CH_2O-\langle\bigcirc\rangle-NHCCH_3$$

phenacetin

15.41 Write flow equations for the following conversions:

(a) [benzene ring with CO_2H and CO_2H] ⟶ [benzene ring with CO_2CH_3 and CO_2H]

(b) [cyclohexane with CH_2CO_2H and CH_2CO_2H] ⟶ [cyclohexane with $CH_2CO_2CH_2C_6H_5$ and CH_2CO_2H]

15.42 Write the IUPAC names of the following esters:

(a) $CH_3CH_2O_2CCH_2-\langle\bigcirc\rangle$ (b) [benzene ring with CO_2CH_3 and NH_2]

odor of honey   odor of grapes

15.43 Write equations that show the syntheses of the esters in the preceding problem from carboxylic acids.

15.44 When 2,4,6-trimethylbenzoic acid is heated with ethanol and a trace of H_2SO_4, no ester is obtained. Suggest a reason and also suggest an alternative route to the ester.

15.45 Predict the products of the reaction of methyl butanoate (odor of apples) with:
(a) hot, dilute aqueous H_2SO_4
(b) concentrated aqueous ammonia
(c) sodium ethoxide
(d) hot, dilute aqueous NaOH
(e) excess phenylmagnesium bromide in diethyl ether
(f) LiAlH$_4$ in THF, followed by an aqueous work-up

15.46 Would acidic hydrolysis or saponification be the method of choice for the conversion of 2-butenyl acetate to acetic acid and 2-buten-1-ol? Why?

15.47 Show by equations how you would convert ethyl acetate to:
(a) acetic acid;   (b) ethanol;
(c) *t*-butyl alcohol;   (d) acetophenone (page 463);
(e) sodium acetate;   (f) *N*-methylacetamide.

15.48 Predict (1) the saponification products (aqueous NaOH), and (2) the products resulting from saponification followed by acidification of the following compounds:

(**a**) cocaine (page 821)   (**b**) nepetalactone (page 658)

15.49 Complete the following equations showing the major products:

(**a**) [cyclic structure] + HBr ⟶

(**b**) $CH_3CH=\overset{\underset{|}{CH_3}}{C}Li$ $\xrightarrow[\text{(2) } H_2O, H^+]{\text{(1) } CH_3CO_2CH_2CH_3}$

(**c**) $ClCH_2CO_2CH_2CH_3 + NH_3$ $\xrightarrow[\text{30 min}]{0°}$

(**d**) $CH_3\overset{\overset{\displaystyle O}{||}}{C}CH_2O\overset{\overset{\displaystyle O}{||}}{C}CH_3 + CH_3OH$ $\xrightarrow{H^+}$

(**e**) $O_2N-\!\!\left\langle\bigcirc\right\rangle\!\!-CO_2CH_3$ $\xrightarrow[\text{(2) } H_2O, H^+]{\text{(1) aqueous NaOH, heat}}$

(**f**) $C_6H_5\overset{\overset{\displaystyle O}{||}}{C}CH_2\overset{\overset{\displaystyle O}{||}}{C}OCH_2CH_3 + C_6H_5NH_2$ ⟶

15.50 Name the following amides, using the IUPAC system:

(**a**) $CH_3(CH_2)_6\overset{\overset{\displaystyle O}{||}}{C}NH_2$ (**b**) $CH_3\overset{\overset{\displaystyle O}{||}}{C}NHC_6H_5$

(**c**) $(CH_3CH_2)_2CH\overset{\overset{\displaystyle O}{||}}{C}NHC_6H_5$ (**d**) $C_6H_5\overset{\overset{\displaystyle O}{||}}{C}N(CH_2CH_3)_2$

15.51 Write equations that show *three* possible reactions leading to *N*-methylhexanamide.

15.52 *Coccidia* are a family of protozoa, some of which can infect the intestines of humans and domestic animals. The anticoccidial drug *Ethopabate* is sometimes used in the treatment of these infections. Suggest a reaction using an acid anhydride that could be used to prepare this compound.

$$CH_3\overset{\overset{\displaystyle O}{||}}{C}NH-\!\!\left\langle\bigcirc\right\rangle\!\!-\overset{\overset{\displaystyle O}{||}}{C}OCH_2CH_3$$

$$OCH_2CH_3$$

Ethopabate

15.53 Predict the products of the hydrolysis of the following compounds (1) in aqueous HCl, and (2) in aqueous NaOH:

(**a**) $HO-\!\!\left\langle\bigcirc\right\rangle\!\!-NH\overset{\overset{\displaystyle O}{||}}{C}CH_3$ (**b**) $Cl-\!\!\left\langle\bigcirc\right\rangle\!\!-NH\overset{\overset{\displaystyle O}{||}}{C}N(CH_3)_2$

acetaminophen   monuron

a pain reliever   *a herbicide*

15.54 Predict the intermediates and products for the following reactions:

(a) $CH_3\overset{O}{\overset{\|}{C}}-\overset{CH_3}{\overset{|}{N}}C_6H_5$ $\xrightarrow[\text{(2) excess } H_2O, H^+]{\text{(1) LiAlH}_4}$ A $\xrightarrow{\text{(3) } H_2O, OH^-}$ B

(91% overall yield)

(b) $\xrightarrow[\text{(2) excess } H_2O, H^+]{\text{(1) excess LiAlH}_4}$ C $\xrightarrow{\text{(3) } H_2O, OH^-}$ D

(90% overall yield)

15.55 Name the following compounds by the IUPAC system:

(a) $(CH_3)_2CHCN$ (b) $CH_3-\bigcirc-CH_2CN$

15.56 Write formulas for   (a) 2-phenylpropanenitrile, and   (b) β-chlorobutyronitrile.

15.57 Suggest syntheses for the following nitriles:

(a) (naphthalene with CN)   (b) $(CH_3)_2C=CHCH_2CN$

15.58 Predict the products of the following reactions:

(a) $NCCH_2CH_2CH_2CN + H_2O \xrightarrow[\text{heat}]{\text{HCl}}$

(b) (benzene ring with CH_3 and CN) $+ H_2O \xrightarrow[\text{heat}]{H_2SO_4}$

(c) (benzene ring with CH_2CN and CO_2H) $+ H_2O \xrightarrow[\text{heat}]{H_2SO_4}$

(d) $NCCH_2CH_2CH_2CN + \text{excess } H_2 \xrightarrow[\text{catalyst}]{\text{heat, pressure}}$

Additional Problems

15.59 Predict the organic products, including stereochemistry where appropriate.

(a) $CH_3CH_2\overset{}{\underset{\underset{CH_3}{|}}{C}}H\overset{O}{\overset{\|}{C}}Br + \bigcirc-CO_2^- \longrightarrow$

(b) $\bigcirc-CO_2H + \bigcirc-OH \xrightarrow[\text{heat}]{H_2SO_4}$

(c) $\bigcirc-CO_2CH_2CH_3 \xrightarrow[\text{(2) } H_2O, H^+]{\text{(1) LiAlH}_4}$

(d) $CH_3CH_2CH_2CO_2H + SOCl_2 \longrightarrow$

(e) $\underset{\substack{|| \ \ || \\ O \ \ O}}{CH_3COCCH_3}$ + (R)-2-octylamine ⟶

(f) $\underset{\substack{|| \\ O}}{CH_3CH_2CCl}$ + LiCu[CH(CH$_3$)$_2$]$_2$ ⟶

(g) $\underset{\substack{|| \\ O}}{CH_3CH_2OCOCH_2CH_3}$ + D$_2$O $\xrightarrow{D^+}$

(h) $\underset{\substack{|| \ \ || \\ O \ \ O}}{C_6H_5COCC_6H_5}$ + (CH$_3$)$_2$CH^{18}OH ⟶

(i) Cl–⟨◯⟩–$\underset{\substack{N \\ }}{\overset{\substack{O \\ || \\ CCH_3}}{|}}$–⟨◯⟩–Cl $\xrightarrow[\text{heat}]{\text{KOH, CH}_3\text{CH}_2\text{OH}}$

(j) $H_3\overset{+}{N}CH_2\underset{\substack{|| \\ O}}{C}O^-$ + (CH$_3$$\underset{\substack{|| \\ O}}{C}$)$_2$O $\xrightarrow{\text{H}_2\text{O}}$

(k) $CH_3(CH_2)_5\underset{\substack{|| \\ O}}{C}Cl$ + ⟨◯⟩$_{N}$ + CH$_3$(CH$_2$)$_5$CO$_2$H $\xrightarrow{\text{benzene}}$

(l) O$_2$N–⟨◯⟩–CH$_2$O$\underset{\substack{|| \\ O}}{C}CH_3$ $\xrightarrow[\text{heat}]{\text{NaOH, H}_2\text{O, CH}_3\text{OH}}$

(m) $CH_3CH_2O\underset{\substack{|| \\ O}}{C}OCH_2CH_3$ $\xrightarrow[\text{(2) H}_2\text{O, H}^+]{\text{(1) excess CH}_3\text{CH}_2\text{MgBr, ether}}$

(n) 1⟨◯⟩–CH$_2$CO$_2$H + 1 Br$_2$ $\xrightarrow[\text{heat}]{\text{trace of PBr}_3\text{, benzene}}$

(o) (CH$_3$)$_3$CCH$_2$CO$_2$H + 2 Br$_2$ $\xrightarrow[\text{heat}]{\text{trace of PBr}_3\text{, benzene}}$

(p) CH$_2$=CHCO$_2$CH$_3$ + HCO$_2$H $\xrightarrow[\text{heat}]{\text{H}_2\text{SO}_4}$

(q) CH$_2$=CHCO$_2$CH$_3$ + CH$_3$CH$_2$OCH$_2$CH$_2$OH $\xrightarrow[\text{heat}]{\text{TsOH}}$

(r) ⟨◯⟩–$\underset{\substack{| \\ OH}}{CH}$–$\underset{\substack{|| \\ O}}{C}$–⟨◯⟩ + (CH$_3$$\underset{\substack{|| \\ O}}{C}$)$_2$O $\xrightarrow{\text{CH}_3\text{CO}_2\text{H}}$

(s) [bicyclic anhydride] + 2 CH$_3$OH $\xrightarrow{\text{TsOH}}$

(t) $CH_3(CH_2)_4$[lactone ring] $\xrightarrow[\text{(2) H}_2\text{O, H}^+]{\text{(1) excess CH}_3\text{MgBr}}$

15.60 Predict the organic products of the LiAlH$_4$ reduction, followed by hydrolysis, of the following compounds: **(a)** dimethyl malonate, CH$_2$(CO$_2$CH$_3$)$_2$; **(b)** propanoyl bromide; **(c)** propanoic anhydride; **(d)** propanenitrile; **(e)** N-methylphenylacetamide.

15.61 Predict the products of catalytic hydrogenation (high temperature and pressure) of:

(a) [structure: cyclohexene ring]—CO_2CH_3 (b) [structure: cyclohexadiene ring]—CO_2H

15.62 *Juvenile hormone* is associated with the pupal development of many insects. Predict the products of hydrolysis in aqueous NaOH of this compound.

[structure of juvenile hormone]

juvenile hormone

15.63 Write an equation for the preparation of each of the following compounds from acetic anhydride and other appropriate reagents:

(a) $CH_3\overset{\overset{O}{\|}}{C}NHCH(CH_3)_2$ (b) [cyclic structure with CH_3CO_2, O_2CCH_3, CH_3CO_2, O, $CH_2O_2CCH_3$]

15.64 Hindered esters and phenyl esters can be synthesized in high yield from acid chlorides and the lithium salt or sodium salt of the phenol. Write equations showing how the following phenyl esters can be synthesized from a carboxylic acid and a phenol using this technique.

(a) H_3C—[benzene ring]—$O\overset{\overset{O}{\|}}{C}C(CH_3)_3$ (b) [benzene ring with two CH_3 groups]—$O\overset{\overset{O}{\|}}{C}CH_2CH_3$

15.65 *m*-Chloroperoxybenzoic acid, used for the epoxidation of carbon–carbon double bonds, is prepared from *m*-chlorobenzoyl chloride and a solution containing NaOH and H_2O_2. Suggest a mechanism that explains the role of the NaOH in the formation of the peroxy acid.

15.66 Predict the structures of the polymers that could be formed from the following monomers:

(a) $HO_2CCH_2CH_2CO_2H + H_2NCH_2CH_2CH_2CH_2NH_2 \xrightarrow{\text{heat}}$ nylon 4,4

(b) $CH_3\overset{\overset{O}{\|}}{CH}\overset{}{\underset{NH_2}{C}}OCH_3 \longrightarrow$ polyalanine

(c) $CH_3O\overset{\overset{O}{\|}}{C}$—[benzene ring]—$\overset{\overset{O}{\|}}{C}OCH_3 + HOCH_2$—[cyclohexane ring]—$CH_2OH \longrightarrow$ Kodel

15.67 The reaction of phosgene ($Cl_2C{=}O$) with 1,2-ethanediol can lead to two products: one is cyclic and one is polymeric. What are the structures of these products?

15.68 Predict the products of the following reactions, and suggest mechanisms for their formation:

OH
|
(a) $CH_3CHCO_2CH_3$ + dil. HCl $\xrightarrow{\text{heat}}$

(b) $(CH_3)_2CHCO_2CH_3$ + dil. NaOH $\xrightarrow{\text{heat}}$

15.69 Suggest a practical method for the synthesis of:

(a) hexanoic acid from 1-bromopentane
(b) 2-hydroxyhexanoic acid from pentanal
(c) β-phenylethylamine ($C_6H_5CH_2CH_2NH_2$) from benzyl bromide
(d) N-cyclohexylacetamide from acetic acid
(e) acetic butanoic anhydride from carboxylic acids
(f) di-t-butyl malonate from malonic acid, $CH_2(CO_2H)_2$
(g) 5-nonanol from compounds containing four or fewer carbons
(h) trans-$CH_3(CH_2)_8CH{=}CHCO_2H$ from $CH_3(CH_2)_8CH_2CH_2CO_2H$
(i) triphenylmethanol from benzoic acid

15.70 Compound A (C_3H_5ON) gives a negative Tollens test, but decolorizes a solution of Br_2 in CCl_4. When A is heated with aqueous NaOH, the fumes have a strong ammoniacal odor.

(a) Suggest a structure for A.
(b) How would you confirm your structure assignment by infrared or nmr spectroscopy? (Tell what absorption you would look for.)

15.71 2,6-Dimethyl-4-methoxy-1-bromobenzene was converted to a Grignard reagent and then was treated with 2,6-dimethyl-4-methoxybenzoyl chloride. Upon work-up, a tertiary alcohol was *not* obtained; however, another product was obtained in 48% yield. Suggest a structure for the product and explain why this reaction proceeds in a different manner from that of the usual reaction of a Grignard reagent and an acid halide.

15.72 Suggest mechanisms for the following reactions:

(a) $CH_2{=}CHCH_2CH_2CO_2H \xrightarrow[\text{H}_2\text{O}]{\text{Br}_2} BrCH_2\text{—}$ $=O$

(b) $\text{—}CH_2\overset{\overset{\displaystyle O}{\|}}{S}CCH_3 \xrightarrow[\text{(2) H}_2\text{O, KOH}]{\text{(1) Br}_2\text{, CCl}_4}$

15.73 o-Phthalic acid (page 593) does not give an acid chloride when treated with $SOCl_2$. Instead, the half-acid chloride undergoes an intramolecular reaction to yield another product. What is this product and how is it formed?

15.74 The acid $H\overset{\overset{\displaystyle O}{\|}}{C}CH_2CH_2CO_2H$ was treated with HCN. The product has the molecular formula $C_5H_5O_2N$. What is this product? Suggest a route for its formation.

15.75 Write a flow equation showing how you would make the following conversion using an amide blocking group:

$\text{—}NH_2 \longrightarrow ClCH_2\overset{\overset{\displaystyle O}{\|}}{C}\text{—}$$\text{—}NH_2$

15.76 The following compound was subjected to heating in aqueous acid to yield compound A, $C_5H_8O_3$. When A was heated with thionyl chloride, compound B, $C_5H_6O_2$, was obtained. What are the structures of A and B?

$$
\begin{array}{c}
\overset{\displaystyle O}{\overset{\|}{CH_3CCHCH_2CO_2CH_3}} \\
\underset{\displaystyle CO_2CH_3}{|}
\end{array}
$$

15.77 When the following compound is treated with silver acetate in acetic acid, an acetate ester A is formed in 98% yield. When A is saponified with aqueous sodium hydroxide, a hemiketal B is obtained in 99% yield. What are the structures of A and B?

15.78 Suggest syntheses for the following compounds:

(a) butanal from compounds containing three or fewer carbons

(b) $NCCH_2CO_2CH_2CH_3$ from acetic acid and other appropriate reagents

(c) from cyclohexanecarboxylic acid

(d) ethyl 4-bromobutanoate from a lactone

(e) $CH_3O_2CCH_2\overset{\displaystyle O}{\overset{\|}{C}}CH_2CO_2H$ from a diacid

(f) from

(g) the local anesthetic *benzocaine* (ethyl p-aminobenzoate) from benzene

(h) from maleic acid (page 1046) and other appropriate reagents

(i) $(CH_3)_2CHCH_2\overset{\displaystyle O}{\overset{\|}{C}}CH_2CH_2CO_2CH_3$ from compounds containing five or fewer carbon atoms

(j) from cyclohexylmethanol as the only organic reactant

(k) from

15.79 Determine the structures of the following compounds:

(a) Compound A ($C_7H_{12}O_4$) shows absorption at 3000 cm^{-1} (3.33 μm) and 1750 cm^{-1} (5.71 μm) in the infrared spectrum. Its proton nmr spectrum shows a triplet ($\delta 1.25$), a singlet ($\delta 3.3$), and a quartet ($\delta 4.1$). The area ratio of these signals is 3:1:2.

(b) Compound B ($C_6H_{10}O_3$) shows its principal infrared absorption at 3000 cm^{-1} (3.33 μm), 1750 cm^{-1} (5.71 μm), and 1725 cm^{-1} (5.79 μm). The proton nmr spectrum shows the following peaks and relative areas: triplet (δ1.25), singlet (δ2.25), singlet (δ3.55), and quartet (δ4.15). The area ratios are 3:3:2:2.

(c) Compound C (C_4H_7NO) shows infrared absorption at 2920 cm^{-1} (3.42 μm), 2230 cm^{-1} (4.48 μm), 1370–1450 cm^{-1} (6.9–7.3 μm), and 1110 cm^{-1} (9.0 μm). The proton nmr spectrum shows two triplets (δ2.6 and 3.55) and a singlet (δ3.35) with area ratios of 2:2:3.

15.80 Explain why the ^{13}C nmr spectrum of *N*-ethylformamide (Figure 15.16) exhibits six peaks instead of three. (See Section 9.12.)

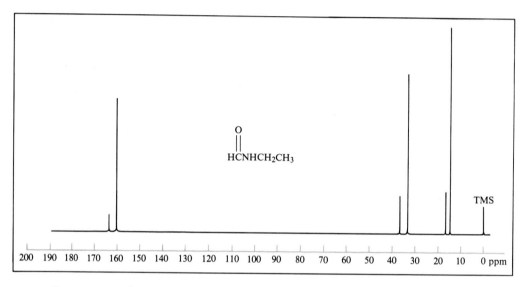

Figure 15.16 ^{13}C nmr spectrum of *N*-ethylformamide for Problem 15.80.

15.81 The proton nmr spectrum of Compound A is shown in Figure 15.17. When A is heated with aqueous acid, the products are acetic acid and acetaldehyde. What is the structure of A?

Figure 15.17 ^1H nmr spectrum for Problem 15.81.

15.82 Compound A ($C_{10}H_{18}O_3$) was treated with dilute aqueous acid to yield the single compound B. When A was heated with ethanol and a trace of H_2SO_4, C was obtained as the only product. The infrared spectra of A and B and the proton nmr spectrum of C are shown in Figure 15.18. What are the structures of A, B, and C?

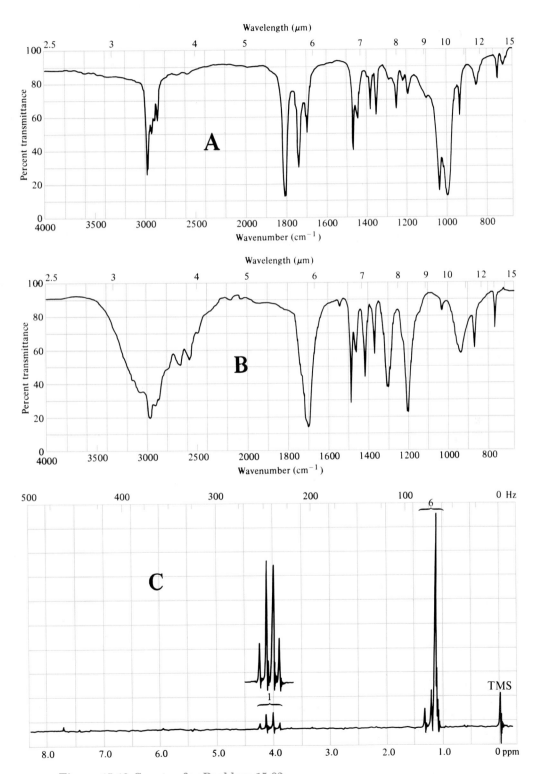

Figure 15.18 Spectra for Problem 15.82.

16 ▼

Enolates and Carbanions: Building Blocks for Organic Synthesis

Nucleophilic reagents react with compounds that contain electron-deficient (partially positive) carbon atoms.

$$\text{Nu:}^- + \overset{\delta+}{\text{R}}-\overset{\delta-}{\text{X}} \longrightarrow \text{Nu}-\text{R} + \text{X}^-$$

$$\text{Nu:}^- + \text{R}-\overset{\overset{\displaystyle O^{\delta-}}{\|}}{\underset{\delta+}{C}}-\text{R} \longrightarrow \text{R}-\overset{\overset{\displaystyle O^-}{|}}{\underset{\underset{\displaystyle \text{Nu}}{|}}{C}}-\text{R}$$

Reagents that contain **nucleophilic carbon atoms**, carbon atoms with carbanion character, also attack partially positive carbon atoms. As an example, a **Grignard reagent**, which has a partially negative carbon atom, attacks carbonyl groups.

$$CH_3\overset{\delta-}{CH_2}-\overset{\delta+}{MgBr} + CH_3\overset{\overset{\displaystyle \ddot{O}:}{\|}}{C}CH_3 \longrightarrow CH_3\underset{\underset{\displaystyle CH_2CH_3}{|}}{\overset{\overset{\displaystyle :\ddot{O}:^- \ ^+MgBr}{|}}{C}}CH_3$$

The attack of one carbon upon another results in a new carbon–carbon bond. Reagents like Grignard reagents with nucleophilic carbon atoms allow us to synthesize compounds with complex carbon skeletons from simple compounds.

Grignard reagents are but one type of many reagents with nucleophilic carbons that are available to the organic chemist. Another versatile class of reagents for building complex molecules are the **enolates**, salts of carbonyl compounds, which also contain nucleophilic carbon atoms.

▼ ## Section 16.1
Acidity of the Alpha Hydrogen

Recall from Section 13.9 that a hydrogen α to a carbonyl group is acidic and can be removed by a strong base. The α hydrogen is acidic primarily because of resonance stabilization of the product enolate ion.

$$CH_3\ddot{O}:^- \ +\overset{\overset{\displaystyle H \quad O}{|\quad\|}}{CH_2CCH_3} \rightleftharpoons CH_3OH + \left[\ ^-\overset{\overset{\displaystyle \ddot{O}:}{\|}}{CH_2-CCH_3} \longleftrightarrow CH_2=\overset{\overset{\displaystyle :\ddot{O}:^-}{|}}{C}CH_3 \right]$$

acetone   resonance structures for
the enolate ion of acetone

Because of resonance stabilization in the enolate ion of acetone, acetone is a far stronger acid than is an alkane. (However, acetone is still a weaker acid than ethanol.)

$$\overset{\overset{\displaystyle H}{|}}{CH_3CH_2} \qquad \overset{\overset{\displaystyle H \quad O}{|\quad\|}}{CH_2CCH_3} \qquad CH_3CH_2-H$$

ethane   acetone   ethanol
$pK_a = 50$ $pK_a = 20$ $pK_a = 15.9$

In an ester, the carbonyl oxygen already carries a partial negative charge from delocalization of the electrons on the alkoxyl oxygen. Therefore, the carbonyl group is somewhat less able to delocalize the anionic negative charge of the enolate. The resonance structures for a typical ester, ethyl acetate, follow:

$$CH_3-\overset{\overset{\displaystyle \ddot{O}:}{\|}}{C}-\ddot{O}CH_2CH_3 \longleftrightarrow CH_3-\overset{\overset{\displaystyle :\ddot{O}:^-}{|}}{C}=\overset{+}{O}CH_2CH_3$$

major contributor

Because the α hydrogen of an ester is less easily removed, a simple ester is less acidic than a ketone.

H   O
| ||
$CH_2COCH_2CH_3$

ethyl acetate
$pK_a = 25$

A hydrogen alpha to a single carbonyl group is less acidic than a hydroxyl hydrogen. Therefore, treatment of an aldehyde, a ketone, or an ester with an alkoxide results in a very low concentration of enolate ions. If we want a reasonably high concentration of the enolate, we must use a much stronger base, such as one of the following:

Some extremely strong bases:

$Na^+ \ {}^-NH_2$ $Na^+ \ H^-$

sodamide   sodium hydride

$CH_3CH_2CH_2CH_2Li$ $(CH_3)_2CHNCH(CH_3)_2 \ Li^+$

n-butyllithium   lithium diisopropylamide (LDA)

$$CH_3\overset{O}{\overset{||}{C}}OCH_2CH_3 + {}^-:\ddot{O}CH_2CH_3 \rightleftharpoons {}^-:CH_2\overset{O}{\overset{||}{C}}OCH_2CH_3 + H\ddot{O}CH_2CH_3$$

ethyl acetate   *not favored*   ethanol
$pK_a = 25$ $pK_a = 16$

$$CH_3\overset{O}{\overset{||}{C}}OCH_2CH_3 + {}^-:\ddot{N}H_2 \rightleftharpoons {}^-:CH_2\overset{O}{\overset{||}{C}}OCH_2CH_3 + :NH_3$$

$pK_a = 25$ *favored* $pK_a = 35$

If a hydrogen is alpha to two carbonyl groups, the negative charge on the anion can be delocalized by both C=O groups. Such a hydrogen is *more acidic than that of an alcohol.* A high concentration of enolate may be obtained by treatment of a β-dicarbonyl compound with an alkoxide. Table 16.1 (page 692) lists the pK_a values for some compounds with hydrogens alpha to one and two carbonyl groups.

$$CH_3\overset{O}{\overset{||}{C}}CH_2\overset{O}{\overset{||}{C}}CH_3 + {}^-OCH_3 \rightleftharpoons CH_3\overset{O}{\overset{||}{C}}CHCH_3\overset{O}{\overset{||}{C}} + CH_3OH$$

2,4-pentanedione   *favored* $pK_a = 15.5$
$pK_a = 9$

Resonance structures of the enolate ion:

$$CH_3\overset{:\ddot{O}:}{\overset{||}{C}}\overset{:\ddot{O}}{\overset{||}{CH}}CCH_3 \longleftrightarrow CH_3\overset{:\ddot{O}:^-}{\overset{|}{C}}=CH\overset{:\ddot{O}}{\overset{||}{C}}CCH_3 \longleftrightarrow CH_3\overset{:\ddot{O}}{\overset{||}{C}}CH=\overset{:\ddot{O}:^-}{\overset{|}{C}}CH_3$$

If an enolate contains more than one alpha carbon atom with ionizable hydrogens, two or more enolate ions can be in equilibrium. For example, in solution, the preceding enolate is in equilibrium with the enolate formed at one

Table 16.1 pK_a values for some carbonyl compounds[a]

Structure	Name	Approx. pK_a
$\overset{O}{\overset{\|}{CH_3C}}\overset{}{\underset{\underset{H}{\|}}{CH}}\overset{O}{\overset{\|}{CCH_3}}$	2,4-pentanedione (acetoacetone)	9
$\overset{O}{\overset{\|}{CH_3C}}\overset{}{\underset{\underset{H}{\|}}{CH}}\overset{O}{\overset{\|}{COCH_2CH_3}}$	ethyl acetoacetate (acetoacetic ester)	11
$\overset{O}{\overset{\|}{CH_3C}}\overset{}{\underset{\underset{R\quad H}{}}{C}}\overset{O}{\overset{\|}{COCH_2CH_3}}$	an alkylacetoacetic ester	13
$CH_3CH_2O\overset{O}{\overset{\|}{C}}\overset{}{\underset{\underset{H}{\|}}{CH}}\overset{O}{\overset{\|}{COCH_2CH_3}}$	diethyl malonate (malonic ester)	13
$\underset{\underset{H}{\|}}{CH_2}\overset{O}{\overset{\|}{CCH_3}}$	acetone	20
$\underset{\underset{H}{\|}}{CH_2}\overset{O}{\overset{\|}{COCH_2CH_3}}$	ethyl acetate	25

[a] For comparison, the pK_a of acetic acid is 4.75; ethanol, 15.9; water, 15.74; and ethane, about 50.

of the terminal carbons. Because of the differences in acidity, however, only the one enolate would be present to any measurable extent.

Equilibrium mixture of enolates:

$$CH_3\overset{O}{\overset{\|}{C}}\overset{..}{\underset{}{CH}}\overset{O}{\overset{\|}{CCH_3}} \quad\longleftrightarrow\quad {}^-\!:CH_2\overset{O}{\overset{\|}{C}}CH_2\overset{O}{\overset{\|}{C}}CH_3$$

more stable enolate favored

Not only a carbonyl group, but any strongly electron-withdrawing group enhances the acidity of an alpha hydrogen. Some other compounds that are more acidic than ethanol are:

$$CH_3CH_2O\overset{O}{\overset{\|}{C}}\underset{\underset{H}{\|}}{CH}CN \qquad NC\underset{\underset{H}{\|}}{CH}CN \qquad \underset{\underset{H}{\|}}{CH_2}NO_2$$

$$\underset{\overset{|}{H}}{\underset{|}{C}}\!H\text{CN} \qquad O_2N\!-\!\!\underset{\overset{|}{H}}{\underset{|}{C}}\!H_2$$

with NO$_2$ group on the ring.

Study Problems

16.1 Which of the indicated hydrogens are acidic?

(a) CH$_3$CH=CHCHO   (b) C$_6$H$_5$CHO   (c) CH$_3$CH$_2$CO$_2$CH$_2$CH$_3$

16.2 Write an equation to show the acid–base reaction, if any, of each of the following compounds with sodium ethoxide:

(a) CH$_3$CH$_2$CHO   (b) (CH$_3$)$_2$CHCO$_2$CH$_2$CH$_3$

(c) (CH$_3$)$_3$CCO$_2$CH$_2$CH$_3$ (d) O$_2$N—⟨NO$_2$ ring⟩—CH$_2$CH$_3$

(e) C$_6$H$_5$CH$_2$NO$_2$ (f) CH$_2$(CO$_2$H)$_2$

▼ Section 16.2
Alkylation of Malonic Ester

One of the more powerful tools at the disposal of the synthetic organic chemist is the reaction of an enolate with an alkyl halide. In this section, we will emphasize the *alkylation of malonic ester*. In general, the end products from alkylation of malonic ester are *α-substituted acetic acids*. In the following example, the R group comes from RX. (In the discussions that follow, we will use —C$_2$H$_5$ to represent the ethyl group.)

General:

$$CH_2(CO_2C_2H_5)_2 \xrightarrow[\text{(2) } \textbf{RX}]{\text{(1) Na}^+ \,{}^-\text{OC}_2\text{H}_5} \textbf{R}CH(CO_2C_2H_5)_2 \xrightarrow[\substack{\text{heat} \\ -CO_2}]{H_2O,\,H^+} \textbf{R}CH_2\overset{\overset{\displaystyle O}{\|}}{C}OH$$

diethyl malonate   a diethyl   an α-substituted
(malonic ester)   alkylmalonate   acetic acid

from RX

$$CH_2(CO_2C_2H_5)_2 \xrightarrow[\text{(2) } CH_3CH_2Br]{\text{(1) Na}^+ \,{}^-\text{OC}_2\text{H}_5} CH_3CH_2CH(CO_2C_2H_5)_2 \xrightarrow[\substack{115° \\ -CO_2}]{H_2O,\,H^+} CH_3CH_2CH_2CO_2H$$

diethyl malonate   diethyl ethylmalonate   butanoic acid (80%)

from RX

A malonic ester alkylation consists of four separate reactions: (1) preparation of the enolate; (2) the actual alkylation; and (3) hydrolysis of the ester, followed by (4) decarboxylation of the resulting β-dicarboxylic acid.

Although there are many chemical reactions involved in this sequence, the laboratory procedure is quite simple because the same reaction vessel can be used for the entire sequence of reactions. (Sometimes the intermediate alkylated ester is purified prior to hydrolysis and decarboxylation in order to simplify the final purification.) Let us discuss each of the steps in this reaction in turn.

A. Formation of the Enolate

The enolate of malonic ester is usually prepared by treatment of the ester with sodium ethoxide, prepared by dissolving sodium metal in anhydrous ethanol (not the common 95% ethanol). (Why not?) Excess ethanol serves as the solvent for the reaction. Diethyl malonate is then added. The ethoxide ion is a stronger base than the enolate ion; therefore, the acid–base equilibrium lies on the side of the resonance-stabilized enolate anion.

$$Na + C_2H_5OH \longrightarrow Na^+ \; ^-:\ddot{O}C_2H_5 + \tfrac{1}{2}H_2\uparrow$$

$$C_2H_5O\overset{\overset{O}{\|}}{C}CH_2\overset{\overset{O}{\|}}{C}OC_2H_5 + \; ^-:\ddot{O}C_2H_5 \;\rightleftharpoons\; C_2H_5O\overset{\overset{O}{\|}}{C}\bar{C}H\overset{\overset{O}{\|}}{C}OC_2H_5 + H\ddot{O}C_2H_5$$

favored

The enolate can also be prepared using NaH and an aprotic solvent, such as benzene, dimethylformamide (DMF), or 1,2-dimethoxyethane.

B. Alkylation

The alkylation reaction is a typical S_N2 displacement by a nucleophile. Methyl and primary alkyl halides give the best yields, with secondary alkyl halides giving lower yields because of competing elimination reactions. (Tertiary alkyl halides give exclusively elimination products, and aryl halides are nonreactive under S_N2 conditions.)

$$(C_2H_5O_2C)_2\underset{\frown}{CH^-} + \underset{\underset{CH_3}{|}}{CH_2}\!\!-\!\!\ddot{Br}: \xrightarrow{S_N2} (C_2H_5O_2C)_2\underset{\underset{CH_3}{|}}{CHCH_2} + :\ddot{Br}:^-$$

Study Problem

16.3 Predict the products of the following reactions:

(a) $CH_2(CO_2C_2H_5)_2 \xrightarrow[\text{(2) } CH_3CH_2CH_2Br]{\text{(1) } Na^+ \; ^-OC_2H_5}$

(b) $CH_3CH(CO_2C_2H_5)_2 \xrightarrow[\text{(2) } CH_3CH_2I]{\text{(1) } Na^+ \; ^-OC_2H_5}$

The product of the alkylation still contains an acidic hydrogen:

$$
\begin{array}{c}
\text{C}_2\text{H}_5\text{OC} \quad \text{CH}_2\text{CH}_3 \\
\overset{\displaystyle \text{O}}{\overset{\|}{}} \quad \\
\text{C} \\
\text{C}_2\text{H}_5\text{OC} \quad \text{H} \quad acidic \\
\overset{\|}{\text{O}}
\end{array}
$$

This second hydrogen can be removed by base, and a *second* R group can be substituted on the malonic ester. This second R group may be the same as, or different from, the first.

$$\text{CH}_3\text{CH}_2\text{CH}(\text{CO}_2\text{C}_2\text{H}_5)_2 \xrightarrow{\text{Na}^+ \, ^-\text{OC}_2\text{H}_5} \text{CH}_3\text{CH}_2\bar{\text{C}}(\text{CO}_2\text{C}_2\text{H}_5)_2 \xrightarrow{\text{CH}_3\text{I}}$$

diethyl ethylmalonate   *an enolate*

$$
\overset{\displaystyle \text{CH}_3}{\underset{}{}}\\
\text{CH}_3\text{CH}_2\overset{|}{\text{C}}(\text{CO}_2\text{C}_2\text{H}_5)_2
$$

diethyl ethylmethylmalonate

The second hydrogen is not as acidic as the first hydrogen. Therefore, the second substitution requires more strenuous reaction conditions than does the first substitution. (For example, a stronger base such as NaH could be used.) Because of the difference in reactivity between the unsubstituted ester and the monoalkyl ester, diethyl malonate can be selectively mono- or dialkylated.

C. Hydrolysis and Decarboxylation

We have mentioned previously that a compound with a carboxyl group beta to a carbonyl group undergoes decarboxylation when heated. The mechanism for decarboxylation was shown in Section 14.9C. If malonic ester (substituted or not) is hydrolyzed in hot acidic solution, the product β-diacid may undergo decarboxylation. (Sometimes decarboxylation does not occur until the diacid is distilled.)

$$\overset{\displaystyle \text{O} \quad \text{O}}{\underset{}{\zeta-\overset{\|}{\text{C}}\text{CH}_2\overset{\|}{\text{C}}\text{OH}}} \xrightarrow{\text{heat}} \overset{\displaystyle \text{O}}{\underset{}{\zeta-\overset{\|}{\text{C}}\text{CH}_3}} + \text{CO}_2$$

Hydrolysis and decarboxylation:

$$
\begin{array}{ccccc}
\overset{\displaystyle \text{O}}{} & & & \overset{\displaystyle \text{O}}{} & \\
\text{R} \quad \overset{\|}{\text{C}}\text{OC}_2\text{H}_5 & & & \text{R} \quad \overset{\|}{\text{C}}\text{OH} & & \text{R} \\
\diagdown \text{C} \diagup & \xrightarrow[\text{heat}]{\text{H}^+, \text{H}_2\text{O}} & & \diagdown \text{C} \diagup & \xrightarrow{-\text{CO}_2} & \diagdown \text{CHCO}_2\text{H} \\
\text{R}' \quad \overset{\|}{\underset{\text{O}}{\text{C}}}\text{OC}_2\text{H}_5 & & & \text{R}' \quad \overset{\|}{\underset{\text{O}}{\text{C}}}\text{OH} & & \text{R}'
\end{array}
$$

an α,α-disubstituted   a β-diacid   an α,α-disubstituted
malonic ester   acetic acid

Preparation of the enolate:

$$CH_2(CO_2C_2H_5)_2 + Na^+ {}^-OC_2H_5 \rightleftharpoons Na^+ {}^-:CH(CO_2C_2H_5)_2 + HOC_2H_5$$

Alkylation:

$$R\!-\!X + {}^-:CH(CO_2C_2H_5)_2 \xrightarrow{S_N2} R\!-\!CH(CO_2C_2H_5)_2 + X^-$$

Hydrolysis and decarboxylation:

$$RCH(CO_2C_2H_5)_2 \xrightarrow[\text{heat}]{H_2O,\ H^+} RCH\begin{matrix} CO_2H \\ \\ CO_2H \end{matrix} \xrightarrow[\text{heat}]{-CO_2} RCH_2CO_2H$$

Figure 16.1 The steps in a malonic ester alkylation. The alkyl halide may be primary, secondary, allylic, or benzylic.

A better yield of carboxylic acid is sometimes obtained when the diester is first saponified and the resulting disodium salt heated with aqueous acid.

$$R_2C(CO_2C_2H_5)_2 \xrightarrow[H_2O]{NaOH} R_2C(CO_2{}^- Na^+)_2 \xrightarrow[\text{heat}]{HCl} R_2CHCO_2H + CO_2$$

What if a chemist does not want a decarboxylation product, but wants a diacid? A diacid can be prepared by *saponification of the diester in base, followed by acidification of the cooled solution.* This way, the dicarboxylic acid itself is not subjected to heat and thus is less likely to undergo decarboxylation.

Saponification and acidification:

$$\begin{matrix} R & CO_2C_2H_5 \\ & \diagdown C \diagup \\ R' & CO_2C_2H_5 \end{matrix} \xrightarrow[\text{heat}]{H_2O,\ OH^-} \begin{matrix} R & CO_2{}^- \\ & \diagdown C \diagup \\ R' & CO_2{}^- \end{matrix} \xrightarrow[\text{cold}]{H^+} \begin{matrix} R & CO_2H \\ & \diagdown C \diagup \\ R' & CO_2H \end{matrix}$$

an α,α-disubstituted
malonic acid

Figure 16.1 summarizes the steps in a malonic ester monoalkylation.

Study Problems

16.4 Write equations for the following reactions:

(a) saponification of diethyl *n*-propylmalonate, followed by treatment with cold HCl
(b) hydrolysis of diethyl dimethylmalonate in hot aqueous acid.

16.5 Give the mechanism for the decarboxylation of methylmalonic acid.

▼ ## Section 16.3
Alkylation of Acetoacetic Ester

Alkylation reactions are not limited to the enolate of diethyl malonate. Other enolates also undergo S_N2 reaction with alkyl halides to yield alkylated products. Another commonly used enolate is that obtained from ethyl acetoace-

Preparation of the enolate:

$$CH_3\overset{O}{\overset{\|}{C}}CH_2\overset{O}{\overset{\|}{C}}OC_2H_5 + Na^+ \ ^-OC_2H_5 \rightleftharpoons Na^+ \ CH_3\overset{O}{\overset{\|}{C}}\overset{..}{\overset{-}{C}}HC\overset{O}{\overset{\|}{C}}OC_2H_5 + HOC_2H_5$$

Alkylation:

$$CH_3\overset{O}{\overset{\|}{C}}\overset{..}{\overset{-}{C}}HC\overset{O}{\overset{\|}{C}}OC_2H_5 + R\overset{\frown}{-}X \xrightarrow{S_N2} CH_3\overset{O}{\overset{\|}{C}}\underset{R}{C}HC\overset{O}{\overset{\|}{C}}OC_2H_5 + X^-$$

Hydrolysis and decarboxylation:

$$CH_3\overset{O}{\overset{\|}{C}}\underset{R}{C}HC\overset{O}{\overset{\|}{C}}OC_2H_5 \xrightarrow[\text{heat}]{H_2O, H^+} CH_3\overset{O}{\overset{\|}{C}}\underset{R}{C}HC\overset{O}{\overset{\|}{C}}OH \xrightarrow[\text{heat}]{-CO_2} CH_3\overset{O}{\overset{\|}{C}}CH_2R$$

Figure 16.2 The steps in an acetoacetic ester alkylation. The alkyl halide may be primary, secondary, allylic, or benzylic.

tate (acetoacetic ester). The end product of alkylation of acetoacetic ester is an α-substituted acetone.

General:

$$CH_3\overset{O}{\overset{\|}{C}}CH_2\overset{O}{\overset{\|}{C}}OC_2H_5 \xrightarrow[\text{(2) RX}]{\text{(1) Na}^+ \ ^-OC_2H_5} CH_3\overset{O}{\overset{\|}{C}}\underset{R}{C}HC\overset{O}{\overset{\|}{C}}OC_2H_5 \xrightarrow[\text{heat} \\ -CO_2]{H_2O, H^+} CH_3\overset{O}{\overset{\|}{C}}CH_2R$$

from RX

ethyl acetoacetate
(acetoacetic ester)

*an ethyl
alkylacetoacetate*

*an α-substituted
acetone*

$$CH_3\overset{O}{\overset{\|}{C}}CH_2\overset{O}{\overset{\|}{C}}OC_2H_5 \xrightarrow[\text{(2) CH}_3CH_2CH_2CH_2Br]{\text{(1) Na}^+ \ ^-OC_2H_5}$$

$$CH_3\overset{O}{\overset{\|}{C}}\underset{CH_2CH_2CH_2CH_3}{C}HC\overset{O}{\overset{\|}{C}}OC_2H_5 \xrightarrow[\text{heat} \\ -CO_2]{H_2O, H^+} CH_3\overset{O}{\overset{\|}{C}}CH_2\overset{\frown}{CH_2CH_2CH_2CH_3}$$

from RX

2-heptanone (60%)

The steps in an acetoacetic ester synthesis are similar to those for a malonic ester synthesis. (See Figure 16.2.)

Study Problems

16.6 Show how the following compounds could be prepared from acetoacetic ester:

(a) $CH_2{=}CHCH_2\underset{CO_2C_2H_5}{\overset{O}{\overset{\|}{C}}HCCH_3}$ (b) $(CH_3CH_2)_2CH\overset{O}{\overset{\|}{C}}CH_3$

16.7 Acetoacetic ester can be dialkylated as well as monoalkylated. Also, other β-keto esters can be used in enolate substitutions. Write equations for the steps of the following reaction sequences:

(a) $\underset{\displaystyle \underset{\displaystyle CH_3}{|}}{CH_3\overset{\displaystyle O}{\overset{\|}{C}}\,\overset{}{C}H\,\overset{\displaystyle O}{\overset{\|}{C}}OC_2H_5}$ $\xrightarrow[\text{(2) } CH_3CH_2CH_2Br]{\text{(1) } Na^+\;{}^-OC_2H_5}$

(b) $C_6H_5\overset{\displaystyle O}{\overset{\|}{C}}CH_2\overset{\displaystyle O}{\overset{\|}{C}}OC_2H_5$ $\xrightarrow[\substack{\text{(2) } CH_3I \\ \text{(3) } H_2O,\,H^+,\,\text{heat}}]{\text{(1) } Na^+\;{}^-OC_2H_5}$

▼ ## Section 16.4
Syntheses Using Alkylation Reactions
A. Malonic Ester and Acetoacetic Ester Syntheses

In general, the products of alkylation reactions of malonic ester or acetoacetic ester are substituted acetic acids or substituted acetones. However, we can also obtain diacids, diesters, keto acids, and keto esters. The various products that can be obtained from alkylation of malonic ester or acetoacetic ester are summarized in Figure 16.3.

It is comparatively easy to predict the products of a reaction when we are given the reactants. It is somewhat more difficult to decide upon specific reactants to use in a synthesis problem. Remember to work the problem backwards: if you are asked to synthesize a compound by an alkylation reaction, first decide what dicarbonyl compound you would need, then pick the alkyl halide.

$CH_3-CH\begin{smallmatrix}CO_2H \\ \\ CO_2H\end{smallmatrix}$

from CH₃X   *from malonic ester*

$\begin{smallmatrix}CH_3 \\ \\ CH_3\end{smallmatrix}CHCO_2H$ *from malonic ester*

from CH₃X

$CH_3CH_2CH_2\,CH\begin{smallmatrix}\overset{\displaystyle O}{\overset{\|}{C}}CH_3 \\ \\ CO_2C_2H_5\end{smallmatrix}$

from CH₃CH₂CH₂X   *from acetoacetic ester*

$CH_3CH_2\,\underset{\displaystyle \underset{CH_3}{|}}{CH}\overset{\displaystyle O}{\overset{\|}{C}}CH_3$ *from acetoacetic ester*

from CH₃X
from CH₃CH₂X

Example

If you were asked to write the equations for the synthesis of 3-methyl-2-pentanone, you would:

1. *write the structure;*
2. *decide what β-dicarbonyl compound you would need; and*

3. *decide what alkyl halides would have to be used for the substitution.*

$$\underset{\underset{\displaystyle CH_3}{|}}{\overset{\displaystyle \overset{O}{\parallel}}{CH_3CCH}}CH_2CH_3 \xleftarrow{\quad\text{from hydrolysis and decarboxylation of}\quad} \underset{\underset{\displaystyle CH_3}{|}}{\overset{\displaystyle \overset{O}{\parallel}\ \ \ CO_2C_2H_5}{CH_3C-\underset{|}{C}-CH_2CH_3}}$$

$CH_2(CO_2C_2H_5)_2$

malonic ester

↓ base
RX

$$RCH(CO_2C_2H_5)_2 \xrightarrow[\text{heat}]{H_2O,\ H^+} RCH(CO_2H)_2 \xrightarrow[\text{heat}]{-CO_2} RCH_2CO_2H$$

a diester　　　　　*a diacid*　　　　　*an acid*

↓ base
R′X

$$\underset{R'}{\overset{R}{\diagdown}}C(CO_2C_2H_5)_2 \xrightarrow[\text{heat}]{H_2O,\ H^+} \underset{R'}{\overset{R}{\diagdown}}C(CO_2H)_2 \xrightarrow[\text{heat}]{-CO_2} \underset{R'}{\overset{R}{\diagdown}}CHCO_2H$$

a diester　　　　　*a diacid*　　　　　*an acid*

$\overset{O}{\overset{\parallel}{CH_3CCH_2CO_2C_2H_5}}$

acetoacetic ester

↓ base
RX

$$\underset{\underset{\displaystyle R}{|}}{\overset{\displaystyle \overset{O}{\parallel}}{CH_3CCHCO_2C_2H_5}} \xrightarrow[\text{heat}]{H_2O,\ H^+} \underset{\underset{\displaystyle R}{|}}{\overset{\displaystyle \overset{O}{\parallel}}{CH_3CCHCO_2H}} \xrightarrow[\text{heat}]{-CO_2} \overset{\displaystyle \overset{O}{\parallel}}{CH_3CCH_2R}$$

a keto ester　　　　*a keto acid*　　　　*a ketone*

↓ base
R′X

$$\underset{\underset{\displaystyle R\ \ \ R'}{|\ \ \ \ |}}{\overset{\displaystyle \overset{O}{\parallel}}{CH_3CCCO_2C_2H_5}} \xrightarrow[\text{heat}]{H_2O,\ H^+} \underset{\underset{\displaystyle R\ \ \ R'}{|\ \ \ \ |}}{\overset{\displaystyle \overset{O}{\parallel}}{CH_3CCCO_2H}} \xrightarrow[\text{heat}]{-CO_2} \underset{\underset{\displaystyle R'}{|}}{\overset{\displaystyle \overset{O}{\parallel}}{CH_3CCHR}}$$

a keto ester　　　　*a keto acid*　　　　*a ketone*

Figure 16.3 Products from the alkylations of malonic ester and acetoacetic ester.

Because a ketone is the product, the β-dicarbonyl starting material is acetoacetic ester. The alkyl halides needed are CH_3X and CH_3CH_2X. Now, equations for the steps in the synthesis can be written:

(1) $\underset{\substack{\| \\ O}}{CH_3CCH_2CO_2C_2H_5}$ $\xrightarrow[\text{(2) } CH_3I]{\text{(1) } NaOC_2H_5}$ $\underset{\substack{\| \quad\quad | \\ O \quad\quad CH_3}}{CH_3CCHCO_2C_2H_5}$

(2) $\underset{\substack{\| \quad\quad | \\ O \quad\quad CH_3}}{CH_3CCHCO_2C_2H_5}$ $\xrightarrow[\text{(2) } CH_3CH_2Br]{\text{(1) } NaOC_2H_5}$ $\underset{\substack{\| \quad | \\ O \quad H_3C \;\; CH_2CH_3}}{CH_3C-C-CO_2C_2H_5}$

(3) $\underset{\substack{\| \quad | \\ O \quad H_3C \;\; CH_2CH_3}}{CH_3C-C-CO_2C_2H_5}$ $\xrightarrow[\substack{\text{heat} \\ -CO_2}]{H^+,\, H_2O}$ $\underset{\substack{\| \quad\quad | \\ O \quad\quad CH_3}}{CH_3CCHCH_2CH_3}$

Sample Problem

Suggest a reaction sequence leading to 3-phenylpropanoic acid.

Solution:

— from benzyl bromide

$\bigcirc$—CH_2 CH_2CO_2H ← *from malonic ester*

(1) $CH_2(CO_2C_2H_5)_2$ $\xrightarrow{Na^+ {}^-OC_2H_5}$ ${}^-CH(CO_2C_2H_5)_2$

(2) $C_6H_5CH_2Br + {}^-CH(CO_2C_2H_5)_2$ $\xrightarrow{-Br^-}$ $C_6H_5CH_2CH(CO_2C_2H_5)_2$

(3) $C_6H_5CH_2CH(CO_2C_2H_5)_2$ $\xrightarrow[\substack{\text{heat} \\ -CO_2}]{H_2O,\, H^+}$ $C_6H_5CH_2CH_2CO_2H$

Study Problem

16.8 Show how you could synthesize the following compounds by alkylation reactions:

(a) $CH_3CH_2CH_2CH_2CH(CO_2H)_2$ (b) $(CH_3)_2CHCO_2H$

(c) $\underset{\substack{\| \quad\quad | \\ O \quad\quad CH_3}}{CH_3CCHCH_2CH_2CH_3}$

(d) $\underset{\substack{| \\ CH_2CO_2C_2H_5}}{CH(CO_2C_2H_5)_2}$

(e) CN / CN

(f) —$CH_2CH(CH_3)_2$

B. Alkylation Reactions of Other Carbonyl Compounds

Ketones and esters with only one carbonyl group and at least one alpha hydrogen can also be alkylated in the alpha position. Because only one activating group is present, a stronger base than an alkoxide is necessary. Lithium diisopropylamide (LDA) is commonly used.

One advantage to reactions with lithium enolates is that they can be carried out at very low temperatures—a reaction condition that minimizes equilibration among different enolates. Thus, these reactions are under kinetic (rate) control and not thermodynamic, or equilibrium, control as are many enolate reactions.

General:

$$\underset{\underset{\underset{H}{\overset{|}{C}}}{\overset{O}{\parallel}}}{RCHCR'} \xrightarrow[- HN[CH(CH_3)_2]_2]{Li^+ \; ^-:\ddot{N}[CH(CH_3)_2]_2, \; cold} \left[\underset{\overset{O}{\parallel}}{R\bar{C}HCR'} \longleftrightarrow \underset{RCH=CR'}{\overset{Li^+ \; O^-}{|}} \right]$$

a lithium enolate

$$\xrightarrow[-LiX]{R''X} \quad \underset{\underset{R''}{\overset{|}{}}}{\overset{O}{\parallel}}{RCHCR'}$$

an α-substituted ketone

a lithium enolate

(93%)

Study Problem

16.9 Suggest alkylation reactions for the following conversions.

(a)

(b)

Section 16.5
Alkylation and Acylation of Enamines

Another type of organic compound containing a nucleophilic carbon that can undergo alkylation reactions is an **enamine**. In Section 13.6B, we discussed the formation of enamines from secondary amines and aldehydes or ketones.

Formation of an enamine:

a ketone   a 2° amine   an enamine

cyclohexanone   piperidine   *an enamine*

The nitrogen of an enamine has an unshared pair of electrons. These electrons are, in a sense, in an *allylic position* and consequently are in conjugation with the double bond. Resonance structures for the enamine show that the carbon β to the nitrogen has a partial negative charge.

resonance structures for an enamine

This β carbon has carbanion character and can act as a nucleophile. For example, when an enamine is treated with an alkyl halide, such as CH_3I, the enamine displaces the halogen of the alkyl halide in an S_N2 reaction. The result is alkylation of the enamine at the position that is β to the nitrogen.

Alkylation:

the enamine

an iminium ion

The product iminium ion is readily hydrolyzed to a ketone. The net result of the entire reaction sequence is alkylation of a ketone in the α position.

Preparation of enamine:

$$R_2CH-\overset{\displaystyle R}{\overset{|}{C}}=O + HN\!\!\bigcirc \;\;\xrightleftharpoons{H^+}\;\; R_2C=\overset{\displaystyle R}{\overset{|}{C}}-N\!\!\bigcirc$$

Substitution reaction:

$$R_2C=\overset{\displaystyle R}{\overset{|}{C}}-N\!\!\bigcirc \;\;\xrightarrow{R'X}\;\; R_2C-\overset{\displaystyle R}{\underset{\underset{\displaystyle R'}{|}}{C}}=\overset{+}{N}\!\!\bigcirc$$

β *to N*

Hydrolysis:

$$R_2\overset{\displaystyle R}{\underset{\underset{\displaystyle R'}{|}}{C}}-C=\overset{+}{N}\!\!\bigcirc + H_2O \;\;\xrightleftharpoons{H^+}\;\; R_2\overset{\displaystyle R}{\underset{\underset{\displaystyle R'}{|}}{C}}-C=O + H_2\overset{+}{N}\!\!\bigcirc$$

α *to C=O*

Figure 16.4 The steps in an enamine alkylation or acylation. The alkyl halide must be a reactive one, such as methyl, benzylic, or allylic.

Hydrolysis:

α *to C=O*

$$\text{(the iminium ion)} + H_2O \;\;\xrightleftharpoons{H^+}\;\; \text{(2-methylcyclohexanone)} \;+\; H_2\overset{+}{N}\!\!\bigcirc$$

the iminium ion   2-methylcyclohexanone   piperidinium ion

a ketone

A generalized sequence for an enamine synthesis is shown in Figure 16.4.

Sample Problem

Give the steps in the enamine synthesis of the following ketone using piperidine as the amine.

$$\text{(cyclopentanone with } CH_2CH=CH_2 \text{ substituent)}$$

Solution:

$$\text{(cyclopentanone)} \;\xrightleftharpoons{\;\;\overset{NH\bigcirc}{}\;\;}\; \text{(enamine)} \;\xrightarrow[-Cl^-]{CH_2=CHCH_2Cl}\; \text{(iminium salt with } CH_2CH=CH_2) \;\xrightleftharpoons{H_2O,\ H^+}\; \text{product}$$

The alkylation step in an enamine synthesis is an S_N2 reaction with a rather weak nucleophile. (Why?) It is not surprising then that only the most reactive halogen compounds are suitable as alkylating agents. These compounds include allylic halides, benzylic halides, α-halocarbonyl compounds, and iodomethane.

Some reactive halides:

CH_3I	$C_6H_5CH_2Cl$	$CH_2{=}CHCH_2Cl$	$CH_3\overset{O}{\overset{\|}{C}}Cl$	$BrCH_2\overset{O}{\overset{\|}{C}}CH_3$
iodomethane	benzyl chloride	allyl chloride	acetyl chloride	bromoacetone

The reactions of enamines with α-halocarbonyl compounds and acid halides follow similar paths to that of alkylation. In each case, the final product (after hydrolysis) is a ketone substituted at the α position.

Figure 16.5 Products from enamine alkylations and acylations.

Figure 16.5 summarizes the products that can be obtained by enamine syntheses.

Sample Problem

How would you prepare the following compound by an enamine synthesis?

$$\underset{\underset{\displaystyle CH_3}{|}}{\overset{\overset{\displaystyle O}{\|}}{CH_3C}}\!-\!\underset{\underset{\displaystyle CH_3}{|}}{C}\!-\!CHO$$

Solution:

Reaction sequence:

Study Problems

16.10 Enamine syntheses are generally carried out with a cyclic amine as the secondary amine. Suggest a reason for this.

Three cyclic amines used in enamine syntheses:

    piperidine   pyrrolidine   morpholine

16.11 Predict the products:

16.12 Show how you would prepare the following compounds by enamine syntheses:

(a) $CH_3CH_2\overset{\overset{\displaystyle O}{\|}}{C}CH(CH_3)_2$ **(b)**

Section 16.6
Aldol Condensations

So far, we have been discussing the displacement of halide ions by nucleophiles. A reagent with a nucleophilic carbon atom can also attack the partially positive carbon of a carbonyl group. The rest of this chapter will be devoted to the reactions of enolates and related anions with carbonyl compounds.

$$Nu:^- + \overset{\overset{\displaystyle \ddot{O}:}{\|}}{-\underset{|}{C}-} \longrightarrow \overset{\overset{\displaystyle :\ddot{O}:^-}{\|}}{-\underset{\underset{\displaystyle Nu}{|}}{C}-}$$

When an aldehyde is treated with a base such as aqueous NaOH, the resulting enolate ion can undergo reaction at the carbonyl group of another molecule of aldehyde. The result is the *addition of one molecule of aldehyde to another.*

$$2\ CH_3\overset{\overset{\displaystyle O}{\|}}{CH} \quad \underset{\longleftarrow}{\overset{OH^-}{\longrightarrow}} \quad CH_3\overset{\overset{\displaystyle OH}{|}}{CH}-CH_2\overset{\overset{\displaystyle O}{\|}}{CH}$$

from one aldehyde

acetaldehyde 3-hydroxybutanal (50%)
(acetaldol or aldol)

This reaction is called an **aldol addition**, or **aldol condensation reaction**. The word "aldol," derived from *ald*ehyde and alcoh*ol*, describes the product, which is a *β-hydroxy aldehyde*. A **condensation reaction** is one in which two or more molecules combine into a larger molecule often with the loss of a small molecule (such as water).

How does an aldol condensation proceed? If acetaldehyde is treated with dilute aqueous sodium hydroxide, a low concentration of enolate ions is formed. The reaction is reversible—as enolate ions undergo reaction, more are formed.

$$CH_3\overset{\overset{\displaystyle O}{\|}}{CH} + OH^- \ \rightleftharpoons \ \left[^-\ddot{C}H_2\overset{\overset{\displaystyle \ddot{O}:}{\|}}{CH} \longleftrightarrow CH_2=\overset{\overset{\displaystyle :\ddot{O}:^-}{|}}{CH} \right] + H_2O$$

resonance structures for the enolate ion

The enolate ion undergoes reaction with another acetaldehyde molecule by adding to the carbonyl carbon to form an alkoxide ion, which abstracts a proton from water to yield the product aldol.

$$CH_3\overset{\overset{\displaystyle \ddot{O}:}{\|}}{CH} + {}^-CH_2\overset{\overset{\displaystyle O}{\|}}{CH} \ \rightleftharpoons \ \left[CH_3\overset{\overset{\displaystyle :\ddot{O}:^-}{|}}{CH}-CH_2\overset{\overset{\displaystyle O}{\|}}{CH} \right] \ \underset{\longleftarrow}{\overset{H_2O}{\longrightarrow}} \ CH_3\overset{\overset{\displaystyle OH}{|}}{CH}CH_2\overset{\overset{\displaystyle O}{\|}}{CH} + OH^-$$

α hydrogens

an alkoxide ion

The starting aldehyde in an aldol condensation must contain a hydrogen α to the carbonyl group so that it can form an enolate ion in base. The aldol product still has a carbonyl group with α hydrogens. Can it undergo further reaction to form trimers? tetramers? polymers? Yes, these materials are by-products of the reaction. For simplicity, we will show only the dimer products and ignore the fact that other higher-molecular-weight products may also be formed.

We have shown the aldol condensation for acetaldehyde. Other aldehydes also undergo this self-addition. Ketones undergo aldol condensations, but the equilibrium does not favor the ketone-condensation product. (Why not?) Although there are a number of laboratory procedures that can be used to induce ketone condensations of the aldol type, the reaction is not as useful with ketones as it is with aldehydes. Therefore, we will concentrate our present discussion on aldehydes. Two other examples of aldol condensations follow:

$$CH_3CH_2CH_2\overset{\overset{\displaystyle O}{\|}}{CH} + CH_3CH_2CH_2\overset{\overset{\displaystyle O}{\|}}{CH} \underset{}{\overset{OH^-}{\rightleftharpoons}} CH_3CH_2CH_2\overset{\overset{\displaystyle OH}{|}}{CH}-\overset{\overset{\displaystyle O}{\|}}{CH}CH$$

with CH_2CH_3 below.

an aldol (75%)

an aldol (76%)

Sample Problem

Show how you could prepare 3-hydroxy-2,2,4-trimethylpentanal by an aldol condensation.

Solution: Write the structure and indicate the carbon–carbon bond formed in the condensation.

the new carbon–carbon bond

Write the equation.

$$2\,(CH_3)_2CHCH\overset{\overset{\displaystyle O}{\|}}{} \overset{OH^-}{\rightleftharpoons} (CH_3)_2CHCH-C(CH_3)_2CH$$

with OH and O substituents.

Study Problems

16.13 Which of the following aldehydes can undergo self-condensations? Explain.

(a) ⬡—CHO   (b) HCHO   (c) cyclopentyl with CH₃ —CHO

(d) $(CH_3)_3CCHO$ **(e)** $(CH_3CH_2)_2CHCHO$

16.14 Predict the product of the self-condensation of: **(a)** acetone;   **(b)** 3-methylbutanal.

A. Dehydration of Aldols

A *β*-hydroxy carbonyl compound, such as an aldol, undergoes dehydration readily because the double bond in the product is in conjugation with the carbonyl group. Therefore, an **α,β-unsaturated aldehyde** can be readily obtained as the product of an aldol condensation.

$$CH_3\overset{\overset{\displaystyle OH}{|}}{C}H-CH_2\overset{\overset{\displaystyle O}{\|}}{C}H \xrightarrow[\text{warm}]{\text{dil. H}^+} CH_3CH=CH\overset{\overset{\displaystyle O}{\|}}{C}H + H_2O$$

3-hydroxybutanal   2-butenal
                                 (crotonaldehyde)

When dehydration leads to a double bond in conjugation with an aromatic ring, dehydration is often spontaneous, even in alkaline solution.

3-hydroxy-3-   3-phenylpropenal
phenylpropanal   (cinnamaldehyde)

B. Crossed Aldol Condensations

An aldehyde with no *α* hydrogens cannot form an enolate ion and thus cannot dimerize in an aldol condensation. However, if such an aldehyde is mixed with an aldehyde that *does* have an alpha hydrogen, a condensation between the two can occur. This reaction is called a **crossed aldol condensation**. A crossed aldol condensation is most useful when only one of the carbonyl compounds has an *α* hydrogen; otherwise, mixtures of products result. Methyl ketones can be used successfully in crossed aldol condensations with aldehydes that contain no *α* hydrogen, as in the second example that follows:

benzaldehyde   acetaldehyde   cinnamaldehyde (90%)

(no α hydrogens)

$$\text{C}_6\text{H}_5\text{-CH} + \text{CH}_3\overset{\text{O}}{\underset{||}{\text{C}}}\text{CH}_3 \xrightarrow{\text{OH}^-} \text{C}_6\text{H}_5\text{-CH}=\text{CH}\overset{\text{O}}{\underset{||}{\text{C}}}\text{CH}_3 + \text{H}_2\text{O}$$

acetone   4-phenyl-3-buten-2-one (90%)
(benzalacetone)

Study Problem

16.15 Predict the major products:

(a) $\text{C}_6\text{H}_5\text{CHO} + \text{CH}_3\text{CH}_2\text{CH}_2\text{CHO} \xrightarrow{\text{OH}^-}$

(b) $\text{C}_6\text{H}_5\text{CHO} + \text{CH}_3\text{CH}_2\overset{\text{O}}{\underset{||}{\text{C}}}\text{CH}_2\text{CH}_3 \xrightarrow{\text{OH}^-}$

C. Syntheses Using Aldol Condensations

In an aldol condensation, two types of product can result: (1) β-hydroxy aldehydes or ketones, and (2) α,β-unsaturated aldehydes or ketones (Figure 16.6). In a synthesis problem, look for these functional groups and decide which aldehydes or ketones must be used for the starting materials.

from acetophenone   *from benzaldehyde*   *from propanal*

Sample Problem

The following ketone can be prepared in 90% yield by a crossed aldol condensation:

$$\text{C}_6\text{H}_5\text{-CH}=\text{CH}\overset{\text{O}}{\underset{||}{\text{C}}}\text{CH(CH}_3)_2$$

What organic reactants are needed?

Solution:

$$\text{C}_6\text{H}_5\text{CH}=\text{CH}\overset{\text{O}}{\underset{||}{\text{C}}}\text{CH(CH}_3)_2$$

from $\text{C}_6\text{H}_5\overset{\text{O}}{\underset{||}{\text{CH}}}$ *from* $\text{CH}_3\overset{\text{O}}{\underset{||}{\text{C}}}\text{CH(CH}_3)_2$

Self-addition (reactant must have an α hydrogen):

$$2\ RCH_2\overset{O}{\overset{\|}{C}}H \underset{}{\overset{OH^-}{\rightleftharpoons}} RCH_2\overset{OH}{\underset{|}{C}}H-\overset{O}{\overset{\|}{C}}H\overset{}{\underset{R}{\overset{|}{C}}}H \underset{heat}{\overset{H^+}{\longrightarrow}} RCH_2CH=\overset{O}{\overset{\|}{C}}\overset{}{\underset{R}{\overset{|}{C}}}H$$

a β-hydroxy   an α,β-unsaturated
aldehyde   aldehyde

Crossed (one reactant must have an α hydrogen):

$$R_3C\overset{O}{\overset{\|}{C}}H + R'CH_2\overset{O}{\overset{\|}{C}}H \underset{}{\overset{OH^-}{\rightleftharpoons}} R_3C\overset{OH}{\underset{|}{C}}H-\overset{O}{\overset{\|}{C}}H\overset{}{\underset{R'}{\overset{|}{C}}}H \underset{heat}{\overset{H^+}{\longrightarrow}} R_3C\overset{O}{\overset{\|}{C}}H=\overset{O}{\overset{\|}{C}}\overset{}{\underset{R'}{\overset{|}{C}}}H$$

a β-hydroxy   an α,β-unsaturated
aldehyde   aldehyde

$$\text{C}_6\text{H}_5-\overset{O}{\overset{\|}{C}}H + RCH_2\overset{O}{\overset{\|}{C}}H \underset{}{\overset{OH^-}{\rightleftharpoons}} \text{C}_6\text{H}_5-CH=\overset{O}{\overset{\|}{C}}\overset{}{\underset{R}{\overset{|}{C}}}H$$

an α,β-unsaturated
aldehyde

Figure 16.6 Products from aldol condensations.

Study Problem

16.16 Suggest syntheses for the following compounds from aldehydes or ketones:

(a) $C_6H_5CH=\underset{\underset{CH_3}{|}}{C}CHO$

(b) [cyclopentene with CH₃ and $-\overset{O}{\overset{\|}{C}}CH_3$ substituents]

(c) [furan]$-CH=CH\overset{O}{\overset{\|}{C}}H$

(d) [cyclohexane with OH and $-\overset{O}{\overset{\|}{C}}H$ substituents]

▼ ■■■ **Section 16.7**
Reactions Related to the Aldol Condensation

A. Knoevenagel Condensation

We have shown aldol condensations and crossed aldol condensations. However, for this type of condensation to occur, all that is needed is one compound with a carbonyl group plus one compound with an acidic hydrogen. The **Knoeve-nagel condensation** is the reaction of an aldehyde with a compound that has a hydrogen α to *two* activating groups (such as C=O or C≡N), usually with ammonia or an amine as the catalyst. Under these conditions, malonic acid itself can be used as a reactant.

Knoevenagel condensations:

$$(CH_3)_2CHCH_2\overset{\overset{\displaystyle O}{\|}}{C}H + CH_2(CO_2C_2H_5)_2 \xrightarrow[\text{heat}]{\overset{\text{piperidine}}{\text{benzene}}} (CH_3)_2CHCH_2CH=C(CO_2C_2H_5)_2 + H_2O$$

3-methylbutanal   diethyl malonate   (78%)

$$\bigcirc\!\!-\overset{\overset{\displaystyle O}{\|}}{C}H + CH_2(CO_2H)_2 \xrightarrow[\text{heat}]{NH_3} \bigcirc\!\!-CH=CHCO_2H + H_2O + CO_2$$

   benzaldehyde   malonic acid   3-phenylpropenoic acid (85%)
                                              (cinnamic acid)

A variation of the Knoevenagel reaction allows the less reactive ketones to undergo condensation with the more acidic ethyl cyanoacetate (pK_a = 9, compared with pK_a = 11 for diethyl malonate).

$$\bigcirc\!\!=\!O + \overset{CN}{\underset{CO_2C_2H_5}{\overset{|}{\underset{|}{CH_2}}}} \xrightarrow[\substack{\text{glacial } CH_3CO_2H \\ \text{benzene} \\ \text{heat}}]{NH_4^+ \ ^-O_2CCH_3} \bigcirc\!\!=\!\overset{CN}{\underset{CO_2C_2H_5}{\overset{|}{\underset{|}{C}}}} + H_2O$$

              ethyl cyanoacetate   (80%)

Sample Problem

How would you prepare the following compound?

$$\bigcirc\!\!-CH=\overset{\overset{\displaystyle CN}{|}}{C}\!\!-\!\bigcirc$$

Solution:

$$\underset{\textit{from } C_6H_5CHO}{\nearrow}\ C_6H_5CH=\overset{\overset{\displaystyle CN}{|}}{C}C_6H_5\ \underset{\textit{from } C_6H_5CH_2CN}{\nwarrow}$$

$$C_6H_5\overset{\overset{\displaystyle O}{\|}}{C}H + \overset{CN}{\underset{}{\overset{|}{CH_2}}}C_6H_5 \xrightleftharpoons{OH^-} C_6H_5CH=\overset{\overset{\displaystyle CN}{|}}{C}C_6H_5 + H_2O$$

Study Problem

16.17 Suggest synthetic routes to the following compounds:

(a) $(CH_3CH_2)_2C=\overset{\overset{\displaystyle O}{\|}}{\underset{\underset{\displaystyle CN}{|}}{C}}COC_2H_5$ **(b)** $\bigcirc\!\!=C(CN)_2$

B. A Biological Aldol-Type Condensation

Esters with the $\overset{\overset{\text{O}}{\|}}{\text{RCS}}$— unit are called **thioesters**. Thioesters undergo reactions, such as hydrolysis, similar to those of ordinary esters. **Acetylcoenzyme A**, which is important in biological reactions, is a thioester.

acetylcoenzyme A

A proton from the acetyl group (at the left in the preceding formula) of acetylcoenzyme A can be removed enzymatically. The resultant enolate ion is used in many biological transformations. For example, this enolate ion can undergo an aldol-type condensation by adding to other carbonyl compounds. The following equations show loss of H^+ from acetylcoenzyme A, nucleophilic attack on the carbonyl group of the oxaloacetate ion, and finally hydrolysis of the thioester to yield the citrate ion and coenzyme A (abbreviated HSCoA).

Enzymatic loss of proton:

acetylcoenzyme A

Nucleophilic attack and thioester hydrolysis:

oxaloacetate
ion

H_2O

CoA—SH + $^-$OCCH$_2$COH

coenzyme A

citrate ion

Section 16.8
Ester Condensations

A. Claisen Condensations

Esters with alpha hydrogens can undergo self-condensation reactions to yield **β-keto esters**. An ester condensation is similar to an aldol condensation; the difference is that the —OR group of an ester can act as a leaving group. The result is therefore *substitution* (whereas aldol condensations are *additions*). Simple ester condensations, such as the following examples, are called **Claisen condensations**. (In the examples below, note the use of *oxo* as a prefix in the name of the β-keto ester to designate the position of the keto group.)

General:

$$\underset{}{RCH_2\overset{O}{\overset{||}{C}}-OR'} \ + \ \underset{R}{H-\overset{O}{\overset{||}{C}}HCOR'} \ \xrightarrow{\text{base}} \ \underset{R}{RCH_2\overset{O}{\overset{||}{C}}-\overset{O}{\overset{||}{C}}HCOR'} \ + \ R'OH$$

a β-keto ester

$$2 \ CH_3\overset{O}{\overset{||}{C}}OC_2H_5 \ \xrightarrow{Na^+ \ ^-OC_2H_5} \ CH_3\overset{O}{\overset{||}{C}}-CH_2\overset{O}{\overset{||}{C}}OC_2H_5 \ + \ C_2H_5OH$$

ethyl acetate   ethyl 3-oxobutanoate (75%)
                          (ethyl acetoacetate)

$$2 \ CH_3CH_2\overset{O}{\overset{||}{C}}OC_2H_5 \ \xrightarrow{Na^+ \ ^-OC_2H_5} \ \underset{CH_3}{CH_3CH_2\overset{O}{\overset{||}{C}}-\overset{O}{\overset{||}{C}}HCOC_2H_5} \ + \ C_2H_5OH$$

ethyl propanoate

ethyl 2-methyl-3-oxopentanoate (45%)

Let us look at the stepwise reaction. First is the *formation of the enolate of the ester* by an acid–base reaction with the alkoxide ion. (An alkoxide, rather than a hydroxide, is used as the base to prevent saponification of the ester.) As in the aldol condensation, a low concentration of enolate is formed because the enolate (with only one carbonyl) is a stronger base than the alkoxide ion.

Enolate formation:

$$CH_3\overset{O}{\overset{||}{C}}OC_2H_5 \ + \ ^-:\ddot{O}C_2H_5 \ \rightleftharpoons \ \left[^-\ddot{C}H_2\overset{:\ddot{O}:}{\overset{||}{C}}OC_2H_5 \ \longleftrightarrow \ CH_2=\overset{:\ddot{O}:^-}{\overset{|}{C}}OC_2H_5 \right] + \ C_2H_5\ddot{O}H$$

ethyl acetate   *resonance structures for the enolate ion*

The nucleophilic carbon then attacks the carbonyl group in a typical carbonyl *addition* reaction. This addition of the enolate is followed by *elimination* of ROH. The entire sequence is thus a typical nucleophilic acyl substitution reaction of a carbonyl compound, similar to those you encountered in Chapter 15.

Attack on carbonyl group:

$$CH_3\overset{\overset{\displaystyle \ddot{O}:}{\|}}{C}OC_2H_5 + {}^-CH_2CO_2C_2H_5 \underset{\text{addition}}{\rightleftharpoons} \left[CH_3\overset{\overset{\displaystyle :\ddot{O}:^-}{|}}{\underset{\underset{\displaystyle CH_2CO_2C_2H_5}{|}}{C}}OC_2H_5 \right]$$

Loss of ROH:

$$\left[CH_3\overset{\overset{\displaystyle :\ddot{O}:^-}{|}}{\underset{\underset{\displaystyle CH_2CO_2C_2H_5}{|}}{C}}\ddot{O}C_2H_5 \right] \underset{\text{elimination}}{\rightleftharpoons} \left[CH_3\overset{\overset{\displaystyle :\ddot{O}}{\|}}{\underset{\underset{\displaystyle CH_2CO_2C_2H_5}{|}}{C}} + {}^-:\ddot{O}C_2H_5 \right] \rightleftharpoons$$

$$CH_3\overset{\overset{\displaystyle O}{\|}}{C}\overset{-}{C}HCO_2C_2H_5 + H\ddot{O}C_2H_5$$

*the enolate of
ethyl acetoacetate*

The product β-keto ester is more acidic than an alcohol because it has hydrogens that are α to two carbonyl groups. Therefore, the product of the condensation is the enolate salt of the β-keto ester. In most Claisen condensations, the formation of this resonance-stabilized enolate ion is the single step that drives the reaction sequence to completion. The β-keto ester is produced when the reaction mixture is acidified with cold dilute mineral acid.

$$CH_3\overset{\overset{\displaystyle O}{\|}}{C}\overset{-}{C}HCO_2C_2H_5 \xrightarrow{H^+} CH_3\overset{\overset{\displaystyle O}{\|}}{C}CH_2CO_2C_2H_5$$

ethyl acetoacetate

a β-keto ester

A β-keto ester can be hydrolyzed by heating in acidic solution, in which case decarboxylation may occur.

Hydrolysis and decarboxylation:

$$CH_3\overset{\overset{\displaystyle O}{\|}}{C}CH_2CO_2C_2H_5 \xrightarrow[-C_2H_5OH]{H^+, H_2O, \text{heat}} CH_3\overset{\overset{\displaystyle O}{\|}}{C}CH_2CO_2H \xrightarrow[\text{heat}]{-CO_2} CH_3\overset{\overset{\displaystyle O}{\|}}{C}CH_3$$

a β-keto ester   *a β-keto acid*   *a ketone*

Sample Problem

Predict the product of the ester condensation of methyl butanoate with sodium methoxide as the base, followed by acidification.

Solution:

1. Write the structure of the starting ester and determine the structure of the enolate ion.

$$CH_3CH_2CH_2\overset{\overset{\displaystyle O}{\|}}{C}OCH_3 + {}^-OCH_3 \rightleftharpoons CH_3CH_2\overset{-}{C}H\overset{\overset{\displaystyle O}{\|}}{C}OCH_3 + CH_3OH$$

2. Write the overall equation for nucleophilic attack on the carbonyl group and loss of ROH.

$$CH_3CH_2CH_2\overset{\displaystyle O}{\overset{\displaystyle \|}{C}}OCH_3 \longrightarrow CH_3CH_2CH_2\overset{\displaystyle O}{\overset{\displaystyle \|}{C}}O \quad + HOCH_3$$

$$\underset{\displaystyle \overset{\displaystyle |}{CH_2CH_3}}{\overset{\displaystyle -\ddot{C}HCOCH_3}{\underset{\displaystyle \|}{\overset{\displaystyle O}{}}}} \qquad \underset{\displaystyle \overset{\displaystyle |}{CH_2CH_3}}{\overset{\displaystyle -:C-\overset{\displaystyle O}{\overset{\displaystyle \|}{C}}OCH_3}{}}$$

3. Acidify:

$$\xrightarrow{\text{H}^+} CH_3CH_2CH_2\overset{\displaystyle O \quad O}{\overset{\displaystyle \| \quad \|}{C}}CH\overset{}{C}OCH_3$$

$$\underset{\displaystyle \overset{\displaystyle |}{CH_2CH_3}}{}$$

4. Write the equation for the combined reactions:

$$2\ CH_3CH_2CH_2\overset{\displaystyle O}{\overset{\displaystyle \|}{C}}OCH_3 \xrightarrow[\text{(2) H}^+]{\text{(1) Na}^+ \ ^-\text{OCH}_3} CH_3CH_2CH_2\overset{\displaystyle O \quad O}{\overset{\displaystyle \| \quad \|}{C}}CHCOCH_3$$

$$\underset{\displaystyle \overset{\displaystyle |}{CH_2CH_3}}{}$$

Study Problem

16.18 Predict the major organic product:

(a) $C_6H_5CH_2CO_2C_2H_5 \xrightarrow[\text{(2) H}^+]{\text{(1) Na}^+ \ ^-\text{OC}_2\text{H}_5}$

(b) [the product from (a)] $\xrightarrow[\text{heat}]{\text{H}_2\text{O, H}^+}$

Sample Problem

A Claisen-like condensation, called a **Dieckmann ring closure**, is used to prepare the following cyclic ketone from a diester. What is the structure of the diester?

Solution: Because the keto group arises from attack of an alpha carbon, the ring must be closed at the following position:

Therefore, the starting diester must be diethyl adipate.

diethyl hexanedioate
(diethyl adipate)

Study Problem

16.19 Write equations for the preparation of the following cyclic compounds from open-chain starting materials:

(a)   (b)

B. Crossed Claisen Condensations

Two different esters can be used in a Claisen condensation. Best results (that is, avoidance of mixtures) are obtained if only one of the esters has an alpha hydrogen.

$$\text{C}_6\text{H}_5\text{COCH}_3 + \text{CH}_3\text{CH}_2\text{CO}_2\text{CH}_3 \xrightarrow[\text{(2) H}^+]{\text{(1) Na}^+ \ ^-\text{OCH}_3} \text{C}_6\text{H}_5\text{CCHCO}_2\text{CH}_3 + \text{CH}_3\text{OH}$$

no α hydrogen

(45%)

Crossed Claisen condensations can be successfully carried out between ketones and esters whether the ester contains an α hydrogen or not. The α hydrogen of the ketone is removed preferentially because ketones are more acidic than esters. For this reason, the crossed Claisen is favored over a self-Claisen condensation of the ester.

ethyl benzoate   acetone   1-phenyl-1,3-butanedione (40%)

$$\text{CH}_3(\text{CH}_2)_4\text{COC}_2\text{H}_5 + \text{CH}_3\text{CCH}_3 \xrightarrow[\text{(2) H}^+]{\text{(1) NaH}} \text{CH}_3(\text{CH}_2)_4\text{CCH}_2\text{CCH}_3 + \text{C}_2\text{H}_5\text{OH}$$

ethyl hexanoate   acetone   2,4-nonanedione (65%)

Sample Problem

A mixture of acetone and diethyl oxalate ($C_2H_5O_2C$—$CO_2C_2H_5$) is added to a mixture of sodium ethoxide in ethanol. After the reaction is completed, the mixture is treated with cold, dilute HCl. A condensation product is isolated from the mixture in 60% yield. What is the product?

Solution:

(1) $\underset{\displaystyle \quad}{CH_3\overset{\displaystyle O}{\overset{\|}{C}}CH_3}$ + $^-OC_2H_5$ $\rightleftharpoons$ $^-CH_2\overset{\displaystyle O}{\overset{\|}{C}}CH_3$ + HOC_2H_5

(2) $C_2H_5O\overset{\displaystyle O}{\overset{\|}{C}}$—$\overset{\displaystyle O}{\overset{\|}{C}}OC_2H_5$ + $^-CH_2\overset{\displaystyle O}{\overset{\|}{C}}CH_3$ $\xrightarrow{\quad}$ $\xrightarrow{\ H^+\ }$

$\underbrace{C_2H_5O\overset{\displaystyle O}{\overset{\|}{C}}\text{—}\overset{\displaystyle O}{\overset{\|}{C}}}_{\textit{from diethyl oxalate} \longrightarrow}\underbrace{\text{—}CH_2\overset{\displaystyle O}{\overset{\|}{C}}CH_3}_{\textit{from acetone}}$ + HOC_2H_5

Study Problem

16.20 Predict the products:

(a) $C_2H_5O_2C$—$CO_2C_2H_5$ + $CH_3CO_2C_2H_5$ $\xrightarrow[\text{(2) H}^+]{\text{(1) Na}^+ \ ^-OC_2H_5}$

(b) $\langle\text{cyclopentane}\rangle=O$ + $CH_3CO_2C_2H_5$ $\xrightarrow[\text{(2) H}^+]{\text{(1) Na}^+ \ ^-OC_2H_5}$

C. Syntheses Using Ester Condensations

Because the product of an ester condensation between two esters is a β-keto ester (or a ketone after hydrolysis and decarboxylation), the decision of which starting materials to use is not difficult. The keto group comes from one starting ester; the ester group with its substituent comes from the other starting ester. The different types of products from ester condensations are summarized in Figure 16.7.

$CH_3CH_2CH_2\overset{\displaystyle O}{\overset{\|}{C}}\text{—}\underset{\displaystyle CH_2CH_3}{CHCO_2C_2H_5}$

from ethyl butanoate

$\langle\text{C}_6\text{H}_5\rangle\text{—}\overset{\displaystyle O}{\overset{\|}{C}}\text{—}\underset{\displaystyle CH_2CH_3}{CHCO_2H}$

from ethyl benzoate   *from ethyl butanoate*

$\langle\text{C}_6\text{H}_5\rangle\text{—}\overset{\displaystyle O}{\overset{\|}{C}}\text{—}CH_2CH_2CH_3$

from ethyl benzoate

from ethyl butanoate (after decarboxylation)

Claisen:

$$2\ RCH_2CO_2C_2H_5 \xrightarrow{\text{base}} RCH_2\overset{\displaystyle O}{\overset{\|}{C}}CHCO_2C_2H_5 \xrightarrow[\text{heat}]{H_2O,\ H^+}$$

with R substituent below the CH

a β-keto ester

$$RCH_2\overset{\displaystyle O}{\overset{\|}{C}}CHCO_2H \xrightarrow[-CO_2]{\text{heat}} RCH_2\overset{\displaystyle O}{\overset{\|}{C}}CH_2R$$

with R substituent below the CH

a β-keto acid   *a ketone*

Dieckmann (for 5- and 6-membered rings):

$$\begin{array}{c} CO_2C_2H_5 \\ (CH_2)_{3\ or\ 4} \\ CH_2CO_2C_2H_5 \end{array}$$

Crossed Claisen:

$$RCO_2C_2H_5\ +\ H_2C\!\!<\quad \xrightarrow{\text{base}}\quad RC\overset{\displaystyle O}{\overset{\|}{}}\!-CH\!<$$

a compound with
an acidic α hydrogen

Figure 16.7 Products of ester condensations.

Sample Problem

What reactants would you need to prepare the following compound by an ester condensation?

Solution:

from (furan)–CO_2C_2H_5

from $CH_3CH_2CO_2C_2H_5$

Study Problems

16.21 Write equations for preparations of the following compounds:

(a) [structure: cyclohexane ring with CH$_3$ at top, =O at right, CO$_2$H at bottom]

(b) [structure: benzene ring attached to $-\overset{\overset{\text{O}}{\|}}{\text{C}}\text{CH}_2\text{CO}_2\text{H}$]

(c) [structure: benzene ring attached to $-\overset{\overset{\text{O}}{\|}}{\text{C}}\text{CH(CN)}_2$]

(d) $(CH_3)_3C\overset{\overset{\text{O}}{\|}}{C}CH_2\overset{\overset{\text{O}}{\|}}{C}CH_3$

16.22 In untreated *diabetes mellitus*, a lack of insulin or an inability to use insulin effectively leads to the poor utilization of glucose (blood sugar) by the cells. One result of this failure to metabolize glucose is that insufficient oxaloacetate ion is formed to break down acetylcoenzyme A (Section 16.7B). Consequently, the level of acetylcoenzyme A increases. In the liver, excess acetylcoenzyme A is converted to β-hydroxybutyrate ions and **ketone bodies**: acetoacetate ion and acetone. (A diabetic's breath and urine may smell of acetone.) Write equations showing how acetylcoenzyme A can be converted to acetoacetate ion and acetone.

▼

Section 16.9
Nucleophilic Addition to α, β-Unsaturated Carbonyl Compounds

A double bond in conjugation with a carbonyl group is susceptible to nucleophilic attack in a 1,4-addition reaction (Section 13.12B).

$$Nu:^- + CH_2{=}CH{-}\overset{\overset{\ddot{O}}{\|}}{C}CH_3 \longrightarrow \left[\overset{\overset{:\ddot{O}:^-}{|}}{\underset{\overset{|}{Nu}}{CH_2{-}CH{=}C}}CH_3 \right] \xrightarrow{H^+} \overset{\overset{\cdot\overset{\cdot}{O}\cdot}{\|}}{\underset{\overset{|}{Nu}}{CH_2CH_2C}}CH_3$$

an enolate

    If an α,β-unsaturated carbonyl compound can undergo nucleophilic attack, we might expect that an enolate ion could add to the double bond. Indeed, it does. The intermediate is the enolate of the 1,4-addition product. The final product is obtained by acidification. This useful synthetic reaction is called a **Michael addition**.

$$CH_2{=}CH{-}\overset{\overset{\ddot{O}}{\|}}{C}H + {}^-\ddot{C}H(CO_2C_2H_5)_2 \rightleftharpoons \left[\overset{\overset{:\ddot{O}:^-}{|}}{\underset{\overset{|}{CH(CO_2C_2H_5)_2}}{CH_2{-}CH{=}CH}} \right] \rightleftharpoons$$

an enolate

$$\left[\overset{\overset{O}{\|}}{\underset{\overset{|}{{}^-:C(CO_2C_2H_5)_2}}{CH_2CH_2CH}} \right] \xrightarrow{H^+} \overset{\overset{O}{\|}}{\underset{\overset{|}{CH(CO_2C_2H_5)_2}}{CH_2CH_2CH}}$$

    more stable enolate   (50%)

$$
\text{C}_6\text{H}_5\text{—CH=CH—}\overset{\displaystyle O}{\overset{\|}{\text{C}}}\text{OC}_2\text{H}_5 + {}^-{:}\text{CH(CO}_2\text{C}_2\text{H}_5)_2 \rightleftharpoons \left[\text{C}_6\text{H}_5\text{—CH—CH=}\overset{\displaystyle O^-}{\overset{|}{\text{C}}}\text{OC}_2\text{H}_5 \atop \overset{|}{\text{CH(CO}_2\text{C}_2\text{H}_5)_2} \right]
$$

$$
\rightleftharpoons \left[\begin{array}{c} \text{C}_6\text{H}_5\text{—CHCH}_2\overset{\displaystyle O}{\overset{\|}{\text{C}}}\text{OC}_2\text{H}_5 \\ \overset{|}{{}^-{:}\text{C(CO}_2\text{C}_2\text{H}_5)_2} \end{array} \right] \xrightarrow{\text{H}^+} \begin{array}{c} \text{C}_6\text{H}_5\text{—CHCH}_2\overset{\displaystyle O}{\overset{\|}{\text{C}}}\text{OC}_2\text{H}_5 \\ \overset{|}{\text{CH(CO}_2\text{C}_2\text{H}_5)_2} \end{array}
$$

more stable enolate   (98%)

The product of the last example shown is a triester. Saponification, followed by acidification, gives a triacid. Under the conditions of acid hydrolysis, however, decarboxylation may occur.

Acid hydrolysis and decarboxylation:

$$
\begin{array}{c} \text{C}_6\text{H}_5\text{—CHCH}_2\text{CO}_2\text{C}_2\text{H}_5 \\ \overset{|}{\text{CH(CO}_2\text{C}_2\text{H}_5)_2} \end{array} \xrightleftharpoons[\text{heat}]{\text{H}_2\text{O, H}^+} \begin{array}{c} \text{C}_6\text{H}_5\text{—CHCH}_2\text{CO}_2\text{H} \\ \overset{|}{\text{CH(CO}_2\text{H)}_2} \end{array} \xrightarrow[\text{heat}]{-\text{CO}_2}
$$

a triester   a triacid

$$
\begin{array}{c} \text{C}_6\text{H}_5\text{—CHCH}_2\text{CO}_2\text{H} \\ \overset{|}{\text{CH}_2\text{CO}_2\text{H}} \end{array}
$$

a diacid

In the decarboxylation of the triacid, it is the malonic acid grouping that loses CO_2. (Why?) This group is highlighted in color in the following equation.

$$
\begin{array}{c} \overset{\textit{this stays}}{\downarrow} \\ \text{C}_6\text{H}_5\text{—CHCH}_2\text{CO}_2\text{H} \\ \overset{|}{\text{CH—CO}_2\text{H}} \\ \overset{|}{\text{CO}_2\text{H}} \end{array} \xrightarrow{\text{heat}} \begin{array}{c} \text{C}_6\text{H}_5\text{—CHCH}_2\text{CO}_2\text{H} \\ \overset{|}{\text{CH}_2\text{CO}_2\text{H}} \end{array} + \text{CO}_2
$$

one of these goes

The types of products that we can obtain from simple Michael additions are shown in Figure 16.8. Michael additions in combination with other condensations are exceedingly useful in laboratory syntheses of complex cyclic compounds, such as steroids (Section 23.7). A portion of one such synthesis is shown. This particular ring-forming sequence (Michael plus aldol) is called a **Robinson annulation**.

$$RCH{=}CHCO_2C_2H_5 + R'CH(CO_2C_2H_5)_2 \xrightarrow{\text{base}}$$

or other compound
with active H

$$\left\{ \begin{array}{l} \underset{|}{RCHCH_2CO_2C_2H_5} \\ R'C(CO_2C_2H_5)_2 \\ \textit{a triester} \end{array} \right.$$

$\Big\downarrow$ H_2O, H^+

$$\begin{array}{l} \underset{|}{RCHCH_2CO_2H} \\ R'C(CO_2H)_2 \\ \textit{a triacid} \end{array}$$

$\Big\downarrow$ heat

$$\begin{array}{l} \underset{|}{RCHCH_2CO_2H} \\ R'CHCO_2H \\ \textit{a diacid} \end{array}$$

Figure 16.8 **Products of Michael addition of a malonic ester with an
α,β-unsaturated ester.**

In an important variation of this sequence, one enantiomer of a chiral
amine is used to catalyze the ring-closure reaction and thus to induce formation
of a single enantiomer of the product.

Sample Problems

Show by equations how you would prepare the following keto acid by a Michael addition:

$$\overset{\overset{\textstyle O}{\|}}{CH_3C}CH_2CH_2CH_2CO_2H$$

Solution: Addition occurs β to the keto group in a 1,4-addition.

β to ketone

$$\overset{\overset{\textstyle O}{\|}}{CH_3C}CH_2CH_2\;\boxed{CH_2CO_2H}$$

from $\overset{\overset{\textstyle O}{\|}}{CH_3C}CH{=}CH_2$ — *from malonic ester*

1. $CH_2(CO_2C_2H_5)_2 \;\underset{\longleftarrow}{\overset{NaOC_2H_5}{\longrightarrow}}\; {}^-{:}CH(CO_2C_2H_5)_2$

2. $CH_3\overset{\overset{\textstyle \ddot{O}:}{\|}}{C}{-}CH{=}CH_2 + {}^-{:}CH(CO_2C_2H_5)_2 \;\rightleftharpoons\; CH_3\overset{\overset{\textstyle :\ddot{O}{:}^-}{|}}{C}{=}CHCH_2CH(CO_2C_2H_5)_2$

$$\underset{\longleftarrow}{\overset{H^+}{}}\; CH_3\overset{\overset{\textstyle O}{\|}}{C}CH_2CH_2CH(CO_2C_2H_5)_2$$

3. $\underset{heat}{\overset{H_2O,\,H^+}{\longrightarrow}}\; CH_3\overset{\overset{\textstyle O}{\|}}{C}CH_2CH_2CH_2CO_2H$

A chemist decided to try the sequence of reactions in the preceding sample problem. After Step 2, the chemist discovered a *cyclic* product, along with the keto diester that had been predicted. What happened?

Solution: The initial product of Step 2 underwent a Dieckmann ring closure. (Note that this intermediate has a nucleophilic carbon that can attack a carbonyl group to yield a six-membered ring.)

$$CH_3\overset{\overset{\textstyle O^-}{|}}{C}{=}CHCH_2CH(CO_2C_2H_5)_2 \;\rightleftharpoons\;$$

$$ {}^-{:}CH_2\overset{\overset{\textstyle O}{\|}}{C}CH_2CH_2\overset{\overset{\textstyle CO_2C_2H_5}{|}}{CH}CO_2C_2H_5 \;\xrightarrow{-C_2H_5OH}\; \xrightarrow{H^+}\;$$

(cyclohexanone ring with $CO_2C_2H_5$ and O)

Study Problem

16.23 If the chemist had subjected the cyclic product in the preceding sample problem to acid hydrolysis and then distilled the organic material, what final product would have been observed?

Sample Problem

How would you prepare the following keto ester by a Michael addition?

Solution: Addition occurs β to the ketone.

$$C_6H_5CH=CHCC_6H_5 + CH_2CO_2C_2H_5 \xrightarrow{Na^+ \ ^-OC_2H_5} \xrightarrow{H^+} \text{product}$$

with C_6H_5 substituent and O on the carbonyl.

Study Problems

16.24 Show by equations how you would prepare the following carboxylic acids by Michael additions:

(a) $(HO_2C)_2\overset{\underset{\displaystyle CH_3}{|}}{C}CHCH_2CO_2H$ with CH_2CH_3

(b) $HO_2CCH_2\overset{\underset{\displaystyle CH_3}{|}}{C}HCH_2CO_2H$

(c)

16.25 Lithium dialkylcuprates (R_2CuLi) undergo 1,4-addition to α,β-unsaturated aldehydes and ketones in a reaction closely related to the Michael addition (Section 13.12B). Predict the product of the following reaction:

$$CH_3-\hspace{-4pt}\bigcirc\hspace{-4pt}=O + (CH_3)_2CuLi \longrightarrow$$

▼ Summary

In this chapter, we have looked at a variety of ways to synthesize compounds with complex carbon skeletons. Each of these reactions is caused by a species with carbanion character.

Alkylations:

$$\diagdown\!CH^- + RX \longrightarrow \diagdown\!CHR + X^-$$

Condensations:

$$\diagdown\!CH^- + R\!-\!\overset{\overset{\textstyle O}{\|}}{C}\!-\!R \longrightarrow R\!-\!\underset{\underset{\textstyle -CH-}{|}}{\overset{\overset{\textstyle OH}{|}}{C}}\!-\!R$$

1,4-Additions:

$$\diagdown\!CH^- + CH_2\!=\!CH\overset{\overset{\textstyle O}{\|}}{C}\!- \longrightarrow CH_2\!-\!\underset{\underset{\textstyle -CH-}{|}}{CH_2}\overset{\overset{\textstyle O}{\|}}{C}\!-$$

Table 16.2 gives an overview of the more important products available from these reactions.

Table 16.2 Major synthetic reactions involving enolates and carbanions

Reaction	Section Reference
Malonic Ester:	
$CH_2(CO_2C_2H_5)_2 \xrightarrow[(2)\ ^-OC_2H_5,\ R'X]{(1)\ ^-OC_2H_5,\ RX} \underset{R'}{\overset{R}{C}}(CO_2C_2H_5)_2$	16.2
Acetoacetic Ester:	
$CH_3\overset{\overset{\textstyle O}{\|}}{C}CH_2CO_2C_2H_5 \xrightarrow[(2)\ ^-OC_2H_5,\ R'X]{(1)\ ^-OC_2H_5,\ RX} CH_3\overset{\overset{\textstyle O}{\|}}{C}\underset{R\ \ R'}{C}CO_2C_2H_5$	16.3
Enamine:	
$R_2CH\overset{\overset{\textstyle O}{\|}}{C}R \xrightarrow[(3)\ H_2O,\ H^+]{\substack{(1)\ R_2NH,\ H^+ \\ (2)\ R'X}} R_2\overset{\overset{\textstyle O}{\|}}{C}\underset{R'}{C}R$	16.5
Aldol Condensation:	
$2\ RCH_2CHO \underset{}{\overset{OH^-}{\rightleftharpoons}} RCH_2\underset{R}{\overset{OH}{C}}HCHCHO$	16.6

(continued)

Table 16.2 (*continued*) **Major synthetic reactions involving enolates and carbanions**

Reaction	Section Reference
Crossed Aldol:	
$RCHO + R'CH_2CHO \xrightleftharpoons{OH^-} RCH\underset{\underset{R'}{|}}{C}HCHCHO$ (with OH on second carbon)	16.6B
Claisen Condensation:	
$2\ RCH_2CO_2C_2H_5 \xrightarrow[(2)\ H^+]{(1)\ ^-OC_2H_5} RCH_2\overset{O}{\overset{\|}{C}}\underset{\underset{R}{|}}{C}HCO_2C_2H_5$	16.8
Crossed Claisen:	
$R\overset{O}{\overset{\|}{C}}OC_2H_5 + R'CH_2CO_2C_2H_5 \xrightarrow[(2)\ H^+]{(1)\ ^-OC_2H_5} R\overset{O}{\overset{\|}{C}}\underset{\underset{R'}{|}}{C}HCO_2C_2H_5$	16.8A
Michael Addition:	
$RCH{=}CHCO_2C_2H_5 + R'CH(CO_2C_2H_5)_2 \xrightarrow[(2)\ H^+]{(1)\ ^-OC_2H_5} \underset{R'C(CO_2C_2H_5)_2}{RCH{-}CH_2CO_2C_2H_5}$	16.9

Study Problems*

16.26 In each structure, indicate the most acidic hydrogen:

(a) $C_6H_5\underset{\underset{CH_3}{|}}{C}HCO_2C_2H_5$ (b) $CH_3\overset{O}{\overset{\|}{C}}CH_2CN$ (c) $H\overset{O}{\overset{\|}{C}}\underset{\underset{CH_3}{|}}{C}HCH{=}CH\overset{O}{\overset{\|}{C}}H$

16.27 Write an equation for the reversible acid–base reaction of each of the following compounds with sodium ethoxide in ethanol:

(a) $C_2H_5O\overset{O}{\overset{\|}{C}}CH_2CN$ (b)   (c)   —CN

16.28 Write resonance structures that show delocalization of the negative charge in the following anions:

(a) $^-{:}CH_2NO_2$ (b) $^-{:}CH_2CH{=}CHCO_2C_2H_5$ (c) $CH_3\overset{O}{\overset{\|}{C}}\bar{\underset{..}{C}}HCN$

---

* For information concerning the organization of the *Study Problems* and *Additional Problems*, see the *Note to student* on page 39.

16.29 Rank the following compounds in order of increasing acid strength (weakest acid first):

(a) [cyclohexyl]—CHO   (b) [cyclohexyl]—CO$_2$C$_2$H$_5$ (c) [2-oxocyclohexyl]—CHO

(d) [2-oxocyclohexyl]—CO$_2$C$_2$H$_5$ (e) C$_2$H$_5$OH   (f) [cyclohexyl]—CO$_2$H

16.30 Complete the following equations for *acid–base reactions*, and indicate by arrow size if each equilibrium is to the right or to the left:

(a) $^-$:CH$_2$CO$_2$C$_2$H$_5$ + CH$_3$CCH$_2$CO$_2$C$_2$H$_5$ $\rightleftharpoons$

(b) CH$_3$CCH$_2$NO$_2$ + OH$^-$ $\rightleftharpoons$

(c) [cyclohexyl]—CO$_2$C$_2$H$_5$ + CH$_3$CCHCCH$_3$ $\rightleftharpoons$

(d) $^-$:CH$_2$CCH$_2$CH $\rightleftharpoons$

16.31 Predict the major organic product:

(a) O=[cyclopentane]=O $\xrightarrow[\text{(2) NaOCH}_3,\ \text{CH}_3\text{I}]{\text{(1) NaOCH}_3,\ \text{CH}_3\text{I}}$

(b) CH$_3$CCH$_2$CO$_2$C$_2$H$_5$ $\xrightarrow[\text{(2) [cyclohexyl]—Br}]{\text{(1) NaOC}_2\text{H}_5}$

(c) [the product from (b)] $\xrightarrow[\text{heat}]{\text{H}_2\text{O, H}^+}$

(d) CH$_3$CCH$_2$CO$_2$C$_2$H$_5$ $\xrightarrow[\text{(2) BrCH}_2\text{CCH}_3]{\text{(1) NaOC}_2\text{H}_5}$

(e) [the product from (d)] $\xrightarrow[\text{heat}]{\text{H}_2\text{O, H}^+}$

(f) O$_2$N—[benzene with NO$_2$]—Cl $\xrightarrow[\text{(2) cold, dilute HCl}]{\text{(1) NCCHCO}_2\text{C}_2\text{H}_5\ \text{Na}^+}$

(*Hint*: See Section 12.4.)

16.32 How would you prepare each of the following compounds from dimethyl malonate or ethyl acetoacetate?

(a) [cyclopentyl]—CH$_2$CH$_2$CO$_2$H   (b) HO$_2$CCH$_2$CHCO$_2$H
                                                                                                                    |
                                                                                                                   CH$_3$

(c) $C_6H_5CH_2CH_2\overset{O}{\underset{\|}{C}}CH_3$

(d)

$$HC\equiv CCH_2$$
$$\overset{}{\underset{}{C}}(CO_2CH_3)_2$$
$$(CH_3)_2C=CHCH_2$$

16.33 Starting with o-bromotoluene, show how you would synthesize 3-(o-bromophenyl)-propanoic acid.

16.34 Predict the major organic products:

(a) $CH_3\overset{O}{\underset{\|}{C}}CH_2\overset{O}{\underset{\|}{C}}OC_2H_5 +$ $\xrightarrow[\text{heat}]{\text{benzene}}$

(b) $+$ $\xrightarrow[\text{toluene, heat}]{p\text{-toluenesulfonic acid}}$

16.35 Predict the hydrolysis products of the following enamines.

(a)

(b) $CH_3\overset{}{C}=CH\overset{O}{\underset{\|}{C}}OCH_2CH_3$

(c) $(CH_3)_2C=CHN$

(d) CO_2CH_3

16.36 Show the mechanism for the formation of the enamine of cyclopentanone and piperidine, followed by reaction with benzoyl chloride (C_6H_5COCl).

16.37 Show how you could synthesize the following ketones from cyclohexanone.

(a)

(b)

(c)

16.38 An aqueous NaOH solution is added to a mixture of acetone and formaldehyde.
   (a) What is the structure of the organic anion formed?
   (b) Would this anion undergo reaction at the faster rate with acetone or with formaldehyde?
   (c) Show each step in the mechanism for the predominant aldol condensation that would occur.

16.39 What aldehydes would be needed to prepare the following aldol condensation products?

$$
\underset{\text{OH}}{\text{(a) } CH_3CH_2\overset{\text{OH}}{\underset{}{C}}HC\overset{\text{O}}{\underset{}{C}}H} \\
\underset{\text{CH}_3}{\underset{\mid}{}}
$$

(a) $CH_3CH_2\underset{\underset{CH_3}{|}}{\overset{\overset{OH}{|}}{C}H}CH\overset{\overset{O}{\parallel}}{C}H$

(b) $C_6H_5CH_2\underset{\underset{C_6H_5}{|}}{\overset{\overset{OH}{|}}{C}H}CH\overset{\overset{O}{\parallel}}{C}H$

(c) $C_6H_5\underset{\underset{C_6H_5}{|}}{\overset{\overset{OH}{|}}{C}H}CH\overset{\overset{O}{\parallel}}{C}H$

(d)

16.40 Predict the major organic product:

(a) $H\overset{\overset{O}{\parallel}}{C}(CH_2)_4\overset{\overset{O}{\parallel}}{C}H$ $\xrightarrow[\text{(2) H}^+\text{, heat}]{\text{(1) OH}^-\text{, H}_2\text{O}}$

(b) $C_6H_5CHO +$ ◯=O $\xrightarrow{\text{OH}^-}$

(c) $\xrightarrow[\text{(2) H}^+\text{, heat}]{\text{(1) OH}^-\text{, H}_2\text{O}}$

(d) $\xrightarrow[\text{(2) H}^+\text{, heat}]{\text{(1) OH}^-\text{, H}_2\text{O}}$

16.41 How would you make the following conversions?

(a) cyclopentanone to ◯=$\underset{\underset{CO_2C_2H_5}{|}}{C}C_6H_5$

(b) ⟨O⟩—CHO to ⟨O⟩—CH=CHNO$_2$

(c) benzaldehyde to $C_6H_5CH=CH\overset{\overset{O}{\parallel}}{C}CH_3$

(d) benzaldehyde to $C_6H_5CH=CH\overset{\overset{O}{\parallel}}{C}CH=CHC_6H_5$

16.42 Predict the major organic products:

(a) $2\ CH_3(CH_2)_3CO_2C_2H_5$ $\xrightarrow[\text{(2) H}_2\text{O, H}^+\text{, heat}]{\text{(1) NaOC}_2\text{H}_5}$

(b) ⬠$\begin{array}{l}CH_2CH_2CO_2C_2H_5\\CH_2CH_2CO_2C_2H_5\end{array}$ $\xrightarrow[\text{(2) H}^+\text{, cold}]{\text{(1) NaOC}_2\text{H}_5}$

(c) $C_6H_5CH_2CO_2C_2H_5 + C_2H_5O\overset{\overset{O}{\parallel}}{C}OC_2H_5$ $\xrightarrow[\text{(2) H}^+\text{, cold}]{\text{(1) NaOC}_2\text{H}_5}$

diethyl carbonate

(d) ⬡$\begin{array}{l}CH_2\overset{\overset{O}{\parallel}}{C}CH_3\\CH_2CO_2C_2H_5\end{array}$ $\xrightarrow[\text{(2) H}^+\text{, cold}]{\text{(1) NaOC}_2\text{H}_5}$

(e) $CH_3\overset{\overset{O}{\parallel}}{C}(CH_2)_4CO_2C_2H_5$ $\xrightarrow[\text{(2) H}^+\text{, cold}]{\text{(1) NaOC}_2\text{H}_5}$

(f) [cyclopentanone] $=O + C_2H_5OCOC_2H_5 \xrightarrow[\text{(2) H}^+,\,\text{cold}]{\text{(1) NaOC}_2\text{H}_5}$

16.43 Suggest synthetic paths leading to the following compounds:

(a) $(CH_3)_3CCCHCO_2C_2H_5$
　　　　　　　$|$
　　　　　　　CH_3

(b) [indandione structure] $-CO_2C_2H_5$

(c) [substituted cyclohexanone] $-CO_2C_2H_5$, with CH_3 substituent

(d) $C_6H_5CCHCO_2C_2H_5$
　　　　　　　　$|$
　　　　　　　　CN

(e) [decalin ketone structure] CO_2H

(f) $C_6H_5CHCCH_3$
　　　　　$|$
　　　　　CN

16.44 Predict the major organic product of each of the following Michael addition reactions:

(a) $1\ CH_2\!=\!CHCCH_2CH_3 + CH_2(CO_2C_2H_5)_2 \xrightarrow[\text{(2) H}^+,\,\text{cold}]{\text{(1) NaOC}_2\text{H}_5}$

(b) $2\ CH_2\!=\!CHCCH_2CH_3 + CH_2(CO_2C_2H_5)_2 \xrightarrow[\text{(2) H}^+,\,\text{cold}]{\text{(1) NaOC}_2\text{H}_5}$

(c) [cyclohexenone] $=O + CH_3CCH_2CCH_3 \xrightarrow[\text{(2) H}^+,\,\text{cold}]{\text{(1) NaOC}_2\text{H}_5}$

(d) $CH_3CH\!=\!CHCO_2C_2H_5 + C_6H_5CH_2CO_2C_2H_5 \xrightarrow[\text{(2) H}^+,\,\text{cold}]{\text{(1) NaOC}_2\text{H}_5}$

16.45 How would you prepare the following compounds by Michael additions?

(a) $CH_3CCHCH_2CH_2CN$
　　　　　　$|$
　　　　　　$CO_2C_2H_5$

(b) [cyclohexanedione] $-CH_2CH_2CO_2C_2H_5$

(c) $CH_3CClCH_2CH_2CN$
　　　　　　$|$
　　　　　　NO_2

(d) [cyclooctanone structure] CCH_3 , $CHCO_2C_2H_5$

Additional Problems

16.46 Predict the major organic products:

(a)

$$\begin{array}{l}(1)\ NaOC_2H_5\\(2)\ H_2O,\ H^+,\ heat\end{array}$$

$\dot{C}H_2CH_2CH_2CO_2C_2H_5$

(b) $CH_3CH_2CHO + C_2H_5O_2CCH_2CN$ $\xrightarrow[\substack{\text{alanine (an amino acid)}\\ \text{heat}}]{\text{benzene, } CH_3CO_2H}$

(c)

$$\begin{array}{l}(1)\ Li^+\ \ \overset{..}{:}\overset{..}{N}[CH(CH_3)_2]_2\\(2)\ BrCH_2CH_2CH_2Cl\end{array}$$

(d)

$-CH_2\overset{\overset{\displaystyle O}{\|}}{C}CH_2CO_2C_2H_5$ $\xrightarrow[\substack{(2)\ H_2O,\ H^+}]{(1)\ K_2CO_3}$

16.47 Other carbon nucleophiles besides enolates can undergo 1,4-addition with α,β-unsaturated carbonyl compounds. One important reaction, called the **Stork enamine reaction**, is the 1,4-addition of an enamine. Predict the major organic product of the following sequence:

16.48 Suggest syntheses for the following compounds from readily available starting materials:

(a) $C_6H_5CH=\overset{\overset{\displaystyle O}{\|}}{\underset{\underset{\displaystyle C_6H_5}{|}}{C}}OC_2H_5$

(b)

$-CH_2CH=CHCH_3$

(c) $(CH_3)_2CHCH(CO_2CH_3)_2$

(d) $C_6H_5CH=C(CN)_2$

(e)

$-CHO$

(f) $C_6H_5\overset{\overset{\displaystyle O}{\|}}{C}CH_2\overset{\overset{\displaystyle O}{\|}}{C}NHC_6H_5$

(g)

$-(CH_2)_6CO_2H$

(h) $C_2H_5O_2C\overset{\overset{\displaystyle O}{\|}}{C}\underset{\underset{\displaystyle CH_2CH_3}{|}}{C}HCO_2C_2H_5$

(i)

$-CH(CO_2C_2H_5)_2$

16.49 An aldehyde with no alpha hydrogen cannot undergo an aldol condensation when treated with aqueous base. However, if such an aldehyde is heated with concentrated NaOH or KOH solution, the aldehyde is converted to a 1:1 mixture of carboxylate and alcohol (**Cannizzaro reaction**). Suggest a mechanism for the following Cannizzaro reaction. (*Hint:* Consider a hydride transfer as part of the mechanism.)

$$2 \text{ HCH} \xrightarrow[\text{heat}]{\text{conc. KOH}} \text{HCO}^- \text{ K}^+ + \text{CH}_3\text{OH}$$

16.50 How would you carry out the following transformations? (Use flow equations.)

(a) $n\text{-C}_7\text{H}_{15}\text{CH}_2\text{CO}_2\text{H}$ from $n\text{-C}_7\text{H}_{15}\text{Br}$

(b) [structure] from [structure]

(c) [structure] from [structure]

(d) [structure] from [structure]

16.51 Propose syntheses for the following compounds:

(a) [structure] from an open-chain compound

(b) [structure] a useful intermediate in prostaglandin syntheses (see Section 23.4), from an open-chain compound

(c) [structure] from 3-methoxy-2-methyl-2-cyclopentenone

(d) [structure] vulpinic acid, a yellow lichen metabolite with anti-inflammatory activity in humans, by an ester condensation of an open-chain compound

16.52 (a) How would you prepare the following compound from 4-methylcyclohexanone?

[structure]

(b) From the dione in (a), show how you could prepare:

[structure]

16.53 A chemist has on his shelf: dilute HCl, dilute NaOH, sodium metal, magnesium metal, anhydrous ethanol, anhydrous ether, bromine, diethyl malonate, a tank of anhydrous HCl, glacial acetic acid, PBr_3, and acetone. (He also has water, heat sources, and solvents for working up his products.) Suggest ways that he could synthesize the following compounds: **(a)** ethyl acetate; **(b)** ethyl acetoacetate; **(c)** 2-pentanone; **(d)** 2-ethylbutanoic acid; **(e)** ethyl α-bromoacetate.

16.54 Show how the following compounds could be synthesized by Robinson annulations:

(a)   **(b)**

16.55 *Dimedone* is a reagent used to separate and identify aldehydes as insoluble condensation products. This reagent can be obtained from *mesityl oxide* by a combination of a Michael condensation and a Dieckmann ring closure. Mesityl oxide, in turn, can be synthesized from acetone. Write a flow equation that shows the reactions in the conversion of acetone to mesityl oxide to dimedone.

$$CH_3\overset{O}{\overset{\|}{C}}CH_3 \longrightarrow (CH_3)_2C\text{=}CH\overset{O}{\overset{\|}{C}}CH_3 \longrightarrow$$

acetone   4-methyl-3-penten-2-one
(mesityl oxide)

5,5-dimethyl-1,3-cyclohexanedione
(dimedone)

16.56 Outline a general synthetic method for the preparation of cycloalkanecarboxylic acids, based upon the alkylation of malonic ester.

16.57 Devise a general method for the synthesis of γ-keto acids from ethyl α-chloroacetate.

16.58 How would you prepare the following carboxylic acids from the *same starting materials?*

(a) —CO_2H **(b)** $HO_2CCH_2(CH_2)_4CH_2CO_2H$

16.59 Write flow equations to show the steps in the following conversions:

(a) excess HCHO + CH_3CHO $\xrightarrow{CO_3^{2-}}$ $HOCH_2\overset{CH_2OH}{\underset{CH_2OH}{C}}CHO$

(b) $\xrightarrow[\text{4 days}]{\text{conc. HCl, heat}}$

(c) $\xrightarrow{NaOC_2H_5}$ $\xrightarrow{H^+}$

16.60 Compound A, which can be obtained from a Diels–Alder reaction of 3-buten-2-one, can be converted to compound B, as shown in the following equation. Compound B, in turn, can be converted to *brevicomin*, the sex attractant of the pine bark beetle, a beetle that spreads tree disease and kills pine forests. Suggest a series of reactions to synthesize B from A and suggest a mechanism for the conversion of compound B to brevicomin.

A

B

p-toluenesulfonic acid
benzene

brevicomin

16.61 Write a flow equation for the steps in the following reaction:

$$C_2H_5OCCH_2{}^{14}CCH_3 + H_2NCCH_2CN \xrightarrow[CH_3OH]{KOH}$$

carbon-14

16.62 Give the structures of compounds **I** and **II**:

$$CH_3CH_2CH_2CO_2C_2H_5 + (CO_2C_2H_5)_2 \xrightarrow{NaOC_2H_5} C_{10}H_{16}O_5 \xrightarrow[heat]{H_2O, H^+} C_5H_8O_3$$

I   **II**

16.63 Write each step in the mechanism of the following reaction sequence:

$$C_2H_5OC(CH_2)_3CO_2C_2H_5 + \left(COC_2H_5\right)_2 \xrightarrow{Na^+ {}^-OC_2H_5}$$

$$\xrightarrow{H^+}$$

16.64 Give the structure of each indicated product:

excess malonic ester $\xrightarrow[BrCH_2CH_2Br]{Na^+ {}^-OC_2H_5} \xrightarrow{H^+}$

$$C_{16}H_{26}O_8 \xrightarrow[BrCH_2CH_2Br]{Na^+ {}^-OC_2H_5} \xrightarrow{H^+} C_{18}H_{28}O_8 \xrightarrow[heat]{H_2O, H^+} C_8H_{12}O_4$$

I   **II**   **III**

16.65 *Pulegone* is a fragrant component of oil of pennyroyal. When this compound is heated in aqueous base, acetone is formed. What other organic product or products could be isolated from the reaction mixture? Explain your answer.

pulegone

16.66 Suggest appropriate reagents for the following conversions, and suggest a mechanism for each reaction.

(a)

(b)

(c) $C_6H_5\overset{O}{\overset{\|}{C}}CH_3$ + HCHO + $(CH_3)_2NH$ $\xrightarrow{H^+}$ $C_6H_5\overset{O}{\overset{\|}{C}}CH_2CH_2N(CH_3)_2$

16.67 Suggest syntheses for the following compounds:

(a) from compounds containing five or fewer carbons

(b) from monocyclic or acyclic starting materials

(c) from acyclic starting materials

(d) from acyclic starting materials

(e) from bromocyclopentane

(f)

$$\underset{\displaystyle \underset{\displaystyle \mathrm{CH_3CCHCO_2C_2H_5}}{|}}{\overset{\displaystyle \overset{\displaystyle O}{\parallel}}{\mathrm{CH_3CCHCO_2C_2H_5}}}$$

from compounds containing six or fewer carbons

(g) H₃C— [structure] starting with diethyl malonate

(h) [structure with H₃C, CH₃, HO₂C] from ethyl acetoacetate

(i) [structure with O and CH₃] from compounds containing six or fewer carbons

(j) [structure with O, CH₂CH₂CH₃, O] from compounds containing six or fewer carbons

(k) [structure with OH, CH₃, =O] from 2-cyclohexenone and other appropriate reagents

(l) [structure with CO₂CH₃, O] from [structure with CO₂CH₃, O]

(m) [structure with CH₃, O, C(CH₃)₃, H] from [structure with O, C(CH₃)₃, H]

16.68 *cis*-Jasmone, a principal odorous component of jasmine oil, can be synthesized by heating the 11-carbon compound *cis*-8-undecen-2,5-dione with basic alumina (Al₂O₃) in benzene. The infrared spectrum of *cis*-jasmone shows principal peaks at 1645 cm⁻¹ (6.08 μm) and 1700 cm⁻¹ (5.88 μm). The molecular weight was determined by mass spectrometry to be 164. The proton nmr spectrum shows the following absorption. What is the structure of *cis*-jasmone?

δ	*splitting*	*relative area*
5.25	multiplet	2
2.85	doublet	2
1.9–2.7	multiplet	6
2.05	singlet	3
0.95	triplet	3

16.69 A chemist subjected ethyl 2,2-dimethylacetoacetate to the following series of reactions. What are the structures of the intermediate products and the two final products?

$$CH_3\overset{\displaystyle O}{\overset{\displaystyle \|}{C}}-\underset{\displaystyle \underset{CH_3}{|}}{\overset{\displaystyle \overset{CH_3}{|}}{C}}-CO_2C_2H_5 \quad \xrightarrow[\text{diethyl ether}]{Na,\ HCO_2C_2H_5} \quad A \quad \xrightarrow{H_2,\ Ni} \quad B \quad \xrightarrow[\text{benzene, heat}]{p\text{-toluenesulfonic acid (trace)}} \quad C + D$$

The δ values and splitting patterns in the proton nmr spectra of C and D follow:

	δ	splitting	relative area
for compound C:	1.21	triplet	3
	1.33	singlet	6
	4.12	quartet	2
	5.61	two doublets	1
	6.30	doublet	1
	6.37	doublet	1
for compound D:	1.35	singlet	6
	2.71	triplet	2
	4.46	triplet	2

17 ▽

Amines

Carbon, hydrogen, and oxygen are the three most common elements in living systems. Nitrogen is fourth. Nitrogen is found in proteins and nucleic acids, as well as in many other naturally occurring compounds of both plant and animal origin. In this chapter, we will discuss the **amines**, organic compounds containing trivalent nitrogen atoms bonded to one or more carbon atoms: RNH_2, R_2NH, or R_3N.

Amines are widely distributed in plants and animals, and many amines have physiological activity. For example, two of the body's natural stimulants of the sympathetic ("fight or flight") nervous system are norepinephrine and epinephrine (adrenaline).

norepinephrine

epinephrine
(adrenaline)

Both norepinephrine and epinephrine are β-phenylethylamines (2-phenyl-ethylamines). A number of other β-phenylethylamines act upon the sympathetic receptors. These compounds are referred to as *sympathomimetic amines* because they "mimic," to an extent, the physiological action of norepinephrine and epinephrine.

Well before the birth of Christ, the compound *ephedrine* was extracted from the *ma-huang* plant in China and used as a drug. Today, it is the active decongestant in nose drops and cold remedies. Ephedrine causes shrinkage of swollen nasal membranes and inhibition of nasal secretions. (Overdoses cause nervousness and sleeplessness.) *Mescaline*, a hallucinogen isolated from the peyote cactus, has been used for centuries by the Indians of the southwestern United States and Mexico in religious ceremonies. *Amphetamine* is a synthetic stimulant that causes sleeplessness and nervousness. Amphetamine is sometimes prescribed for obesity, because it is also an appetite depressant. Like many other sympathomimetic amines, amphetamine contains a chiral carbon and has a pair of enantiomers. The more active enantiomer of amphetamine (the dextrorotatory one) is called *dexedrine*.

ephedrine

a decongestant

mescaline

a hallucinogen

amphetamine

a stimulant

Single enantiomers of chiral amines are common in plants. Because of their basicity, some of these amines find use in the resolution of racemic carboxylic acids. (See Section 4.10.) Strychnine, isolated from the seeds of the Asiatic tree *Strychnos nux-vomica*, is an example of such an amine. Strychnine is a highly toxic stimulant of the central nervous system. Besides its laboratory use, it is used to kill rodents and predatory animals.

strychnine

▼ ## Section 17.1
Classification and Nomenclature of Amines

Amines may be classified as **primary**, **secondary**, or **tertiary**, according to the number of alkyl or aryl substituents bonded to the nitrogen.

one C attached   *two C's attached*   *three C's attached*

CH_3NH_2 —NHCH$_3$ $(CH_3CH_2)_3N$

a 1° alkylamine   *a 2° arylalkylamine*   *a 3° trialkylamine*

This classification is different from that of alkyl halides or alcohols. Their classification is based on the number of groups attached to the carbon that has the halide or hydroxyl group.

three C's attached to head C

$$CH_3-\overset{\overset{\displaystyle CH_3}{|}}{\underset{\underset{\displaystyle CH_3}{|}}{C}}-OH \qquad CH_3-\overset{\overset{\displaystyle CH_3}{|}}{\underset{\underset{\displaystyle CH_3}{|}}{C}}-NH_2$$

one C attached to N

t-butyl alcohol   *t*-butylamine

a 3° alcohol   *a 1° amine*

An amine nitrogen can have *four* groups or atoms bonded to it, in which case the nitrogen is part of a positive ion. These ionic compounds fall into two categories. If one or more of the attachments is H, the compound is an **amine salt**. If all four groups are alkyl or aryl (no H's on the N), the compound is a **quaternary ammonium salt**.

Amine salts:

$(CH_3)_2NH_2{}^+ \ Cl^-$

dimethylammonium chloride

salt of a 2° amine

$$\overset{CH_3}{\underset{H}{N^+}} \qquad Br^-$$

N-methylpiperidinium bromide

salt of a 3° amine

Quaternary ammonium salts:

$(CH_3)_4N^+ \ Cl^-$ $CH_3CO_2CH_2CH_2\overset{+}{N}(CH_3)_3 \ Cl^-$

tetramethylammonium   acetylcholine chloride
chloride

involved in the transmission of nerve impulses

Study Problem

17.1 Classify each of the following compounds as a 1°, 2°, or 3° amine; as a salt of a 1°, 2°, or 3° amine; or as a quaternary ammonium salt:

(a) $CH_3CH_2\overset{\overset{\displaystyle CH_3}{|}}{C}HNH_2$ **(b)** ⬡NH

(c) $(CH_3)_3NH^+ NO_3^-$ **(d)** $(CH_3CH_2)_3CNH_2$

(e) $\overset{+}{N}(CH_3)_2\ Cl^-$ **(f)** NCH₃

Simple amines are usually named by the functional-group system. The alkyl or aryl group is named, then the ending **-amine** is added.

$CH_3CH_2CH_2NH_2$ —NH₂ $(CH_3CH_2)_2NH$

propylamine   cyclohexylamine   diethylamine

Diamines are named from the name of the parent alkane (with appropriate prefix numbers) followed by the ending **-diamine**.

$H_2NCH_2CH_2CH_2NH_2$

1,3-propanediamine

If more than one type of alkyl group is attached to the nitrogen, the largest alkyl group is considered the parent. A subsidiary alkyl group is designated by an **N-alkyl-** prefix.

$$\underset{\text{N-methyl-2-propylamine}}{\overset{\overset{\displaystyle CH_3}{|}}{CH_3CHNHCH_3}} \qquad \underset{\text{N,N-dimethyl-2-propylamine}}{\overset{\overset{\displaystyle CH_3}{|}}{CH_3CHN(CH_3)_2}}$$

If a functional group of higher nomenclature priority is present, an **amino-** prefix is used. (See the appendix for a list of nomenclature priorities.)

$$H_2NCH_2CH_2OH \qquad \underset{\text{2-(N-methylamino)propanoic acid}}{\overset{\overset{\displaystyle NHCH_3}{|}}{CH_3CHCO_2H}}$$

2-aminoethanol

The chemistry of the **nonaromatic heterocyclic amines** is similar to that of their open-chain counterparts. The commonly encountered compounds of this class generally have individual names:

pyrrolidine   piperidine   piperazine   morpholine

In the numbering of heterocyclic rings, the heteroatom is considered position 1. Oxygen has priority over nitrogen.

2-methylpyrrolidine   3,5-dimethylmorpholine

Study Problem

17.2 Give each of the following compounds an acceptable name:

(a) $H_2NCH_2CH_2CH_2CH_2NH_2$ (commonly called *putrescine*, an odorous compound found in decaying flesh)

(b) $H_2NCH_2(CH_2)_3CH_2NH_2$ (*cadaverine*, origin similar to that of putrescine)

(c) $(CH_3)_2CHCH_2NH_2$ (d) $CH_3CH_2N(CH_3)_2$

(e) Cl—⬡—$NHCH_3$ (Name as a derivative of aniline.)

Section 17.2
Bonding in Amines

The bonding in an amine is directly analogous to that in ammonia: an sp^3-hybridized nitrogen atom bonded to three other atoms or groups (H or R) and with a pair of unshared valence electrons in the remaining sp^3 orbital (see Figure 2.19, page 66).

H—N̈—H CH_3—N̈—CH_3 ⬡N̈
| | |
H CH_3 H

ammonia   trimethylamine   piperidine

In an amine salt or a quaternary ammonium salt, the unshared pair of electrons forms the fourth sigma bond. The cations are analogous to the ammonium ion.

H CH_3 CH_3
|+ |+   /
H—N—H Cl⁻ CH_3—N—CH_3 Cl⁻   ⬡N⁺ ⁻O_2CCH_3
| | \
H CH_3 H

ammonium   tetramethylammonium   *N*-methylpiperidinium
chloride   chloride   acetate

An amine molecule with three different groups bonded to the nitrogen is chiral. However, enantiomers of most amines cannot be isolated because rapid inversion between mirror images occurs at room temperature. The inversion proceeds by way of a planar transition state (sp^2 nitrogen). The result is that the nitrogen pyramid flips inside out, much as an umbrella in a strong wind. The energy required for this inversion is about 6 kcal/mol, about twice the energy required for rotation around a carbon–carbon sigma bond.

The mirror images are interconvertible:

sp^2

R_1—N̈⁗R_3 ⇌ [R_1—N ⟨R_3, R_2] ⇌ R_1—N̈⟨R_3, R_2
   |
   R_2

transition state with
two e⁻ in p orbital

If an amine nitrogen has three different substituents and interconversion between the two mirror-image structures is restricted, then a pair of enantiomers can be isolated. **Tröger's base** is an example of such a compound. The methylene bridge between the two nitrogens prevents interconversion between the mirror images; Tröger's base can be separated into a pair of enantiomers.

Tröger's base

Some quaternary ammonium salts can also exist as isolable enantiomers. These compounds are structurally similar to compounds containing sp^3-hybridized carbon atoms. If four different groups are bonded to the nitrogen, the ion is chiral and the salt can be separated into enantiomers.

A pair of enantiomers:

Study Problem

17.3 Which of the following structures could exist as isolable enantiomers?

(a) $[(CH_3)_2CH]_2\overset{+}{N}(CH_2CH_2Cl)_2\ Cl^-$ (b) $CH_3NHCH_2CH_2Cl$

(c)

(d)

▼ ## Section 17.3
Physical Properties of Amines

As we described in Chapter 1, amines undergo hydrogen bonding. The N---HN hydrogen bond is weaker than the O---HO hydrogen bond because N is less electronegative than O and therefore the NH bond is less polar. This weaker hydrogen bonding between amine molecules results in boiling points that fall between those for nonhydrogen-bonded compounds (like alkanes or ethers) and those for strongly hydrogen-bonded compounds (like alcohols) of comparable molecular weight. (See Table 17.1.)

$$CH_3CH_2OCH_2CH_3 \qquad (CH_3CH_2)_2NH \qquad CH_3CH_2CH_2CH_2OH$$

bp 34.5°	bp 56°	bp 117°

Table 17.1 Physical properties of some amines

Name	Structure	Bp, °C	Solubility in Water
methylamine	CH_3NH_2	−6.3	∞
dimethylamine	$(CH_3)_2NH$	7.5	∞
trimethylamine	$(CH_3)_3N$	3	∞
ethylamine	$CH_3CH_2NH_2$	17	∞
benzylamine	$C_6H_5CH_2NH_2$	185	∞
aniline	$C_6H_5NH_2$	184	3.7 g/100 mL

Because they do not have an NH bond, tertiary amines in the pure liquid state cannot undergo hydrogen bonding. The boiling points of tertiary amines are lower than those for comparable primary or secondary amines and are closer to the boiling points of alkanes of similar molecular weight.

no hydrogen bonding   *hydrogen bonding*

$(CH_3)_3N$ $(CH_3)_3CH$ $CH_3CH_2CH_2NH_2$

bp 3°   bp −10°   bp 48°

Amines of low molecular weight are soluble in water because they can undergo hydrogen bonding with water. Tertiary amines, as well as primary and secondary amines, can undergo this type of hydrogen bonding because they have unshared pairs of electrons that can be used to form hydrogen bonds with water.

$$(CH_3)_3N\text{:---}H\text{---}O\text{---}H$$

Study Problem

17.4 Show all types of hydrogen bonding that could exist in **(a)** pure dimethylamine and **(b)** aqueous dimethylamine.

Volatile amines have very distinctive and usually offensive odors. Methylamine has an odor similar to that of ammonia. Trimethylamine smells like dead salt-water salmon. Piperidine smells like dead fresh-water fish. Arylamines are not as unpleasant smelling as alkylamines. However, arylamines such as aniline are toxic and are especially insidious because they can be absorbed through the skin. Some, like β-naphthylamine, are carcinogenic.

2-naphthylamine (β-naphthylamine)

a carcinogenic arylamine

Amine salts and quaternary ammonium salts behave physically like inorganic salts—high-melting, water-soluble, and odorless.

▼ Section 17.4
Spectral Properties of Amines

A. Infrared Spectra

The bonds that give rise to infrared absorption characteristic of amines are the C—N and N—H bonds (Table 17.2). All aliphatic amines show C—N stretching in the fingerprint region. However, only primary and secondary amines show the distinctive NH stretching absorption, which is observed to the left of CH absorption in the spectrum. This is the same region where OH absorption is observed. However, the two can often be differentiated because the OH absorption is usually broader and stronger than NH absorption. The stronger absorption by an OH bond is due to the greater polarity and hydrogen bonding of this group.

In Chapter 9, we described how primary amines show two NH absorption peaks, secondary amines show one NH peak, and tertiary amines show no

Table 17.2 Characteristic infrared absorption for amines

	Position of Absorption	
Type of Absorption	*cm^{-1}*	*μm*
1° amines:		
N—H stretching (pure liquid)	3250–3400 (2 peaks)	2.9–3.1
C—N stretching	1020–1250	8.0–9.8
2° amines:		
N—H stretching (pure liquid)	3330	3.0
C—N stretching	1020–1250	8.0–9.8
3° amines:		
C—N stretching	1020–1250	8.0–9.8

Figure 17.1 ^{1}H nmr spectrum of *n*-butylamine.

absorption in this region. Spectra of the three types of amines are shown in Figure 9.10 (page 333).

B. Proton Nmr Spectra

The NH absorption in the ^{1}H nmr spectrum is generally a sharp singlet, not split by adjacent protons. In this respect, NH absorption is similar to OH absorption (Section 9.10E). Aliphatic amines show NH absorption at δ values of about 1.0–2.8 ppm, while arylamines absorb at about $\delta2.6$–4.7. (The exact position depends upon the solvent used.) The α protons are somewhat deshielded by the electronegative nitrogen. The chemical shift for these protons is in the range of $\delta2.2$–2.8. (See Figure 17.1.)

Section 17.5
Preparation of Amines

Techniques for the preparation of amines fall into three general categories. We will discuss each category in turn.

Nucleophilic substitution:

$$\text{RX} \quad + \text{NH}_3 \xrightarrow{\text{S}_\text{N}2} \text{RNH}_3^+ \text{ X}^- \xrightarrow{\text{OH}^-} \text{RNH}_2$$

an alkyl halide

Reduction:

$$\overset{\displaystyle O}{\overset{\|}{\text{RCNH}_2}} \quad \text{or} \quad \text{RCN} \xrightarrow{\text{[H]}} \text{RCH}_2\text{NH}_2$$

an amide or nitrile

Amide rearrangement:

$$\overset{\displaystyle O}{\overset{\|}{\text{RCNH}_2}} \xrightarrow{\text{Br}_2,\ \text{OH}^-} \text{RNH}_2$$

an amide

A. Synthesis by Substitution Reactions

Reaction of amines and alkyl halides. Ammonia or an amine carries an unshared pair of electrons and can act as a nucleophile in a substitution reaction with an alkyl halide. The reaction of a nitrogen nucleophile is similar to the reaction of any other nucleophile with RX. The product of the reaction with ammonia or an amine is an amine salt. The free amine can be obtained by the treatment of this amine salt with a base such as NaOH.

S_N2 reaction:

$$\text{H}_3\text{N:} \quad + \quad \overset{\displaystyle \text{CH}_3}{\overset{|}{\text{CH}_2}}\!-\!\ddot{\text{Br}}: \xrightarrow{\text{S}_\text{N}2} \quad \overset{\displaystyle \text{CH}_3}{\overset{|}{\text{H}_3\overset{+}{\text{N}}\!-\!\text{CH}_2}} \quad :\ddot{\text{Br}}:^-$$

ammonia   bromoethane   ethylammonium bromide

an amine salt

Treatment with base:

$$CH_3CH_2NH_3^+ \ Br^- + \ :\overset{..}{O}H^- \longrightarrow CH_3CH_2\overset{..}{N}H_2 + H_2\overset{..}{\underset{..}{O}}: + Br^-$$

ethylamine

The order of reactivity of alkyl halides is typical for S_N2 reactions: $CH_3X > 1° > 2°$. Tertiary alkyl halides do not undergo substitution reactions with ammonia or amines; elimination products are obtained.

The principal disadvantage of this route to amines is that the product amine salt can exchange a proton with the starting ammonia or amine.

$$CH_3CH_2NH_3^+ \ Br^- + \ NH_3 \ \rightleftharpoons \ CH_3CH_2NH_2 \ + NH_4^+ \ Br^-$$

also a nucleophile

This proton exchange results in two or more nucleophiles competing in the reaction with the alkyl halide. For this reason, a mixture of mono-, di-, and trialkylamines and the quaternary ammonium salt is frequently obtained from the reaction of ammonia with an alkyl halide.

$$NH_3 \ \xrightarrow{RX} \ RNH_2 \ \xrightarrow{RX} \ R_2NH \ \xrightarrow{RX} \ R_3N \ \xrightarrow{RX} \ R_4N^+ \ X^-$$

If the amine is very inexpensive or if ammonia is used, a large excess can be used to favor monoalkylation. In this case, RX is more likely to collide with the molecules of the desired reactant and less likely to collide with those of the alkylated product. In the following example, an excess of ammonia favors the primary amine product.

$$CH_3CH_2CH_2CH_2Br + \text{excess } NH_3 \ \longrightarrow \ \xrightarrow{OH^-} \ CH_3CH_2CH_2CH_2NH_2 + Br^-$$

1-bromobutane   *n*-butylamine (45%)

If the quaternary ammonium salt is desired, an excess of the alkyl halide would be used.

$$(CH_3CH_2)_2NH + \text{excess } CH_3CH_2I \ \longrightarrow \ (CH_3CH_2)_4N^+ \ I^-$$

Gabriel phthalimide synthesis.   A reaction sequence that yields primary amines without secondary and tertiary amines is the Gabriel phthalimide synthesis. The first step in the sequence is an S_N2 reaction with the phthalimide anion as the nucleophile. Hydrolysis of the substituted phthalimide yields the amine.

potassium
phthalimide

N-ethylphthalimide

$$+ \ H_2NCH_2CH_3$$

ethylamine

*free of 2° and 3°
amines*

Phthalimide is prepared by heating phthalic anhydride with ammonia. The potassium salt is made by treating phthalimide with KOH. Usually, a proton cannot be removed from an amide nitrogen so easily. However, like other β-dicarbonyl compounds, imides are acidic because the anion is resonance stabilized.

Preparation of phthalimide anion:

phthalic anhydride   phthalimide   *resonance-stabilized*
                  (pK_a = 8.3)

Study Problem

17.5 Write resonance structures that show the delocalization of the negative charge in the phthalimide anion.

Treatment of the potassium phthalimide with an alkyl halide yields an *N*-alkylphthalimide.

Attack on RX:

an N-alkylphthalimide

Finally, the alkylphthalimide is hydrolyzed. This reaction is simply the hydrolysis of an amide (Section 15.8C).

Hydrolysis:

half-hydrolyzed

Study Problem

17.6 List the sequence of reagents that you would add to potassium phthalimide to prepare: **(a)** *n*-propylamine; **(b)** allylamine; **(c)** benzylamine.

An ingenious variation of the Gabriel phthalimide synthesis is used to prepare α-amino acids, the building blocks of proteins. This sequence is: (1) treatment of potassium phthalimide with diethyl bromomalonate; (2) treatment of the imide-malonate with base to remove the α hydrogen; and (3) treatment with RX, which gives a typical malonic ester alkylation reaction.

1. Reaction with bromomalonic ester:

$$\text{N}^- + \text{BrCH(CO}_2\text{C}_2\text{H}_5)_2 \xrightarrow{-\text{Br}^-} \text{NCH(CO}_2\text{C}_2\text{H}_5)_2$$

diethyl
bromomalonate

the imide-malonate

2. Treatment with base:

$$\text{NC(CO}_2\text{C}_2\text{H}_5)_2 \xleftarrow{-\text{OC}_2\text{H}_5} \text{N}\bar{\text{C}}(\text{CO}_2\text{C}_2\text{H}_5)_2 + \text{C}_2\text{H}_5\text{OH}$$

an enolate ion

3. Reaction with RX:

$$\text{N}\bar{\text{C}}(\text{CO}_2\text{C}_2\text{H}_5)_2 + \text{R}-\text{X} \xrightarrow[\text{S}_\text{N}2]{-\text{X}^-} \text{NC(CO}_2\text{C}_2\text{H}_5)_2$$

Acid hydrolysis results in hydrolysis of both imide and diester plus decarboxylation of the diacid. The product is the protonated amino acid.

hydrolyzed

*hydrolyzed and
decarboxylated*

$$\text{N}-\text{C(CO}_2\text{C}_2\text{H}_5)_2 \xrightarrow[\text{heat}]{\text{H}_2\text{O, H}^+} \text{(}_{\text{CO}_2\text{H}}^{\text{CO}_2\text{H}}\text{)} + \text{H}_3\overset{+}{\text{N}}\text{CHCO}_2\text{H} + \text{C}_2\text{H}_5\text{OH} + \text{CO}_2$$

*a protonated
α-amino acid*

Sample Problem

How would you prepare phenylalanine, $\text{C}_6\text{H}_5\text{CH}_2\text{CHCO}_2\text{H}$, by a phthalimide synthesis?

$$\underset{\text{NH}_2}{|}$$

Solution:

1. Determination of reagents needed:

$$\text{C}_6\text{H}_5\text{CH}_2\text{CHCO}_2\text{H}$$

$$\underset{\text{NH}_2}{|}$$

from $\text{C}_6\text{H}_5\text{CH}_2\text{X}$

*from phthalimide
and the bromomalonate*

2. *Steps in the synthesis:*

Study Problem

17.7 How would you prepare the following amino acid by a phthalimide synthesis?

$$(CH_3)_2CHCH_2\underset{\underset{NH_2}{|}}{C}HCO_2H$$

leucine

B. Synthesis by Reduction

Reduction reactions often provide convenient routes to amines. The reduction of *aromatic nitro compounds* to arylamines was discussed in Section 12.3A.

2,4-dinitrotoluene   2,4-toluenediamine (75%)

Some arylamines can be synthesized by the reaction of ammonia or amines with activated aryl halides (Section 12.4).

2,4-dinitroaniline (70%)

Nitriles undergo catalytic hydrogenation or reduction with LiAlH$_4$ to yield primary amines of the type RCH$_2$NH$_2$ in yields of approximately 70%. Nitriles are available from alkyl halides. Therefore, a nitrile synthesis is a technique for lengthening a carbon chain as well as for preparing an amine.

$$(CH_3)_2CHCH_2Br \xrightarrow[-Br^-]{CN^-} (CH_3)_2CHCH_2CN \xrightarrow[(2) H_2O, H^+]{(1) LiAlH_4} (CH_3)_2CHCH_2CH_2NH_2$$

1-bromo-2-methylpropane   3-methylbutanenitrile   (3-methyl-1-butyl)amine

a 1° alkyl halide

Amides also yield amines when treated with reducing agents:

$$CH_3(CH_2)_{10}\overset{\overset{\displaystyle O}{\|}}{C}NHCH_3 \xrightarrow[(2) H_2O, H^+]{(1) LiAlH_4} CH_3(CH_2)_{10}CH_2NHCH_3$$

N-methyldodecanamide   *N*-methyldodecylamine (95%)

Reductive amination, a reaction that converts ketones or aldehydes to primary amines, was discussed in Section 13.7D. This reaction is much better for synthesizing an amine of the type R_2CHNH_2 than is the reaction of R_2CHBr and NH_3 because the latter reaction may lead to elimination products. Secondary and tertiary amines can also be synthesized by reductive amination if a primary or secondary amine is used instead of ammonia.

$$\text{benzaldehyde} \xrightarrow[-H_2O]{NH_3} [\text{an imine}] \xrightarrow[60°, 90 \text{ atm}]{H_2, Ni} \text{benzylamine (85\%)}$$

benzaldehyde   *an imine*   benzylamine (85%)

$$HOCH_2CH_2NH_2 + CH_3\overset{\overset{\displaystyle O}{\|}}{C}CH_3 \xrightarrow{-H_2O} [HOCH_2CH_2N\!=\!C(CH_3)_2]$$

2-aminoethanol   acetone   *an imine*
(ethanolamine)

$$\xrightarrow{H_2, Pt} HOCH_2CH_2NHCH(CH_3)_2$$

2-(*N*-isopropylamino)ethanol (95%)

Study Problems

17.8 Show with flow equations how you would carry out the following syntheses:

(a) cyclohexylamine from cyclohexanone
(b) $CH_2\!=\!CHCH_2CH_2CH_2NH_2$ from 4-bromo-1-butene
(c) *N*,*N*-dimethylbenzylamine from a carboxylic acid

17.9 Suggest two techniques for preparing *sec*-butylamine free of higher alkylation products.

C. Amide Rearrangement

When an unsubstituted amide ($RCONH_2$) is treated with an alkaline, aqueous solution of bromine, it undergoes rearrangement to yield an amine. This reaction is called the **Hofmann rearrangement**. The carbonyl group is lost as CO_3^{2-}. Therefore, the amine contains one less carbon than the starting amide.

$$CH_3(CH_2)_4\overset{\overset{\displaystyle O}{\|}}{C}NH_2 + 4\ OH^- + Br_2 \xrightarrow{H_2O} CH_3(CH_2)_4NH_2 + CO_3^{2-} + 2\ H_2O + 2\ Br^-$$

hexanamide   *n*-pentylamine (85%)

Because the carbonyl group appears to be lost from the interior of the molecule, let us consider the mechanism of the Hofmann rearrangement. The reaction proceeds by a series of discrete steps. *Step 1* is bromination at the nitrogen. *Step 2* is loss of a proton from the nitrogen and results in an unstable anion. The rearrangement step is *Step 3* of the sequence. Note that this rearrangement is a *1,2-shift* very similar to those in carbocation rearrangements (Section 5.6F). The product of the rearrangement is an isocyanate, stable under some conditions, but not in aqueous base. In aqueous base, the isocyanate undergoes hydrolysis (*Step 4*) to an amine and the carbonate ion.

Step 1 (bromination of N):

$$R\overset{\overset{\displaystyle O}{\|}}{C}\overset{..}{N}H + OH^- \underset{-H_2O}{\rightleftharpoons} \left[R\overset{\overset{\displaystyle O}{\|}}{C}\overset{..}{\underset{..}{N}}H \right] \xrightarrow{Br_2} R\overset{\overset{\displaystyle O}{\|}}{C}\overset{..}{N}H + Br^-$$

$pK_a = \sim 15$

Step 2 (extraction of H$^+$ by OH$^-$):

$$R\overset{\overset{\displaystyle O}{\|}}{C}\overset{..}{N}H + {}^-OH \rightleftharpoons \left[R\overset{\overset{\displaystyle O}{\|}}{C}\overset{..}{\underset{..}{N}}Br \right] + H_2O$$

                           unstable

Step 3 (displacement of Br$^-$ by R—, a 1,2-shift):

$$\left[\overset{\overset{\displaystyle O}{\|}}{\underset{R}{C}}\overset{..}{N}{-}Br \right]^- \longrightarrow R{-}\overset{..}{N}{=}C{=}O + Br^-$$

                           an isocyanate

Step 4 (hydrolysis of the isocyanate):

$$RN{=}C{=}O \xrightarrow[H_2O]{OH^-} \left[RN H\overset{\overset{\displaystyle O}{\|}}{C}O^- \right] \xrightarrow[H_2O]{OH^-} RNH_2 + CO_3^{2-}$$

                           the amine

It has been found that the Hofmann rearrangement proceeds with *retention of configuration* at the α carbon of the amide. This evidence leads us to believe that the rearrangement step (Step 3) has a bridged transition state.

$$CH_3CH_2{-}\overset{\overset{\displaystyle CH_3}{|}}{\underset{\underset{\displaystyle H}{|}}{C}}{-}\overset{\overset{\displaystyle O}{\|}}{C}\underset{NH_2}{} \xrightarrow{Br_2,\ OH^-} CH_3CH_2{-}\overset{\overset{\displaystyle CH_3}{|}}{\underset{\underset{\displaystyle H}{|}}{C}}{-}NH_2$$

(*R*)-2-methylbutanamide   (*R*)-2-butylamine

A bridged transition state in Step 3:

$$CH_3CH_2-\underset{\underset{H}{|}}{\overset{\overset{CH_3}{|}}{C}}-\overset{\overset{O}{\|}}{C}\underset{\overset{|}{Br}}{\overset{..}{N}}^- \longrightarrow \left[CH_3CH_2-\underset{\underset{H}{|}}{\overset{\overset{CH_3}{|}}{C}}\cdots\overset{\overset{O}{\|}}{\underset{\underset{\cdot Br^{\delta-}}{N^{\delta-}}}{C}} \right] \xrightarrow{-Br^-}$$

$$CH_3CH_2-\underset{\underset{H}{|}}{\overset{\overset{CH_3}{|}}{C}}-\overset{..}{N}=C=O$$

The advantage of the Hofmann rearrangement is that yields of pure primary amines are good. This would be the best route to a primary amine containing a tertiary alkyl group, such as $(CH_3)_3CNH_2$.

Study Problem

17.10 Predict the major organic products when the following compounds are treated with aqueous alkaline Br_2:

(a) (R)-⟨◯⟩—$CH_2\underset{\underset{CH_3}{|}}{\overset{\overset{O}{\|}}{CH}C}NH_2$ **(b)** $H_2N\overset{\overset{O}{\|}}{C}CH_2CH_2CH_2\overset{\overset{O}{\|}}{C}NH_2$

D. Summary of Amine Syntheses

We have shown several routes to amines. By one or another of these routes, shown in the scheme below, a chemist may synthesize:

1. an amine with the same number of carbons as the starting material;
2. an amine with one additional carbon; or
3. an amine with one less carbon.

The specific reactions are summarized in Table 17.3.

General routes to amines:

$$RCH_2OH \xrightarrow{[O]} \overset{\overset{O}{\|}}{R}COH \xrightarrow{SOCl_2} \overset{\overset{O}{\|}}{R}CCl \xrightarrow{NH_3} \overset{\overset{O}{\|}}{R}CNH_2 \xrightarrow{Br_2,\ OH^-} RNH_2$$

one less carbon

$$RCH_2OH \xrightarrow{PX_3} RCH_2X \xrightarrow{CN^-} RCH_2CN \xrightarrow{[H]} RCH_2CH_2NH_2$$

one more carbon

$$RCH_2X \xrightarrow{Gabriel} RCH_2NH_2$$

same number of carbons

Table 17.3 Summary of laboratory syntheses of amines[a]

Reaction	Section Reference

Primary Amines:

substitution

$$1° \text{ RX } \xrightarrow[\text{(2) OH}^-]{\text{(1) excess NH}_3} \text{ RNH}_2 \qquad 17.5A$$

$$\text{RX } \xrightarrow[\substack{\text{(2) H}_2\text{O, H}^+ \\ \text{(3) OH}^-}]{\text{(1) K}^+ \text{ phthalimide}} \text{ RNH}_2 \qquad 17.5A$$

reduction

$$\text{ArNO}_2 \xrightarrow[\text{(2) OH}^-]{\text{(1) Fe, HCl}} \text{ ArNH}_2 \qquad 12.3A$$

$$\text{RCN } \text{ or } \underset{\substack{\| \\ O}}{\text{RCNH}_2} \xrightarrow[\text{(2) H}_2\text{O, H}^+]{\text{(1) LiAlH}_4} \text{ RCH}_2\text{NH}_2 \qquad 17.5B$$

$$\text{R}_2\text{C}=\text{O} \xrightarrow{\text{NH}_3, \text{ H}_2, \text{ Ni}} \text{ R}_2\text{CHNH}_2 \qquad 13.7D, 17.5B$$

rearrangement

$$\underset{\substack{\| \\ O}}{\text{RCNH}_2} \xrightarrow{\text{X}_2, \text{ OH}^-} \text{ RNH}_2 \qquad 17.5C$$

Secondary and Tertiary Amines:

reduction

$$\underset{\substack{\| \\ O}}{\text{RCNR}'_2} \xrightarrow[\text{(2) H}_2\text{O}]{\text{(1) LiAlH}_4} \text{ RCH}_2\text{NR}'_2 \qquad 17.5C$$

$$\text{R}_2\text{C}=\text{O} + \text{R}'_2\text{NH} \xrightarrow{\text{H}_2, \text{ Ni}} \text{ R}_2\text{CHNR}'_2 \qquad 13.7D, 17.5B$$

[a] Some arylamines can be synthesized by nucleophilic aromatic substitution. See Section 12.4.

Section 17.6
Basicity of Amines

The unshared pair of valence electrons of ammonia or an amine can be donated to an electron-deficient atom, ion, or molecule. In water solution, an amine is a *weak base* and accepts a proton from water in a reversible acid–base reaction.

$$(\text{CH}_3)_3\text{N:} \quad + \text{H} \overset{\frown}{\ddot{\text{O}}}\text{H} \rightleftharpoons (\text{CH}_3)_3\overset{+}{\text{N}}\text{H} + \text{:}\ddot{\text{O}}\text{H}^-$$

trimethylamine

Table 17.4 pK_b values for some amines and pK_a values for their conjugate acids

Structure	pK_b	Structure	pK_a
NH_3	4.75	NH_4^+	9.25
CH_3NH_2	3.34	$CH_3NH_3^+$	10.66
$(CH_3)_2NH$	3.27	$(CH_3)_2NH_2^+$	10.73
$(CH_3)_3N$	4.19	$(CH_3)_3NH^+$	9.81
⬡NH	2.88	⬡NH_2^+	11.12
⬡$-NH_2$	9.37	⬡$-NH_3^+$	4.63

The calculation of basicity constants and pK_b values for weak bases was discussed in Section 1.10. Table 17.4 lists a few amines along with their pK_b values. (Recall that decreasing values for pK_b indicate increasing base strength.) We have also included the pK_a values for the conjugate acids of these amines. (Note that pK_a = 14 − pK_b.) An amine that is a stronger base has a *weaker* conjugate acid with a *higher* pK_a.

The same structural features that affect the relative acid strengths of carboxylic acids and phenols (Section 14.5) affect the relative base strengths of amines.

1. *If the free amine is stabilized relative to the cation, the amine is less basic.*
2. *If the cation is stabilized relative to the free amine, the amine is a stronger base.*

*if the free amine
is stabilized, R_3N
is a weaker base*

*if the cation is
stabilized, R_3N is
a stronger base*

$$R_3N\text{:} + H_2O \rightleftharpoons R_3NH^+ + OH^-$$

An *electron-releasing group*, such as an alkyl group, on the nitrogen increases basicity by dispersing the positive charge in the cation. (This dispersal of positive charge is analogous to that in carbocations, Section 5.6E.) By dispersal of the positive charge, the cation is stabilized relative to the free amine. Therefore, base strength increases in the series NH_3, CH_3NH_2, and $(CH_3)_2NH$.

$$NH_3 \qquad CH_3NH_2 \qquad CH_3NHCH_3$$

ammonia   methylamine   dimethylamine

increasing basicity

$$CH_3 \!\rightarrow\! \ddot{N}H_2 + H_2O \rightleftharpoons CH_3 \!\rightarrow\! NH_3^+ + OH^-$$

*stabilized by
dispersal of
positive charge*

The cation is also stabilized by *increasing solvation*. In this case, the solvent helps disperse the positive charge. Dimethylamine (pK_b = 3.27) is a slightly stronger base than methylamine. However, trimethylamine (pK_b = 4.19) is a *weaker base* than dimethylamine. The reason is that trimethylamine is more hindered, and the cation is less stabilized by solvation. These arguments explain why the nonaromatic heterocyclic amines (with their alkyl groups "tied back" away from the unshared electrons of the nitrogen) are more basic than comparable open-chain secondary amines.

$$\overset{\displaystyle CH_3}{\underset{}{CH_3-N-CH_3}}$$ is a weaker base than $$\overset{\displaystyle CH_3}{\underset{}{CH_3-N-H}}$$

trimethylamine   dimethylamine
pK_b = 4.19   pK_b = 3.27

$CH_3CH_2NHCH_2CH_3$ is a weaker base than

diethylamine
$pK_b = 3.01$

pyrrolidine
$pK_b = 2.73$

Hybridization of the nitrogen atom in a nitrogen compound also affects the base strength. An sp^2 orbital contains more s character than an sp^3 orbital. A molecule with an sp^2-hybridized nitrogen is less basic because its unshared electrons are more tightly held and thus the free nitrogen compound is stabilized instead of the cation.

pyridine
$pK_b = 8.75$

piperidine
$pK_b = 2.88$

Resonance also affects the base strength of an amine. Cyclohexylamine is a far stronger base than is aniline.

aniline
$pK_b = 9.37$

cyclohexylamine
$pK_b = 3.3$

The reason for the low basicity of aniline is that the positive charge of the anilinium ion cannot be delocalized by the aromatic pi cloud. However, the pair of electrons of the free amine are delocalized by the ring. The result is that the free amine is stabilized in comparison with the conjugate acid (the cation).

aniline

anilinium ion

*resonance-stabilized
and favored*

*no resonance stabilization
of positive charge*

Resonance structures of aniline:

Sample Problem

Explain why piperidine is a stronger base than morpholine.

piperidine
$pK_b = 2.88$

morpholine
$pK_b = 5.67$

Solution: The oxygen atom in morpholine is *electron-withdrawing*, making the nitrogen more positive. The cation is not stabilized relative to the free morpholine, but is *de-stabilized*:

$$O\text{—}NH + H_2O \rightleftharpoons O\text{—}NH_2^+ + OH^-$$

less stable
because of less
dispersal of + charge

Piperidine has no such destabilizing effect. Its cation is stabilized by electron release of the attached CH_2 groups:

$$NH + H_2O \rightleftharpoons NH_2^+ + OH^-$$

stabilized

Study Problems

17.11 Which would you expect to be more basic:
(a) piperazine (page 740) or piperidine?   (b) piperazine or hexamethylenetetramine?

hexamethylenetetramine

17.12 Explain the following trend in pK_b values:

$$O_2N\text{—}\langle\bigcirc\rangle\text{—}NH_2 \qquad \langle\bigcirc\rangle\text{—}NH_2 \qquad CH_3\text{—}\langle\bigcirc\rangle\text{—}NH_2$$

$pK_b = 13.0$ $pK_b = 9.37$ $pK_b = 8.9$

▼ **Section 17.7**
Amine Salts

The reaction of an amine with a mineral acid (such as HCl) or a carboxylic acid yields an **amine salt**. The salts are commonly named in one of two ways: as **substituted ammonium salts** or, by an older system, as **amine-acid complexes**.

$$(CH_3)_3N\!: \ + HCl \longrightarrow (CH_3)_3NH^+ \ Cl^-$$

trimethylamine   trimethylammonium chloride
                                              (trimethylamine hydrochloride)

$$CH_3CH_2\ddot{N}H_2 + CH_3CO_2H \longrightarrow CH_3CH_2\overset{+}{N}H_3 \ ^-O_2CCH_3$$

ethylamine   acetic acid   ethylammonium acetate
                                                            (ethylamine acetate)

Because of its ability to form salts, an amine that is insoluble in water can be made soluble by treatment with dilute acid. In this fashion, compounds containing amino groups can be separated from water- and acid-insoluble materials. Naturally occurring amines in plants, called **alkaloids** (Section 19.3), can be extracted from their sources, such as leaves or bark, by aqueous acid. Many compounds containing amino groups are used as drugs. These drugs are often administered as their water-soluble salts rather than as water-insoluble amines.

novocaine
water-insoluble

novocaine hydrochloride
water-soluble

Study Problem

17.13 *Piperazine citrate* is a crystalline solid used in the treatment of pinworms and roundworms. Write an equation that shows the formation of the piperazine citrate formed from one molecule of each reactant.

citric acid   piperazine

A free amine can be regenerated from one of its salts by treatment with a strong base, usually NaOH. Quaternary ammonium salts, which have no acidic protons, do not undergo this reaction.

$$RNH_3^+ \ Cl^- + OH^- \longrightarrow R\ddot{N}H_2 \ + H_2O + Cl^-$$

an amine salt   *an amine*

$$R_4N^+ \ Cl^- \ + OH^- \longrightarrow \text{no reaction}$$

a quaternary
ammonium salt

Because of the nitrogen's ionic charge, quaternary ammonium salts have some interesting applications. For example, quaternary ammonium salts with long hydrocarbon chains are used as detergents. The combination of a long, hydrophobic hydrocarbon tail with an ionic hydrophilic head results in two types of interactions with other substances. One part of the molecule is soluble in nonpolar organic solvents, fats, and oils, while the other part is soluble in water. Soaps (Section 23.2) exhibit similar behavior.

In Section 8.6, we discussed how crown ethers can act as *phase-transfer agents* by carrying inorganic ions into an organic solution. Quaternary ammonium salts can also act as phase-transfer agents. For example, let us say that we wish to carry out a substitution reaction using an alkyl halide and NaCN. When an aqueous NaCN solution is mixed with a solution of RX in a water-insoluble organic solvent, two layers result. Reaction can occur only at the interface of these layers. A quaternary ammonium salt can be used to transfer the CN⁻ ions to the organic solution so that reaction can occur in the organic solution as well as at the interface. Also, recall that nucleophiles such as CN⁻ are more nucleophilic and more reactive when they are not solvated by water. The result is that reaction is much faster.

The catalytic action of $R_4N^+ \ X^-$ arises from the fact that it is water-soluble and also slightly soluble in organic solvents. If $R_4N^+ \ X^-$ is dissolved in the aqueous phase of the two-phase reaction mixture, some of the salt also becomes dissolved in the organic layer. However, if the aqueous layer contains an excess of CN⁻ ions, the salt that is transferred is mainly $R_4N^+ \ CN^-$, not $R_4N^+ \ X^-$.

Anion exchange in the aqueous phase:

$$(n\text{-}C_4H_9)_4N^+ \ Cl^- \ + \ CN^- \ \rightleftharpoons \ (n\text{-}C_4H_9)_4N^+ \ CN^- + \ Cl^-$$

tetrabutylammonium   *excess*   *migrates to organic*
chloride   *phase*

S_N2 reaction in the organic phase:

$$CH_3(CH_2)_7Cl + (n\text{-}C_4H_9)_4N^+ \ CN^- \longrightarrow CH_3(CH_2)_7CN + \ (n\text{-}C_4H_9)_4N^+ \ Cl^-$$

1-chlorooctane   nonanenitrile   *returns to aqueous phase*
                                                    (90%)   *for additional exchange*

▼ Section 17.8
Substitution Reactions with Amines

We have already mentioned a variety of substitution reactions with amines. The problems of the reaction of an amine with an alkyl halide were discussed earlier in this chapter (Section 17.5).

$$RNH_2 + R'Cl \xrightarrow{\ S_N2\ } \underset{\underset{R'}{|}}{R\overset{+}{N}H_2} \ Cl^- \quad \text{and also} \quad R\overset{+}{N}HR'_2 \ Cl^- \quad \text{and} \quad R\overset{+}{N}R'_3 \ Cl^-$$

In Chapter 15, we discussed the acylation of amines as a technique for the synthesis of amides. For example:

$$CH_3CCl + CH_3NH_2 \longrightarrow \left[CH_3C\overset{\cdot\cdot}{\underset{\underset{H}{\overset{+}{N}HCH_3}}{\overset{:\overset{\cdot\cdot}{O}:^-}{C}}}Cl \right] \xrightarrow{-HCl} CH_3C\underset{:NHCH_3}{\overset{:\overset{\cdot\cdot}{O}}{\|}}$$

N-methylacetamide

The utility of this reaction is that amines can be used to synthesize other amines by conversion to the amide, followed by reduction.

$$\underset{\textit{an acid chloride}}{R\overset{O}{\underset{\|}{C}}Cl} + \underset{\textit{an amine}}{R'_2NH} \longrightarrow \underset{\textit{an amide}}{R\overset{O}{\underset{\|}{C}}NR'_2} \xrightarrow[\text{(2) H}_2\text{O, H}^+]{\text{(1) LiAlH}_4} \underset{\textit{a new amine}}{RCH_2NR'_2}$$

In Section 15.10, the preparation of sulfonamides was discussed. The reaction of amines with benzenesulfonyl chloride is sometimes used to test whether an amine is primary, secondary, or tertiary. This test is called the **Hinsberg test**.

$$1° \ RNH_2 \xrightarrow[\text{OH}^-]{\text{Cl}-\text{SO}_2-\text{C}_6\text{H}_5} \left[\underset{\cdot\cdot}{RN}\overset{\overset{\text{— }acidic}{H}}{-}SO_2-\hspace{-2mm}\bigcirc \right] \xrightarrow[-\text{H}_2\text{O}]{\text{OH}^-}$$

$$R\overset{\cdot\cdot}{\underset{\cdot\cdot}{N}}-SO_2-\hspace{-2mm}\bigcirc \xrightarrow{\text{H}^+} RNH-SO_2-\hspace{-2mm}\bigcirc \downarrow$$

anion of a sulfonamide:
soluble in base   a sulfonamide:
insoluble in acid

$$2° \ R_2NH \xrightarrow[\text{OH}^-]{\text{Cl}-\text{SO}_2-\text{C}_6\text{H}_5} R_2\overset{\overset{\text{— }no\ acidic\ H}{}}{N}-SO_2-\hspace{-2mm}\bigcirc \downarrow$$

a sulfonamide:
insoluble in acid or base

$$3° \ R_3N \xrightarrow[\text{OH}^-]{\text{Cl}-\text{SO}_2-\text{C}_6\text{H}_5} \left[R_3\overset{+}{N}-SO_2-\hspace{-2mm}\bigcirc \right] \longrightarrow$$

unstable in base:
no change observed

$$R_3N + {}^-OSO_2-\hspace{-2mm}\bigcirc \xrightarrow{\text{H}^+} R_3\overset{+}{N}H + HOSO_2-\hspace{-2mm}\bigcirc$$

soluble in acid

Amines also react with aldehydes and ketones to yield imines and enamines. (For the mechanisms, see Section 13.6.)

Section 17.9
Reactions of Amines with Nitrous Acid

In Section 12.3, we discussed the formation of benzenediazonium chloride ($C_6H_5N_2^+$ Cl^-) by the treatment of aniline with cold aqueous nitrous acid, HNO_2 (prepared *in situ* from $NaNO_2$ and HCl). Recall that the aryldiazonium salts are stable at $0°$ and are useful synthetic intermediates because of the excellent leaving ability of N_2.

The treatment of a *primary alkylamine* with $NaNO_2$ and HCl also results in a diazonium salt, but an alkyldiazonium salt is unstable and decomposes to a mixture of alcohol and alkene products along with N_2. The decomposition proceeds by way of a carbocation.

When treated with $NaNO_2$ and HCl, *secondary amines* (alkyl or aryl) yield **N-nitrosoamines**, compounds containing the N—N=O group. Many N-nitrosoamines are carcinogenic.

Tertiary amines are not entirely predictable in their reactions with nitrous acid. A tertiary arylamine usually undergoes ring substitution with —NO because of the ring activation by the —NR_2 group. A tertiary alkylamine (and sometimes tertiary arylamines, too) may lose an R group and form an N-nitroso derivative of a secondary amine.

Study Problems

17.14 When *n*-butylamine is treated with a cold aqueous solution of HCl and NaNO$_2$, the following products are obtained: 1-chlorobutane, 2-chlorobutane, 1-butanol, 2-butanol, 1-butene, 2-butene, and nitrogen gas. Suggest a mechanism that accounts for each of these products.

17.15 Suggest a reason why benzenediazonium chloride is more stable than ethyldiazonium chloride.

Section 17.10
Hofmann Elimination

Quaternary ammonium hydroxides (R$_4$N$^+$ OH$^-$) are amine derivatives that are used in structure-determination studies because they undergo elimination reactions to yield alkenes and amines. We will look briefly at how these compounds are prepared, then at their elimination reactions, and finally at their utility in structure studies.

A. Formation of Quaternary Ammonium Hydroxides

When a quaternary ammonium halide is treated with aqueous silver oxide, the quaternary ammonium hydroxide is obtained:

$$2\,R_4N^+\,X^- \;+\; Ag_2O + H_2O \longrightarrow \;2\,R_4N^+\,OH^- \;+\; 2\,AgX\downarrow$$

*a quaternary
ammonium halide*   *a quaternary
ammonium hydroxide*

N,N-dimethylpiperidinium
chloride   *N,N*-dimethylpiperidinium
hydroxide

A quaternary ammonium hydroxide cannot be obtained by the ionic reaction of R$_4$N$^+$ X$^-$ with aqueous NaOH because the reactants and products are all water-soluble ionic compounds. If such a reaction were attempted, a mixture of R$_4$N$^+$ OH$^-$ and R$_4$N$^+$ Cl$^-$ (along with NaOH and NaCl) would result. However, silver hydroxide, which is formed *in situ* from moist silver oxide (Ag$_2$O + H$_2$O → 2AgOH), removes the halide ion as an AgX precipitate. Removal of the AgX by filtration, followed by evaporation of the water, yields the pure quaternary ammonium hydroxide.

B. The Elimination

When a quaternary ammonium hydroxide (as a solid) is heated, an elimination reaction called a **Hofmann elimination** occurs. This reaction is an E2 reaction in which the leaving group is an amine.

E2 transition state

This elimination generally yields the *Hofmann product*, the alkene with *fewer alkyl groups* on the pi-bonded carbons. The formation of the less substituted, less stable alkene can be attributed to steric hindrance in the transition state (Section 5.9E) due to the bulky R_3N^{+-} group.

$$(CH_3)_3\overset{+}{N}-\underset{\underset{CH_2CH_3}{|}}{\overset{\overset{CH_3}{|}}{CH}} \ OH^- \xrightarrow{\text{heat}}$$

sec-butyltrimethylammonium hydroxide

$$(CH_3)_3N + CH_2{=}CHCH_2CH_3 + CH_3CH{=}CHCH_3 + H_2O$$

1-butene (95%)   2-butene (5%)

Study Problem

17.16 Predict the major organic products when the following compounds are heated:

(a) $CH_3CH_2CH_2CH_2\overset{\overset{CH_3}{|}}{\underset{\underset{CH_3}{|}}{N^+}}{-}CH_2CH_3 \ OH^-$ (b) $C_6H_5CH_2CH_2\overset{\overset{CH_3}{|}}{\underset{\underset{CH_3}{|}}{N^+}}{-}CH_2CH_3 \ OH^-$

C. Exhaustive Methylation

Many compounds in nature contain heterocyclic nitrogen rings. A quaternary ammonium hydroxide of a heterocyclic ring undergoes elimination in the same fashion as an open-chain amine. When the nitrogen atom is part of a ring, fragmentation does not occur. Instead, the amino group and alkenyl group both remain in the same molecule.

N,N-dimethylpiperidinium   *N,N*-dimethyl-4-penten-1-amine
hydroxide

In the determination of the structure of a compound, the goal is often to degrade the compound into small identifiable fragments. To show how the Hofmann elimination can do this, let us consider the simple ring system of piperidine. The quaternary ammonium hydroxide is prepared by treatment with CH_3I (S_N2 reaction) followed by reaction with Ag_2O. Heating results in elimination to yield an alkenylamine, as shown in the preceding equation.

Because the product of this elimination still contains an amino group, it can undergo another reaction with CH_3I and Ag_2O to yield a new quaternary ammonium hydroxide. Heating this product results in a new alkene, while the nitrogen is expelled from the molecule as trimethylamine. This series of reactions is called **exhaustive methylation**.

$$\text{(ring)}-N(CH_3)_2 \xrightarrow[\text{(2) Ag}_2\text{O}]{\text{(1) CH}_3\text{I}} \text{(ring)}-\overset{+}{N}(CH_3)_3 \ OH^- \xrightarrow{\text{heat}} \text{(ring)} + (CH_3)_3N$$

trimethylamine

Study Problems

17.17 Fill in the blanks:

(a) $\underset{\underset{H}{N}}{\text{(ring)}}-CH_3 \xrightarrow[\text{(2) Ag}_2\text{O}]{\text{(1) excess CH}_3\text{I}} \underline{\quad} \xrightarrow{\text{heat}} \underline{\quad} \xrightarrow[\substack{\text{(2) Ag}_2\text{O} \\ \text{(3) heat}}]{\text{(1) CH}_3\text{I}} \underline{\quad}$

(b) $\text{(ring)}-NH_2 \xrightarrow[\text{(2) Ag}_2\text{O}]{\text{(1) excess CH}_3\text{I}} \underline{\quad} \xrightarrow{\text{heat}} \underline{\quad}$

17.18 *Coniine*, $C_8H_{17}N$, is a toxic constituent of poison hemlock (*Conium maculatum*), the extract of which is believed to have killed Socrates. The proton nmr spectrum of coniine shows no doublets. Coniine reacts with two equivalents of CH_3I. Reaction with Ag_2O followed by pyrolysis yields an intermediate ($C_{10}H_{21}N$) which, upon further methylation followed by conversion to the hydroxide and pyrolysis, yields trimethylamine, 1,5-octadiene, and 1,4-octadiene. What are the structures of coniine and the intermediate compound?

▼ Summary

An **amine** is a compound that contains a trivalent nitrogen that has from one to three alkyl or aryl groups attached: RNH_2, R_2NH, or R_3N. A compound with four groups attached to the nitrogen is an **amine salt** ($R_3NH^+ \ X^-$) or a **quaternary ammonium salt** ($R_4N^+ \ X^-$).

Amines can be prepared by *substitution reactions*, by *reduction reactions*, or by *rearrangement*. These synthetic reactions are summarized in Table 17.3 (page 753).

Because the nitrogen of an amine has a pair of unshared valence electrons, amines are *weak bases*. The base strength is affected by *hybridization* ($sp^3 > sp^2 > sp$), by *electron-withdrawing groups* (base-weakening), by *electron-releasing groups* (base-strengthening), and by *conjugation* (base-weakening).

$$\text{(ring)}NH > \text{(ring)}N \qquad CH_3{\rightarrow}NH_2 > NH_3$$

hybridization   *electron-release*

$$CH_3CH_2NH_2 > HO \leftarrow CH_2CH_2NH_2 \qquad \langle \ \rangle -NH_2 > \bigcirc -NH_2$$

electron-withdrawal   *conjugation*

Amines undergo reaction with acids to yield amine salts:

$$R_3N \underset{OH^-}{\overset{HX}{\rightleftarrows}} R_3NH^+ \ X^-$$

an amine salt

Most amines are nucleophiles and can displace good leaving groups or can attack carbonyl groups. When primary amines are treated with cold nitrous acid, diazonium salts are formed. Alkyldiazonium salts are unstable, but aryldiazonium salts can be used to prepare a variety of substituted aromatic compounds. The reactions of amines are summarized in Table 17.5.

Quaternary ammonium hydroxides, when heated, lose water and an amine. The least substituted alkene is usually formed. This reaction is called the **Hofmann elimination**.

$$CH_3CH_2\overset{+}{N}(CH_3)_2 \ OH^- \xrightarrow{\text{heat}} CH_2{=}CH_2 + N(CH_3)_2 \qquad + H_2O$$
$$\underset{CH_2CH_2CH_3}{|} \qquad\qquad\qquad \underset{CH_2CH_2CH_3}{|}$$

Table 17.5 Summary of reactions of amines

Reaction		Product	Section Reference
$R_3N + R'X$	$\longrightarrow$ $R_3\overset{+}{N}R' \ X^-$	amine salt or quaternary ammonium salt	17.5A
$R_2NH + R'\overset{O}{\overset{\|}{C}}Cl$	$\longrightarrow$ $R_2N\overset{O}{\overset{\|}{C}}R'$	**amide**[a]	15.3C, 17.8
$1° \ RNH_2 + R'_2C{=}O$	$\xrightarrow{H^+}$ $RN{=}CR'_2$	**imine**	13.6A
$2° \ R_2NH + R'_2CH\overset{O}{\overset{\|}{C}}R'$	$\xrightarrow{H^+}$ $R'_2C{=}\overset{NR_2}{\overset{\|}{C}}R'$	**enamine**[b]	13.6B
$ArNH_2$	$\xrightarrow[0°]{\overset{NaNO_2}{HCl}}$ $ArN_2^+ \ Cl^-$	**aryldiazonium salt**[c,d]	12.3, 17.9
$R_2CH\overset{NR'_2}{\overset{\|}{C}}R_2$	$\xrightarrow[\text{(3) heat}]{\overset{\text{(1) } CH_3I}{\text{(2) } Ag_2O, H_2O}}$ $R_2C{=}CR_2$	**alkene**	17.10

[a] Other routes to amides are the similar reactions of amines with acid anhydrides (Section 15.4C) and with esters (Section 15.5C).

[b] Enamines can be converted to α-substituted aldehydes or ketones (see Section 16.5).

[c] Aryldiazonium salts can be converted to aryl halides, nitriles, etc. (see Section 12.3).

[d] Secondary amines (alkyl or aryl) yield *N*-nitrosoamines when treated with HNO_2.

Study Problems*

17.19 Write formulas for the following compounds:

(a) phenylalanine (2-amino-3-phenylpropanoic acid), an amino acid constituent of proteins

(b) *N*-methyl-1-cyclohexyl-2-propylamine, a vasoconstrictor

(c) *N,N*-dimethylaniline, a solvent and industrial intermediate

(d) 1,2-propanediamine, used with $CuSO_4$ to test for mercury

17.20 Name the following compounds:

(a) $C_6H_5CH_2NCH_3$
   $\quad\quad\quad\quad | $
   $\quad\quad\quad CH_2CH_3$

(b)

(c) $(C_6H_5)_2CHN(CH_2CH_3)_2$

(d) $CH_3CHCH_2CH_2CH$
   $\quad\quad |\quad\quad\quad\quad\| $
   $\quad\quad NH_2\quad\quad\quad O$

17.21 Which of the following structures has enantiomers, ge   isomers, both, or neither? [*Hint:* In (f), consider the hybridization of the N and t   tant geometry.]

(a) $C_6H_5CHNH_2$
   $\quad\quad\quad | $
   $\quad\quad CH_2CH_3$

(b) $CH_3NCH_2CH_3$
   $\quad\quad\quad | $
   $\quad\quad\quad C_6H_5$

(c)
```
      CH3
       \  +
        N        Cl⁻
       /
      CH2CH3
```

(d)
```
        CH3
        /
   [ring]N⁺       Cl⁻
        \
  |      CH2CH3
  CH3
```

(e) $CH_3NCH_2CH_3$
   $\quad\quad | $
   $\quad\quad C_6H_5$
   (with O^- on N⁺)

(f) $C_6H_5CH{=}NO$

17.22 Explain the following observations:

(a) Cyclohexylamine is more water soluble than cyclohexanol.

(b) Trimethylamine has a lower boiling point than dimethylamine.

(c) Ethylamine has a higher boiling point than dimethylamine.

17.23 Which of the following species can act as a nucleophile? Explain.

(a) $(CH_3)_2NH$ (b) $(CH_3)_3N$ (c) $H_2N{-}NH_2$

(d) [ring]N (e) $HN\quad NCH_3$ (f) [ring]$N{-}CH_3$ (with ⁺)

17.24 Suggest syntheses for the following compounds, starting with or a   al gen compounds or alcohols:

(a) [cyclopentyl]$-CH_2NH_2$

(b) [cyclohexyl with] OH and CH_2NH_2

(c) $HO_2CCH_2CHCO_2H$
   $\quad\quad\quad\quad\quad | $
   $\quad\quad\quad\quad\quad NH_2$

(d) $(CH_3)_2CHNH_2$

(e) [cyclohexyl]$-NHCH_2CH_2CH_3$

(f) $CH_3CH_2CH_2CH_2N(CH_3)_2$

---

* For information concerning the organization of the *Study Problems* and *Additional Problems*, see the *Note to student* on page 39.

17.25 How would you convert 1-pentanol to: **(a)** *n*-pentylamine (free of higher alkylated amines)? **(b)** *n*-hexylamine? **(c)** *n*-butylamine?

17.26 Suggest a way to make each of the following conversions: **(a)** benzene to aniline; **(b)** benzamide to aniline; **(c)** aniline to acetanilide ($C_6H_5NHCOCH_3$); **(d)** (*R*)-2-butanol to (*S*)-2-butylamine; **(e)** toluene to benzylamine; **(f)** acetic acid to acetamide.

17.27 Which is more basic? **(a)** aniline or *p*-bromoaniline; **(b)** trimethylamine or tetramethylammonium hydroxide; **(c)** *p*-nitroaniline or 2,4-dinitroaniline; **(d)** ethylamine or ethanolamine ($HOCH_2CH_2NH_2$); **(e)** *p*-toluidine (*p*-methylaniline) or *p*-(trichloromethyl)aniline. Explain your answers.

17.28 Complete the following equations:

(a)

(b)

(c)

(d) $(CH_3)_3NH^+$ +

(e)

17.29 List each of the following groups of cations in order of increasing acidity (weakest acid first):

(a) (1)   (2)   (3)

(b) (1)   (2)   (3)

(c) (1)   (2)   (3)

(d) (1) H_3O^+ (2)

17.30 Name the following amine salts:

(a)   **(b)**

(c) $(CH_3)_3\overset{+}{N}CH_2CH_2Cl\ Cl^-$

a height shortener for wheat

17.31 Predict the major product of the reaction of pyrrolidine (page 740) with: **(a)** benzoyl chloride; **(b)** acetic anhydride; **(c)** CH_3I (excess); **(d)** phthalic anhydride; **(e)** benzenesulfonyl chloride ($C_6H_5SO_2Cl$); (**f**) acetyl chloride, followed by $LiAlH_4$ (then hydrolysis); (**g**) cold nitrous acid; (**h**) dilute HCl; (**i**) acetone + H^+.

17.32 Suggest techniques for separating the following mixtures:

(**a**) cyclohexanol, cyclohexylamine, and cyclohexanecarboxylic acid

(**b**) hexanamide and *n*-hexylamine

17.33 Predict the major organic products when each of the following compounds is heated:

(**a**) OH^- (**b**) OH^-

17.34 The following compound is a venomous constituent from cryptic thief ants collected in Florida. Predict the major products of its exhaustive methylation.

17.35 An amine ($C_8H_{15}N$) is subjected to two stages of exhaustive methylation (two rounds of treatment with CH_3I, Ag_2O, and heat). The final products are trimethylamine and the following diene:

What is a likely structure for the original amine?

Additional Problems

17.36 Complete the following equations:

(**a**) $(CH_3CH_2)_2NH + ClCH_2CH_2OH \xrightarrow{\text{NaOH}}$

(**b**) $CH_3NH_2 \xrightarrow[\text{(2) } H^+]{\text{(1) excess } ClCH_2CO_2^- \ Na^+, \text{NaOH}}$

(**c**) $\xrightarrow[\text{(2) } H_2O, H^+]{\text{(1) } LiAlH_4}$

(**d**) $CH_3NHCO_2C_2H_5 \xrightarrow[\text{HCl}]{\text{NaNO}_2}$

(**e**) $C_6H_5CH_2\overset{\overset{\displaystyle NH_2}{|}}{C}HCO_2CH_3 \xrightarrow[\text{CH}_3\text{CH}_2\text{OH}]{C_6H_5CHO, (CH_3CH_2)_3N}$

(**f**) + 1 HCl $\longrightarrow$

nicotine

(g) 2 [pyrrolidine structure with N-H] + BrĊHCO$_2$CH$_2$CH$_3$ (with CH$_3$ on carbon) ⟶

17.37 Predict the organic products when 1.0 mol of the local anesthetic lidocaine (Xylocaine) is treated with: **(a)** 1.0 mol of cold dilute H$_2$SO$_4$; **(b)** excess dilute HCl and heat; **(c)** excess dilute NaOH and heat.

$$(CH_3CH_2)_2NCH_2\overset{O}{\overset{\|}{C}}NH-\text{[2,6-dimethylphenyl ring with CH}_3\text{ groups]}$$

lidocaine

17.38 What chemical reaction could you use to distinguish between:
   (a) aniline and n-hexylamine?
   (b) n-octylamine and octanamide?
   (c) triethylammonium chloride and tetraethylammonium chloride?

17.39 Suggest synthetic routes to the following compounds:
   (a) isoleucine (2-amino-3-methylpentanoic acid)
   (b) β-phenylethylamine from toluene
   (c) 2-methyl-1-phenyl-1,3-pentanedione from benzoic acid
   (d) aniline from benzoic acid
   (e) the plant product *ubine* (2-dimethylamino-1-phenylethanol) from an alkene

   (f) H$_2$NCH$_2$—[benzene ring]—CH$_2$NH$_2$ from toluene

   (g) 3,5-dimethoxyaniline from a nitrile

17.40 Suggest synthetic routes to the following compounds:
   (a) the pain reliever *phenacetin* (p-ethoxyacetanilide) from p-nitrophenol. (For the structure of acetanilide, see Table 11.1, page 463.)
   (b) methyl orange (page 866) from substituted benzenes

   (c) [cyclohexane ring]—C(=O)—[cyclohexanone ring with =O]   from compounds containing six or fewer carbon atoms

17.41 The treatment of (cyclopentylmethyl)amine with nitrous acid resulted in a 76% yield of cyclohexanol. Another alcohol and three alkenes were also present in the product mixture.

   (a) Give a plausible mechanism for the formation of cyclohexanol.
   (b) What are the likely structures of the other products?

17.42 A chemist attempted to carry out a Hofmann rearrangement of butanamide with bromine and potassium hydroxide in methanol, rather than in water. Instead of n-propylamine, the chemist obtained a carbamate, CH$_3$CH$_2$CH$_2$NHCO$_2$CH$_3$. Explain how this product was formed.

17.43 When the following compound is subjected to catalytic hydrogenation, a compound with the formula C$_9$H$_{17}$N is produced. What is the structure of this product?

$$CH_2CH_2CN$$

(structure: cyclohexanone ring with CH$_2$CH$_2$CN substituent)

17.44 A tertiary amine is oxidized by peroxides such as H_2O_2 to an **amine oxide**, a compound containing the —NO group. An amine oxide with a β hydrogen undergoes elimination when heated. (This reaction is called a **Cope elimination**.) Suggest a mechanism for the following elimination reaction of an amine oxide.

$$C_6H_5CH-CHN(CH_3)_2 \longrightarrow \underset{H_3C}{\overset{C_6H_5}{}}C=C\underset{CH_3}{\overset{H}{}} + (CH_3)_2NOH$$

$$\underset{(S)}{CH_3}\ \ \underset{(R)}{CH_3}$$

(E)
major alkene

17.45 Suggest a mechanism for the following conversion:

$$C_6H_5CH_2CHCO_2H \xrightarrow[\text{(2) }(CH_3C)_2O]{\text{(1) NaOH, }C_6H_5CCl}$$

$$\overset{+}{N}H_3\ Cl^-$$

phenylalanine
hydrochloride

(product structure with $C_6H_5CH_2$, ring containing N and O, and C_6H_5)

(95%)

17.46 4-Chloro-1-phenyl-1-butanone was heated with 1,2-ethanediol, *p*-toluenesulfonic acid, and benzene. The product A was heated with potassium phthalimide in dimethylformamide to yield B in 57% overall yield. Finally, B was heated with KOH in ethanol to yield C ($C_{10}H_{11}N$). What are the structures of A, B, and C?

17.47 A chemist treated 1-bromobutane with ammonia and isolated products A and B. When A was treated with acetic anhydride, C was obtained. When B was treated with acetic anhydride, D was obtained. The infrared spectra of C and D are shown in Figure 17.2. Identify A, B, C, and D.

Figure 17.2 Infrared spectra for Problem 17.47 (*continued on page 770*).

Figure 17.2 (*continued*) Infrared spectra for Problem 17.47.

17.48 The infrared spectrum for Compound A ($C_8H_{11}N$) is given in Figure 17.3. A is soluble in dilute acid. Oxidation of A with hot $KMnO_4$ yields benzoic acid. What are the two possible structures for A? How could you distinguish these two possibilities by nmr spectroscopy?

Figure 17.3 Infrared spectrum for Compound A in Problem 17.48.

17.49 A compound $C_{10}H_{13}NO$ (A) has the proton nmr spectrum shown in Figure 17.4. A is insoluble in dilute aqueous acid. Heating A with aqueous NaOH, followed by acidification, results in two compounds: acetic acid and an amine salt. The free amine (B) from this salt has the formula $C_8H_{11}N$.

Compound B was subjected to high-pressure hydrogenation to yield C ($C_8H_{17}N$). Compound C was subjected to one stage of exhaustive methylation (in which it took up 2 mols of CH_3I) and yielded trimethylamine and 3-methylcyclohexene. What are the structures of A, B, and C?

Figure 17.4 ¹H nmr spectrum of Compound A in Problem 17.49.

Polycyclic and Heterocyclic Aromatic Compounds

In Chapters 11 and 12, we discussed benzene and substituted benzenes. However, benzene is but one member of a large number of aromatic compounds. Many other aromatic compounds can be grouped into two classes: **polycyclic compounds and heterocyclic compounds**. The polycyclic aromatic compounds are also referred to as *polynuclear*, *fused-ring*, or *condensed-ring* aromatic compounds. These aromatic compounds are characterized by rings that jointly share carbon atoms and by a common aromatic pi cloud.

Some polycyclic aromatic compounds:

naphthalene

anthracene

phenanthrene

top view   side view

Figure 18.1 The structure of graphite.

The polycyclic aromatic hydrocarbons and most of their derivatives are solids. Naphthalene has been used as mothballs and flakes, and derivatives of naphthalene are used in motor fuels and lubricants. The most extensive use made of the polycyclic aromatics is that of synthetic intermediates, for example, in the manufacture of dyes. (See Section 21.4B.)

Graphite is one of the more interesting polycyclic compounds. The structure of graphite consists of planes of fused benzene rings (Figure 18.1). The distance (3.5 Å) between any pair of planes is believed to be the thickness of the pi system of benzene. The "slipperiness" of graphite is due to the ability of these planes to slide across each other. Because of this property, graphite is a valuable lubricant that can be used even in outer space where ordinary oils and greases would solidify. Because of its mobile pi electrons, graphite can conduct electricity and finds use where an inert electrode is needed. Flashlight batteries, for example, contain graphite electrodes.

A **heterocyclic compound** is a cyclic compound in which the ring atoms are of two more different elements. Heterocyclic rings can be aromatic, just as carbon rings can. Approximately a third of the organic chemical literature deals with heterocyclic compounds. Many **alkaloids**, such as morphine (Section 19.3); the **nucleic acids**, the carriers of the genetic code; and many other compounds of biological importance are examples of heterocyclic compounds.

Two structurally similar heterocycles:

tryptophan

an amino acid

psilocin

a hallucinogen in Psilocybe mushrooms

Section 18.1
Nomenclature of Polycyclic Aromatic Compounds

The ring systems of common polycyclic aromatic compounds have individual names. Unlike the numbering of benzene or a cycloalkane ring, which starts at the position of a substituent, the numbering of a polycyclic ring is fixed by convention and does not change with the position of a substituent.

naphthalene   anthracene   phenanthrene

The position of a substituent in a monosubstituted naphthalene is often designated by a Greek letter. The positions adjacent to the ring-junction carbons are called α positions, while the next positions are β positions. By this system, 1-nitronaphthalene is called α-nitronaphthalene, while 2-nitronaphthalene is called β-nitronaphthalene. Naphthalene itself has four equivalent α positions and four equivalent β positions. (Only number designations are used in the anthracene and phenanthrene systems.)

1-nitronaphthalene   2-nitronaphthalene
(α-nitronaphthalene)   (β-nitronaphthalene)

▼ ## Section 18.2
Bonding in Polycyclic Aromatic Compounds

For a monocyclic ring system to be aromatic, it must meet three criteria:

1. Each atom in the ring system must have a *p* orbital available for bonding.
2. The ring system must be planar.
3. There must be $(4n + 2)$ pi electrons in the ring system (Hückel rule).

These criteria were discussed in Section 11.7. The aromatic system of a polycyclic aromatic compound must also contain atoms with *p* orbitals available for bonding, and the entire aromatic ring system must be planar. The Hückel rule, which was devised for monocyclic systems, also is applicable to polycyclic systems in which the sp^2-hybridized carbons are peripheral, or on the outside edge of the ring system. In polycyclic compounds, the number of pi electrons is easily counted when Kekulé formulas are used.

10 pi electrons   *14 pi electrons*   *14 pi electrons*
($n = 2$)   ($n = 3$)   ($n = 3$)

Like benzene, the polycyclic aromatic systems are more stable than the corresponding hypothetical polyenes with localized pi bonds. The energy differences between the hypothetical polyenes and the real compounds (that is, the *resonance energies*) have been calculated from heats of combustion and hydrogenation data.

resonance energy
(kcal/mol):   36   61   84   92

The resonance energy for a polycyclic aromatic compound is less than the sum of the resonance energies for a comparable number of benzene rings. Although the resonance energy for benzene is 36 kcal/mol, that for naphthalene is only 61 kcal/mol (about 30 kcal/mol for each ring).

In benzene, all carbon–carbon bond lengths are the same. This fact leads us to believe that there is an equal distribution of pi electrons around the benzene ring. In the polycyclic aromatic compounds, the carbon–carbon bond lengths are *not* all the same. For example, the distance between carbons 1 and 2 (1.36 Å) in naphthalene is smaller than the distance between carbons 2 and 3 (1.40 Å).

C—C in ethane: 1.54 Å

C=C in ethylene: 1.34 Å

C—C in benzene: 1.40 Å

From these measurements, we conclude that there is *not* an equal distribution of pi electrons around the naphthalene ring. From a comparison of bond lengths, we would say that the carbon 1–carbon 2 bond of naphthalene has more double-bond character than the carbon 2–carbon 3 bond. The resonance structures for naphthalene also indicate that the carbon 1–carbon 2 bond has more double-bond character.

two out of three resonance structures
show a carbon 1–carbon 2 double bond

Because all the carbon–carbon bonds in naphthalene are not the same, many chemists prefer to use Kekulé-type formulas for this compound instead of using circles to represent the pi cloud. We will use Kekulé-type formulas in our discussion of the reactions of naphthalene.

Phenanthrene shows similar differences among its bonds. The double-bond character of the 9,10-bond of phenanthrene is particularly evident in its chemical reactions. These positions of the phenanthrene ring system undergo addition reactions that are typical of alkenes but are not typical for benzene.

double-bond character

Study Problem

18.1 Write the Kekulé resonance structures for phenanthrene. On the basis of these structures, explain why the 9,10-bond has double-bond character.

▼ Section 18.3
Oxidation of Polycyclic Aromatic Compounds

The polycyclic aromatic compounds are more reactive toward oxidation, reduction, and electrophilic substitution than is benzene. The reason for the greater reactivity is that the polycyclic compounds can undergo reaction at one ring and still have one or more intact benzenoid rings in the intermediate and in the product. Less energy is required to overcome the aromatic character of a single ring of the polycyclic compounds than is required for benzene.

Benzene is not easily oxidized; however, naphthalene can be oxidized to products in which much of the aromaticity is retained. Phthalic anhydride is prepared commercially by the oxidation of naphthalene. This reaction probably proceeds by way of *o*-phthalic acid.

Under controlled conditions, 1,4-naphthoquinone can be isolated from an oxidation of naphthalene (although yields are usually low).

Anthracene and phenanthrene can also undergo oxidation to quinones:

anthracene

9,10-anthraquinone

phenanthrene

9,10-phenanthraquinone (48%)

Study Problem

18.2 From the following observations, tell whether chromate oxidation of naphthalene derivatives involves an initial *electrophilic attack* or *nucleophilic attack*. Explain.

1-nitronaphthalene

3-nitro-1,2-phthalic acid

1-naphthylamine

o-phthalic acid

▼ Section 18.4
Reduction of Polycyclic Aromatic Compounds

Unlike benzene, the polycyclic compounds can be partially hydrogenated without heat and pressure, or they can be reduced with sodium and ethanol.

tetralin

9,10-dihydroanthracene

The partially reduced ring systems still contain one or more benzenoid rings. Most of the aromatic character of the original ring systems has been retained in these partially reduced products. To hydrogenate the polycyclic aromatics completely would, of course, require heat and pressure, just as it does for benzene.

naphthalene

tetralin

decalin

Study Problem

18.3 Predict the product of the reaction of phenanthrene with sodium and ethanol.

Section 18.5
Electrophilic Substitution Reactions of Naphthalene

The polycyclic aromatic ring systems are more reactive toward electrophilic attack than is benzene. Naphthalene undergoes electrophilic aromatic substitution reactions predominantly at the 1-position. The reasons for the enhanced reactivity and for this position of substitution will be discussed shortly.

1-bromonaphthalene

1-nitronaphthalene

1-naphthalenesulfonic acid

1-acetylnaphthalene

Anthracene, phenanthrene, and larger fused-ring compounds are even more reactive than naphthalene toward electrophilic substitution. However, these reactions are not as important as those of naphthalene because mixtures of isomers (which are often difficult to separate) are obtained. Phenanthrene, for example, undergoes mononitration at each available position to yield five nitrophenanthrenes.

A. Position of Substitution of Naphthalene

The mechanism for naphthalene substitution is similar to that for benzene substitution. Let us look at the stepwise bromination reaction to see why substitution at the 1-position is favored and why this reaction occurs more readily than the bromination of benzene.

1-Substitution (favored):

intermediate

Resonance structures for the 1-substitution intermediate:

major contributors

The resonance structures of the intermediate for substitution at the 1-position show four contributors in which the benzene ring is intact. Because of aromatic resonance-stabilization, these structures are of lower energy than the other resonance structures and are major contributors to the real structure of the intermediate. This is the reason that naphthalene undergoes electrophilic substitution more readily than benzene. For benzene to go to a benzenonium ion requires the loss of aromaticity, about 36 kcal/mol. For naphthalene to go to its intermediate requires only partial loss of aromaticity, about 25 kcal/mol (the difference in resonance energy between naphthalene and benzene). Because of the lower E_{act} leading to the intermediate, the rate of bromination of naphthalene is faster than that of benzene.

requires 36 kcal/mol to destroy aromaticity

requires only 25 kcal/mol

Why is 1-substitution favored over 2-substitution for naphthalene? Inspect the resonance structures for the intermediate leading to 2-substitution:

2-Substitution (not favored):

intermediate

Resonance structures for the 2-substitution intermediate:

major contributors

The intermediate leading to the 2-substituted isomer has only two contributing resonance structures in which a benzenoid ring is intact, while the 1-intermediate shows four such resonance structures. The 1-intermediate is more stabilized by resonance, and its transition state is of lower energy. For this reason, the E_{act} is lower and the 1-intermediate is formed faster.

The sulfonation of naphthalene, which is a reversible reaction (Section 11.9F), is more complex than bromination. At 80°, the expected 1-naphthalenesulfonic acid is the product. However, at higher temperatures (160–180°), the product is 2-naphthalenesulfonic acid. At low temperatures, the reaction is under **kinetic control**—that is, the relative rates of reaction determine the product ratio. At high temperatures, the reaction is under **thermodynamic**, or **equilibrium, control**—the relative stabilities of the products determine the product ratio.

We have already seen why the 1-substitution of naphthalene is faster than 2-substitution. At less than 80°, the rate of formation of either naphthalenesulfonic acid is relatively slow. The reaction proceeds through the lower-energy 1-intermediate just as it does for bromination.

at 80°:   1-naphthalenesulfonic acid (91%)

2-naphthalenesulfonic acid

Even though 1-naphthalenesulfonic acid is formed at low temperatures, this isomer is less stable than the 2-isomer because of repulsions between the —SO₃H group and the hydrogen at position 8.

1-naphthalenesulfonic acid

less stable

2-naphthalenesulfonic acid

more stable

At a higher temperature, the rates of both forward reactions and the rates of both reverse reactions are all increased. Although the 1-product may be formed more readily, it can revert quickly to naphthalene. The 2-product is formed more slowly, but the rate of its reverse reaction is even slower because the 2-product is more stable and of lower energy. At higher temperatures, the 2-product accumulates in the reaction mixture and is the observed product. (See Figure 18.2.)

Figure 18.2 Energy diagram for the sulfonation of naphthalene.

Study Problems

18.4 On the basis of what you learned in Chapter 11 about activating and deactivating groups and the position(s) to which they direct an incoming substituent, predict the major organic products of aromatic nitration of the following compounds:

18.5 Write equations showing how you could prepare the following compounds from naphthalene.

Section 18.6
Nomenclature of Aromatic Heterocyclic Compounds

Because of their widespread occurrence in nature, the aromatic heterocycles are of more general interest to chemists than are the polycyclic compounds containing only carbon atoms in their rings. Like the polycyclic aromatic compounds, the aromatic heterocycles generally have individual names. The names and structures of some of the more important members of this class of compounds are listed in Table 18.1.

The numbering of three representative heterocycles follows:

pyridine   thiazole*   imidazole*

When a heterocycle contains only one heteroatom, Greek letters may also be used to designate ring position. The carbon atom adjacent to the heteroatom is the α carbon. The next carbon is the β carbon. The next carbon in line, if any,

---

* A circle in the ring is not a proper representation of an aromatic pi cloud in a five-membered heterocycle. How the aromatic pi cloud is formed in these compounds will be discussed in Section 18.9.

Table 18.1 Some important aromatic heterocyclic compounds

Structure	Name	Structure	Name
	pyrrole		pyrimidine
	furan		quinoline
	thiophene		isoquinoline
	imidazole		indole
	thiazole		purine
	pyridine		

is γ. Pyridine has two α positions, two β positions, and one γ position. Pyrrole has two α and two β positions.

pyridine

pyrrole

Section 18.7
Pyridine, a Six-Membered Aromatic Heterocycle

Of the common six-membered heterocycles, only the nitrogen heterocycles are stable aromatic compounds.

pyridine
or
pyridine

aromatic

sp^3-hybridized carbon

pyran

not aromatic

Pyridine has a structure similar to that of benzene. Pyridine contains a planar, six-membered ring consisting of five carbons and one nitrogen. Each of these ring atoms is sp^2-hybridized and has one electron in a *p* orbital that

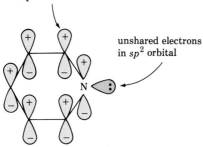

Each ring atom contributes
one pi electron

unshared electrons
in sp^2 orbital

Figure 18.3 The lowest-energy π molecular orbital in pyridine.

contributes to the aromatic pi cloud (six pi electrons). Figure 18.3 shows the lowest-energy π molecular orbital of pyridine.

Note the differences between benzene and pyridine. Benzene is symmetrical and nonpolar, but pyridine contains an electronegative nitrogen and therefore is *polar*.

E^+ *attack on ring*
not favored

dipole moment $\mu = 2.26$ D

Because the nitrogen is more electronegative than carbon, the rest of the pyridine ring is electron-deficient. An electron-deficient ring means that the carbon atoms in the ring carry a *partial positive charge*. A pyridine ring therefore has a low reactivity toward electrophilic substitution compared with benzene. Besides the electronegative nitrogen rendering the ring partially positive, pyridine forms a cation with many Lewis acids. Cation formation renders the ring even more electron-deficient.

e^-*-deficient* $\longrightarrow$ $\qquad$ $\longleftarrow$ *more* e^-*-deficient*

$^-$FeBr$_3$

Pyridine does not undergo Friedel–Crafts alkylations or acylations, nor does it undergo coupling with diazonium salts. Bromination proceeds only at high temperatures in the vapor phase and probably proceeds by a free-radical path. When substitution does occur, it occurs at the 3-position.

$\xrightarrow[300°]{Br_2}$

pyridine $\qquad$ 3-bromopyridine (37%) $\qquad$ 3,5-dibromopyridine (26%)

Another major difference between pyridine and benzene is that the nitrogen in pyridine contains an unshared pair of electrons in an sp^2 orbital. This pair of electrons can be donated to a hydrogen ion. Like amines, pyridine is basic. The basicity of pyridine ($pK_b = 8.75$) is less than that of aliphatic amines ($pK_b = \sim4$) because the unshared electrons are in an sp^2 orbital instead of an sp^3 orbital. Nonetheless, pyridine undergoes many reactions typical of amines.

Like that of benzene, the aromatic ring of pyridine is resistant to oxidation. Side chains can be oxidized to carboxyl groups under conditions that leave the ring intact.

A. Nucleophilic Substitution on the Pyridine Ring

When a benzene ring is substituted with electron-withdrawing groups such as —NO_2, *aromatic nucleophilic substitutions* can take place (Section 12.4).

The nitrogen in pyridine withdraws electron density from the rest of the ring. It is not surprising, then, that nucleophilic substitution also occurs with pyridine. Substitution proceeds most readily at the 2-position, followed by the 4-position, but not at the 3-position.

2-bromopyridine   2-aminopyridine

4-chloropyridine   4-aminopyridine

Let us look at the mechanisms for substitution at the 2- and 3-positions to see why the former reaction proceeds more readily.

2-Substitution (favored):

resonance structures for intermediate

3-Substitution (not favored):

resonance structures for intermediate

The intermediate for 2-substitution is especially stabilized by the contribution of the resonance structure in which nitrogen carries the negative charge. Substitution at the 3-position goes through an intermediate in which the nitrogen cannot help stabilize the negative charge. The intermediate for 3-substitution is of higher energy; the rate of reaction going through this intermediate is slower.

Study Problem

18.6 Give the resonance structures for the intermediate in the reaction of 4-chloropyridine with ammonia.

Benzene itself (with no substituents) does not undergo nucleophilic substitution. This reaction does occur with pyridine if an extremely strong base such as a lithium reagent or amide ion (NH_2^-) is used.

$$\text{benzene} + \ddot{:}NH_2 \xrightarrow{100°} \text{no reaction}$$

$$\text{pyridine} + \ddot{:}NH_2 \xrightarrow[-H_2]{100°} \xrightarrow{H_2O} + OH^-$$

2-aminopyridine (70%)

$$\text{pyridine} + \text{phenyl-Li} \xrightarrow{100°} + LiH$$

2-phenylpyridine (50%)

In the reaction of pyridine with NH_2^-, the initial product is the anion of 2-aminopyridine. The free amine is obtained by treatment with water.

Step 1 (attack of NH_2^- and loss of H_2):

resonance structures for intermediate

the anion of
2-aminopyridine

Step 2 (treatment with H_2O):

$$+ H_2O \longrightarrow + OH^-$$

2-aminopyridine

▼ Section 18.8
Quinoline and Isoquinoline

Quinoline is a fused-ring heterocycle that is similar in structure to naphthalene, but with a nitrogen at position 1. Isoquinoline is the 2-isomer. (Note that the numbering of isoquinoline starts at a carbon, not at the nitrogen.)

quinoline   isoquinoline

Both quinoline and isoquinoline contain a pyridine ring fused to a benzene ring. The nitrogen ring in each of these two compounds behaves somewhat like the pyridine ring. Both quinoline and isoquinoline are weak bases (pK_b = 9.1 and 8.6, respectively). Both compounds undergo *electrophilic substitution* more easily than pyridine, but in positions 5 and 8 (on the benzenoid ring, not on the deactivated nitrogen ring). The positions of substitution are determined by intermediates similar to those in naphthalene substitution reactions.

quinoline   →   5-nitroquinoline (52%)   +   8-nitroquinoline (48%)

isoquinoline   →   5-nitroisoquinoline (90%)   +   8-nitroisoquinoline (10%)

Study Problem

18.7 Draw resonance structures of the intermediates for nitration at the 5- and 6-positions of quinoline to show why 5-nitroquinoline is formed preferentially.

Like pyridine, the nitrogen-containing ring of either quinoline or isoquinoline can undergo **nucleophilic substitution**. The position of attack is α to the nitrogen in either ring system, just as it is in pyridine.

quinoline   (1) NH_2^- (2) H_2O 2-aminoquinoline

isoquinoline   (1) CH_3Li (2) H_2O 1-methylisoquinoline

▼ Section 18.9
Pyrrole, a Five-Membered Aromatic Heterocycle

For a five-membered-ring heterocycle to be aromatic, the heteroatom must have *two electrons* to donate to the aromatic pi cloud. Pyrrole, furan, and thiophene all meet this criterion and therefore are aromatic. We will emphasize pyrrole in our discussion of five-membered aromatic heterocycles because it is typical in terms of both bonding and chemical reactivity.

pyrrole   furan   thiophene

Unlike pyridine and the amines, pyrrole ($pK_b = \sim 14$) is *not basic* under the usual conditions.

$$+ \; H^+ \longrightarrow \text{no stable cation}$$

To see why pyrrole is not basic, we must consider the electronic structure of pyrrole. We know that pyrrole is aromatic because (1) its heat of combustion is about 25 kcal/mol less than that calculated for a diene structure; (2) pyrrole undergoes aromatic substitution reactions; and (3) the protons of pyrrole absorb in the aromatic region of the nmr spectrum (Figure 18.4). (Recall that aryl protons absorb downfield from most other protons because they are de-shielded by the effects of the ring current. See Section 9.7B.)

In a five-membered ring, the minimum number of pi electrons needed for aromaticity is six ($4n + 2$, where $n = 1$). The four carbons of pyrrole each contribute one electron; therefore, the nitrogen atom of pyrrole must contribute *two* electrons (not just one as it does in pyridine). In addition to contributing two electrons to the pi molecular orbitals, the nitrogen in pyrrole shares three electrons in sigma bonds to two ring carbons and to a hydrogen. Consequently, all five bonding electrons of the nitrogen are used in

Figure 18.4 ^{1}H nmr spectrum of pyrrole, C_4H_5N. (The NH absorption is a low, broad band near 8 ppm.)

p orbitals in π_1

Figure 18.5 **The lowest-energy molecular orbital in pyrrole.**

bonding. The pyrrole nitrogen does not have unshared electrons and is not basic. Figure 18.5 shows the orbital picture for the lowest-energy π molecular orbital.

Study Problem

18.8 If pyrrole did form a cation when treated with HCl, what would its structure be? (Show the *p* orbitals.) Would the pyrrole cation be aromatic or not?

Because the nitrogen atom in pyrrole contributes two electrons to the aromatic pi cloud, the nitrogen atom is electron-deficient and therefore not basic. The pyrrole ring, however, has six pi electrons for only five ring atoms. The ring is electron-rich and therefore partially negative. The dipole moments reflect this: nitrogen is usually the negative end of a dipole, but the nitrogen in pyrrole is the positive end of the molecule.

μ: 2.26 D   1.81 D

A. Electrophilic Substitution on the Pyrrole Ring

Because the ring carbons are the negative part of the pyrrole molecule, these carbons are *activated toward electrophilic attack, but deactivated to nucleophilic attack.* (This reactivity is opposite to that of pyridine.) The principal chemical characteristic of pyrrole and the other five-membered aromatic heterocycles is the ease with which they undergo electrophilic substitution.

$$\underset{\text{pyrrole}}{\overset{\displaystyle\text{(pyrrole ring, N—H)}}{}}$$

$\xrightarrow[\text{pyridine}]{SO_3}$ (pyrrole ring)—SO_3H

2-pyrrolesulfonic acid (90%)

$\xrightarrow[\substack{(CH_3CO)_2O \\ 5°}]{HNO_3}$ (pyrrole ring)—NO_2

2-nitropyrrole (80%)

Electrophilic substitution occurs principally at the 2-position of the pyrrole ring. A look at the resonance structures for the intermediates of 2- and 3-nitration shows why. Three resonance structures can be drawn for the 2-intermediate, while only two can be drawn for the 3-intermediate. There is greater delocalization of the positive charge in the intermediate leading to 2-nitration than there is in that leading to 3-nitration.

2-Nitration (favored):

resonance structures for the intermediate

3-Nitration (not favored):

resonance structures for the intermediate

Study Problem

18.9 Predict the major organic monosubstitution products:

(a) furan + $(CH_3CO)_2O$ $\xrightarrow{BF_3}$

(b) thiophene + H_2SO_4 $\xrightarrow{25°}$

(c) pyrrole + $C_6H_5N_2^+$ Cl^- $\xrightarrow{25°}$

(d) furan + Br_2 $\xrightarrow[25°]{dioxane}$

B. Porphyrins

The **porphyrin ring system** is a biologically important unit found in *heme*, the oxygen-carrying component of hemoglobin; in *chlorophyll*, a plant pigment; and in the *cytochromes*, compounds involved in utilization of O_2 by animals. Structures of these compounds are shown in Figure 18.6.

Note that the porphyrin ring system is composed of four pyrrole rings joined by $=CH-$ groups. The entire ring system is aromatic.

highlighted hydrogen atoms can be replaced by metal ions

porphyrin

Figure 18.6 Some biologically important porphyrins.

The pyrrole hydrogens in the porphyrin ring system can be replaced by a variety of metal ions. The product is a **chelate** (Greek, *chele,* "crab's claw"), a compound or ion in which a metal ion is held by more than one bond from the original molecule. With porphyrin, the chelate is planar around the metal ion, and resonance results in four equivalent bonds from the nitrogen atoms to the metal.

*planar and
resonance-stabilized*

▼ Summary

Three common **polycyclic aromatic compounds** are **naphthalene**, **anthracene**, and **phenanthrene**. These compounds exhibit less resonance energy per ring than benzene, and certain C—C bonds have more double-bond character than others. These compounds may be oxidized to quinones or may be partially hydrogenated, as shown in Table 18.2.

Naphthalene undergoes electrophilic substitution at the 1-position.

The six-membered aromatic heterocycle **pyridine** is a weak base and has a partially positive ring. Compared to benzene, pyridine is deactivated toward electrophilic substitution, but activated toward nucleophilic substitution.

Quinoline and **isoquinoline** undergo electrophilic substitution on the benzenoid ring, but nucleophilic substitution on the nitrogen ring.

Pyrrole is an aromatic five-membered nitrogen heterocycle. It is not basic. Its ring is partially negative and is activated toward electrophilic substitution, but deactivated toward nucleophilic substitution.

electron pair in aromatic pi system

The order of reactivity of the heterocycles and benzene toward electrophilic and nucleophilic substitution follows:

Electrophilic substitution:

Nucleophilic substitution:

Table 18.2 Summary of the reactions of naphthalene, pyridine, and pyrrole

Reaction	Section Reference

Naphthalene:

[O] → or ... 18.3

[H] → or ... 18.4

E⁺ → ... 18.5

Pyridine:

HX → ... 18.7

RX → ... 18.7

(1) ⁻:N̈H₂ (2) H₂O → ... 18.7A

RLi → ... 18.7A

Substituted Pyridine:

Nu:⁻ → ... 18.7A

where Cl is 2- or 4-

Quinoline:

E⁺ → ... 18.8

Nu:⁻ → ... 18.8

where Nu:⁻ is ⁻NH₂ or R⁻ from RLi

(continued)

Table 18.2 (*continued*) **Summary of the reactions of naphthalene, pyridine, and pyrrole**

Reaction	Section Reference
Isoquinoline:	

18.8

18.8

where Nu:⁻ is ⁻NH₂ or R⁻ from R Li

Pyrrole:

18.9A

ᵃ At 160°, sulfonation yields 2-naphthalenesulfonic acid.

Study Problems*

18.10 Name the following compounds:

(a)

*used as an antioxidant
in lubricating oils*

(b)

*1,6-Cleve's acid
used in the dye industry*

(c)

naproxen

an anti-inflammatory agent

(d)

juglone

*a colored compound
in walnut shells*

18.11 Give structures for each of the following names:

(a) 3-hydroxy-2-naphthoic acid, used to stabilize the B-vitamin riboflavin

---

* For information concerning the organization of the *Study Problems* and *Additional Problems*, see the *Note to student* on page 39.

(b) 1-naphthaleneacetic acid, a plant growth hormone that induces flowering and thinning of fruit

(c) sodium 3,6-di-*t*-butyl-1-naphthalenesulfonate, one of the isomeric compounds in "dibunate sodium," a cough suppressant

18.12 How many pi electrons does each of the following structures have? Which structures are *fully aromatic*, the aromatic pi cloud involving the entire ring system?

(a)

(b)

(c)

18.13 Predict the principal organic products, including stereochemistry where appropriate:

(a)

$$\xrightarrow[\text{heat}]{\text{CrO}_3, \text{H}_2\text{SO}_4}$$

(b)

$$+ \text{H}_2 \xrightarrow[\text{no heat or pressure}]{\text{Cu, CrO}_3}$$

(c)

$$+ \text{ excess H}_2 \xrightarrow[\text{heat, pressure}]{\text{Rh}/\text{Al}_2\text{O}_3, \text{CH}_3\text{CH}_2\text{OH}}$$

18.14 Predict the major organic products. (If no reaction occurs, write *no reaction*.)

(a)

$$+ \text{C}_6\text{H}_5\overset{O}{\underset{\|}{\text{C}}}\text{Cl} \xrightarrow{\text{AlCl}_3}$$

(b)

$$\xrightarrow[\text{(2) H}_2\text{O, H}^+]{\text{(1) C}_6\text{H}_5\text{MgBr}}$$

(c)

$$+ \text{HNO}_3 \longrightarrow$$

(d)

$$+ \text{HNO}_3 \xrightarrow{\text{H}_2\text{SO}_4}$$

(e)

$$\xrightarrow[\text{(2) HNO}_3]{\text{(1) CH}_3\overset{O}{\underset{\|}{\text{C}}}\text{Cl}}$$

18.15 Show the stepwise mechanism for the following reaction. (Include resonance structures of the intermediate in your answer.)

$$+ \text{HNO}_3 \xrightarrow{\text{H}_2\text{SO}_4}$$

18.16 Propose syntheses for the following compounds from naphthalene:

(a)

(b)

18.17 How would you explain the following observations? Aromatic nitration of 2-methyl-naphthalene yields 75% 2-methyl-1-nitronaphthalene, but aromatic sulfonation of 2-methylnaphthalene yields 80% 6-methylnaphthalene-2-sulfonic acid.

18.18 Name the following compounds:

(a)

(b)

(c)

*a metabolic product
from pyridoxine*

(d)

chloroxine

used in dandruff treatment

18.19 Draw structures for: **(a)** 2,4-dinitrothiophene;   **(b)** *N*-phenylpyrrole;   **(c)** furan-2-carboxylic acid;   **(d)** 1,8-dimethylisoquinoline.

18.20 Complete the following equations. (If no reaction occurs, write *no reaction.*)

(a) + HBr ⟶

(b) + CH_3CH_2Br ⟶

(c) + $(CH_3)_3NH^+ Cl^-$ ⟶

18.21 Which reaction would you expect to have the faster rate and why: **(a)** the reaction of pyridine and sodamide ($NaNH_2$) or   **(b)** the reaction of 2-chloropyridine and sodamide?

18.22 Predict the major organic products:

(a)
(1) C_6H_5MgBr
(2) H_2O, H^+

(b) + Br_2 $\xrightarrow{FeBr_3}$

(c)
(1) $CH_3(CH_2)_3Li$
(2) H_2O

18.23 Draw the *p*-orbital components in the lowest-energy pi molecular orbitals for the following compounds (see Figure 18.3). Indicate the number of *p* electrons contributed by each atom and any unshared electrons.

(a) thiazole   **(b)** pyrimidine   **(c)** purine   **(d)** thiophene   **(e)** pyran

18.24 Show the direction of the dipoles in **(a)** 2-methylpyridine; **(b)** 2-ethylpyrrole; **(c)** isoquinoline.

18.25 Predict the major organic products:

(a) [structure: furan] + CH_3CCl (with C=O) $\xrightarrow{SnCl_4}$ **(b)** [structure: 3-bromothiophene] + HNO_3 $\xrightarrow{CH_3CO_2H}$

(c) [structure: thiophene-3-carboxylic acid, CO_2H] + Br_2 $\xrightarrow{CH_3CO_2H}$

18.26 Suggest a reason why the rate of aromatic bromination of furan-2-carboxylic acid (2-furoic acid) is less than that of furan.

18.27 Although pyrrole is not basic, thiazole is. Explain.

18.28 Pyrrole ($pK_a = \sim 15$) is a weak acid: it can lose a proton and form an anion. Suggest a reason for the stability of this anion compared to that of $(CH_3CH_2)_2\ddot{N}:^-$.

18.29 List the following compounds in order of increasing rates of reaction (slowest first) toward acetic anhydride with $AlCl_3$ catalyst:

(a) 3-nitrofuran   **(b)** 2,5-dinitrofuran   **(c)** 3,4-dinitrofuran   **(d)** 3-methylfuran

18.30 Using an unsubstituted porphyrin ring system with a chelated magnesium ion, show four resonance structures (there are more) that illustrate four equivalent bonds to the Mg.

Additional Problems

18.31 Predict the major organic products:

(a) [structure: 4-chloropyridine] + Na^+ $^-OCH_2CH_3$ $\xrightarrow{CH_3CH_2OH}$ **(b)** [structure: 2-nitronaphthalene, NO_2] $\xrightarrow[160°]{H_2SO_4 + SO_3}$

(c) [structure: 2-methylthiophene, H_3C] + HNO_3 $\xrightarrow{(CH_3CO)_2O}$ **(d)** [structure: 1-methylnaphthalene, CH_3] + H_2SO_4 $\xrightarrow{40°}$

(e) [structure: 1-methylnaphthalene, CH_3] + Br_2 $\xrightarrow{h\nu}$ **(f)** [structure: 4-methylquinoline, CH_3] + HCl $\longrightarrow$

(g) [structure: pyrrole, N–H] + CH_3CH_2MgBr $\longrightarrow$

(h) [structure: furan with H_3C, CH_3O_2C, and $C(CH_2)_3CH_3$ with C=O] $\xrightarrow[(2)\ H_2O,\ H^+]{(1)\ NH_2NH_2,\ KOH,\ HOCH_2CH_2OH,\ heat}$

18.32 Propose syntheses for the following compounds:

(a) 2-naphthoxyacetic acid, a plant growth hormone and a sedative, from 2-naphthol
(b) 9,10-phenanthrenediol from phenanthrene
(c) 4-amino-1-naphthalenesulfonic acid from naphthalene

(d)   from naphthalene and furan

(e)   from an open-chain compound and any other reagents needed

(f)   from a monocyclic compound

(g)   from

benz[j]aceanthrylene

a carcinogen

CH₂CO₂H — CH_2CO_2H

(h)   from

benz[k]acephenanthrylene

a carcinogen

(i)   from 4-methylpyridine

isoniazid

*used in the treatment
of tuberculosis*

(j)   from

(k)   from

(l)   from

cyclopenta[*cd*]pyrene   4-bromopyrene

*a carcinogen in carbon black
and auto exhaust*

18.33 Only one of the three isomeric monohydroxypyridines (OH on a ring carbon) exhibits the chemical characteristics of a phenol. Explain.

18.34 When pyrrole is treated with strong acid, a polymer called *pyrrole red* is formed. Suggest a mechanism for the first step in the polymerization.

18.35 What is the most likely position of substitution by Br_2 on the indole ring system? Explain your answer by showing the mechanism.

indole

18.36 Predict the position of aromatic electrophilic and nucleophilic substitution of pyrimidine (Table 18.1). Would you expect pyrimidine to be more or less reactive than pyridine toward electrophilic substitution? Toward nucleophilic substitution?

18.37 *Imidazole* (Table 18.1) is found in proteins. The equilibrium between imidazole ($pK_b = 7$) and the imidazolium ion (imidazole—H^+) helps buffer proteins in biological systems. Use *p*-orbital pictures to explain why imidazole and the imidazolium ion are both aromatic.

18.38 Substituted naphthalene compounds can be prepared by Friedel–Crafts acylations of substituted benzenes with succinic anhydride. Tell what intermediate and final products would be obtained in the following sequence. [*Hints*: HF catalyzes aromatic acylation with a carboxylic acid, and Pd/heat causes catalytic dehydrogenation (loss of H_2) leading to aromatization of nonaromatic rings.]

toluene + succinic anhydride $\xrightarrow{\text{AlCl}_3}$ (a) $\xrightarrow[\text{HCl}]{\text{Zn(Hg)}}$ (b) $\xrightarrow{\text{HF}}$ (c) $\xrightarrow[\text{heat}]{\text{Pd}}$ (d)

18.39 The **Skraup synthesis** is the classical method for the synthesis of quinoline and substituted quinolines. In this reaction sequence, a mixture of glycerol and aniline is heated in the presence of concentrated H_2SO_4 and a mild oxidizing agent, such as nitrobenzene. The principal steps in the sequence are shown. Give mechanisms for steps 1 and 2 leading to 1,2-dihydroquinoline.

glycerol   *not isolated*

(2) [aniline structure] + [CHO/CH/CH₂ structure] $\xrightarrow{\text{H}^+}$ [1,2-dihydroquinoline structure] + H_2O

aniline   1,2-dihydroquinoline
                                not isolated

(3) 3 [quinoline-H structure] + [nitrobenzene structure]—NO_2 $\xrightarrow{\text{H}^+}$ 3 [quinoline structure] + [benzene]—NH_2 + 2 H_2O

                    nitrobenzene   quinoline
                                                        (85%)

18.40 Predict the product of the reaction of aniline and 3-buten-2-one in the presence of a Lewis-acid catalyst.

18.41 The aromatic dianion A of *sym*-dibenzocyclooctatetraene can be prepared by the treatment of the hydrocarbon with potassium metal. When A is treated with excess dry ice, followed by acidification, two products are formed: B ($C_{18}H_{12}O_3$) in 22% yield and C ($C_{18}H_{14}O_4$) in 51% yield. The proton nmr spectrum (in $CDCl_3$) of B showed a multiplet at $\delta 7.20$, a singlet at $\delta 6.98$, and a singlet at $\delta 4.64$ (area ratio 4:1:1). What are the structures of A, B, and C?

[structure of sym-dibenzocyclooctatetraene]

sym-dibenzocyclooctatetraene

18.42 Acid hydrolysis of the carbohydrates in oat hulls or corncobs gives Compound A, $C_5H_4O_2$, in almost 100% yield. Catalytic hydrogenation of A with CuO—Cr_2O_3 catalyst at 175° and 100 atm gives Compound B, $C_5H_6O_2$. The infrared spectrum of B and proton nmr spectra of A and B are shown in Figure 18.7. What are the structures of A and B?

Figure 18.7 Spectra for Problem 18.42.

19 ▼

Natural Products: Studies in Organic Synthesis

Identification of organic compounds and understanding of their chemical and physical properties are major goals of organic chemists. The synthesis of organic compounds is an equally important goal. In synthesis, the goal is to prepare a particular pure compound, with defined stereochemistry, by a short series of organic reactions.

Synthetic studies are frequently aimed at preparing natural products. Naturally occurring organic compounds often have unique structures as well as properties. They are often found in nature as a single enantiomer or diastereomer. The synthesis of one of these naturally occurring compounds can be a great challenge.

It is not uncommon for a synthesis of a natural product to begin as an intellectual challenge. However, such academic studies sometimes result in important chemical discoveries and new commercial products. Many natural products have potent biological properties and some have unusual physical properties. For example, plants biosynthesize a class of nitrogen-containing organic compounds called alkaloids. Morphine (Section 19.3), a painkiller, is an alkaloid obtained from the poppy *Papaver somniferum*. Once an alkaloid with biological properties such as morphine is synthesized, variations on the synthesis can be carried out and structural analogs of the natural product prepared. Such studies have led to the development of numerous commercially important pharmaceuticals.

We will consider three syntheses and briefly look at the class of natural products to which the target compound belongs. The first synthesis is of abscisic acid, a terpene—a class of naturally occurring compounds composed of isoprene (2-methylbutyl) skeletal units. Abscisic acid is involved in the biological control of the dormancy of plants.

The second synthesis is of (*S*)-4-methyl-3-heptanone, an alarm pheromone of the leaf-cutting ant (*Atta texana*) and other insects. The (*S*) enantiomer is 400 times more active than the (*R*) enantiomer. Therefore, we will discuss methods by which a single stereoisomer of such a chiral compound can be synthesized.

The third synthesis is of (+)-retronecine, which is part of a pyrrolizidine alkaloid that is a cumulative plant poison of grazing animals.

| (+)-abscisic acid | (*S*)-4-methyl-3-heptanone | (+)-retronecine |
| a terpenoid | a pheromone | an alkaloid component |

▼ Section 19.1
Terpenes

Essential oils are the odorous plant components that can be separated from other plant material by steam distillation. Many essential oils, such as those from flowers, are used in perfumes. Quite early in the history of organic chemistry, these essential oils, especially the oils of turpentine, attracted the attention of chemists. It was soon discovered that the ratio of the number of carbon atoms to the number of hydrogen atoms in turpentine is 5:8. Other natural products with a carbon-to-hydrogen ratio of 5:8 were then classified as **terpenes**. Later, chemists learned that the terpenes were composed of compounds containing a *head-to-tail* joining of isoprene skeletal units. (The *head* is the end closer to the methyl branch.)

The original definition of terpenes was then expanded to include all compounds containing isoprene skeletal units. To emphasize this relationship to isoprene, terpenes are also called **isoprenoids**. Terpenes may contain two, three, or more isoprene units. Their molecules may be open-chain or cyclic. They may contain double bonds, hydroxyl groups, carbonyl groups, or other functional groups. A terpene-like structure that contains elements other than C and H is called a **terpenoid**.

head   CH$_3$ *tail*

$\overset{\curvearrowright}{}$ CH$_2$=CCH=CH$_2$ $\overset{\curvearrowleft}{}$

isoprene

tail   *head*

CH$_3$ ↓   ↓   CH$_2$

CH$_3$C=CHCH$_2$—CH$_2$CCH=CH$_2$ or   *a terpene*

tail   *head*

CH$_3$ ↓   ↓   CH$_3$

CH$_3$C=CHCH$_2$—CH$_2$C=CHCH$_2$OH   or   *a terpenoid*

CH$_2$OH

CH$_3$

CH

head → CH$_2$ CH$_2$

tail → CH$_2$ CHOH   or

CH

CH

H$_3$C   CH$_3$

OH

Although the idea is appealing, naturally occurring terpenes do not arise from the polymerization of isoprene. The first step in terpene biosynthesis is an enzymatic ester condensation of the acetyl portions of acetylcoenzyme A. Intermediates in the formation of terpenes are the pyrophosphates (diphosphates) of mevalonic acid and a pair of isopentenyl alcohols. An abbreviated biosynthetic route to terpenes and steroids is shown in Section 23.7C.

A. Classification of Terpenes

Terpenes are classified by the *number of isoprene units* they contain:

monoterpenes: two isoprene units (10 carbons)
sesquiterpenes: three isoprene units (15 carbons)
diterpenes: four isoprene units (20 carbons)
triterpenes: six isoprene units (30 carbons)
tetraterpenes: eight isoprene units (40 carbons)

Monoterpenes, with skeletons that contain only two isoprene units, are the simplest of the terpenes. Yet, even monoterpenes exhibit a variety of structures. (Although some terpenes and terpenoids, such as geraniol, are found in a variety of plants and animals, we have indicated sources especially rich in the following compounds.) Note that the trivial names of terpenes often are derived from the original sources of these compounds.

Acyclic monoterpenes:

geraniol	citral (geranial)
in roses	*in lemongrass*

Cyclic monoterpenes:

limonene	menthol	camphor	α-pinene
in citrus fruits	*in mint*	*from camphor trees*	*in turpentine (from pine trees)*

Carrots contain an orange-colored tetraterpene called *carotene*. (If a person eats too many carrots, the deposition of carotene will color his or her skin orange. However, time is a cure for this condition.) Carotene can be cleaved enzymatically into two units of vitamin A. (The role of vitamin A in vision will be discussed in Chapter 21.)

β-carotene

an orange tetraterpene

$$\downarrow [O]$$

all-*trans*-vitamin A

Study Problem

19.1 Show the isoprene units in menthol and α-pinene.

B. Synthesis of Abscisic Acid, a Terpenoid

Abscisic acid, a sesquiterpenoid, is widely distributed in the plant kingdom. It is involved in the regulation of the dormant state of plants, allowing them to survive under adverse environmental conditions.

Several different syntheses of abscisic acid have been reported. The one we describe (see Figure 19.1) begins with isophorone, a commercially available, achiral cyclohexenone synthesized by an aldol-dehydration reaction of acetone.

Comparison of the structures of abscisic acid and isophorone reveals that the keto group, the double bond, and the methyl groups needed are present in the starting isophorone. The synthetic work, therefore, involves building the allylic dienoic acid side chain and the insertion of the allylic tertiary hydroxyl group.

Retrosynthetic analysis (working backwards from the final, or target, compound) shows that one possible synthetic approach could be (1) the

Figure 19.1 An outline of the synthesis of abscisic acid from isophorone. (M. G. Constantino, P. M. Donate, and N. Petragnani, *J. Org. Chem.* 1986, *51*, 253.)

formation of the target tertiary alcohol and insertion of the side chain by the reaction of a Grignard reagent or related organometallic reagent with a ketone, preceded by (2) formation of a keto group at the key position.

abscisic acid   a diketone   isophorone

Study Problem

19.2 Write an equation for the reaction of the intermediate diketone shown with excess CH_3MgI, followed by hydrolysis.

Part 1. Synthesis of the ketone. Because the synthesis hinges on the reaction of only one of two keto groups, the existing keto group in the starting isophorone must be blocked. The blocking group chosen was a cyclic ketal, a functional group inert to Grignard-type reactions, but readily reconverted to a keto group by acidic hydrolysis (Section 13.5B). Reaction of isophorone with ethylene glycol using *p*-toluenesulfonic acid as catalyst gave the ketal in 88.5% yield.

isophorone   a ketal (88.5%)

So that the ketal formation would proceed to completion, the water formed by the cyclization was removed as a toluene–water azeotrope. Under these reaction conditions, acid and heat, the double bond of isophorone equilibrates, forming a 70:30 mixture of two double-bond isomers. These isomers were separated by fractional distillation.

(30%)   (70%)

used for the next step

Alkaline permanganate oxidation of the double bond of the ketal gave the α-hydroxy ketone in 72% yield. The alkene was *cis*-hydroxylated by permanganate to yield a diol with secondary and tertiary hydroxyl groups. Continued oxidation of this diol converted only the secondary —OH group to a carbonyl.

Tertiary hydroxyl groups are inert to oxidation under alkaline conditions (Section 7.9C).

(72%)

A mixture of methanesulfonyl chloride and pyridine was used to dehydrate the tertiary alcohol. Under these alkaline conditions, the ketal group remains intact.

(63%)

A possible pathway for this alkaline dehydration is an initial formation of a methylsulfonate ester that decomposes into a carbocation. (Recall from Section 7.8B that sulfonate groups are good leaving groups.) The most substituted conjugated alkene forms from the carbocation.

a methanesulfonate
ester

Study Problem

19.3 Write an equation showing the dehydration of the preceding hydroxy–ketal–ketone under acidic conditions.

Part 2. Synthesis of the final product. Insertion of the side chain was accomplished by reaction of the ketone with an acetylenic lithium reagent at low temperature. The reagent was obtained by treating methyl 2-penten-4-ynoate with lithium diisopropylamide (LDA), a hindered (and thus non-nucleophilic) strong base, at −78°C. A rapid acid–base reaction between the lithium amide and the acetylenic proton yielded the organolithium reagent.

methyl (Z)-3-methyl-2-penten-4-ynoate   lithium diisopropylamide (LDA)

a lithium acetylide   diisopropylamine

Immediately following the formation of the lithium acetylide, the cyclic ketone was added to the reaction mixture and the mixture maintained at a low temperature ($-78°$ to $-30°$). (At these low temperatures, the ester group does not undergo reaction.) The product lithium salt of a tertiary alcohol was treated with water to yield the alcohol.

($\sim 100\%$)

The final steps in the synthesis were (1) reduction of the triple bond to a *trans* double bond, (2) removal of the ketal blocking group, and (3) saponification of the ester.

The reduction was carried out with a reducing agent, an aqueous-dimethylformamide solution of chromous sulfate ($CrSO_4$), that allows the thermodynamically more-stable *trans* isomer to form. During work-up, the ketal blocking group was hydrolyzed to a ketone. This hydrolysis reaction was catalyzed by ammonium ions, which are sufficiently acidic to hydrolyze the ketal group.

(35%)

Saponification, followed by acidification, provided the final abscisic acid in 80% yield from the methyl ester.

$$\text{(1) NaOH, CH}_3\text{OH, H}_2\text{O} \quad \xrightarrow{\text{(2) H}^+}$$

(±)-abscisic acid (80%)

In designing and executing this synthesis of abscisic acid, the organic chemists had to consider the numerous possible side reactions that the three different oxygen-containing functional groups and the three differently substituted carbon–carbon double bonds could undergo. Note that this particular route to abscisic acid was not designed to be stereoselective. It produced the natural product as the racemic mixture.

Study Problem

19.4 Suggest reasons why acidic hydrolysis of methyl abscisate would be a poor idea.

▼ ## Section 19.2
Pheromones: Synthesis of Chiral Natural Products

Humans communicate by talking, using sign language, drawing pictures, or writing. Insects and some animals communicate, in part, by chemicals. A chemical or a mixture of chemicals secreted by one individual of a species that brings forth a response in another individual of the same species is called a **pheromone** (from the Greek *phero*, "carrier").

Insect pheromones are generally classified by the response they elicit. For example, *alarm pheromones* signify danger, and *sex attractants* help the different sexes of the same species to find one another. Other pheromones are used to recruit. For example, a pheromone secreted by one bee helps alert other bees to the location of a food source.

Extremely small quantities of a pheromone can elicit the desired response. A typical female insect may carry only 10^{-8} gram of sex attractant, yet that is enough to attract over a billion males from miles away! A male gypsy moth can "smell" a female at a distance of 7 miles.

The biological activity of a pheromone is usually highly dependent upon stereochemistry—one enantiomer or geometric isomer is active and all other stereoisomers inactive.

Many of the known insect pheromones are not complex in structure. Geraniol and citral, both terpenes, are recruiting pheromones for honeybees, while isoamyl acetate (not a terpene) is a bee alarm pheromone. (Isoamyl acetate is also the principal odorous component of banana oil.)

$$CH_3COCH_2CH_2CH(CH_3)_2$$

3-methylbutyl acetate
(isoamyl acetate)

geraniol   citral

The following compounds have sex-attractant activity for different species of insects.

$$cis\text{-}(CH_3)_2CH(CH_2)_4\overset{\displaystyle O}{\overset{\displaystyle \triangle}{C}}HCH(CH_2)_9CH_3$$

gypsy moth

$$cis\text{-}CH_3(CH_2)_3CH{=}CH(CH_2)_6\overset{\displaystyle O}{\overset{\displaystyle \|}{O}}CCH_3$$

cabbage looper

$$CH_2OH$$

boll weevil

$$cis\text{-}CH_3(CH_2)_{12}CH{=}CH(CH_2)_7CH_3$$

house fly

An example of a three-component pheromone mixture is the sex attractant from the male *Ips paraconfusus* (an engraver pine beetle). To elicit response, all three compounds (ipsenol, ipsdienol, and *cis*-verbenol) must be present. We say that the mixture is *synergistic*. It is postulated that the synergism has the effect of increasing specificity because only the correct combination functions as the pheromone. Thus, this synergism may allow the pheromone to be distinguished from the multitude of background chemicals found in a forest.

HO   "ipsenol"   HO   "ipsdienol"   —OH   H   *cis*-verbenol

Pheromone chemistry is an exciting and vigorous field of research that includes isolation, structure determination, synthesis, and biological studies. It is an area of study that also is of immediate practical value. For example, sex-attractant pheromones have been used for insect control. In some cases, male insects may be lured by a sex-attractant pheromone, trapped, and then sterilized and released to mate unproductively with females.

The research effort in this field has been directed toward the elucidation and synthesis of single compounds as pheromones. As more information becomes available, it may turn out that a single compound believed to be the active pheromone is actually only one component of a complex mixture of interacting substances, such as the three-component system shown.

A. Synthesis of Chiral Compounds

It is not possible to synthesize only one enantiomer of a chiral compound from an achiral starting material using achiral reagents and catalysts. When such a reaction is carried out, both configurations, (R) and (S), of the product are formed; that is, the product is always racemic.

To synthesize one stereoisomer of a chiral compound, chemists must use one of two synthetic approaches. The first is to synthesize the compound as a racemic mixture and then separate the enantiomers. The separation of enantiomers, called **resolution**, was discussed in Section 4.10. The second approach is to synthesize the compound with a chiral auxiliary—a chiral catalyst, reagent, solvent, or intermediate. This technique, called **asymmetric synthesis**, will be discussed here.

Because many asymmetric syntheses are not 100% efficient, the terms **optical purity** and **enantiomeric excess** are often used to describe the products.

$$\text{optical purity} = \frac{\text{observed } [\alpha]_D^T}{[\alpha]_D^T \text{ of pure enantiomer}} \times 100\%$$

$$\text{enantiomeric excess} = \frac{\text{excess of desired enantiomer}}{\text{total yield}} \times 100\%$$

$$\text{or } \frac{A - B}{A + B} \times 100\%$$

Thus, a chemist might report a 61% yield with a 94% enantiomeric excess.

Study Problem

19.5 **(a)** Calculate the enantiomeric excess of A in a mixture of enantiomers that contains 120 mg of A and 20 mg of B. **(b)** Calculate the enantiomeric excess of A in a mixture of 50 mg of A and 50 mg of B (a racemic mixture).

Chiral catalyst.   Enzymes are chiral catalysts. Because of their asymmetric surfaces, enzymes are able to hold an achiral substrate so that only one face of the molecule undergoes reaction. The reduction of ketones, hydroxy ketones, and keto esters with baker's yeast is an example of an enzymatic asymmetric synthesis.

$$\underset{\text{1-hydroxy-2-heptanone}}{CH_3(CH_2)_4\overset{\overset{\displaystyle O}{\|}}{C}CH_2OH} \quad \xrightarrow[\substack{\text{sucrose} \\ H_2O}]{\text{baker's yeast}} \quad \underset{\substack{(R)\text{-1,2-heptanediol (56\%)} \\ (\text{enantiomeric excess, } 100\%)}}{\underset{CH_3(CH_2)_4 \quad CH_2OH}{\overset{\overset{\displaystyle OH}{|}}{C}\cdots H}}$$

Nonenzymatic chiral catalysts have also been developed. These catalysts function in the same manner as their enzymatic counterparts. An example is a chiral hydrogenation catalyst that can induce chirality during hydrogenation (Section 10.13A).

$$\underset{CH_3\overset{\overset{\displaystyle O}{\|}}{C}NH-\overset{\overset{\displaystyle CO_2H}{|}}{C}H=CH_2}{} + H_2 \quad \longrightarrow \quad \underset{\substack{CH_3\overset{\overset{\displaystyle O}{\|}}{C}NH \quad CH_3 \\ (\text{enantiomeric excess, } 90\%)}}{\overset{\overset{\displaystyle CO_2H}{|}}{\underset{}{C}}\cdots H \; (R)}$$

Chiral reagent.   If a chiral reagent is used to form a chiral center in an achiral starting material, one enantiomer will usually be formed in excess. For example, a single enantiomer can be isolated from the hydroboration–oxidation of an achiral diene when a chiral dialkylborane is used in the hydroboration step.

$$(21–31\%)$$

Chiral solvent. We can often enhance the formation of one enantiomer by using a chiral solvent for the reaction of two achiral reagents.

(enantiomeric excess, 33%)

Chiral intermediate. Another approach to asymmetric syntheses is the use of chiral intermediates. In this procedure, a chiral auxiliary reagent is used to prepare a chiral intermediate from an achiral substrate. The intermediate is then subjected to a standard chemical reaction that forms a chiral center. The bonded chiral auxiliary allows only one face or side of the substrate to undergo reaction. After the reaction, the chiral auxiliary is removed and the chiral product is released. This approach is used in the pheromone synthesis that follows.

B. Synthesis of (S)-4-Methyl-3-heptanone, a Pheromone

The principal alarm pheromone of the leaf-cutting ant (*Atta texana*) is (S)-4-methyl-3-heptanone, a compound with a relatively simple structure containing one chiral carbon. The racemic form of this pheromone has been synthesized by a number of different routes, and the enantiomers have been separated by resolution.

We will discuss an asymmetric synthesis of the (S) enantiomer that starts with 3-pentanone, an achiral compound.

The synthetic pathway is outlined in Figure 19.2. The chiral auxiliary reagent is a hydrazine, (S)-1-amino-2-methoxymethylpyrrolidine (abbreviated SAMP). This optically active hydrazine is obtained from (S)-proline, a naturally occurring amino acid, by a multistep synthesis, which we will not

Figure 19.2 Synthesis of (S)-(+)-4-methyl-3-heptanone, a pheromone.
(D. Enders, H. Kipphardt, and P. Fey, *Org. Syn.* 1987, *65*, 183.)

discuss. However, it is important to recognize that the chirality of the chiral auxiliary reagent originates with a naturally occurring compound and is not created from achiral compounds by a laboratory procedure.

(S)-proline

naturally occurring

(S)-1-amino-2-methoxymethylpyrrolidine
(SAMP)
(50% overall)

The first step in the asymmetric synthesis of the ant pheromone was conversion of 3-pentanone to a chiral hydrazone. This step created the molecular chiral environment that allowed only one configuration at the new chiral center later in the synthesis.

hydrazine group

a hydrazone

The alpha hydrogens of hydrazones, like those of aldehydes and ketones, are slightly acidic (Chapter 16). In this synthesis, lithium diisopropylamide (LDA) was used to convert the optically active hydrazone to the lithium salt, a lithium aza-enolate.

Enolate formation:

$$\overset{\overset{O}{\|}}{R\,C\,CH_2R} + base{:}^- \longrightarrow \overset{\overset{O}{\|}}{R\,C\,\overset{..}{C}HR} + base{-}H$$

$$\overset{\overset{NNR_2}{\|}}{CH_3CH_2\,C\,CH_2CH_3} + Li^+\ {}^-{:}\overset{..}{N}[CH(CH_3)_2]_2 \longrightarrow \overset{\overset{NNR_2}{\|}}{CH_3CH_2\,C\,\overset{-}{C}HCH_3}\ Li^+ + H\overset{..}{N}[CH(CH_3)_2]_2$$

a lithium aza-enolate

The lithium aza-enolate is stabilized by forming a complex involving the lithium cation, the alpha carbanion, and the chiral auxiliary. This chiral chelated carbanion is a nucleophile that will react with alkyl halides in nucleophilic substitution reactions. Treatment of the complex with 1-iodopropane at $-110°$ resulted in alkylation. The chiral pyrrolidine portion of the hydrazone allowed only one of two diastereomers, (S,S) and not (R,S), to form because of the chiral environment at the site of reaction. This type of reaction is called **asymmetric induction**.

Nmr spectroscopy revealed that the (Z) isomer of the alkylated hydrazone was initially formed and that this isomer slowly isomerized to the more thermodynamically stable (E) isomer. The fact that the (Z) isomer formed first shows that the chelated face of the molecule is the one that undergoes alkylation. The equilibration between the (Z) and (E) hydrazones did not pose a problem in the overall synthesis because either hydrazone would yield the (S) ketone after hydrolysis.

Simple alkaline hydrolysis would convert the hydrazone group to the desired keto group, but would also cause racemization at the alpha carbon by way of the symmetrical enol tautomer. For this reason, ozone without a subsequent water work-up was used to oxidize the pi bond of the hydrazone and to form the keto group. (In another study, a two-phase system of 3 M HCl and pentane was used.)

an (S)-N-nitrosoamine

Study Problem

19.6 Write an equation showing the racemization of (S)-4-methyl-3-heptanone in aqueous sodium hydroxide. Include the structure of the intermediate in your answer.

▼ Section 19.3
Alkaloids

A. Some Common Alkaloids

Primitive people have often used extracts of roots, bark, leaves, flowers, berries, and seeds as drugs. This use of plants for medicinal purposes was not necessarily based on superstition or wishful thinking. Many plants contain compounds that have profound physiological impact. The active agents in many of these plant substances have been isolated and have been found to be heterocyclic nitrogen compounds.

Many of the nitrogen compounds in plants contain basic nitrogen atoms and thus can be extracted from the bulk of the plant material by dilute acid. These compounds are called **alkaloids**, which means "like an alkali." After extraction, the free alkaloids can be regenerated by subsequent treatment with aqueous base.

Extraction:

$$R_3N: + HCl \xrightarrow{H_2O} R_3NH^+ \; Cl^-$$

Regeneration:

$$R_3NH^+ \; Cl^- + OH^- \longrightarrow R_3N: + H_2O + Cl^-$$

Alkaloids vary from simple to complex in their structures. One of the simplest in structure, but not in its physiological effects, is *nicotine*.

nicotine

Nicotine is highly toxic and has been used as an insecticide. In very small doses (such as a smoker obtains from cigarettes), nicotine acts by stimulating the autonomic (involuntary) nervous system. If small doses are continued, nicotine can depress this same nervous system into less than normal activity.

The first isolation of an alkaloid in the pure state was reported in 1805. This alkaloid was *morphine* (from the Greek *Morpheus*, the god of dreams), one of many alkaloids to come from the gum and seeds of the opium poppy, *Papaver somniferum.*

morphine   codeine   heroin

Codeine is the methoxy derivative of morphine (at the phenolic group), while *heroin* is the diacetyl derivative. Codeine, like morphine, is a powerful analgesic and occurs naturally in the seeds of the opium poppy. Codeine is also an excellent cough suppressant that is sometimes used in prescription cough medicines. In recent years, it has been largely replaced by *dextromethorphan*, a nonaddictive, synthetic drug that is an equally effective cough suppressant. (Note the similarities in structure.)

dextromethorphan

Heroin does not occur naturally, but may be synthesized from morphine in the laboratory. Heroin, like morphine and codeine, is a powerful analgesic. In some parts of the world, heroin is used to relieve pain in terminal cancer patients. Because it is even more addictive than morphine, its medicinal use is prohibited in the United States.

A large number of physiologically active alkaloids contain the **tropane ring system**:

tropane

One of the tropane alkaloids is *atropine*, found in *Atropa belladonna* and other members of the nightshade plant family. Atropine is used in eye drops to dilate the pupils. *Scopolamine* (a so-called "truth" serum) is used as a preoperative sedative. Chemically, it is the epoxide of atropine. *Cocaine*, an addictive stimulant and pain reliever, also contains the tropane ring system.

atropine   scopolamine   cocaine

Study Problem

19.7 When atropine is subjected to acid hydrolysis, two products can be isolated: *tropine*, which is not optically active, and *tropic acid*, which is obtained as a racemic mixture. What are the structures of these two compounds?

B. Synthesis of Retronecine, an Alkaloid

Alkaloids that contain a pyrrolizidine ring system are called **pyrrolizidine alkaloids**.

pyrrolizidine

These alkaloids are found worldwide in a variety of plants, such as asters and ragworts. They are of interest because they are cumulative poisons to grazing animals, causing liver cirrhosis (leading to a condition called "horse staggers"), liver tumors, and death.

About 100 different pyrrolizidine alkaloids have been identified. Most of these are found in nature as esters that can be hydrolyzed to a carboxylic acid, called a *necic acid*, and an amino alcohol (a substituted pyrrolizidine), called a *necine*.

monocrotalic acid   retronecine

a typical necic acid   *the most widespread necine*

Although a large number of necines are known, the most common is retronecine, which was first isolated in 1909 and was synthesized in the laboratory in 1962. Figure 19.3 shows the reaction sequence used in this stereospecific synthesis.

The first step in the synthesis is a Michael addition (Section 16.9). Under the conditions of this addition, the product tetraethoxy compound **A** undergoes a Dieckmann ring closure (Section 16.8) to yield a β-keto ester **B**. Acid hydrolysis of this keto ester yields a β-keto acid (not isolated), which undergoes decarboxylation. The conditions of the hydrolysis and decarboxylation

Figure 19.3 Synthesis of retronecine, a component of several widely spread plant toxins. (T.A. Geissman and A. C. Waiss, Jr., *J. Org. Chem.* 1962, *27,* 139.)

(overnight at room temperature in 12 M HCl) are sufficiently mild that the carbon–nitrogen bond of the carbamate group is not cleaved.

The decarboxylated product is then re-esterified with ethanol to yield compound **C**. Reduction of the keto group in **C** with sodium borohydride yields a *cis* hydroxy acid, which undergoes lactonization to yield compound **D**. (The reason for the preferential formation of the *cis* acid instead of the *trans* isomer is discussed in Section 13.7B.)

Hydrolysis of the carbon–nitrogen bond in the carbamate group occurs when compound **D** is heated at reflux in aqueous hydroxide for 16 hours. These reaction conditions also open the lactone ring. Upon acidification, the lactone ring re-forms and the amino nitrogen atom is protonated to yield salt **E**. This salt is neutralized with sodium carbonate and alkylated (a nucleophilic substitution reaction between an amine and a reactive organohalogen compound). Treatment of the product ester **F** with potassium ethoxide results in transesterification of the lactone followed by another Dieckmann ring closure.

The ketone group of keto ester **G** is catalytically hydrogenated. The resulting hydroxyl group is eliminated during the alkaline saponification of dihydroxy ester **H** (not the usual conditions for alcohol dehydration reactions). The remainder of the synthetic sequence is the esterification of the carboxylic acid and the reduction of the product ester to yield racemic retronecine, as shown in Figure 19.3.

This synthesis leads to the *trans* isomer of retronecine with the correct relative configurations of its two chiral carbons. The product, however, is racemic. The natural product, (+)-retronecine, was separated from this racemic mixture by a resolution (see Section 4.10) using (+)-camphoric acid.

Study Problem

19.8 Write equations showing the mechanism of the following conversion, shown at the start of the synthetic sequence in Figure 19.3.

$$C_2H_5O_2CCH_2CH_2NHCO_2C_2H_5 + \textit{trans-}C_2H_5O_2CCH{=}CHCO_2C_2H_5 \longrightarrow B$$

▼ Summary

Terpenes and **terpenoids**, found in both plants and animals, have carbon skeletons of diisoprene, triisoprene, etc.

A **pheromone** is a chemical excreted by an individual of a species (notably insects) that elicits a response in another individual of the same species. Generally of simple structure, pheromones are used to signify danger or food, or to act as sex attractants.

An **asymmetric synthesis** is the synthesis of a chiral compound from an achiral starting material. This type of synthesis requires a chiral catalyst, reagent, solvent, or auxiliary.

Alkaloids are acid-soluble, nitrogen-containing plant materials. Typical alkaloids are nicotine, morphine, and atropine.

Study Problems*

19.9 Indicate the isoprene units and classify each structure as a monoterpene, sesquiterpene, or diterpene:

(a)

caryophyllene

in oil of cloves

(b)

in oil of balsam

(c)

vitamin A

19.10 A line formula for squalene is shown below.

(a) Is squalene a mono-, di-, tri-, or tetraterpene?

(b) Designate the link between each pair of isoprene units as being head-to-head, head-to-tail, or tail-to-tail.

(c) Write the IUPAC name of squalene, designating those double bonds that have stereoisomers as *E* or *Z*. (*Hint*: See Table A1 in the appendix.)

(d) Write an equation showing the enzymatic formation of squalene from farnesol (Figure 23.4, page 952).

* For information concerning the organization of the *Study Problems* and *Additional Problems*, see the *Note to student* on page 39.

19.11 Write equations, showing starting materials and reagents, for the following conversions:

(a) an aldehyde ⟶ CH₃CH₂CH〈O—O〉

(b) a hydrocarbon ⟶ [cyclohexanone with CH₂CH₃ and CH₃ substituents]

(c) an alkyne ⟶ CH₃CH₂C≡C:⁻ Li⁺

19.12 The following carbonyl compounds yield (*S*) products when the keto group is reduced with baker's yeast. What are the structures of the products?

(a) CH₃CCH₂COCH₂CH₃   (b) [benzene ring]—CCH₂CH₂CH₃

19.13 We have shown a synthesis in which chirality is induced by a reagent called SAMP (Section 19.2B). Another compound used to induce chirality is called RAMP. Suggest a complete name and formula for RAMP.

19.14 Give the structures of the products:
   (a) morphine + cold, dilute HCl ⟶
   (b) morphine + cold, dilute NaOH ⟶
   (c) codeine + cold, dilute NaOH ⟶
   (d) psilocin (page 774) + cold, dilute HCl ⟶

Additional Problems

19.15 The structure of the aggregating pheromone of the bark beetle follows:

What are the organic products of its reaction with: (a) excess Br₂ in CCl₄;   (b) excess gaseous HBr (no O₂ or peroxides);   (c) hot KMnO₄ solution?

19.16 Many terpenes undergo rearrangements in the presence of acid. Suggest mechanisms for the following terpene rearrangements:

19.17 A sex attractant pheromone of the Douglas fir tussock moth was synthesized by the following route. The 11-carbon compound (*Z*)-5,10-undecadien-1-ol was oxidized to compound A with pyridinium chlorochromate, a reagent that oxidizes alcohols to aldehydes or ketones without oxidizing carbon–carbon double bonds. Compound A was treated with *n*-decylmagnesium bromide, followed by acidic hydrolysis to yield B. Compound B was subjected to pyridinium chlorochromate oxidation to yield the pheromone. Write a flow equation for the synthesis, showing the structures of A, B, and the pheromone.

19.18 *Citronellal* $(C_{10}H_{18}O)$ is a terpenoid that undergoes reaction with Tollens reagent to yield citronellic acid $(C_{10}H_{18}O_2)$. Chromate oxidation of citronellal yields acetone and $HO_2CCH_2CH(CH_3)CH_2CH_2CO_2H$. What is the structure of citronellal?

19.19 The synthesis of the acyclic monoterpene *alloocimene* was accomplished by the following route. What is the structure of alloocimene? (Show the structures of the intermediate products in your answer.)

$$\underset{\displaystyle CH_3\overset{\displaystyle O}{\overset{\|}{C}}CH_2CH_3}{} \xrightarrow{\ NaC\equiv CH\ } A \xrightarrow{\ C_2H_5MgBr\ } B \xrightarrow[\text{(2) } H_2O]{\text{(1) } (CH_3)_2CHCHO}$$

$$C \xrightarrow[\text{poisoned Pd}]{H_2} D \xrightarrow[\text{heat}]{H^+} \text{alloocimene}$$

19.20 The terpenoid *pinol* $(C_{10}H_{16}O)$ can be obtained by the treatment of the dibromide of α-terpineol with $NaOC_2H_5$. What is the structure of pinol? (*Hint*: Consider the stereochemistry of the intermediates.)

α-terpineol

19.21 Suggest a synthesis for α-terpineol from readily available, open-chain starting materials.

20 ▼

Pericyclic Reactions

Except for S_N2 and E2 reactions, most of the organic reactions we have discussed so far proceed *stepwise* by way of intermediates such as carbocations or free radicals. A large number of reactions of conjugated polyenes, called **pericyclic reactions** (from *peri*, "around" or "about"), proceed by concerted (*single-step*) mechanisms just as an S_N2 reaction does. That is, old bonds are broken as new bonds are formed, all in one step. Pericyclic reactions are characterized by a cyclic transition state involving the pi bonds.

The energy of activation for pericyclic reactions is supplied by heat (**thermal induction**) or by ultraviolet light (**photo-induction**). (Solvents and electrophilic or nucleophilic reagents have little or no effect on the course of a pericyclic reaction.) Pericyclic reactions are generally stereospecific, and it is not uncommon that the two modes of induction yield products of opposite stereochemistry. For example, a thermally induced pericyclic reaction might yield a *cis*-product, while the photo-induced reaction of the same reactant yields the *trans*-product.

There are three principal types of pericyclic reactions:

1. **Cycloaddition reactions**, in which two molecules combine to form a ring. In these reactions two pi bonds are converted to two sigma bonds. The best-known example of a cycloaddition reaction is the Diels–Alder reaction, discussed in Section 10.16. "Line" formulas, such as those in the following equation, were also discussed in that section. Recall that a reactant must be in the s-*cis* (not s-*trans*) form to undergo cycloaddition.

1,3-butadiene   ethylene   cyclohexene

2. **Electrocyclic reactions**, reversible reactions in which a compound with conjugated double bonds undergoes cyclization. In the cyclization, two pi electrons are used to form a sigma bond.

3. **Sigmatropic rearrangements**, concerted intramolecular rearrangements in which an atom or group of atoms shifts from one position to another.

For many years, a theoretical understanding of the mechanisms of pericyclic reactions eluded chemists. However, since 1960, several theories have been developed to rationalize these reactions. R. B. Woodward of Harvard University and R. Hoffmann of Cornell University have proposed explanations based upon the symmetry of the molecular orbitals of the reactants and products. Hoffmann received a Nobel Prize in 1981 for this work. (Woodward received a Nobel Prize in 1965 for his work on organic synthesis of complex compounds, including chlorophyll.) A similar treatment of pericyclic reactions has been developed by K. Fukui (Nobel Prize, 1981) of Kyoto University in Japan. In this text, we will emphasize Fukui's approach, which is called the **frontier orbital method** of analyzing pericyclic reactions.

Before we discuss the mechanisms of pericyclic reactions, we will introduce some features of the molecular orbitals of conjugated systems. We suggest that you first review Sections 2.1–2.3 to refresh your understanding of bonding and antibonding molecular orbitals.

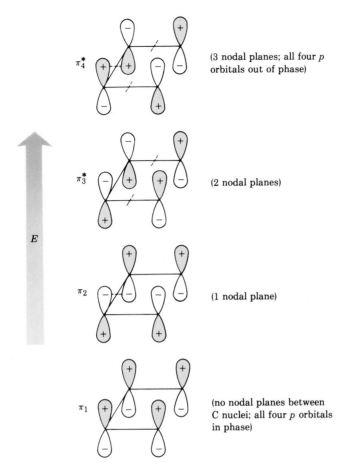

π_4^* (3 nodal planes; all four p orbitals out of phase)

π_3^* (2 nodal planes)

E

π_2 (1 nodal plane)

π_1 (no nodal planes between C nuclei; all four p orbitals in phase)

Figure 20.1 The bonding and antibonding π molecular orbitals of 1,3-butadiene, $CH_2=CHCH=CH_2$. The π_1 and π_2 orbitals are bonding orbitals; π_3^* and π_4^* are antibonding orbitals.

Section 20.1
Molecular Orbitals of Conjugated Polyenes

A conjugated polyene contains either $4n$ or $(4n + 2)$ pi electrons, where n is an integer, in its conjugated system. The simplest $4n$ system is represented by 1,3-butadiene, where $n = 1$. Any conjugated diene contains π molecular orbitals similar to those of 1,3-butadiene. Therefore, we can use 1,3-butadiene as a model for all conjugated dienes.

In 1,3-butadiene, four p orbitals are used in the formation of the π molecular orbitals; thus, four π molecular orbitals result. In this system, π_1 and π_2 are the bonding orbitals and π_3^* and π_4^* are the antibonding orbitals. Figure 20.1 depicts these orbitals in terms of increasing energy. Note that the higher-energy molecular orbitals are those with a greater number of nodes between nuclei.

In the ground state, 1,3-butadiene has its four pi electrons in the two orbitals of lowest energy: π_1 and π_2. In this case, π_2 is the **Highest Occupied**

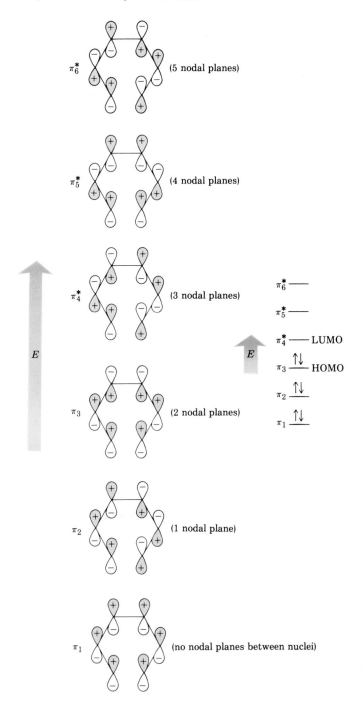

Figure 20.2 The bonding and antibonding π molecular orbitals of 1,3,5-hexa-triene, CH_2=$CHCH$=$CHCH$=CH_2.

Molecular Orbital, or **HOMO**, and π_3^* is the **Lowest Unoccupied Molecular Orbital**, or **LUMO**. The HOMO and LUMO are referred to as **frontier orbitals** and are the orbitals used in the frontier orbital method of analyzing pericyclic reactions.

Ground state of 1,3-butadiene:

$$\pi_4^* \underline{\quad\quad}$$

$$\pi_3^* \underline{\quad\quad} \longleftarrow \textit{the LUMO}$$

$$\pi_2 \underline{\underset{\uparrow\downarrow}{\quad\quad}} \longleftarrow \textit{the HOMO}$$

$$\pi_1 \underline{\underset{\uparrow\downarrow}{\quad\quad}}$$

E

When 1,3-butadiene absorbs a photon of the proper wavelength, an electron is promoted from the HOMO to the LUMO, which then becomes the new HOMO.

Excited state of 1,3-butadiene:

$$\pi_4^* \underline{\quad\quad}$$

$$\pi_3^* \underline{\underset{\downarrow}{\quad\quad}} \longleftarrow e^- \textit{ promoted to LUMO;}$$

$$\pi_2 \underline{\underset{\uparrow}{\quad\quad}} \qquad \pi_3^* \textit{ is now the HOMO}$$

$$\pi_1 \underline{\underset{\uparrow\downarrow}{\quad\quad}}$$

E

Aside from ethylene ($n = 0$), the simplest ($4n + 2$) system is represented by a conjugated triene ($n = 1$), such as 1,3,5-hexatriene. Because a triene contains a pi system formed from six p orbitals, a total of six π molecular orbitals results. These are depicted in Figure 20.2, along with the π orbital diagram of the ground state.

Study Problems

20.1 Draw the π orbital diagram for the lowest-energy *excited* state of 1,3,5-hexatriene.

20.2 Draw the π orbital diagram for the ground state of the nonconjugated diene 1,4-pentadiene. (Be careful!)

▼ ## Section 20.2
Cycloaddition Reactions

A **cycloaddition reaction** is a reaction in which two unsaturated molecules undergo an addition reaction to yield a cyclic product. For example,

two π electrons   *two π electrons*

$$\begin{array}{ccc} CH_2 & CH_2 & \\ \| & \| & \xrightarrow{h\nu} \\ CH_2 & CH_2 & \end{array} \qquad \begin{array}{c} CH_2-CH_2 \\ | \qquad | \\ CH_2-CH_2 \end{array}$$

ethylene   cyclobutane

The cycloaddition of ethylene or any two simple alkenes is called a **[2 + 2] cycloaddition**, because *two pi electrons + two pi electrons* are involved.

The Diels–Alder reaction is an example of a **[4 + 2] cycloaddition**. The diene contains four pi electrons that are used in the cycloaddition, while the dienophile contains two. (The carbonyl pi electrons in the following example are not used in bond formation in the reaction and therefore are not included in the number classification of this cycloaddition.)

the diene   *the dienophile*
(4 π electrons)   *(2 π electrons)*

Study Problem

20.3 Classify the following cycloaddition reaction by the number of pi electrons involved:

(95%)

Cycloaddition reactions are concerted, stereospecific reactions. (See Section 10.16 for a discussion of the stereochemistry of the Diels–Alder reaction.) Also, any particular cycloaddition reaction is either thermally induced or photo-induced, but not both.

A. [2 + 2] Cycloaddition

Cycloaddition reactions of the [2 + 2] type proceed readily in the presence of light of the proper wavelength, but not when the reaction mixture is heated. This behavior is readily explained by the frontier orbital theory: by assuming that electrons "flow" from the HOMO of one molecule to the LUMO of the other.

Let us consider the [2 + 2] cycloaddition of ethylene to yield cyclobutane. Ethylene has two π molecular orbitals: π_1 and π_2^*. In the ground state, π_1 is the bonding orbital and the HOMO, while π_2^* is the antibonding orbital and the LUMO.

Ethylene in the ground state:

Sample Problem

Draw the orbital diagram and the *p* orbitals for the lowest-energy *excited* state of ethylene, and indicate the HOMO.

Solution:

π_2^* ——— HOMO ↓

π_1 ——— ↑

In a cycloaddition reaction, the HOMO of one molecule must overlap with the LUMO of the second molecule. (It cannot overlap with the HOMO of the second molecule because that orbital is already occupied.) Simultaneously with the merging of the π orbitals, these orbitals also undergo hybridization to yield the new sp^3 sigma bonds.

When ethylene is heated, its π electrons are not promoted, but remain in the ground state, π_1. If we examine the phases of the ground-state HOMO of one ethylene molecule and the LUMO of another ethylene molecule, we can see why cyclization does not occur by thermal induction.

HOMO, π_1

$\xrightarrow{\text{heat}}$ no reaction

phases wrong for overlap, symmetry-forbidden

LUMO, π_2^*

For bonding to occur, the phases of the overlapping orbitals must be the same. This is not the case for the ground-state HOMO and LUMO of two ethylene molecules or any other [2 + 2] system. Because the phases of the orbitals are incorrect for bonding, a thermally induced [2 + 2] cycloaddition is said to be a **symmetry-forbidden reaction**. A symmetry-forbidden reaction may occur under some circumstances, but the energy of activation would be very high—possibly so high that other reactions, such as free-radical reactions, would occur first.

When ethylene is irradiated with ultraviolet light, a pi electron is promoted from the π_1 to the π_2^* orbital in some, but not all, of the molecules. The result is a mixture of the ground-state and excited-state ethylene molecules. If we examine the HOMO of an excited molecule (π_2^*) and the LUMO of a ground-state molecule (also π_2^*), we see that the phases are now correct for bonding. Such a reaction has a relatively low energy of activation, and is said to be **symmetry-allowed**.

excited state HOMO, π_2*

phases correct for bonding,
symmetry-allowed

ground-state LUMO, π_2*

transition state
in which π orbitals
are being rehybri-
dized to σ orbitals

Although the cycloaddition of ethylene itself proceeds in poor yield, other photo-induced [2 + 2] cycloadditions do find synthetic utility. Probably the widest use of this type of reaction is with intramolecular cyclizations, which can yield some very unusual "cage" structures.

(74%)   (80%)

Study Problem

20.4 Suggest synthetic routes to the following compounds from open-chain starting materials.

(a)   (b)

B. [4 + 2] Cycloadditions

As we have mentioned, the Diels–Alder reaction is the best-known [4 + 2] cycloaddition. The examples on page 442 illustrate the versatility of this reaction. Note that the Diels–Alder reaction requires heat, not ultraviolet light, for its success. This experimental condition is different from that required for a [2 + 2] cycloaddition. To see why this is so, we will examine the HOMO–LUMO interactions of *only the p-orbital components that will form the new sigma bonds* in a [4 + 2] cycloaddition. We will compare the HOMO–LUMO interactions for the ground state (for a thermally induced reaction) and those for the excited state (for an attempted photo-induced reaction). Based upon experimental observations, we would expect to find that the HOMO–LUMO interactions of the thermally induced reaction are symmetry-allowed and those of the photo-induced reaction are symmetry-forbidden.

We will use the simplest [4 + 2] system: the cycloaddition of 1,3-butadiene (the diene) and ethylene (the dienophile). The frontier orbital pictures may be extrapolated to other [4 + 2] cycloadditions. In the thermally induced reaction, we can visualize the pi electrons "flowing" from the HOMO (π_2) of the diene (Figure 20.1) to the LUMO (π_2^*) of the dienophile. Note the phases of the orbitals that lead to the thermally induced reaction. This reaction is symmetry-allowed.

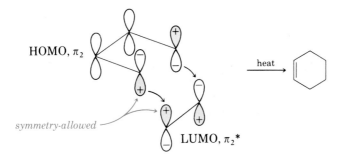

When a diene is excited by light, its HOMO becomes the π_3^* orbital, and this molecular orbital cannot overlap with the LUMO of the dienophile. The photo-induced [4 + 2] cyclization is therefore symmetry-forbidden.

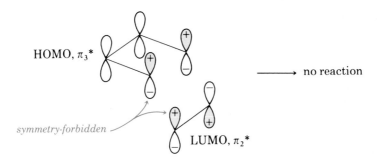

Study Problems

20.5 Predict whether a [4 + 2] cycloaddition could be photo-induced if the dienophile, instead of the diene, were the excited reactant. Explain your answer.

20.6 Is the cycloaddition reaction in Problem 20.3 (page 832) a thermally induced or photo-induced reaction? Explain.

Section 20.3
Electrocyclic Reactions

An **electrocyclic reaction** is the concerted interconversion of a conjugated polyene and a cycloalkene. We will discuss primarily the cyclization. The reverse reaction, ring opening, proceeds by the same mechanism, but in reverse.

Electrocyclic reactions are induced either thermally or photochemically:

1,3-butadiene   cyclobutene   1,3,5-hexatriene   1,3-cyclohexadiene

An intriguing feature about electrocyclic reactions is that the stereochemistry of the product is dependent on whether the reaction is thermally induced or photo-induced. For example, when (2*E*, 4*Z*, 6*E*)-octatriene is heated, the *cis*-dimethylcyclohexadiene is the product. When the triene is irradiated with ultraviolet light, however, the *trans*-dimethylcyclohexadiene is formed. We will discuss the reasons for this in Section 20.3B.

cis-5,6-dimethyl-1,3-cyclohexadiene

(2*E*,4*Z*,6*E*)-octatriene

trans-5,6-dimethyl-1,3-cyclohexadiene

A. Cyclization of 4n Systems

A conjugated polyene yields a cycloalkene by the end-to-end overlap of its *p* orbitals and the simultaneous rehybridization of the carbon atoms involved in bond formation. 1,3-Butadiene, which has 4*n* pi electrons, is the simplest polyene. Therefore, we will introduce the mechanism with this compound. Because cyclobutanes are strained, the reverse reaction (ring opening) is usually favored. However, the mechanism for ring opening is simply the reverse of that for ring closing.

The two lobes of each *p* orbital that will form the new sigma bond in cyclization are either *in phase* or *out of phase* with each other:

in phase   *out of phase*

To form a new sigma bond, the existing C—C sigma bonds must rotate so that the *p* orbitals can undergo end-to-end overlap. For this to occur, the existing pi bonds must be broken. The energy for the pi-bond breakage and the bond rotation is supplied by the heat or ultraviolet light. To form a sigma bond, the pair of overlapping lobes of the two *p* orbitals must be *in phase* after rotation.

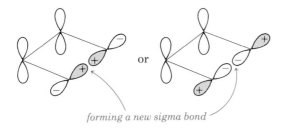

forming a new sigma bond

There are two different ways in which the existing C—C sigma bonds can rotate in order to position the *p* orbitals for overlap. (1) The two C—C sigma bonds can rotate in the *same direction* (either both clockwise or both counterclockwise). This type of rotation is referred to as **conrotatory motion**. (2) The two C—C sigma bonds can rotate in *different* directions, one clockwise and one counterclockwise. This type of rotation is **disrotatory motion**.

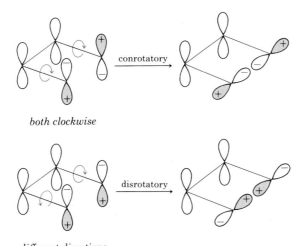

both clockwise

different directions

Note that in the two preceding equations, the phases of the *p* orbitals in the two starting dienes are different. Therefore, the direction of rotation for symmetry-allowed overlap depends upon the phases of the *p*-orbitals just prior to cyclization. If the *p* orbitals are out of phase before rotation, then conrotatory motion brings them into phase after rotation. If the *p* orbitals are in phase before rotation, then disrotatory motion is required. To determine which diene system is present just prior to reaction, we must consider the phases of the *p* orbitals in the ground state and the excited state of the diene.

When 1,3-butadiene is *heated*, reaction takes place from the ground state. The electrons that are used for sigma-bond formation are in the HOMO (π_2, in this case). In Figure 20.1 (page 829), it can be seen that the pertinent *p* orbitals in this HOMO are out of phase with each other. For the new sigma bond to form, rotation must be *conrotatory*. Only in this way are the in-phase lobes allowed to overlap. (Disrotatory motion would not place the in-phase lobes together.)

π_2
thermal
(out of phase)

In *photo-induced* cyclization, the phases of the *p* orbitals of the HOMO (now π_3^*) are the reverse of that in thermal cyclization (see Figure 20.1). Therefore, the symmetry-allowed rotation is *disrotatory* instead of conrotatory.

π_3^*
photo
(*in phase*)

Study Problem

20.7 Draw structures showing *conrotatory* motion of 1,3-butadiene in the excited state (π_3^*). Are the potential bonding *p* orbitals in a symmetry-allowed or symmetry-forbidden orientation?

B. Stereochemistry of a 4*n* Electrocyclization

If (2*E*,4*Z*)-hexadiene were cyclized, the *cis*-dimethylcyclobutane would result from heating. In thermal cyclization, conrotatory motion is required for bond formation. Both methyl groups rotate in the same direction. As a result, they end up on the same side of the ring, or *cis*, in the product.

Just the reverse would occur in photochemical cyclization. In the disrotatory motion, one of the methyl groups rotates up and the other rotates down. The result is that the two methyl groups are *trans* in the product.

Sample Problems

Would the photochemical electrocyclic reaction of (2*E*,4*E*)-hexadiene yield *cis*- or *trans*-3,4-dimethylcyclobutene?

Solution: 2,4-Hexadiene is a 4*n* polyene. Therefore, the photochemical electrocyclic reaction takes place by disrotatory motion.

(2*E*,4*E*)-hexadiene   *cis*-3,4-dimethylcyclobutene

What is the structure and expected stereochemistry of the ring-opened product when *trans*-3,4-dimethylcyclobutene is heated?

Solution:

(2*E*,4*E*)-hexadiene

C. Cyclization of (4*n* + 2) Systems

Figure 20.2 (page 830) shows the π orbitals of 1,3,5-hexatriene, a (4*n* + 2) polyene. In the HOMO of the ground state (π_3), the *p* orbitals that form the sigma bond in the cyclization are in phase. Therefore, the thermal cyclization proceeds by *disrotatory motion*.

When an electron of 1,3,5-hexatriene is promoted by photon-absorption, π_4^* becomes the HOMO and thus the p orbitals in question become out of phase. Therefore, photo-induced cyclization proceeds by *conrotatory motion*. The symmetry-allowed reactions of this $(4n + 2)$ system are just the opposite of those for 1,3-butadiene, a $4n$ system.

A summary of the types of motion to be expected from the different types of polyenes under the influence of heat and ultraviolet light is shown in Table 20.1.

Study Problem

20.8 Without referring to the start of this section, predict the stereochemistry of the products.

(a)

(b)

Table 20.1 Types of electrocyclic reactions

Number of Pi Electrons	Reaction	Motion
$4n$	thermal	conrotatory
$4n$	photochemical	disrotatory
$(4n + 2)$	thermal	disrotatory
$(4n + 2)$	photochemical	conrotatory

The thermally induced electrocyclic reactions of (2*E*,4*Z*,6*Z*,8*E*)-deca-tetraene provide elegant examples of electrocyclic reactions. The starting tetraene forms a cyclooctatriene near room temperature. The tetraene is a 4*n* polyene. Therefore, a conrotatory motion is the expected mode of cyclization. Indeed, the *trans*-dimethylcyclooctatriene is the product of this initial cyclization. When this cyclooctatriene is heated to a slightly higher temperature, another electrocyclic ring closure occurs. However, the cyclooctatriene is a (4*n* + 2) polyene. Therefore, this thermally induced electrocyclic reaction proceeds with disrotatory motion, and a *cis* ring junction is formed.

trans methyl groups   *cis ring junction*

Section 20.4
Sigmatropic Rearrangements

A **sigmatropic rearrangement** is a concerted intramolecular shift of an atom or a group of atoms. Two typical examples of sigmatropic rearrangements are:

Cope rearrangement:

1,5-heptadiene   *transition state*   3-methyl-1,5-hexadiene

Claisen rearrangement:

allyl phenyl ether   *transition state*

keto form   *o*-allylphenol

                              enol form

A. Classification of Sigmatropic Rearrangements

Sigmatropic rearrangements are classified by a double numbering system that refers to the relative positions of the atoms involved in the migration. This method of classification is different from those for cycloadditions or electrocyclic reactions, which are classified by the number of π electrons involved in the cyclic transition state.

The method used in classifying sigmatropic reactions is best explained by example. Consider the following rearrangement:

numbering of the migrating group

numbering of the alkenyl chain

Both the alkenyl chain and the migrating group are numbered *starting at the position of their original attachment*, not necessarily at a carbon atom. (Note that these numbers are not related to nomenclature numbers.) In the example, atom 1 of the migrating group ends up on atom 3 of the alkenyl chain. Therefore, this sigmatropic rearrangement would be classified as a [1,3] sigmatropic rearrangement.

In a similar manner, the following reaction would be classified as a [1,7] sigmatropic shift. (In this example, there is no atom 2 in the migrating group.)

It is not always the first atom of the migrating group that becomes bonded to the alkenyl chain in the rearrangement. Consider the following example. In this case, atom 3 of the migrating group becomes bonded to atom 3 of the alkenyl chain. This is an example of a [3,3] sigmatropic rearrangement.

the migrating group

Atom 3 of the migrating group becomes bonded to atom 3 of the alkenyl chain

the alkenyl chain

Study Problem

20.9 Classify the following rearrangements by the preceding technique:

(a) the Claisen rearrangement

(b) $CH_3CH_2CH{=}CHCH{=}CDCH_3 \longrightarrow CH_3CH{=}CHCH{=}CHCHDCH_3$

B. Mechanism of Sigmatropic Rearrangements

Sigmatropic rearrangements of the [1,3] type are relatively rare, while [1,5] sigmatropic rearrangements are fairly common. We can use the frontier orbital approach to analyze these reactions and see why this is so. Let us first consider the following thermally induced sigmatropic rearrangement, which is a [1,3] shift:

$$\overset{\overset{\displaystyle H}{\underset{\displaystyle |}{}}}{CH_2CH}=CD_2 \xrightarrow{\text{difficult}} \overset{\overset{\displaystyle H}{\underset{\displaystyle |}{}}}{CH_2=CHCD_2}$$

For the purpose of analyzing the orbitals, it is assumed that the sigma bond connecting the migrating group to its original position (the CH bond in our example) undergoes homolytic cleavage to yield two radicals. This is *not* how the reaction takes place (the reaction is concerted), but this assumption does allow analysis of the molecular orbitals.

$$\overset{\overset{\displaystyle H}{\underset{\displaystyle |}{}}}{CH_2CH}=CD_2 \xrightarrow[\text{homolytic cleavage}]{\text{hypothetical}} \underset{\text{allyl radical}}{H\cdot \quad \cdot CH_2CH=CD_2}$$

The products of the hypothetical cleavage are a hydrogen atom and an allyl radical, which contains three pi electrons and thus three π molecular orbitals. The π molecular orbitals of the allyl radical are shown in Figure 20.3.

The actual shift of the H· could take place in one of two directions. In the first case, the migrating group could remain on the same side of the π orbital system. Such a migration is termed a **suprafacial process**. As you can see, in this system a suprafacial migration is geometrically feasible but symmetry-forbidden.

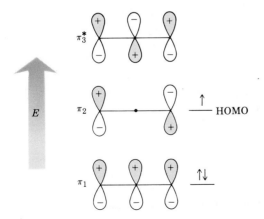

Figure 20.3 The three π molecular orbitals of an allyl radical. (Note that π_2 contains one node at carbon 2.)

Suprafacial migration:

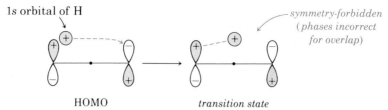

HOMO   *transition state*

Let us consider the second mode of migration. For a symmetry-allowed [1,3] sigmatropic shift to occur, the migrating group (H· in our example) must shift by an **antarafacial process**—that is, it must migrate to the *opposite face* of the orbital system.

Antarafacial migration:

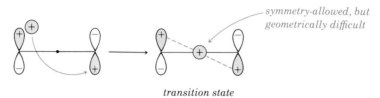

transition state

While symmetry-allowed, a [1,3] antarafacial sigmatropic rearrangement of H is not geometrically favorable. Our conclusion is that [1,3] sigmatropic shifts should not occur readily. This conclusion is in agreement with experimental facts. As we have mentioned, [1,3] sigmatropic rearrangements are rare.

By contrast, [1,5] sigmatropic shifts are quite common. A simple example follows:

$$\overset{\overset{\displaystyle H}{|}}{CH_2}CH{=}CHCH{=}CD_2 \xrightarrow{\text{[1,5]}} CH_2{=}CHCH{=}\overset{\overset{\displaystyle H}{|}}{CH}CD_2$$

If we again assume a homolytic bond cleavage for purposes of analysis, we must consider the π molecular orbitals of a pentadienyl radical, which contains five pi electrons. These orbitals are depicted in Figure 20.4.

$$\overset{\overset{\displaystyle H}{|}}{CH_2}CH{=}CHCH{=}CH_2 \xrightarrow[\text{homolytic cleavage}]{\text{hypothetical}} \overset{\overset{\displaystyle H\cdot}{|}}{\cdot CH_2}CH{=}CHCH{=}CH_2$$

pentadienyl radical

Considering the HOMO of this radical and the orbital symmetry, we can see that the [1,5] shift is both symmetry-allowed and suprafacial.

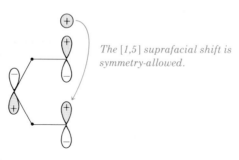

The [1,5] suprafacial shift is symmetry-allowed.

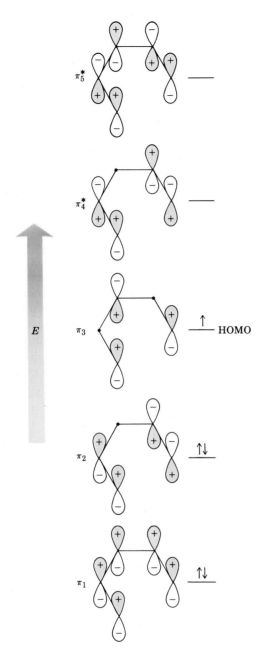

Figure 20.4 The five π molecular orbitals of the pentadienyl radical.

Study Problem

20.10 Which of the following known sigmatropic rearrangements would proceed readily and which slowly? Explain your answers.

(a)

(b)

Section 20.5
Pericyclic Reactions Leading to Vitamin D

Pericyclic reactions are not merely laboratory curiosities; they are also observed in natural processes. As one example, let us briefly consider some of the transformations that occur in the vitamin D group of compounds.

Vitamin D is essential for the proper growth of bones. Lack of vitamin D results in defective bone growth, a condition called *rickets*. Humans can obtain vitamin D by a number of routes. One is by the action of sunlight on a particular steroid, 7-dehydrocholesterol, which is found in the skin. (Steroids are discussed in Section 23.7.) This steroid undergoes a photo-induced electrocyclic ring opening to yield a triene. The triene, previtamin D, then undergoes a thermally induced [1,7] sigmatropic shift to yield vitamin D_3. (The subscript 3 is used to differentiate this vitamin D from other structurally similar compounds with vitamin D activity.)

7-dehydrocholesterol

in the skin

previtamin D

vitamin D_3

Another source of vitamin D is *irradiated ergosterol*, which is commonly added to milk. The conversion of ergosterol to vitamin D_2 proceeds by the same series of reaction steps as the conversion of 7-dehydrocholesterol.

ergosterol

*in yeast, soybean oil,
and ergot (a fungus)*

vitamin D_2
(calciferol)

Study Problem

20.11 Previtamin D can undergo both thermally induced and photo-induced electrocyclic ring-closure reactions. What will be the stereochemistry at positions 9 and 10 of the resulting products? (See page 947 for the numbering of the steroid ring.)

▼ Summary

Pericyclic reactions are concerted, thermally induced or photo-induced reactions with cyclic transition states. Three types of pericyclic reactions are:

Cycloaddition:

Electrocyclization:

Sigmatropic rearrangement:

In the **frontier orbital method** of analyzing cycloaddition reactions, electrons are assumed to flow from the **HOMO** of one molecule to the **LUMO** of the other. If the phases of these orbitals are the same, the reaction is **symmetry-allowed**. If the orbital phases are opposite and show antibonding character, the reaction is **symmetry-forbidden**. Symmetry-allowed [2 + 2] cycloadditions are photo-induced, while [4 + 2] cycloadditions are thermally induced.

In electrocyclic reactions, *p*-orbital components of the HOMO undergo end-to-end overlap to form the new sigma bond. To do this, they must undergo **conrotatory** or **disrotatory** motion, which, in turn, determines the stereochemistry. A summary of the types of motion to be expected is shown in Table 20.1 (page 840).

Sigmatropic rearrangements occur **suprafacially** or **antarafacially**, depending on the phases of the interacting orbitals in the HOMO of a hypothetical radical system. The geometry of the transition state determines whether the rearrangement proceeds readily or not. The classification of sigmatropic rearrangements is discussed in Section 20.4A.

Study Problems*

20.12 Identify each of the following pericyclic reactions as being   (1) a cycloaddition;   (2) an electrocyclic reaction; or   (3) a sigmatropic rearrangement:

(a) 2 $\xrightarrow{\text{heat}}$

(b) $CH_2{=}CHCCH_2CH{=}CH_2$ (with CO_2CH_3 above and CO_2CH_3 below) $\xrightarrow{\text{heat}}$ $CH_2{=}CHCH_2CH_2CH{=}C$ (with CO_2CH_3 and CO_2CH_3)

(c) $\xrightarrow{\text{heat}}$ $CH_2{=}CHCH{=}C$ (with CH_3 and CH_3)

(d) $\xrightarrow{\text{heat}}$

(e) $\xrightarrow{h\nu}$

20.13 (1) Draw the p-orbital arrays in the π molecular orbitals of the following ions.   (2) Draw diagrams showing the occupied orbitals of the ground states and indicate the HOMO's. (See Figure 20.2, page 830, for an example.)

(a) $CH_2{=}CH\overset{+}{C}HCH_3$ (b) $CH_2{=}CH\overset{\cdot}{C}HCH_3$
(c) $CH_3CH{=}CHCH{=}CH\overset{\cdot\cdot}{C}HCH_3$

20.14 Classify each of the following cycloadditions as [2 + 2], [4 + 2], etc.:

(a) $2\ CH_2{=}C{=}CH_2 \longrightarrow$

(b)

* For information concerning the organization of the *Study Problems* and *Additional Problems*, see the *Note to student* on page 39.

(c) =CH$_2$ + CH$_3$O$_2$CC≡CCO$_2$CH$_3$ ⟶

20.15 Which of the following types of cycloadditions would you predict to proceed easily upon heating?

(a) [6 + 2]   (b) [6 + 4]   (c) [8 + 2]   (d) [8 + 4]

20.16 1,2-Diphenylcyclobutene undergoes a photochemical dimerization. What is the structure of the dimer? (Include the stereochemistry in your answer.)

20.17 Suggest synthetic routes to the following compounds from monocyclic or acyclic starting materials:

(a)

(b)

(c)

(d)

20.18 (1) Tell whether each of the following dienes or trienes would undergo conrotatory or disrotatory motion in a cyclization reaction. (2) What product would be observed? (Include stereochemistry where appropriate.)

(a) $\xrightarrow{\text{heat}}$

(b) $\xrightarrow{hv}$

(c) $\xrightarrow{hv}$

(d) $\xrightarrow{\text{heat}}$

20.19 Predict the electrocyclic product and its stereochemistry:

(a) $\xrightarrow{hv}$

(b) $\xrightarrow{\text{heat}}$

(c) $\xrightarrow{\text{heat}}$

(d) $\xrightarrow{hv}$

(e) $\xrightarrow{\text{heat}}$

20.20 The following compounds both undergo photo-induced electrocyclic reactions. What are the structures and stereochemistry of the products?

(a)

(b)

20.21 How would you carry out the following conversions?

(a)

(b)

20.22 Classify each of the following sigmatropic rearrangements as [1,3], [3,3], etc.

(a) $\xrightarrow{\text{heat}}$

(b) $\xrightarrow{\text{heat}}$

(c) $\xrightarrow{\text{heat}}$

20.23 Predict the product or products of the sigmatropic rearrangement of 7,7-dideuterio-1,3,5-cycloheptatriene.

20.24 Explain the following observation:

20.25 Should you use thermal induction or photo-induction to carry out the following reactions?

(a)

(b) $+ CH_3CH_2C \equiv CCH_2CH_3 \longrightarrow$

(c) $CH_2 = CHCHCH_2CH = CH_2 \longrightarrow CH_3CH = CHCH_2CH_2CH = CH_2$

with CH_3 above the third carbon.

20.26 When a mixture of quinone (page 507) and cyclopentadiene is heated, the two compounds undergo an addition reaction. When the product of this addition is exposed to sunlight, it undergoes isomerization. What are the structures of the addition product and its isomer?

Additional Problems

20.27 Predict the structures and stereochemistry of the major organic products:

(a) CH_2=$\overset{\underset{|}{CH_3}}{C}CH$=$CH_2$ + C_6H_5CH=CH_2 $\xrightarrow{\text{heat}}$

(b) (E,E)-CH_3OCH=$CHCH$=$CHOCH_3$ + CH_3O_2CC≡CCO_2CH_3 $\xrightarrow{\text{heat}}$

(c)
$\xrightarrow{\text{heat}}$

(d) $trans$-C_6H_5CH=CHC_6H_5 + $trans$-CH_3CH=$CHCO_2CH_3$ $\xrightarrow{hv}$

(e)
$\xrightarrow{hv}$

(f) [the product from (e)] $\xrightarrow{\text{heat}}$

20.28 What two products could be formed by a [4 + 2] cycloaddition dimerization of 3-buten-2-one?

20.29 Suggest synthetic routes to the following compounds from acyclic or monocyclic starting materials:

(a)

(b)

(c)

(d)

(e)

(f)

(g)

(h)

20.30 Suggest mechanisms for the following reactions:

(a)
$\xrightarrow{\text{heat}}$

(b)

$$CH_3CH_2CH_2CO_2H + CH_3CC \equiv CCH_2CH_2CH_3 \xrightarrow{h\nu}$$

with the $C=O$ drawn above the CC as:

$$\overset{O}{\overset{\|}{C}}$$

(c) $CH_3CH_2CH_2CO_2H + CH_3\overset{\overset{\displaystyle O}{\|}}{C}C \equiv CCH_2CH_2CH_3 \xrightarrow{h\nu}$

$$CH_3CH_2CH_2\overset{\overset{\displaystyle O}{\|}}{C}\overset{}{C}H\overset{\overset{\displaystyle O}{\|}}{C}CH_2CH_2CH_3$$
$$\underset{\underset{\displaystyle CH_3C=O}{|}}{}$$

20.31 Complete the following equations, showing the stereochemistry of the products:

(a) $2 \cdot$ $\xrightarrow{h\nu}$

(b)

$+ \; CH_2=C=CH_2 \xrightarrow{heat}$

(c) $+ \; CH_2=CH_2 \xrightarrow{heat}$

(d) $+ \; CH_3O_2CC \equiv CCO_2CH_3 \xrightarrow{heat}$

21 ▽

Spectroscopy II: Ultraviolet Spectra, Color and Vision, Mass Spectra

In Chapter 9, we discussed the absorption of infrared and radiofrequency radiation by organic compounds and how the absorption can be used in structure identification. In this chapter, we will consider the absorption of **ultraviolet (uv) and visible light** by organic compounds. Ultraviolet and visible spectra are also used in structure determination. More important, the absorption of visible light results in vision. We will also discuss this topic, along with colors and dyes. Last, we will introduce **mass spectra**, which arise from fragmentation of molecules when they are bombarded with high-energy electrons.

▼ ## Section 21.1
Ultraviolet and Visible Spectra

The wavelengths of uv and visible light are substantially shorter than the wavelengths of infrared radiation (see Figure 9.2, page 324). The unit we will use to describe these wavelengths is the *nanometer* (1 nm = 10^{-7} cm). The visible spectrum spans from about 400 nm (violet) to 750 nm (red), while the ultraviolet spectrum ranges from 100 to 400 nm.

The quantity of energy absorbed by a compound is inversely proportional to the wavelength of the radiation:

$$\Delta E = h\nu = \frac{hc}{\lambda}$$

where ΔE = energy absorbed, in ergs

h = Planck's constant, 6.6×10^{-27} erg-sec

ν = frequency, in Hz

c = speed of light, 3×10^{10} cm/sec

λ = wavelength, in cm

Infrared radiation is relatively low-energy radiation. Absorption of infrared radiation by a molecule leads to increased vibrations of covalent bonds. Molecular transitions from the ground state to an excited vibrational state require about 2–15 kcal/mol.

Both uv and visible radiation are of higher energy than infrared radiation. Absorption of ultraviolet or visible light results in **electronic transitions**, promotion of electrons from low-energy ground-state orbitals to higher-energy excited-state orbitals. These transitions require about 40–300 kcal/mol. The energy absorbed is subsequently dissipated as heat, as light, or in chemical reactions (such as isomerization or free-radical reactions).

The wavelength of uv or visible light absorbed depends on the ease of electron promotion. Molecules that require *more energy* for electron promotion absorb at *shorter wavelengths*. Molecules that require *less energy* absorb at *longer wavelengths*. Compounds that absorb light in the visible region (that is, colored compounds) have more-easily promoted electrons than compounds that absorb at shorter uv wavelengths.

absorption at 100 nm (uv) $\longrightarrow$ 750 nm (visible)

increasing ease of electronic transition

Study Problem

21.1 Which would have the more-easily promoted electrons, anthracene (colorless) or Tyrian purple?

anthracene   Tyrian purple

A uv or visible spectrophotometer has the same basic design as an infrared spectrophotometer (see Figure 9.5, page 328). Absorption of radiation by a

Figure 21.1 Ultraviolet spectrum of mesityl oxide, 9.2×10^{-5} M, 1.0-cm cell. (The $n \rightarrow \pi^*$ transition, at about 325 nm, is too weak to be detected at this solution concentration. The different types of transitions are described in Section 21.3.)

sample is measured at various wavelengths and plotted by a recorder to give the spectrum. (Figure 21.1 shows a typical uv spectrum.)

Because energy absorption by a molecule is quantized, we might expect that the absorption for electronic transitions would be observed at discrete wavelengths as a spectrum of lines or sharp peaks. This is not the case. Instead, a uv or visible spectrum consists of broad absorption bands over a wide range of wavelengths. The reason for the broad absorption is that the energy levels of both the ground state and the excited state of a molecule are subdivided into *rotational and vibrational sublevels*. Electronic transitions may occur from any one of the sublevels of the ground state to any one of the sublevels of an excited state (Figure 21.2). Since these various transitions differ slightly in energy, their wavelengths of absorption also differ slightly and give rise to the broad band observed in the spectrum.

Figure 21.2 Schematic representation of electronic transitions from a low energy level to a high energy level.

▼   ## Section 21.2
Expressions Used in Ultraviolet Spectroscopy

Figure 21.1 shows the uv spectrum of a dilute solution of mesityl oxide (4-methyl-3-penten-2-one). The spectrum shows the scan from 200 to 400 nm. (Because absorption by atmospheric carbon dioxide becomes significant below 200 nm, the 100–200 nm region is usually not scanned.) The wavelength of absorption is usually reported as λ_{max}, the wavelength at the highest point of the curve. The λ_{max} for mesityl oxide is 232 nm.

The absorption of energy is recorded as **absorbance** (not transmittance as in infrared spectra). The absorbance at a particular wavelength is defined by the equation:

$$A = \log \frac{I_0}{I}$$

where A = absorbance

I_0 = intensity of the reference beam

I = intensity of the sample beam

The absorbance by a compound at a particular wavelength increases with an increasing number of molecules undergoing transition. Therefore, the absorbance depends on the electronic structure of the compound and also upon the concentration of the sample and the length of the sample cell. For this reason, chemists report the energy absorption as **molar absorptivity ε** (sometimes called the *molar extinction coefficient*) rather than as the actual absorbance. Often, uv spectra are replotted to show ε or $\log \varepsilon$ instead of A as the ordinate. The $\log \varepsilon$ value is especially useful when values for ε are very large.

$$\varepsilon = \frac{A}{cl}$$

where ε = molar absorptivity

A = absorbance

c = concentration, in M

l = cell length, in cm

The molar absorptivity (usually reported at the λ_{max}) is a reproducible value that takes into account concentration and cell length. Although ε has the units $M^{-1}\,cm^{-1}$, it is usually shown as a unitless quantity. For mesityl oxide, the ε_{max} is $1.2 \div (9.2 \times 10^{-5} \times 1.0)$, or 13,000 (values taken from Figure 21.1).

Sample Problem

A flask of cyclohexane is known to be contaminated with benzene. At 260 nm, benzene has a molar absorptivity of 230, and cyclohexane has a molar absorptivity of zero. A uv spectrum of the contaminated cyclohexane (1.0-cm cell length) shows an absorbance of 0.030. What is the concentration of benzene?

Solution:

$$c = \frac{A}{\varepsilon l} = \frac{0.030}{230 \times 1.0} = 0.00013M$$

Section 21.3
Types of Electron Transitions

Let us consider the different types of electron transitions that give rise to ultraviolet or visible spectra. The ground state of an organic molecule contains valence electrons in three principal types of molecular orbitals: **sigma (σ) orbitals; pi (π) orbitals; and filled, but nonbonded, orbitals (n).**

$$\overset{\sigma\ electrons}{\underset{}{\text{H:CH}_3}} \qquad \overset{\pi\ electrons}{\underset{}{\text{CH}_2::\text{CH}_2}} \qquad \overset{n\ electrons}{\underset{}{\text{CH}_3\ddot{\text{O}}\text{H}}}$$

Both σ and π orbitals are formed from the overlap of two atomic or hybrid orbitals. Each of these molecular orbitals therefore has an antibonding σ^* or π^* orbital associated with it. An orbital containing n electrons does not have an antibonding orbital (because it was not formed from two orbitals). Electron transitions involve the promotion of an electron from one of the three ground state orbitals (σ, π, or n) to one of the two excited state orbitals (σ^* or π^*). There are six possible transitions; the four important transitions and their relative energies are shown in Figure 21.3.

The most useful region of the uv spectrum is at wavelengths longer than 200 nm. The following transitions give rise to absorption in the nonuseful 100–200 nm range: $\pi \rightarrow \pi^*$ for an isolated double bond and $\sigma \rightarrow \sigma^*$ for an ordinary carbon–carbon bond. The useful transitions (200–400 nm) are $\pi \rightarrow \pi^*$ for compounds with conjugated double bonds, and some $n \rightarrow \sigma^*$ and $n \rightarrow \pi^*$ transitions.

A. Absorption by Polyenes

Less energy is required to promote a π electron of 1,3-butadiene than is needed to promote a π electron of ethylene. The reason is that the energy difference between the HOMO (Highest Occupied Molecular Orbital) and the LUMO (Lowest Unoccupied Molecular Orbital) for conjugated double bonds is less than the energy difference for an isolated double bond.

Figure 21.3 Energy requirements for important electronic transitions. (The corresponding wavelengths are in parentheses.)

For $CH_2=CH_2$:   For $CH_2=CHCH=CH_2$:

Because less energy is needed for a $\pi \rightarrow \pi^*$ transition of 1,3-butadiene, this diene absorbs uv radiation of longer wavelengths than does ethylene. As more conjugated double bonds are added to a molecule, the energy required to reach the first excited state decreases. Sufficient conjugation shifts the absorption to wavelengths that reach into the visible region of the spectrum; a compound with sufficient conjugation is colored. For example, lycopene, the compound responsible for the red color of tomatoes, has eleven conjugated double bonds. (Line formulas, such as the following one, are described in Section 10.16A.)

lycopene
$\lambda_{max} = 505$ nm

Table 21.1 lists the λ_{max} values for $\pi \rightarrow \pi^*$ transitions of a series of aldehydes with increasing conjugation. Inspection of the table reveals that the position of absorption is shifted to longer wavelengths as the extent of the conjugation increases. Generally, this increase is about 30 nm per conjugated double bond in a series of polyenes.

Study Problems

21.2 List the following all-*trans*-polyenes in order of increasing λ_{max}:

(a) $CH_3(CH=CH)_{10}CH_3$ (b) $CH_3(CH=CH)_9CH_3$
(c) $CH_3(CH=CH)_8CH_3$

21.3 The λ_{max} for Compound (a) in Problem 21.2 is 476 nm. Predict the λ_{max} values for Compounds (b) and (c)

Table 21.1 Ultraviolet absorption for some unsaturated aldehydes

Structure	λ_{max}, nm
$CH_3CH=CHCHO$	217
$CH_3(CH=CH)_2CHO$	270
$CH_3(CH=CH)_3CHO$	312
$CH_3(CH=CH)_4CHO$	343
$CH_3(CH=CH)_5CHO$	370

B. Absorption by Aromatic Systems

Benzene and other aromatic compounds exhibit more-complex spectra than can be explained by simple $\pi \to \pi^*$ transitions. The complexity arises from the existence of *several* low-lying excited states. Benzene absorbs strongly at 184 nm ($\varepsilon = 47{,}000$) and at 202 nm ($\varepsilon = 7{,}000$) and has a series of absorption bands between 230–270 nm. A value of 260 nm is often reported as the λ_{max} for benzene because this is the position of strongest absorption above 200 nm. Solvents and substituents on the ring alter the uv spectra of benzene compounds.

The absorption of uv radiation by aromatic compounds composed of fused benzene rings is shifted to longer wavelengths as the number of rings is increased because of increasing conjugation and greater resonance-stabilization of the excited state.

benzene
$\lambda_{max} = 260$ nm

naphthalene
$\lambda_{max} = 280$ nm

phenanthrene
$\lambda_{max} = 350$ nm

anthracene
$\lambda_{max} = 375$ nm

naphthacene
$\lambda_{max} = 450$ nm
(yellow)

pentacene
$\lambda_{max} = 575$ nm
(blue)

coronene
$\lambda_{max} = 400$ nm
(yellow)

Study Problem

21.4 Suggest a reason why coronene absorbs at a *shorter* wavelength than does naphthacene.

C. Absorption Arising from Transitions of *n* Electrons

Compounds that contain nitrogen, oxygen, sulfur, phosphorus, or one of the halogens all have unshared *n* electrons. If the structure contains no π bonds, these *n* electrons can undergo only $n \to \sigma^*$ transitions. Because the *n* electrons are of higher energy than either the σ or π electrons, less energy is required to promote an *n* electron, and transitions occur at longer wavelengths than $\sigma \to \sigma^*$ transitions. Note that some of these values are within the usual uv spectral range of 200–400 nm (Table 21.2). The π^* orbital is of lower energy than the σ^* orbital. Consequently, $n \to \pi^*$ transitions require less energy than $n \to \sigma^*$ transitions and often are in the range of a normal instrument scan.

The *n* electrons are in a different region of space from σ^* and π^* orbitals, and the probability of an *n* transition is low. Since molar absorptivity depends

Table 21.2 Ultraviolet absorption arising from $n \rightarrow \sigma^*$ transitions

Structure	λ_{max}, nm	ε	Structure	λ_{max}, nm	ε
CH$_3$ÖH	177	200	CH$_3$CH$_2$CH$_2$B̈r:	208	300
(CH$_3$)$_3$N̈	199	3950	CH$_3$Ï:	259	400
CH$_3$C̈l:	173	200			

on the number of electrons undergoing transition, ε values for n transitions are low, in the 10–100 range (compared to about 10,000 for a $\pi \rightarrow \pi^*$ transition).

A compound such as acetone that contains both a π bond and n electrons exhibits both $\pi \rightarrow \pi^*$ and $n \rightarrow \pi^*$ transitions. Acetone shows absorption at 187 nm ($\pi \rightarrow \pi^*$) and 270 nm ($n \rightarrow \pi^*$).

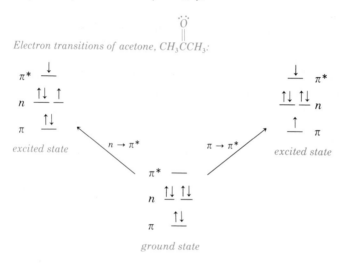

Electron transitions of acetone, CH$_3$CCH$_3$:

Section 21.4
Color and Vision

Color has played a significant role in human society ever since people first learned to color clothes and other articles. Color is the result of a complex set of physiological and psychological responses to wavelengths of light of 400–750 nm striking the retina of the eye. If all wavelengths of visible light strike the retina, we perceive white. If none of them do, we perceive black or darkness. If a small range of wavelengths hits the eye, then we observe individual colors. Table 21.3 lists the wavelengths of the visible spectrum with their corresponding colors and complementary colors, which we will discuss shortly.

Our perception of color arises from a variety of physical processes. A few examples of how light of a particular wavelength may be directed to the eye follow. (1) The yellow-orange color of a sodium flame results from the **emission of light** with a wavelength of 589 nm. The emission is caused by excited electrons returning to lower-energy orbitals. (2) A prism causes a **diffraction of light** that varies with the wavelength. We observe the separated wavelengths as a rainbow pattern. (3) **Interference** results from light being reflected from two surfaces of a very thin film (e.g., soap bubbles or bird feathers). The light wave

Table 21.3 Colors in the visible spectrum

Wavelength, nm	Color	Complementary (Subtraction) Color
400–424	violet	green-yellow
424–491	blue	yellow
491–570	green	red
570–585	yellow	blue
585–647	orange	green-blue
647–700	red	green

reflected from the farther surface is reflected out of phase with the reflection from the nearer surface, resulting in wave interference and cancellation of some wavelengths. Hence, we see color instead of white.

The fourth, and most common, process that leads to color is the **absorption of light of certain wavelengths** by a substance. Organic compounds with extensive conjugation absorb certain wavelengths of light because of $\pi \to \pi^*$ and $n \to \pi^*$ transitions. We do not observe the color absorbed, but we see its **complement**, which is reflected. A complementary color, sometimes called a **subtraction color**, is the result of subtraction of some of the visible wavelengths from the entire visual spectrum. For example, pentacene (page 859) absorbs at 575 nm, in the yellow portion of the visible spectrum. Thus, pentacene absorbs the yellow light (and, to a lesser extent, that of the surrounding wavelengths) and reflects the other wavelengths. Pentacene has a blue color, which is the complement of yellow.

Some compounds appear yellow even though their λ_{max} is in the ultraviolet range of the spectrum (for example, coronene, page 859). In such a case, the tail of the absorption band extends from the ultraviolet region into the visible region and absorbs the violet to blue wavelengths. Figure 21.4 depicts the spectrum of such a compound.

A. Mechanism of Vision

The human eye is an amazingly intricate organ that converts photons of light into nerve pulses that travel to the brain and result in vision. The mechanism of the eye is remarkably sensitive. About one quantum of light energy is all that is

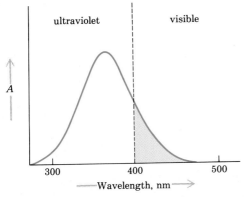

Figure 21.4 A compound with a λ_{max} in the uv region may also absorb light in the visible region.

necessary to trigger the mechanism resulting in a visual nerve pulse. We can detect as few as 100 photons of light. (For comparison, a typical flashlight bulb radiates about 2×10^{18} photons per second.)

The retina of the eye contains two types of photoreceptors—the **rods** and the **cones**. The cones are responsible for color vision and for vision in bright light. Animals that lack cones are colorblind. The rods are responsible for black and white photoreception and for vision in very dim light. While more is known about rods than cones, much is still to be learned, for example, how the nerve impulse is generated.

In the rods, light is detected by a reddish-purple pigment called **rhodopsin**, or **visual purple** ($\lambda_{max} \cong 500$ nm). Rhodopsin is formed from an aldehyde called *11-cis-retinal* and a protein called *opsin*. These two components of rhodopsin are bonded together by a protonated imine link between the aldehyde group of 11-*cis*-retinal and an amino group of a lysine residue in opsin. As is frequently the case with protein complexes, opsin has a shape that fits around 11-*cis*-retinal and holds it in a pocket. Compounds with other shapes do not fit into this pocket. In the combined form, the imine bond joining 11-*cis*-retinal and opsin is protected by the rest of the opsin molecule and is not readily hydrolyzed.

11-*cis*-retinal   rhodopsin

When a photon of light (*hν*) is absorbed by rhodopsin, the 11-*cis* double bond of the retinal portion is isomerized to a *trans* double bond. The product is a high-energy intermediate that undergoes a series of transformations. Finally, because all-*trans*-retinal does not fit into opsin's pocket, the now exposed iminium ion link is hydrolyzed and all-*trans*-retinal is released (see Figure 21.5).

rhodopsin $\longrightarrow$ bathorhodopsin $\longrightarrow$

lumirhodopsin $\longrightarrow$ metarhodopsin I $\longrightarrow$ metarhodopsin II $\xrightarrow{\text{H}_2\text{O}}$

all-*trans*-retinal

Figure 21.5 When 11-*cis*-retinal undergoes isomerization to all-*trans*-retinal, its shape changes so that it no longer fits into its pocket. Thus, the iminium ion link becomes exposed and can be hydrolyzed. Note that the isomerization also separates the iminium ion's positive charge from its balancing negative charge in opsin; this charge separation is one reason for the high energy contained in the photoisomerization product.

In the process of the hydrolysis, enzymes are activated that change the ionic permeability of the photoreceptor cell and thus change its electrical character. These changes are responsible for the generation of the nerve impulse.

Vitamin A is important in the human diet partly because it is the precursor of 11-*cis* retinal.

B. Colored Compounds, Dyes, and Indicators

Nature abounds with color. Some colors, such as those of hummingbird or peacock feathers, arise from light diffraction by the unique structure of the feathers. However, most of Nature's colors are due to the absorption of certain wavelengths of visible light by organic compounds.

Before the theories of electronic transition were developed, it was observed that some types of organic structures give rise to color, while others do not. The partial structures necessary for color (unsaturated groups that can undergo $\pi \to \pi^*$ and $n \to \pi^*$ transitions) were called **chromophores**, a term coined in 1876 (Greek *chroma*, "color," and *phoros*, "bearing").

Some chromophores:

It was also observed that the presence of some other groups caused an intensification of color. These groups were called **auxochromes** (Greek *auxanein*, "to increase"). We now know that auxochromes are groups that cannot undergo $\pi \to \pi^*$ transitions, but can undergo transitions of n electrons.

Some auxochromes:

—OH　　—OR　　—NH$_2$　　—NHR　　—NR$_2$　　—X

Naphthoquinones and **anthraquinones** are common natural coloring materials. *Juglone* is a naphthoquinone that is partly responsible for the coloring of walnut hulls. *Lawsone* is similar in structure to juglone; it is found in Indian henna, which is used as a red hair dye. A typical anthraquinone, carminic acid, is the principal red pigment of *cochineal*, a ground-up insect (*Coccus cacti L.*) that is used as a red dye in food and cosmetics.

juglone　　　　lawsone　　　　　　　carminic acid

A **dye** is a colored organic compound that is used to impart color to an object or a fabric. The history of dyes goes back to prehistoric times. *Indigo* (page 865), the oldest known dye, was used by the ancient Egyptians to dye mummy cloths. *Tyrian purple* (page 854), which was obtained from *Murex* snails found near the city of Tyre, was used by the Romans to dye the togas of the emperors.

To be useful as a dye, a compound must be *fast* (remain in the fabric during washing or cleaning). To be fast, a dye must, in one way or another, be bonded to the fabric. A fabric composed of fibers of polypropylene or a similar hydrocarbon is difficult to dye because it has no functional groups to attract dye molecules. Successful dyeing of these fabrics can be accomplished, however, by the incorporation of a metal–dye complex into the polymer. Dyeing of cotton (cellulose) is easier because hydrogen bonding between hydroxyl groups of the glucose units and groups of the dye molecule hold the dye to the cloth. Polypeptide fibers, such as wool or silk, are the easiest fabrics to dye because they contain numerous polar groups that can interact with dye molecules.

A **vat dye** is a dye that is applied to fabric (in a vat) in a soluble form and then is allowed to undergo reaction to an insoluble form. The blue coats supplied by the French to the Americans during the American Revolution were dyed with *indigo*, a typical vat dye. Indigo was obtained by a fermentation of the woad plant (*Isatis tinctoria*) of Western Europe or of plants of the *Indigofera* species, found in tropical countries. Both types of plants contain the glucoside *indican*, which can be hydrolyzed to glucose and *indoxyl*, a colorless precursor of indigo. Fabrics were soaked in the fermentation mixture containing indoxyl, then were allowed to air dry. Air oxidation of indoxyl yields the blue, insoluble indigo. Indigo is deposited in the *cis* form, which undergoes spontaneous isomerization to the *trans* isomer.

indoxyl   *cis*-indigo   *trans*-indigo

Azo dyes are the largest and most important class of dyestuffs, their numbers running into the thousands. In azo-dyeing, the fabric is first impregnated with an aromatic compound activated toward electrophilic substitution, then is treated with a diazonium salt to form the dye.

Direct Blue 2B

an azo dye

An **acid–base indicator** is an organic compound that changes color with a change in pH. These compounds are most frequently encountered as titration endpoint indicators. Test papers, such as litmus paper, are impregnated with one or more of these substances.

Two typical indicators are *methyl orange* and *phenolphthalein*. Methyl orange is red in acidic solutions that have pH values less than 3.1. It is yellow in solutions with pH values greater than 4.4. Phenolphthalein, on the other hand, changes color on the alkaline side of the pH range. Up to pH 8.3, phenolphthalein is colorless. At pH 10, it is red. In strongly alkaline solutions, it again becomes colorless.

Indicators change color because the chromophoric system is changed by an acid–base reaction. In acidic solution, methyl orange exists as a resonance hybrid of a protonated azo structure; this resonance hybrid is red. The azo

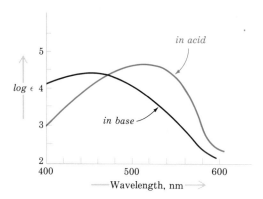

Figure 21.6 Visible spectra of methyl orange in acidic and alkaline solutions (pH 1 and 13, respectively).

nitrogen is not strongly basic, and the protonated azo group loses the hydrogen ion at about pH 4.4. The loss of the proton changes the electronic structure of the compound, resulting in a change of color from red to yellow. Figure 21.6 shows spectra of methyl orange at two different pH values.

$$^-O_3S-\bigcirc-\ddot{N}=\ddot{N}-\bigcirc-\ddot{N}(CH_3)_2$$

methyl orange

yellow in base

$$OH^- \updownarrow H^+$$

$$^-O_3S-\bigcirc-\overset{+}{\underset{H}{N}}=\ddot{N}-\bigcirc-\ddot{N}(CH_3)_2 \longleftrightarrow {}^-O_3S-\bigcirc-\underset{H}{\ddot{N}}-\ddot{N}=\bigcirc=\overset{+}{N}(CH_3)_2$$

red in acid

The commercial value of phenolphthalein is that it serves as the active ingredient in "candy" and "gum" laxatives. However, phenolphthalein is also one of the best-known titration indicators. In acidic solution, phenolphthalein is a colorless lactone. In the lactone, the center carbon is in the sp^3-hybrid state; consequently, the three benzene rings are isolated, not conjugated.

At pH values greater than 8.3 (alkaline solution), a phenolic hydrogen is removed from phenolphthalein, the lactone ring opens, and the center carbon becomes sp^2-hybridized. In this form, the benzene rings are in conjugation, and the extensive pi system gives rise to the red color that is observed in mildly alkaline solution. Figure 21.7 shows these reactions.

In strongly alkaline solution, the center carbon of phenolphthalein is hydroxylated and is converted to the sp^3-hybrid state. This reaction isolates the three pi systems again. At high pH values, phenolphthalein is colorless.

red in base

H⁺ ‖ OH⁻

phenolphthalein

colorless in acid

colorless in strong base

Figure 21.7 The acid–base reactions of phenolphthalein.

Study Problem

21.5 One of the following indicators is blue-green at pH 7; the other is violet. Which is which? Explain your answer.

(a)

(b)

Section 21.5 Photochemistry

The electrons of most stable compounds are paired and their spins (represented as $+1/2$ or $-1/2$) cancel each other. If the electron spins of a molecule cancel, regardless of whether the electrons are all paired in orbitals, the molecule is said to be in the **singlet state**. If the electrons are paired in their lowest-energy orbitals, the molecule is in the lowest-energy singlet state, or ground state, S_0.

When a molecule absorbs a photon of ultraviolet or visible light, an electron is promoted from a ground-state orbital to an excited-state orbital. This promotion of an electron is extremely rapid (about 10^{-15} second). The spin state of the electron that absorbs the energy does not change. Therefore, the molecule is still in the singlet state, now an **excited singlet state**: S_1, S_2, or higher excited state, depending on the energy of the excited state.

Immediately after promotion (in about 10^{-11} second), the electron drops to the lowest-energy excited singlet state (S_1) by a process called **internal conversion**. The energy lost during internal conversion is transformed into heat and molecular motion. The lifetime of an electron in the S_1 state is in the range of 10^{-8} to 10^{-7} second.

A **triplet state (T)** is an excited state in which the spin states of the electrons of the molecule do not cancel because the spin state of one electron in the molecule has been changed.

$$\pi_2^* \;\underline{\;\downarrow\;}\qquad\qquad\qquad \pi_2^* \;\underline{\;\uparrow\;}$$

$$\pi_1 \;\underline{\;\uparrow\;}\qquad\qquad\qquad \pi_1 \;\underline{\;\uparrow\;}$$

a pi-bond excited singlet state;　　　*a pi-bond triplet state;*
electrons of opposite spin　　　*electrons with the same spin*

The average energy of the lowest-energy triplet excited state T_1 is generally greater than that of S_0, but less than that of S_1. The process by which an electron in S_1 is transferred to T_1 is called **intersystem crossing**. An S_1 or T_1 electron must lose its excess energy and return to the ground state. The energy can be lost as heat, as light, or by a chemical reaction. In this section, we will briefly discuss two of these processes—photon emission (fluorescence) and photochemical reactions.

Figure 21.8 summarizes the relative energy relationships of the various energy states.

A. Fluorescence

Besides intersystem crossing, one way in which an excited molecule with electrons in the S_1 state can return to the ground state S_0 is to lose its energy as light, a process called **fluorescence**. This process is fast (10^{-7} second). The energy lost by this emission of light is slightly *less* than the energy initially absorbed because of internal conversion. Consequently, the wavelength of light emitted is slightly longer than that absorbed.

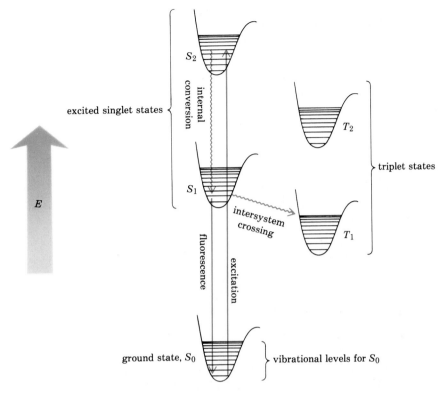

Figure 21.8 Simplified energy relationships of the ground state (S_0), the singlet excited states S_1 and S_2, and the triplet excited states T_1 and T_2. Each state contains various vibrational levels, as shown in the diagram and labeled for S_0. Energy can also be lost by excited species in chemical reactions and phosphorescence (emission of light when an electron in the T_1 state returns to the S_0 state).

A compound that absorbs light in the visible range appears colored. When the same compound emits light of a different wavelength, it appears two-colored, or fluorescent. An example of a fluorescent compound is **fluorescein**, which has been used as a marker for airplanes downed at sea. In aqueous solution and in the presence of light, fluorescein appears red with an intense yellow-green fluorescence.

fluorescein

Some fluorescent compounds, called **optical bleaches**, are used as fabric whiteners. These are colorless compounds that absorb ultraviolet light just out of the visible range, then emit blue-violet light at the edge of the visible spectrum. This blue-violet color masks yellowing of the fabric.

Two optical bleaches:

Blankophor R

Calcofluor SD

B. Photochemical Reactions

We have already discussed some types of photochemical reactions. The photolysis, or cleavage caused by light, of Cl_2 into $2 \, Cl\cdot$ as the initiation step in free-radical halogenation (Chapter 6) is one example. Many pericyclic reactions (Chapter 20) are also photochemical reactions. Let us introduce a few other types of these reactions.

Photoisomerization.   Isomerization of *cis* and *trans* isomers is possible under proper experimental conditions. A *cis*-to-*trans* isomerization is usually an exothermic process that occurs when the two isomers are allowed to equilibrate. The difference in energy between the *cis* and *trans* isomers determines the position of the equilibrium and, thus, the extent of the isomerization.

Isomerization of a *trans* isomer to a *cis* isomer is usually endothermic and, thus, equilibration does not favor the *cis* isomer. However, a *trans* isomer can be converted to *cis* by irradiation with a frequency that promotes a pi electron of the *trans* system, but not a pi electron of the *cis* system.

When a pi electron is excited, the pi bond of either the *cis* or *trans* isomer can assume a twisted shape. However, if the frequency of radiation is chosen properly, the *cis* isomer cannot absorb radiation, and its reaction leading back to the *trans* isomer cannot occur.

| *trans* | twisted S_1 or T_1 | *cis* |

cis
cannot absorb hv
and thus accumulates

An example of a photoisomerization reaction follows:

trans-dibenzoylethene   *cis*-dibenzoylethene

Photodimerization.   Irradiation of an α,β-unsaturated carbonyl compound or a conjugated polyene can lead to dimerization with the formation of four-membered rings (cycloaddition; see Section 20.2). For example, cyclopentenone undergoes dimerization to yield the following diones.

Photolysis.   A diatomic molecule, such as bromine or chlorine, can dissociate into a pair of radicals after the absorption of a photon of light. Ultraviolet light can also cause aldehydes and ketones to undergo photolytic reactions. Photolytic reactions take place by two principal pathways—the **Norrish type I** and the **Norrish type II**, named to honor R. G. W. Norrish, who received the 1967 Nobel Prize in Chemistry for his work on photochemical reactions.

Norrish type I reactions, which apply mainly to ketones in the gaseous state, involve the homolysis of the sigma bond joining the carbonyl carbon to the alpha carbon and subsequent free-radical reactions. The products depend upon the structures of the radicals. An example of a Norrish type I reaction is the reaction caused by the irradiation of gaseous acetone at 313 nm. The primary photochemical reaction is the formation of a methyl radical ($\cdot CH_3$) and an ethanoyl radical ($CH_3CO\cdot$). The products result from the free-radical reactions of these two radicals.

Norrish type I fragmentation:

Subsequent free-radical reactions:

A *Norrish type II* pathway involves an initial excitation of an aldehyde or a ketone by ultraviolet light followed by an internal abstraction of a gamma hydrogen. The resulting diradical can fragment, cyclize, or form other radical products.

A Norrish type II reaction:

Subsequent free-radical reactions:

an enol

Photoreduction of aryl ketones. Irradiation of an aryl ketone, such as benzophenone, in the presence of a hydrogen donor, such as a secondary alcohol, causes a reductive coupling reaction.

The initial photo product is an excited singlet state, which then goes to a triplet state. The triplet contains two unpaired electrons and thus has diradical character. This radical can abstract a hydrogen atom from some other molecule in solution. In the photoreduction, the diradical abstracts a hydrogen atom from the secondary alcohol and forms two radicals that can dimerize to yield the observed product.

Step 1 (formation of T_1):

benzophenone S_1: electrons paired T_1: a diradical

Step 2 (hydrogen abstraction):

a hydrogen donor

Step 3 (dimerization):

"pinacol"

Section 21.6
Mass Spectrometry

Most of the spectral techniques we have discussed arise from absorption of energy by molecules. **Mass spectrometry** is based on different principles. In a mass spectrometer, a sample in the gaseous state is bombarded with electrons of sufficient energy to exceed the first ionization potential of the compound. (The ionization potentials of most organic compounds are in the range of 185–300 kcal/mol.) Collision between an organic molecule and one of these high-energy electrons results in the loss of an electron from the molecule and the formation of an organic ion. The organic ions that result from this high-energy electron bombardment are unstable and fly apart into smaller fragments, both radicals and other ions. In a typical mass spectrometer, the *positively charged fragments* are detected. The **mass spectrum** is a plot of **abundance** (the relative amounts of the different positively charged fragments) versus the **mass-to-charge ratio** *(m/e,* or *m/z)* of the fragments. The ionic charge of most particles detected in a mass spectrometer is +1. The *m/e* value for such an ion is equal to its mass. Consequently, from a practical standpoint, the mass spectrum is a record of particle mass versus relative abundance of the particles.

How a molecule or ion breaks into fragments depends upon the carbon skeleton and functional groups present. Therefore, the structure and mass of the fragments give clues about the structure of the parent molecule. Also, it is frequently possible to determine the molecular weight of a compound from its mass spectrum.

Let us introduce mass spectrometry using methanol as an example. When methanol is bombarded with high-energy electrons, one of the valence electrons is lost. The result is an **ion radical**, a species with one unpaired electron and a charge of +1. An ion radical is symbolized by $\overset{+}{\cdot}$. The ion radical that results from abstraction of one electron of a molecule is called the **molecular ion** and is symbolized $M^{\overset{+}{\cdot}}$. The mass of the molecular ion is the molecular weight of the compound. The molecular ion of methanol has a mass of 32 and a charge of +1. Its mass-to-charge ratio (*m/e*) is 32. (In the following example, the half-headed arrow signifies the loss of one electron from a methanol molecule.)

$$e^- + CH_3\overset{..}{\underset{..}{O}}H \xrightarrow{-2e^-} CH_3\overset{.}{O}H^+, \quad \text{usually written} \quad [CH_3OH]^{\overset{+}{\cdot}}$$

the molecular ion of methanol, m/e = 32

A molecular ion can undergo fragmentation after it has been formed. In the case of $[CH_3OH]^{\overset{+}{\cdot}}$, the ion radical can lose a hydrogen atom and become a cation: $[CH_2{=}OH]^+$. This fragment has a *m/e* of 31. In the mass spectrum of methanol (Figure 21.9), peaks for the particles with *m/e* values of 31 and 32 are evident. (The fragments that give rise to the other peaks in this mass spectrum will be discussed in Section 21.11.)

$$\left[\begin{matrix} H \\ H{:}\overset{..}{C}{:}\overset{..}{\underset{..}{O}}{:}H \\ H \end{matrix} \right]^{+} \longrightarrow H\cdot + \left[\begin{matrix} H \\ \overset{..}{C}{::}\overset{..}{\underset{..}{O}}{:}H \\ H \end{matrix} \right]^{+}$$

a cation, m/e = 31
(not an ion radical because
all electrons are paired)

As can be seen in Figure 21.9, a mass spectrum is presented as a bar graph. Each peak in the spectrum represents a fragment of the molecule. The fragments are scanned so that the peaks are arranged by increasing *m/e* from

Figure 21.9 Mass spectrum of methanol, CH_3OH.

left to right in the spectrum. The intensities of the peaks are proportional to the relative abundance of the fragments, which in turn depends on their relative stabilities. By convention, the tallest peak in a spectrum, called the **base peak**, is given the intensity value of 100%. Lesser peaks are reported as 20%, 30%, or whatever their value is relative to the base peak. The base peak sometimes arises from the molecular ion, but often it arises from a smaller fragment.

Study Problems

21.6 Write the formula for each product and tell whether it is a cation, an ion-radical, or a radical:

(a) CH_4 minus one e^- (b) $[CH_4]^+$ minus H·
(c) $[CH_3CH_2]^+$ minus H·   (d) $[CH_3CH_2]^+$ minus H·

21.7 Write the formulas for the molecular ions of   (a) CH_4 and   (b) $CH_3CH_2CH_3$.

21.8 Give the m/e value for each of the following particles:

(a) $[CH_4]^+$ (b) $[(CH_3)_2CH]^+$ (c) $[O_2]^+$ (d) $[H_2O]^+$

▼ Section 21.7
The Mass Spectrometer

A diagram of a common type of mass spectrometer is shown in Figure 21.10. The sample is introduced, vaporized, and allowed to feed in a continuous stream to the *ionization chamber*. The ionization chamber (as well as the entire instrument) is kept under vacuum to minimize collisions and reactions between radicals, air molecules, and so forth. In this chamber, the sample passes through a stream of high-energy electrons, which causes the ionization of some of the sample molecules into their molecular ions.

After its formation, a molecular ion can undergo fragmentation and rearrangement. These processes are extremely rapid ($10^{-10} - 10^{-6}$ sec). The longer-lived particles can be detected by the ion collector, but a shorter-lived particle may not have a sufficient lifetime to reach the ion collector. In some cases, the molecular ion is too short-lived to be detected, and only its fragmentation products exhibit peaks.

As the ion radicals and other particles are formed, they are fed past two electrodes, the *ion accelerator plates*, which accelerate the positively charged

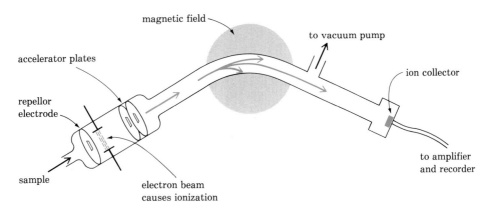

Figure 21.10 Diagram of a mass spectrometer.

particles. (The neutral and negatively charged particles are not accelerated and are removed continuously by the vacuum pumps.) From the accelerator plates, the positively charged particles pass into the *analyzer tube*, where they are deflected into a curved path by a magnetic field.

The radius of the curved path depends upon the particle's velocity, which in turn is dependent on the magnetic field strength, the accelerating voltage, and the m/e of the particle. At the same field strength and voltage, the particles of higher m/e have a path with a wider radius, while the lower-m/e particles have a path of smaller radius (see Figure 21.11). The continuous flow of positively charged particles through the analyzer tube therefore forms a pattern: higher-m/e particles with a larger radius and lower-m/e particles with a smaller radius. If the accelerating voltage is slowly and continuously decreased, the velocities of all the particles decrease, and the radii of the paths of all the particles also decrease. By this technique, particles of successively higher m/e are allowed to strike the detector. Figure 21.11 shows the effects of decreasing the accelerating voltage on the paths of particles with three different m/e values. (The same effect can be obtained by increasing the magnetic field strength instead of decreasing the accelerating voltage.)

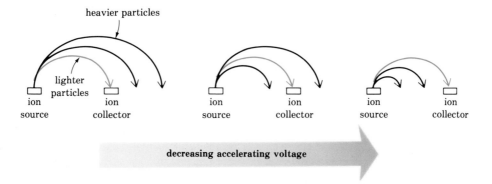

Figure 21.11 As the accelerating voltage is decreased, particles of successively higher m/e hit the ion collector.

Figure 21.12 Mass spectrum of bromomethane, CH_3Br.

▼ ## Section 21.8
Isotopes in Mass Spectra

A mass spectrometer is so sensitive that particles differing by 1.0 mass unit give separate signals. The molecular weight of CH_3Br is 94.9 (15.0 atomic mass units for CH_3 and 79.9 for Br). However, the mass spectrum of this compound (Figure 21.12) does not show one molecular ion peak at $m/e = 94.9$. Instead, two peaks are observed, one at $m/e = 94$ and the other at $m/e = 96$. The reason for the two peaks is that naturally occurring Br exists in two isotopic forms, one with atomic mass 79 and the other with atomic mass 81. When we calculate the molecular weight of a bromine compound, we use a weighted average of these two isotopic masses (79.9). Because a mass spectrometer detects particles containing each of these isotopes as individual species, we cannot use average atomic masses when dealing with mass spectra, as we do when calculating the stoichiometry of a chemical reaction.

Table 21.4 lists the common natural isotopes and their relative abundance. Naturally occurring bromine exists as a 50.5%–49.5% mixture of bromine-79 and bromine-81, respectively. Particles of the same structure containing Br give a pair of peaks of approximately the same intensity that are 2.0 mass units apart. *The particle containing the lower-mass isotope is considered the molecular ion.* The peak for the other particle is called the **M + 2 peak** (molecular ion plus two mass units).

Table 21.4 Natural abundance of some isotopes

Isotope	Abundance, %	Isotope	Abundance, %	Isotope	Abundance, %
^1H	99.985	^2H	0.015		
^{12}C	98.89	^{13}C	1.11		
^{14}N	99.63	^{15}N	0.37		
^{16}O	99.76	^{17}O	0.04	^{18}O	0.20
^{32}S	95.0	^{33}S	0.76	^{34}S	4.2
^{19}F	100				
^{35}Cl	75.5			^{37}Cl	24.5
^{79}Br	50.5			^{81}Br	49.5
^{127}I	100				

Naturally occurring chlorine is a mixture of 75.5% chlorine-35 and 24.5% chlorine-37. The particle containing chlorine-35 is considered the molecular ion, while the particle containing chlorine-37 gives rise to the M + 2 peak, which has an intensity approximately one-third that of the molecular ion peak.

Most elements common in organic compounds (except for Cl and Br) exist in nature as predominantly one isotope. For example, carbon is 98.89% carbon-12, and hydrogen is 99.985% hydrogen-1. For this reason, we generally assume for mass spectral purposes that all of C is carbon-12 and ignore the tiny proportion that is carbon-13. The presence of the isotopes of the common elements explains the multitude of small peaks surrounding a large peak in a mass spectrum (for example, the small peaks around the peaks at $m/e = 94$ and $m/e = 96$ in Figure 21.12).

Study Problems

21.9 Calculate the m/e value for the molecular ion of each of the following compounds: **(a)** ethane; **(b)** 1,2-dichloroethane; **(c)** ethanol; **(d)** *p*-bromophenol.

21.10 Figure 21.13 contains mass spectra of four compounds. Which of these compounds contains Br? Which contains Cl?

Figure 21.13 Mass spectra for Problem 21.10.

Section 21.9
Ionization and Fragmentation in Mass Spectra

In the mass spectrometer, the first reaction of a molecule is the initial ionization—the abstraction of a single electron. The loss of an electron gives rise to the molecular ion. From the peak for this ion radical, which is usually the peak that is farthest to the right in the spectrum, the molecular weight of the compound may be determined. (Remember, it is the exact molecular weight for a molecule containing single isotopes and not an average molecular weight.)

The question arises, "Which type of electron is lost from the molecule?" This question cannot be answered with accuracy. It is believed that the electron in the highest-energy orbital (the "loosest" electron) is the first to be lost. If a molecule has n (unshared) electrons, one of these is lost. If there are no n electrons, then a pi electron is lost. If there are neither n electrons nor pi electrons, the molecular ion is formed by loss of a sigma electron.

$$CH_3\ddot{O}H \longrightarrow [CH_3OH]^{\dot{+}} + n\,e^-$$

$$CH_3CH\!::\!CH_2 \longrightarrow [CH_3CH\!-\!CH_2]^{\dot{+}} + \pi\,e^-$$

$$H\!:\!CH_3 \longrightarrow [CH_4]^{\dot{+}} + \sigma\,e^-$$

After the initial ionization, the molecular ion undergoes fragmentation, a process in which free radicals or small neutral molecules are lost from the molecular ion. A molecular ion does not fragment in a random fashion, but tends to form the *most stable fragments possible*. In equations showing fragmentation, it is common practice not to show the free-radical fragments because they are not detected by the mass spectrometer.

Let us reconsider the mass spectrum of methanol in Figure 21.9. The spectrum consists of three principal peaks at $m/e = 29$, 31, and 32. The structures of the fragments can often be deduced from their masses. The $M^{\dot{+}}$ peak (at 32) of methanol arises from loss of one electron. The peak at 31 must arise from loss of an H atom (which has a mass of 1.0). The peak at 29 must arise from an ion that has lost two more H atoms. What about the minor peak at 15? This peak arises from the loss of $\cdot OH$ from the molecular ion.

$$CH_3OH \xrightarrow{-e^-} [CH_3OH]^{\dot{+}}$$

$$m/e = 32$$

with branches:
$$\xrightarrow{-H\cdot} [CH_2\!=\!OH]^+ \xrightarrow{-H_2} [CH\!\equiv\!O]^+$$
$$m/e = 31 \qquad\qquad m/e = 29$$
$$\xrightarrow{-\cdot OH} [CH_3]^+$$
$$m/e = 15$$

Could other fragmentation patterns occur? For example, could the molecular ion lose H^+ to become $[CH_3O]^{\cdot}$? It possibly could, but we do not know, because only the positively charged particles are accelerated and detected.

A. Effect of Branching

Branching in a hydrocarbon chain leads to fragmentation primarily at the branch because secondary ion radicals and secondary carbocations are more stable than primary ion radicals and primary carbocations. Carbocation stability is a more important factor than free-radical stability. For example, a methylpropane molecular ion yields predominantly an isopropyl cation and a methyl radical (not the reverse).

$$\left[\begin{array}{c} CH_3 \\ | \\ CH_3CH \cdot CH_3 \end{array}\right]^{\dot{+}} \longrightarrow CH_3\overset{CH_3}{\underset{|}{CH^+}} + \cdot CH_3 \quad \text{and very little} \quad CH_3\overset{CH_3}{\underset{|}{\underset{\cdot}{CH}}} + {}^+CH_3$$

Study Problem

21.11 What would be the molecular ion and the principal positively charged fragments arising from ionization of: **(a)** 2-methylpentane; **(b)** 2,2-dimethylpropane; **(c)** 1-pentene?

B. Effect of a Heteroatom or Carbonyl Group

Consider another spectrum, that of *N*-ethylpropylamine (Figure 21.14). The molecular ion has $m/e = 87$. Fragmentation of this molecular ion takes place alpha to the nitrogen atom and yields fragments with $m/e = 58$ (loss of an ethyl group) and 72 (loss of a methyl group). This type of fragmentation is called **α-fission** and is common in both amines and ethers.

fission here

$$[CH_3\text{—}CH_2NHCH_2\text{—}CH_2CH_3]^{\dot{+}}$$
$$m/e = 87$$

$$\xrightarrow{-\cdot C_2H_5} [CH_3CH_2NH{=}CH_2]^+$$
$$m/e = 58$$

$$\xrightarrow{-\cdot CH_3} [CH_2{=}NHCH_2CH_2CH_3]^+$$
$$m/e = 72$$

The reason for α-fission is that the cation formed in this reaction is resonance stabilized:

$$[R\text{—}CH_2\text{—}\overset{\cdot\cdot}{N}HR]^{\dot{+}} \xrightarrow{-R\cdot} [CH_2{=}\overset{+}{N}HR \longleftrightarrow \overset{+}{C}H_2\text{—}\overset{\cdot\cdot}{N}HR]$$

A similar fragmentation occurs at a bond adjacent to a carbonyl group (or α to the oxygen). Again, the resultant cation is resonance stabilized.

$$\left[\begin{array}{c} \overset{\cdot\cdot}{O}: \\ || \\ RC\text{—}R \end{array}\right]^{\dot{+}} \xrightarrow{-R\cdot} [RC{\equiv}\overset{+}{O}: \longleftrightarrow R\overset{+}{C}{=}\overset{\cdot\cdot}{O}:]$$

Figure 21.14 Mass spectrum of *N*-ethylpropylamine, $CH_3CH_2NHCH_2CH_2CH_3$.

Study Problem

21.12 The mass spectrum of an ether is shown in Figure 21.15. What is the structure of the ether?

Figure 21.15 Mass spectrum for Problem 21.12.

C. Loss of a Small Molecule

Small stable molecules, such as H_2O, CO_2, CO, and C_2H_4, can be lost from a molecular ion. An alcohol, for example, readily loses H_2O and shows a peak at 18 mass units less than the peak of the molecular ion. This peak is referred to as the **M–18 peak**. In many alcohols, elimination of H_2O is so facile that the molecular-ion peak is not even observed in the spectrum. The spectrum of 1-butanol (Figure 21.16) is a typical mass spectrum of an alcohol.

$$[CH_3CH_2CH_2CH_2OH]^{\ddagger} \xrightarrow{-H_2O} [CH_3CH_2CHCH_2]^{\ddagger}$$
$$M^{\ddagger}, m/e = 74 \qquad\qquad M - 18, m/e = 56$$

D. McLafferty Rearrangement

When there is a hydrogen atom γ to a carbonyl group in the molecular ion, a **McLafferty rearrangement** may occur. In this rearrangement, an alkene is lost from the molecular ion.

Figure 21.16 Mass spectrum of 1-butanol.

γ hydrogen

$$-\overset{\overset{\displaystyle H}{|}}{\underset{|}{C}}CH_2CH_2\overset{\overset{\displaystyle O}{\|}}{C}-$$

Rearrangement:

$$\left[\begin{array}{c} H_2C \overset{H}{\diagup} \quad \overset{\cdot\cdot}{O}\cdot \\ \quad \| \\ H_2C \diagdown \diagup CH \\ \quad CH_2 \end{array}\right]^{\ddagger} \longrightarrow \begin{array}{c} H_2C \\ \| \\ H_2C \end{array} + \left[\begin{array}{c} H\overset{\cdot\cdot}{O}\cdot \\ | \\ CH \\ \diagup\diagup \\ CH_2 \end{array}\right]^{\ddagger}$$

$$m/e = 72 \qquad\qquad m/e = 44$$

Study Problem

21.13 Predict the m/e values for the products of the McLafferty rearrangement of the following compounds:

(a) $CH_3CH_2CH_2\overset{\overset{\displaystyle O}{\|}}{C}CH_3$

(b) $CH_3\overset{\overset{\displaystyle CH_3}{|}}{C}HCH_2\overset{\overset{\displaystyle O}{\|}}{C}H$

(c) $CH_3\overset{\overset{\displaystyle O}{\|}}{C}OCH_2CH_2CH_3$

(d) $(CH_3)_2CHCH_2\overset{\overset{\displaystyle O}{\|}}{C}OCH_2CH_3$

▼ Summary

Absorption of ultraviolet (200–400 nm) or visible (400–750 nm) light results in **electronic transitions**, promotion of electrons from the ground-state orbitals to orbitals of higher energy. The wavelength λ of absorption is inversely proportional to the energy required. The uv or visible spectrum is a plot of **absorbance A** or **molar absorptivity** ε vs λ, where $\varepsilon = A/cl$. The position of maximum absorption is reported as λ_{max}.

The important electronic transitions are $\pi \rightarrow \pi^*$ for conjugated systems and $n \rightarrow \pi^*$. Increasing amounts of conjugation result in shifts of λ_{max} toward longer wavelengths. Compounds that absorb at wavelengths longer than 400 nm are colored; the apparent color is the **complementary color** of the wavelength absorbed.

Vision is made possible by the conversion of 11-*cis*-retinal in rhodopsin to all-*trans*-retinal.

Dyes are colored compounds that adhere to fabric or other substance. An **acid–base indicator** is a compound that undergoes a color change in a reaction with acid or base. The color change arises from a change in the conjugated system and thus in the wavelength of absorption.

Photochemistry is the study of physical changes, such as fluorescence, and chemical changes caused by the absorption of ultraviolet or visible light. Photochemical reactions include *cis-trans* isomerization, dimerization of polyunsaturated compounds, photolysis of carbonyl compounds, and photoreduction of aryl ketones.

Table 21.5 Summary of some fragmentation patterns in mass spectra

Fragmentation	*Reaction Type*
Alkanes:	
$[R_2CH \dot{+} CH_3]^{\ddot{+}} \xrightarrow{-\cdot CH_3} R_2\overset{+}{C}H$	σ-bond fission to most stable carbocation
Amines and Ethers:	
$[R \dot{+} CH_2 - NR'_2]^{\ddot{+}} \xrightarrow{-R\cdot} CH_2 = \overset{+}{N}R'_2$	α-fission
$[R \dot{+} CH_2 - OR']^{\ddot{+}} \xrightarrow{-R\cdot} CH_2 = \overset{+}{O}R'$	α-fission
Carbonyl Compounds:	
$\left[\overset{\overset{O}{\parallel}}{RC} \dot{+} Y \right]^{\ddot{+}} \xrightarrow{-Y\cdot} R\overset{+}{C} = O$	α-fission
$\left[\overset{\overset{H}{\mid}}{R_2CCH_2} \dot{+} \overset{\overset{O}{\parallel}}{CH_2CY} \right]^{\ddot{+}} \xrightarrow{-R_2C=CH_2} \left[CH_2 = \overset{\overset{OH}{\mid}}{CY} \right]^{\ddot{+}}$	McLafferty rearrangement[a]
where Y = H, R', OH, OR', etc.	
Alcohols:	
$\left[\overset{\overset{OH}{\mid}}{R_2CHCR_2} \right]^{\ddot{+}} \xrightarrow{-H_2O} [R_2C=CR_2]^{\ddot{+}}$	loss of H_2O

[a] If a ketone or ester undergoes rearrangement and if Y contains a γ hydrogen, two types of rearrangements may be observed.

A **mass spectrum** is a graph of **abundance** versus **mass-to-charge ratio** (m/e) of positively charged particles that result from bombardment of a compound with high-energy electrons. Removal of one electron from a molecule of the compound results in the **molecular ion**. The molecular ion can lose atoms, ions, radicals, and small molecules to yield a variety of fragmentation products. Table 21.5 summarizes some of the fragmentation patterns.

Study Problems*

21.14 A chemist prepared 1,3,5-hexatriene and 1,3,5,7-octatetraene, placed them in separate flasks, but neglected to label them. How could the two compounds be differentiated by uv spectroscopy?

21.15 Which of the following pairs of compounds could probably be distinguished from each other by their uv spectra? Explain.

(a) $CH_3CO_2CH_2CH_3$, $CH_3CH_2CO_2CH_3$

(b)

(c) $CH_3CH=CHCH_2\overset{\overset{O}{\parallel}}{C}CH_3$, $CH_3CH_2CH=CHCCH_3$

---

* For information concerning the organization of the *Study Problems* and *Additional Problems*, see the *Note to student* on page 39.

(d) ,

21.16 Calculate the molar absorptivities of the following compounds at the specified wavelengths:

(a) adenine (9.54×10^{-5} M solution, 1.0-cm cell), absorbance of 1.25 at 263 nm
(b) cyclohexanone (0.038 M solution, 1.0-cm cell), absorbance of 0.75 at 288 nm

21.17 3-Buten-2-one shows uv absorption maxima at 219 nm and 324 nm. **(a)** Why are there two maxima? **(b)** Which has the greater ε_{max}?

21.18 What types of electronic transitions would give rise to uv absorption in each of the following compounds?

(a) 2,4-octadiene **(b)** 2-cyclohexenone **(c)** aniline **(d)** formaldehyde

21.19 Rank the following compounds in the order of increasing λ_{max}: **(a)** benzene; **(b)** biphenyl (C_6H_5—C_6H_5); **(c)** styrene (C_6H_5CH=CH_2); **(d)** stilbene (C_6H_5CH=CHC_6H_5).

21.20 A chemist has a liquid compound of unknown structure with the formula C_5H_8; the uv spectrum shows a λ_{max} at about 220 nm. What are the likely possible structures of this compound?

21.21 Suggest a reason why the λ_{max} for *trans*-stilbene (C_6H_5CH=CHC_6H_5)is 295 nm, while that for *cis*-stilbene is 280 nm.

21.22 *Indophenol blue* is a vat dye that is oxidized to an insoluble blue dye after application to fabric. Give the structure of the oxidized form.

reduced form of Indophenol blue

21.23 *Phenol red* is an acid–base indicator that is yellow at pH 6 but red at pH 9. Draw the structures (and resonance structures) for the two forms of this compound.

Phenol red

21.24 Each of the following reactions proceeds by either a Norrish type I path or a Norrish type II path. State which path, and suggest a mechanism for each.

(a) $\xrightarrow{h\nu}$ + CO

(b) $C_6H_5\overset{O}{\overset{\|}{C}}CH_2CH_2CH_3$ $\xrightarrow{h\nu}$ C_6H_5 + $C_6H_5\overset{O}{\overset{\|}{C}}CH_3$ + CH_2=CH_2

(c) $CH_3\overset{\overset{\displaystyle O}{\|}}{C}CH_3 \xrightarrow{h\nu} CH_3\overset{\overset{\displaystyle O}{\|}}{C}-\overset{\overset{\displaystyle O}{\|}}{C}CH_3 + CH_3CH_3$

21.25 Predict the products of each of the following photo-induced reactions:

(a) $\xrightarrow{h\nu}$

(b) $\xrightarrow{h\nu}$

(c) $\underset{C_6H_5}{\overset{C_6H_5}{\diagdown}}N=N\diagdown_{C_6H_5} \xrightarrow{h\nu}$

21.26 Give the structures of the products:

(a) $[CH_3CH_2CH_2OH]^{\ddagger} \xrightarrow{-H_2O}$ (b) $\xrightarrow{-e^-}$

(c) $\left[\bigcirc-OH\right]^{\ddagger} \xrightarrow{-H\cdot}$ (d) $[(CH_3)_2CHCl]^{\ddagger} \xrightarrow{-Cl\cdot}$

21.27 For each of the following compounds, predict the structures and m/e values for the molecular ion and likely positively charged fragmentation products:

  (a) ethyl isopropyl ether   (b) ethyl isobutyl ether
  (c) 2-chloropropane   (d) 2,5-dimethylhexane
  (e) 2-propanol   (f) 4-cyclopentylbutanal (McLafferty fragment only)

21.28 Suggest structures and fragmentation patterns that account for the following observed peaks in the mass spectra:

  (a) *n*-butane, $m/e = 58, 57, 43, 29, 15$
  (b) benzamide, $m/e = 121, 105, 77$
  (c) 1-bromopropane, $m/e = 124, 122, 43, 29, 15$
  (d) $C_6H_5CH_2OCH_3$, $m/e = 122, 121, 91, 77$
  (e) 5-methyl-2-hexanone, $m/e = 71, 58, 43$

Additional Problems

21.29 What reactants would you need to prepare the yellow azo dye *Crysamine G*?

$HO-\bigcirc-N=N-\bigcirc-\bigcirc-N=N-\bigcirc-OH$

Crysamine G

21.30 Suggest a synthesis for the indicator methyl orange (page 866) from benzene.

21.31 The mass spectrum of nicotine shows principal peaks at $m/e = 162\,(M^+)$, 161, 133, and 84. Write equations to show the fragments giving rise to these peaks.

21.32 *Phosphorescence* is the emission of light by a molecule returning from the T_1 excited state to the ground state. For the same compound, would you expect phosphorescence to be at lower or higher frequencies than fluorescence? Explain.

21.33 In uv spectra, the presence of an additional double bond in conjugation adds about 30 nm to the λ_{max}. From the following observed values for λ_{max}, state how the *degree of substitution* on the sp^2 carbons of a polyene affects the position of the λ_{max}.

Structure	λ_{max}, nm
$CH_2=CHCH=CH_2$	217
$CH_3CH=CHCH=CH_2$	223
$CH_3CH=CHCH=CHCH_3$	227
$CH_2=CCH=CHCH_3$ with CH_3 substituent	227

21.34 From your answer in Problem 21.33, predict the λ_{max} values for the following polyenes:

(a)

(b)

21.35 Predict the λ_{max} values where indicated:

Structure	λ_{max}, nm	Structure	λ_{max}, nm
$CH_2=CHCCH_3$ ($\overset{O}{\|}$)	219	$(CH_3)_2C=CHCCH_3$ ($\overset{O}{\|}$)	(a)
$CH_3CH=CHCCH_3$ ($\overset{O}{\|}$)	224	$(CH_3)_2C=CCCH_3$ ($\overset{O}{\|}$) with CH_3	(b)
		$CH_2=CHCH=CHCCH_3$ ($\overset{O}{\|}$)	(c)

21.36 A ketone ($C_6H_{12}O$) gives a positive iodoform test and shows principal peaks in the mass spectrum at the following m/e values: 100, 85, 57, and 43. Which of the following compounds are compatible with these data?

(a) $CH_3CH_2CCH_2CH_2CH_3$ ($\overset{O}{\|}$)

(b) $CH_3CCH_2CH_2CH_2CH_3$ ($\overset{O}{\|}$)

(c) $CH_3CCHCH_2CH_3$ ($\overset{O}{\|}$) with CH_3

(d) $CH_3CCH_2CH(CH_3)_2$ ($\overset{O}{\|}$)

21.37 A compound contains only C, H, and O. The infrared spectrum shows strong absorption at 1724 cm^{-1} (5.8 μm), 1388 cm^{-1} (7.2 μm), and 1231 cm^{-1} (8.1 μm) (plus other minor absorption). The ^{1}H nmr spectrum shows only one singlet at $\delta 2.1$. The mass spectrum has principal peaks at 58 m/e and 43 m/e. What is the structure of the compound?

21.38 The infrared, ^{1}H nmr, and mass spectra for Compounds A through C are shown in Figures 21.17 through 21.19. From the spectra, deduce the structure of each compound.

21.39 Compound A is a terpenoid with one chiral carbon. The (S)-enantiomer is an essential oil in rose oil and geranium oil. Spectra for Compound A are shown in Figure 21.20. What is the structure of the (S)-enantiomer of Compound A? (*Hint*: What is the significance of the fact that Compound A is a terpenoid?)

Figure 21.17 Spectra for Compound A, Problem 21.38.

Figure 21.18 Spectra for Compound B, Problem 21.38.

Figure 21.19 Spectra for Compound C, Problem 21.38.

Figure 21.20 Spectra for Compound A, Problem 21.39 (*continued on page 890*).

Figure 21.20 (*continued*) **Spectra for Compound A, Problem 21.39.**

22 ▼

Carbohydrates

Carbohydrates are naturally occurring compounds of carbon, hydrogen, and oxygen. Many carbohydrates have the empirical formula CH_2O. For example, the molecular formula for glucose is $C_6H_{12}O_6$ (six times CH_2O). These compounds were once thought to be "hydrates of carbon," hence the name carbohydrates. In the 1880s, it was recognized that the idea was a misconception and that carbohydrates are actually polyhydroxy aldehydes and ketones or their derivatives.

Carbohydrates vary dramatically in their properties. For example, *sucrose* (table sugar) and *cotton* are both carbohydrates. One of the principal differences between various types of carbohydrates is the size of the molecules. The **monosaccharides** (often called *simple sugars*) are the simplest carbohydrate units. They cannot be hydrolyzed to smaller carbohydrate molecules. Figure 22.1 shows dimensional formulas and Fischer projections for five of the most important monosaccharides. (You may find it helpful to review Fischer projections in Section 4.6C.)

①CHO   CHO   CH$_2$OH   CH$_2$OH
②H—C—OH   H——OH   C=O   C=O
③HO—C—H   HO——H   HO—C—H   HO——H
      or   or
④H—C—OH   H——OH   H—C—OH   H——OH
⑤H—C—OH   H——OH   H—C—OH   H——OH
⑥CH$_2$OH   CH$_2$OH   CH$_2$OH   CH$_2$OH

D-glucose   D-fructose

CHO   CHO
H—C—OH   H——OH
HO—C—H   HO——H
      or
HO—C—H   HO——H
H—C—OH   H——OH
CH$_2$OH   CH$_2$OH

D-galactose

CHO   CHO   CHO   CHO
H—C—OH   H——OH   CH$_2$ CH$_2$
H—C—OH   H——OH   H—C—OH   H——OH
   or   or
H—C—OH   H——OH   H—C—OH   H——OH
CH$_2$OH   CH$_2$OH   CH$_2$OH   CH$_2$OH

D-ribose   2-deoxy-D-ribose

Figure 22.1 Some important monosaccharides.

Monosaccharides can be bonded together to form dimers, trimers, etc., and, ultimately, polymers. The dimers are called **disaccharides**. Sucrose is a disaccharide that can be hydrolyzed to one unit of glucose plus one unit of fructose. The monosaccharides and disaccharides are soluble in water and are generally sweet-tasting.

$$1 \text{ sucrose} \xrightarrow[\text{heat}]{\text{H}_2\text{O, H}^+} 1 \text{ glucose } + 1 \text{ fructose}$$

a disaccharide

Carbohydrates composed of two to eight units of monosaccharide are referred to as **oligosaccharides** (Greek *oligo*-, "a few"). If more than eight units of monosaccharide result from hydrolysis, the carbohydrate is a **polysaccharide**. Examples of polysaccharides are *starch*, found in flour and cornstarch, and *cellulose*, a fibrous constituent of plants and the principal component of cotton.

$$\text{starch or cellulose} \xrightarrow[\text{heat}]{H_2O,\ H^+} \text{many units of glucose}$$

polysaccharides

In this chapter, we will consider first the monosaccharides and the conventions used by carbohydrate chemists. Then we will discuss some disaccharides, and finally, a few polysaccharides.

Section 22.1
Some Common Monosaccharides

Glucose, the most important monosaccharide, is sometimes called *blood sugar* (because it is found in the blood), *grape sugar* (because it is found in grapes), or *dextrose* (because it is dextrorotatory). Mammals can convert sucrose, lactose (milk sugar), maltose, and starch to glucose, which is then used by the organism for energy or stored as *glycogen* (a polysaccharide). When the organism needs energy, the glycogen is again converted to glucose. Excess carbohydrates can be converted to fat; therefore, a person can become obese on a fat-free diet. Carbohydrates can also be converted to steroids (such as cholesterol) and, to a limited extent, to protein. (A source of nitrogen is also needed in protein synthesis.) Conversely, an organism can convert proteins and fats to carbohydrates.

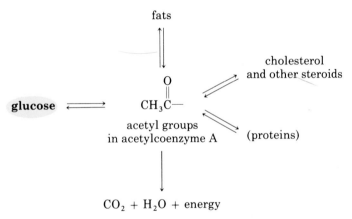

Fructose, also called *levulose* because it is levorotatory, is the sweetest-tasting sugar. It occurs in fruit and honey. Combined with glucose, it is also found in sucrose. **Galactose** is found, bonded to glucose, in the disaccharide lactose. **Ribose** and **deoxyribose** form part of the polymeric backbones of nucleic acids. The prefix *deoxy-* means "minus an oxygen"; the structures of ribose and deoxyribose (Figure 22.1) are the same except that deoxyribose lacks an oxygen at carbon 2.

Section 22.2
Classification of the Monosaccharides

The suffix **-ose** is used to designate a carbohydrate. All the monosaccharides and many oligosaccharides and polysaccharides have names ending in *-ose* (for example, sucrose and cellulose).

Monosaccharides that contain aldehyde groups are referred to as **aldoses** (*ald*ehyde plus -*ose*). Glucose, galactose, ribose, and deoxyribose are all aldoses. Monosaccharides, such as fructose, with ketone groups are called **ketoses** (*ket*one plus -*ose*).

The number of carbon atoms in a monosaccharide (usually three to seven) may be designated by tri-, tetr-, etc. For example, a **triose** is a three-carbon monosaccharide, while a **hexose** is a six-carbon monosaccharide. Glucose is an example of a hexose. These terms may be combined. Glucose is an **aldohexose** (six-carbon aldose), while ribose is an **aldopentose** (five-carbon aldose). Ketoses are often given the ending -**ulose**. Fructose is an example of a **hexulose** (six-carbon ketose), or **ketohexose**.

Study Problem

22.1 Classify each of the following monosaccharides by the preceding system:

$$
\begin{array}{cc}
& \text{CH}_2\text{OH} \\
\text{CHO} & | \\
& \text{C}=\text{O} \\
\textbf{(a)} \ \text{H}{-}\text{OH} & \textbf{(b)} \ \text{H}{-}\text{OH} \\
\text{H}{-}\text{OH} & \text{H}{-}\text{OH} \\
\text{CH}_2\text{OH} & \text{CH}_2\text{OH} \\
\text{erythrose} & \text{ribulose}
\end{array}
$$

(c) galactose (Figure 22.1)

▼ Section 22.3
Configurations of the Monosaccharides

Monosaccharides are quite similar to one another in structure. Some monosaccharides are structurally different. For example, glucose is an aldehyde and fructose is a ketone. Other common monosaccharides are diastereomers (nonenantiomeric stereoisomers) of one another. For example, glucose and galactose are **epimers**, diastereomers that differ in configuration at only one of their chiral carbon atoms.

A. The D and L System

In the late 19th century, it was determined that the configuration of the last chiral carbon in each of the naturally occurring monosaccharides is the same as that of (+)-glyceraldehyde. Today we call that configuration the (*R*)-configuration, but chemists at that time had no way to determine the absolute configuration around a chiral carbon. Instead, chemists devised the D and L system for designating relative configurations. (Do not confuse D and L with *d* and *l*, sometimes used to refer to the direction of rotation of the plane of polarization of plane-polarized light; see Section 4.7A.)

In the D and L system, (+)-glyceraldehyde was arbitrarily assigned the configuration with its OH on carbon 2 to the right in the Fischer projection (an assumption later shown to be correct). A monosaccharide is a member of the **D-series** if the hydroxyl group on the chiral carbon farthest from carbon 1 is also *on the right* in the Fischer projection. (Almost all naturally occurring carbohydrates are members of the D-series.) In addition, each monosaccharide was given its own name. For example, the following two diastereomeric aldopentoses are named D-lyxose and D-ribose.

```
                              CHO              CHO
         CHO           HO——H          H——OH
    H——OH             HO——H          H——OH
         CH₂OH          H——OH          H——OH
                              CH₂OH          CH₂OH
   D-(+)-glyceraldehyde    D-lyxose        D-ribose
```

If the OH on the last chiral carbon is projected to the *left*, then the compound is a member of the L-series. The following two examples are the enantiomers of D-lyxose and D-ribose.

```
         CHO              CHO
    H——OH          HO——H
    H——OH          HO——H
   HO——H          HO——H
         CH₂OH          CH₂OH
   L-lyxose        L-ribose
```

Study Problem

22.2 Assign each chiral carbon atom an (*R*) or (*S*) designation and tell whether each compound is D or L:

```
         CO₂H              CO₂H
(a) H——OH      (b) HO——H
         CH₃              CH₃

         CH₃
         C=O
(c) HO——H
    H——OH
         CH₂OH
```

B. Relating Configurations

We have mentioned that early chemists could not determine the absolute configurations of chiral carbons. Instead, configurations relative to that of (+)-glyceraldehyde were determined. How are other compounds related to glyceraldehyde? One example of determining a relative configuration follows. If the aldehyde group of D-glyceraldehyde is oxidized to a carboxylic acid, the product, glyceric acid, necessarily has the same configuration around the chiral carbon as that in D-glyceraldehyde. The product, even though levorotatory, is still a member of the D-series.

D-(+)-glyceraldehyde   D-(−)-glyceric acid

The configurations of the tartaric acids relative to D-glyceraldehyde were established in 1917 by the sequence shown in Figure 22.2, which produced two of the three isomers of tartaric acid.

In the first step of the sequence, D-glyceraldehyde is treated with HCN to yield a mixture of cyanohydrins. A new site of chirality is introduced in this step, and both diastereomers are formed. The diastereomers are separated; then, in the second step, each diastereomeric cyanohydrin is hydrolyzed.

In the third step, the terminal CH_2OH group is oxidized to yield two tartaric acids. Carbon 3 of each of these tartaric acids has the same configuration as carbon 2 of D-glyceraldehyde because the series of reactions has not affected the configuration around that carbon. However, the configurations around carbon 2 in the tartaric acids are different. One of the tartaric acids obtained from this synthesis does not rotate plane-polarized light. This is the *meso* isomer, the one with an internal plane of symmetry. The other tartaric acid obtained from the synthesis rotates plane-polarized light to the left. It must have the second structure as follows:

meso-tartaric acid   D-(−)-tartaric acid

Study Problem

22.3 Starting with L-(−)-glyceraldehyde, what tartaric acid(s) would be produced in the preceding sequence?

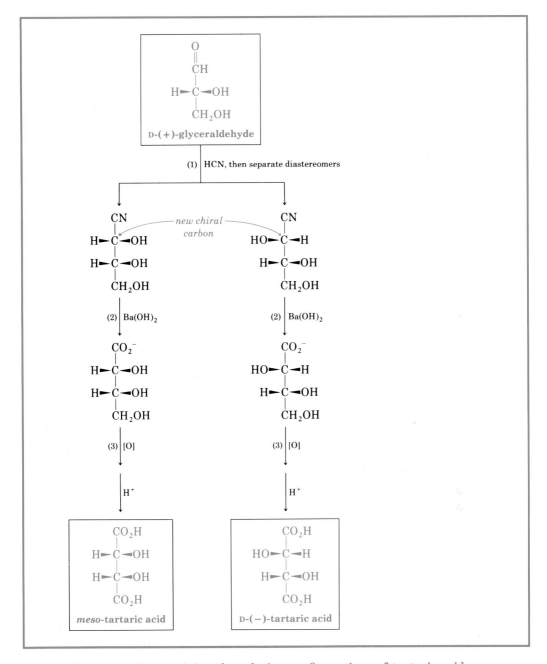

Figure 22.2 Determining the relative configurations of tartaric acids.

C. Configurations of the Aldohexoses

Glucose has six carbon atoms, four of which are chiral (carbons 2, 3, 4, and 5). Because the terminal carbon atoms of glucose have different functional groups, there can be no internal plane of symmetry; therefore, this compound has 2^4, or sixteen, stereoisomers. Only half of these sixteen stereoisomers belong to the D-series; of these, only D-glucose, D-galactose, and D-mannose occur in abundance.

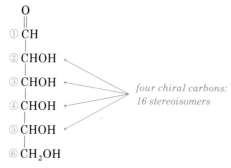

The Fischer projections of all the D-aldoses, from D-glyceraldehyde through the D-aldohexoses, are shown in Figure 22.3. Going from the triose, D-glyceraldehyde, to the tetroses, one carbon is added to the "top" of the molecule in the Fischer projection. The addition of one more carbon creates a new chiral carbon in each step down in the figure. Therefore, D-glyceraldehyde leads to a pair of tetroses, each tetrose leads to a pair of pentoses, and each pentose leads to a pair of hexoses.

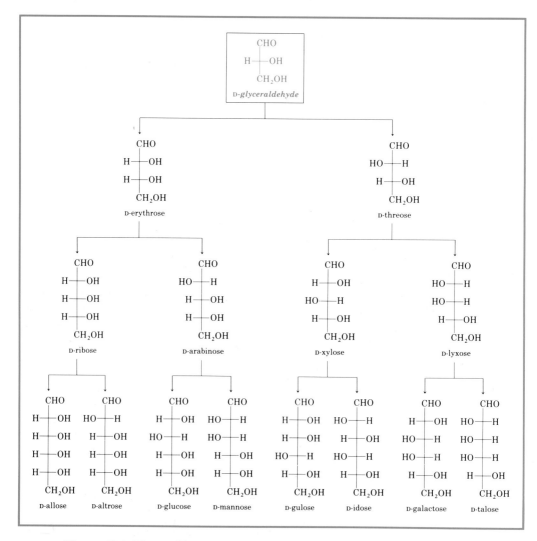

Figure 22.3 The D-aldoses.

Study Problems

22.4 There are eight D-aldohexoses and also eight L-aldohexoses. Write the Fischer projections of   **(a)** L-glucose, and   **(b)** L-mannose.

22.5 Although most naturally occurring sugars belong to the D series, L-arabinose is more common in plants than D-arabinose is. Write the Fischer projection for L-arabinose.

22.6 What is the name of the aldohexose in which only the OH at carbon 5 has the opposite configuration from that of D-glucose?

▼ ## Section 22.4
Cyclization of the Monosaccharides

Glucose has an aldehyde group at carbon 1 and hydroxyl groups at carbons 4 and 5 as well as at carbons 2, 3, and 6. A general reaction of alcohols and aldehydes is that of *hemiacetal formation* (see Section 13.5B).

$$\underset{\text{RCH}}{\overset{\text{O}}{\parallel}} + \text{R'OH} \;\underset{}{\overset{\text{H}^+}{\rightleftharpoons}}\; \underset{\text{RCH}}{\overset{\text{OH}}{\mid}}\!\!-\!\text{OR'}$$

a hemiacetal

In water solution, glucose can undergo an intramolecular reaction to yield *cyclic hemiacetals*. Either five-membered ring hemiacetals (using the hydroxyl group at carbon 4) or six-membered ring hemiacetals (using the hydroxyl group at carbon 5) can be formed.

Although Fischer projections are useful in discussions of open-chain carbohydrates, they are awkward for cyclic compounds. Let us rewrite the formula for glucose to show the more important of the cyclization reactions, the reaction leading to the six-membered rings.

D-glucose

open-chain aldehyde form

a pair of diastereomers;
hemiacetal groups highlighted

Carbon 1 (the aldehyde carbon), which is not chiral in the open-chain structure, becomes chiral in the cyclization. Therefore, a pair of diastereomers results from the cyclization. Because all the hemiacetal structures are in equilibrium with the aldehyde in water solution, they are also in equilibrium with each other.

5-membered cyclic hemiacetal $\rightleftharpoons$ glucose $\rightleftharpoons$ 6-membered cyclic hemiacetal

(two diastereomers)   (open-chain)   *(two diastereomers)*

A. Furanose and Pyranose Rings

A monosaccharide in the form of a five-membered ring hemiacetal is called a **furanose**. *Furan-* is from the name of the five-membered oxygen heterocycle *furan*. Similarly, the six-membered ring form is called a **pyranose** after *pyran*. The terms furanose and pyranose are often combined with the name of the monosaccharide. For example, D-**glucopyranose** is used for the six-membered ring of D-glucose, or D-**fructofuranose** for the five-membered ring of fructose.

furan   pyran

Of the two ring systems for glucose, the six-membered cyclic hemiacetal, or glucopyranose, is favored; we will emphasize this ring size in our discussion. Part of the reason that glucose preferentially forms the six-membered ring in solution is that the bond angles and staggering of attached groups are favorable in the chair form of this ring. Even though the pyranose ring of a mono-saccharide may predominate in equilibrium in water, it may be the furanose ring that is incorporated enzymatically into natural products. For example, in ribonucleic acids, ribose is found as a furanose, and not as a pyranose.

B. Haworth and Conformational Formulas

To represent the cyclic structures of sugars, **Haworth perspective formulas** were developed. By convention, a Haworth formula is drawn with the ring oxygen on the far side of the ring and carbon 1 on the right. The terminal —CH₂OH group is positioned above the plane of the ring in the D-series and below the plane of the ring in the L-series. The hydrogen atoms on the ring carbons are not usually shown in Haworth projections.

up = D $\rightarrow$ CH$_2$OH   *down* = L   *hydrogen atom*

O

CH$_2$OH

O

Using Fischer and Haworth formulas, we can rewrite the equations showing the cyclization of glucose. *Note that any group that is to the* right *in the*

Fischer projection is down *in the Haworth projection, and any group that is to the* left *in the Fischer projection is* up *in the Haworth formula.*

Fischer

Haworth

The flat Haworth formula is not an entirely correct representation of a pyranose ring (although it is fairly correct for the more planar furanose ring). A pyranose, like cyclohexane, exists primarily in the chair form, as the following conformational formula shows. In this chapter, we will use both Haworth formulas and conformational formulas.

α-D-glucopyranose

If an OH is down in a Haworth formula, it is also down (below the plane of the ring) in the conformational formula. Similarly, if an OH is up in the Haworth formula, it is also up in the conformational formula. As for any substituted six-membered ring, the ring assumes the conformation in which the majority of the groups are equatorial.

Study Problem

22.7 Draw a conformational formula of α-D-glucopyranose in which the CH_2OH is in an axial position.

C. Anomers

To yield a pyranose, the hydroxyl group at carbon 5 of glucose attacks the aldehyde carbon, carbon 1. A hemiacetal group is formed. Two of the most important consequences of this cyclization reaction are that a new chiral carbon is formed (carbon 1) and that a pair of diastereomers results. These diastereomers, monosaccharides that differ only in the configuration at carbon 1, are called **anomers** of each other. The carbonyl carbon in any monosaccharide is the **anomeric carbon**. This is the carbon that becomes chiral in the cyclization reaction.

In a Haworth formula of a D-sugar, the structure in which the anomeric OH is projected *down* (*trans* to the terminal CH_2OH) is called the **α-anomer**, while the structure in which the anomeric OH is projected *up* (*cis* to the terminal CH_2OH) is called the **β-anomer**.

Thus, the two anomeric D-glucoses can be called **α-D-glucopyranose** (or simply α-D-glucose) and **β-D-glucopyranose** (or β-D-glucose).

α-D-glucopyranose   β-D-glucopyranose

Sample Problem

Although fructose can form a six-membered cyclic hemiketal, in sucrose it is found in the furanose form. Draw Haworth formulas for α- and β-D-fructofuranose.

Solution:

Study Problem

22.8 Draw the Haworth and conformational formulas for the anomers of D-galactopyranose.

D. Mutarotation

Pure glucose exists in two crystalline forms: α-D-glucose and β-D-glucose. Pure α-D-glucose has a melting point of 146°C. The specific rotation of a freshly prepared solution is +112°. Pure β-D-glucose has a melting point of 150°C and a specific rotation of +18.7°. The specific rotation of a solution of either α- or β-D-glucose changes slowly until it reaches an equilibrium value of +52.6°. This slow spontaneous change in optical rotation, first observed in 1846, is called **mutarotation**.

Mutarotation occurs because, in solution, either α- or β-D-glucose undergoes a slow equilibration with the open-chain form and with the other anomer. Regardless of which anomer is dissolved, the result is an equilibrium mixture of 64% β-D-glucose, 36% α-D-glucose, and 0.02% of the aldehyde form of D-glucose. The final specific rotation is that of the equilibrium mixture.

α-D-glucose (36%)   D-glucose (0.02%)   β-D-glucose (64%)

Note that the equilibrium mixture of the anomers of D-glucose contains a greater percentage of the β-anomer than of the α-anomer. The reason is that the β-anomer is the more stable of the two. From our discussion of conformational analysis in Chapter 4, this is the expected result. The hydroxyl group at carbon 1 is *equatorial* in the β-anomer, but *axial* in the α-anomer.

α-D-glucose   β-D-glucose

Other monosaccharides also exhibit mutarotation. In water solution, the other aldoses with a 5-hydroxyl group also exist primarily in the pyranose forms. However, the percentages of the various species involved in the equilibrium may vary. For example, the equilibrium mixture of D-ribose in water is 56% β-pyranose, 20% α-pyranose, 18% β-furanose, 6% α-furanose, and a trace of the open-chain, aldehyde form.

Although the β-anomer of the pyranose rings is generally the more stable anomer, this is not always the case. For example, α-D-mannose is more stable than its β-anomer and predominates in an equilibrium mixture. This apparent anomaly, termed the **anomeric effect**, arises from interactions between the polar substituents on the ring.

β-D-ribopyranose (56%)   α-D-ribopyranose (20%)

β-D-ribofuranose (18%)   α-D-ribofuranose (6%)

Because of the facile conversion in water of the hemiacetal OH group between α and β, it often is not possible to specify the configuration at this carbon. For this reason, we will sometimes represent the hemiacetal OH bond with a squiggle, which means the structure may be α or β or a mixture.

α or β or a mixture

▼ ## Section 22.5
Glycosides

When a hemiacetal is treated with an alcohol, an acetal is formed (Section 13.5B). The acetals of monosaccharides are called **glycosides** and have names ending in **-oside**.

a hemiacetal   an acetal

β-D-glucopyranose   methyl β-D-glucopyranoside

a glycoside

The glycoside carbon (carbon 1 in an aldose) is easy to recognize because it has two OR groups attached.

two OR groups

OCH$_3$

CH$_3$CHOCH$_3$

an acetal

two OR groups

$\sim$OCH$_3$

a glycoside

Although a hemiacetal of a monosaccharide is in equilibrium with the open-chain form and with its anomer in water solution, an acetal is stable in neutral or alkaline solution. Therefore, a glycoside is not in equilibrium with the aldehyde or with its anomer in water solution. However, glycosides can be hydrolyzed to the hemiacetal (and aldehyde) forms by treatment with aqueous acid. This reaction is simply the reverse of glycoside formation.

CH$_2$OH

OCH$_3$

OH

HO

OH

methyl β-D-glucopyranoside

$+ H_2O \xrightleftharpoons{H^+}$

CH$_2$OH

O

OH

HO

$\sim$OH $+ CH_3OH$

OH

D-glucopyranose

Disaccharides and polysaccharides are glycosides; we will discuss these compounds later in this chapter. Other types of glycosides are also common in plants and animals. *Amygdalin* and *laetrile* (Section 13.5B) are glycosides found in the kernels of apricot pits and bitter almonds. *Vanillin* (used as vanilla flavoring) is another example of a structure found in nature as a glycoside, in this case, as a β-D-glucoside. In these types of glycosides, the nonsugar portion of the structure is called an **aglycone**. Vanillin is the aglycone in the following example.

CH$_3$O

HO

O

CH

vanillin

HO

HO

CH$_2$OH

O

OH

O

CH$_3$O

O

CH

vanillin β-D-glucoside
(glucovanillin)

Section 22.6
Oxidation of Monosaccharides

An aldehyde group is easily oxidized to a carboxyl group. Chemical tests for aldehydes depend upon this ease of oxidation (Section 13.8). Sugars that can be oxidized by such mild oxidizing agents as Tollens reagent, an alkaline solution of $Ag(NH_3)_2{}^+$, are called **reducing sugars** (because the inorganic oxidizing agent is *reduced* in the reaction). The cyclic hemiacetal forms of all aldoses are readily oxidized because they are in equilibrium with the open-chain aldehyde form.

D-glucopyranose

a reducing sugar

Although fructose is a ketone, it is also a reducing sugar.

D-fructose

a reducing sugar

The reason that fructose can be oxidized so readily is that, in alkaline solution, fructose is in equilibrium with two diastereomeric aldehydes through an enediol tautomeric intermediate. (In Section 13.10A, a similar enzymatic isomerization was discussed.)

a ketose   an enediol   an aldose
                  intermediate

In glycosides, the carbonyl group is blocked. Glycosides are **nonreducing sugars**.

A. Aldonic Acids

The product of oxidation of the aldehyde group of an aldose is a polyhydroxy carboxylic acid called an **aldonic acid**. Although Tollens reagent can effect the conversion, a more convenient and less expensive reagent for the synthetic reaction is a buffered solution of bromine.

D-glucose   D-gluconic acid

an aldonic acid

In alkaline solution, the aldonic acids exist as open-chain carboxylate ions. Upon acidification, they form lactones (cyclic esters), just as any γ- or δ-hydroxy acid would (Section 15.6). Most aldonic acids have both γ and δ hydroxyl groups, and either a five- or a six-membered ring could be formed. The five-membered rings (γ-lactones) are favored in these cases.

D-gluconic acid   *a lactone*

Study Problem

22.9 Predict the product (if any) of bromine oxidation of each of the following compounds:

B. Aldaric Acids

Vigorous oxidizing agents oxidize the aldehyde group and also the terminal hydroxyl group (a primary alcohol) of a monosaccharide. The product is a polyhydroxy dicarboxylic acid called an **aldaric acid**. Aldaric acids also form lactones readily.

D-glucose   D-glucaric acid

an aldaric acid

Study Problem

22.10 Which of the aldohexoses form *meso*-aldaric acids upon oxidation with hot nitric acid?

C. Uronic Acids

Although it is not easy to do in the laboratory, in biological systems the terminal CH_2OH group can be oxidized enzymatically without oxidation of the aldehyde group. The product is called a **uronic acid**.

D-glucose   D-glucuronic acid

a uronic acid

Glucuronic acid is important in animal systems because many toxic substances are excreted in the urine as **glucuronides**, derivatives of this acid. Also, in plant and animal systems, D-glucuronic acid can be converted to L-gulonic acid, which is used to biosynthesize L-ascorbic acid (vitamin C). (This last conversion does not take place in primates or guinea pigs, which require a dietary source of vitamin C.) The fact that a compound of the D-series becomes a compound of the L-series is not due to a biochemical change in configuration; rather, the change arises from the change in the numbering of the carbons, as may be seen in the following equation.

D-glucuronic acid   L-gulonic acid

L-ascorbic acid
(vitamin C)

D. Periodic Acid Oxidation

The **periodic acid oxidation** is a test for 1,2-diols and for 1,2- or α-hydroxy aldehydes and ketones. A compound containing such a grouping is oxidized and cleaved by periodic acid (HIO_4). In the case of a simple 1,2-diol, the products are two aldehydes or ketones.

The periodic acid reaction goes through a cyclic intermediate, a fact that explains why isolated hydroxyl groups are not oxidized.

Study Problem

22.11 Predict the products of the periodic acid oxidation of the following compound:

In the reaction of an α-hydroxy aldehyde or ketone, the carbonyl group is oxidized to a carboxyl group, while the hydroxyl group is again oxidized to an aldehyde or ketone.

$$to -CO_2H \qquad to -CHO$$

$$\underset{\text{RC}}{\overset{O}{\|}} | \underset{\text{CHR}}{\overset{OH}{|}} \xrightarrow{HIO_4} \underset{\text{RCOH}}{\overset{O}{\|}} + \underset{\text{HCR}}{\overset{O}{\|}} + HIO_3$$

$$\underset{\text{HC}}{\overset{O}{\|}} \underset{\text{CHR}}{\overset{OH}{|}} \xrightarrow{HIO_4} \underset{\text{HCOH}}{\overset{O}{\|}} + \underset{\text{HCR}}{\overset{O}{\|}} + HIO_3$$

The periodic acid oxidation is used in carbohydrate analysis. In carbohydrates, the oxidation of the interior hydroxyl groups proceeds further than it does with a simple 1,2-diol. For example, the products of the periodic acid oxidation of erythrose are formaldehyde and formic acid in the molar ratio of 1:3.

oxidized to HCO_2H
oxidized to HCHO

erythrose

This oxidation of the CHOH groups in erythrose to formic acid can be rationalized by considering the reaction to be stepwise.

Step 1:

Step 2:

Step 3:

Study Problem

22.12 Predict the periodic acid oxidation products of glucose.

▼ ## Section 22.7
Reduction of Monosaccharides

Both aldoses and ketoses can be reduced by carbonyl reducing agents, such as hydrogen and catalyst or a metal hydride, to polyalcohols called **alditols**. The suffix for the name of one of these polyalcohols is **-itol**. The product of reduction of D-glucose is called D-*glucitol*, or *sorbitol*.

D-glucose

D-glucitol
(sorbitol)

Natural D-glucitol has been isolated from many fruits (for example, cherries, plums, apples, pears, and mountain ash berries) and from algae and seaweed. Synthetic D-glucitol is used as an artificial sweetener.

▼ ## Section 22.8
Reactions at the Hydroxyl Groups

The hydroxyl groups in carbohydrates behave in a manner similar to that of other alcohol groups. They can be esterified by either carboxylic acids or inorganic acids, and they can be converted to ethers. Carbohydrates can also act as diols and form cyclic acetals or ketals with aldehydes or ketones. These reactions were discussed in Chapters 7 and 13.

A. Acetate Formation

A common reagent for esterification of alcohols is acetic anhydride, with either sodium acetate or pyridine as an alkaline catalyst. If the reaction is carried out below 0°C, the acylation reaction is faster than the $\alpha-\beta$ anomeric interconversion. Under these conditions, either α- or β-D-glucose yields its corresponding pentaacetate. At higher temperatures, a mixture of the α- and β-pentaacetates is formed, with the β-pentaacetate predominating.

β-D-glucopyranose

penta-*O*-acetyl-β-D-glucopyranose

B. Ether Formation

Dimethyl sulfate is an inorganic ester with an excellent leaving group. This compound is used to form methyl ethers.

$$\overset{..}{\underset{..}{RO}}:^- + CH_3 - \overset{\overset{O}{\|}}{\underset{\underset{O}{\|}}{OSOCH_3}} \xrightarrow{S_N2} \overset{..}{\underset{..}{RO}} - CH_3 + :\overset{..}{\underset{..}{OSOCH_3}}$$

dimethyl sulfate   *a methyl ether*

When a monosaccharide is treated with an excess of dimethyl sulfate and NaOH, all the hydroxyl groups (including a hemiacetal or hemiketal OH group) are converted to methoxyl groups.

D-glucose $\xrightarrow[\text{NaOH}]{(CH_3O)_2SO_2}$ methyl tetra-*O*-methyl-D-glucopyranoside (55 %)

In a typical Williamson ether synthesis (RO⁻ + RX → ROR + X⁻; Section 8.3B), the alkoxide must be prepared with a stronger base than NaOH. In the case of the carbohydrates, NaOH is a sufficiently strong base to yield alkoxide ions. (The inductive effect of the electronegative oxygens on adjacent carbons renders each hydroxyl group more acidic than a hydroxyl group in an ordinary alcohol.) Because the acetal bond is stable in base, the configuration at the anomeric carbon of a glycoside is not changed in this methylation reaction.

Study Problem

22.13 Give the structure of the product of the treatment of methyl 2-deoxy-α-D-ribofuranoside with: **(a)** acetic anhydride; **(b)** an alkaline solution of dimethyl sulfate.

C. Cyclic Acetal and Ketal Formation

Because carbohydrates contain numerous OH groups, it is sometimes desirable to block some of them so that selective reactions can be carried out on the other hydroxyl groups. Acetals and ketals are two common blocking groups (Section 13.5B). For example, an aldehyde, such as benzaldehyde, reacts with 1,3-diol groupings in sugar molecules. Other aldehydes and ketones can react

Figure 22.4 The conversion of L-sorbose to L-ascorbic acid (vitamin C).

preferentially at different diol grouping. (In some cases, the product is the furanose, rather than the pyranose, ring.) The different products arise because of subtle (generally unpredictable) steric and electronic effects.

In the commercial conversion of L-sorbose to vitamin C, acetone is used to block four hydroxyl groups so that a single CH_2OH group can be oxidized. This conversion is outlined in Figure 22.4.

Figure 22.5 summarizes the important reactions of monosaccharides.

Section 22.9
The Structure Determination of Glucose

In 1888, it was known that glucose is an aldohexose. The question was, Which of the 16 possible stereoisomeric aldohexoses is it? In 1891, the German chemist Emil Fischer reported the structure of the open-chain aldehyde form of D-glucose. For this work, he received the Nobel Prize in 1902.

The determination of configuration of a compound with four chiral carbons might seem an overwhelming task, but Fischer accomplished this task

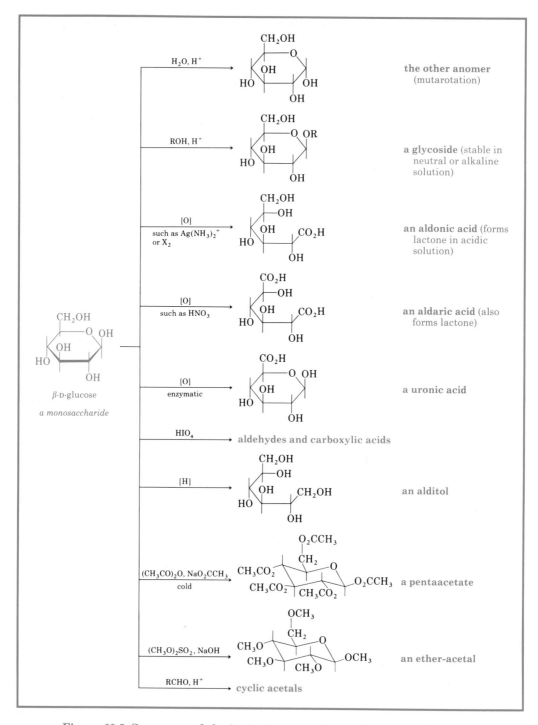

Figure 22.5 Summary of the important reactions of monosaccharides.

in a series of simple reactions. From his data, it was possible to determine only the *relative configuration* of glucose, not the absolute configuration, which was finally determined 50 years later by x-ray diffraction. Therefore, Fischer made the assumption that the OH on carbon 2 in D-(+)-glyceraldehyde and thus the OH on carbon 5 in D-(+)-glucose are projected to the right in the Fischer projections. (It turned out later that his assumption was correct; see Sections 4.8 and 22.3B.) This assumption narrowed the choices for the configuration of glucose to the eight D-aldohexoses shown in Figure 22.3, page 898.

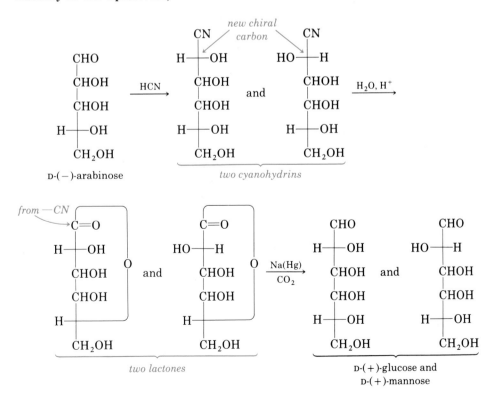

Fact 1: It was known that the aldopentose (−)-arabinose could be converted to the aldohexoses (+)-glucose and (+)-mannose. Heinrich Kiliani discovered the chain-lengthening step in 1886, and Fischer completed the synthesis by reduction of the resulting lactone to yield the aldohexoses in 1890. The following sequence is consequently known as the **Kiliani–Fischer synthesis**. (We show the intermediate lactones here as Fischer projections for consistency in the equations.)

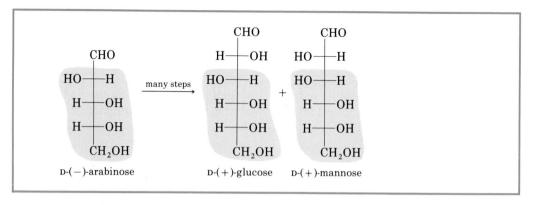

Figure 22.6 D-(+)-Glucose and D-(+)-mannose have the same configuration at the last three chiral carbons.

Because (−)-arabinose yields both (+)-glucose and (+)-mannose, all three of these sugars have *the same configuration at the last three chiral carbons* (carbons 3, 4, and 5 of glucose and mannose). (+)-Glucose and (+)-mannose must differ *only* by the configuration at carbon 2. These conclusions are pictured in Figure 22.6 with the structures as they are now known.

Fact 2: Fischer found that the oxidation of both end groups of (−)-arabinose yielded an optically active diacid, and not the *meso*-diacid.

$$
\begin{array}{ccc}
\text{CHO} & & \text{CO}_2\text{H} \\
| & & | \\
\text{CHOH} & & \text{CHOH} \\
| & \xrightarrow[\text{heat}]{\text{HNO}_3} & | \\
\text{CHOH} & & \text{CHOH} \\
\text{H——OH} & & \text{H——OH} \\
| & & | \\
\text{CH}_2\text{OH} & & \text{CO}_2\text{H}
\end{array}
$$

(−)-arabinose   *not meso*

Therefore, Fischer concluded that carbon 2 in (−)-arabinose must have OH on the *left*. If it were on the right, the *meso*-diacid would result.

left →

$$
\begin{array}{cc}
\text{CO}_2\text{H} & \text{CO}_2\text{H} \\
\text{HO——H} & \text{H——OH} \\
| & | \\
\text{CHOH} & \text{CHOH} \\
\text{H——OH} & \text{H——OH} \\
| & | \\
\text{CO}_2\text{H} & \text{CO}_2\text{H}
\end{array}
$$

this diacid would be meso, regardless of configuration at carbon 3

only possibility   *not this*

From the data presented so far, it is possible to write almost complete structures for (−)-arabinose, (+)-glucose, and (+)-mannose. Figure 22.7 shows these data with the known structures.

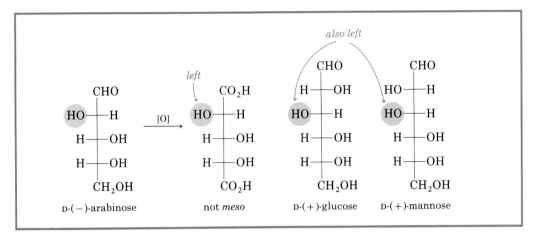

Figure 22.7 Because D-(−)-arabinose yields an optically active aldaric acid, the OH at carbon 2 must be projected on the left.

CHO
HO—H
CHOH
H—OH
CH₂OH

D-(−)-arabinose

CHO
H—OH
HO—H
CHOH
H—OH
CH₂OH

CHO
HO—H
HO—H
CHOH
H—OH
CH₂OH

one is D-(+)-glucose;
one is D-(+)-mannose

Fact 3: Fischer observed that both (+)-glucose and (+)-mannose are oxidized to optically active diacids. This means that the OH at carbon 4 of both of these monosaccharides is *on the right*. (If it were on the left, one of the two sugars would yield a *meso*-diacid.) Figure 22.8 shows these reactions with the known structures.

CHO
CHOH
HO—H
H—OH   *right*
H—OH
CH₂OH

$\xrightarrow{\text{HNO}_3}$
heat

CO₂H
CHOH
HO—H
H—OH
H—OH
CO₂H

*not meso,
regardless of
configuration
at carbon 2*

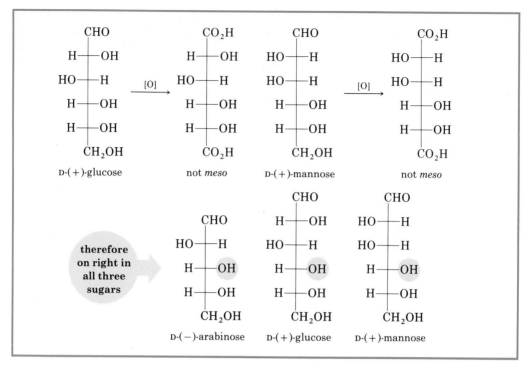

Figure 22.8 Because both D-(+)-glucose and D-(+)-mannose yield optically active aldaric acids, the 4-hydroxyl must be on the right.

If OH at carbon 4 were on the left:

Now a complete structure for (−)-arabinose can be written. All that is needed is to differentiate between (+)-glucose and (+)-mannose.

Fact 4: The sugar (+)-gulose (another aldohexose) and (+)-glucose both give the *same diacid* when oxidized. Of the possible diacids that can be obtained from the two structures that are (+)-glucose and (+)-mannose, *only one* can come from two *different* sugars.

$$
\begin{array}{c}
\text{CHO} \\
\text{H}\!-\!\!-\!\text{OH} \\
\text{HO}\!-\!\!-\!\text{H} \\
\text{H}\!-\!\!-\!\text{OH} \\
\text{H}\!-\!\!-\!\text{OH} \\
\text{CH}_2\text{OH}
\end{array}
\xrightarrow{\text{HNO}_3}
\begin{array}{c}
\text{CO}_2\text{H} \\
\text{H}\!-\!\!-\!\text{OH} \\
\text{HO}\!-\!\!-\!\text{H} \\
\text{H}\!-\!\!-\!\text{OH} \\
\text{H}\!-\!\!-\!\text{OH} \\
\text{CO}_2\text{H}
\end{array}
\qquad
\begin{array}{c}
\text{CHO} \\
\text{HO}\!-\!\!-\!\text{H} \\
\text{HO}\!-\!\!-\!\text{H} \\
\text{H}\!-\!\!-\!\text{OH} \\
\text{H}\!-\!\!-\!\text{OH} \\
\text{CH}_2\text{OH}
\end{array}
\xrightarrow{\text{HNO}_3}
\begin{array}{c}
\text{CO}_2\text{H} \\
\text{HO}\!-\!\!-\!\text{H} \\
\text{HO}\!-\!\!-\!\text{H} \\
\text{H}\!-\!\!-\!\text{OH} \\
\text{H}\!-\!\!-\!\text{OH} \\
\text{CO}_2\text{H}
\end{array}
$$

I · **II**

can arise from two aldohexoses   ·   *can arise from only one aldohexose*

Let us look at diacid II first. The two potential aldohexoses that can lead to diacid II follow:

$$
\begin{array}{c}
\text{CHO} \\
\text{HO}\!-\!\!-\!\text{H} \\
\text{HO}\!-\!\!-\!\text{H} \\
\text{H}\!-\!\!-\!\text{OH} \\
\text{H}\!-\!\!-\!\text{OH} \\
\text{CH}_2\text{OH}
\end{array}
\xrightarrow{\text{rotate } 180^\circ}
\begin{array}{c}
\text{CH}_2\text{OH} \\
\text{HO}\!-\!\!-\!\text{H} \\
\text{HO}\!-\!\!-\!\text{H} \\
\text{H}\!-\!\!-\!\text{OH} \\
\text{H}\!-\!\!-\!\text{OH} \\
\text{CHO}
\end{array}
$$

both are D-(+)-*mannose*

But these two aldohexoses are the *same*. If we rotate either projection 180° in the plane of the paper, the structure is the same as the other. Only this one aldohexose can yield diacid II. This aldohexose must be (+)-mannose.

Now let us look at diacid I. It can arise from two possible aldohexoses that are *not the same*: D-glucose and L-gulose. (Refer to Figure 22.3, page 898, for the structure of D-gulose.)

$$
\begin{array}{c}
\text{CHO} \\
\text{H}\!-\!\!-\!\text{OH} \\
\text{HO}\!-\!\!-\!\text{H} \\
\text{H}\!-\!\!-\!\text{OH} \\
\text{H}\!-\!\!-\!\text{OH} \\
\text{CH}_2\text{OH}
\end{array}
\qquad
\begin{array}{c}
\text{CH}_2\text{OH} \\
\text{H}\!-\!\!-\!\text{OH} \\
\text{HO}\!-\!\!-\!\text{H} \\
\text{H}\!-\!\!-\!\text{OH} \\
\text{H}\!-\!\!-\!\text{OH} \\
\text{CHO}
\end{array}
$$

D-(+)-glucose   ·   L-(+)-gulose

rotating one by 180°
does not give the other

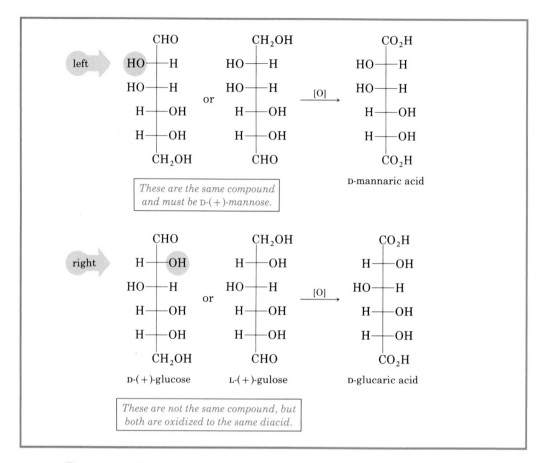

Figure 22.9 Because the aldaric acid from D-(+)-glucose can also be formed from L-(+)-gulose, D-(+)-glucose has the OH at carbon 2 on the right. Because the aldaric acid from D-(+)-mannose can come *only* from D-(+)-mannose, this sugar is the one with the OH at carbon 2 on the left.

If these two sugars give the same diacid (that is, diacid I), then the left-hand structure is (+)-glucose, the right-hand structure is (+)-gulose, and the aldohexose that leads to diacid II is (+)-mannose. See Figure 22.9 for these reactions with known structures.

A. Determination of Ring Size

The cyclic structures for glucose were postulated in 1895, but it was not shown until 1926 that glucose forms six-membered cyclic hemiacetals and glycosides. The reactions that were used to determine the ring size are, again, not complex.

A monosaccharide undergoes reaction with dimethyl sulfate to yield a completely methylated structure. In acidic solution, the methylated methyl glycoside can be hydrolyzed and the ring opened. The methoxyl groups, which are ethers, are not affected by this reaction. The hydrolyzed acetal therefore has only one hydroxyl group. Determination of the position of this hydroxyl group provides the information needed to know the ring size of the original acetal.

ring opening   *one OH is formed*

$$\text{methylated ring structure} \xrightarrow[\text{H}_2\text{O, H}^+]{} \text{open-chain structure}$$

The position of the OH group was determined by vigorous oxidation in which the —CHO group is oxidized to —CO_2H and the single CHOH group is oxidized to a keto group. Under these oxidizing conditions, cleavage of the molecule occurs adjacent to the ketone group (on either side) to yield two dicarboxylic acids. (See Section 13.10 for a discussion of this type of oxidation.)

$$\text{methylated D-glucose} \xrightarrow[\text{heat}]{\text{HNO}_3} \text{products}$$

methylated D-glucose

From methylated glucose, the two methoxy diacids shown in the preceding equation were observed as principal products (along with methoxyacetic acid and CO_2). The structures of the two diacids were determined by comparing their physical properties with those of methoxy diacids of known structure. Because these product diacids are methoxy derivatives of *four-* and *five-carbon* diacids, it was concluded that carbon 5 holds the OH group in the hydrolyzed glycoside and that the original glycoside ring is a six-membered ring.

carbon 5 —

four carbons   *five carbons*

Study Problem

22.14 What products would be observed from methylation, hydrolysis, and oxidation of a methyl D-glucofuranoside?

Section 22.10
Disaccharides

A **disaccharide** is a carbohydrate composed of two units of monosaccharide joined together by a glycoside link from carbon 1 of one unit to an OH of the other unit. A common mode of attachment is an α or β glycoside link from the first unit to the 4-hydroxyl group of the second unit. This link is called a 1,4'-α or a 1,4'-β link, depending on the stereochemistry at the glycoside carbon.

a 1,4'-β link
(conformational)

a 1,4'-β link
(Haworth)

Let us look at the preceding structures more closely. Unit 1 (the left-hand unit in each structure) has a β-glycoside link to unit 2. In aqueous solution, this glycoside link is fixed. It is not in equilibrium with the anomer. However, unit 2 (the right-hand unit in each structure) contains a hemiacetal group. In aqueous solution, this particular group is in equilibrium with the open-chain aldehyde form and with the other anomer.

A. Maltose

The disaccharide **maltose** is used in baby foods and malted milk. It is the principal disaccharide obtained from the hydrolysis of starch. Starch is broken down into maltose in an apparently random fashion by an enzyme in saliva called *α-1,4-glucan 4-glucanohydrolase*. The enzyme *α-1,4-glucan maltohydrolase*, found in sprouted barley (*malt*), converts starch specifically into maltose units. In beer-making, malt is used for the conversion of starches from corn or other sources into maltose. An enzyme in yeast (*α-glucosidase*) catalyzes the hydrolysis of the maltose into D-glucose, which is acted upon by other enzymes from the yeast to yield ethanol. One molecule of maltose yields two molecules of D-glucose, regardless of whether the hydrolysis takes place in a laboratory flask, in an organism, or in a fermentation vat.

$$\text{starch} \xrightarrow[\text{H}^+ \text{ or enzymes}]{\text{H}_2\text{O}} \text{maltose} \xrightarrow[\text{H}^+ \text{ or enzymes}]{\text{H}_2\text{O}} \text{D-glucose} \xrightarrow[\text{enzymes}]{} \underset{\text{ethanol}}{\text{CH}_3\text{CH}_2\text{OH}}$$

A molecule of maltose contains two units of D-glucopyranose. The first unit (shown on the left) is in the form of an α-glycoside. This unit is attached to the oxygen at carbon 4′ in the second unit by a 1,4′-α link.

conformational formula for maltose
4-O-(α-D-glucopyranosyl)-D-glucopyranose

The anomeric carbon of the second unit of glucopyranose in maltose is part of a hemiacetal group. As a result, there are two forms of maltose (α- and β-maltose), which are in equilibrium with each other in solution. Maltose undergoes mutarotation, is a reducing sugar, and can be oxidized to the carboxylic acid **maltobionic acid** by a bromine–water solution.

Study Problem

22.15 Give the structures of the products:

(a) α-maltose $\xrightarrow{\text{H}_2\text{O, H}^+}$ (b) β-maltose $\xrightarrow{\text{Br}_2, \text{H}_2\text{O}}$

B. Cellobiose

The disaccharide obtained from the partial hydrolysis of cellulose is called **cellobiose**. Like maltose, cellobiose is composed of two glucopyranose units joined together by a 1,4′-link. Cellobiose differs from maltose in that the 1,4-linkage is β rather than α.

cellobiose
4-*O*-(β-D-glucopyranosyl)-D-glucopyranose

Chemical hydrolysis of cellobiose with aqueous acid yields a mixture of α- and β-D-glucose, the same products that are obtained from maltose. Cellobiose can also be hydrolyzed with the enzyme *β-glucosidase* (also called *emulsin*), but not by *α-glucosidase*, which is specific for the α link (that is, the link in maltose).

Study Problems

22.16 Give the structures of the products:

(a) α-cellobiose $\xrightarrow{\text{H}_2\text{O, H}^+}$ (b) β-cellobiose $\xrightarrow{\text{H}_2\text{O, H}^+}$

(c) α-cellobiose $\xrightarrow{\text{Br}_2,\ \text{H}_2\text{O}}$ (d) α-cellobiose $\xrightarrow{\text{Tollens reagent}}$

(e) α-cellobiose $\xrightarrow{\text{β-glucosidase}}$

22.17 Which would you expect to be more stable, β-maltose or β-cellobiose? Why?

C. Lactose

The disaccharide **lactose** (milk sugar) is different from maltose or cellobiose in that it is composed of two different monosaccharides, D-glucose and D-galactose.

glucopyranose
β-galactopyranose

lactose
4-*O*-(β-D-galactopyranosyl)-D-glucopyranose

Lactose is a naturally occurring disaccharide found only in mammals. Cow's milk and human milk contain about 5% lactose. Lactose is obtained commercially as a by-product in the manufacture of cheese.

In normal human metabolism, lactose is hydrolyzed enzymatically to D-galactose and D-glucose; then the galactose is converted to glucose, which can undergo metabolism. A condition called **galactosemia** that affects some infants is caused by lack of the enzyme used to convert galactose to glucose. Galactosemia is characterized by high levels of galactose in the blood and urine. Symptoms range from vomiting to mental and physical retardation and sometimes death. Treatment consists of removing milk and milk products from the diet. (An artificial milk made from soybeans may be substituted.)

D. Sucrose

The disaccharide **sucrose** is common table sugar. Sugar cane was grown domestically as early as 6000 B.C. in India. (The words "sugar" and "sucrose" come from the Sanskrit word *sarkara*.) Sucrose was encountered by the soldiers of Alexander the Great, who entered India in 325 B.C. In later centuries, the use of sucrose was spread by the Arabs and the Crusaders. Sugar cane was introduced into the New World by Columbus, who brought some to Santo Domingo in 1493. In the 1700s, it was discovered that certain beets also contain high levels of sucrose. The discovery meant that sugar could be obtained from plants grown in temperate climates as well as from sugar cane grown in the tropics.

sucrose
β-D-fructofuranosyl α-D-glucopyranoside

Whether it comes from beets or sugar cane, the chemical composition of sucrose is the same: one unit of fructose joined to one unit of glucose. The glycoside link joins the ketal and acetal carbons, β from fructose and α from glucose. Note the difference between sucrose and the other disaccharides we have discussed: in sucrose, *both* anomeric carbon atoms (not just one) are used in the glycoside link. In sucrose, neither fructose nor glucose has a hemiacetal group; therefore, sucrose in water is not in equilibrium with an aldehyde or keto form. Sucrose does not exhibit mutarotation and is not a reducing sugar.

Invert sugar is a mixture of D-glucose and D-fructose obtained by the acidic or enzymatic hydrolysis of sucrose. The enzymes that catalyze the hydrolysis of sucrose, called *invertases*, are specific for the β-D-fructofuranoside link and are found in yeast and in bees. (Honey is primarily invert sugar.) Because of the presence of free fructose (the sweetest sugar), invert sugar is sweeter than sucrose. A synthetic invert sugar called *Isomerose* is prepared by the enzymatic isomerization of glucose in corn syrup. It has commercial use in the preparation of ice cream, soft drinks, and candy.

The name "invert sugar" is derived from inversion in the sign of the specific rotation when sucrose is hydrolyzed. Sucrose has a specific rotation of $+66.5°$, a *positive* rotation. The mixture of products (glucose, $[\alpha] = +52.7°$, and fructose, $[\alpha] = -92.4°$) has a net *negative* rotation.

▼ # Section 22.11
Polysaccharides

A **polysaccharide** is a compound in which the molecules contain many units of monosaccharide joined together by glycoside links. Upon complete hydrolysis, a polysaccharide yields monosaccharides.

Polysaccharides serve three purposes in living systems: structural, nutritional, and as specific agents. Typical architectural polysaccharides are *cellulose*, which gives strength to the stems and branches of plants, and *chitin*, the structural component of the exoskeletons of insects. Typical nutritional polysaccharides are *starch* (as is found in wheat and potatoes) and *glycogen*, an animal's internal store of readily available carbohydrate. *Heparin*, an example of a specific agent, is a polysaccharide that prevents blood coagulation.

$$\text{CO}_2\text{H} \qquad\qquad \text{CH}_2\text{OSO}_3\text{H}$$

heparin

Polysaccharides can also be bonded to other types of molecules, as in *glycoproteins* (polysaccharide–protein complexes) and *glycolipids* (polysaccharide–lipid complexes).

A. Cellulose

Cellulose is the most abundant organic compound on earth. It has been estimated that about 10^{11} tons of cellulose are biosynthesized each year, and that cellulose accounts for about 50% of the bound carbon on earth! Dry leaves contain 10–20% cellulose; wood, 50%; and cotton, 90%. The most convenient laboratory source of pure cellulose is filter paper.

Cellulose forms the fibrous component of plant cell walls. The rigidity of cellulose arises from its overall structure. Cellulose molecules are chains, or microfibrils, of up to 14,000 units of D-glucose that occur in twisted rope-like bundles held together by hydrogen bonding.

A single molecule of cellulose is a linear polymer of 1,4'-β-D-glucose. Complete hydrolysis in 40% aqueous HCl yields only D-glucose. The disaccharide isolated from partially hydrolyzed cellulose is cellobiose, which can be further hydrolyzed to D-glucose with an acidic catalyst or with the enzyme emulsin. Cellulose itself has no hemiacetal carbon—it cannot undergo muta-rotation or be oxidized by such test reagents as Tollens reagent. (There may be a hemiacetal at one end of each cellulose molecule. However, this is but a small portion of the whole and does not lead to observable reaction.)

cellulose

Study Problem

22.18 Predict the major product (and give a structure or partial structure) when cellulose is treated with: (**a**) an excess of hot aqueous H_2SO_4, then water;   (**b**) hot water;   (**c**) hot aqueous NaOH solution;   (**d**) an excess of NaOH and dimethyl sulfate.

Although mammals do not produce the proper enzymes for breaking down cellulose into glucose, certain bacteria and protozoa do have these enzymes. Grazing animals are capable of using cellulose as food only indirectly. Their stomachs and intestines support colonies of microorganisms that live and reproduce on cellulose. The animals use these microorganisms and their by-products as food.

B. Starch

Starch is the second most abundant polysaccharide. Starch can be separated into two principal fractions based upon solubility when triturated (pulverized) with hot water: about 20% of starch is **amylose** (soluble) and the remaining 80% is **amylopectin** (insoluble).

Amylose.   Complete hydrolysis of amylose yields only D-glucose; partial hydrolysis yields maltose as the only disaccharide. We conclude that amylose is a linear polymer of 1,4′-linked α-D-glucose. The difference between amylose and cellulose is the glycoside link: β in cellulose, α in amylose. This difference is responsible for the different properties of these two polysaccharides.

amylose

There are 250 or more glucose units per amylose molecule; the exact number depends upon the species of animal or plant. (Measurement of chain length is complicated by the fact that natural amylose degrades into smaller chains upon separation and purification.)

Amylose molecules form helices or coils around I_2 molecules. A deep blue color arises from electronic interactions between the two. This color is the basis of the **iodine test for starch**, in which a solution of iodine is added to an unknown as a test for the presence of starch.

Amylopectin, a much larger polysaccharide than amylose, contains 1000 or more glucose units per molecule. Like the chain in amylose, the main chain of amylopectin contains 1,4′-α-D-glucose. Unlike amylose, amylopectin is *branched* so that there is a terminal glucose about every 25 glucose units (Figure 22.10). The bonding at the branch point is a 1,6′-α-glycosidic bond.

amylopectin

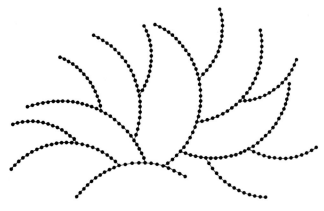

Figure 22.10 A representation of the branched structure of amylopectin. Each ─•─ represents a glucose molecule.

Complete hydrolysis of amylopectin yields only D-glucose. However, incomplete hydrolysis yields a mixture of the disaccharides maltose and isomaltose, the latter arising from the 1,6'-branching. The oligosaccharide mixture obtained from the partial hydrolysis of amylopectin, which is referred to as **dextrins**, is used to make glue, paste, and fabric sizing.

$$\text{amylopectin} \xrightarrow{\ H_2O\ } \text{dextrins} \xrightarrow{\ H_2O\ } \text{maltose + isomaltose} \xrightarrow{\ H_2O\ } \text{D-glucose}$$

isomaltose
6-*O*-(α-D-glucopyranosyl)-D-glucopyranose

Glycogen is a polysaccharide that is used as a storehouse (primarily in the liver and muscles) for glucose in an animal system. Structurally, glycogen is related to amylopectin. It contains chains of 1,4'-α-linked glucose with branches (1,6'-α). The difference between glycogen and amylopectin is that glycogen is more branched than amylopectin.

C. Chitin

The principal structural polysaccharide of the arthropods (for example, crabs and insects) is **chitin**. It has been estimated that 10^9 tons of chitin are biosynthesized each year! Chitin is a linear polysaccharide consisting of β-linked

N-acetyl-D-glucosamine. Upon hydrolysis, chitin yields 2-amino-2-deoxy-D-glucose. (The acetyl group is lost in the hydrolysis step.) In nature, chitins are bonded to nonpolysaccharide material (proteins and lipids).

chitin

Study Problem

22.19 Give the structures for the major organic products when chitin is treated with: **(a)** hot dilute aqueous HCl; **(b)** hot dilute aqueous NaOH.

▼ Summary

Carbohydrates are polyhydroxy aldehydes and ketones or their derivatives. A **monosaccharide** is the smallest carbohydrate; it does not undergo hydrolysis to smaller units. Monosaccharides may be classified as to the number of carbons and to the principal functional group:

an aldotriose   *a tetrulose*

Epimers are diastereomers that differ in configuration at only one chiral carbon atom.

Natural monosaccharides generally belong to the **D-series**. Because of the presence of both hydroxyl and carbonyl groups, monosaccharides that can form **furanose** or **pyranose** hemiacetal or hemiketal rings undergo cyclization. The cyclization creates a new chiral carbon and therefore gives rise to a pair of diastereomers called **α** and **β anomers**. In solution, the anomers are in equilibrium with each other.

CHO
H——OH
HO——H
H——OH
H——OH
CH₂OH

Because of the equilibrium, a monosaccharide undergoes reactions typical of aldehydes as well as reactions at the hydroxyl groups. The reactions of monosaccharides are summarized in Figure 22.5, page 914.

Disaccharides are composed of two monosaccharide units joined by a glycoside link from one unit to an OH group of the second unit. **Maltose** is composed of two D-glucopyranose units joined by a 1,4'-α link. **Cellobiose** is composed of two D-glucopyranose units joined by a 1,4'-β link. **Lactose** is composed of β-D-galactopyranose joined to the 4-position of D-glucopyranose. **Sucrose** is composed of α-D-glucopyranose and β-D-fructofuranose joined by a 1,2' link.

A **polysaccharide** is composed of many monosaccharide units joined by glycosidic links:

cellulose:	1,4'-β-D-glucopyranose
amylose:	1,4'-α-D-glucopyranose
amylopectin:	1,4'-α-D-glucopyranose with 1,6'-α-branching

Study Problems*

22.20 Draw the Fischer projections for the open-chain forms of all the isomeric D-hexuloses (ketohexoses) and indicate the chiral carbons in each.

22.21 Classify the following monosaccharides (as an aldohexose, for example):

(a)
CHO
HO——H
H——OH
CH₂OH

(b)
CH₂OH
C=O
CH₂OH

(c)
CHO
CH₂
HO——H
HO——H
CH₂OH

22.22 State whether each of the monosaccharides in the preceding problem belongs to the D series or to the L series, if either.

22.23 (a) When (R)-2-hydroxypropanal is treated with HCN, what products are obtained?
(b) Are those products optically active? Explain.

* For information concerning the organization of the *Study Problems* and *Additional Problems*, see the *Note to student* on page 39.

22.24 A chemist has a single stereoisomer of 1,2,3,4-butanetetraol of unknown configuration. He also has a flask of D-glyceraldehyde. Suggest a reaction sequence that could be used to relate the configuration of the tetraol to that of D-glyceraldehyde.

22.25 Match each of the following classes of compound with a structure on the right:

(a) a hexulose   (1)

(b) a pentopyranose   (2)

(c) a pentofuranose   (3)

(d) a pentofuranoside   (4)

22.26 Label each of the structures in Problem 22.25 as D or L.

22.27 Give structures for (and label as α or β) the anomers of compounds (1) and (4) in Problem 22.25.

22.28 Draw Haworth formulas for the following monosaccharides. [For (c) and (d), refer to Figure 22.3, page 898.]

(a)   (b)

(c) β-D-altropyranose   (d) α-D-lyxofuranose

22.29 Write equations (using Haworth formulas) that illustrate:

(a) the mutarotation of pure β-D-arabinofuranose in water
(b) the conversion of β-D-fructofuranose to β-D-fructopyranose
(c) the mutarotation of β-maltose (Section 22.10A)

22.30 What would you expect to be the most stable conformation for: (a) β-D-galactopyranose, and   (b) α-D-idopyranose? Give structures and reasons for your answers.

22.31 Give the Haworth formulas for the major organic products:

(a) D-glucose + $(CH_3)_2CHOH$ $\xrightarrow{H^+}$

(b) D-galactose + CH_3CH_2OH $\xrightarrow{H^+}$

(c) methyl α-D-ribofuranoside $\xrightarrow[\text{heat}]{H_2O,\ H^+}$

22.32 The oxidation of D-fructose with Tollens reagent yields a mixture of anions of D-mannonic acid and D-gluconic acid. Explain.

22.33 Which of the following compounds would *not* undergo mutarotation? Why?

(e) 6-*O*-(α-D-galactopyranosyl)-β-D-glucopyranose
(f) α-D-glucopyranosyl α-D-glucopyranoside

22.34 Which of the compounds in Problem 22.33 are nonreducing sugars?

22.35 Give the structure and the name of the organic product obtained when D-galactose is treated with: **(a)** aqueous Br_2; **(b)** hot dilute HNO_3; **(c)** Tollens reagent; **(d)** $NaBH_4$, followed by aqueous HCl.

22.36 *Algin* is a polysaccharide obtained from seaweed. It is hydrolyzed to D-mannuronic acid. What is the structure of this acid? (Use a Fischer projection.)

22.37 Fill in the blanks (give *all* possibilities):

(a) $\underset{\text{(a D-tetrose)}}{\underline{\hspace{2cm}}} \xrightarrow{\text{hot } HNO_3}$ *meso*-tartaric acid

(b) $\underline{\hspace{2cm}} \xrightarrow[\text{cold}]{\text{dil. HCl}}$ [structure: HOCH₂ ring with =O]

(c) $\underset{\text{(a D-aldohexose)}}{\underline{\hspace{2cm}}} \xrightarrow[\text{(2) } H_2O, H^+]{\text{(1) } NaBH_4}$ a *meso*-alditol

22.38 In Section 22.6B, we stated that aldaric acids can form lactones. Write formulas for the lactones that could be formed by D-glucaric acid.

22.39 Three compounds (A, B, and C) were subjected to oxidation with HIO_4. The following products were obtained from each reaction, respectively. What are the structures of A, B, and C?

(a) $CH_3\overset{O}{\overset{||}{C}}OH + H\overset{O}{\overset{||}{C}}CH_2CH_3$ **(b)** $CH_3\overset{O}{\overset{||}{C}}CH_2CH_2CH_2\overset{O}{\overset{||}{C}}H$ **(c)** $H\overset{O}{\overset{||}{C}}H + CH_3\overset{O}{\overset{||}{C}}CH_3$

22.40 Write equations showing each step in the periodic acid oxidation of arabinose.

22.41 Write equations to represent the reaction of β-D-ribofuranose with **(a)** an excess of diethyl sulfate plus NaOH, and **(b)** an excess of acetic anhydride.

22.42 Methyl β-D-gulopyranoside is treated with: (1) excess dimethyl sulfate plus NaOH; (2) H_2O, H^+; and then (3) hot HNO_3. Write the equations that illustrate the steps in this reaction sequence.

22.43   Write an equation that shows the acid-catalyzed reaction of D-mannopyranose with benzaldehyde.

22.44   **(a)** Write the sequence of reactions by which D-ribose is converted into D-allose.
      (b) Show how you would determine that your product in **(a)** is indeed D-allose and not D-altrose.

22.45   Predict the major organic product of the treatment of amylose with dimethyl sulfate and NaOH, followed by hydrolysis in dilute HCl.

22.46   How many possible disaccharides could be formed from just D-glucopyranose? (The α and β links form *different* disaccharides.)

22.47   *Trehalose* is a nonreducing sugar with the formula $C_{12}H_{22}O_{11}$. Upon hydrolysis, only D-glucose is obtained. What are the possible structures of trehalose?

22.48   If a polysaccharide were composed of 1,4'-α-linked D-glucopyranose with 1,3'-α-linked branches, what are the possible disaccharides that would be obtained upon partial hydrolysis?

Additional Problems

22.49   Predict the major organic product for the reaction of D-mannose with: **(a)** $Br_2 + H_2O$;   **(b)** HNO_3;   **(c)** ethanol + H^+;   **(d)** [the product from **(c)**] plus excess dimethyl sulfate and NaOH;   **(e)** [the product from **(c)**] plus CH_3I and Ag_2O;   **(f)** acetic anhydride;   **(g)** acetyl chloride + pyridine;   **(h)** $NaBH_4$;   **(i)** HCN followed by H_2O and HCl;   **(j)** $LiAlH_4$ followed by aqueous acid;   **(k)** H_2 and Ni catalyst;   **(l)** [the product from **(c)**] plus H_2O, HCl.

22.50   How would you distinguish chemically between   **(a)** maltose and sucrose?   **(b)** D-lyxose and D-xylose?   **(c)** D-mannose and D-mannitol?   **(d)** D-mannose and D-mannonic acid?   **(e)** D-glucose and D-galactose?   **(f)** D-glucose and methyl α-D-glucoside?

22.51   List some properties of D-glucose that are consistent with a cyclic hemiacetal structure but that are inconsistent with an open-chain aldehyde form.

22.52   Suggest a synthesis of 2,3-di-*O*-methylglucose from glucose. (Ignore the stereochemistry at the anomeric carbon.)

22.53   Today, the ring size of a glycoside is determined by HIO_4 oxidation of the glucoside (Section 22.6D). Predict the products of the HIO_4 oxidation of methyl D-glucopyranoside and methyl D-glucofuranoside.

22.54   A carbohydrate A $(C_{12}H_{22}O_{11})$ was treated with   (1) CH_3OH, H^+, and   (2) excess CH_3I and Ag_2O. The product B was hydrolyzed to 2,3,4,6-tetra-*O*-methyl-D-galactose and 2,3,6-tri-*O*-methyl-D-glucose. When A was treated with aqueous acid, the products were D-galactose and D-glucose in equal amounts. When A was treated with aqueous Br_2, a carboxylic acid C was isolated. Hydrolysis of C with aqueous HCl resulted in D-gluconic acid as the only acid. What are the structures of A, B, and C?

22.55   *Raffinose* is a trisaccharide found in beets. Complete hydrolysis of raffinose yields D-fructose, D-glucose, and D-galactose. Partial enzymatic hydrolysis of raffinose with invertase yields D-fructose and the disaccharide *melibiose*. Partial hydrolysis of raffinose with an α-glycosidase yields D-galactose and sucrose. Methylation of raffinose followed by hydrolysis yields 2,3,4,6-tetra-*O*-methylgalactose; 2,3,4-tri-*O*-methylglucose; and 1,3,4,6-tetra-*O*-methylfructose. What are the structures of raffinose and melibiose?

22.56   *Bleomycin* is an antitumor antibiotic active against squamous cell carcinomas. Part of the structure of this compound is a reducing disaccharide A with the formula $C_{12}H_{22}O_{11}$.

Compound A is hydrolyzed in acidic solution to D-mannose and L-gulose. The hydrolysis is also catalyzed by enzymes specific for α-mannose glycosides. When A is subjected to treatment with (1) excess dimethyl sulfate + NaOH and then (2) acidic hydrolysis, the products are the 2,3,4,6-tetramethyl ether of D-mannose and the 3,4,6-trimethyl ether of L-gulose. What is the structure of A?

22.57 *Linustatin* ($C_{16}H_{27}NO_{11}$) is a disaccharide in defatted meal from flaxseed (*Linum usitatissimum*) that can protect livestock against the toxic effects of selenium compounds in forage plants. Acidic hydrolysis of linustatin yields D-glucose as the only monosaccharide. Enzymatic hydrolysis with β-glucosidase yields two compounds: D-glucose and a glycoside. Comparison of the ^{13}C nmr spectrum of linustatin with those of known compounds indicates the presence of two CH_3 groups, one quaternary C (no hydrogens bonded to it), and one unsubstituted —CH_2OH group, as well as other carbon atoms. The infrared spectrum of linustatin shows strong absorption at 3400 cm^{-1} (2.94 μm) and weak absorption at 2240 cm^{-1} (4.46 μ). No carbonyl absorption is apparent. What is the structure of linustatin?

23 ▽

Lipids

A lipid is defined as a naturally occurring organic compound that is insoluble in water, but soluble in nonpolar organic solvents such as a hydrocarbon or diethyl ether. This definition sounds as if it might include many types of compounds, and indeed it does. The various classes of lipids are related to one another by this shared physical property; but their chemical, functional, and structural relationships, as well as their biological functions, are diverse. We will discuss here the classes usually thought of as lipids: fats and oils, steroids, and a few other compounds of interest. The class of lipids called terpenes is discussed in Chapter 19.

a fat:
a triglyceride,
or triacylglycerol

cholesterol

a steroid

Section 23.1
Fats and Oils

Fats and oils are **triglycerides**, or **triacylglycerols**; both terms mean triesters of glycerol. The distinction between a fat and an oil is arbitrary: at room temperature a fat is solid and an oil is liquid. Most glycerides in animals are fats, while those in plants tend to be oils; hence the terms *animal fats* (bacon fat, beef fat) and *vegetable oils* (corn oil, safflower oil).

The carboxylic acid obtained from the hydrolysis of a fat or oil, called a **fatty acid**, generally has a long, unbranched hydrocarbon chain. Fats and oils are often named as derivatives of these fatty acids. For example, the tristearate of glycerol is named tristearin, and the tripalmitate of glycerol is named tripalmitin. A fat or oil can also be named as an ester in the usual manner, for example, glyceryl tristearate and glyceryl tripalmitate.

$$\begin{array}{l} CH_2O_2C(CH_2)_{16}CH_3 \\ CHO_2C(CH_2)_{16}CH_3 \\ CH_2O_2C(CH_2)_{16}CH_3 \end{array} + 3\,H_2O \xrightarrow{\;H^+\;} \begin{array}{l} CH_2OH \\ CHOH \\ CH_2OH \end{array} + 3\,CH_3(CH_2)_{16}CO_2H$$

glyceryl tristearate
(tristearin)

a typical fat

glycerol
(glycerine)

stearic acid

a fatty acid

Because they can be hydrolyzed to smaller molecules, fats and oils are said to be **complex lipids**. Lipids such as cholesterol that cannot be hydrolyzed are called **simple lipids**.

Fatty acids can also be obtained from **waxes**, such as beeswax. In these cases, the fatty acid is esterified with a simple long-chain alcohol.

$$C_{25}H_{51}CO_2C_{28}H_{57} \qquad C_{27}H_{55}CO_2C_{32}H_{65} \qquad C_{15}H_{31}CO_2C_{16}H_{33}$$

in beeswax   *in carnauba wax*   cetyl palmitate

in spermaceti

Most naturally occurring fats and oils are *mixed* triglycerides; that is, the three fatty-acid portions of the glyceride are not the same. Table 23.1 lists some representative fatty acids, and Table 23.2 shows the fatty-acid composition of some plant and animal triglycerides.

Table 23.1 Selected fatty acids and their sources[a]

Name of Acid	Structure	Source
Saturated:		
butyric	$CH_3(CH_2)_2CO_2H$	milk fat
palmitic	$CH_3(CH_2)_{14}CO_2H$	animal and plant fat
stearic	$CH_3(CH_2)_{16}CO_2H$	animal and plant fat
Unsaturated:		
palmitoleic	$CH_3(CH_2)_5CH{=}CH(CH_2)_7CO_2H$	animal and plant fat
oleic	$CH_3(CH_2)_7CH{=}CH(CH_2)_7CO_2H$	animal and plant fat
linoleic[b]	$CH_3(CH_2)_4CH{=}CHCH_2CH{=}CH(CH_2)_7CO_2H$	plant oils
linolenic[b]	$CH_3CH_2CH{=}CHCH_2CH{=}CHCH_2CH{=}CH(CH_2)_7CO_2H$	linseed oil
arachidonic[b]	$CH_3(CH_2)_4(CH{=}CHCH_2)_4(CH_2)_2CO_2H$	plant oils

[a] The carbon–carbon double bonds in naturally occurring unsaturated fatty acids are *cis*.
[b] Essential fatty acids that must be present in the human diet and that are used for the synthesis of prostaglandins (see Section 23.4).

Almost all naturally occurring fatty acids have an *even* number of carbon atoms because they are biosynthesized from the two-carbon acetyl groups in acetylcoenzyme A.

$$8\ \ CH_3\overset{\displaystyle O}{\overset{\|}{C}}{-}SCoA \xrightarrow{\text{many steps}} CH_3(CH_2)_{14}CO_2H$$

two carbons   *an even number of carbons*

acetylcoenzyme A   palmitic acid

The hydrocarbon chain in a fatty acid may be saturated or it may contain double bonds. The most widely distributed fatty acid in nature, oleic acid, contains one double bond. Fatty acids with more than one double bond are not uncommon, particularly in vegetable oils. These oils are the so-called *polyunsaturates*.

The configuration around any double bond in a naturally occurring fatty acid is *cis*, a configuration that results in the low melting points of oils. A

Table 23.2 Approximate fatty-acid composition of some common fats and oils

Source	Composition (%)[a]			
	Palmitic	*Stearic*	*Oleic*	*Linoleic*
corn oil	10	5	45	38
soybean oil	10	—	25	55
lard	30	15	45	5
butter	25	10	35	—
human fat	25	8	46	10

[a] Other fatty acids are also found in lesser amounts.

saturated fatty acid forms zigzag chains that can fit compactly together, resulting in high van der Waals attractions. Therefore, saturated fats are solids. If a few *cis* double bonds are present in the chains, the molecules cannot form neat, compact lattices, but tend to coil. Polyunsaturated triglycerides tend to be oils. Figure 23.1 shows models of the two types of chains.

A constant fluidity of cell membranes in different environments is maintained by the ratio of saturated to unsaturated fatty acids in the membranes. For example, bacteria grown at higher temperatures contain more saturated fatty acids in their cell membranes than do bacteria grown at lower temperatures.

A saturated triglyceride:

can fit into a solid lattice

An unsaturated triglyceride:

cannot fit as well into a lattice

Figure 23.1 The shapes of saturated and unsaturated triglycerides. Adapted from William H. Brown, *Introduction to Organic and Biochemistry, 4th ed.* (Copyright 1987 by Wadsworth, Inc. Used by permission of Brooks/Cole Publishing Co.)

▼ ## Section 23.2
Soaps and Detergents

Soaps are the alkali metal salts (usually sodium salts) of fatty acids. Soaps contain primarily C_{16} and C_{18} salts, but may also contain some lower-molecular-weight carboxylates.

Soaps were probably discovered by the ancient Egyptians several thousand years ago. Soap making by the Teutonic tribes was reported by Julius Caesar. The technique of soap making was lost in many parts of Europe during the Dark Ages, but was rediscovered during the Renaissance. The use of soap did not become widespread until the 18th century.

Soap is manufactured today by practically the same techniques as were used in ages past. Molten tallow (beef fat) or other fat is heated with lye (sodium hydroxide) and is thus saponified to glycerol and sodium salts of fatty acids. In the past, wood ashes (which contain bases such as potassium carbonate) were used instead of lye.

Saponification:

$$CH_2O_2C(CH_2)_{16}CH_3$$
$$CHO_2C(CH_2)_{16}CH_3 \ + \ 3\,NaOH \ \xrightarrow{\ heat\ } \ CHOH \ + \ 3\,CH_3(CH_2)_{16}CO_2^- \ Na^+$$
$$CH_2O_2C(CH_2)_{16}CH_3$$

glyceryl tristearate   glycerol   sodium stearate

                                                a soap

Once the saponification is complete, salt is added to help precipitate the soap, the water layer containing glycerol is drawn off, and the glycerol is recovered by distillation. Glycerol is used as a moisturizer in tobacco, pharmaceuticals, and cosmetics. (The moisturizing properties arise from the hydroxyl groups, which can hydrogen-bond with water and prevent its evaporation.) The soap is purified by boiling in fresh water to leach out excess lye, NaCl, and glycerol. Additives such as pumice, dyes, and perfume are added. The solid soap is then melted and poured into a mold.

A molecule of a soap contains a long hydrocarbon chain plus an ionic end. The hydrocarbon portion of the molecule is hydrophobic and soluble in nonpolar substances, while the ionic end is hydrophilic and water-soluble. Because of the hydrocarbon chain, a soap molecule as a whole is not truly soluble in water. However, soap is readily suspended in water because it forms **micelles**, clusters of 50–150 soap molecules with their hydrocarbon chains grouped together and with their ionic ends facing the water (see Figure 23.2).

The value of a soap is that it can emulsify oily dirt so that it can be rinsed away. This ability to act as an emulsifying agent arises from two properties of the soap. First, the hydrocarbon chain of a soap molecule dissolves in nonpolar substances, such as droplets of oil. Second, the anionic end of the soap molecule, which is attracted to water, is repelled by the anionic ends of soap molecules protruding from other drops of oil. Because of these repulsions between the soap–oil droplets, the oil cannot coalesce, but remains suspended.

water
and Na$^+$ ions

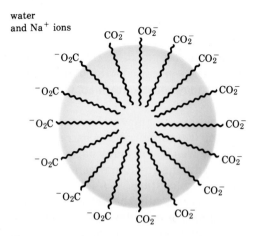

Figure 23.2 A micelle of the alkylcarboxylate ions of a soap.

*in soapy water, oil droplets repel each other because
of similar charges of soap's carboxylate groups*

The principal disadvantage of soaps is that they precipitate with hard water (water that contains Ca^{2+}, Mg^{2+}, Fe^{3+}, etc.) and leave a residue variously called scum, bathtub ring, or tattletale gray.

$$2\,RCO_2^- + Ca^{2+} \longrightarrow (RCO_2)_2Ca\downarrow$$

insoluble

After World War II, **synthetic detergents** were developed. Detergents are long-chain sulfonate or sulfate salts of sodium ($RSO_3^-\ Na^+$ and $ROSO_3^-\ Na^+$). Detergents have the advantage of not precipitating with the metal ions in hard water.

$$CH_3(CH_2)_{11}\!-\!\!\left\langle\!\!\bigcirc\!\!\right\rangle\!\!-\!SO_3^-\ Na^+ \qquad CH_3(CH_2)_{11}OSO_3^-\ Na^+$$

sodium *p*-dodecylbenzenesulfonate

sodium dodecyl sulfate
(sodium lauryl sulfate)

Study Problem

23.1 Write equations for a reaction sequence that would convert glyceryl trioleate into a sodium alkyl sulfate detergent.

Soaps and detergents belong to a general class of compounds called **surfactants** (from *surface-active agents*), which are compounds that can lower the surface tension of water. Any surfactant molecule contains a hydrophobic end (one or more hydrocarbon chains) and a hydrophilic end (usually, but not necessarily, ionic). The hydrocarbon portion of a surfactant molecule must contain 12 or more carbon atoms to be effective.

General symbol for a surfactant:

hydrophobic tail   *hydrophilic head*

Surfactants can be classed as *anionic, cationic,* or *neutral*, depending on the nature of the hydrophilic group. Soaps, with their carboxylate groups, are anionic surfactants. The antibacterial "benzalkonium" chlorides (*N*-benzyl quaternary ammonium chlorides) are examples of cationic surfactants. A neutral surfactant contains nonionic, but polar, groups that can hydrogen bond with water. These are becoming increasingly popular because they are superior for cleaning synthetic fabrics.

$$\overset{O}{\overset{\|}{C_{11}H_{23}C}}O^- \ Na^+ \qquad C_{12}H_{25}OSO_3^- \ Na^+$$

anionic surfactants

$$\bigcirc\!\!-CH_2-\overset{\overset{\displaystyle CH_3}{\displaystyle |}}{\underset{\underset{\displaystyle CH_3}{\displaystyle |}}{N^+}}-C_{12}H_{25} \ Cl^-$$

a cationic surfactant
(a benzalkonium chloride)

$$\overset{O}{\overset{\|}{RC}}OCH_2CH_2O-(CH_2CH_2O)_2-CH_2CH_2OH$$

a neutral surfactant

Surfactants lower the surface tension of water by disrupting the hydrogen bonding on the water's surface. They cause this disruption by positioning their hydrophilic heads on the surface of the water and their hydrophobic tails extending away from the water's surface.

water surface

▼ Section 23.3
Phospholipids

Phospholipids are lipids that contain phosphate ester groups. **Phosphoglycerides**, one type of phospholipid, are closely related to the fats and oils. These compounds usually contain fatty-acid esters at two positions of glycerol with a phosphate ester at the third position. Phosphoglycerides are distinctive because their molecules contain *two* long hydrophobic tails and a highly polar hydrophilic group—a dipolar-ion group. Phosphoglycerides are, therefore, neutral surfactants. They are excellent emulsifying agents. In mayonnaise, the phosphoglycerides of the egg yolk keep the oil emulsified in the vinegar.

Two types of phosphoglycerides:

a lecithin,
or phosphatidylcholine

a cephalin,
or phosphatidylethanolamine

General symbol for a phosphoglyceride:

hydrophobic tails   *dipolar hydrophilic head*

Lecithins and **cephalins** are two types of phosphoglyceride that are found principally in the brain, nerve cells, and liver of animals and are also found in egg yolks, wheat germ, yeast, soybeans, and other foods. These two types of compounds are similar to each other in structure. Lecithins are derivatives of choline chloride, $HOCH_2CH_2N(CH_3)_3{}^+ Cl^-$, which is involved in the transmission of nerve impulses. Cephalins are derivatives of ethanolamine, $HOCH_2CH_2NH_2$.

a plasmalogen

sphingomyelin

a sphingolipid

Other classes of phospholipids are represented by **plasmalogens**, which have vinyl ether groups instead of ester groups at carbon 1 of glycerol, and

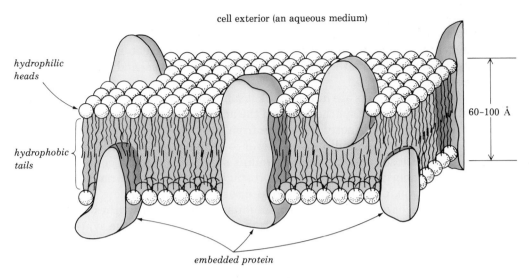

cell exterior (an aqueous medium)

hydrophilic heads

60–100 Å

hydrophobic tails

embedded protein

cell interior (an aqueous medium)

Figure 23.3 The bilayer of phospholipids in a cell membrane.

sphingolipids, of which sphingomyelin is an example. Sphingomyelin is a phosphate ester, not of glycerol, but of a long-chain allylic alcohol with an amide side chain.

Study Problem

23.2 Hydrolysis of sphingomyelin yields phosphoric acid, choline, a 24-carbon fatty acid, and *sphingosine*. What is the structure of sphingosine?

Phospholipids are important in cell membranes. These membranes are formed primarily from proteins associated with a bilayer, or double layer, of phosphoglyceride molecules with their hydrophobic ends pointing inward and their hydrophilic ends pointing outward. The hydrocarbon portion of the membrane does not allow passage of water, ions, or polar molecules. The function of the embedded proteins is to pass water, ions, and other substances *selectively* in and out of the cell (see Figure 23.3).

It is thought that the sphingolipids such as sphingomyelin contribute strength to the myelin (nerve cell) sheath by the intertwining of their hydrocarbon chains. Phospholipids are also thought to act as electrical insulation for the nerve cells. The myelin sheaths of people with multiple sclerosis (and some other diseases that affect this membrane) are deficient in the long hydrocarbon chains.

Section 23.4
Prostaglandins

One of the more exciting areas of biochemical research today is that of **prostaglandins**. These compounds were isolated and studied by the Swedish scientists Sune Bergstrom and Bengt Samuelsson, who received a 1982 Nobel prize for their work in this area. The prostaglandins were first discovered in semen, and it was recognized that they were synthesized in the prostate gland (hence the name). We now know that prostaglandins are found throughout the body and are also synthesized in the lungs, liver, uterus, and other organs and tissues.

Prostaglandins are hormone-like compounds that alter the activity of the cells where they are synthesized and the activity of adjoining cells. The nature of a prostaglandin's activities varies from one type of cell to another and from one type of prostaglandin to another. For example, administration of remarkably small doses of some prostaglandins stimulates uterine contractions and can cause abortions. Imbalances in prostaglandins can lead to nausea, diarrhea, inflammation, pain, fever, menstrual disorders, asthma, ulcers, hypertension, drowsiness, or blood clots. Aspirin inhibits prostaglandin synthesis by acetylating and thus deactivating *cyclooxygenase*, a key enzyme in the synthesis.

The prostaglandins are 20-carbon carboxylic acids that contain cyclopentane rings. They are biosynthesized from the 20-carbon unsaturated fatty acids.

(8Z,11Z,14Z)-eicosatrienoic acid
(also called homo-γ-linolenic acid)

PGE$_1$

PGF$_{1\alpha}$

(5Z,8Z,11Z,14Z)-eicosatetraenoic acid
(arachidonic acid)

PGE$_2$

PGF$_{2\alpha}$

There are several known prostaglandins, but the four shown as products in the preceding equations are the most common ones. Although similar to each other in structure, the prostaglandins differ in (1) the number of double bonds, and (2) whether the cyclopentane portion is a diol or a keto alcohol. The terms

PGE_1, $PGF_{1\alpha}$, PGE_2, and $PGF_{2\alpha}$ are symbols for these compounds. PG means *prostaglandin*, E means the *keto alcohol*, and F means the *diol*. The subscript numbers refer to the number of double bonds. (For example, the subscript 1 means *one* double bond.) The subscript α refers to the configuration of the OH at carbon 9 (*cis* to the carboxyl side chain).

▼ Section 23.5
Steroids

We have mentioned steroids several times. Now let us take a more detailed look at these compounds. A **steroid** is a compound that contains the following ring system. The four rings are designated *A*, *B*, *C*, and *D*. The carbons are numbered as shown, starting with ring *A*, progressing to ring *D*, then to the angular (bridgehead) methyl groups, and finally to a side chain if it is present.

cholestane

Many steroids may be named as derivatives of this structure, which is called **cholestane**. (Steroid nuclei with different side chains also have names; however, we will not present them here.)

4-cholesten-3-one

a steroid with a cholestane skeleton

Steroids are found in almost all types of living systems. In animals, many steroids act as hormones. These, as well as synthetic steroids, find wide use as pharmaceuticals. In this section, we will briefly consider the conformation of the steroid ring system and then turn our attention to some important steroids. Finally, we will briefly discuss the biosynthesis of these compounds.

A. Conformation of Steroids

Recall from Section 4.5B that the more stable isomer of a 1,2-dialkylcyclo-hexane is the one with both substituents *equatorial*. A *trans*-1,2-disubstituted cyclohexane can exist in a preferred conformation in which both substituents are equatorial, while the *cis*-isomer must have one axial substituent and one equatorial substituent. The *trans*-isomer is therefore the more stable one.

trans (e,e)   *more stable than*   *cis (a,e)*

One ring fused to another in the 1,2-positions can be *cis* or *trans*. The following structures represent *cis-* and *trans*-decalin.

cis-decalin: or   Note cis H's as well as cis R groups.

trans-decalin: or   Note trans H's.

As can be seen in the formulas for *trans-* and *cis*-decalin, the *trans* ring juncture is *e,e*, while the *cis* ring junture is *a,e*. The *trans*-isomer is more stable than the *cis*-isomer by about 3 kcal/mol. The steroid nucleus contains three ring junctures (*A/B*, *B/C*, and *C/D*). In nature, these are usually the more stable *trans* ring junctures (but we will encounter an exception later in this chapter).

trans ring junctures

Groups substituted on a steroid ring system can be *below* or *above* the plane of the ring as the ring system is generally drawn. A group that is below the plane (*trans* to the angular methyl groups) is called an α group, while one that is above the plane (*cis* to the angular methyl groups) is called a β group.

angular methyl groups

β: up (cis to angular methyl groups)

The terms α and β can be used in the names of steroids to designate the stereochemistry of substituents. In the name, α or β immediately follows the position number for the substituent.

17α-ethynyl-17β-hydroxy-4-estren-3-one

an oral contraceptive, usually called norethindrone

Study Problems

23.3 Draw conformational formulas for:

(a) cholestane   (b) 3α-cholestanol   (c) norethindrone

23.4 Which is the more stable isomer, 3α-cholestanol or 3β-cholestanol? Why?

B. Some Important Steroids

Cholesterol, the most widespread animal steroid, is found in almost all animal tissues. Human gallstones and egg yolks are especially rich sources of this compound. Cholesterol is an important component of cell membranes of higher animals and is a necessary intermediate in the biosynthesis of the steroid hormones. However, because it can be synthesized from acetylcoenzyme A, it is not a dietary necessity. High levels of blood cholesterol are associated with arteriosclerosis (hardening of the arteries), a condition in which cholesterol and other lipids coat the insides of the arteries.

cholesterol
(5-cholesten-3β-ol)

A steroid related to cholesterol, *7-dehydrocholesterol*, which is found in the skin, is converted to *vitamin D* when irradiated with ultraviolet light. This reaction is discussed in Section 20.5.

Cortisone and **cortisol** (hydrocortisone) are two of 28 or more hormones secreted by the adrenal cortex. They are widely used to treat inflammation due to allergies or rheumatoid arthritis. Many related steroids with a carbonyl or hydroxyl group at carbon 11 have similar activity.

cortisone   cortisol

Sex hormones are produced primarily in the testes or the ovaries. Their production is regulated by pituitary hormones. The sex hormones impart secondary sex characteristics and regulate the sexual and reproductive functions. Male hormones are collectively called **androgens**; female hormones, **estrogens**; and pregnancy hormones, **progestins**.

In pregnant females, the presence of progesterone suppresses ovulation and menstruation. Synthetic progestins, such as norethynodrel (Enovid), are used to suppress ovulation as a method of birth control.

Androgens:

testosterone   androsterone

the principal male hormone   *a metabolized form of testosterone*

Estrogens:

estradiol   estrone

Progestins:

progesterone   norethynodrel

suppresses ovulation   *a synthetic progestin*

The tissue-building steroids are closely related to the androgens. Note the similarity in structure between methandrostenolone and testosterone.

methandrostenolone

a tissue-building steroid

Bile acids are found in bile, which is produced in the liver and stored in the gall bladder. The structure of cholic acid, the most abundant bile acid, follows. Cholic acid, as well as other bile acids, has a *cis A/B* ring juncture instead of the usual *trans A/B* ring juncture.

cholic acid

Bile acids are secreted into the intestines in combination with sodium salts of either glycine or taurine ($H_2NCH_2CH_2SO_3H$). The bile acid–amino acid link is an amide link between the carboxyl group of the bile acid and the amino group of the amino acid. In this combined form, the bile acid–amino acid acts to keep lipids emulsified in the intestines, thereby promoting their digestion.

$$RCO_2H \; + \; H_2NCH_2CO_2^- \; Na^+ \quad \xrightarrow[-H_2O]{\text{enzymes}} \quad \overset{\displaystyle O}{\overset{\displaystyle \|}{R}}\text{C}NHCH_2CO_2^- \; Na^+$$

cholic acid   sodium salt of   *an emulsifying agent*
                 glycine

C. Biosynthesis of Steroids

The biosynthesis of steroids proceeds by way of *terpenes* (Section 19.1). An abbreviated biosynthetic path to terpenes is shown in Figure 23.4. The first

Figure 23.4 Generalized biosynthetic pathway leading to the terpenes. (To emphasize the terpene portions of the compounds, the pyrophosphate groups are not shown.)

step is an enzymatic ester condensation of the acetyl portions of acetylcoenzyme A. Intermediates in the formation of terpenes are the pyrophosphates (diphosphates) of mevalonic acid and a pair of isopentenyl alcohols. Figure 23.4 shows the formation of the sesquiterpene *farnesol*, which is used to synthesize the higher terpenes.

The precursor for steroids is the triterpene squalene, which arises from the condensation of two farnesol molecules. The sequence of reactions leading from squalene to lanosterol, an intermediate that can be converted to cholesterol and other steroids, is shown in Figure 23.5. The first step is an enzymatic epoxide formation. Next, the epoxide is protonated and then a series of electron shifts results in the closure of four rings and the formation of the steroid ring system. The intermediate product of the ring-closure step is a carbocation. The final steps are a series of 1,2-shifts involving hydrides and methyl groups. Finally, a proton is lost and a stable alkene is generated. It is amazing that this whole sequence is catalyzed by only one enzyme, squalene oxidocyclase.

Figure 23.5 The conversion of squalene to lanosterol.

▼ Summary

Fats and **oils** are triglycerides, or triacylglycerols; that is, triesters of glycerol and long-chain fatty acids. Generally, oils (liquid) contain more unsaturation than fats (solid). Fatty acids contain even numbers of carbon atoms and *cis* double bonds.

A **soap** is the alkali-metal salt of a fatty acid. A **detergent** is a salt of a sulfonate or sulfate that contains a long hydrocarbon chain. These compounds are **surfactants**. Naturally occurring waxes are esters of fatty acids and long-chain alcohols.

Phosphoglycerides, such as lecithins and cephalins, generally contain glycerol esterified with two fatty acids and a phosphatidylamine. Other phospholipids may differ in structure, but all contain long hydrocarbon chains plus a phosphate group and an amino group.

Prostaglandins are 20-carbon carboxylic acids containing a cyclopentane ring, hydroxyl groups, one or more carbon–carbon double bonds, and sometimes a keto group. They are biosynthesized from unsaturated 20-carbon fatty acids.

Cholesterol, cortisone, the sex hormones, and the bile acids are all **steroids**, compounds that contain the following ring system:

Study Problems*

23.5 Hydrolysis of *trimyristin*, a fat obtained from nutmeg, yields only one fatty acid, myristic acid. This same acid can be obtained from 1-bromododecane (the 12-carbon, continuous-chain alkyl bromide) and diethyl malonate in a malonic ester synthesis. What are the structures of myristic acid and trimyristin?

23.6 A fat of unknown structure is found to be optically active. Saponification, followed by acidification, yields two equivalents of palmitic acid and one equivalent of oleic acid. What is the structure of the fat? Write an equation for its saponification, followed by acidification.

23.7 Starting with glyceryl tristearate as the only organic reagent, show by equations how you would prepare a wax.

23.8 Write formulas for the products of complete hydrolysis of each of the following phosphoglycerides with dilute aqueous NaOH.

(a)
$$CH_3(CH_2)_{16}CO_2CH_2$$
$$CH_3(CH_2)_7CH{=}CH(CH_2)_7CO_2CH \quad O^-$$
$$CH_2OPOCH_2CH_2\overset{+}{N}(CH_3)_3 \ Cl^-$$
$$\overset{\parallel}{O}$$

(b)
$$CH_3(CH_2)_{16}CO_2CH_2$$
$$CH_3(CH_2)_7CH{=}CH(CH_2)_7CO_2CH \quad O^-$$
$$CH_2OPOCH_2CH_2\overset{+}{N}H_3 \ Cl^-$$
$$\overset{\parallel}{O}$$

---

* For information concerning the organization of the *Study Problems* and *Additional Problems*, see the *Note to student* on page 39.

23.9 *Cerebrosides* are glycosphingolipids that are found primarily in the sheaths of nerve cells. What fatty acid and sugar would be isolated from hydrolysis of the following cerebroside in acidic solution?

23.10 Tell whether the substituents or the ring junctures in the following compounds are *cis* or *trans*:

23.11 Which structure of each of the following pairs is more stable? Why?

23.12 Draw the structure of each of the following steroids. (*Acetoxy-* means CH_3CO_2—.)

   (**a**) 5-cholesten-3α-ol
   (**b**) 1-cholesten-3-one
   (**c**) 4,6-cholestadien-3β-ol
   (**d**) 3β-acetoxy-6α,7β-dibromocholestane

23.13 The structure of the bile acid *desoxycholic acid* is shown below. Redraw the structure showing the conformation of the ring system.

Additional Problems

23.14 Show how you could convert palmitoleic acid to: (**a**) ethyl palmitoleate; (**b**) ethyl palmitate; (**c**) nonanedioic acid; (**d**) 2-chloropalmitic acid; (**e**) heptanal; (**f**) 2-methyl-10-heptadecen-2-ol, $CH_3(CH_2)_5CH{=}CH(CH_2)_7C(OH)(CH_3)_2$.

23.15 A mixed triglyceride contains two units of stearic acid and one unit of palmitoleic acid. What are the major organic products when this triglyceride is treated with:

   (**a**) an excess of dilute aqueous NaOH and heat?
   (**b**) H_2, copper chromite catalyst, heat, and pressure?
   (**c**) bromine in CCl_4?

23.16 What are the major organic products of ozonolysis, followed by oxidative work-up, of each of the following fatty acids? (**a**) palmitic acid; (**b**) palmitoleic acid; (**c**) linoleic acid; (**d**) linolenic acid.

23.17 How would you distinguish chemically between: (**a**) tripalmitin and tripalmitolein; (**b**) beeswax and beef fat; (**c**) beeswax and paraffin wax; (**d**) linoleic acid and linseed oil; (**e**) sodium palmitate and sodium *p*-decylbenzenesulfonate; (**f**) a vegetable oil and a motor oil?

23.18 How would you separate a mixture of estradiol and testosterone?

23.19 When cholic acid is treated with acetic anhydride, one monoacetate is formed preferentially. What is the structure of this monoacetate and why is it formed preferentially?

23.20 A biochemist prepared the following samples of mevalonic acid labeled with carbon-14. The three mevalonic acids were fed to a series of plants. Later, the isopentenyl alcohols from the plants were isolated. In which position would the carbon-14 be found in each of the product isopentenyl alcohols?

23.21 In a plant, two isopentenyl alcohol molecules combine to yield geraniol (see Figure 23.4). Identify the position of carbon-14 in geraniol arising from the labeled mevalonic acid in Problem 23.20(b).

23.22 *Prostacyclin*, also called prostaglandin I$_2$, dilates blood vessels and inhibits blood platelet aggregation (the start of the blood-clotting mechanism). This prostaglandin is unstable in aqueous acid and undergoes the following reaction. Suggest a mechanism.

prostacyclin

24 ▼

Amino Acids and Proteins

Proteins are among the most important compounds in an animal organism. Appropriately, the word protein is derived from the Greek *proteios*, which means "primary." Proteins are *polyamides*, and hydrolysis of a protein yields *amino acids*.

$$\{-NHCHC\overset{O}{\underset{R}{\|}}-NHCHC\overset{O}{\underset{R'}{\|}}-\} \xrightarrow[\text{heat}]{H_2O, H^+} H_2NCHCO_2H + H_2NCHCO_2H \quad \text{etc.}$$

a protein   *amino acids*

Only twenty amino acids are commonly found in plant and animal proteins, yet these twenty amino acids can be combined in a variety of ways to form muscles, tendons, skin, fingernails, feathers, silk, hemoglobin, enzymes, antibodies, and many hormones. We will first consider the amino acids and then discuss how their combinations can lead to such diverse products.

Section 24.1
The Structures of Amino Acids

The amino acids found in proteins are **α-aminocarboxylic acids**. Variation in the structures of these monomers occurs in the side chain.

$$\text{α amino group} \curvearrowright \quad H_2N{-}\underset{\underset{R}{|}}{\overset{\overset{CO_2H}{|}}{C}}{-}H \quad \text{variation in structure occurs in the side chain}$$

The simplest amino acid is aminoacetic acid ($H_2NCH_2CO_2H$), called *glycine*, which has no side chain and consequently does not contain a chiral carbon. All other amino acids have side chains, and therefore their α carbons are chiral. Amino acids from proteins belong to the L-series; that is, the groups around the α carbon have the same configuration as L-glyceraldehyde. (It is interesting that *racemic* α-amino acids have been detected in certain carbonaceous meteorites.)

$$HO{-}\underset{\underset{CH_2OH}{|}}{\overset{\overset{CHO}{|}}{C}}\blacktriangleleft H \qquad H_2N\blacktriangleright\underset{\underset{R}{|}}{\overset{\overset{CO_2H}{|}}{C}}\blacktriangleleft H$$

L-glyceraldehyde   *an L-amino acid*

Study Problem

24.1 While naturally occurring amino acids belong to the L-series, not all have an (*S*) configuration. Which L-amino acid has an (*R*) configuration? (See Table 24.1.)

Amino acids do not always behave like organic compounds. For example, they have melting points of over 200°, whereas most organic compounds of similar molecular weight are liquids at room temperature. Amino acids are soluble in water and other polar solvents, but insoluble in nonpolar solvents such as diethyl ether or benzene. Amino acids have large dipole moments. Also, they are less acidic than most carboxylic acids and less basic than most amines.

$$\underset{pK_a\ =\ \sim 5}{RCO_2H} \qquad \underset{pK_b\ =\ \sim 4}{RNH_2} \qquad H_2N\underset{\underset{R}{|}}{\overset{\overset{CO_2H}{|}}{C}}H$$

$$pK_a = \sim 10$$
$$pK_b = \sim 12$$

Why do amino acids exhibit such unusual properties? The reason is that an amino acid contains a basic amino group and an acidic carboxyl group in the same molecule. An amino acid undergoes an internal acid–base reaction to yield a **dipolar ion**, also called a **zwitterion** (from German *zwitter*, "hybrid"). Because of the resultant ionic charges, an amino acid has many properties of a salt. Furthermore, the pK_a of an amino acid is not the pK_a of a —CO_2H group, but that of an —NH_3^+ group. The pK_b is not that of a basic amino group, but that of the very weakly basic —CO_2^- group.

$$\overset{\curvearrowright}{H_2\overset{..}{N}}-\underset{R}{\overset{\overset{\displaystyle CO_2H}{|}}{C}}-H \;\rightleftharpoons\; H_3\overset{+}{N}-\underset{R}{\overset{\overset{\displaystyle CO_2^-}{|}}{C}}-H$$

a dipolar ion

Table 24.1 contains a complete list of the twenty amino acids commonly found in proteins. Even though amino acids exist as dipolar ions, their structures are commonly represented in the nonionic form, as shown in the table. Also included in this table are the three-letter abbreviations for the amino acid names. The use of these abbreviations will be discussed later in this chapter.

Table 24.1 The common amino acids found in proteins

Name	Abbreviation	Structure
alanine	ala	$CH_3\underset{\underset{\displaystyle NH_2}{|}}{C}HCO_2H$
arginine*	arg	$H_2N\underset{\underset{\displaystyle NH}{\|}}{C}NHCH_2CH_2CH_2\underset{\underset{\displaystyle NH_2}{|}}{C}HCO_2H$
asparagine	asn	$H_2N\overset{\overset{\displaystyle O}{\|}}{C}CH_2\underset{\underset{\displaystyle NH_2}{|}}{C}HCO_2H$
aspartic acid	asp	$HO_2CCH_2\underset{\underset{\displaystyle NH_2}{|}}{C}HCO_2H$
cysteine	cyS	$HSCH_2\underset{\underset{\displaystyle NH_2}{|}}{C}HCO_2H$
glutamic acid	glu	$HO_2CCH_2CH_2\underset{\underset{\displaystyle NH_2}{|}}{C}HCO_2H$
glutamine	gln	$H_2N\overset{\overset{\displaystyle O}{\|}}{C}CH_2CH_2\underset{\underset{\displaystyle NH_2}{|}}{C}HCO_2H$
glycine	gly	$\underset{\underset{\displaystyle NH_2}{|}}{C}H_2CO_2H$
histidine*	his	imidazole ring—$CH_2\underset{\underset{\displaystyle NH_2}{|}}{C}HCO_2H$

(continued)

Table 24.1 (*continued*) The common amino acids found in proteins

Name	Abbreviation	Structure
isoleucine*	ile	$$\overset{\displaystyle CH_3}{\underset{\displaystyle NH_2}{CH_3CH_2\overset{\mid}{C}H\overset{\mid}{C}HCO_2H}}$$
leucine*	leu	$(CH_3)_2CHCH_2\underset{\displaystyle NH_2}{\overset{\mid}{C}HCO_2H}$
lysine*	lys	$H_2NCH_2CH_2CH_2CH_2\underset{\displaystyle NH_2}{\overset{\mid}{C}HCO_2H}$
methionine*	met	$CH_3SCH_2CH_2\underset{\displaystyle NH_2}{\overset{\mid}{C}HCO_2H}$
phenylalanine*	phe	$\bigcirc\!\!-CH_2\underset{\displaystyle NH_2}{\overset{\mid}{C}HCO_2H}$
proline	pro	$\underset{\overset{\mid}{H}}{\underset{N}{\diagup}}\!\!-CO_2H$
serine	ser	$HOCH_2\underset{\displaystyle NH_2}{\overset{\mid}{C}HCO_2H}$
threonine*	thr	$$\overset{\displaystyle OH}{\underset{\displaystyle NH_2}{CH_3\overset{\mid}{C}H\overset{\mid}{C}HCO_2H}}$$
tryptophan*	trp	indole$-CH_2\underset{\displaystyle NH_2}{\overset{\mid}{C}HCO_2H}$
tyrosine	tyr	$HO\!-\!\bigcirc\!\!-CH_2\underset{\displaystyle NH_2}{\overset{\mid}{C}HCO_2H}$
valine*	val	$(CH_3)_2CH\underset{\displaystyle NH_2}{\overset{\mid}{C}HCO_2H}$

* Essential amino acid.

A. Essential Amino Acids

Some amino acids can be synthesized by an organism from its "pool" of organic compounds. One such mode of synthesis is the conversion of an amino acid that is present in excess to a desired amino acid by a **transamination reaction**. The mechanism for this reaction was presented in Section 13.6.

Circled groups exchanged:

$$\underset{\text{old amino acid}}{\overset{\displaystyle\text{CO}_2\text{H}}{\underset{\displaystyle\text{R}}{\text{H}_2\text{N}\,\text{CH}}}} \quad+\quad \underset{\text{old keto acid}}{\overset{\displaystyle\text{CO}_2\text{H}}{\underset{\displaystyle\text{R}'}{\text{C}=\text{O}}}} \quad\xrightarrow[\text{(many steps)}]{\overset{\text{transaminase}}{\text{enzymes}}}\quad \underset{\text{new keto acid}}{\overset{\displaystyle\text{CO}_2\text{H}}{\underset{\displaystyle\text{R}}{\text{C}=\text{O}}}} \quad+\quad \underset{\text{new amino acid}}{\overset{\displaystyle\text{CO}_2\text{H}}{\underset{\displaystyle\text{R}'}{\text{H}_2\text{N}\,\text{CH}}}}$$

Not all amino acids can be obtained by interconversions from other amino acids or by synthesis from other compounds in the animal system. Amino acids that are required for protein synthesis and cannot be synthesized by the organism must be present in the diet. Such compounds are referred to as **essential amino acids**. Which amino acids are essential depends upon the species of animal and even upon individual differences.

The amino acids usually considered essential for humans are starred in Table 24.1. Of these, tryptophan, phenylalanine, methionine, and histidine can be enzymatically converted from (R) to (S); thus, a racemic mixture of these amino acids can be totally utilized. The other essential amino acids must be provided in the diet in their (S)-configurations to be used in protein biosynthesis.

As an example of individual differences in requirements for amino acids, let us look at *tyrosine*. This amino acid is not considered an essential amino acid because, in most people, tyrosine can be synthesized from phenylalanine. A small percent of individuals who have inherited a condition known as **phenylketonuria (PKU)**, which means "phenyl ketones in the urine," do not have the enzyme *phenylalanine hydroxylase* necessary for this conversion. In these people, the phenylalanine is converted to phenylpyruvic acid. The diet of a person with PKU must contain some tyrosine and must also be limited in its quantity of phenylalanine. Otherwise, excessive amounts of phenylpyruvic acid accumulate in the brain, causing mental retardation.

B. Importance of Side-Chain Structure

How can polymers composed of twenty similar amino acids have such a wide variety of properties? Part of the answer lies in the nature of the side chains in the amino acids. Note in Table 24.1 that some amino acids have side chains that contain carboxyl groups; these are classified as **acidic amino acids**. Amino acids containing side chains with amino groups are classified as **basic amino acids**. These acidic and basic side chains help determine the structure and reactivity of the proteins in which they are found. The rest of the amino acids are classified as **neutral amino acids**. The side chains of neutral amino acids are also important. For example, some of these side chains contain —OH, —SH, or other polar groups that can undergo hydrogen bonding, which we will find is an important feature in overall protein structure.

$$
\begin{array}{cccc}
\text{CO}_2\text{H} & \text{CO}_2\text{H} & \text{CO}_2\text{H} & \text{CO}_2\text{H} \\
| & | & | & | \\
\text{H}_2\text{NCH} & \text{H}_2\text{NCH} & \text{H}_2\text{NCH} & \text{H}_2\text{NCH} \\
| & | & | & | \\
\text{CH}_2\text{CH}_2\text{CO}_2\text{H} & (\text{CH}_2)_4\text{NH}_2 & \text{CH(CH}_3)_2 & \text{CH}_2\text{OH}
\end{array}
$$

glutamic acid	lysine	valine	serine
an acidic amino acid	*a basic amino acid*	*a neutral amino acid with a nonpolar side chain*	*a neutral amino acid with a polar side chain*

The characteristics of a protein are changed if a carboxylic acid group in a side chain is converted to an amide. Note the difference in the side chains of glutamic acid and glutamine.

$$
\begin{array}{ll}
\text{CO}_2\text{H} & \text{CO}_2\text{H} \\
| \qquad\qquad\quad \nearrow acidic & | \qquad\qquad\qquad \nearrow neutral,\ polar \\
\text{H}_2\text{NCH} & \text{H}_2\text{NCH} \qquad\ \ \text{O} \\
| & | \qquad\qquad\qquad \| \\
\text{CH}_2\text{CH}_2\text{CO}_2\text{H} & \text{CH}_2\text{CH}_2\text{CNH}_2
\end{array}
$$

glutamic acid	glutamine

The —SH group in cysteine plays a unique role in protein structure. Thiols can undergo oxidative coupling to yield disulfides (Section 8.7). This coupling between two units of cysteine yields a new amino acid called cystine and provides a means of cross-linking protein chains. Giving hair a "permanent wave" involves breaking some existing S—S cross-links by reduction and then reforming new S—S links in other positions of the protein chains.

$$
\begin{array}{l}
\qquad\qquad\qquad\qquad\qquad\quad \text{O} \qquad\qquad\qquad\qquad\qquad \text{O} \\
\qquad\qquad\qquad\qquad\qquad\quad \| \qquad\qquad\qquad\qquad\qquad\ \| \\
\text{first protein chain:}\ \ \sim\!\!-\text{NHCHC}\!-\!\sim \qquad \sim\!\!-\text{NHCHC}\!-\!\sim \\
\qquad\qquad\qquad\qquad\qquad\quad | \qquad\qquad\qquad\qquad\qquad\ | \\
\qquad\qquad\qquad\qquad\qquad\ \text{CH}_2\text{SH} \qquad\qquad\qquad\ \text{CH}_2 \\
\qquad\qquad\qquad\qquad\qquad\qquad\qquad\qquad\qquad\qquad\qquad\ | \\
\qquad\qquad\qquad\qquad\qquad\qquad\qquad\qquad\qquad\qquad\qquad \text{S} \\
\qquad\qquad\qquad\qquad\qquad\quad \xrightarrow[\text{[H]}]{\text{[O]}} \qquad\qquad\quad | \\
\qquad\qquad\qquad\qquad\qquad\qquad\qquad\qquad\qquad\qquad\qquad \text{S} \\
\qquad\qquad\qquad\qquad\qquad\qquad\qquad\qquad\qquad\qquad\qquad\ | \\
\qquad\qquad\qquad\qquad\qquad\ \text{CH}_2\text{SH} \qquad\qquad\qquad\ \text{CH}_2 \\
\qquad\qquad\qquad\qquad\qquad\quad | \qquad\qquad\qquad\qquad\qquad\ | \\
\text{second protein chain:}\ \ \sim\!\!-\text{NHCHC}\!-\!\sim \qquad \sim\!\!-\text{NHCHC}\!-\!\sim \\
\qquad\qquad\qquad\qquad\qquad\quad \| \qquad\qquad\qquad\qquad\qquad\ \| \\
\qquad\qquad\qquad\qquad\qquad\quad \text{O} \qquad\qquad\qquad\qquad\qquad \text{O}
\end{array}
$$

cross-link between chains

The order in which amino acids are found in a protein molecule determines the relationship of the side chains to one another and consequently determines how the protein interacts with itself and with its environment. For example, a hormone or other water-soluble protein contains many amino acids with polar side chains, while an insoluble muscle protein contains a greater proportion of amino acids with nonpolar side chains.

The importance of the side chains of amino acids is illustrated by the condition known as **sickle-cell anemia**. The difference between normal hemoglobin and sickle-cell hemoglobin is that, in a protein molecule of 146 amino acid units, one single unit has been changed from glutamic acid (with an acidic side chain) to valine (with a nonpolar side chain). This one small error in the protein renders the affected hemoglobin less soluble and thus less able to perform its prescribed task of carrying oxygen to the cells of the body.

Section 24.2
Synthesis of Amino Acids

The common amino acids are relatively simple compounds, and racemic mixtures of most amino acids can be synthesized by standard techniques. The racemic mixtures can then be resolved to yield the pure enantiomeric amino acids.

The **Strecker synthesis** of amino acids, developed in 1850, is a two-step sequence. The first step is the reaction of an aldehyde with a mixture of ammonia and HCN to yield an aminonitrile. Hydrolysis of the aminonitrile results in the amino acid.

Step 1:

$$\underset{\text{acetaldehyde}}{\overset{\overset{\text{O}}{\|}}{CH_3CH}} \;\underset{}{\overset{NH_3}{\rightleftharpoons}}\; \underset{}{\overset{\overset{\text{OH}}{|}}{CH_3CHNH_2}} \;\underset{}{\overset{-H_2O}{\rightleftharpoons}}\; CH_3CH{=}NH \;\overset{HCN}{\longrightarrow}\; \underset{\substack{\text{2-amino-}\\\text{propanenitrile}}}{\overset{\overset{\text{NH}_2}{|}}{CH_3CHCN}}$$

Step 2:

$$\underset{}{\overset{\overset{\text{NH}_2}{|}}{CH_3CHCN}} \;\xrightarrow[\text{(2) neutralize}]{\text{(1) }H_2O,\ H^+}\; \underset{(R)(S)\text{-alanine (60\%)}}{\overset{\overset{\text{NH}_2}{|}}{CH_3CHCO_2H}}$$

Another synthetic route to amino acids is by the **amination of an α-halo acid** with an excess of ammonia. (An excess of NH_3 must be used to neutralize the acid and to minimize overalkylation. See Section 17.5A.)

$$\underset{}{\overset{\overset{\text{Br}}{|}}{(CH_3)_2CHCHCO_2H}} \;\xrightarrow[\text{(2) neutralize}]{\text{(1) excess }NH_3}\; \underset{(R)(S)\text{-valine (48\%)}}{\overset{\overset{\text{NH}_2}{|}}{(CH_3)_2CHCHCO_2H}} + NH_4Br$$

The **Gabriel phthalimide synthesis** (Section 17.5A) is a more elegant route to amino acids. The advantage of this synthesis over direct amination is that overalkylation cannot occur.

Reductive amination of an α-keto acid is another procedure used to

Table 24.2 Summary of synthetic routes to α-amino acids

Reaction	Section Reference

Substitution:

$$\underset{\text{RCHCO}_2\text{H}}{\overset{\text{X}}{|}} \xrightarrow[\text{(2) H}^+]{\text{(1) NH}_3} \underset{\text{RCHCO}_2\text{H}}{\overset{\text{NH}_2}{|}}$$ 17.5A

$$\text{XCH(CO}_2\text{C}_2\text{H}_5)_2 \xrightarrow[\substack{\text{(2) NaOC}_2\text{H}_5,\ \text{RX} \\ \text{(3) H}_2\text{O, H}^+,\ \text{heat}}]{\text{(1) K}^+\ \text{phthalimide}} \underset{\text{RCHCO}_2\text{H}}{\overset{\text{NH}_2}{|}}$$ 17.5A

Strecker Synthesis:

$$\underset{\text{RCH}}{\overset{\text{O}}{\|}} \xrightarrow[\substack{\text{(2) HCN} \\ \text{(3) H}_2\text{O, H}^+}]{\text{(1) NH}_3,\ -\text{H}_2\text{O}} \underset{\text{RCHCO}_2\text{H}}{\overset{\text{NH}_2}{|}}$$ 24.2

Reductive Amination:

$$\underset{\text{RCCO}_2\text{H}}{\overset{\text{O}}{\|}} \xrightarrow{\text{H}_2,\ \text{NH}_3,\ \text{Pd}} \underset{\text{RCHCO}_2\text{H}}{\overset{\text{NH}_2}{|}}$$ 13.7D

obtain racemic amino acids. (Remember, carboxyl groups are not easily reduced.)

$$\underset{\text{CH}_3\text{CCO}_2\text{H}}{\overset{\text{O}}{\|}} \xrightarrow{\text{H}_2,\ \text{NH}_3,\ \text{Pd}} \underset{\text{CH}_3\text{CHCO}_2\text{H}}{\overset{\text{NH}_2}{|}}$$

an α-keto acid   *(R)(S)-alanine*

The reactions leading to α-amino acids are summarized in Table 24.2.

Study Problem

24.2 A synthetic procedure suitable for the incorporation of short-lived ^{11}C (half-life, 20 minutes) from ^{11}CN$^-$ into amino acids follows. What are the structures of the intermediates A, B, and C and the structure of the amino acid D?

$$(\text{CH}_3)_2\text{CHCHO} \xrightarrow[\text{HCN}]{\text{CN}^-} \text{A} \xrightarrow{\text{SOCl}_2} \text{B} \xrightarrow{\text{liquid NH}_3} \text{C} \xrightarrow{\text{OH}^-} \text{D}$$

anion of an amino acid

▼ ## Section 24.3
Reactions of Amino Acids

A. Amphoterism of Amino Acids

An amino acid contains both a carboxylate ion ($-\text{CO}_2^-$) and an ammonium ion ($-\text{NH}_3^+$) in the same molecule. Therefore, an amino acid is **amphoteric**: it can undergo reaction with either an acid or a base to yield a cation or an anion, respectively.

In acid:

$$\underset{\underset{R}{|}}{\overset{\overset{CO_2^-}{|}}{H_3\overset{+}{N}-C-H}} + \overset{+}{H^+} \; \rightleftharpoons \; \underset{\underset{R}{|}}{\overset{\overset{CO_2H}{|}}{H_3\overset{+}{N}-C-H}}$$

a cation

In base:

$$\underset{\underset{H \quad R}{| \quad |}}{\overset{\overset{CO_2^-}{|}}{H_2\overset{+}{N}-C-H}} + OH^- \; \rightleftharpoons \; \underset{\underset{R}{|}}{\overset{\overset{CO_2^-}{|}}{H_2\overset{..}{N}-C-H}} + H_2O$$

an anion

Study Problems

24.3 When each of the following amino acids is dissolved in water, would the solution be acidic, basic, or near-neutral? (Refer to Table 24.1 for the structures.)

(**a**) glutamic acid   (**b**) glutamine   (**c**) leucine   (**d**) lysine   (**e**) serine

24.4 Monosodium glutamate (MSG) is widely used as a condiment. What is the most likely structure for this compound? (*Hint*: Which carboxyl group in glutamic acid is more acidic?)

24.5 Predict the product of reaction of (**a**) proline and (**b**) tyrosine with an excess of aqueous HCl and with an excess of aqueous NaOH.

Isoelectric points. You might think that an aqueous solution of a so-called neutral amino acid would be neutral. However, aqueous solutions of neutral amino acids are slightly *acidic* because the $-NH_3^+$ group is a stronger acid than $-CO_2^-$ is a base. The result of the difference in acidity and basicity is that an aqueous solution of alanine contains more amino acid anions than cations. We can say that alanine carries a *net negative charge* in aqueous solution.

At pH 7, alanine carries a net negative charge:

weaker base

$$\underset{\underset{CH_3}{|}}{\overset{\overset{CO_2^-}{|}}{H_3\overset{+}{N}-C-H}} + H_2O \; \rightleftharpoons \; \underset{\underset{CH_3}{|}}{\overset{\overset{CO_2^-}{|}}{H_2N-C-H}} + H_3O^+$$

stronger acid

no net charge   *net negative charge*

If a small amount of HCl or other acid is added to an alanine solution, the acid–base equilibrium is shifted so that the net charge on the alanine ions becomes zero. The pH at which an amino acid carries no net ionic charge is defined as the **isoelectric point** of that amino acid. The isoelectric point of alanine is 6.0.

At pH 6, alanine carries no net charge:

Isoelectric points can be determined by **electrophoresis**, a process of measuring the migration of ions in an electric field. This is accomplished by placing an aqueous solution of an amino acid on an adsorbent between a pair of electrodes. In this cell, anions migrate toward the positive electrode and cations migrate toward the negative electrode. If alanine or other neutral amino acid is dissolved in plain water, there is a net migration of amino acid ions toward the positive electrode. At its isoelectric point, an amino acid exhibits no net migration toward either electrode in an electrophoresis cell.

Isoelectric points can also be determined by titration. Figure 24.1 shows the titration curve for the titration of the cationic form of glycine, $H_3N^+{-}CH_2CO_2H$, with base. As base is added, the completely protonated ion is converted to the neutral dipolar ion, $H_3N^+{-}CH_2CO_2^-$. When one half of the cationic form has been neutralized, the pH equals the pK_1 for the reaction.

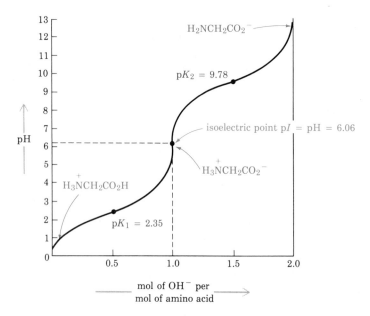

Figure 24.1 Titration curve for glycine hydrochloride.

$$H_3\overset{+}{N}CH_2CO_2H \rightleftharpoons H_3\overset{+}{N}CH_2CO_2^- + H^+$$

$$K_1 = \frac{[H_3\overset{+}{N}CH_2CO_2^-][H^+]}{[H_3\overset{+}{N}CH_2CO_2H]}$$

When $[H_3\overset{+}{N}CH_2CO_2^-] = [H_3\overset{+}{N}CH_2CO_2H]$,

$$K_1 = [H^+] \quad \text{and therefore} \quad pK_1 = pH$$

As more base is added, all the cationic form is converted to the neutral dipolar ion. The pH at which this occurs is the isoelectric point. (Again, refer to Figure 24.1.) As yet more base is added, the dipolar ion is converted to an anion. At the halfway point, the pH equals the pK_2.

$$H_3\overset{+}{N}CH_2CO_2^- \rightleftharpoons H^+ + H_2NCH_2CO_2^-$$

$$K_2 = \frac{[H^+][H_2NCH_2CO_2^-]}{[H_3\overset{+}{N}CH_2CO_2^-]}$$

When $[H_2NCH_2CO_2^-] = [H_3\overset{+}{N}CH_2CO_2^-]$,
$$K_2 = [H^+] \quad \text{and therefore} \quad pK_2 = pH$$

The isoelectric point is shown on the titration curve. It can also be calculated as the average of pK_1 and pK_2.

$$\text{isoelectric point } pI = \frac{2.35 + 9.78}{2} = 6.06$$

The isoelectric point of a given amino acid is a physical constant. The value varies from amino acid to amino acid, but falls into one of three general ranges. For a *neutral amino acid*, the isoelectric point, which depends primarily on the relative pK_a and pK_b of the $-NH_3^+$ and $-CO_2^-$ groups, is around 5.5–6.0.

The second carboxyl group in an *acidic amino acid* means that there is another group that can interact with water. A water solution of an acidic amino acid is distinctly acidic, and the amino acid ion carries a net negative charge.

$$\underset{acidic}{\overset{CO_2^-}{\underset{CH_2CH_2CO_2H}{H_3\overset{+}{N}-\overset{|}{\underset{|}{C}}-H}}} + H_2O \rightleftharpoons \underset{net\ negative\ charge}{\overset{CO_2^-}{\underset{CH_2CH_2CO_2^-}{H_3\overset{+}{N}-\overset{|}{\underset{|}{C}}-H}}} + H_3O^+$$

A *greater concentration of H^+* is required to bring an acidic amino acid to the isoelectric point than is needed for a neutral amino acid. The isoelectric points for acidic amino acids are near pH 3.

A *basic amino acid* has a second amino group that undergoes reaction with water to form a positive ion. Hydroxide ions are needed to neutralize a basic

$$\underset{basic}{\overset{CO_2^-}{\underset{(CH_2)_4\ \ddot{N}H_2}{H_3\overset{+}{N}-\overset{|}{\underset{|}{C}}-H}}} + H_2O \rightleftharpoons \underset{net\ positive\ charge}{\overset{CO_2^-}{\underset{(CH_2)_4\overset{+}{N}H_3}{H_3\overset{+}{N}-\overset{|}{\underset{|}{C}}-H}}} + OH^-$$

Table 24.3 Isoelectric points for some amino acids

Name	Structure	Isoelectric Point
Neutral:		
alanine	CH_3CHCO_2H $\quad\quad\vert$ $\quad\quad NH_2$	6.00
glutamine	$\quad\quad O$ $\quad\quad\Vert$ $H_2NCCH_2CH_2CHCO_2H$ $\quad\quad\quad\quad\quad\quad\vert$ $\quad\quad\quad\quad\quad\quad NH_2$	5.65
Acidic:		
glutamic acid	$HO_2CCH_2CH_2CHCO_2H$ $\quad\quad\quad\quad\quad\vert$ $\quad\quad\quad\quad\quad NH_2$	3.22
aspartic acid	$HO_2CCH_2CHCO_2H$ $\quad\quad\quad\quad\vert$ $\quad\quad\quad\quad NH_2$	2.77
Basic:		
lysine	$H_2N(CH_2)_4CHCO_2H$ $\quad\quad\quad\quad\quad\vert$ $\quad\quad\quad\quad\quad NH_2$	9.74
arginine	$H_2NCNH(CH_2)_3CHCO_2H$ $\quad\quad\Vert\quad\quad\quad\quad\vert$ $\quad\quad NH\quad\quad\quad NH_2$	10.76

amino acid and bring it to its isoelectric point. For basic amino acids, we would expect the isoelectric points to be above pH 7, and that indeed is the case. These isoelectric points are in the range of 9–10. Isoelectric points of some representative amino acids are found in Table 24.3.

Study Problem

24.6 Suggest a reason for the fact that the isoelectric point of lysine is 9.74, but that for tryptophan (Table 24.1) is only 5.89. (*Hint*: Think of a reason why the heterocyclic N in tryptophan is not basic.)

B. Acylation

The amino group of an amino acid can be readily acylated with either an acid halide or an acid anhydride to yield an amide group. Because an amide nitrogen is not basic, an acylated amino acid does not form a dipolar ion. For this reason, acylated amino acids exhibit physical properties typical of organic compounds instead of those of salts. In the synthesis of peptides (small proteins), the *N*-acyl group is used as a blocking group (see Section 24.8).

$$
\underset{\text{acetic anhydride}}{CH_3\overset{O}{\overset{\Vert}{C}}O\overset{O}{\overset{\Vert}{C}}CH_3} + \underset{\underset{CH_2CH(CH_3)_2}{\vert}}{\overset{+}{N}H_3CHCO_2^-} \longrightarrow \underset{\underset{CH_2CH(CH_3)_2}{\vert}}{CH_3\overset{O}{\overset{\Vert}{C}}-NHCHCO_2H} + CH_3\overset{O}{\overset{\Vert}{C}}OH
$$

acetic anhydride   leucine   *N*-acetylleucine (85%)

C. Reaction with Ninhydrin

α-Amino acids react with ninhydrin to form a blue-violet product called **Ruhemann's purple**. The reaction is used as a spot test to detect the presence of amino acids on chromatography paper. Because the reaction is quantitative, it also finds use in automated amino acid analyzers, instruments that determine the percentages of the various amino acids present in a sample.

ninhydrin

Ruhemann's purple

(*blue-violet*)

Study Problem

24.7 Unlike the other amino acids, proline does not form a blue-violet product when treated with ninhydrin. Explain.

Figure 24.2 summarizes the reactions of amino acids.

Figure 24.2 Summary of reactions of amino acids.

▼ ## Section 24.4
Peptides

A **peptide** is an amide formed from two or more amino acids. The amide link between an α-amino group of one amino acid and the carboxyl group of another amino acid is called a **peptide bond**. The following example of a peptide formed from alanine and glycine, called *alanylglycine*, illustrates the formation of a peptide bond.

— peptide bond

$$
\underset{\substack{\text{alanine}}}{\underset{\substack{|\\ \text{CH}_3}}{\text{H}_2\text{NCHCOH}}} + \underset{\substack{\text{glycine}}}{\text{H}_2\text{NCH}_2\text{COH}} \xrightarrow{-\text{H}_2\text{O}} \underset{\substack{\text{alanylglycine}}}{\underset{\substack{|\\ \text{CH}_3}}{\text{H}_2\text{NCHC}-\text{NHCH}_2\text{COH}}}
$$

a dipeptide

Each amino acid in a peptide is called a **unit**, or a **residue**. Alanylglycine has two residues: the alanine residue and the glycine residue. Depending upon the number of amino acid residues in the molecule, a peptide may be referred to as a **dipeptide** (two residues), a **tripeptide** (three residues), and so forth. A **polypeptide** is a peptide with a large number of amino acid residues. What is the difference between a polypeptide and a protein? None, really. Both are polyamides constructed from amino acids. By convention, a polyamide with fewer than 50 amino acid residues is classified as a polypeptide, while a larger polyamide is considered to be a protein.

In the dipeptide alanylglycine, the alanine residue has a free amino group and the glycine unit has a free carboxyl group. However, alanine and glycine could be joined another way to form *glycylalanine*, in which glycine has the free amino group and alanine has the free carboxyl group.

Two different dipeptides from alanine and glycine:

$$
\underset{\substack{\text{alanylglycine}}}{\underset{\substack{|\\ \text{CH}_3}}{\text{H}_2\text{NCHC}-\text{NHCH}_2\text{COH}}} \qquad \underset{\substack{\text{glycylalanine}}}{\underset{\substack{|\\ \text{CH}_3}}{\text{H}_2\text{NCH}_2\text{C}-\text{NHCHCOH}}}
$$

The greater the number of amino acid residues in a peptide, the greater is the number of structural possibilities. Glycine and alanine can be bonded together in two ways. In a tripeptide, three amino acids can be joined in six different ways. Ten different amino acids could lead to over four trillion decapeptides!

For purposes of discussion, we must represent peptides in a systematic manner. The amino acid with the free amino group is usually placed at the left end of the structure. This amino acid is called the **N-terminal amino acid**. The amino acid with the free carboxyl group is placed at the right and is called the **C-terminal amino acid**. The name of the peptide is constructed from the names of the amino acids as they appear left-to-right, starting with the *N*-terminal amino acid.

N-terminal on left

C-terminal on right

$$\underset{\underset{CH_3}{|}}{H_2NCHC}\overset{\overset{O}{||}}{—}NHCH\overset{\overset{O}{||}}{C}\underset{\underset{CH_2-\bigcirc-OH}{|}}{—}NHCH_2\overset{\overset{O}{||}}{C}OH$$

alanyltyrosylglycine

a tripeptide

For the sake of convenience and clarity, the names of the amino acids are often abbreviated. We have shown the abbreviations of the twenty common amino acids in Table 24.1. Using the abbreviated names, alanyltyrosylglycine becomes *ala–tyr–gly*.

Sample Problem

What is the structure of leu–lys–met?

Solution:

$$\underset{\underset{(CH_3)_2CHCH_2}{|}}{H_2NCHC}\overset{\overset{O}{||}}{—}\underset{\underset{(CH_2)_4NH_2}{|}}{NHCHC}\overset{\overset{O}{||}}{—}\underset{\underset{CH_2CH_2SCH_3}{|}}{NHCHCO_2H}$$

leu   lys   met

Study Problem

24.8 Give all possibilities for the structure of a tripeptide consisting of ala, gly, and phe. (Use abbreviated names rather than structural formulas.)

▼ ## Section 24.5
Bonding in Peptides

As we mentioned in Section 15.8C, an amide bond has some double-bond character due to partial overlap of the *p* orbitals of the carbonyl group with the unshared electrons of the nitrogen.

$$\underset{}{—\overset{\overset{\ddot{O}:}{||}}{C}—\ddot{N}<} \longleftrightarrow —\overset{\overset{^-:\ddot{O}:}{|}}{C}=\overset{+}{N}<$$

Evidence for the double-bond character of a peptide bond is found in bond lengths. The bond length of the peptide bond is shorter than that of the usual C—N single bond: 1.32 Å in the peptide bond versus 1.47 Å for a typical C—N single bond in an amine.

Because of the double-bond character of the peptide bond, rotation of groups around this bond is somewhat restricted, and the atoms attached to the carbonyl group and to the N all lie in the same plane. X-ray analysis shows that the amino acid side chains around the plane of the peptide bond are in a *trans* type of relationship. This stereochemistry minimizes steric hindrance between side chains.

Section 24.6
Some Interesting Peptides

Although most proteins are relatively large molecules, a peptide need not contain thousands of amino acid residues for biological activity. Consider the following tripeptide:

pyroglutamylhistidylprolinamide

This tripeptide is a **thyrotropic-hormone releasing factor** (TRF) that has been isolated from the hypothalamus glands of hogs and cattle. Administered to humans (intravenously or orally), it stimulates the secretion of thyrotropin, which in turn stimulates the secretion of thyroid hormones, the regulators of the body's metabolism.

The structure of TRF exemplifies some commonly encountered structural variations in peptides and proteins. The *N*-terminal amino acid is a derivative of glutamic acid. The side-chain carboxyl group has joined with the free amino group to form a lactam (a cyclic amide).

$$O \qquad\qquad O$$
$$\parallel \qquad\qquad \parallel$$
$$HO\overset{\parallel}{C}CH_2CH_2CHCOH \quad \xrightarrow{-H_2O} \quad O=C \qquad CH-\overset{\parallel}{C}OH$$

glutamic acid   pyroglutamic acid

a lactam

The *C*-terminal residue of TRF is an amide of proline. Amides of free carboxyl groups are not at all uncommon in natural protein structures. These are generally denoted in the abbreviated structures by adding NH_2 to the *C*-terminal. For example, *gly* means $-NHCH_2CO_2H$, but *gly*–NH_2 means

$$-NHCH_2\overset{\parallel}{C}NH_2.$$
$$O$$

Study Problem

24.9 Give the structural formula for pro–leu–ala–NH_2.

Oxytocin, a pituitary hormone that causes uterine contractions during parturition, is another important peptide. Note that oxytocin is a cyclic peptide joined by a S—S cystine link. Figure 24.3 shows the structural formula.

Enkephalins, the body's own painkillers, are brain peptides that contain as few as five amino acid residues. It is thought that the opiates (morphine, *etc.*)

Figure 24.3 The structure of oxytocin, gln-asn-cyS-pro-leu-gly-NH_2.

act as analgesics by mimicking the shapes and polarities of the enkephalins and attaching themselves to the pain-relief reception sites in the brain. It is theorized that "jogger's high" (a feeling of well-being generated by jogging or running) is caused by the release of enkephalins in the brain.

Two enkephalins:

tyr–gly–gly–phe–met   tyr–gly–gly–phe–leu

▼ Section 24.7
Determination of Peptide Structure

The determination of a peptide structure is not an easy task. Complete hydrolysis in acidic solution yields the individual amino acids. These may be separated and identified by such techniques as chromatography or electrophoresis. The molecular weight of the peptide can be determined by physical–chemical methods. With this information, the chemist can determine the number of amino acid residues, the identity of the amino acid residues, and the number of residues of each amino acid in the original peptide. But this information reveals nothing about the *sequence* of the amino acids in the peptide. Several techniques have been developed to determine this sequence. The first is **terminal-residue analysis**.

A. Terminal-Residue Analysis

The Edman reagent. Analysis for the *N*-terminal can be accomplished by treating the peptide with phenyl isothiocyanate (the Edman reagent, named after the Swedish biochemist P. Edman, who developed it). The isothiocyanate undergoes reaction with the free amino group in a series of steps that eventually results in cleavage of the *N*-terminal amino acid from the peptide and in the formation of a phenylthiohydantoin (PTH), a derivative of an amino acid that can be isolated and characterized. The cleavage reaction is carried out by heating the intermediate adduct with acid in an anhydrous solvent such as nitromethane. (If water were present, the entire peptide would be hydrolyzed.)

phenyl isothiocyanate   *N-terminal*
(Edman reagent)   *residue*

a phenylthiohydantoin

Why cannot the chemist continue to treat the peptide stepwise with phenyl isothiocyanate? In each step the *N*-terminal residue could be broken off until the entire peptide was degraded and its order of amino acids determined. To a certain extent, this can be done; however, each cycle of thiourea formation and hydrolysis results in some internal hydrolysis of the remaining peptide. After about 20–30 cycles, the hydrolysis of the peptide is sufficient to produce many smaller peptides, each of which has an *N*-terminal residue. The phenylthiohydantoin is thus no longer that of a single end group, but a mixture from a variety of end groups. In recent years, automated peptide analyzers called *sequenators* have been used to degrade and analyze peptides of 20–60 amino acid residues.

The Sanger reagent.   Another reagent useful for determining the *N*-terminal residue is the Sanger reagent, 1-fluoro-2,4-dinitrobenzene. The fluoro group of the Sanger reagent can undergo aromatic nucleophilic substitution with amines. The substitution is facile because the intermediate carbanion is stabilized by the nitro groups (Section 12.4).

1-fluoro-2,4-dinitrobenzene

a 2,4-dinitrophenylamine

The Sanger reagent reacts readily with the *N*-terminal amino acid of a peptide and converts the amino group to an arylamino group. After complete hydrolysis of the treated peptide, the *N*-terminal amino acid remains bonded to the 2,4-dinitrophenyl group and thus can be separated from the other amino

from the N-terminal amino acid

acids and identified. The principal disadvantage of using the Sanger reagent is that a peptide cannot be degraded one amino acid at a time, as in an Edman degradation.

The Sanger reagent was developed by Sir Frederick Sanger, who received the 1958 Nobel Prize for being the first to establish the complete amino acid sequence of a protein (insulin).

The *C*-terminal amino acid residue can be determined enzymatically. *Carboxypeptidase* is a pancreatic enzyme that specifically catalyzes hydrolysis of the *C*-terminal amino acid, but not that of other peptide bonds.

$$\text{rest of peptide} -\underset{\underset{R}{|}}{\text{NHCHCO}_2\text{H}} + \text{H}_2\text{O} \xrightarrow{\text{carboxypeptidase}} \text{rest of peptide} + \underset{\underset{R}{|}}{\text{H}_2\text{NCHCO}_2\text{H}}$$

C-terminal
amino acid

The principal difficulty with this method of analysis is that after the original *C*-terminal amino acid has been cleaved, the remaining portion of the peptide has another *C*-terminal residue, which in turn can undergo cleavage. Very careful experimental work is required to determine which amino acid in the resulting mixture was the *C*-terminal residue of the starting peptide.

B. Internal Sequence of Amino Acids

A large polypeptide must usually be hydrolyzed into smaller fragments for determination of the internal amino acid sequence. The hydrolysis mixture is separated and the order of amino acid residues in each fragment determined (by end-group analysis, for example). The structures of the fragments are then pieced together like a jigsaw puzzle to give the structure of the entire peptide.

Sample Problem

What is the structure of a pentapeptide that yields the following tripeptides when partially hydrolyzed?

gly–glu–arg,   glu–arg–gly,   arg–gly–phe

Solution: Fit the pieces together:

gly–glu–arg

    glu–arg–gly

        arg–gly–phe

The pentapeptide is gly–glu–arg–gly–phe.

In theory, partial hydrolysis can be achieved by heating the polypeptide with water and acid or base. In practice, proteolytic (peptide-hydrolyzing) enzymes or chemical reagents are used. These enzymes and reagents have the advantage that they cleave the polypeptides at specific peptide bonds. Some examples are listed in Figure 24.4.

$$\overset{\underset{\displaystyle}{\text{O}}}{\underset{|}{\text{—NHCHC}}}\overset{\underset{\displaystyle}{\text{O}}}{\underset{|}{\text{—NHCHC—}}}$$

| | |
| R | R' |

site of cleavage

chymotrypsin: R = tyr, phe, trp
pepsin: R' = tyr, phe, trp
trypsin: R = lys, arg
cyanogen bromide (BrCN):   R = met
thermolysin: R' = ile, leu, val

Figure 24.4 Specificities of proteolytic enzymes and reagents used to cleave internal peptide bonds.

A new and indirect way of determining the sequence of amino acids in a protein molecule is by isolating the portion of a DNA molecule that is responsible for the biosynthesis of that protein and then determining the base sequence in the DNA fragment (Chapter 25).

Sample Problem

List the amino acids and peptides formed when the following peptide is treated with chymotrypsin:

    tyr–ala–leu–tyr

Solution: Chymotrypsin cleaves the amide bond on the carbonyl side of tyrosine:

    tyr⁄ala–leu–tyr ⟶ tyr + ala–leu–tyr

Study Problem

24.10 List the amino acids and peptides formed when the following polypeptide is treated with: **(a)** trypsin;   **(b)** chymotrypsin;   **(c)** pepsin:

    trp–met–gly–gly–phe–arg–phe–trp–val–lys–ala–gly–ser

▼ Section 24.8
Synthesis of Peptides

The first peptide was synthesized by Emil Fischer, who in 1902 also put forth the idea that proteins were polyamides.

The synthesis of ordinary amides from acid chlorides and amines is a straightforward reaction (Section 15.3C):

    $\text{RCOCl} + \text{R'NH}_2 \longrightarrow \text{RCONHR'}$

However, the synthesis of peptides or proteins by this route is not straightforward. The principal problem is that there is more than one way in which the amino acids may join.

$$\text{gly + ala} \longrightarrow \text{gly–ala or ala–gly or gly–gly or ala–ala}$$

$$\text{gly–ala + phe} \longrightarrow \text{phe–gly–ala or gly–ala–phe or}$$

$$\text{gly–ala–gly–ala or phe–phe}$$

To prevent unwanted reactions, every other reactive group, including reactive groups in side chains, must be blocked. By leaving only the desired amino group and carboxyl group free, the chemist can control the positions of reaction. (See Figure 24.5.)

The criteria for a good blocking group are (1) that it be inert to the reaction conditions needed for forming the desired amide link, and (2) that it be readily removable when the synthesis is complete. One such blocking group is a *carbamate group*—inert to the amide-formation reaction, but easily removed in a later step without disturbing the rest of the molecule. This approach to peptide synthesis was developed in 1932.

The carbamate group formed from benzyl chloroformate, as in the following equation, is called the *carbobenzoxy* group, frequently abbreviated CBz or simply Z.

Preparation of carbamate to protect an amino group:

benzyl chloroformate   glycine   carbamate group

from $Cl_2C{=}O + C_6H_5CH_2OH$ benzyloxycarbonylglycine (80%)

Figure 24.5 A general procedure for synthesizing peptides in the laboratory.

The amino-blocked glycine could be treated with $SOCl_2$ to form the acid chloride and then treated with a new amino acid to form an amide. However, acid chlorides are highly reactive and unwanted side reactions may occur despite blocking. To circumvent this problem, the amino-blocked glycine is usually treated with ethyl chloroformate to yield an **activated ester**.

$$Z-NHCH_2\overset{\overset{O}{\|}}{C}OH + Cl\overset{\overset{O}{\|}}{C}OC_2H_5 \xrightarrow{-HCl} Z-NHCH_2\underbrace{\overset{\overset{O}{\|}}{C}O\overset{\overset{O}{\|}}{C}OC_2H_5}$$

| amino-blocked glycine | ethyl chloroformate | *"activated ester group" (an anhydride-carbonate)* |

Like an acid chloride, this activated ester can undergo reaction with an amino group of another amino acid to give the desired dipeptide.

$$Z-NHCH_2\overset{\overset{O}{\|}}{C}O\overset{\overset{O}{\|}}{C}OC_2H_5 + \underset{\underset{CH_2C_6H_5}{|}}{H_2NCHCO_2H} \xrightarrow[-C_2H_5OH]{-CO_2} Z-NHCH_2\overset{\overset{O}{\|}}{C}-\underset{\underset{CH_2C_6H_5}{|}}{NHCHCO_2H}$$

| | phenylalanine or its ethyl ester | amino-blocked gly–phe |

At this point, the sequence can be repeated to add a third amino acid. When the peptide synthesis is complete, the carbamate group is cleaved by reduction to yield the free peptide.

Removal of carbamate blocking group:

$$\bigcirc\!\!-CH_2\overset{\overset{O}{\|}}{-O\!-\!C}-NHCH_2\overset{\overset{O}{\|}}{C}-\underset{\underset{CH_2C_6H_5}{|}}{NHCHCO_2H} \xrightarrow{H_2,\ Pd}$$

$$\bigcirc\!\!-CH_3 + CO_2 + H_2NCH_2\overset{\overset{O}{\|}}{C}-\underset{\underset{CH_2C_6H_5}{|}}{NHCHCO_2H}$$

gly–phe

Study Problem

24.11 Write the equations illustrating the addition of alanine to the amino-blocked gly–phe.

A. Solid-Phase Peptide Synthesis

New and better methods of synthesizing peptides are always under investigation. One relatively new technique is called **solid-phase peptide synthesis**, or the **Merrifield peptide synthesis** (after Bruce Merrifield at Rockefeller University, who developed the technique and was awarded a Nobel prize in 1984 for his work). In this type of synthesis, resins hold the *C*-terminal amino acid by the carboxyl group as the peptide is being synthesized. The resin is a polystyrene that contains about 1% *p*-(chloromethyl)styrene units.

a polystyrene resin
containing *p*-(chloromethyl)styrene units

The amino group of the first amino acid is initially blocked, often as a *t*-butoxycarbonyl ("Boc") group. This amino-blocked amino acid, as the carboxylate, reacts with the benzylic chloride groups of the resin to form ester groups (a typical substitution reaction between a carboxylate and a benzylic halide).

Step 1 (formation of Boc derivative):

Step 2 (reaction with benzylic chloride group):

The amino-blocking group is removed by treatment with an anhydrous acid, such as HCl in acetic acid. Then, a second amino-blocked amino acid (with an activated carbonyl group) is added.

A common technique for activating the $-CO_2H$ group (so that it will undergo reaction with the amine) is by the addition of dicyclohexylcarbodiimide to the carboxylic acid. This compound reacts with the carboxylic acid to yield an intermediate with a leaving group that can be displaced by the amine in a typical nucleophilic acyl substitution reaction. The product is the amide.

dicyclohexylcarbodiimide (DCC)

good leaving group

the amide

dicyclohexylurea

In the solid-phase peptide synthesis, the product of this reaction is the amino-blocked dipeptide, still bonded to the resin at the carboxyl end. The amino-blocking group is removed, and the process is repeated until the peptide synthesis is complete. A controlled cleavage of the resin–ester bond with an acid such as anhydrous HF releases the peptide and removes the final amino-blocking group.

The classical type of peptide synthesis is tedious because intermediate peptides must be isolated and purified. However, in a solid-phase peptide synthesis, most impurities can simply be washed away from the resin after each step. This technique is successful enough that commercial **automated peptide synthesizers** have been developed.

Unfortunately, solid-phase peptide synthesis techniques are not practical for synthesizing large protein molecules in a pure state and in high yields. Poor-quality products may result because the resin tends to hold some impurities. However, recent advances in "gene-splicing" have provided routes to high yields of pure peptides and proteins.

Section 24.9
Classification of Proteins

Proteins may be roughly categorized by the type of function they perform. These classes are summarized in Table 24.4. **Fibrous proteins** (also called **structural proteins**), which form skin, muscles, the walls of arteries, and hair, are composed of long thread-like molecules that are tough and insoluble.

Another functional type of protein is the class of **globular proteins**. These are small proteins, somewhat spherical in shape because of folding of the protein chains upon themselves. Globular proteins are water-soluble and

Table 24.4 Classes of proteins

Class	Comments
Fibrous, or Structural (Insoluble):	
collagens	form connective tissue; comprise 30% of mammalian protein; lack cysteine and tryptophan; rich in hydroxyproline
elastins	form tendons and arteries
keratins	form hair, quills, hoofs, nails; rich in cysteine and cystine
Globular (Soluble):	
albumins	examples are egg albumin and serum albumin
globulins	an example is serum globulin
histones	occur in glandular tissue and with nucleic acids; rich in lysine and arginine
protamines	associated with nucleic acids; contain no cysteine, methionine, tyrosine, or tryptophan; rich in arginine
Conjugated (Combined with Other Substances):	
nucleoproteins	combined with nucleic acids
mucoproteins	combined with $>4\%$ carbohydrates
glycoproteins	combined with $<4\%$ carbohydrates
lipoproteins	combined with lipids, such as phosphoglycerides or cholesterol

perform various functions in an organism. For example, *hemoglobin* transports oxygen to the cells. *Insulin* aids in carbohydrate metabolism. *Antibodies* render foreign protein inactive. *Fibrinogen* (soluble) can form insoluble fibers that result in blood clots. *Hormones* carry messages throughout the body.

Conjugated proteins, proteins connected to a nonprotein moiety such as a sugar, perform various functions throughout the body. A common mode of linkage between the protein and nonprotein is by a functional side chain of the protein. For example, an acidic side chain of the protein can form an ester with an —OH group of a sugar molecule.

▼ ■ **Section 24.10**
Higher Structures of Proteins

The sequence of amino acids in a protein molecule is called the **primary structure** of the protein. However, there is much more to protein structure than just the primary structure. Many of the properties of a protein are due to the orientation of the molecule as a whole. The shape (such as a helix) into which a protein molecule arranges its backbone is called the **secondary structure**. Further interactions, such as folding of the backbone upon itself to form a sphere, result in the **tertiary structure**. Interactions between certain protein subunits, such as between the globins in hemoglobin (page 987), result in the **quaternary structure**. The secondary, tertiary, and quaternary structures are collectively referred to as the **higher structure** of the protein.

One protein that has been well-studied in terms of its secondary structure is **keratin**, found in fur and feathers. Each protein molecule in keratin has the shape of a spiral, called a *right-handed α-helix* (Figure 24.6). "Right-handed" refers to the direction of the turns in the helix; the mirror image is a left-handed helix. In the mid-1930s, the term "α" was coined to differentiate the x-ray pattern of keratin from that of some other proteins.

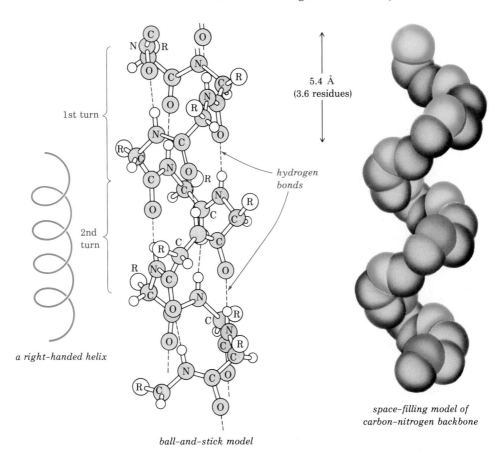

5.4 Å
(3.6 residues)

1st turn

hydrogen bonds

2nd turn

a right-handed helix

space-filling model of carbon-nitrogen backbone

ball-and-stick model

Figure 24.6 The protein chains in keratin form right-handed α-helices. The ball-and-stick model is adapted from C. B. Anfinsen, *The Molecular Basis of Evolution* (John Wiley and Sons, Inc., New York, 1964). The space-filling model is adapted from W. H. Brown, *Introduction to Organic and Biochemistry, 4th ed.* (Copyright 1987 by Wadsworth, Inc. Used by permission of Brooks/Cole Publishing Co.)

In keratin, each turn of the helix contains 3.6 amino acid residues. The distance from one coil to the next is 5.4 Å. The helix is held in its shape primarily by hydrogen bonds between one amide–carbonyl group and an NH group that is 3.6 amino acid units away (Figure 24.7). The helical shape gives a strong, fibrous, flexible product.

Hydrogen bonding between an α-amino group and a carbonyl group is one contributor to the shape of a protein molecule. Other inter- and intramolecular interactions also contribute to the higher structure. Some of these interactions are hydrogen bonding between side chains, S—S cross-links, and *salt bridges* (ionic bonds such as RCO_2^- ^+H_3NR between side chains). The most stable higher structure is the one with the greatest number of stabilizing interactions. (Each hydrogen bond lends ∼5 kcal/mol of stability.) Given a particular primary structure, a protein naturally assumes its most stable higher structure.

Figure 24.7 The helix in keratin is held in its shape by hydrogen bonds. (Nonparticipating R's and H's have been omitted.) From R. J. Fessenden and J. S. Fessenden, *Chemical Principles for the Life Sciences* (Allyn and Bacon, Inc., Boston, 1976).

Let us look at some other types of protein structure. **Collagen** is a general classification of a tough, strong protein that forms cartilage, tendons, ligaments, and skin. Collagen derives its strength from its higher structure of "super-helices": three right-handed α-helical polypeptides entwined to form a triple left-handed helical chain. The entwined molecules are collectively called a *tropocollagen molecule.* One of these tropocollagen molecules is about 15 Å in diameter and 2800 Å long. The tropocollagen triple helix, like a single helix, is held together by hydrogen bonding.

Gelatin is obtained by boiling collagen-containing animal material. However, gelatin is not the same type of protein as collagen. It has been found that the molecular weight of gelatin is only one-third that of collagen. Presumably, in gelatin formation, the tropocollagen molecule unravels and the single strands form hydrogen bonds with water, resulting in the characteristic gel-formation.

Helical structures are not the only type of secondary structure of proteins. Another type of structure, referred to as a *β-sheet*, or *pleated sheet*, is found in silk fibroin. The pleated sheet is an arrangement in which single protein molecules are lined up side by side and held there by hydrogen bonds between the chains (Figure 24.8).

The protein chains in silk fibroin are not simply stretched-out zigzag chains. Analysis by x-ray diffraction shows repetitive units every 7.0 Å. This repetition probably arises from a puckering (or "pleating") in the chains that alleviates steric hindrance. It is interesting that silk fibroin contains 46% glycine (no side chain) and 38% of a mixture of alanine and serine (small side chains). The lack of bulky R groups in these amino acids allows the side-by-side arrangement of protein chains in the fibroin structure.

7.0 Å per pleat

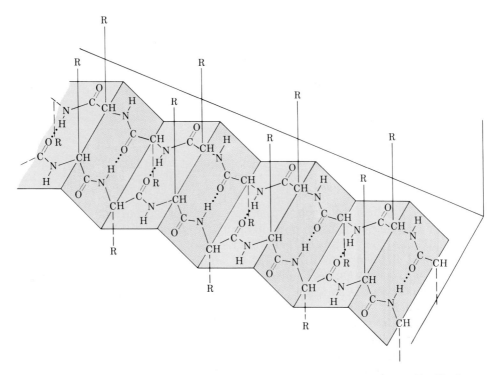

Figure 24.8 The pleated sheet structure of silk fibroin. From P. Karlson, *Introduction to Modern Biochemistry, 3rd ed.* (Georg Thieme Verlag, Stuttgart, 1968).

A *globular protein* depends on a tertiary structure to maintain its intricately folded, globular shape, which is necessary to maintain solubility. In a globular protein, polar hydrophilic side chains are situated on the outside of the sphere (to increase water-solubility) and nonpolar hydrophobic side chains are arranged on the interior surface, where they may be used to catalyze nonaqueous reactions. The unique surface of each globular protein enables it to "recognize" certain complementary organic molecules. This recognition allows enzymes to catalyze reactions of particular molecules, but not others.

Hemoglobin, the portion of erythrocytes (red blood cells) that is responsible for the transport of oxygen in the bloodstream, is a good example of a globular protein. One unit of hemoglobin, which has a "molecular weight" of about 65,000, contains four protein molecules called *globins*. Each globin is folded in such a way that (1) it fits perfectly with the three other globins to maintain the hemoglobin entity and (2) it forms a molecular crevice of just the right size and shape to hold a unit of heme along with its O_2 molecule.

Heme (page 793) is a **prosthetic group**, a nonprotein organic molecule held firmly by a protein. It is composed of a porphyrin ring system with a chelated ferrous ion in the center. Each heme unit is bound to its globin by a coordinate bond from the ferrous ion to the nitrogen in an imidazole ring of histidine, one of the amino acids that forms the globin. The ferrous ion also can form a coordinate bond with an O_2 molecule.

For heme to function as a transporter of oxygen to the cells, the Fe^{2+} ion must release the oxygen readily. Such a release could not occur if Fe^{2+} were oxidized to Fe^{3+} and oxygen were reduced. The nonpolar hydrophobic environment of the protein around the O_2 binding site assures that electron-transfer from Fe^{2+} to O_2 does not occur. (Remember that nonpolar solvents cannot support ions in solution.) A similar phenomenon is observed in the laboratory when Fe^{2+} ions are embedded in polystyrene. Oxygen is attracted, but no oxidation–reduction reaction takes place as it would in an ionic medium.

Carbon monoxide poisoning is the result of CO molecules taking the place of O_2 molecules in hemoglobin. The CO molecules are firmly held by the iron and are not as readily released as oxygen molecules.

Figure 24.9 lists definitions and examples of the higher structures of proteins.

Section 24.11
Denaturation of Proteins

Denaturation of a protein is the loss of its higher structural features by disruption of hydrogen bonding and other secondary forces that hold the molecule together. The result of denaturation is the loss of many of the biological properties of the protein.

One of the factors that can cause denaturation of a protein is a *change in temperature.* Cooking an egg white is an example of an irreversible denaturation. An egg white is a colorless liquid containing albumins, which are soluble globular proteins. Heating the egg whites causes the albumins to unfold and precipitate. The result is a white solid.

primary:	amino acid sequence
secondary:	shape of backbone
	Examples: α-helix, β-pleated sheet
tertiary:	folding of helix
	Example: folded helix in a globular protein
quaternary:	interactions between two or more protein molecules
	Example: the association of four globins in hemoglobin

Figure 24.9 Summary of the higher structures of proteins.

A change in pH can also cause denaturation. When milk sours, the change in pH arising from lactic acid formation causes *curdling*, or precipitation of soluble proteins. Other factors that can cause denaturation are detergents, radiation, oxidizing or reducing agents (which can alter S—S links), and changes in the type of solvent.

Some proteins (those in skin and the lining of the gastrointestinal tract, for example) are quite resistant to denaturation, while other proteins are more susceptible. Denaturation may be reversible if a protein has been subjected to only mild denaturing conditions, such as a slight change in pH. When this protein is replaced in its natural environment, it may resume its natural higher structure in a process called **renaturation**. Unfortunately, renaturation may be very slow or may not occur at all. One of the problems in protein research is how to study proteins without disrupting the higher structures.

Section 24.12
Enzymes

The word **enzyme** means "in yeast." Even without any knowledge of their structures or functions, humans have used enzymes since prehistoric times in the production of wine, vinegar, and cheese. Pasteur thought that living yeast cells were necessary for fermentation processes. We now know that a living cell is not necessary. The proper enzymes, plus reaction conditions that do not cause denaturation, are all that are needed for enzymatic reactions.

An enzyme is a *biological catalyst*. A higher animal contains thousands of enzymes. Virtually every biochemical reaction is catalyzed by an enzyme. Even the equilibrium $CO_2 + H_2O \rightleftarrows H_2CO_3$ is enzyme-catalyzed because the rate of the uncatalyzed equilibration does not produce carbonic acid fast enough for an animal's needs.

Enzymes are more efficient catalysts than most laboratory or industrial catalysts (such as Pd in a hydrogenation reaction). Biological reactions in humans occur at 37°C and in aqueous media. High temperature, high pressure, or very reactive reagents (such as NaOH or $LiAlH_4$) are not available to an organism. Enzymes also allow a selectivity of reactants and a control over reaction rate that can be obtained with no other class of catalyst.

All enzymes are proteins. Some are relatively simple in structure. However, most are complex. The structures of many enzymes are still unknown. For biological activity, some enzymes require prosthetic groups, or **cofactors**. These cofactors are nonprotein portions of the enzyme. A cofactor may be a simple *metal ion*; for example, copper ion is the cofactor for the enzyme *ascorbic acid oxidase*. Other enzymes contain *nonprotein organic molecules* as cofactors. An organic prosthetic group is frequently referred to as a **coenzyme**. (One coenzyme, coenzyme A, was mentioned in Section 16.7B.)

If an organism cannot synthesize a necessary cofactor, the cofactor must be present in small amounts in the diet. The active units of many cofactors are vitamins. Table 24.5 shows a few cofactors and the corresponding vitamins.

A. Naming Enzymes

Most enzymes are named after the reactions that they catalyze. The ending for an enzyme name is usually **-ase**. The name may be *general* and refer to a class of

Table 24.5 Some cofactors that contain vitamins

Name of Cofactor	Vitamin Needed	Structure of Vitamin
vitamin C (ascorbic acid)	vitamin C	
vitamin B_1 (thiamine)	vitamin B_1	
biotin	biotin	
coenzyme A	pantothenic acid	
NAD^{+a}	nicotinic acid (niacin) or nicotinamide (niacinamide)	
pyridoxyl phosphate	pyridoxine	

[a] Nicotinamide adenine dinucleotide, a biological oxidizing agent.

enzymes that catalyze a general type of reaction. For example, a **polymerase** is any enzyme that catalyzes a polymerization reaction, and a **reductase** is any enzyme that catalyzes a reduction reaction. An enzyme name may also refer to a *specific* enzyme: **ascorbic acid oxidase** is the enzyme that catalyzes the oxidation of ascorbic acid, while **phosphoglucoisomerase** catalyzes the isomerization of glucose 6-phosphate to fructose 6-phosphate.

Study Problem

24.12 Suggest the function of each of the following enzymes: **(a)** an acetyltransferase; **(b)** phenylalanine hydroxylase; **(c)** pyruvate dehydrogenase.

B. How Enzymes Work

Some enzymes have been studied in detail, yet there is still much to learn about even the well-known enzymes. It is believed that an enzyme fits itself around the substrate (the molecule to be acted upon) to form an **enzyme–substrate complex**. The bonds of the substrate may be strained by attractions between itself and the enzyme. Strained bonds are of higher energy and are more easily broken. Therefore, the desired reaction proceeds easily and yields an **enzyme–product complex**.

In many cases, the product is not the same shape as the reacting substrate; thus the fit between the product and the enzyme is no longer perfect. The altered shape of the product causes a dissociation of the complex, and the enzyme surface is ready to accept another molecule of substrate. This theory of enzyme activity is called the **induced-fit theory**.

$$E \;+\; S \;\longrightarrow\; E\text{—}S \;\longrightarrow$$

enzyme   substrate   enzyme–substrate
complex

$$E\text{—}P \;\longrightarrow\; E \;+\; P$$

enzyme–product   enzyme   product
complex

Enzymes have molecular weights of 12,000–120,000 and higher. Most substrates (for example, an amino acid or a unit of glucose) are much smaller molecules. The specific location of the large enzyme structure where reaction occurs is called the **active site**. This site is where the prosthetic group (if any) is located. Metallic prosthetic groups are thought to serve as electrophilic agents and, in this way, catalyze the desired reactions. In NAD^+, the active site is the nicotinamide end of the cofactor. NAD^+ is readily reduced and therefore catalyzes oxidation reactions.

$$R_2CHOH + \text{NAD}^+ \longrightarrow R_2C{=}O + \text{NADH} + H^+$$

NAD⁺   NADH

oxidized   *the reduced form of NAD⁺*

The rest of the enzyme molecule is not simply excess molecular weight! It is believed that this portion of the enzyme recognizes its substrate and holds it in place. It was suggested in the 1890s by Emil Fischer that enzymes are chiral molecules and that reactants must complement this chirality in order to undergo reaction. Fischer compared the fitting together of the substrate structure and the enzyme structure to a key fitting into a lock (Figure 24.10).

Recognition may occur by a series of dipole–dipole interactions, by hydrogen bonding, or by covalent bonding, in which the stereochemistry must be just right. In some cases, the rest of the enzyme molecule is folded to form a

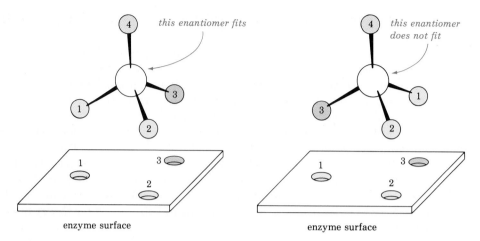

Figure 24.10 One enantiomer fits on the enzyme surface; its mirror image does not.

hydrophobic pocket that holds a nonpolar portion of the substrate in place. (We mentioned this type of structure for hemoglobin.) If the nonpolar end of a potential substrate does not fit in the pocket correctly, enzyme catalysis diminishes or is nonexistent. Therefore, the functional group to be acted upon must fit the active site on the enzyme, and the rest of the substrate molecule must fit together with other portions of the enzyme molecule for reaction to proceed. This dual type of recognition is the basis of the unique specificity of most enzymes.

Both the active site and the rest of an enzyme are important in enzyme activity. Let us look at one reaction in which the active site seems to be the more important factor in substrate recognition. The enzyme *succinate dehydrogenase* catalyzes the dehydrogenation of succinic acid to the *trans*-diacid fumaric acid. (The *cis*-isomer, maleic acid, is not produced in this reaction.) The oxidizing agent in this reaction is *flavin adenine dinucleotide* (FAD), which is reduced by a 1,4-addition of two hydrogen atoms (plus two electrons). (We show only the functional portion of FAD here.)

$$HO_2CCH_2CH_2CO_2H \; + \quad\quad\quad\quad\quad \xrightleftharpoons[\text{dehydrogenase}]{\text{succinate}}$$

succinic acid   FAD

fumaric acid   FADH$_2$

the reduced form of FAD

Other diacids, such as oxalic acid, malonic acid, and glutaric acid, inhibit the dehydrogenation of succinic acid.

Inhibitors of succinate dehydrogenase:

$$HO_2C-CO_2H \qquad HO_2CCH_2CO_2H \qquad HO_2CCH_2CH_2CO_2H$$

oxalic acid   malonic acid   glutaric acid

Of these diacids, malonic acid has the greatest inhibiting effect. Malonic acid is very similar in structure to succinic acid, but is structurally incapable of undergoing dehydrogenation. If succinic acid contains only 2% malonic acid, the rate of enzymatic dehydrogenation of the succinic acid is halved! The probability is excellent that malonic acid competes with succinic acid for a position on the active site and that malonic acid is attracted and held there preferentially. The presence of malonic acid on the active site thus blocks the approach of succinic acid.

▼ Summary

A **protein** is a polyamide. Hydrolysis yield α-amino acids of (L)-configuration at the α carbon. Amino acids undergo an internal acid–base reaction to yield **dipolar ions**.

an α-amino acid   *a dipolar ion*

Essential amino acids are those that cannot be synthesized by an organism and must be present in the diet. **Acidic amino acids** are those with a carboxyl group in the side chain (R in the preceding equation). **Basic amino acids** contain an amino group in the side chain. **Neutral amino acids** contain neither $-CO_2H$ nor $-NH_2$ in the side chain, but may contain OH, SH, or other polar group. **Cross-linking** in proteins may be provided by the SH group in cysteine, which can link with another SH in an oxidation reaction: $2\,RSH \rightarrow RSSR + 2\,H$.

Racemic amino acids may be synthesized by a variety of routes, which are summarized in Table 24.2, page 966.

The **isoelectric point** of an amino acid is the pH at which the dipolar ion is electrically neutral and does not migrate toward an anode or cathode. The isoelectric point depends on the acidity or basicity of the side chain.

in H₂O, a net positive charge   *at pH 9.74, no net charge*

Reactions of amino acids include reactions with acid and base, acylation, and conversion to Ruhemann's purple.

A **peptide** is a polyamide of fewer than 50 amino acid residues. The *N*-**terminal amino acid** is the amino acid with a free α-amino group, while the *C*-**terminal amino acid** has a free carboxyl group at carbon 1. **Terminal residue analysis** (using such reagents as the Edman reagent, the Sanger reagent, and hydrolytic enzymes) to determine the *N*- and *C*-terminals and **partial hydrolysis** to smaller peptides are two techniques for peptide structure determination.

In the synthesis of a peptide, reactive groups (except for the groups desired to undergo reaction) must be blocked. A **carbamate group** may be used to protect an amino group. A **solid-phase peptide synthesis** provides a blocking group for the *C*-terminal carboxyl group.

Proteins are polyamides of more than 50 amino acid residues. The order of side chains in a protein determines its **higher structures**, which arise from internal and external hydrogen bonding, van der Waals forces, and other interactions between side chains. The higher structures of proteins give them a variety of physical and chemical properties so that they may perform a variety of functions. **Denaturation** is the disruption of hydrogen bonds and thus the disruption of the higher structure of the protein.

Enzymes are proteins that catalyze biochemical reactions. Enzymes are efficient and specific in their catalytic action. The specificity is provided for by the unique shape and by the polar (or nonpolar) groups contained within the enzyme structure. Some enzymes work in conjunction with a nonprotein **cofactor**, which may be organic or inorganic.

Study Problems*

24.13 Label each of the following amino acids as *acidic*, *basic*, or *neutral*: **(a)** isoleucine; **(b)** aspartic acid; **(c)** asparagine; **(d)** serine; **(e)** histidine; **(f)** glutamine.

24.14 Which amino acids have more than one chiral carbon (see Table 24.1)?

24.15 Show the possible oxidative coupling products between 1.0 mol each of cysteine and the following compound:

$$\underset{\underset{CH_2SH}{|}}{HSCH_2\overset{\overset{O}{\|}}{C}NHCHCO_2H}$$

24.16 Outline a Strecker synthesis for: **(a)** phenylalanine, and **(b)** isoleucine. **(c)** What would be the configuration of the chiral carbons in the products?

24.17 Suggest a synthesis for valine from 3-methylbutanoic acid. Give the stereochemistry of the product.

24.18 Suggest a route to alanine from (*R*)-lactic acid (2-hydroxypropanoic acid) using a reductive amination. What is the stereochemistry of the product?

---

* For information concerning the organization of the *Study Problems* and *Additional Problems*, see the *Note to student* on page 39.

24.19 Predict the major organic products:

(a) $CH_3CH_2CO_2H$ $\xrightarrow[\text{(2) excess NH}_3]{\text{(1) Br}_2, \text{PBr}_3}$

(b)

$\xrightarrow[\text{(2) HCl, H}_2\text{O, heat}]{\text{(1) ClCH}_2\text{CO}_2\text{C}_2\text{H}_5}$

(c)

$\xrightarrow[\text{(3) HCl, H}_2\text{O, heat}]{\substack{\text{(1) Na}^+ \ ^-\text{OC}_2\text{H}_5 \\ \text{(2) C}_6\text{H}_5\text{CH}_2\text{Cl}}}$

(d) $(CH_3)_2CHCH_2\overset{\displaystyle O}{\overset{\displaystyle \|}{C}}H + NH_3 + HCN \longrightarrow$

(e) $C_6H_5CH_2\underset{\underset{\displaystyle NH_2}{|}}{C}HCO_2H + CH_3OH + HCl \xrightarrow{\text{heat}}$

24.20 Predict the products of the reaction of alanine with: (a) dilute aqueous HCl; (b) dilute aqueous KOH; (c) methanol + H_2SO_4 with heat; (d) an equimolar quantity of acetic anhydride.

24.21 Predict the major products:

(a) $1\ H_2N\underset{\underset{\displaystyle CH(CH_3)_2}{|}}{C}HCO_2^- + 1\ HCl \longrightarrow$

(b) $1\ H_2N\underset{\underset{\displaystyle (CH_2)_4NH_2}{|}}{C}HCO_2^- + 1\ HCl \longrightarrow$

(c) $1\ ^+H_3N\underset{\underset{\displaystyle CH(CH_3)_2}{|}}{C}HCO_2H + 1\ NaOH \longrightarrow$

(d) $1\ ^+H_3N\underset{\underset{\displaystyle (CH_2)_4NH_3^+}{|}}{C}HCO_2H + 1\ NaOH \longrightarrow$

24.22 Write equations to represent: (a) dipolar-ion formation of histidine; (b) the equilibria between histidine and water; (c) the reaction of histidine with dilute aqueous HCl; (d) the reaction of histidine with dilute aqueous NaOH.

24.23 Explain the following observations:

(a) Although saturated carboxylic acids absorb at about 1720 cm^{-1} (5.81 μm) in the infrared region, amino acids do not absorb at this position.

(b) If a neutral solution of an amino acid is acidified, the infrared spectrum then shows absorption at 1720 cm^{-1}.

24.24 Without referring to the text, match the amino acid with its isoelectric point.

(a) cysteine   (1) 10.76
(b) aspartic acid   (2) 6.30
(c) proline   (3) 5.07
(d) arginine   (4) 2.77

24.25 Without referring to the text, predict the approximate isoelectric point of each of the following amino acids: **(a)** serine; **(b)** histidine; **(c)** glutamic acid; **(d)** glutamine; **(e)** lysine.

24.26 Complete the following equations:

(a) $H_2NCCH_2CH_2CHCO_2H$ + 2 [structure] $\longrightarrow$

(b) $H_2NCCH_2CH_2CHCO_2H$ + CH_3COCCH_3 $\longrightarrow$

(c) tyrosine + excess acetic anhydride $\longrightarrow$

(d) serine + excess acetic anhydride $\longrightarrow$

(e) alanine + ninhydrin $\longrightarrow$

24.27 In the following structure, label: **(a)** the peptide bond(s); **(b)** the *N*-terminal amino acid; **(c)** the *C*-terminal amino acid. **(d)** Is this structure a dipeptide, tripeptide, or tetrapeptide? **(e)** Would this peptide be considered acidic, basic, or neutral?

$$(CH_3)_2CHCH_2CHCNHCHCNHCH_2CO_2H$$
$$\quad\quad\quad NH_2 \quad CH_2CH_2SCH_3$$

24.28 Write the structures for **(a)** glycylglycine, and **(b)** alanylleucylmethionine.

24.29 Write the full name and the abbreviated name for each of the following peptides:

(a) $H_2NCHCNHCH_2CNHCHCO_2H$
$\quad\quad CH_2OH \quad\quad\quad CH_2CH(CH_3)_2$

(b) [structure] $-CNHCHC-NHCHCO_2H$ $\longrightarrow$
$\quad N \quad HOCHCH_3 \quad CH_2CH_2SCH_3$ $\longrightarrow$
$\quad H$

24.30 Give the structures of the principal organic ionic species in the following solutions:

(a) glycyllysine in dilute aqueous HCl
(b) glycylglutamic acid in dilute aqueous NaOH
(c) glycyltyrosine in dilute aqueous NaOH

24.31 A globular protein is hydrolyzed by aqueous acid. When the reaction mixture is neutralized, ammonia is given off. What can you deduce about the structure of the protein from this observation?

24.32 Each of the following peptides is subjected to reaction with an alkaline solution of phenyl isothiocyanate, followed by acid hydrolysis. Give the structures of the products.

(a) gly–ala   **(b)** ala–gly   **(c)** ser–phe–met

24.33 Write the structures of the products of the reaction of 1-fluoro-2,4-dinitrobenzene, followed by hydrolysis, with: **(a)** valine; **(b)** alanylvaline; **(c)** glutamylglycine.

24.34 A pentapeptide obtained from treatment of a protein with trypsin contains arginine, aspartic acid, leucine, serine, and tyrosine. To determine the amino acid sequence, the peptide was treated with Edman reagent three times. The composition of the peptide remaining after each treatment was: (1) arginine, aspartic acid, leucine, serine; (2) arginine, aspartic acid, serine; and (3) arginine, serine. What is the amino acid sequence of the pentapeptide?

24.35 Each of the following peptides is treated with trypsin. What fragments will be formed?
   (a) lys–asp–gly–ala–ala–glu–ser–gly
   (b) trp–cyS–lys–ala–arg–arg–gly
   (c) ala–ala–his–arg–glu–lys–phe–ile–gly–glu–gly–glu

24.36 The product fragments in Problem 24.35(a) are treated with 1-fluoro-2,4-dinitrobenzene, followed by hydrolysis. List the products.

24.37 Each of the peptides in Problem 24.32 is treated with the enzyme carboxypeptidase. What products are formed?

24.38 *Bradykinin* is a pain-causing nonapeptide that is released by globulins in blood plasma as a response to toxins in wasp stings. Partial hydrolysis of bradykinin results in the following tripeptides:

   ser–pro–phe   gly–phe–ser   pro–phe–arg   arg–pro–pro

   pro–gly–phe   pro–pro–gly   phe–ser–pro

   What is the amino acid sequence in bradykinin?

24.39 A chemist decided to prepare the dipeptide val–ala by the following sequence of reactions. Using structural formulas, rewrite each set of reactants and tell what possible products were obtained in each case.
   (a) val + SOCl$_2$ **(b)** [product(s) from (a)] + ala

24.40 Starting with monomeric amino acids, show how to prepare ala–gly and phe–val and then show how these dipeptides can be joined to yield ala–gly–phe–val. (Do not use a solid-phase synthesis.)

24.41 Which of the following structural features would contribute to water-solubility of a protein?
   (a) rich in glutamic acid
   (b) rich in valine
   (c) formation of bundles of helices
   (d) formation of a helix folded into a sphere
   (e) combination with glucose
   (f) combination with cholesterol (page 949)

24.42 Summarize the nature of each type of bond listed below and its role in protein structure:
   (a) peptide bond;   **(b)** disulfide bond;   **(c)** hydrogen bond;   **(d)** ionic salt-bridge bond.

24.43 Meat can be tenderized by soaking it overnight in a marinade of vinegar, spices, and sugar.
   (a) Why is the meat tenderized in this process?
   (b) Would a water solution of sucrose alone have the same tenderizing effect?

24.44 Match each of the following enzymes (or class of enzymes) with a reaction on the right that it might catalyze. (There may be more than one correct answer.)

(a) a phosphatase

(b) a glycosidase

(c) α-1,4-glucan 4-glucanohydrolase

(d) an aminopeptidase

(e) an oxidase

(f) phosphohexose isomerase

(1) $Fe^{2+} \longrightarrow Fe^{3+}$

(2) cleavage of the *N*-terminal amino acid from a protein

(3) sucrose $\longrightarrow$ fructose + glucose

(4) glucose 6-phosphate $\longrightarrow$ glucose

(5) cellulose $\longrightarrow$ glucose

(6) amylose $\longrightarrow$ glucose

(7) glucose 6-phosphate $\longrightarrow$ fructose 6-phosphate

24.45 Which of the following compounds would be most likely to inhibit the enzymatic incorporation of nicotinic acid (3-pyridinecarboxylic acid) into NAD^+?

(a) pyridine   (b) 3-pyridinesulfonic acid

(c) 3-methylpyridine   (d) adenine

(e) acetic acid   (f) malonic acid

Additional Problems

24.46 The tripeptide pyroglutamylhistidylprolinamide (page 974) is subjected to complete hydrolysis with HCl and H_2O. What is the structure of each product?

24.47 A new amino acid, γ-carboxyglutamic acid (gla) was discovered in 1974 as one of the amino acid residues in blood coagulation proteins and in proteins from calcified tissues. (The presence of this residue allows the protein to chelate metal ions.) The reason for the late discovery of this amino acid was that, in structure determinations, proteins are often hydrolyzed to smaller fragments by heating with acid.

(a) Write an equation to show what happens to γ-carboxyglutamic acid when heated in aqueous acid.

(b) Write an equation to show its probable reaction when heated in aqueous base.

$$\overset{\displaystyle NH_2}{\underset{\displaystyle |}{(HO_2C)_2CHCH_2CHCO_2H}}$$

γ-carboxyglutamic acid (gla)

24.48 A peptide containing one equivalent each of tyr, ile, gly, arg, and cyS had no *C*-terminal amino acid and no *N*-terminal amino acid. Explain.

24.49 Suggest a reason for the differences in the isoelectric points of   (a) lysine (9.74) and histidine (7.59), and   (b) lysine and arginine (10.76).

24.50 (a) When the following peptide is treated with 1-fluoro-2,4-dinitrobenzene, followed by hydrolysis, *two* dinitrophenylamino acids are isolated. What are they?

(b) Would the presence of the second product interfere with the determination of the peptide structure? Explain.

ala–lys–ala–gly

24.51 An amino acid with an (*R*), or D, configuration at the α carbon cannot be enzymatically incorporated into proteins. Many D-amino acids are oxidized in a reaction catalyzed by **D-amino acid oxidase**.

$$\underset{\displaystyle R}{\overset{\displaystyle CO_2H}{H{-}C{-}NH_2}} + \tfrac{1}{2}O_2 \longrightarrow \underset{\displaystyle R}{\overset{\displaystyle CO_2H}{C{=}O}}$$

This enzyme-catalyzed reaction is very slow for L-amino acids, D-glutamic acid, D-lysine, and D-aspartic acid. The reaction is fast for D-proline, D-alanine, D-methionine, and D-tyrosine.

(a) Would you say that the differences in reaction rate are due to steric hindrance or to electronic effects?

(b) Would you predict that the binding site on the enzyme is polar or nonpolar?

(c) Which would be likely to undergo this enzymatic oxidation at a faster rate, D-cysteine or D-arginine?

24.52 The complete hydrolysis of an acyclic nonapeptide yields a mixture of ala, asp, glu, gly, leu, lys, phe, tyr, and val. Terminal-residue analysis shows the *N*-terminal amino acid to be val and the *C*-terminal amino acid to be gly. Partial enzymatic hydrolysis with chymotrypsin yields a pentapeptide and a tetrapeptide, among other products. The tetrapeptide is partially hydrolyzed to three dipeptides. One dipeptide contains ala and gly; the second contains asp and tyr; and the third contains asp and ala. Partial hydrolysis of the nonapeptide with trypsin yields a pentapeptide and a tetrapeptide. The amino acid content of the tetrapeptide is glu, leu, lys, and val. What are the two possible structures of the nonapeptide? How could they be differentiated?

24.53 In mammals, arginine undergoes enzymatic hydrolysis to urea $(H_2N)_2C=O$, which is excreted in the urine, and a basic amino acid called *ornithine*. What is the structure of ornithine?

24.54 Treatment of ornithine (see Problem 24.53) with an alkaline solution of $H_2N—CN$ yields arginine. Suggest a mechanism for this reaction. (*Hint*: Your first step should be to assign $\delta+$ and $\delta-$ charges to the atoms.)

24.55 *Glutathione* is a tripeptide found in most living cells. Partial hydrolysis yields cyS, glu, gly, glu–cyS, and cyS–gly.

(a) What is the amino acid sequence in glutathione?

(b) It has been discovered that glutamic acid forms a peptide link in glutathione with the *side chain* carboxyl group, rather than the carboxyl group adjacent to the amino group. What is the structural formula of glutathione?

25 ▼

Nucleic Acids

The study of nucleic acids is a fascinating area of modern research. These polymers carry the genetic code and are responsible for its expression by protein synthesis. The two types of nucleic acids are **deoxyribonucleic acids (DNA)** and **ribonucleic acids (RNA)**. In all cells, DNA is the repository of the genetic code and RNA is involved in protein synthesis. Viruses, however, contain their genetic code in either DNA or RNA and use the host cells that they have infected for protein synthesis.

The component parts of nucleic acids can be obtained by direct or enzymatic hydrolysis (Figure 25.1). Complete hydrolysis yields a sugar (ribose from RNA or deoxyribose from DNA), purine and pyrimidine heterocyclic bases, and inorganic phosphate ions. Partial hydrolysis of a nucleic acid yields **nucleotides** (sugar bonded to a base and to a phosphate group). Nucleotides are the monomeric units of nucleic acids. Hydrolysis of the nucleotides results in **nucleosides** (sugar bonded to base).

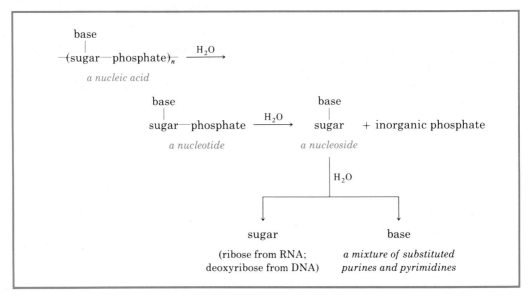

Figure 25.1 Hydrolysis of the nucleic acids.

In this chapter, we will first discuss the hydrolysis fragments. Next, we will consider DNA itself—its primary and higher structures—and some related topics. We will then consider RNA. We will conclude with an overview of the role of RNA in protein synthesis.

Section 25.1
Hydrolysis Products of the Nucleic Acids

In this section, we will discuss the complete hydrolysis products of the nucleic acids (the phosphate, the sugars, and the bases), then the nucleosides, and finally the nucleotides.

A. The Inorganic Phosphate

The inorganic phosphate generated by the complete hydrolysis of the nucleic acids is either $H_2PO_4^-$ or HPO_4^{2-}, depending on the pH and the medium used for hydrolysis.

In the nucleotides, the phosphate group is bonded to a hydroxyl group of the sugar. Because the protonated phosphate group is acidic, protonated nucleotides are also acidic.

$$\underset{\substack{\text{protonated phosphate} \\ \text{group in nucleotides}}}{HO-\overset{\overset{O}{\|}}{\underset{\underset{OH}{|}}{P}}-O-\xi} \quad \underset{(pK_1 \cong 1)}{\overset{-H^+}{\rightleftharpoons}} \quad \underset{10\%}{HO-\overset{\overset{O}{\|}}{\underset{\underset{O^-}{|}}{P}}-O-\xi} \quad \underset{(pK_2 \cong 6)}{\overset{-H^+}{\rightleftharpoons}} \quad \underset{90\%}{{}^-O-\overset{\overset{O}{\|}}{\underset{\underset{O^-}{|}}{P}}-O-\xi}$$

at physiological pH

B. The Sugars

The sugar in RNA, as the name ribonucleic acid implies, is β-D-ribose, which is in the furanose form. The sugar in DNA (deoxyribonucleic acid) is 2-deoxy-β-D-ribose, also in the furanose form. The suffix 2-deoxy in this name means that the structure lacks an oxygen at position 2.

D-ribofuranose   2-deoxy-D-ribofuranose

C. The Bases

Complete hydrolysis of nucleic acids yields several heterocyclic ring systems, usually called **bases**. These compounds are substituted pyrimidines and purines.

purine   pyrimidine

the parent ring system   *the parent ring system*
of adenine and guanine   *of thymine, uracil, and cytosine*

The four principal bases obtained from DNA are adenine, guanine, cytosine, and thymine. These bases are usually identified in abbreviated formulas by the first letters of their names (A, G, C, and T).

The four principal bases found in DNA:

purine bases   pyrimidine bases

adenine (A)   guanine (G)   thymine (T)   cytosine (C)

The four principal bases obtained from RNA are adenine, guanine, cytosine, and uracil. Three of these bases are the same as those found in DNA, but uracil (U) in RNA replaces thymine (T) in DNA.

In RNA:

uracil (U)

In DNA:

thymine (T)

Table 25.1 A few of the minor bases and nucleosides found in the nucleic acids

Name	Formula	Where Found
5-methylcytosine		many types of DNA
N^6-methyladenine		many types of DNA and RNA
dihydrouracil		tRNA
1-methylguanine		tRNA
ribothymidine		tRNA
pseudouridine (Ψ-uridine)		tRNA
1-methylinosine		tRNA

Although the bases we have described are the most common, a variety of so-called minor bases have also been isolated. In some cases, their concentrations are not really minor. For example, some plant DNA contains up to 7% 5-methylcytosine. These minor bases result from enzymatic modification of a principal base after the nucleic acid has been biosynthesized instead of being incorporated during the biosynthesis.

In Table 25.1, the structures of some of the minor bases in DNA and RNA are listed. Transfer RNA (tRNA) is a particularly rich source for these minor bases.

Tautomers.   The pyrimidine and purine bases can undergo lactam–lactim tautomerization, a type of keto–enol tautomerization. In neutral and acidic media, the lactam form of these bases predominates.

Examples of lactam–lactim tautomerization:

lactim form of uracil   *a lactam form of uracil*

lactim form of guanine   *a lactam form of guanine*

Study Problems

25.1 There are two lactam–lactim tautomers of uracil in addition to those shown above. Draw their structures.

25.2 Show the possible lactam–lactim structures for thymine.

Acid–base reactions.   Some of the bases from the nucleic acids are amphoteric. They can be protonated to form cationic salts or can lose a proton to form an anion. For example,

Acid–base reactions for uracil:

uracil

The ionic form of the bases depends on the pH of the solution. At physiological pH (near 7), all the bases are in a neutral nonionic form. The ionic forms of the bases, however, play a major role in the mechanism of denaturation of the nucleic acids by acids and bases (Section 25.2D).

Pyrimidine is a relatively weak base; its pK_b is 12.7. Uracil and thymine, both pyrimidine bases, are both far less basic than even pyrimidine. The reason for this low basicity is that the nitrogens in uracil and thymine are in amide groups in their lactam forms. The unshared pair of electrons of an amide nitrogen is delocalized by the carbonyl groups (Section 15.8C). The delocalized electrons are not available for bonding and, thus, these so-called bases are not basic.

Study Problem

25.3 Using formulas in your answer, explain why cytosine is far more basic than uracil or thymine. (*Hint*: Consider the resonance stabilization of the cationic salt.)

A purine ring system contains imidazole and pyrimidine fused rings. The five-membered imidazole ring is basic, while the pyrimidine ring is only weakly basic.

Guanine, like the parent purine, is protonated at nitrogen 7. The two ring systems have similar pK_b values (11.0 for guanine and 11.7 for purine). Nitrogen 1 of guanine is an amide nitrogen and, therefore, not basic.

The amino group of adenine changes the position of protonation of the purine ring. The amino group stabilizes the adjacent positive charge at nitrogen 1 by resonance, but it cannot stabilize a positive charge at position nitrogen 7. Therefore, protonation of adenine occurs at nitrogen 1.

protonated at position 1

adenine

D. The Nucleosides

A nucleoside contains a purine or pyrimidine base bonded to a sugar by a β-glycosidic linkage.

a nucleoside from DNA   *a nucleoside from RNA*

These glycosides are often called **N-glycosides** to differentiate them from the common carbohydrate O-glycosides. The chemical properties of the N- and O-glycosides are similar.

Because the bond is glycosidic, the anomeric carbon of the sugar is locked into one configuration (beta for all known nucleosides).

β O-glycosidic group   *β N-glycosidic group*

The structures of the four common nucleosides that can be isolated from DNA hydrolysis are shown in Figure 25.2. The principal nucleosides obtained from RNA are the same as those obtained from DNA with two exceptions: the sugar is ribose instead of deoxyribose, and uracil replaces thymine as a base.

When numbering the two ring systems of a nucleoside, we use primes for the sugar's positions.

Figure 25.2 The principal nucleosides from the hydrolysis of DNA. (Deoxy-thymidine is sometimes called thymidine because thymine is not a principal base in nucleic acids other than DNA.)

The N-glycosidic bond of nucleosides is stable in base, but readily cleaved in acid. The rate of acidic cleavage depends on the structure of the heterocyclic base. The purine nucleosides are rapidly hydrolyzed by aqueous acid to a heterocyclic base and a sugar. The pyrimidine nucleosides are cleaved only after prolonged treatment with concentrated acid.

E. The Nucleotides

A nucleotide contains a base, a sugar, and a phosphate group. The nucleotides are monomers for the nucleic acid polymers.

base
|
phosphate—sugar

a nucleotide

base
|
$\{$ —phosphate—sugar—$\}$

the nucleic acids

Deoxyribonucleotides with a phosphate group at position 3′ or position 5′ are known. The 5′-phosphates are more common and we will emphasize them in our discussions. Figure 25.3 shows the structures of the principal deoxyribonucleotides.

A ribose residue in a nucleoside has *three* hydroxyl groups (at positions 2′, 3′, and 5′) that can form phosphate esters. Alkaline hydrolysis of RNA yields ribonucleotides with the phosphate group at position 2′ or 3′. Enzymatic hydrolysis of RNA yields nucleotides with either a 3′-phosphate or a 5′-phosphate depending on the enzyme used. The 3′- and 5′-phosphates are analogous to those obtained from DNA. The 2′-phosphate arises from a cyclic phosphate produced during the alkaline hydrolysis.

deoxyadenosine 5′-phosphate
(5′-deoxyadenylic acid,
deoxyadenosine monophosphate,
5′-dAMP)

deoxyguanosine 5′-phosphate
(deoxyguanylic acid,
deoxyguanosine monophosphate,
5′-dGMP)

deoxycytidine 5′-phosphate
(deoxycytidylic acid,
deoxycytidine monophosphate,
5′-dCMP)

deoxythymidine 5′-phosphate
deoxythymidylic acid,
deoxythymidine monophosphate,
5′-dTMP)

Figure 25.3 Formulas and names for the principal deoxyribonucleotides.

Table 25.2 pK_a values of the principal ribonucleotides

Nucleotide	First Phosphate Proton	Second Phosphate Proton	Base
adenosine 5′-phosphate	0.9	6.1	3.8
uridine 5′-phosphate	1.0	6.4	9.5
cytidine 5′-phosphate	0.8	6.3	4.5
guanosine 5′-phosphate	0.7	6.1	2.4, 9.4

a cyclic phosphate

The structures of the principal ribonucleotides with a phosphate group at the 5′ position are the same as the corresponding deoxyribonucleotides with two exceptions: ribose replaces deoxyribose as the sugar and uracil replaces thymine as a base.

The nucleotides are amphoteric. They contain a phosphoric acid ester group with two acidic protons as well as the basic heterocyclic ring. Table 25.2 lists the pK_a values of the principal ribonucleotides.

Section 25.2
Structure and Properties of DNA

DNA is a giant, even among macromolecules. It is difficult to isolate DNA and obtain accurate measurements of its mass because such large molecules fragment readily. Estimates of the size, however, can be made. The chromosome of the intestinal bacterium *Escherichia coli* (*E. coli*) contains DNA with a molecular weight of about 2.6×10^9. DNA molecules from higher animals are larger and more complex than those from bacteria.

A. Primary Structure of DNA

The DNA polymer consists of nucleoside residues linked by phosphate groups. The phosphate forms an inorganic ester link between the 3' hydroxyl of one sugar residue and the 5' hydroxyl of the next sugar (Figure 25.4). A linear DNA polymer, therefore, has a free 5'-hydroxyl group at one end and a free 3'-hydroxyl group at the other end.

A DNA molecule contains a backbone of sugar–phosphate groups bearing a series of bases (A, G, C, and T). The order in which the bases occur is called the **base sequence**. A number of shorthand formulas have been developed to represent the base sequence of a single DNA strand. The simplest method is to list the letters for the base sequence starting at the 5'-hydroxyl end of the chain (on the left) and proceeding toward the 3'-hydroxyl end (on the right)—for example, A—C—A—.

B. Secondary Structure of DNA: The Double Helix

How does the DNA polymer with its base sequence carry the genetic code? In 1953, J. D. Watson and F. H. C. Crick proposed a model for DNA that accounts

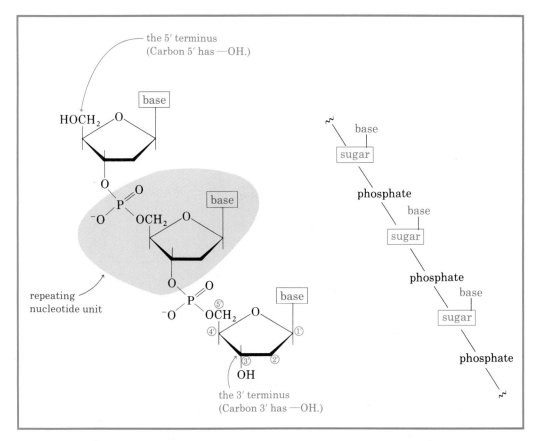

Figure 25.4 Primary structure of DNA. Each base is one of four heterocycles—cytosine, thymine, adenine, or guanine.

for its behavior. In 1962, these two men, along with Maurice Wilkins, who provided evidence for the structure of the model by x-ray analysis, were awarded the Nobel Prize for their work.

The Watson–Crick model of DNA is a double helix of two long antiparallel DNA molecules held together by hydrogen bonds. In this model, antiparallel means that two linear DNA molecules are parallel but aligned in opposite directions. Each end of the double helix thus consists of a 5′ terminus (from one molecule) and a 3′ terminus (from the other molecule).

$$5' \text{ end} \longleftarrow \text{A—G—T—C—A—A—G—T—G—G—C—C} \longrightarrow 3' \text{ end}$$
$$3' \text{ end} \longleftarrow \text{T—C—A—G—T—T—C—A—C—C—G—G} \longrightarrow 5' \text{ end}$$

The hydrogen bonds between the DNA strands are not random, but are specific between pairs of bases: guanine is hydrogen bonded to cytosine, and adenine to thymine.

Why are the hydrogen bonds specific—thymine to adenine and cytosine to guanine but not other combinations? Thymine and adenine can be joined by *two* hydrogen bonds (approximate total strength, 10 kcal/mol). Cytosine and guanine can be joined by *three* hydrogen bonds (approximate total strength, 17 kcal/mol). No other pairing of the four bases leads to such strong hydrogen bonding. Figure 25.5 shows the structures, dimensions, and hydrogen bonds of the bases.

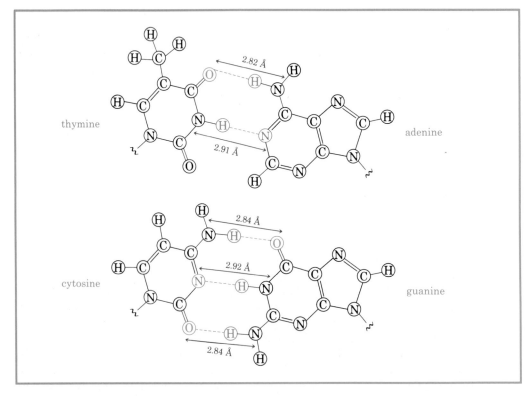

Figure 25.5 Hydrogen bonds and dimensions in base pairs of DNA: (a) thymine to adenine and (b) cytosine to guanine.

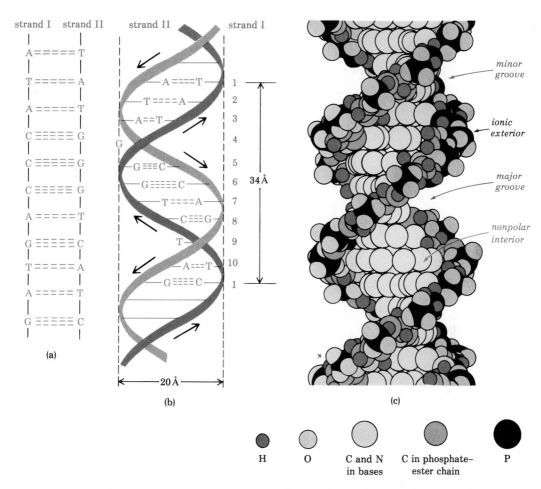

strand I   strand II

A ===== T

T ===== A

A ===== T

C ===== G

C ===== G

C ===== G

A ===== T

G ===== C

T ===== A

A ===== T

G ===== C

(a)

strand II   strand I

A ==== T   1

T ==== A   2

A == T   3

G

G === C   5

G ===== C   6   34Å

T ==== A   7

C === G   8

T   9

A === T   10

G ==== C   1

← 20Å →

(b)

minor groove

ionic exterior

major groove

nonpolar interior

(c)

H   O   C and N in bases   C in phosphate-ester chain   P

Figure 25.6 Three representations of the double-helical model of DNA. (a) Two uncoiled DNA molecules are joined by hydrogen bonds between complementary base pairs. (Not shown are the sugar and phosphate units.) (b) The DNA strands are coiled in a double helix, with ten base pairs for every complete turn of the helix. (c) A "space-filling" model of DNA. (Adapted from William H. Brown, *Introduction to Organic and Biochemistry, 4th ed.* **Copyright 1987 by Wadsworth, Inc. Used by permission of Brooks/Cole Publishing Co.)**

Picture the double-stranded helix of DNA held together by a series of particular hydrogen-bonded pairs, as shown in Figure 25.6(a). Wherever an adenine (A) appears in one strand, a thymine (T) appears opposite it in the other strand. The two strands are completely complementary in this respect. Figure 25.6 shows the base-pairing and also shows a space-filling model of the appearance of the DNA double helix.

In the DNA double helix, the purine and pyrimidine bases are on the *inside* of the helix, and the deoxyribose and phosphate units are on the *outside*. In addition to hydrogen bonding between the stacked purines and pyrimidines, other forces, such as induced dipole attractions, contribute to the maintenance of the two-stranded structure. The energy of these attractions is

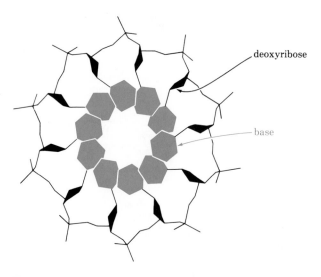

deoxyribose

base

Figure 25.7 Diagram of DNA viewed down the helical axis.

called the **stacking energy**. Figure 25.7 shows a representation of the DNA helix viewed down the axis.

Study Problem

25.4 With formulas, show why thymine and guanine cannot form as strong hydrogen bonds as thymine and adenine.

C. Denaturation and Renaturation

Disruption of the secondary structure of a polymer is called *denaturation* (Section 24.11). DNA, like proteins, can be denatured by changes in pH, increases in temperature, and some reagents. Loss of the base-pair hydrogen bonding causes the double-stranded helix to unwind and separate.

Strong acid or base causes the normally neutral heterocyclic bases in DNA to acquire ionic charges. These charges affect the bases' ability to form effective hydrogen bonds. Changes in hydrogen bonding between the bases is easily detected by ultraviolet spectroscopy. Because of the pi-electron systems, all the heterocyclic bases absorb ultraviolet radiation of characteristic wavelengths and with characteristic absorptivities. When the bases are strongly hydrogen bonded, the absorbance is diminished. As denaturation occurs, an increase in the absorbance occurs (Figure 25.8).

When denatured DNA is held at its **annealing temperature**, a temperature just below its denaturation temperature, or "melting point," the double-stranded helix can re-form. **Renaturation** is the reassembly of the original two strands that were separated.

Renaturation is a two-step process. The first step is **nucleation**—the joining of the two strands by a few base pairings. Nucleation is generally slow because the process is second order, but circular DNA nucleates rapidly be-

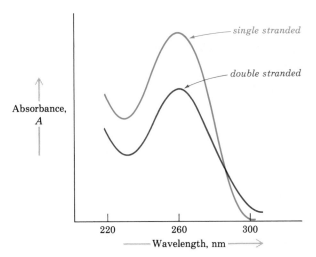

Figure 25.8 The absorbance of DNA at a wavelength of 260 nm increases when the double helix is unraveled into single strands.

cause the process is first order. Once the two strands are loosely joined, the second step, a rapid **zippering**, the pairing of the remaining bases, occurs.

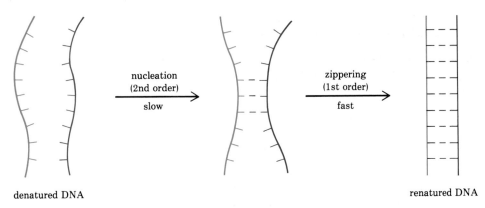

In addition to concentration, the rate of renaturation is dependent on the complexity of the DNA. The synthetic nucleic acids polyadenine and polyuracil undergo rapid renaturation because almost any collision between the two strands results in base pairing. Mammalian DNA, on the other hand, renatures very slowly because few collisions are fruitful.

D. Replication

In living cells, *the genetic code is contained by the DNA and consists of a particular sequence of bases that specifies the sequence of amino acids to be incorporated into any one protein* (a particular enzyme, for example). For passing the genetic code from one cell to its offspring (mother cell to daughter cell), the DNA duplicates itself exactly in a process called **replication**.

In replication, the double helix of DNA becomes enzymatically unraveled. Each strand serves as a template, or pattern, for the synthesis of a new complementary chain. When the syntheses are complete, two double-stranded helices exist where only one did before. Thus, after cell division, both cells contain identical DNA.

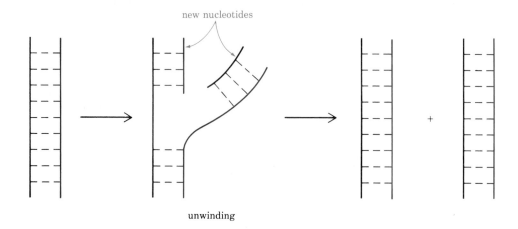

unwinding

Replication is a complex series of reactions that requires many different enzymes. Addition of new nucleotides to the growing DNA chain, which is catalyzed by the enzyme *DNA polymerase*, occurs by the substitution of a nucleotide 5'-triphosphate at the free 3'-hydroxyl group of the growing chain. A diphosphate ion is the leaving group; one monophosphate group becomes part of the chain. Figure 25.9 illustrates the addition of a nucleotide.

Both DNA strands are synthesized in the 5' → 3' direction. Because the two existing chains are aligned in opposite directions, one strand must have a 3'-hydroxyl group near the **replication fork**, or site of unraveling, while the other strand must have a free 5'-hydroxyl group. The complement of the 3' → 5' strand is synthesized as a continuous single chain, but the other new strand must be synthesized discontinuously, as a series of short chains. These pieces of this new DNA strand are then bonded together by other enzymes.

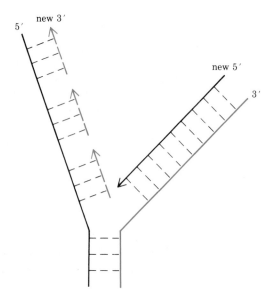

5′ terminus

new chain

3′ terminus

O

$^-O-P=O$

OCH_2

$C\!\equiv\!\equiv\!\equiv\!G-$

replication
proceeding in
this direction

leaving group

$:\!O-H$

O^-

new nucleotide
being added

$^-O-P-O-P-O-P=O$

CH_2

$T\!=\!=\!=\!A-$

OH

to 3′ terminus

5′ terminus
of original strand

Figure 25.9 Addition of a nucleotide to a growing DNA chain. The nucleotide triphosphate loses a diphosphate ion in the enzyme-catalyzed reaction.

E. Conformers of DNA

Because of the chain of single bonds in its phosphodiester backbone, DNA can flex into several conformations. The three most important conformers are the B-form, the A-form, and the Z-form. B-DNA, also called the Watson–Crick B-form, is the naturally occurring form of DNA. A-DNA is derived from the B-form by dehydration. The formation of Z-DNA occurs only when the DNA chains have specific base sequences.

The major structural features of the DNA molecule affected by conversion of one form to the other are (1) the conformation of the sugar, (2) the projection of the plane of the base around the glycosidic bond, and (3) the angle at which the bases are stacked in the helix in reference to the helical axis.

Sugar conformation.   The two major conformations of 2-deoxyribose are the 2′-*endo* conformer and the 3′-*endo* conformer. The 2′-*endo* conformer has carbon 2′ puckered toward the base whereas the 3′-*endo* conformer has the 3′ carbon puckered in that direction.

carbon 2'

carbon 3'

| HOCH₂—O—base | HOCH₂ ... base (2' endo) | HOCH₂ ... base (3' endo) |

Haworth projection   *2' endo*   *3' endo*

conformational formulas

The 2'-endo conformation of deoxyribose is found in B-DNA, while the 3'-endo conformer occurs in A-DNA. The conformation of a Z-DNA sugar depends on the base to which it is bonded. If the base is cytosine, the sugar is in the 2'-endo conformation. If the base is guanine, the sugar is in the 3'-endo conformation.

Glycosidic-bond conformers.   Rotation about the N-glycosidic bond allows the base to be oriented toward the sugar (*syn*) or away from it (*anti*).

syn   *anti*

In B-DNA and A-DNA, all bases are in the *anti* orientation. In Z-DNA, the pyrimidine bases are *anti* and the purine bases are *syn*.

Stacking angle.   In B-DNA, the planes of the bases are stacked nearly parallel to one another and at approximately right angles to the axis passing through the center of the helix. In A-DNA, the bases are tilted about 19° from the per-

a base pair   ~90°   9° tilt   19° tilt

helical axis

B–DNA   Z–DNA   A–DNA

pendicular to the helical axis. The tilt of the Z-DNA bases (9°) is intermediate between that in A-DNA and B-DNA.

Properties of the conformers.   **B-DNA**: The natural form of DNA is the B-form. This form is characterized by a polar exterior (sugar–phosphate) and a relatively nonpolar interior (bases). The external spiral of anionic phosphate groups requires neutralization with cations, commonly protonated proteins. The nonpolar bases exclude most water molecules from the interior of the DNA and, therefore, they form hydrogen-bonded pairs with maximum strength.

The B-DNA helix is characterized by a major groove and a minor groove that spiral along the surface of the helix. (See Figure 25.6, page 1013.) In spite of the nonregularity of bases in the sequence, the major groove is lined with carbon 6, carbon 8, and nitrogen 7 of the purines and carbons 4–6 of the pyrimidines. The narrower minor groove is paved with carbon 2 and nitrogen 3 of the purine bases and carbon 2 of the pyrimidine bases.

B-DNA is a right-handed helix. The number of bases per turn of the helix is 10–10.4 and the pitch (rise) per turn of the helix is 34 Å.

A-DNA is formed from B-DNA by dehydration. The double helix is compressed, resulting in a greater number of bases per turn (about 10.7–11). The bases are tilted, and the sugar is in the 3'-*endo* conformation. Like B-DNA, A-DNA is a right-handed helix but the major groove is wider and deeper and the minor groove is narrower and shallower than in B-DNA.

Z-DNA is a left-handed helix. However, it is not a left-handed version of either A-DNA or B-DNA. Z-DNA can be formed only by DNA containing alternating pyrimidine and purine bases. Z-DNA can be formed from synthetic polynucleotides or from segments of natural B-DNA containing the required base sequence. The formation of a Z-DNA segment occurs when base pairs in the chain are flipped over, as shown in Figure 25.10. Methylation of cytosine enhances the tendency of a segment of B-DNA to form a Z-DNA segment.

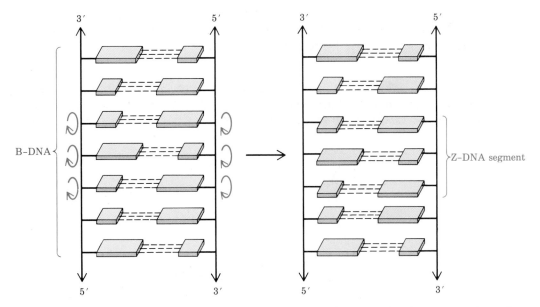

Figure 25.10 Diagram showing how bases can rotate in the conversion of a segment of B-DNA to Z-DNA.

F. Base Sequences in DNA

Exons and introns.   Not all bases in DNA are involved in the informational portion of the genetic code. For example, many of the genes in eukaryotic DNA (DNA from cells containing well-defined nuclei) are segmented. That is, the gene has portions that contain information, called **exons**, and portions that do not contain genetic information, called **introns**. The function of the introns is not yet known.

Palindromes.   Sequences of paired bases with the same order in both strands as we proceed from the 5′ direction to the 3′ end are called **palindromes**, a term describing a word or phrase that reads the same forward or backward (for example, radar).

A palindrome base sequence:

$$5' \longleftarrow \text{G—A—A—T—T—C} \longrightarrow 3'$$
$$3' \longleftarrow \text{C—T—T—A—A—G} \longrightarrow 5'$$

Restriction endonucleases.   Often called **restriction enzymes**, restriction endonucleases catalyze the hydrolysis, or cutting, of DNA at certain locations. These enzymes, found in bacteria, cut only specific base sequences, commonly palindrome sequences. The function of restriction enzymes is the cleavage of foreign DNA that enters the bacterial cell. Associated with these enzymes are *modifying enzymes* that modify the bacteria's own DNA (by methylation of certain bases, for example) to prevent its cleavage by the restriction enzyme.

More than 500 restriction enzymes have been isolated. They are used to cleave DNA at specific locations in genetic engineering and in gene-mapping of chromosomes. These enzymes are also used in the fingerprinting of DNA, a technique used in criminology laboratories, for example, to establish the individual source of body fluids (blood, semen) or to determine parentage.

Cleavage of a palindromic sequence by a restriction enzyme obtained from a strain of E. coli:

cleave here

$$5' \longleftarrow \text{G—A—A—T—T—C} \longrightarrow 3'$$
$$3' \longleftarrow \text{C—T—T—A—A—G} \longrightarrow 5'$$

and cleave here

G. Species Variations in DNA

Because the base sequence of DNA determines the types of protein in an organism, the DNA of each species of plant or animal has a different base sequence. DNA molecules in different types of organisms also differ in size and shape. In general, the size of a DNA molecule is proportional to the amount of information it contains. Viruses have the smallest DNA molecules (about 5000 base pairs), and humans have the largest (about 2.9×10^9 base pairs). A DNA molecule of *E. coli* contains about 4.3×10^6 base pairs, which contain the code for 2000–3000 different proteins.

In eukaryotic cells, the long-chain DNA double-stranded helix is found in the cell nucleus, where it is complexed with proteins to form the chromosomes. The nucleus in each cell of an organism contains DNA of the same structure. Therefore, each cell contains all the genetic information required for the entire organism! Exceptions are cells that produce antibodies. The DNA molecules in these cells contain variable regions so that these cells can "custom-tailor" antibodies to fight specific antigens (invaders). Other exceptions are the germ cells produced by meiosis for the purpose of sexual reproduction.

In prokaryotic cells (cells that lack true nuclei, such as those of bacteria and blue-green algae), DNA forms circular, compactly folded molecules in the cytoplasm (cellular fluid). Some bacteria contain small circular double-stranded DNA molecules called **plasmids** in addition to their main DNA molecules. The genetic information in plasmids is independent of that in the principal bacterial DNA and is instrumental in the development of bacterial resistance to antibiotics. Because of their small size and relative lack of complexity, plasmids are used in genetic research and engineering.

Viruses are small particles composed of either DNA or RNA (not both) protected by a protein coating. Viruses can be simple in structure because they use cells of higher life forms, called *host cells*, to reproduce viral nucleic acids and proteins.

Section 25.3
Structure and Properties of RNA

RNA and DNA are similar in primary structure, but differ in a few key features. (1) The sugar in RNA is ribose while in DNA it is deoxyribose. (2) RNA contains uracil in place of thymine. (3) RNA molecules are single stranded (while DNA molecules are double stranded) and are in the A-form (Section 25.2E). Finally, (4) RNA molecules are generally smaller than DNA. For example, human DNA contains about 46 million base pairs, and the largest RNA contains only about 50,000.

The three major classes of RNA found in cells are:

Messenger RNA (mRNA): As its name implies, mRNA carries genetic information from DNA to the site of protein synthesis, a portion of the cell containing the ribosomes.
Ribosomal RNA (rRNA): Together with the ribosomal proteins, rRNA serves as the structural material for the ribosome.
Transfer RNA (tRNA): Molecules of tRNA carry the amino acids to the ribosome for incorporation into a growing polypeptide chain.

In addition to these three classes, a number of other types of RNA exist. Within the cell, for example, small RNA molecules serve specialized functions, such as mRNA processing. Also, the genetic codes for a number of viruses are contained in RNA instead of DNA. For brevity, we will limit our discussion to the principal RNA classes.

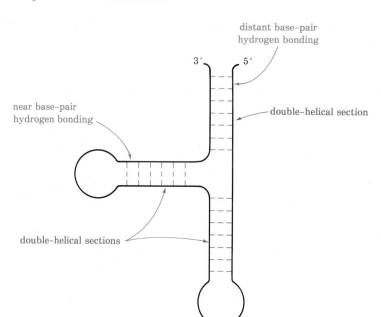

distant base–pair
hydrogen bonding

3′   5′

near base–pair
hydrogen bonding

double–helical section

double–helical sections

Figure 25.11 Typical structural features of RNA.

A. General Structure of RNA

The primary structure of RNA is similar to that of DNA—a series of sugar units (*ribose*, in this case) linked together by $5' \rightarrow 3'$ phosphodiester bonds. Each ribose residue is bonded to a base at the anomeric carbon. The principal RNA bases are *adenine, guanine, cytosine,* and *uracil* (replacing thymine). Uracil, like thymine, forms strong hydrogen bonds with adenine. These structural features were discussed in Section 25.1.

Although an RNA molecule is single stranded, it is neither linear nor a random coil. Its secondary structure is an A helix with base pairing between distant bases of the same strand. In addition, base pairing between near bases of the same strand gives hairpin-loop structures (Figure 25.11). The tertiary structure of RNA involves folding and base stacking.

Although the major classes of RNA are single stranded, double-stranded RNA molecules do occur in certain viruses. In these double-stranded molecules, the 2′-hydroxyl group of the ribose forces the chain into a conformation similar to that of the A-form of DNA.

B. Messenger RNA (mRNA)

Messenger RNA is only a small percentage of the total RNA in the cell. The reason is, in part, that mRNA molecules have a rather short lifetime of 1–3 minutes in prokaryotic cells and 2–24 hours in eukaryotic cells.

Cells synthesize various types of protein from the genetic code in DNA. Protein synthesis is accomplished in two stages. First, a portion of the DNA unwinds. In eukaryotic (nucleated) cells, the sequence of bases in one **gene**, the part of the DNA that codes for one type of protein, is used as a template for the synthesis of messenger RNA (mRNA). Thus, the mRNA is the complement of the portion of the DNA strand that synthesized it. This synthesis of mRNA is called **transcription**, a term referring to the transcription of the genetic code from DNA to RNA.

In mRNA, a particular series of three bases in a row, called a **codon**, specifies a particular amino acid. For example, C—C—U specifies proline. We will discuss codons in greater detail in Section 25.4A.

Because each type of RNA molecule carries information for one gene from DNA to the ribosomes, each mRNA contains a different base sequence. Consequently, mRNA is a heterogeneous group of compounds that is difficult to study.

In eukaryotic cells, the DNA base sequence for a single gene, consisting of both introns and exons (Section 25.2), is first transcribed from one portion of one DNA strand onto a nascent mRNA, called **hnRNA** (heterogeneous nuclear RNA), which is then modified, or processed, in the nucleus to mRNA. The first steps in processing involve removing the introns and splicing the exons together (Figure 25.12).

Before the mRNA leaves the nucleus, the 5′ end of its chain is capped by the substitution of a 7-methylguanylate group and the methylation of the 2′-hydroxyl groups of the first two ribose residues. The 3′ end of the RNA chain is tailed by the addition of up to 200 adenyl residues.

Capping of the 5′ terminus:

7-methylguanylate

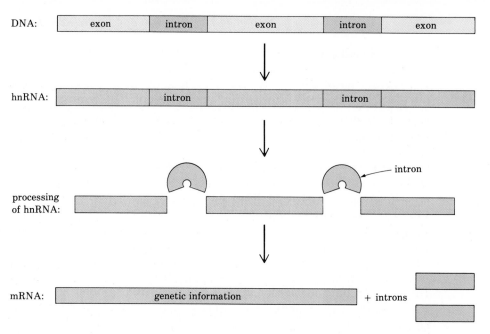

Figure 25.12 Diagram of the conversion of hnRNA to mRNA by the enzymatic removal of introns and the splicing of exons.

Figure 25.13 Representations of a typical tRNA molecule: (a) diagram showing cloverleaf structure; (b) diagram emphasizing the folded L shape; (c) three-dimensional representation.

C. Transfer RNA (tRNA)

The tRNA molecules comprise about 10% of the total RNA of a cell. There are 50–60 different types of tRNA in a cell, with at least one tRNA for each amino acid codon (three-base sequence in mRNA). For convenience, the different types of tRNA are designated by the amino acid they carry (tRNA^{ala} is tRNA that carries alanine).

All types of tRNA have similar chemical and physical properties. Each tRNA consists of a single RNA strand folded into a four-loop cloverleaf structure. The loops are labeled I through IV. Loop I is called the *dihydro-U loop* or D loop because it contains dihydrouridine (see Table 25.1). Loop II is called the *anticodon loop* because it contains a series of three bases (the **anticodon**) complementary to the codon in mRNA for its amino acid. Loop III is called the *variable loop* and contains 3–21 bases depending upon the tRNA. Loop IV is called the *pseudouridine* (Ψ) *loop* because it contains a pseudouridine base (Table 25.1).

In its tertiary structure, the tRNA molecule folds back upon itself to form an L-shaped structure with the amino acid at one end of the L and the anticodon at the other end of the L. This folding allows the bases at the 5′ and 3′ ends of the molecule to pair and gives rise to a region called the **acceptor stem** (see Figure 25.13). The last three bases at the 3′ end of all tRNA molecules are C–C–A; the amino acid is bonded to the 3′-hydroxyl group of the terminal ribose.

Bonding of the amino acid to the terminal adenosine nucleoside at the 3′ end of the chain:

amino acid ester at
3′ hydroxyl of ribose

D. Ribosomal RNA (rRNA) and the Ribosomes

The most abundant ribonucleoproteins (ribonucleic acid–protein complexes) in the cytoplasm are the ribosomes. They are composed of ribosomal RNA (rRNA) and numerous proteins.

Ribosomes are isolated by the centrifugation of ruptured cells. They are identified by **Svedberg (S) units**, with 1S equal to 10^{-13} second, a unit used in determining the rate of sedimentation of compounds in an ultracentrifuge, a high-speed analytical centrifuge.

Prokaryotic cells contain 70S ribosomes, while eukaryotic cells contain larger 80S ribosomes. The 70S ribosome of *E. coli* has been extensively studied. It can be dissociated into two subunits, the 50S and 30S subunits. (Because they represent a physical property, not a structural feature, the S units are not additive.) The 50S subunit contains 34 proteins and two RNA

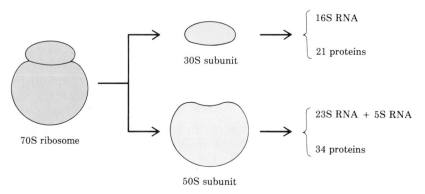

Figure 25.14 Dissociation of the prokaryotic 70S ribosome of *E. coli* into its components.

molecules, 23S and 5S RNA. The 23S RNA contains 2904 nucleotides, and the 5S RNA contains 120 nucleotides. The 30S subunit has 21 proteins and one RNA, 16S RNA, which contains 1542 nucleotides. The 70S ribosome thus contains a total of 3 RNA molecules and 55 protein residues (Figure 25.14).

Section 25.4
The Role of RNA in Protein Biosynthesis

After transcription and processing, the mRNA is used as a template for the synthesis of the protein. This process, called **translation**, refers to the translation of the nucleotide code into a sequence of amino acids in a peptide molecule. Translation encompasses the integrated actions of mRNA, many types of transfer RNA (which deliver appropriate amino acids to mRNA), ribosomes (particles that provide the sites of the synthesis), and many enzymes.

replication

$$\text{DNA} \xrightarrow{\text{transcription}} \text{mRNA} \xrightarrow{\text{translation}} \text{peptide or protein}$$

tRNA

nucleotides   amino acids

A. Reading the Codons

When mRNA leaves the nucleus of a eukaryotic cell, ribosomes, which are also partly formed in the nucleus, become attached to the 5′ end of its chain. The ribosomes are the actual sites of protein synthesis.

Once the mRNA molecule is bonded to ribosomes, the bases are read *three at a time* as the ribosomes progress along the chain toward the 3′ end. Each set of three bases, a *codon*, signals the incorporation of a single amino acid into the growing peptide chain. The codons for the amino acids, which are the same

Table 25.3 Codons for the amino acids

Amino Acid	*Codons*[a]
phenylalanine (phe)	UUU, UUC
serine (ser)	UCU, UCC, UCA, UCG, AGU, AGC
tyrosine (tyr)	UAU, UAC
cysteine (cys)	UGU, UGC
tryptophan (trp)	UGG
leucine (leu)	CUU, CUC, CUA, CUG, UUA, UUG
proline (pro)	CCU, CCC, CCA, CCG
histidine (his)	CAU, CAC
glutamine (gln)	CAA, CAG
arginine (arg)	CGU, CGC, CGA, CGG, AGA, AGG
lysine (lys)	AAA, AAG
asparagine (asn)	AAU, AAC
isoleucine (ile)	AUU, AUC, AUA
methionine (met) or *N*-formylmethionine (fmet)	AUG
threonine (thr)	ACU, ACC, ACA, ACG
valine (val)	GUU, GUC, GUA, GUG
alanine (ala)	GCU, GCC, GCA, GCG
aspartic acid (asp)	GAU, GAC
glutamic acid (glu)	GAA, GAG
glycine (gly)	GGU, GGC, GGA, GGG

[a] When preceded by an initiator region, the codon AUG signals "Start a new peptide molecule beginning with *N*-formylmethionine, or fmet." The codons UAA, UAG, and UGA signal termination of the synthesis.

in all known life forms, have been determined experimentally. They are listed in Table 25.3.

Bases are read three at a time;
each codon (set of three) designates one amino acid.

$5' \longleftarrow \{ \text{—C—C—U—} \, \text{—C—U—C—} \, \text{—G—C—U—} \} \longrightarrow 3'$

a segment of mRNA

codon for	*codon for*	*codon for*
proline	*leucine*	*alanine*

In Table 25.3, note that more than one codon can signal a particular amino acid to be incorporated into a protein. In addition, some codons serve special functions. For example, the codon AUG serves two functions: (1) as an initiator codon signaling for the start of synthesis of a peptide, and (2) as a codon for the incorporation of methionine into the growing peptide chain. Other special-purpose codons are UAA, UAG, and UGA, all of which signal STOP. When the ribosomal synthesis site encounters one of these stop codons, the peptide chain is released from the ribosome and assumes its secondary and tertiary structures.

B. Anticodon of tRNA

The amino acids are carried to the mRNA–ribosome by tRNA molecules. One key feature of the tRNA structure is its anticodon, a series of three bases that is complementary to a particular codon. For example, if mRNA contains the codon GCC (for the amino acid alanine), the tRNA that carries alanine would contain the anticodon CGG. The anticodon allows a tRNA to recognize the correct location on the mRNA molecule.

C. Steps in Protein Biosynthesis

Figure 25.15 diagrams the overall process of protein biosynthesis. The three principal steps in this process are *initiation, chain elongation,* and *termination.*

Initiation. In the presence of certain protein initiator factors, mRNA becomes bound to a ribosome by a leader sequence near the 5′ end of the mRNA. In bacterial cells, the START codon (AUG) after the leader sequence calls for

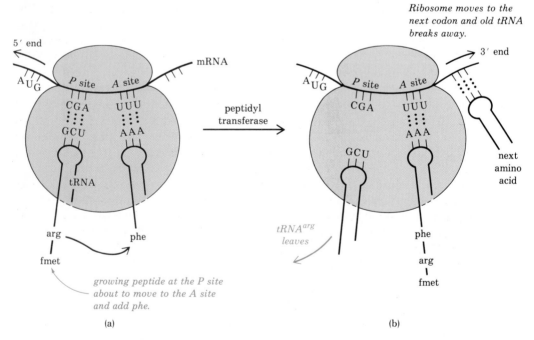

Figure 25.15 Biosynthesis of a peptide. (a) *fMet,* or *N*-formylmethionine (the initiating amino acid, with the codon AUG), has become bonded to *arg* (codon CGA). The next amino acid to be added will be *phe* (codon UUU). (b) The growing peptide chain has been transferred to phe. The ribosome moves along the mRNA chain (to the right) so that the peptide chain can be transferred to the next amino acid.

N-formylmethionine (fmet) as the first amino acid of the peptide to be synthesized. Depending on the protein's structure, the formyl group or even the methionine residue may be hydrolyzed from the final protein. However, most proteins in a bacterium like *E. coli* have methionine as their *N*-terminal amino acid.

formyl group

N-formylmethionine

The codon AUG for "start with *N*-formylmethionine" is the same as the codon for the incorporation of methionine itself in the interior of the peptide chain. However, the tRNA that brings the *N*-formylmethionine to start a peptide chain differs from the tRNA that brings methionine itself. Initiation factors on the mRNA chain also help differentiate *N*-formylmethionine from methionine.

Chain elongation.   (1) The tRNA holding the second amino acid becomes hydrogen bonded to the mRNA codon and positions its amino acid at the **amino acid binding site (acyl site)**, or **A site**, on the ribosome. (2) A peptide bond is formed between the *N*-formylmethionine and the incoming amino acid. (3) The dipeptide is translocated to the **peptide binding site (peptidyl site)**, or **P site**, of the ribosome. (4) The ribosome moves along the mRNA chain so that the next tRNA with its amino acid can move in to the A site, and the process is repeated for the next amino acid.

Termination.   The peptide is complete when the ribosome encounters one of the STOP codons. These codons terminate the sequence because no tRNA molecules correspond to them. Protein-release factors catalyze the hydrolysis of the peptide so that it and the ribosome are released from each other and from the mRNA.

Although one mRNA directs the synthesis of a specific protein, it can direct the synthesis of many molecules of this protein. As one ribosome, holding a growing peptide, moves along the mRNA chain, other ribosomes become successively attached to the 5′ end of the mRNA to follow along behind the first tRNA. Thus, several peptide molecules can be synthesized, one after the other, in assembly-line fashion.

▼ Summary

Deoxyribonucleic acids (DNA), which are the carriers of the genetic code, and **ribonucleic acids (RNA)**, which implement the code, are polymers. The backbones consist of sugar molecules (deoxyribose in DNA and ribose in RNA) linked together by phosphate units. Each sugar unit is also bonded to a heterocyclic base. Nucleic acids can be hydrolyzed to **nucleotides** (phosphate–sugar–base) and to **nucleosides** (sugar–base).

The bases in DNA are adenine (A), guanine (G), thymine (T), and cytosine (C). In RNA, uracil (U) replaces thymine.

Hydrogen bonding between specific base pairs (G and C, T and A) and stacking energy hold DNA in an antiparallel double-stranded helix (Watson–Crick model). DNA can undergo **replication** because the base pairs are specific. The three principal conformers of DNA are the **B-form**, the **A-form**, and the **Z-form**. The B-form of DNA is the naturally occurring form.

The portion of a DNA strand specific for the coding of the amino acid sequence of one protein is called a **gene**. This portion of DNA synthesizes a complementary **mRNA (transcription)**, which leaves the nucleus to direct the synthesis of that protein molecule (**translation**) with the aid of **ribosomes**.

The bases of mRNA are read three at a time. Each three-base sequence (a **codon**) designates a particular amino acid to be incorporated into a growing peptide chain. The **tRNA** molecules carry amino acids to the site of protein synthesis and recognize their proper locations because of **anticodons**.

Study Problems*

25.5 Draw formulas for **(a)** 2-deoxy-β-D-ribose 3-phosphate and **(b)** 2-deoxy-β-D-ribose 5-phosphate.

25.6 Draw the formula for a tautomer of uracil that shows its close relationship to pyrimidine.

25.7 Using uracil, β-D-ribose, and a phosphate group, draw formulas for **(a)** a nucleoside and **(b)** a nucleotide.

25.8 Write equations for the complete alkaline hydrolysis of 2-deoxycytidine 5′-phosphate.

25.9 Adenosine 5′-triphosphate (ATP) is important in many biological schemes. What is its structure?

25.10 Why are guanine and cytosine paired in the DNA helix, but not adenine and cytosine? Use formulas to show the hydrogen bonding.

25.11 Write formulas showing the hydrogen bonds between nucleotides containing the following pairs:

(a) uracil and adenine
(b) uracil and guanine
(c) uracil and cytosine

25.12 In the preceding problem, state which pairs of bases are paired in nucleic acids. Explain why.

25.13 Why does the melting point (denaturation temperature) of DNA increase with increasing guanosine–cytosine content?

---

* For information concerning the organization of the Study Problems and Additional Problems, see the Note to student on page 39.

25.14 List the following compounds in order of increasing molecular weight (lowest first).

(a) mRNA   (b) tRNA   (c) DNA

25.15 Define the following terms:

(a) replication   (b) transcription   (c) translation

25.16 (a) A portion of a DNA molecule has the following base sequence. What is the base sequence of the complementary strand?

5′ end ← A T T C G G T A T → 3′ end

(b) What is the base sequence in an RNA molecule synthesized by the DNA shown above?

25.17 Three different codons of mRNA result in the incorporation of isoleucine in a protein: AUU, AUC, and AUA.

(a) What sequences of bases in DNA give rise to these codons?
(b) What are the corresponding anticodons in tRNA?

Additional Problems

25.18 (a) *Theophylline*, which occurs in tea and is also synthesized industrially, is a cardiac stimulant, diuretic (urine-secretion stimulant), and muscle relaxant. The structure of theophylline is a purine ring system with carbonyl groups at positions 2 and 6 and methyl groups at nitrogens 1 and 3. The protonated nitrogen is at position 7. What is the structure of theophylline?

(b) *Caffeine* has the same structure as theophylline with an additional *N*-methyl group at position 7. What is its structure?

25.19 What tripeptides would be formed by the sequence of codons in (a) and by the mutated sequence in (b)?

(a) mRNA: 5′ ← UUU—UAU—AGU → 3′
(b) mRNA: 5′ ← UUU—UAC—AGU → 3′

25.20 In sickle-cell anemia, valine replaces a glutamic acid unit of normal hemoglobin. Referring to Table 25.3, postulate the change occurring in the genetic code for hemoglobin that causes sickle-cell anemia.

25.21 Write letters for all possible base sequences in mRNA that would produce the dipeptide glu–arg.

▼ **Appendix**

Nomenclature of Organic Compounds

A complete discussion of definitive rules of organic nomenclature would require more space than can be allotted in this text. We will survey some of the more common nomenclature rules, both IUPAC and trivial.* The following references contain more detail:

IUPAC Nomenclature of Organic Chemistry, Sections A, B, & C, 2nd Ed., Butterworths, London, 1971. Now available from Pergamon Press, Elmsford, New York. [Also may be found in *Pure Appl. Chem. II* (1–2) (1965).]

Chemical Abstracts Service, *Naming and Indexing of Chemical Substances for* Chemical Abstracts *during the Ninth Collective Period (1972–1976) (January–June, 1972)*, American Chemical Society, Columbus, Ohio, 1973.

---

* *Chemical Abstracts* has its own nomenclature system, which we will not cover here. To find a *Chemical Abstracts* index name, refer to one of the references listed or consult a recent *Index Guide.*

Chemical Abstracts Service, *Combined Introductions to the Indexes to Volume 66 (January–June, 1967)*, American Chemical Society, Columbus, Ohio, 1968.

A. M. Patterson, L. T. Capell, and D. F. Walker, *The Ring Index, 2nd Ed.*, American Chemical Society, Washington, D.C., 1960; *Supplement I (1957–1959)*, 1963; *Supplement II (1960–1961)*, 1964; *Supplement III (1962–1963)*, 1965.

R. S. Cahn and O. C. Dermer, *Introduction to Chemical Nomenclature, 5th Ed.*, Butterworths, London, 1979.

Alkanes

The names for the first thirty continuous-chain alkanes are listed in Table A1.

Branched alkanes. In the name of an alkane with alkyl substituents, the longest continuous chain is considered the parent. The parent is numbered from one end to the other, the direction being chosen to give the lower numbers to the substituents. (When considering complex cases in which a question about numbering arises, the lower individual numbers up to the first point of difference are the correct numbers. The numbers with the lower sum may *not* be the correct ones. Therefore, the prefix numbers 1,7,7- would be preferred to 2,2,8-.) The entire name of the structure is composed of (1) the numbers of the positions of the substituents; (2) the names of the substituents; and (3) the name of the parent.

Alkyl substituents. The names of the alkyl substituents (also called *branches*, or *radicals*) are derived from the names of their corresponding alkanes with the ending changed from *-ane* to *-yl*. For example, CH_3CH_2— is ethyl (from ethane). Multiple substituents are placed in alphabetical order,

$$CH_2CH_3$$
$$|$$
$$CH_3CH_2CHCH_2CH_2CH_3$$
① ② ③ ④ ⑤ ⑥

3-ethylhexane

$$CH_3 \quad CH_2CH_3$$
$$| \qquad |$$
$$CH_3CHCH_2CHCH_2CHCH_2CH_3$$
① ② ③ ④ ⑤ ⑥ ⑦ ⑧
$$|$$
$$CH_3$$

4-ethyl-2,6-dimethyloctane

Table A1 Names of some continuous-chain alkanes

Molecular Formula	Name	Molecular Formula	Name
CH_4	methane	$C_{16}H_{34}$	hexadecane
C_2H_6	ethane	$C_{17}H_{36}$	heptadecane
C_3H_8	propane	$C_{18}H_{38}$	octadecane
C_4H_{10}	butane	$C_{19}H_{40}$	nonadecane
C_5H_{12}	pentane	$C_{20}H_{42}$	eicosane
C_6H_{14}	hexane	$C_{21}H_{44}$	heneicosane
C_7H_{16}	heptane	$C_{22}H_{46}$	docosane
C_8H_{18}	octane	$C_{23}H_{48}$	tricosane
C_9H_{20}	nonane	$C_{24}H_{50}$	tetracosane
$C_{10}H_{22}$	decane	$C_{25}H_{52}$	pentacosane
$C_{11}H_{24}$	undecane	$C_{26}H_{54}$	hexacosane
$C_{12}H_{26}$	dodecane	$C_{27}H_{56}$	heptacosane
$C_{13}H_{28}$	tridecane	$C_{28}H_{58}$	octacosane
$C_{14}H_{30}$	tetradecane	$C_{29}H_{60}$	nonacosane
$C_{15}H_{32}$	pentadecane	$C_{30}H_{62}$	triacontane

Table A2 Trivial names for some common alkyl groups

Structure	Name
$CH_3CH_2CH_2—$	*normal-*, or *n*-propyl[a]
$(CH_3)_2CH—$	isopropyl
$CH_3CH_2CH_2CH_2—$	*n*-butyl[a]
$(CH_3)_2CHCH_2—$	isobutyl
$CH_3CH_2CH(CH_3)—$	*secondary-*, or *sec*-butyl
$(CH_3)_3C—$	*tertiary-*, *tert-*, or *t*-butyl
$CH_3CH_2CH_2CH_2CH_2—$	*n*-pentyl (or *n*-amyl)[a]
$(CH_3)_2CHCH_2CH_2—$	isopentyl (or isoamyl)
$(CH_3)_3CCH_2—$	neopentyl

[a] The use of *n*- (to denote a continuous chain) is optional.

each preceded by its number and like substituents grouped together. Some common branched alkyl substituents have trivial names (see Table A2).

Alkenes and Alkynes

Unbranched hydrocarbons having one double bond are named in the IUPAC system by replacing the ending *-ane* of the alkane name with *-ene*. With two or more double bonds, the ending is *-adiene*, *-atriene*, etc. The chain is numbered to give the lowest numbers to the double bonds. (The *lower* number of the two carbons joined by the double bond is used to give the position.)

$$\overset{①}{C}H_3\overset{②}{C}H=\overset{③}{C}H\overset{④}{C}H_2\overset{⑤}{C}H_3 \qquad \overset{①}{C}H_2=\overset{②}{C}H\overset{③}{C}H=\overset{④}{C}H\overset{⑤}{C}H_3$$

2-pentene   1,3-pentadiene

Unbranched hydrocarbons having one triple bond are named by replacing the ending *-ane* of the alkane name with *-yne*. If there are two or more triple bonds, the ending is *-adiyne, -atriyne*, etc. The chain is numbered to give the lowest numbers to the triple bonds. Again, the lower number is used to give the position. For example, $CH_3CH_2C≡CH$ is 1-butyne.

Unbranched hydrocarbons with both double and triple bonds are named by replacing the ending *-ane* of the alkane with the ending *-enyne*. When necessary, position numbers are used. The position numbers are chosen to be as low as possible. For example, $CH≡CCH=CHCH_3$ is 3-penten-1-yne, not 2-penten-4-yne. If there is a choice, the double bond receives the lower number. For example, $CH≡CCH=CH_2$ is 1-buten-3-yne, not 3-buten-1-yne. Trivial names for some alkenes and alkynes are listed in Table A3.

Table A3 Trivial names for some alkenes and alkynes

Structure	Name
$CH_2=CH_2$	ethylene
$CH≡CH$	acetylene
$CH_2=C=CH_2$	allene
$CH_2=CHCH_3$	propylene
$CH_3C≡CH$	methylacetylene
$(CH_3)_2C=CH_2$	isobutylene
$CH_2=C(CH_3)CH=CH_2$	isoprene

Branched alkenes and alkynes.   In the IUPAC name of a branched alkene or alkyne, the parent chain is the longest chain that contains the maximum number of double or triple bonds. (This may or may not be the longest continuous chain in the structure.)

$$CH_3CH_2CH_2$$
$$CH_2=CHC=CCH=CHCH_3$$
$$CH_2CH_2CH_3$$

3,4-dipropyl-1,3,5-heptatriene

Several unsaturated substituents have trivial names; some of these are listed in Table A4, along with their systematic names.

In complex compounds, the symbol Δ is sometimes used to denote double bonds.

is $\Delta^{4,6}$-cholestadiene

Geometric isomers.   There are two methods for naming geometric isomers. In one, the prefixes *cis-* (same side) and *trans-* (opposite sides) are used.

cis-2-butene   *trans*-2-butene

The other method uses (E) (groups with higher priority, according to the Cahn–Ingold–Prelog system, on opposite sides) or (Z) (groups with higher priority on the same side) for designation of geometric isomers. The priority rules are listed in Section 4.1B (page 119).

Table A4 IUPAC names for some unsaturated groups

Structure	Name
$CH_2=$	methylene
$CH_2=CH-$	ethenyl (vinyl)[a]
$CH_3CH=$	ethylidene
$CH\equiv C-$	ethynyl
$CH_2=CHCH_2-$	2-propenyl (allyl)[a]
$CH_3CH=CH-$	1-propenyl
$(CH_3)_2C=$	isopropylidene
	2-cyclopenten-1-yl

[a] Names in parentheses are trivial.

(Z)-3-methyl-2-pentene   (E)-3-methyl-2-pentene

ethyl has priority over methyl; methyl has priority over H

Cyclic Hydrocarbons

Cycloalkanes and cycloalkenes. The names of saturated monocyclic hydro-carbons are formed by prefixing *cyclo-* to the name of the alkane with the same number of carbon atoms.

is cyclohexane

A ring is considered the parent unless there is a longer chain attached.

propylcyclohexane   1-cyclohexylheptane

Alkyl substituents are named as prefixes and are given the lowest possible position numbers.

  cis-1,2-dimethylcyclopentane

Unsaturated monocyclic hydrocarbons are named by changing the ending from *-ane* to *-ene* (*-adiene*, etc.). The ring is numbered to give the lowest numbers possible to the double bonds. A ring with alkyl substituents and double bonds is numbered so that the double bonds receive the lowest possible numbers.

1,3-cyclohexadiene   5-methyl-1,3-cyclohexadiene

Some common terpene ring structures have individual names.

menthane   pinane   bornane   norbornane

Table A5 Names for some arenes and aryl groups

Structure	Name	Structure	Name
Arenes:		**Aryl Groups:**	
benzene ring	benzene	C_6H_5-	phenyl-
naphthalene (numbered 1–8)	naphthalene	$C_6H_5CH_2-$	benzyl-
anthracene (numbered 1–10)	anthracene	$CH_3-\bigcirc-$	*p*-tolyl-
phenanthrene (numbered 1–10)	phenanthrene	1-naphthyl structure	1-naphthyl- (*α*-naphthyl-)
$\bigcirc-CH_3$	toluene	$C_6H_5CH{=}CHCH_2-$	cinnamyl-
$\bigcirc-CH(CH_3)_2$	cumene		
$\bigcirc-CH{=}CH_2$	styrene		
benzene with two CH_3 (ortho)	*o*-xylene		
benzene with three CH_3	mesitylene		

Aromatic hydrocarbons.   Many aromatic hydrocarbons are referred to by their trivial names (see Table A5). Systems composed of five or more linear fused benzene rings are named by a Greek-number prefix followed by *-cene*. (The prefix denotes the number of fused rings.)

pentacene structure

pentacene

The aromatic system is the parent unless a longer chain is attached.

$$C_6H_5CH_2CH_2CH_3 \qquad C_6H_5CH_2(CH_2)_5CH_3$$

propylbenzene   1-phenylheptane

The positions of two substituents on a benzene ring are designated either by prefix numbers or by o-, m-, or p- (*ortho-*, *meta-*, or *para-*). For more than two substituents, numbers must be used.

o-dibromobenzene   m-dibromobenzene   p-dibromobenzene   1,2,3-triethylbenzene

The principal group, or a group that is part of the parent (for example, the CH_3 in toluene), is always considered to be bonded to position 1 on the ring. The numbers of substituents are chosen to be as low as possible.

m-chlorophenol   p-nitrotoluene   3,5-dimethylstyrene

Heterocycles

Some common heterocycles are listed in Table A6. In monocyclic heterocycles with only one heteroatom, that atom is considered position 1. Other ring systems are numbered by convention (see Table A6).

To name monocyclic compounds with one or more heteroatoms, prefixes may be used: *oxa-* (—O—), *aza-* (—NH—), *thia-* (—S—). For unsaturated rings, the ring size is designated by suffixes: *-ole* means five, and *-ine* means six (used only with rings that contain nitrogen). *Example*: An *oxazole* is a five-membered ring containing O and N. In numbering, O has priority over N.

1,3-oxazole   1,3,5-triazine

Heteroatoms in Chains

The *oxa-, aza-* system may also be used for naming aliphatic compounds. This method is called *replacement nomenclature*. All atoms in the chain are numbered so that the heteroatoms receive the lowest possible numbers. The parent is the

Table A6 Names of some common heterocycles

Structure	Name	Structure	Name
	furan		pyrimidine
	4H-pyran (γ-pyran)		morpholine
	pyrrole		thiophene
	pyrazole		indole
	imidazole		carbazole
	pyridine		purine
	piperidine		quinoline
	piperazine		isoquinoline

alkane that has the same number of carbon atoms as the total number of atoms in the continuous chain (heteroatoms, but not hydrogens, included).

$$\overset{①}{CH_3}\overset{②③}{OCH_2CH_2}\overset{④}{O}\overset{⑤⑥}{CH_2CH_2}\overset{⑦}{O}\overset{⑧⑨}{CH_2CH_2}\overset{⑩}{CH_3}$$

2,5,8-trioxadecane

Putting Names Together

The prefixes.   In the IUPAC system, alkyl and aryl substituents and many functional groups are named as prefixes to the parent name (for example,

Table A7 Some common functional groups named as prefixes

Structure	Name	Structure	Name
—OR	alkoxy-[a]	—F	fluoro-
—NH$_2$	amino-	—H	hydro-[b]
—N=N—	azo-	—I	iodo-
—Br	bromo-	—NO$_2$	nitro-
—Cl	chloro-	—NO	nitroso-

[a] *methoxy-, ethoxy-*, etc., depending upon R.
[b] *Hydro-* is a prefix used to designate a hydrogenated derivative of an unsaturated parent. *Perhydro-* means completely hydrogenated.

4a,8a-dihydronaphthalene   perhydrophenanthrene

iodomethane). Some common functional groups named as prefixes are listed in Table A7. Other prefix names are sometimes used for carbonyl groups, hydroxyl groups, etc. These are mentioned under their specific headings.

Like treatment of like things.   Names should be as simple as possible and as consistent as possible. Two identical substituents should be treated alike, even though a few other rules may be broken. Although C$_6$H$_5$CH$_3$ would be methylbenzene or toluene (rather than phenylmethane), (C$_6$H$_5$)$_2$CH$_2$ is called diphenylmethane.

Prefixes to designate like things.   In simple compounds, the prefixes *di-*, *tri-*, *tetra-*, *penta-*, *hexa-*, etc. (not italicized) are used to indicate the number of times a substituent is found in the structure: e.g., dimethylamine for (CH$_3$)$_2$NH or dichloromethane for CH$_2$Cl$_2$.

In complex structures, the prefixes *bis-*, *tris-*, and *tetrakis-* (not italicized) are used: *bis-* means two of a kind; *tris-*, three of a kind; and *tetrakis-*, four of a kind.

[(CH$_3$)$_2$N]$_2^-$ is bis(dimethylamino)- and not di(dimethylamino)-

The prefix *bi-* is used for (1) "double" molecules, and (2) bridged hydrocarbons.

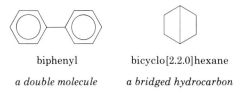

biphenyl   bicyclo[2.2.0]hexane

a double molecule   *a bridged hydrocarbon*

Order of prefixes.   Prefixes are listed in *alphabetical order* (ethylmethyl-). In alphabetizing, a prefix denoting the number of times a substituent is found (di-, tri-, etc.) is disregarded. Ethyldimethyl- is the correct alphabetical order, even though *d* comes before *e*.

When to drop a vowel. In *conjunctive names* (names that are formed by combining two names), vowels are not elided, but are maintained. For example, the *e* in indoleacetic acid is retained (not indolacetic acid).

If there are two successive suffixes, the vowel at the end of the first suffix is dropped (propenoic acid, not propeneoic acid), unless it is followed by a consonant (propanediol, not propandiol).

Nomenclature priority of functional groups. The various functional groups are ranked in priority as to which receives the suffix name and the lowest position number. A list of these priorities is given in Table A8. These are *not* the same priorities that are used for (E) and (Z) or (R) and (S).

$$\underset{\underset{NH_2}{|}}{CH_3\overset{\overset{O}{\|}}{C}CHCH_3} \qquad \text{is 3-amino-2-butanone (not 2-amino-3-butanone)}$$

The principal functional group is indicated at the end of the name. A name with only one functional ending is preferred. *Example*: $HOCH_2CH_2CO_2H$ is 3-hydroxypropanoic acid. In names that must have two endings, the terminal ending refers to the principal functional group (see Table A8). *Example*: $CH_3CH{=}CHCO_2H$ is 2-butenoic acid. An exception to this rule is found in the naming of -enynes.

Numbering of the parent. The parent is numbered so that the principal function receives the lowest number (see preceding section). Greek letters are reserved for trivial names and may not correspond directly to the IUPAC

Table A8 Nomenclature priority[a]

Structure	*Name*
—N(CH$_3$)$_3$$^+$ (as one example)	onium ion
—CO$_2$H	carboxylic acid
—SO$_3$H	sulfonic acid
—CO$_2$R	ester
—COX	acid halide
—CONR$_2$	amide
—CN	nitrile
—CHO	aldehyde
—CO—	ketone
ROH	alcohol
ArOH	phenol
—SH	thiol
—NR$_2$	amine
—O—O—	peroxide
—MgX (as one example)	organometallic
$\diagup C{=}C\diagdown$	alkene
—C≡C—	alkyne
R—, X—, etc.	other substituents

[a] Highest nomenclature priority is at the top.

numbers. *Alpha* (α, the first letter of the Greek alphabet) means *on the nearest carbon*, which is often number 2 in the systematic name.

$$\overset{\displaystyle Br}{\underset{}{|}}$$
$$CH_3CHCO_2H$$

2-bromopropanoic acid (IUPAC)
α-bromopropionic acid (trivial)

Numbering of substituents and use of parentheses. If two numbering systems are required for identification of all atoms in the molecule, primes (′) are often used for one of the systems to prevent confusion.

Alkyl substituents are numbered separately from the parent chain, beginning at the point of attachment. In these cases, a prime is not necessary provided that parentheses are used to enclose complex prefixes.

p-(2-bromoethyl)aniline

2-(2-chloropropyl)-1,3,5-hexatriene

Configuration around a chiral carbon. Four different substituents around a chiral carbon can have either an (*R*) or (*S*) configuration. For designation of the configuration as (*R*) or (*S*), the structure is placed with the lowest-priority group in the rear. (See Section 4.1B for determination of priority, which is *not the same* as nomenclature priority.) The direction from the highest-priority group to the second-highest-priority group is then determined. If the direction is *clockwise*, the chiral carbon is (*R*); if it is *counterclockwise*, then the carbon is (*S*). If the structure has only one chiral carbon, (*R*) or (*S*) is used as the first prefix in the name. If the molecule has more than one chiral carbon, the designation of each chiral carbon and its position number is enclosed in parentheses in the prefix—e.g., (2*R*,3*R*)-dibromopentane.

(*R*)-2-butanol

(*S*)-2-butanol

Carboxylic Acids

Four principal types of names for carboxylic acids are: (1) IUPAC; (2) trivial; (3) carboxylic acid; and (4) conjunctive. In addition, it is sometimes necessary to name a carboxyl group with a prefix.

IUPAC names.   Except for acids of one to five carbons and some fatty acids, aliphatic monocarboxylic acids are named by the IUPAC system. The longest chain containing the $-CO_2H$ group is chosen as the parent, and the chain is numbered starting with the carbon of the $-CO_2H$ group as position 1. The name is taken from the name of the alkane with the same number of carbons, with the final -*e* replaced by -*oic acid*.

$$CH_3CH_2CH_2CH_2CH_2CO_2H$$

hexanoic acid

Substituents are designated by prefixes. A double bond is designated as a suffix preceding -oic acid.

$$\overset{\displaystyle CH_3}{\underset{\displaystyle |}{CH_2=CHCH_2CHCH_2CO_2H}}$$

3-methyl-5-hexenoic acid

Trivial names.   Table A9 gives a list of commonly encountered trivial names for carboxylic acids.

Carboxylic acid names.   A carboxylic acid name is used when a $-CO_2H$ group is attached to a ring. The name is a combination of the name of the ring system with the suffix -*carboxylic acid*. The carboxyl group is considered bonded to position 1 of the ring unless the ring system has its own unique numbering system. (The carbon of the $-CO_2H$ group is not numbered as it is in an IUPAC name.)

cyclohexanecarboxylic acid   2-pyridinecarboxylic acid

Conjunctive names.   Conjunctive names are combinations of two names: in the following examples, the name of the ring plus the name of the acid.

cyclohexaneacetic acid   indole-2-acetic acid

Table A9 Trivial names of some monocarboxylic acids

Structure	Name of Acid	Structure	Name of Acid
Saturated Chain:		**Other Functionality[b]:**	
HCO_2H	formic	$CH_3COCH_2CO_2H$	acetoacetic
CH_3CO_2H	acetic	$CH_3CH(OH)CO_2H$	lactic
$CH_3CH_2CO_2H$	propionic	CH_3COCO_2H	pyruvic
$CH_3(CH_2)_2CO_2H$	butyric	$CH_3COCH_2CH_2CO_2H$	levulinic
$CH_3(CH_2)_3CO_2H$	valeric		
$CH_3(CH_2)_{10}CO_2H$	lauric	**On Rings:**	
$CH_3(CH_2)_{12}CO_2H$	myristic	$C_6H_5CO_2H$	benzoic
$CH_3(CH_2)_{14}CO_2H$	palmitic		
$CH_3(CH_2)_{16}CO_2H$	stearic		

For the "On Rings" entries with structures:

salicylic (o-hydroxybenzoic acid structure, OH and CO₂H on benzene ring)

2-naphthoic (naphthalene with CO₂H at position 2)

nicotinic (pyridine with CO₂H at position 3)

Structure	Name of Acid
Unsaturated Chain[a]:	
$CH_2{=}CHCO_2H$	acrylic
$CH_2{=}C(CH_3)CO_2H$	methacrylic
$trans\text{-}CH_3CH{=}CHCO_2H$	crotonic

[a] For unsaturated fatty acids, see Table 23.1, page 939.
[b] For α-amino acids, see Table 24.1, page 961.

Diacids and polycarboxylic acids. Diacids may be named systematically as *-dioic acids.* For example, $HO_2CCH_2CO_2H$ is propanedioic acid. Trivial names are commonly used. Some of these are listed in Table A10.

Aliphatic polycarboxylic acids containing more than two carboxyl groups are named by the carboxylic-acid system. The longest chain with the greatest number of carboxyl groups is chosen as the parent. If there is unsaturation, the double or triple bonds are included in the chain if possible.

$$\overset{③}{HO_2C}CH_2\overset{②}{CH}CH_2\overset{①}{CO_2H} \quad \text{is 1,2,3-propanetricarboxylic acid}$$
$$|$$
$$CO_2H$$

Prefix names. The following example shows the use of the *carboxy-* prefix in naming a compound impossible to name as a trioic acid.

$$HO_2CCH_2CHCH_2CO_2H$$
$$|$$
$$CH_2CO_2H$$

3-(carboxymethyl)pentanedioic acid

Sulfonic acids. Sulfonic acids are named by adding the ending *-sulfonic acid* to the name of the rest of the structure.

$$CH_3CH_2SO_3H \qquad C_6H_5SO_3H$$

ethanesulfonic acid   benzenesulfonic acid

Table A10 Trivial names of some diacids

Structure	Name of Acid
Aliphatic:	
HO_2CCO_2H	oxalic
$HO_2CCH_2CO_2H$	malonic
$HO_2C(CH_2)_2CO_2H$	succinic
$HO_2C(CH_2)_3CO_2H$	glutaric
$HO_2C(CH_2)_4CO_2H$	adipic
$HO_2C(CH_2)_5CO_2H$	pimelic
$HO_2C(CH_2)_6CO_2H$	suberic
$HO_2C(CH_2)_7CO_2H$	azelaic
$HO_2C(CH_2)_8CO_2H$	sebacic
cis-$HO_2CCH{=}CHCO_2H$	maleic
$trans$-$HO_2CCH{=}CHCO_2H$	fumaric
$HO_2CCH(OH)CH(OH)CO_2H$	tartaric

Aromatic:

phthalic

isophthalic

terephthalic

Acid Anhydrides

Acid anhydrides are named from the names of the component acid or acids with the word *acid* dropped and the word *anhydride* added. (Either IUPAC or trivial acid names may be used.)

benzoic anhydride

acetic propionic anhydride

Cyclic anhydrides are named from the parent diacid.

maleic acid   maleic anhydride

Acid Halides

Acid halides are named by changing the ending of the carboxylic acid name from *-ic acid* to *-yl* plus the name of the halide. An ending of *-yl halide* in the name of a *diacid* implies that both carboxyl groups are acid-halide groups.

$$\underset{\text{acetyl chloride}}{CH_3\overset{\displaystyle O}{\overset{\|}{C}}Cl} \qquad \underset{\text{benzoyl bromide}}{C_6H_5\overset{\displaystyle O}{\overset{\|}{C}}Br} \qquad \underset{\text{succinyl chloride}}{Cl\overset{\displaystyle O}{\overset{\|}{C}}CH_2CH_2\overset{\displaystyle O}{\overset{\|}{C}}Cl}$$

Alcohols

The names of alcohols may be (1) IUPAC; (2) trivial; or, occasionally, (3) conjunctive. *IUPAC names* are taken from the name of the alkane with the final *-e* changed to *-ol*. For polyols, the prefix, *di-*, *tri-*, etc. is placed just before *-ol*, with the position numbers placed at the start of the name, if possible.

$$CH_3CH_2CH_2CH_2CH_2OH \qquad HO-\hexagon-OH$$

1-pentanol   1,4-cyclohexanediol

In cases where confusion is possible, the number precedes *-ol. Example*: $CH_3CH{=}CHCH_2OH$ is 2-buten-1-ol.

In a molecule containing a functional group with higher nomenclature priority and in complex molecules, the prefix *hydroxy-* is used. For example, $CH_3CH(OH)CH_2CO_2H$ is 3-hydroxybutanoic acid.

Trivial names are generally composed of the name of the alkyl group plus the word *alcohol*.

$$(CH_3)_3COH \qquad\qquad C_6H_5CH_2OH$$

tert-butyl alcohol   benzyl alcohol

Conjunctive names are used principally with structures in which ring systems are bonded to an aliphatic alcohol.

pyridine-3-methanol

Polyols. Structures with two OH groups on adjacent carbons (1,2-diols) are sometimes given trivial *glycol* names: the name of the *alkene* (not the alkane) from which the diol could be formed and the word *glycol*. *Glycerol* and *glycerin* are trivial names for 1,2,3-propanetriol.

$$\underset{\substack{\text{1,2-propanediol (IUPAC)}\\\text{propylene glycol (trivial)}}}{CH_3\overset{\displaystyle OH}{\overset{|}{C}}HCH_2OH} \qquad\qquad \underset{\substack{\text{1,2,3-propanetriol (IUPAC)}\\\text{glycerol (trivial)}}}{HOCH_2\overset{\displaystyle OH}{\overset{|}{C}}HCH_2OH}$$

Phenols

Phenols are compounds in which an OH is attached directly to an arene ring. In these cases, phenol (or naphthol, etc.) is considered the parent.

p-nitrophenol

1-naphthol (IUPAC)   2-naphthol (IUPAC)
α-naphthol (trivial)   β-naphthol (trivial)

Many phenols and substituted phenols have trivial names.

o-cresol   pyrocatechol   resorcinol   hydroquinone

Aldehydes

Aldehydes may be named by the IUPAC system or by trivial aldehyde names. In the IUPAC system, the *-oic acid* ending of the corresponding carboxylic acid is changed to *-al*.

$$CH_3CH_2CH_2CH_2CH_2CHO$$

hexanal

In trivial names, the *-ic* (or *-oic*) *acid* ending is changed to *-aldehyde*.

$$CH_3CHO \qquad C_6H_5CHO$$

acetaldehyde   benzaldehyde   cyclohexanecarbaldehyde

Some aldehydes have specific trivial names:

$HOCH_2CH(OH)CHO$

2-furaldehyde   glyceraldehyde
or furfural

Nitriles

The functional group of a nitrile is the *cyano group*, $-C\equiv N$. In the IUPAC system, nitriles are named from the alkane parent that contains the same number of carbons (including the C in the $-C\equiv N$ group) and the suffix *-nitrile* to denote $\equiv N$.

carbon 1

3 carbons
$$CH_3CH_2C\equiv N \qquad (CH_3)_2CHCN$$

propanenitrile   2-methylpropanenitrile

When the cyano group is bonded to a ring, the ending *-carbonitrile* (to denote the entire $-C\equiv N$ group) is used.

$-CN$

cyclohexanecarbonitrile   1-pyrrolecarbonitrile

If groups of higher priority are present, the prefix *cyano-* (to denote the entire $-C\equiv N$ group) is used.

$$\overset{O}{\overset{\|}{HOCCH_2CH_2C}}\equiv N$$

3-cyanopropanoic acid

In trivial names, the *-ic acid* or *-oic acid* ending of the corresponding carboxylic acid name is replaced by *-onitrile*.

from benzoic acid
from acetic acid
$$CH_3C\equiv N \qquad \langle\bigcirc\rangle - C\equiv N$$

acetonitrile   benzonitrile

Occasionally, nitriles are named as cyanides, such as phenyl cyanide for C_6H_5-CN. These names are also trivial.

Amides

In both the IUPAC and trivial systems, an amide is named by dropping the *-ic* (or *-oic*) *acid* ending of the corresponding acid name and adding *-amide*.

$$\overset{O}{\overset{\|}{CH_3(CH_2)_4C}}NH_2 \qquad \overset{O}{\overset{\|}{CH_3C}}NH_2$$

hexanamide (IUPAC)   acetamide (trivial)

Substituents on the amide nitrogen are named as prefixes preceded by *N*- or *N,N*-. $C_6H_5CONHCH_3$ is *N*-methylbenzamide, and $C_6H_5CON(CH_3)_2$ is *N,N*-dimethylbenzamide. *N*-Phenylamides have trivial names of *anilides*.

$$\underset{\text{acetanilide}}{CH_3\overset{\displaystyle O}{\overset{\|}{C}}NHC_6H_5} \qquad \underset{\text{benzanilide}}{C_6H_5\overset{\displaystyle O}{\overset{\|}{C}}NHC_6H_5}$$

Sulfonamides are named by attaching the ending *-sulfonamide* to the name for the rest of the structure. *Example*: $C_6H_5SO_2NH_2$ is benzenesulfonamide.

$$H_2N-\underset{5\quad 6}{\overset{3\quad 2}{\underset{4}{\bigcirc}}}{}^{1}-SO_2NH_2$$

p-aminobenzenesulfonamide
(sulfanilamide)

Amines

The preferred method for naming simple amines is by using the name of the hydrocarbon portion of the structure, omitting the final *-e* before another vowel, and adding the ending *-amine*. Note that the largest alkyl group bonded to the *N* is considered the parent alkane.

$$\underset{\text{ethanamine}}{CH_3CH_2NH_2} \qquad \underset{\text{1,2-ethanediamine}}{H_2NCH_2CH_2NH_2} \qquad \underset{\text{N-methylethanamine}}{CH_3NHCH_2CH_3}$$

larger group

The more commonly encountered names, also recognized by IUPAC, are formed by listing the alkyl groups, not the alkanes, before the ending *-amine*.

$$\underset{\text{ethylamine}}{CH_3CH_2NH_2} \qquad \underset{\text{N-methylethylamine}}{CH_3NHCH_2CH_3}$$

If functionality of higher nomenclature priority is present, the prefix *amino-* is used.

$$\underset{\text{2-amino-1-propanol}}{CH_3\overset{\displaystyle NH_2}{\overset{|}{C}H}CH_2OH} \qquad \underset{\text{2-(methylamino)-1-propanol}}{CH_3\overset{\displaystyle NHCH_3}{\overset{|}{C}H}CH_2OH}$$

Some arylamines have their own names.

$$\underset{\text{aniline}}{C_6H_5NH_2} \qquad \underset{\text{p-toluidine}}{CH_3-\bigcirc-NH_2}$$

The prefix *aza-* is sometimes used to identify a nitrogen in a chain or ring; see page 1039.

Amine salts.   Amine salts are named as *ammonium salts* or (in simple cases) as amine hydrochlorides, etc.

$$(CH_3)_3NH^+ \ Cl^-$$

trimethylammonium chloride
(trimethylamine hydrochloride)

Cyclic salts often are named as *-inium salts.*

$$C_6H_5NH_3^+ \ Br^-$$

anilinium bromide
(aniline hydrobromide)

pyridinium acetate

Esters and Salts of Carboxylic Acids

Esters and salts of carboxylic acids are named with two words. The first word is the name of the substituent on the oxygen. The second word is derived from the name of the parent carboxylic acid with the ending changed from *-ic acid* to *-ate.*

$$CH_3(CH_2)_4CO_2CH_3$$ methyl hexanoate (IUPAC)
$$CH_3CO_2CH_2CH_3$$ ethyl acetate (trivial)
$$CH_3CH_2CO_2Na$$ sodium propanoate (or sodium propionate)

$$CH_3 - \bigcirc - SO_3Na$$ sodium *p*-toluenesulfonate (or sodium tosylate)

Ethers

Ethers are usually named with the names of the alkyl or aryl groups bonded to the O followed by the word *ether.* (These are trivial names.)

$$CH_3CH_2OCH_2CH_3$$ $$CH_3O - \bigcirc$$

diethyl ether   cyclohexyl methyl ether

In the names of more complex ethers, an *alkoxy-* prefix may be used. (This is the IUPAC preference.)

$$\overset{\displaystyle OCH_3}{\underset{\displaystyle}{CH_3CH_2CH_2CHCH_2CO_2H}}$$

3-methoxyhexanoic acid

Sometimes the prefix *oxa-* is used; see page 1039.

Ketones

In the *systematic* names for ketones, the *-e* of the parent alkane name is dropped and *-one* is added. A prefix number is used if necessary.

$$
\begin{matrix} & O & & & O & & & O & O \\ & \| & & & \| & & & \| & \| \end{matrix}
$$

CH$_3$CCH$_2$CH$_2$CH$_3$ CH$_3$CCH=CHCH$_3$ CH$_3$CCH$_2$CCH$_3$

2-pentanone   3-penten-2-one   2,4-pentanedione

— e retained before consonant

In a complex structure, a keto group may be named in the IUPAC system with the prefix *oxo-*. (The prefix *keto-* is also sometimes encountered.) Contrast the use of *oxo-* with that of *oxa-* (an ether).

$$
\begin{matrix} O \\ \| \end{matrix}
$$

CH$_3$CCH$_2$CO$_2$H   is 3-oxobutanoic acid

In *trivial* names, the alkyl groups bonded to the C=O group are named, and the word *ketone* is added. Methyl ethyl ketone (CH$_3$COCH$_2$CH$_3$) is a common example. The CH$_3$CO— group is sometimes called the *aceto-*, or *acetyl*, group, while the C$_6$H$_5$CO— group is the *benzo-*, or *benzoyl*, group. Also encountered are *-phenone*, *-naphthone*, or *-acetone* endings, where one of these groups is an attachment to the ketone carbonyl group.

$$
\begin{matrix} O & O & & O & & O \\ \| & \| & & \| & & \| \end{matrix}
$$

CH$_3$CCH$_2$CCH$_3$ CH$_3$CC$_6$H$_5$ C$_6$H$_5$CC$_6$H$_5$

acetylacetone   acetophenone   benzophenone

Some ketones also have specific trivial names: CH$_3$COCH$_3$ is called acetone.

Organometallics and Metal Alkoxides

Organometallic compounds, those with C bonded to a metal, are named by the name of the alkyl or aryl groups plus the name of the metal.

(CH$_3$CH$_2$)$_3$Al   C$_6$H$_5$MgBr

triethylaluminum   phenylmagnesium bromide

Organosilicon or boron compounds are often named as derivatives of the metalloid hydrides. SiH$_4$ is silane; (CH$_3$)$_4$Si is tetramethylsilane. BH$_3$ is borane; (CH$_3$CH$_2$)$_3$B is triethylborane.

Sodium or potassium alkoxides are named as salts: the name of the cation plus the name of the alcohol with the *-anol* ending changed to *-oxide*. Salts of phenols are called *phenoxides*. (An alternate ending is *-olate*.)

CH$_3$CH$_2$ONa   C$_6$H$_5$OK

sodium ethoxide   potassium phenoxide
(sodium ethanolate)   (potassium phenolate)

Glossary of some prefix symbols used in organic chemistry

(+)   dextrorotatory

(−)   levorotatory

(±)   racemic

α-   alpha: (1) on the adjacent carbon;   (2) refers to configuration of carbon 1 in sugars;   (3) refers to configuration of substituents on steroid ring systems

aldo-   aldehyde

allo-   closely related

andro-   relating to male

anhydro-   denoting abstraction of water

antho-   relating to flowers

anthra-   relating to coal or to anthracene

anti-   on opposite faces or sides

β-   beta: (1) opposite to that of α in configuration;   (2) second carbon removed from a functional group or a heteroatom

bi-   twice or double

bisnor-   indicating removal of two carbons

chromo-   color or colored

cis-   on the same side of a double bond or ring

cyclo-   cyclic

Δ-   double bond

D-   on the right in the Fischer projection (see Section 22.3A)

d-   dextrorotatory

de-   removal of something, such as hydrogen (*dehydro-*) or oxygen (*deoxy-*)

dextro-   to the right, as in dextrorotatory

dl-   racemic

(E)-   on the opposite sides of a double bond

(continued)

Glossary of some prefix symbols used in organic chemistry (continued)

---

endo-   (1) on or in the ring and not on a side chain;   (2) opposite the bridge side of a ring system:

(3) attached as a bridge within a ring, as 1,4-*endo*-methyleneanthracene

epi-   (1) epimeric;   (2) a bridge connection on a ring, as 9,10-epidioxyanthracene

erythro-   related in configuration to erythrose

exo-   (1) on a side chain attached to a ring;   (2) on the bridge side of a ring system (see *endo-*)

gem-   attached to the same atom

hemi-   one-half

hydro-   (1) denotes presence of H;   (2) sometimes relating to water

hypo-   indicating a low, lower, or lowest state of oxidation

i-, iso-   (1) methyl branch at the end of the chain;   (2) isomeric;   (3) occasionally *i-* is used for *inactive*

L-   on the left in the Fischer projection (see Section 22.3A)

l-   levorotatory

leuco-   colorless or white

levo-   to the left, as in levorotatory

m-, meta-   (1) 1,3 o benzene;   (2) closely related compound, as metaldehyde

meso-   (1) with a plane of symmetry and optically inactive;   (2) middle position of certain cyclic organic compounds

n-   *normal*: continuous chain

neo-   one C connected to four other C's

nor-   (1) removal of one or more C's (with H's);   (2) structure isomeric to that of root name, as norleucine.

o-, ortho-   1,2 on benzene

oligo-   few (units of)

(continued)

Glossary of some prefix symbols used in organic chemistry (continued)

---

p-, para-   (1) 1,4 on benzene;   (2) polymeric, as paraformaldehyde

per-   saturated with, as in *perhydro-*, or *peroxy-*

peri-   (1) 1,8 on naphthalene;   (2) fusion of ring to two or more adjoining rings, as perixanthenoxanthene.

pheno-   relating to phenyl or benzene

poly-   many (units of)

Ψ-, pseudo-   has a resemblance to

pyro-   indicating formation by heat

(R)-   clockwise configuration around a chiral carbon

(R)(S)-   racemic

s-   abbreviation for *secondary-* or *symmetrical-*

(S)-   counterclockwise configuration around a chiral carbon

seco-   denoting ring cleavage

sec-   abbreviation for *secondary-*

sym-   symmetrical

syn-   on the same face or side

t-, tert-   abbreviation for *tertiary-*

threo-   related in configuration to threose

trans-   on opposite sides of a double bond or ring

uns-, unsym-   unsymmetrical

v-, vic-   vicinal: on adjacent C's

(Z)-   on the same side of a double bond

▼ Answers to Problems

The answers to the problems within the textual material are given here. For the answers to the chapter-end problems, see the *Solutions Manual* that accompanies this text.

▼ Chapter 1

1.1 (a) :Cl:C:Cl: with :O: above (b) Cl₂C::CCl₂ structure (c) H₂C::CHBr structure (d) H:C⋮⋮C:F:

1.2 (a) neutral molecule, no formal charges   (b) ion: $\overset{0}{C}H_3\overset{+1}{N}H_3$

(c) neutral molecule: :$\overset{-1}{C}$⋮⋮$\overset{+1}{O}$:

1.3 (a) all atoms 0;   (b) $CH_2=\overset{+1}{N}=\overset{-1}{N}$;   (c) $CH_3O\overset{+2}{S}$—OH, other atoms 0 with O^{-1} above and O^{-1} below

1.4 (a) $(CH_3)_2CHCH_2Cl$;   (b) CH_3CHCl_2;   (c) $CH_3(CH_2)_3CHClCH_2Cl$;
(d) $(CH_3)_2C=C(CH_3)_2$;   (e) $CH_2(C\equiv N)_2$

1.5 (a) structural formula   (b) structural formula   (c) structural formula

1.6 (a) pentagon   (b) fused bicyclic   (c) oxygen ring   (d) cyclobutane with three CH₃   (e) cyclopentene

1.7 (a) liberates more energy

1.8 (a) $\overset{\delta+}{C}H_3-\overset{\delta-}{Br}$;   (b) $CH_3\overset{}{C}-\overset{\delta-}{O}-\overset{\delta+}{H}$ with $\overset{\delta-}{O}$ double-bonded above, $\overset{\delta+}{}$ below C

1057

1.9 (a) CH_3OH;   (b) $CH_3O—H$;   (c) CH_3Cl

1.10 (a) $CH_3\overset{+}{C}\overset{\longrightarrow}{\equiv}N$;   (b) $CH_3\overset{O\uparrow}{\underset{\|+}{\overset{\|}{C}}}CH_3$;   (c) none;   (d) [cyclohexane ring]$\overset{+\longrightarrow}{=}O$

1.11 (a) $CH_3CH_2CH_2\overset{\overset{H}{|}}{\underset{\underset{H}{|}}{N}}H———:\overset{\overset{H}{|}}{N}CH_2CH_2CH_3$

(b) $CH_3\overset{..}{\underset{..}{O}}H———:\overset{\overset{H}{|}}{\underset{..}{O}}CH_3$, $CH_3\overset{..}{\underset{..}{O}}H———:\overset{\overset{H}{|}}{\underset{..}{O}}H$, $CH_3\overset{..}{\underset{..}{O}}:———H\overset{\overset{H}{|}}{\underset{..}{O}}:$, $H_2\overset{..}{O}:———H\overset{\overset{H}{|}}{\underset{..}{O}}:$

(c) none   (d) $(CH_3)_2\overset{..}{O}:———H\overset{\overset{H}{|}}{\underset{..}{O}}:$, $H\overset{..}{O}:———H\overset{\overset{H}{|}}{\underset{..}{O}}:$

1.12 (a) base: $^-:\overset{..}{N}H_2 + H_2\overset{..}{O}: \rightleftharpoons :NH_3 + {}^-:\overset{..}{O}H$

(b) acid: $CH_3CH_2CH_2\overset{\overset{\cdot\overset{\cdot}{O}\cdot}{\|}}{C}OH + H_2\overset{..}{O}: \rightleftharpoons CH_3CH_2CH_2\overset{\overset{\cdot\overset{\cdot}{O}\cdot}{\|}}{C}\overset{..}{O}:^- + H_3O:^+$

(c) base: $^-:\overset{..}{O}CH_2CH_3 + H_2\overset{..}{O}: \rightleftharpoons H\overset{..}{O}CH_2CH_3 + {}^-:\overset{..}{O}H$

(d) acid: [cyclohexane ring]$\overset{}{N}H_2^+ + H_2\overset{..}{O}: \rightleftharpoons$ [cyclohexane ring]$:NH + H_3O:^+$

(e) base: [cyclohexane ring]$:NH + H_2\overset{..}{O}: \rightleftharpoons$ [cyclohexane ring]$NH_2^+ + {}^-:\overset{..}{O}H$

1.13 (a) $CH_3CH_2\overset{\overset{\cdot\overset{\cdot}{O}\cdot}{\|}}{C}\overset{..}{O}⌒H + H\overset{..}{O}H \rightleftharpoons CH_3CH_2\overset{\overset{\cdot\overset{\cdot}{O}\cdot}{\|}}{C}\overset{..}{O}:^- + H\overset{\overset{H}{\underset{+}{|}}}{O}H$

(b) [cyclohexane ring]$—\overset{\overset{\cdot\overset{\cdot}{O}\cdot}{\|}}{C}\overset{..}{O}⌒H + {}^-:\overset{..}{O}H \longrightarrow$ [cyclohexane ring]$—\overset{\overset{\cdot\overset{\cdot}{O}\cdot}{\|}}{C}\overset{..}{O}:^- + H\overset{..}{O}H$

(c) $(CH_3)_2\overset{}{N}H + H⌒\overset{..}{O}H \rightleftharpoons (CH_3)_2\overset{+}{N}H_2 + {}^-:\overset{..}{O}H$

(d) [cyclohexane ring]$\overset{}{\underset{\underset{H}{|}}{N}}: + H⌒\overset{..}{O}\overset{\overset{\cdot\overset{\cdot}{O}\cdot}{\|}}{C}CH_3 \rightleftharpoons$ [cyclohexane ring]$\overset{\overset{H}{|}}{\underset{\underset{H}{|}}{N}}{}^+ + {}^-:\overset{..}{O}\overset{\overset{\cdot\overset{\cdot}{O}\cdot}{\|}}{C}CH_3$

1.14 (a) $HO—H$ is a stronger acid than $H—CH_3$ because O is more electronegative than C and thus is more able to carry a negative ionic charge.

$$H\overset{..}{\underset{..}{O}}⌒H + B:^- \rightleftharpoons H\overset{..}{\underset{..}{O}}:^- + HB$$

$$H—CH_3 + B:^- \rightleftharpoons {}^-:CH_3 + HB$$

(b) $H—NH_2$ is a stronger acid than $H—CH_3$ because N is more electronegative than C.

(c) $HS—H$ is a stronger acid than $HO—H$ because S is below O in the periodic table. Thus, S contains one more electron shell than O, is larger than O, and is better able to carry a negative charge.

$$H\overset{..}{\underset{..}{S}}—H + B:^- \rightleftharpoons H\overset{..}{\underset{..}{S}}:^- + HB$$

$$H\overset{..}{\underset{..}{O}}—H + B:^- \rightleftharpoons H\overset{..}{\underset{..}{O}}:^- + HB$$

1.15 (a) $CH_3\overset{\displaystyle O}{\overset{\|}{C}}CH_3$, Lewis base; H^+, Lewis acid

(b) $(CH_3)_3C^+$, Lewis acid; Cl^-, Lewis base

(c) $CH_3\overset{\displaystyle O}{\overset{\|}{C}}OCH_3$, Lewis acid; $^-OCH_3$, Lewis base

1.16 $pH = pK_a + \log\dfrac{[A^-]}{[HA]}$

$[A^-] = [HA]$

therefore, $pH = pK_a = 2.85$

1.17 (a) < **(b)** < **(c)**

1.18 (c) < **(b)** < **(a)**

▼ Chapter 2

2.1 (a) [structure] **(b)** [structure] All C—C bonds are sp^3-sp^3 and all C—H bonds are sp^3-s.

2.2 (a) [structure] sp^3-s for three C—H bonds; sp^2-sp^3; sp^3-s for three C—H bonds

(b) [structure] sp^2-sp^2; All C—H bonds are sp^2-s.

(c) [structure] sp^3-sp^2; sp^2-s; sp^2-sp^2 and $p-p$; sp^2-s; sp^3-sp^2; All other C—C bonds are sp^3-sp^3 and all other C—H bonds are sp^3-s.

2.3 All C and H atoms are in one plane; the pi bonds are above and below this plane.

2.4

2.5   **(a)** (2), (1);   **(b)** (2), (1);   **(c)** (1), (2), (3)

2.6   **(a)**   **(b)** ⟨hexane⟩—NH₂ ;   **(c)**

2.7   RCO₂H

2.8   **(a)**   H—C—N——C—N—H   All C—H and N—H bonds are sp^3–s,
                                     and all C—N bonds are sp^3–sp^3

with H substituents and H—C—H branch

(b) sp^2–sp^2 and p–p sp^2–s
           H—N—C—N—C—C≡N: sp^3–sp   All unmarked C—H and N—H bonds are sp^3–s.
                            sp–sp and two p–p
           sp^3–sp^2 sp^3–sp^3

(c)   All C—C and C—N bonds are sp^3–sp^3;
          all C—H and N—H bonds are sp^3–s

2.9   The following structures are not the only possible answers:

(a) CH₃CH₂CH₂CH with =O   **(b)** CH₃CCH₂CH₃ with =O   **(c)** CH₃CH₂CH₂COH with =O   **(d)** CH₃CCH₂COH with two =O

2.10   Five conjugated double bonds, no isolated double bonds.

2.11   **(a)** CH₃CH=CHCH=CHCH=CHCH₃   **(b)** CH₂=CHCH₂CH=CHCH=CHCH₃
            (c) CH₂=CHCH₂CH=CHCH₂CH=CH₂

2.12   **(a)**   **(b)**

2.13   CH₂=CHCCH₃ with =O

2.14

2.15 The right-hand structure is the major contributor because each atom has an octet.

2.16 (a) $CH_3CH{=}CH{-}\overset{+}{C}HCH_3 \longleftrightarrow CH_3\overset{+}{C}H{-}CH{=}CHCH_3$; *equivalent*

(b)

major

(c) $CH_3\overset{O}{\overset{||}{C}}{-}\overset{-}{C}H_2{}^{-} \longleftrightarrow CH_3\overset{:O:^-}{\overset{|}{C}}{=}CH_2$

major

(d) $^{-}\overset{\frown}{C}H_2{-}C{\equiv}N{:} \longleftrightarrow CH_2{=}C{=}\ddot{N}{:}^{-}$

major

(e)

major

(f)

All four structures are equivalent.

In (f), note that the negative charge is not delocalized by the benzene ring. In any additional resonance structures we can draw, the resonance stabilization of the ring is lost and the $-CO_2{}^-$ group carries two negative charges. For example:

▼ **Chapter 3**

3.1 (a) $CH_3CH_2\overset{O}{\overset{||}{C}}CH_2CH_3$ (b)

$-CH_2OH$ (c) $CH_3\overset{O}{\overset{||}{C}}CH_3$

3.2 (a) HOCH$_2$CH (with =O above CH); **(b)** [cyclohexane]—CH$_2$OH; **(c)** CH$_3$CH$_2$CH$_2$CH$_2$CH (with =O above final CH)

3.3 (a) C$_n$H$_{2n+2}$ is an open-chain alkane.
  (b) C$_n$H$_{2n}$ contains either one double bond or one ring.
  (c) C$_n$H$_{2n-2}$ contains either one triple bond, one double bond plus one ring, two double bonds, or two rings.

3.4 (a) CH$_3$CH$_2$CH$_2$—[cyclohexane]; **(b)** (CH$_3$)$_2$CHCH$_2$—[cyclohexane]; **(c)** CH$_3$CH$_2$CH$_2$CHCH$_2$CH$_2$CH$_2$CH$_3$ with C(CH$_3$)$_3$ substituent

3.5 (a) Cl$_2$CHCH$_2$Cl; **(b)** Cl—[cyclohexane with Cl]—NO$_2$

3.6 (a) propene; **(b)** cyclohexene; **(c)** 1,3-cyclohexadiene;
  (d) 1-chloropropene or 1-chloro-1-propene

3.7 (a) [cyclopentene] **(b)** CH≡CC≡CCH$_3$ **(c)** [cyclobutane with CH$_3$]

3.8 CH$_3$CH$_2$CH$_2$CH$_2$CH (=O); CH$_3$CCH$_2$CH$_2$CH$_3$ (=O) or CH$_3$CH$_2$CCH$_2$CH$_3$ (=O); CH$_3$CH$_2$CH$_2$CH$_2$COH (=O)

  pentanal   2-pentanone   3-pentanone   pentanoic acid

3.9 (a) 7-nitro-3-heptene; **(b)** 3-bromo-1-phenyl-1-propanol

▼ Chapter 4

4.1 (a) (E)-2-chloro-1-fluoro-1-butene; **(b)** (Z)-2,3-dichloro-2-butene

4.2

[structures: C=C with CO$_2$H, CO$_2$H, H, H — *maleic acid*; C=C with H, CO$_2$H, CO$_2$H, H — *fumaric acid*]

4.3 (a) (E): **(b)** (Z): **(c)** (E): **(d)** (Z)

4.4

[four cyclohexane ring structures with CH$_3$, OH, and CH(CH$_3$)$_2$ substituents; the last labeled *menthol*]

4.5 (a) [Newman projections: Br, H, H, H, Cl — *anti*; Br, H, Cl, H, H — *gauche*] **(b)** [Newman projections: CO$_2$H, H, H, OH — *anti*; CO$_2$H, H, OH, H — *gauche*]

4.6   Only (c), because thujone contains one three-membered ring. (a) and (b) contain only six-membered rings.

4.7   (**a**) *trans e,e*;   (**b**) *cis a,e*;   (**c**) *trans e,e*;   (**d**) *trans e,e*

4.8   (**a**) none;   (**b**) CH₃CH₂ĊHBr;   (**c**) C₆H₅CH₂ĊHC₆H₅

with CH₃ above (b) and F above (c), both with * below the chiral carbon.

$$\text{(a)} \quad CH_3 \quad (b) \quad F$$

4.9   (**a**) C₆H₅━Ċ◄Br   Br━Ċ◄C₆H₅   (**b**) H━Ċ◄OH   HO━Ċ◄H

with CH₃ below each in (a) and CH₂CH₃ below each, CH₃ above each in (b)

dimensional   *dimensional*

4.10 (**a**)

CO₂H
Cⁱⁱⁱⁱ‚H
(CH₃)₂CH   NH₂

(**b**)

CO₂H
Cⁱⁱⁱⁱ‚H
HO₂CCH₂   NH₂

4.11 (**a**) H₂N─┼─H   (**b**) H₂N─┼─H

with CO₂H above and CH₃ below (a); CO₂H above and CH₂SH below (b)

4.12 (**a**) (*R*);   (**b**) (*S*);   (**c**) (*R*);   (**d**) (*S*)

4.13 (**a**)

Br
Cⁱⁱⁱⁱ‚H
CH₃CH₂ (CH₂)₃CH₃

(**b**)

OH
Cⁱⁱⁱⁱ‚H
CH₃CH₂CH₂ CH₃

4.14 (**a**) four;   (**b**) two;   (**c**) eight

4.15

CH₃   CH₃   CH₃
H━C◄Cl   Cl━C◄H   H━C◄Cl
H━C◄Cl   H━C◄Cl   Cl━C◄H
CH₃   CH₃   CH₃

meso   ‾‾‾‾‾‾ *enantiomers* ‾‾‾‾‾‾

4.16 Make models of the (*R,R*) and (*S,S*) enantiomers to verify the mirror-image relationships of the conformers and to verify the nonsuperimposability of the enantiomers.

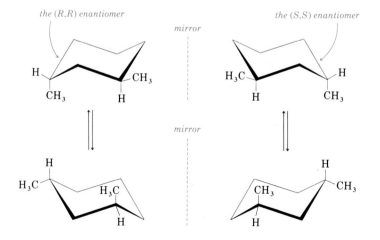

the (R,R) enantiomer   *mirror*   *the (S,S) enantiomer*

mirror

4.17 (a)

A (1e, 2a) ⇌ B (1a, 2e)

no change   *mirror*   *rotate model*

A   B (nonsuperposable on A)

(b) a boat conformation (*meso*)

plane of symmetry   *plane of symmetry*

CH₃   or   CH₃   CH₃

Neither structure has a *meso* form.

4.18 (a)   *the cis enantiomers*   *the trans enantiomers*   **(b)** *enantiomers*

4.19 (a) (*R*): H—NH₂   (*S*): H₂N—H

CH₂C₆H₅   CH₂C₆H₅
CH₃   CH₃

(b) CH₃CHCO₂H + H₂N—C—H ⟶

OH   CH₂C₆H₅
(*R*)(*S*)   CH₃
(*S*)

H—C—OH H₃N—C—H + HO—C—H H₃N—C—H

CO₂⁻   CH₂C₆H₅   CO₂⁻   CH₂C₆H₅
CH₃   CH₃   CH₃   CH₃
(*R,S*)   (*S,S*)

▼ Chapter 5

5.1   **(a)** 2-iodo-2-methylpropane,   *t*-butyl iodide   **(b)** 2-chloropropane,   isopropyl chloride
(c) 1-iodo-2-methylpropane,   isobutyl iodide

5.2   **(a)** Br₂CHCH₂CH₂CH₃   **(b)** CH₃CHClCH=CH₂   **(c)** FCH₂CH₂OH

5.3 NC⁻ +

$$CH_3CH_2\overset{H}{\underset{CH_3}{\overset{|}{\underset{|}{C}}}}Br \overset{(S)}{\longrightarrow} NC-\overset{H}{\underset{CH_3}{\overset{|}{\underset{|}{C}}}}CH_2CH_3 + Br^-$$

5.4 (a) [cyclohexyl]—CH₂Cl   **1°**

(b) (CH₃)₂CHCH₂CHClCH₃   *less hindered*

(c) [benzene ring]—CH₂Cl

The other compound is aryl and nonreactive.

5.5 (a) [cyclohexyl]—CH₂Cl + NaCN ⟶ [cyclohexyl]—CH₂CN + NaCl

(b) (CH₃)₂CHCH₂CHCH₃ (with Cl) + NaCN ⟶ (CH₃)₂CHCH₂CHCH₃ (with CN) + NaCl

(c) [benzene ring]—CH₂Cl + NaCN ⟶ [benzene ring]—CH₂CN + NaCl

5.6 (a) (CH₃CH₂)₃COH + HI   (b) [cyclohexane with CH₃ and OCH₃] + HCl

5.7 (b), (a), (c)

5.8 (a) no observed rearrangement products

(b) (CH₃)₂C⁺CHCH₂CH₃ (with H) ⟶ (CH₃)₂C⁺CHCH₂CH₃ (with H)

5.9 (CH₃)₂CHCHCH₂CH₃ (with OCH₂CH₃) + (CH₃)₂CCH₂CH₂CH₃ (with OCH₂CH₃) + HCl

5.10 (a) allylic;   (b) vinylic;   (c) aryl;   (d) benzylic

5.11 (a) [benzene ring]—CHCH=CH₂ (with OH) + [benzene ring]—CH=CHCH₂OH

(b) CH₃—[cyclohexene]—OH + CH₃—[cyclohexene with HO]

5.12 (CH₃)₂CCH₂CH₃ (with Br) $\xrightarrow[-\text{HBr}]{\text{CH}_3\text{CH}_2\text{OH}}$ (CH₃)₂CCH₂CH₃ (with OCH₂CH₃) + (CH₃)₂C=CHCH₃ + CH₂=CCH₂CH₃ (with CH₃)

5.13 (a) CH₃CH=CHCH₂CH₂CH₃ + CH₃CH₂CH=CHCH₂CH₃

   cis and *trans*   *cis* and *trans*

(b) [cyclohexene]—CH₃ + [cyclohexane]=CH₂

5.14 [cyclohexene]—CH₃

5.15

(Z)-alkene

5.16

(1R,2S)   (1S,2R)

5.17 The enantiomers yield (*Z*)-1-bromo-1,2-diphenylethene (*trans*-phenyls); the *meso* form yields the (*E*)-isomer (*cis*-phenyls).

5.18

Only one H can be abstracted.   only E2 product

5.19 (a) $(CH_3)_2CHCHBrCH_2CH_3$ $\xrightarrow{-HBr}$ $(CH_3)_2C=CHCH_2CH_3$ + $(CH_3)_2CHCH=CHCH_3$

Saytzeff   *Hofmann*

(b)

Saytzeff   *Hofmann*

5.20 (a) $CH_3CH_2CH_2CH=CH_2$ + $CH_3CH_2CH=CHCH_3$;   **(b)** *trans*-$CH_3CH=CHCH_3$

mainly trans

5.21 CH_3S^- is the better nucleophile because S is larger and more polarizable than O.

5.22 (a) $CH_3(CH_2)_5\overset{\displaystyle OH}{\underset{|}{C}}HCH_3$ + $CH_3(CH_2)_5\overset{\displaystyle OCH_2CH_3}{\underset{|}{C}}HCH_3$ + *trans*-$CH_3(CH_2)_4CH=CHCH_3$

The two substitution products are formed by an S_N1 path, and the alkene by an E1 path. The alkyl halide is secondary. A weak nucleophile, but polar solvent, is used. These conditions favor carbocation reactions, while the higher temperature favors elimination.

(b) $HOCH_2\overset{\displaystyle OH}{\underset{|}{C}}HCH_2O$—⟨benzene ring⟩ + NaCl

The organic halide is primary. A reasonably strong nucleophile (but weak base) is used. We would expect reaction to occur by an S_N2 path.

(c) $CH_3CH_2OCH_2CH_2CN$ + NaCl

The halide is primary and ^-CN is a strong nucleophile. We would expect substitution by an S_N2 path.

$$\underset{\text{(d)}}{\overset{\displaystyle \text{OCH}_2\text{CH}_3}{\text{CH}_3\text{CHCH}=\text{CH}_2}} + \text{CH}_3\text{CH}=\text{CHCH}_2\text{OCH}_2\text{CH}_3 + \text{HCl}$$

(d)

The halide is allylic. The solvent is polar and weakly nucleophilic. We would expect substitution by an S_N1 path and the possibility of allylic rearrangement.

5.23 (a) ⬡—O^- + $\text{CH}_3\text{CH}_2\text{CH}_2\text{I}$ ⟶

(b) $\text{C}_6\text{H}_5\text{CH}_2\text{Br} + \text{CH}_3\text{S}^-$ or $\text{C}_6\text{H}_5\text{CH}_2\text{S}^- + \text{CH}_3\text{I}$ ⟶

(c) $\text{CH}_3\text{CH}_2\text{Br} + {}^-\text{OCH}_2\text{CH}_2\text{CH}_3$ or $\text{CH}_3\text{CH}_2\text{CH}_2\text{Br} + {}^-\text{OCH}_2\text{CH}_3$ ⟶

(d) $\text{C}_6\text{H}_5\text{CO}_2^- + \text{CH}_2=\text{CHCH}_2\text{Br}$ ⟶

(e) $\text{C}_6\text{H}_5\text{O}^- + \text{CH}_3\text{CH}_2\text{Br}$ ⟶

(f) $(\text{CH}_3)_2\text{CHCH}_2\text{CHBrCH}_3 \xrightarrow[\text{CH}_3\text{OH}]{{}^-\text{OCH}_3}$ or $(\text{CH}_3)_2\text{CHCHBrCH}_2\text{CH}_3 \xrightarrow[(\text{CH}_3)_3\text{COH}]{{}^-\text{OC(CH}_3)_3}$

▼ **Chapter 6**

6.1 (a) $\text{H}:\overset{\cdot\cdot}{\underset{\cdot\cdot}{\text{O}}}\cdot$ **(b)** $\text{H}:\overset{\text{H}}{\underset{\text{H}}{\overset{\cdot\cdot}{\text{C}}}}:\overset{\cdot\cdot}{\underset{\cdot\cdot}{\text{O}}}\cdot$ **(c)** $\text{H}:\overset{\text{H}}{\underset{\text{H}}{\text{C}}}:\overset{\text{H}}{\underset{\text{H}}{\text{C}}}\cdot$

6.2 *initiation:* $\text{Cl}_2 \xrightarrow{hv} 2\,\text{Cl}\cdot$

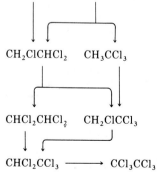

propagation: ⬡ + Cl· ⟶ ⬡· + HCl

⬡· + Cl_2 ⟶ ⬡—Cl + Cl·

termination: ⬡· + Cl· ⟶ ⬡—Cl   or other combination of two radicals

6.3 $\text{CHCl}_3 + \text{Cl}\cdot \longrightarrow \cdot\text{CCl}_3 + \text{HCl}$
$\cdot\text{CCl}_3 + \text{Cl}_2 \longrightarrow \text{CCl}_4 + \text{Cl}\cdot$

6.4 nine

$\text{CH}_3\text{CH}_3 \longrightarrow \text{CH}_3\text{CH}_2\text{Cl} \longrightarrow \text{CH}_2\text{ClCH}_2\text{Cl} + \text{CH}_3\text{CHCl}_2$

$\text{CH}_2\text{ClCHCl}_2 \quad \text{CH}_3\text{CCl}_3$

$\text{CHCl}_2\text{CHCl}_2 \quad \text{CH}_2\text{ClCCl}_3$

$\text{CHCl}_2\text{CCl}_3 \longrightarrow \text{CCl}_3\text{CCl}_3$

6.5 not racemic, $\underset{\underset{\underset{(R)(S)}{\text{CH}_3}}{\big|}}{\overset{\overset{\text{Cl}}{\big|}}{\text{CH}_3\text{CHCHCH}_2\text{Cl}}}$ (R)

6.6 9 to 1

6.7 (a) CH_2=CH—$\dot{C}H_2$ ⟷ $\dot{C}H_2$—CH=CH_2

(b)

6.8 (b), (c), (a), (d)

6.9 (a) $C_6H_5CHBrCH_2CH_2CH_3$; **(b)** C_6H_5CH=$CHCH_2Br$ (conjugated)

6.10 The reaction is faster because a $3°$ hydrogen is abstracted more readily than a $2°$ hydrogen.

$CH_3CH_2OCH_2CH_3$ $(CH_3)_2CHOCH(CH_3)_2$

6.11 $(CH_3)_2C$—N=N—$C(CH_3)_2$ $\xrightarrow{\text{heat}}$ $2(CH_3)_2\dot{C}\cdot$ + :N≡N:

The product radical is $3°$ and "allylic" and is thus relatively stable. The other product is N_2, which is easily lost from a molecule because of its stability.

6.12 R· + ⟶ RH +

relatively stable free radical
(resonance stabilized by both the
phenyl and naphthyl groups)

6.13 $+CF_2CF_2\rightarrow_x$ **6.14** CH_2=CHCN

▼ Chapter 7

7.1 (a) 2,4-dimethyl-3-pentanol;   **(b)** 2,3-dimethyl-2,3-butanediol

7.2 (a)   **(b)**

7.3 (a) 3-cyclopentenol;   **(b)** 4-bromocyclohexanol;   **(c)** 2-bromoethanol

7.4 (a) $2°$, benzylic;   **(b)** $3°$, but *not* allylic;   **(c)** $2°$, allylic

7.5 (a)

(b)

7.6 (a), (d), and (f), which contain acidic hydrogens that would destroy any Grignard reagent formed.

7.7 (structure) —Br $\xrightarrow{\text{Mg}}$ (structure) —MgBr $\xrightarrow{\text{D}_2\text{O}}$ product

7.8 (structure) —Br $\xrightarrow{\text{Mg}}$ (structure) —MgBr $\xrightarrow[\text{(2) H}_2\text{O, H}^+]{\text{(1) CH}_3\text{CHO}}$ (structure with OH) CHCH$_3$

7.9 (a) $C_6H_5Br \xrightarrow[\text{ether}]{\text{Mg}} C_6H_5MgBr \xrightarrow[\text{(2) H}_2\text{O, H}^+]{\text{(1) HCHO}} C_6H_5CH_2OH$

(b) $CH_3(CH_2)_3Cl \xrightarrow{\text{OH}^-} CH_3(CH_2)_3OH$

(c) $CH_3CH_2Br \xrightarrow[\text{ether}]{\text{Mg}} CH_3CH_2MgBr \xrightarrow[\text{(2) H}_2\text{O, H}^+]{\text{(1) C}_6\text{H}_5\text{CHO}} C_6H_5\overset{\text{OH}}{\underset{|}{\text{C}}}HCH_2CH_3$

7.10 (a) (structure) CH$_2$OH + PBr$_3$ $\longrightarrow$ (structure) CH$_2$Br + H$_3$PO$_3$

Although HBr could also be used for this conversion, some ring opening would occur (see Section 8.4).

(b) (structure) CH$_2$OH + SOCl$_2$ $\xrightarrow{\text{pyridine}}$ (structure) CH$_2$Cl + SO$_2$ + HCl

PCl$_3$ or HCl (ZnCl$_2$ catalyst) would also be suitable reagents.

7.11 (structure) $\xrightarrow[-\text{HCl}]{\text{SOCl}_2}$ [structure OSOCl] $\xrightarrow{\text{Cl}^-}$ (structure Cl) + SO$_2$ + Cl$^-$

trans D and Cl

7.12 *trans*, because it is more stable

7.13 $(CH_3)_2\overset{\text{CH}_3}{\underset{|}{\text{C}}}-\overset{|}{\underset{:\ddot{O}H}{\text{C}}}HCH_3 \xrightarrow{\text{H}^+} (CH_3)_2\overset{\text{CH}_3}{\underset{|}{\text{C}}}-\overset{|}{\underset{:\overset{+}{O}H_2}{\text{C}}}HCH_3 \xrightarrow{-\text{H}_2\text{O}}$

$(CH_3)_2\overset{\text{CH}_3}{\underset{|}{\text{C}}}-\overset{+}{\text{C}}HCH_3 \xrightarrow{\text{1,2 methyl shift}} (CH_3)_2\overset{+}{\text{C}}-\overset{\text{CH}_3}{\underset{\text{H}}{\text{CCH}_3}} \xrightarrow{-\text{H}^+} (CH_3)_2C=C(CH_3)_2$

7.14 (structure CHCH$_3$, CH$_3$) $\xrightarrow{\text{H}^+}$ (structure $\overset{+}{O}H_2$ CHCH$_3$, CH$_3$) $\xrightarrow{-\text{H}_2\text{O}}$ (structure $\overset{+}{}$CHCH$_3$, CH$_3$) $\xrightarrow{\text{rearrangement}}$

(structure $\overset{+}{}$CHCH$_3$, CH$_3$) ≡ (structure CH$_3$, H, CH$_3$) $\xrightarrow{-\text{H}^+}$ (structure CH$_3$, CH$_3$)

7.15 (a) $CH_3CH_2CH_2OH + TsCl \xrightarrow{-\text{HCl}} CH_3CH_2CH_2OTs$

(b) (R)-$CH_3\overset{\text{OH}}{\underset{|}{\text{C}}}H(CH_2)_3CH_3 + TsCl \xrightarrow{-\text{HCl}}$ (R)-$CH_3\overset{\text{OTs}}{\underset{|}{\text{C}}}H(CH_2)_3CH_3$

7.16

7.17 (*S*)-CH$_3$CH(CH$_2$)$_5$CH$_3$ with OH   **7.18** (b), (c), (a)   **7.19** (b), (a), (c)

7.20 (a) =O;   **(b)** C$_6$H$_5$CO$_2$H

7.21 CH$_3$CH$_2$OH $\xrightarrow{\text{H}_2\text{CrO}_4}$ CH$_3$CO$_2$H $\xrightarrow[\text{heat}]{\text{CH}_3\text{CH}_2\text{OH, H}^+}$ CH$_3$CO$_2$CH$_2$CH$_3$

7.22 (a) (CH$_3$)$_2$CHCH$_2$Br $\xrightarrow[\text{ether}]{\text{Mg}}$ (CH$_3$)$_2$CHCH$_2$MgBr $\xrightarrow[\text{(2) H}_2\text{O, H}^+]{\text{(1) CH}_3\overset{\text{O}}{\overset{\|}{\text{C}}}\text{CH}_2\text{CH}_3}$ product

(b) —Br $\xrightarrow[\text{ether}]{\text{Mg}}$ —MgBr $\xrightarrow[\text{(2) H}_2\text{O, H}^+]{\text{(1) HCHO}}$ product

(e) CH$_2$=CHCH$_2$Br $\xrightarrow[\text{ether}]{\text{Mg}}$ CH$_2$=CHCH$_2$MgBr

(*E*) isomer

Chapter 8

8.1 (a) CH$_3$CH$_2$OC$_6$H$_5$;   **(b)** (CH$_3$)$_2$CHO——OCH(CH$_3$)$_2$ **(c)** (CH$_3$)$_2$C—CH$_2$

8.2 (a)   **(b)** CH$_3$CH$_2$O$\overset{\text{O}}{\underset{\text{O}}{\overset{\|}{\underset{\|}{\text{S}}}}}OCH_2CH_3$ **(c)** CH$_3$CH$_2$O$\overset{\text{O}}{\underset{\text{O}}{\overset{\|}{\underset{\|}{\text{S}}}}}$OH   **(d)** —ONO$_2$

8.3 (a) + CH$_3$I $\longrightarrow$;   **(b)** (CH$_3$)$_2$CHO$^-$ + CH$_3$CH$_2$Br $\longrightarrow$

8.4 + C$_6$H$_5$CO$_3$H $\longrightarrow$

cis

$\xrightarrow{\text{aqueous KOH}}$

(1*R*,2*R*) or (1*S*,2*S*)

8.5

$$\text{(trimethoxybenzene)} \xrightarrow[\text{heat}]{3\text{ HI}} \text{(trihydroxybenzene)} + 3\text{ CH}_3\text{I} \xrightarrow[\text{H}_2\text{O}]{3\text{ AgNO}_3} 3\text{ CH}_3\text{OH} + 3\text{ AgI}\downarrow + 3\text{ HNO}_3$$

8.6 (a)

(b)

8.7 Dimethyl sulfoxide contains only two hydrophobic methyl groups and contains a very polar S=O group, which can hydrogen-bond with water.

▼ **Chapter 9**

9.1 **(a)** 1600 cm^{-1}; **(b)** 200 nm; **(c)** 60,004 Hz

9.2 **(a)** 1667 cm^{-1}; **(b)** 12.5 μm; **(c)** 1.5 μm

9.3 I (d); II (b)

9.4 The structure of the compound is CH$_3$CH$_2$OCH$_2$CH$_3$.

$$\text{CH}_3\text{CH}_2\text{OCH}_2\text{CH}_3 + \text{HI} \longrightarrow \text{CH}_3\text{CH}_2\text{I} + \text{CH}_3\text{CH}_2\text{OH}$$

$$\text{CH}_3\text{CH}_2\text{OH} + \text{HI} \longrightarrow \text{CH}_3\text{CH}_2\text{I} + \text{H}_2\text{O}$$

Assignment of infrared absorption:

2850–2975 cm^{-1} (3.4–3.5 μm): sp^3 C—H stretch
1380 cm^{-1} (7.25 μm): sp^3 C—C stretch
1110 cm^{-1} (9.0 μm): C—O stretch

9.5 Compound I is an ester while compound II is a ketone. Although both spectra show C=O absorption, the spectrum of compound I also shows strong absorption in the 1230 cm^{-1} (8.1 μm) region (C—O stretch).

9.6 Calculate the ratio of protons.

$$\frac{62.5\text{ mm}}{20.8\text{ mm}} = 3.00 \qquad \frac{20.8\text{ mm}}{20.8\text{ mm}} = 1.00$$

The proton ratio is 3:1. Compound **(a)** would exhibit only a singlet and, therefore, does not fit the data. Compound **(b)**, CH$_3$CH$_2$CH$_3$, has a proton ratio of 6:2, or 3:1, and thus is compatible with the spectrum. Compound **(c)**, CH$_3$CH$_2$CH$_2$CH$_3$, has a proton ratio of 6:4, or 3:2, and is not compatible.

9.7 **(a)** six neighboring protons: 7 peaks   **(b)** nine neighboring protons: 10 peaks
(c) one neighboring proton: 2 peaks   **(d)** one neighboring proton: 2 peaks

9.8 **(a), (c), (d), (e)**   **9.9** **(a)** one at room temperature;   **(b)** one

9.10 **(a)** two for each
(b) two for the group arising from the CH_2 protons and three in the group arising from the other proton
(c) two in the group arising from the CH_3 protons and four in the group arising from the other proton

9.11 Cl—⟨O⟩—$\overset{\overset{\displaystyle O}{\|}}{C}CH_3$

9.12 Two Cl atoms have a greater electron-withdrawing power than one Cl atom.

9.13   *triplet, 2.9* *triplet, 4.3*
$C_6H_5CH_2CH_2O_2CCH_3$ **9.14** I (e); II (d); III (a)
singlet, 7.15 *singlet, 2.0*

9.15

(a)   (b)   (c)

9.16

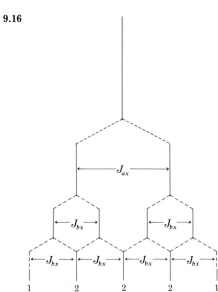

9.17 The compound contains an ester group. A partial structure can be written.

$$\begin{matrix} O \\ \| \\ -CO- \end{matrix} + C_4H_7Br$$

Saturated noncyclic alkyl groups have the general formula C_nH_{2n+1} or, if containing Br, $C_nH_{2n}Br$. The C_4H_7Br grouping must thus contain one double bond or be part of a ring.

From the ^{13}C nmr spectra, we deduce the following groups, in order of their appearance in the spectra:

$$\begin{matrix} O & Br \\ \| & | \\ -C- & CH= \end{matrix} \qquad -CH= \qquad -OCH_3 \qquad -CH_2-$$

not bonded   *not bonded to O*
to O or Br

Putting the pieces together, we arrive at the structure:

$$\begin{matrix} O \\ \| \\ BrCH=CHCH_2COCH_3 \end{matrix} \quad (cis \text{ or } trans)$$

9.18 $(CH_3)_2CHOH$ **9.19** $C_6H_5CH_2OH$

9.20 $CH_3CH_2CHClCO_2H$ **9.21**

$$\text{(benzene ring)}-\overset{\displaystyle O}{\overset{\displaystyle \|}{C}}CH_2CH_2CH_3$$

▼ Chapter 10

10.1 (a)

$$\begin{matrix} CH_3 & CH_2CH_3 \\ & \diagdown \quad \diagup \\ & C=C \\ & \diagup \quad \diagdown \\ ClCH_2CH_2 & CH_3 \end{matrix}$$

(b)

$$\begin{matrix} CH_2=CH & CH_3 \\ \diagdown \quad \diagup \\ C=C \\ \diagup \quad \diagdown \\ H & H \end{matrix}$$

(c) (cyclohexyl)$-C\equiv CH$ (d) $C_6H_5C\equiv CC_6H_5$

10.2 (a) 2-methyl-1,4-pentadiene;   (b) 2-propynol or 2-propyn-1-ol

10.3

J_{ab} very small and
peaks overlap

10.4 (a) (cyclohexene) ;   (b) $CH_2=C(CH_3)_2$;   (c) $CH_3CH=C(CH_3)_2$

10.5 (a) $CH_3C{\equiv}CH$ $\xrightarrow{NaNH_2}$ $CH_3C{\equiv}C{:^-}\ Na^+$ $\xrightarrow{CH_3CH_2Br}$ $CH_3CH_2C{\equiv}CCH_3$

(b) $CH_3C{\equiv}C{:^-}$ $\xrightarrow{C_6H_5CH_2CH_2Br}$ $C_6H_5CH_2CH_2C{\equiv}CCH_3$

10.6 $CH{\equiv}CH$ $\xrightarrow[-CH_3CH_3]{1\ CH_3CH_2MgBr}$ $CH{\equiv}CMgBr$ $\xrightarrow[(2)\ H_2O,\ H^+]{(1)\ CH_3CH_2\overset{\displaystyle O}{\overset{\|}{C}}H}$ $CH_3CH_2\overset{\displaystyle OH}{\underset{|}{C}}HC{\equiv}CH$

10.7 (a) $CH_3CH_2CH_2CH_2C{\equiv}CH + 1\ HBr \longrightarrow CH_3CH_2CH_2CH_2CBr{=}CH_2$

(b) $CH_3CH_2CH_2CH_2C{\equiv}CH + 2\ HBr \longrightarrow CH_3CH_2CH_2CH_2CBr_2CH_3$

10.8 (a) $CH_3CH_2\overset{+}{C}HCH_3$ $\xrightarrow{Br^-}$ $CH_3CH_2\overset{\displaystyle Br}{\underset{|}{C}}HCH_3$

(c) $(CH_3)_2\overset{+}{C}CH_2CH_3$ $\xrightarrow{Br^-}$ $(CH_3)_2\overset{\displaystyle Br}{\underset{|}{C}}CH_2CH_3$

10.9 $CH_3CH{=}CH_2$ $\xrightarrow{H^+}$ $[CH_3\overset{+}{C}HCH_3]$ $\xrightarrow{CH_3CH_2\ddot{O}H}$ $CH_3CH_2\overset{+}{\underset{\displaystyle CH_3\overset{\displaystyle}{\underset{|}{C}}HCH_3}{\ddot{O}H}}$ $\underset{-H^+}{\rightleftharpoons}$ $CH_3CH_2\ddot{O}CH(CH_3)_2$

10.10 (a) $(CH_3)_3CCH{=}CH_2$ $\xrightarrow{HI}$ $(CH_3)_3CCHICH_3 + (CH_3)_2CICH(CH_3)_2$

(b) $C_6H_5CH{=}CH_2$ $\xrightarrow{HCl}$ $C_6H_5CHClCH_3$

(c) $(CH_3)_2CHCH{=}CH_2$ $\xrightarrow{HCl}$ $(CH_3)_2CHCHClCH_3 + (CH_3)_2CClCH_2CH_3$

(d) $(CH_3)_3CCH{=}CHCH_3$ $\xrightarrow{HCl}$ $(CH_3)_3CCH_2CHClCH_3$
$+ (CH_3)_3CCHClCH_2CH_3 + (CH_3)_2CCl\overset{\displaystyle}{\underset{\displaystyle CH_3}{\underset{|}{C}}}HCH_2CH_3$

10.11 (a) $(CH_3)_2CHCH_2Br$;   (b) $(CH_3)_2CBrCH_3$

10.12 $CH_3CH_2CH{=}CH_2$ $\xrightarrow{H^+}$ $[CH_3CH_2\overset{+}{C}HCH_3]$ $\xrightarrow{HSO_4^-}$ $CH_3CH_2\overset{\displaystyle OSO_3H}{\underset{|}{C}}HCH_3$

10.13 (a) $CH_3CH_2\overset{\displaystyle OH}{\underset{|}{C}}HCH_3$;   (b) $(CH_3)_2\overset{\displaystyle OH}{\underset{|}{C}}CH(CH_3)_2$;

(c)   (by way of the allylic cation)

10.14

10.15 (a) $(CH_3)_3CCH{=}CH_2$ $\xrightarrow[-\ ^-O_2CCH_3]{Hg(O_2CCH_3)_2}$ $\left[(CH_3)_3C\overset{+}{C}H\!-\!\overset{HgO_2CCH_3}{\underset{}{C}}H_2 \right]$ $\xrightarrow[-H^+]{H_2O}$

$(CH_3)_3CCH\!-\!\overset{\displaystyle HgO_2CCH_3}{\underset{\displaystyle OH}{\underset{|}{C}}}H_2$ $\xrightarrow{NaBH_4}$ $(CH_3)_3C\overset{\displaystyle}{\underset{\displaystyle OH}{\underset{|}{C}}}CHCH_3$

(b) $(CH_3)_3CCH=CH_2 + H_2O \xrightarrow{H^+} (CH_3)_2CCH(CH_3)_2$ *rearranged*

with OH above the product carbon.

10.16 $CH_2=CCH_2CH_3$ (with CH_3) $\xrightarrow[-\ ^-O_2CCH_3]{Hg(O_2CCH_3)_2}$ $CH_2-CCH_2CH_3$ (with CH_3CO_2Hg and CH_3, cationic) $\xrightarrow[-H^+]{CH_3OH}$

$\xrightarrow{} CH_2-CCH_2CH_3$ (with CH_3CO_2Hg, CH_3, OCH_3) $\xrightarrow{NaBH_4} CH_3CCH_2CH_3$ (with CH_3, OCH_3)

10.17

trans transition state
(cannot yield a *cis* alcohol)

10.18 (a)

with OH and CH_2CH_3

trans (and racemic)

(b)

(2R,3R)-3-methyl-2-pentanol
and its (2S,3S) enantiomer

10.19 (a) $(CH_3)_2CBrCH_2Br$;   **(b)** $(CH_3)_2CBrCHBrCH_3$

10.20

$\xrightarrow[anti\text{-addition}]{Br_2}$

(2S,3S)   (2R,3R)

10.21 (a)

$\xrightarrow{Br_2}$

(E)-1,2-dideuterioethene   *meso*-1,2-dibromo-1,2-dideuterioethane

(b)

$\xrightarrow{Br_2}$

(Z)-1,2-dideuterioethene   (1S,2S)-1,2-dibromo-1,2-dideuterioethane   (1R,2R)-1,2-dibromo-1,2-dideuterioethane

10.22 No, because the intermediate contains Br.

10.23

10.24 $CHBr_3 + (CH_3)_3CO^- \longrightarrow :CBr_2 + (CH_3)_3COH + Br^-$

dibromocarbene

probable by-product   (63%)

10.25 (a) The difference in energy is greater between *cis*- and *trans*-$(CH_3)_3CCH{=}CHCH_2CH_3$ because of the greater steric hindrance in the *cis*-isomer. **(b)** The difference in energy is greater between *cis*- and *trans*-ClCH=CHCl because the *cis*-isomer contains greater dipole–dipole repulsions between the two chlorine atoms.

10.26 (a)

(b)   Both products are racemic.

10.27 (a) in either case

(b) from reductive,   from oxidative

(c) OHC—CHO + OHC$(CH_2)_4$CHO from reductive

$HO_2C{-}CO_2H + HO_2C(CH_2)_4CO_2H$ from oxidative

(d) OHC$(CH_2)_4$CHO from reductive, $HO_2C(CH_2)_4CO_2H$ from oxidative

10.28 $CH_3CHBrCHBrCH_3$; $CH_3CHClCHClCHClCHClCH_3$

10.29 $[CH_3CH_2\overset{+}{C}H{-}CH{=}CHCH_3 \longleftrightarrow CH_3CH_2CH{=}CH\overset{+}{C}HCH_3] \longrightarrow$

$CH_3CH_2CHICH{=}CHCH_3 + CH_3CH_2CH{=}CHCHICH_3$

10.30 (a)

resonance structures for the intermediate

(b)

resonance structures for the intermediate

10.31 the *trans*-isomer, because it is more stable

10.32 A, BrCH$_2$CH=CHCH$_2$Br; **B**, BrCH$_2$CHBrCH=CH$_2$

10.33

10.34 (a) s-*trans*; **(b)** s-*cis*; **(c)** s-*trans*. Only **(a)** is interconvertible.

10.35 (a)

(b)

10.36

10.37

10.38

and

10.39 (a)

(b) CH$_3$CH$_2$CH=CH$_2$ $\xrightarrow{\text{Br}_2}$ CH$_3$CH$_2$CHBrCH$_2$Br $\xrightarrow[\text{(2) H}^+]{\text{(1) NaNH}_2}$ product

(c)

▼ **Chapter 11**

11.1 (a)

—CH$_2$CH$_3$ **(b)**

(c) CH$_3$CH$_2$—

—OH

(d)

—CH$_2$CH$_2$OH **(e)**

—CH$_2$Br

11.2 A, *p*-iodoanisole; B, *p*-chloroaniline

11.3 not aromatic

11.4 Compound **(a)** is not aromatic because one ring carbon is sp^3 hybridized. Ion **(b)** contains sp^2-hybridized ring atoms, but eight ($4n$) pi electrons, and thus is not aromatic. Ion **(c)** contains sp^2-hybridized ring atoms and six ($4n + 2$) pi electrons; the cycloheptatrienyl cation is aromatic.

11.5 (a) C$_6$H$_5$CHCH$_2$CH$_3$;   **(b)** C$_6$H$_5$C(CH$_3$)$_3$;   **(c)** C$_6$H$_5$C(CH$_3$)$_2$;   **(d)** (C$_6$H$_5$)$_2$CH$_2$

11.6

11.7 **(a)** $C_6H_5CCH_2CH_3$ with two CH_3 groups **(b)** $C_6H_5CH_2CHCH_2CH_3$ with CH_3

11.8 Two parts *o*-, two parts *m*-, and one part *p*-.

11.9 **(a)** **(b)**

11.10

11.11 The unshared electrons of the nitrogen are delocalized by the carbonyl group and are less available for donation to the ring. Because the amide nitrogen is partially positive, it also exerts a greater electron-withdrawal than an amine nitrogen.

11.12 **(a)**

(b)

(c)

(d) Relative rates: (b) > (c) > (d) > (a).

11.13 **(a)**

(b)

(c)

because the right-hand ring is activated and the left-hand ring is deactivated

▼ **Chapter 12**

12.1

12.2

12.3 (a) and **(b)**, $C_6H_5O^- + Na^+$

12.4 The nitro group helps share the negative charge in the anion.

12.5 (a)

(b)

12.6 (a)

(b)

(c) C_6H_6 $\xrightarrow[\text{H}_2\text{SO}_4]{\text{HNO}_3}$ $C_6H_5NO_2$ $\xrightarrow[\text{(2) OH}^-]{\text{(1) Fe, HCl}}$ $C_6H_5NH_2$ $\xrightarrow[0°]{\overset{\text{NaNO}_2}{\text{HCl}}}$

$C_6H_5N_2^+$ Cl^- $\xrightarrow[\text{heat}]{\text{H}_2\text{O, H}^+}$ C_6H_5OH $\xrightarrow[\text{(2) CH}_3\text{I}]{\text{(1) OH}^-}$ $C_6H_5OCH_3$ $\xrightarrow{C_6H_5N_2^+ Cl^-}$ product

12.7

12.8 (a)

(50%)   (50%)

the observed products

(b)

(100%)

not observed

12.9 (a) C_6H_6 $\xrightarrow[\text{H}_2\text{SO}_4]{\text{HNO}_3}$ $C_6H_5NO_2$ $\xrightarrow[\text{FeCl}_3]{\text{Cl}_2}$

(b) C_6H_6 + $\overset{\text{O}}{\overset{\|}{\text{ClC}}}(CH_2)_3CH_3$ $\xrightarrow{\text{AlCl}_3}$ $C_6H_5\overset{\text{O}}{\overset{\|}{\text{C}}}(CH_2)_3CH_3$ $\xrightarrow[\text{HCl}]{\text{Zn/Hg}}$ $C_6H_5(CH_2)_4CH_3$

(c) excess C_6H_6 + $\text{Cl}\overset{\text{O}}{\overset{\|}{\text{CCH}}}_2C_6H_5$ $\xrightarrow{\text{AlCl}_3}$ $C_6H_5\overset{\text{O}}{\overset{\|}{\text{CCH}}}_2C_6H_5$ $\xrightarrow[\text{HCl}]{\text{Zn/Hg}}$ $C_6H_5CH_2CH_2C_6H_5$

(d) excess C_6H_6 + $CHCl_3$ $\xrightarrow{\text{AlCl}_3}$ $(C_6H_5)_3CH$

▼ ## Chapter 13

13.1 (b)   **13.2** C_6H_5CHO

13.3 (a) because of electron-withdrawal by the Cl atoms

13.4 (a)

hemiketal

(b)

ketal

(c)

hemiacetal

13.5 (a)

=O + HOCH_2CH_2OH

(b)

=O + HOCH_2CH_2OH

(c)

—OH + $\overset{\text{O}}{\overset{\|}{\text{H}}}\text{CCH}_3$

13.6 (a)

(b)

13.7

13.8 (a)

(b) (1) Extract the ether solution with an aqueous, alkaline solution to remove the carboxylic acid as RCO_2^- Na^+.

(2) Extract the ether solution with aqueous $NaHSO_3$ solution to remove the aldehyde as $RCHSO_3^-$ Na^+.
$\quad\;$|
$\quad$ OH

(3) Regenerate the acid by acidification of its solution. Regenerate the aldehyde by treatment of its solution with acid or base. The ketone is still in the ether solution.

13.9 (a) C_6H_5MgBr; (b) $CH_3CH_2CH_2MgBr$

13.10 (a) and (b) contain acidic protons relative to RMgX.

13.11 An arylamine yields a product in which the double bond is in conjugation with the aromatic ring.

13.12 A pair of stereoisomers:

planar with   *cis* CH_3 groups
trans CH_3 groups

13.13 (a)

(b)

(c)

(d)

13.14 (a) (b) (c)

13.15

double bond in conjugation with benzene ring

13.16 (a) $CH_3\overset{O}{\overset{\|}{C}}CH_2CH_3$ **(b)** [cyclohexane]=NNH—[2,4-dinitrophenyl]   **(c)** $C_6H_5\overset{NNH_2}{\overset{\|}{C}}CH_3$

13.17 There are two possible Wittig routes to each product.

(a) (1) $CH_3CH_2CH_2Br \xrightarrow{(C_6H_5)_3P} (C_6H_5)_3\overset{+}{P}—CH_2CH_2CH_3 \ Br^- \xrightarrow{n\text{-}C_4H_9Li}$

$(C_6H_5)_3P{=}CHCH_2CH_3 \xrightarrow{C_6H_5CHO} C_6H_5CH{=}CHCH_2CH_3$

(2) $C_6H_5CH_2Br \xrightarrow{(C_6H_5)_3P} (C_6H_5)_3\overset{+}{P}—CH_2C_6H_5 \ Br^- \xrightarrow{n\text{-}C_4H_9Li}$

$(C_6H_5)_3P{=}CHC_6H_5 \xrightarrow{CH_3CH_2CHO} C_6H_5CH{=}CHCH_2CH_3$

(b) (1) [cyclopentyl]—Br $\xrightarrow{(C_6H_5)_3P}$ $(C_6H_5)_3\overset{+}{P}$—[cyclopentyl] $Br^- \xrightarrow{n\text{-}C_4H_9Li}$

$(C_6H_5)_3P{=}$[cyclopentylidene] $\xrightarrow{\text{[cyclopentyl]}CHO}$ [cyclopentylidene]$=CH$—[cyclopentyl]

(2) [cyclopentyl]$CH_2Br \xrightarrow{(C_6H_5)_3P} (C_6H_5)_3\overset{+}{P}—CH_2$[cyclopentyl] $Br^- \xrightarrow{n\text{-}C_4H_9Li}$

$(C_6H_5)_3P{=}CH$[cyclopentyl] $\xrightarrow{\text{[cyclopentanone]}}$ product

(c) (1) $BrCH_2CH_2CH_2CH_2Br \xrightarrow{2(C_6H_5)_3P} (C_6H_5)_3\overset{+}{P}—CH_2CH_2CH_2CH_2—\overset{+}{P}(C_6H_5)_3 + 2\ Br^- \xrightarrow{2n\text{-}C_4H_9Li}$

$(C_6H_5)_3P{=}CHCH_2CH_2CH{=}P(C_6H_5)_3 \xrightarrow{\text{[benzene-1,2-dicarbaldehyde]}} \text{product}$

(2) [benzene, ortho CH_2Br, CH_2Br] $\xrightarrow{2(C_6H_5)_3P}$ [benzene, ortho $CH_2—\overset{+}{P}(C_6H_5)_3 \ Br^-$, $CH_2—\overset{+}{P}(C_6H_5)_3 \ Br^-$] $\xrightarrow{2n\text{-}C_4H_9Li}$

[benzene, ortho $CH{=}P(C_6H_5)_3$, $CH{=}P(C_6H_5)_3$] $\xrightarrow{OHCCH_2CH_2CHO} \text{product}$

13.18 $CH_3(CH_2)_4CHO \xrightarrow[(2)\ H_2O,\ H^+]{(1)\ CH_3MgI}$

$CH_3CHO \xrightarrow[(2)\ H_2O,\ H^+]{(1)\ CH_3(CH_2)_3CH_2MgBr}$ → $CH_3(CH_2)_4\overset{OH}{\overset{|}{C}}HCH_3$

$CH_3(CH_2)_4\overset{O}{\overset{\|}{C}}CH_3 \xrightarrow[\text{heat, pressure}]{H_2,\ Ni}$

13.19 (a) $CH_3CH_2CHO \xrightarrow[(2)\ H_2O,\ H^+]{(1)\ NaBH_4}$

(b) [cyclohexenone] $\xrightarrow[(2)\ H_2O,\ H^+]{(1)\ NaBH_4}$ **(c)** [cyclohexyl]$\overset{O}{\overset{\|}{C}}CH_3 \xrightarrow[(2)\ H_2O,\ H^+]{(1)\ NaBH_4}$

13.20

$$
\begin{array}{c}
CH_2OH \\
H{-}C{-}OH \\
HO{-}C{-}H \\
H{-}C{-}OH \\
H{-}C{-}OH \\
CH_2OH
\end{array}
$$

13.21 (a) $C_6H_5\overset{\overset{\ddot{O}:}{\|}}{C}{-}\overset{\cdot\cdot}{C}H{-}\overset{:\ddot{O}:}{\overset{\|}{C}}CH_3 \longleftrightarrow C_6H_5\overset{\overset{\ddot{O}:}{|}}{C}{-}CH{=}\overset{:\ddot{O}:\bar{}}{C}CH_3 \longleftrightarrow C_6H_5\overset{:\ddot{O}:^-}{C}{=}CH{-}\overset{:\ddot{O}}{\overset{\|}{C}}CH_3$

(b) $CH_3\overset{\overset{\ddot{O}:}{\|}}{C}{-}\overset{\cdot\cdot}{C}H{-}\overset{:\ddot{O}:}{\overset{\|}{C}}CH_3 \longleftrightarrow CH_3\overset{\overset{\ddot{O}:}{|}}{C}{-}CH{=}\overset{:\ddot{O}:\bar{}}{C}CH_3 \longleftrightarrow CH_3\overset{:\ddot{O}:^-}{C}{=}CHCCH_3$

13.22 Enol form of ethyl acetoacetate:

$$
CH_3C{\underset{\underset{H}{\|}}{C}}\diagdown{C}{\diagup}{\underset{}{\diagdown OCH_2CH_3}}
$$
(with O⋯H⋯Ö hydrogen bond ring)

13.23 $\underset{\begin{array}{c}| \\ CH_2OH\end{array}}{\overset{\begin{array}{c}CH_2OPO_3H^- \\ |\end{array}}{C{=}O}} \rightleftharpoons \left[\underset{\begin{array}{c}\| \\ CHOH\end{array}}{\overset{\begin{array}{c}CH_2OPO_3H^- \\ |\end{array}}{COH}} \right] \rightleftharpoons \underset{\begin{array}{c}| \\ CHO\end{array}}{\overset{\begin{array}{c}CH_2OPO_3H^- \\ |\end{array}}{HCOH}}$

intermediate enediol

13.24 Ketone and OH⁻, second-order kinetics

13.25 (1) $\overset{\overset{O}{\|}}{CH_3CCH_2Br} + OH^- \underset{}{\overset{-H_2O}{\rightleftharpoons}} \left[CH_3\overset{:\ddot{O}:}{C}{-}\overset{\cdot\cdot}{C}HBr \longleftrightarrow CH_3\overset{:\ddot{O}:^-}{C}{=}CHBr \right]$

stabilized

(2) $\overset{\overset{O}{\|}}{CH_3C}\overset{\cdot\cdot}{C}HBr + :\overset{\cdot\cdot}{Br}{-}\overset{\cdot\cdot}{Br}: \longrightarrow CH_3\overset{\overset{O}{\|}}{C}CHBr_2 + :\overset{\cdot\cdot}{Br}:^-$

The anion in step (1) is stabilized by electron withdrawal by the electronegative Br atom.

13.26 (a) $CH_3\overset{\overset{+OH}{\|}}{C}CH_3$

(b) The rates would be the same because neither Br_2 nor I_2 is involved in the rate-determining step.

13.27 (a), (c)

13.28 (a)

(b) $(CH_3)_2CH\overset{\overset{O}{\|}}{C}CH_3$ **(c)** $CH_3\overset{\overset{O}{\|}}{C}$—⟨benzene ring⟩—$\overset{\overset{O}{\|}}{C}CH_3$

13.29 (a)

(b)

(c)

(d)

(e)

13.30 (a) $CH_3\overset{\overset{O}{\|}}{C}CH_3 \xrightarrow[\text{(2) } H_2O, H^+]{\text{(1) } C_6H_5MgBr} CH_3\underset{\underset{C_6H_5}{|}}{\overset{\overset{OH}{|}}{C}}CH_3 \xrightarrow[\text{heat}]{H^+} \text{product}$

(b) $\xrightarrow{\text{HCN}}$ $-\text{CN}$ $\xrightarrow{\text{H}_2\text{O, H}^+}$ product

(c) $\text{CH}_3(\text{CH}_2)_4\text{CHO}$ $\xrightarrow[\text{(2) H}_2\text{O, H}^+]{\text{(1)} \; \langle\;\rangle-\text{MgBr}}$ $\text{CH}_3(\text{CH}_2)_4\overset{\overset{\displaystyle \text{OH}}{|}}{\text{CH}}-\langle\;\rangle$ $\xrightarrow[\text{heat}]{\text{H}_2\text{CrO}_4}$

$\text{CH}_3(\text{CH}_2)_4\overset{\overset{\displaystyle O}{\|}}{\text{C}}-\langle\;\rangle$ $\xrightarrow{(\text{C}_6\text{H}_5)_3\text{P}=\text{CH}_2}$ product

▼ Chapter 14

14.1 **(a)** propenoic acid;   **(b)** pentanedioic acid;   **(c)** 2-bromopropanoic acid

14.2 **(a)** α-bromopropionic acid;   **(b)** β-hydroxypropionic acid

14.3 **(a)**

(b) and plus hydrogen bonds between molecules of *o*-hydroxybenzoic acid, as in **(a)**

(c) $\text{C}_6\text{H}_5\text{C}$ plus hydrogen bonds within and between benzoic acid molecules and between methanol molecules.

14.4 **(a)** $\text{HOCH}_2\text{CH}_2\text{Cl}$ $\xrightarrow{\text{KCN}}$ $\text{HOCH}_2\text{CH}_2\text{CN}$ $\xrightarrow{\text{H}_2\text{O, H}^+}$ $\text{HOCH}_2\text{CH}_2\text{CO}_2\text{H}$

(b) $(\text{CH}_3)_3\text{CCH}_2\text{Cl}$ $\xrightarrow[\text{ether}]{\text{Mg}}$ $(\text{CH}_3)_3\text{CCH}_2\text{MgCl}$ $\xrightarrow[\text{(2) H}_2\text{O, H}^+]{\text{(1) CO}_2}$ $(\text{CH}_3)_3\text{CCH}_2\text{CO}_2\text{H}$

(c) $\xrightarrow[\text{ether}]{\text{Mg}}$ $\xrightarrow[\text{(2) H}_2\text{O, H}^+]{\text{(1) CO}_2}$

(d) $\text{C}_6\text{H}_5\text{CH}_2\text{Br}$ $\xrightarrow{\text{KCN}}$ $\text{C}_6\text{H}_5\text{CH}_2\text{CN}$ $\xrightarrow{\text{H}_2\text{O, H}^+}$ $\text{C}_6\text{H}_5\text{CH}_2\text{CO}_2\text{H}$

or $\text{C}_6\text{H}_5\text{CH}_2\text{Br}$ $\xrightarrow[\text{ether}]{\text{Mg}}$ $\text{C}_6\text{H}_5\text{CH}_2\text{MgBr}$ $\xrightarrow[\text{(2) H}_2\text{O, H}^+]{\text{(1) CO}_2}$ $\text{C}_6\text{H}_5\text{CH}_2\text{CO}_2\text{H}$

14.5 **(a)** bromoacetic acid, because Br is more strongly electron withdrawing than a phenyl group
(b) dibromoacetic acid, because two Br are more strongly electron withdrawing than one is
(c) 2-iodopropanoic acid, because the I is one carbon closer to the carboxylate group and thus exhibits a stronger inductive effect.

14.6 **(a)** weaker;   **(b)** 3,5-dinitrobenzoic acid

14.7

stabilized by   *not stabilized by*
hydrogen bonding   *hydrogen bonding*

14.8 (d), (c), (a), (b)

14.9 Dissolve the aldehyde in diethyl ether, extract the ether solution with dilute aqueous $NaHCO_3$ to remove the acid, and then evaporate the ether to recover the aldehyde.

$$\underbrace{RCHO + RCO_2H}_{\text{ether-soluble}} \xrightarrow[\text{H}_2\text{O}]{\text{NaHCO}_3} \underbrace{RCHO}_{\text{ether-soluble}} + \underbrace{RCO_2^- Na^+}_{\text{water-soluble}} + CO_2\uparrow$$

Separate in a separatory funnel

14.10 Extract with aqueous $NaHCO_3$, which removes the acid. Extract with aqueous NaOH, which removes the naphthol. The octanol remains in the ether. The acid and phenol can be regenerated by acidification.

$$\underbrace{RCO_2H + ArOH + ROH}_{\text{ether-soluble}} \xrightarrow[\text{H}_2\text{O}]{\text{NaHCO}_3} \underbrace{RCO_2^- Na^+}_{\text{water-soluble}} + \underbrace{ArOH + ROH}_{\text{ether-soluble}} \xrightarrow[\text{H}_2\text{O}]{\text{NaOH}}$$

$$\underbrace{ArO^- Na^+}_{\text{water-soluble}} + \underbrace{ROH}_{\text{ether-soluble}}$$

14.11

14.12 (a) CH_3—⟨○⟩—$CO_2CH(CH_3)_2$ (b) $CH_3CH_2O_2C$—⟨○⟩—$CO_2CH_2CH_3$

(c) (R)-$CH_3CO_2\underset{\underset{CH_3}{|}}{C}HCH_2CH_3$

14.13

14.14 (a) ⟨○⟩—CH_2OH (b) HO—⟨○⟩—CH_2OH

cis and *trans*

14.15 (a)   (b)

14.16

14.17 (a) CH_3CO_2H (b) $CH_3\overset{O}{\overset{||}{C}}CH_2CH_3$ (c) no reaction

14.18 $CH_3CH_2CH_2CO_2H + CH_3OH \xrightarrow[\text{heat}]{H^+}$

$CH_3CH_2CH_2CO_2H \xrightarrow[\text{(2) } CH_3I]{\text{(1) } OH^-}$ $\longrightarrow CH_3CH_2CH_2CO_2CH_3$

14.19 (a) $CH_3CH_2CO_2H \xrightarrow[\text{(2) } H_2O]{\text{(1) } LiAlH_4} CH_3CH_2CH_2OH \xrightarrow{HBr} CH_3CH_2CH_2Br \xrightarrow[\text{ether}]{Mg}$

$CH_3CH_2CH_2MgBr \xrightarrow[\text{(2) } H_2O, H^+]{\text{(1) } CO_2} \text{product}$

(b) $ClCH_2CH_2CO_2H \xrightarrow[\text{(2) } KCN]{\text{(1) neutralize with } OH^-} NCCH_2CH_2CO_2^- \xrightarrow{H_2O, H^+} \text{product}$

(c) $C_6H_5\!-\!\!\langle\!\!\rangle \xrightarrow[\text{(2) } H_2O_2, H^+]{\text{(1) } O_3} C_6H_5CH\!\!\begin{array}{c} CO_2H \\ \\ CO_2H \end{array} \xrightarrow[-CO_2]{\text{heat}} \text{product}$

▼ Chapter 15

15.1 Because of both resonance stabilization and the inductive effect of the more electronegative oxygen, the carbonyl carbon is more positive than the vinyl carbon.

$$CH_3\overset{\overset{\textstyle\ddot O:}{\|}}{C}Cl \longleftrightarrow CH_3\overset{\overset{\textstyle:\ddot O:^-}{|}}{\underset{+}{C}}Cl$$

15.2 $CH_3O\overset{\overset{\textstyle O}{\|}}{C}CH_2CH_2CN$

15.3 (a) $CH_3CH_2CO_2H + SOCl_2 \longrightarrow CH_3CH_2\overset{\overset{\textstyle O}{\|}}{C}Cl + HCl + SO_2$

(b) $3\, HO_2C\!-\!CO_2H + 2\, PCl_3 \longrightarrow 3\, Cl\overset{\overset{\textstyle O}{\|}}{C}\!-\!\overset{\overset{\textstyle O}{\|}}{C}Cl + 2\, H_3PO_3$

(c) $3\, C_6H_5CH_2CO_2H + PCl_3 \longrightarrow 3\, C_6H_5CH_2\overset{\overset{\textstyle O}{\|}}{C}Cl + H_3PO_3$
(*Note*: Either $SOCl_2$ or PCl_3 could be used in each case.)

15.4 $CH_3CH_2CH_2\overset{\overset{\textstyle:\ddot O}{\|}}{C}Cl + :\ddot OH \longrightarrow \left[CH_3CH_2CH_2\overset{\overset{\textstyle:\ddot O:^-}{|}}{\underset{:\ddot OH}{C}}\!\!-\!\!\ddot Cl: \right] \xrightarrow{-Cl^-}$

$CH_3CH_2CH_2\overset{\overset{\textstyle:\ddot O}{\|}}{\underset{:\ddot OH}{C}} \xrightarrow{OH^-} CH_3CH_2CH_2\overset{\overset{\textstyle:\ddot O}{\|}}{C}\ddot O:^- + H_2O$

15.5 Either or both positions in the carboxyl group:

(a) $\langle\!\!\rangle\!\!-\!\!\overset{\overset{\textstyle O}{\|}}{C}\!-\!Cl + H_2^{18}O \longrightarrow \langle\!\!\rangle\!\!-\!\!\overset{\overset{\textstyle O}{\|}}{C}\!-\!^{18}OH$

(b)

(c)

15.6 $CH_3CH_2CH_2\overset{O}{\underset{}{C}}Cl \xrightarrow{C_6H_5\ddot{O}H} \left[CH_3CH_2CH_2\overset{:\ddot{O}:^-}{\underset{\overset{+}{C_6H_5\ddot{O}H}}{C}}\ddot{C}l: \right] \xrightarrow{-HCl} CH_3CH_2CH_2\overset{:\ddot{O}}{\underset{}{C}}\ddot{O}C_6H_5$

15.7 $CH_3CO_2^- Na^+ + Cl^- +$ unreacted CH_3NH_2

15.8 $C_6H_5\overset{:\ddot{O}}{\underset{}{C}}Cl + {}^-:\ddot{O}\overset{O}{\underset{}{C}}CH_3 \longrightarrow \left[C_6H_5\overset{:\ddot{O}:^-}{\underset{:\ddot{O}CCH_3}{\underset{\underset{O}{\|}}{C}}}\ddot{C}l: \right] \xrightarrow{-Cl^-} C_6H_5\overset{O}{\underset{}{C}}O\overset{O}{\underset{}{C}}CH_3$

15.9 $C_6H_6 \xrightarrow[AlCl_3]{\overset{O}{\underset{}{ClCCH(CH_3)_2}}} C_6H_5\overset{O}{\underset{}{C}}CH(CH_3)_2 \xrightarrow[HCl]{Zn/Hg} C_6H_5CH_2CH(CH_3)_2$

15.10 (a) $(CH_3)_2CHCl \xrightarrow[\text{(2) CuI}]{\text{(1) Li}} LiCu[CH(CH_3)_2]_2 \xrightarrow{C_6H_5\overset{O}{\underset{}{C}}Cl}$ product

(b) $CH_3I \xrightarrow[\text{(2) CuI}]{\text{(1) Li}} LiCu(CH_3)_2 \xrightarrow{\overset{O}{\underset{}{C}}Cl}$ product

or $\xrightarrow[\text{(2) CuI}]{\text{(1) Li}} LiCu(C_6H_{11})_2 \xrightarrow{CH_3\overset{O}{\underset{}{C}}Cl}$ product

15.11 (a) $C_6H_5CHBrCO_2H$ **(b)** $ClCH_2CHClCO_2H$

15.12 (a) pentanoic anhydride
(b) acetic benzoic anhydride or benzoic ethanoic anhydride

15.13 (a)

[structure: benzene ring with OCCH₃ (ester, O=C) group and OCH₃ group] + CH_3CO_2H **(b)** [naphthalene with O_2CCH_3 group] + $CH_3CO_2^-$

(c) [cyclohexane with $-OCCH_3$ (O=C)] $+ CH_3CO_2H$ **(d)** [benzene ring with CO_2H and $CO_2CH_2CH_3$ groups]

15.14 (a)

[benzene ring with CH_3 and $-NH_2$] $\xrightarrow{(CH_3CO)_2O}$ [benzene ring with CH_3 and $-NHCCH_3$ (O=C)] $\xrightarrow{HNO_3}$

[benzene ring with CH_3, $-NHCCH_3$ (O=C), and NO_2] $\xrightarrow{H_2O,\ OH^-}$ [benzene ring with CH_3, $-NH_2$ and NO_2] $+ CH_3CO_2^-$

(b) [benzene ring with CH_3 and $-NHCCH_3$ (O=C)] $\xrightarrow{HNO_3}$ O_2N-[benzene ring with CH_3 and $-NHCCH_3$ (O=C)] $\xrightarrow{H_2O,\ OH^-}$ O_2N-[benzene ring with CH_3 and $-NH_2$]

(c) The nitric acid would form a salt with the amine. The resulting $-\overset{+}{N}H_3$ group is strongly deactivating. (Then, oxidation of the $-\overset{+}{N}H_3$ group occurs.)

[benzene ring with CH_3 and $-NH_2$] $+ HNO_3 \longrightarrow$ [benzene ring with CH_3 and $-\overset{+}{N}H_3\ NO_3^-$]

15.15 (a) $CH_3CH_2CH_2CO_2H + CH_3OH \xrightarrow{H^+} CH_3CH_2CH_2CO_2CH_3 + H_2O$

$CH_3CH_2CH_2CO_2H \xrightarrow{SOCl_2} CH_3CH_2CH_2\overset{O}{\overset{||}{C}}Cl \xrightarrow[\text{pyridine}]{CH_3OH} CH_3CH_2CH_2CO_2CH_3$

(b) $\overset{\bullet}{CH_3}CO_2H + HOCH_2CH=C(CH_3)_2 \xrightarrow{H^+} CH_3CO_2CH_2CH=C(CH_3)_2 + H_2O$

$CH_3\overset{O}{\overset{||}{C}}O\overset{O}{\overset{||}{C}}CH_3 + HOCH_2CH=C(CH_3)_2 \longrightarrow CH_3CO_2CH_2CH=C(CH_3)_2 + CH_3CO_2H$

(Acetyl chloride could also be used.)

15.16 $CH_3CO_2H + H^{18}OCH_2CH_3$

15.17 $CH_3CO_2^- + H^{18}OCH_2CH_3$

15.18 (a) $C_6H_5CO_2CH_2CH_3 + OH^- \longrightarrow C_6H_5CO_2^- + HOCH_2CH_3$

(b) [six-membered ring with two O and two C=O] $+ 2\ OH^- \longrightarrow HOCH_2CH_2OH + {}^-O\overset{O}{\overset{||}{C}}-\overset{O}{\overset{||}{C}}O^-$

(c) [bicyclic lactone structure] $+ 2\ OH^- \longrightarrow$ [benzene ring with $CH_2CH_2CO_2^-$ and O^-] $+ H_2O$

15.19 (a)

$$CH_3\overset{:O:}{\overset{\|}{C}}OCH_2CH_3 \;\underset{\longleftarrow}{\overset{H^+}{\rightleftharpoons}}\; \left[CH_3\overset{\overset{+}{\ddot{O}}H}{\overset{\|}{C}}OCH_2CH_3 \right] \;\xrightarrow{CH_3\ddot{O}H}$$

$$\left[\begin{array}{c} :\ddot{O}H \\ | \\ CH_3\overset{|}{\underset{|}{C}}OCH_2CH_3 \\ | \\ CH_3\overset{+}{\underset{..}{O}}H \end{array} \right] \;\underset{\text{transfer}}{\overset{\text{proton}}{\rightleftharpoons}}\; \left[\begin{array}{c} :\ddot{O}{-}H \\ | \\ CH_3\overset{|}{\underset{|}{C}}{-}\ddot{O}CH_2CH_3 \\ | \\ CH_3\overset{|}{\underset{|}{O}}\;\;H \end{array} \right] \;\overset{-H^+}{\rightleftharpoons}\; CH_3\overset{O}{\overset{\|}{C}}OCH_3 + HOCH_2CH_3$$

(b)

$$CH_3\overset{:\ddot{O}:}{\overset{\|}{C}}OCH_2CH_3 \;\overset{{}^-:\ddot{O}CH_3}{\rightleftharpoons}\; \left[\begin{array}{c} :\ddot{O}:^- \\ | \\ CH_3\overset{|}{\underset{|}{C}}{-}\ddot{O}CH_2CH_3 \\ | \\ :\ddot{O}CH_3 \end{array} \right] \;\rightleftharpoons\; CH_3\overset{\ddot{O}:}{\overset{\|}{C}}OCH_3 + {}^-:\ddot{O}CH_2CH_3$$

15.20 When the hydroxy acid is treated with a chlorinating agent, the hydroxyl group would be converted to a chloro substituent.

$$\underset{\underset{OH}{|}}{CH_3CHCO_2H} \;\xrightarrow{SOCl_2}\; \underset{\underset{Cl}{|}}{CH_3\overset{O}{\overset{\|}{CHC}}Cl}$$

Treatment of a carboxylic ester of a hydroxy acid with ammonia would yield the hydroxy amide because NH_3 does not replace OH^-.

$$\underset{\underset{OH}{|}}{CH_3CHCO_2H} \;\xrightarrow{CH_3OH,\,H^+}\; \underset{\underset{OH}{|}}{CH_3CHCO_2CH_3} \;\xrightarrow{NH_3}\; \underset{\underset{OH}{|}}{CH_3\overset{O}{\overset{\|}{CHC}}NH_2}$$

15.21

$$\underset{\underset{CH_2O_2C(CH_2)_{14}CH_3}{|}}{\overset{\overset{CH_2O_2C(CH_2)_{14}CH_3}{|}}{CHO_2C(CH_2)_{14}CH_3}} \;\xrightarrow[\text{(2) } H_2O,\, H^+]{\text{(1) LiAlH}_4}\; \underset{\underset{CH_2OH}{|}}{\overset{\overset{CH_2OH}{|}}{CHOH}} + 3\,HOCH_2(CH_2)_{14}CH_3$$

15.22 (a)

$$CH_3CH_2CO_2CH_3 \;\xrightarrow[\text{(2) } H_2O,\, H^+]{\text{(1) 2 CH}_3\text{MgI}}\; \underset{\underset{OH}{|}}{CH_3CH_2C(CH_3)_2}$$

(b)

$$CH_3CH_2CO_2CH_3 \;\xrightarrow[\text{(2) } H_2O,\, H^+]{\text{(1) 2 CH}_3\text{CH}_2\text{MgBr}}\; \underset{\underset{OH}{|}}{CH_3CH_2C(CH_2CH_3)_2}$$

15.23

$$HC\overset{O}{\overset{\|}{O}}OCH_2CH_3 \;\xrightarrow[\text{(2) } H_2O,\, H^+]{\text{(1) 2 CH}_3\text{CH}_2\text{MgBr}}\; HOCH(CH_2CH_3)_2$$

15.24

[structure: a six-membered ring with CH₃ and O substituents, two ring oxygens, two carbonyl O's and a CH₃]

15.25 (a)

$$\underset{\underset{{}^+NH_3\,Cl^-}{|}}{HO_2CCH_2CHCO_2H} + \underset{\underset{{}^+NH_3\,Cl^-}{|}}{C_6H_5CH_2CHCO_2H} + CH_3OH$$

(b)

$$\underset{\underset{NH_2}{|}}{Na^+\,{}^-O_2CCH_2CHCO_2{}^-\,Na^+} + \underset{\underset{NH_2}{|}}{C_6H_5CH_2CHCO_2{}^-\,Na^+} + CH_3OH$$

15.26 (a)

+ NH_4^+ **(b)** + $HN(CH_2CH_3)_2$

(c) $^-O_2C(CH_2)_3NHCH_3$

15.27 (a)

(b)

15.28

15.29 (a) CH_3CH_2CN;   **(b)** $CH_3CH_2CH_2CN$

15.30 (a)

(b) $(CH_3)_2CHCH_2CO_2CH_3$ $\xrightarrow[\text{(2) H}_2\text{O, H}^+]{\text{(1) 2 CH}_3\text{CH}_2\text{CH}_2\text{MgBr}}$ product

(c) $CH_3CH_2CH_2OH$ $\xrightarrow{\text{H}_2\text{CrO}_4}$ $CH_3CH_2CO_2H$ $\xrightarrow{\text{SOCl}_2}$ $CH_3CH_2\overset{\overset{\displaystyle O}{\|}}{C}Cl$

$CH_3CH_2CH_2OH$ $\xrightarrow{\text{HBr}}$ $CH_3CH_2CH_2Br$ $\xrightarrow{\text{KCN}}$ $CH_3CH_2CH_2CN$

$\xrightarrow[\text{(2) H}_2\text{O, H}^+]{\text{(1) LiAlH}_4}$ $CH_3CH_2CH_2CH_2NH_2$ $\xrightarrow{\text{CH}_3\text{CH}_2\overset{\overset{\displaystyle O}{\|}}{C}Cl}$ product

▼ Chapter 16

16.1   (a), because its conjugate base is resonance stabilized.

16.2   (a) $CH_3CH_2CHO + CH_3CH_2\overset{\cdot\cdot}{\underset{\cdot\cdot}{O}}:^-$ $\rightleftharpoons$ $CH_3\overset{-}{\underset{\cdot\cdot}{C}}HCHO + CH_3CH_2\overset{\cdot\cdot}{\underset{\cdot\cdot}{O}}H$

(b) $(CH_3)_2CHCO_2CH_2CH_3 + CH_3CH_2O^-$ $\rightleftharpoons$ $(CH_3)_2\overset{-}{\underset{\cdot\cdot}{C}}CO_2CH_2CH_3 + CH_3CH_2OH$

(c) no reaction

(d) $+ CH_3CH_2O^-$ $\rightleftharpoons$ $+ CH_3CH_2OH$

(e) $C_6H_5CH_2NO_2 + CH_3CH_2O^-$ $\rightleftharpoons$ $C_6H_5\overset{-}{\underset{\cdot\cdot}{C}}HNO_2 + CH_3CH_2OH$

(f) $CH_2(CO_2H)_2 + 2\,CH_3CH_2O^-$ $\rightleftharpoons$ $CH_2(CO_2^-)_2 + 2\,CH_3CH_2OH$

16.3   (a) $CH_3CH_2CH_2CH(CO_2C_2H_5)_2$ **(b)** $CH_3CH_2\overset{\overset{\displaystyle CH_3}{|}}{C}(CO_2C_2H_5)_2$

16.4 (a) $CH_3CH_2CH_2CH(CO_2C_2H_5)_2 \xrightarrow[\text{heat}]{H_2O,\ OH^-}$

$CH_3CH_2CH_2CH(CO_2{}^-)_2 \xrightarrow{H^+} CH_3CH_2CH_2CH(CO_2H)_2$

(b) $(CH_3)_2C(CO_2C_2H_5)_2 \xrightarrow[\text{heat}]{H_2O,\ H^+} [(CH_3)_2C(CO_2H)_2] \xrightarrow[-CO_2]{\text{heat}} (CH_3)_2CHCO_2H$

16.5 $\xrightarrow{-CO_2}$ $\rightleftharpoons$

an enol

16.6 (a) $CH_3\overset{O}{\overset{\|}{C}}CH_2\overset{O}{\overset{\|}{C}}OC_2H_5 \xrightarrow[\text{(2) } CH_2=CHCH_2Br]{\text{(1) } NaOC_2H_5}$

(b) $CH_3\overset{O}{\overset{\|}{C}}CH_2\overset{O}{\overset{\|}{C}}OC_2H_5 \xrightarrow[\text{(2) } NaOC_2H_5, \text{ then } CH_3CH_2Br]{\text{(1) } NaOC_2H_5, \text{ then } CH_3CH_2Br} CH_3\overset{O}{\overset{\|}{C}}\underset{CO_2C_2H_5}{C}(CH_2CH_3)_2 \xrightarrow[\substack{\text{heat}\\-CO_2}]{H_2O,\ H^+}$

16.7 (a) $CH_3\overset{O}{\overset{\|}{C}}\underset{CH_3}{CH}\overset{O}{\overset{\|}{C}}OC_2H_5 \xrightarrow[-C_2H_5OH]{Na^+\ {}^-OC_2H_5} CH_3\overset{O}{\overset{\|}{C}}-\underset{CH_3}{\overset{\cdot\cdot}{C}}-\overset{O}{\overset{\|}{C}}OC_2H_5\ Na^+ \xrightarrow[-Na^+,\ -Br^-]{CH_3CH_2CH_2Br} CH_3\overset{O}{\overset{\|}{C}}-\underset{CH_3}{\overset{CH_2CH_2CH_3}{C}}-CO_2C_2H_5$

(b) $C_6H_5\overset{O}{\overset{\|}{C}}CH_2\overset{O}{\overset{\|}{C}}OC_2H_5 \xrightarrow[-C_2H_5OH]{Na^+\ {}^-OC_2H_5} C_6H_5\overset{O}{\overset{\|}{C}}\overset{\cdot\cdot}{C}H\overset{O}{\overset{\|}{C}}OC_2H_5\ Na^+ \xrightarrow{CH_3I}$

$C_6H_5\overset{O}{\overset{\|}{C}}\underset{CH_3}{CH}\overset{O}{\overset{\|}{C}}OC_2H_5 \xrightarrow[\text{heat}]{H_2O,\ H^+} C_6H_5\overset{O}{\overset{\|}{C}}\underset{CH_3}{CH}CO_2H \xrightarrow[-CO_2]{\text{heat}} C_6H_5\overset{O}{\overset{\|}{C}}CH_2CH_3$

16.8 (a) $CH_2(CO_2C_2H_5)_2 \xrightarrow[\text{(2) } CH_3(CH_2)_3Br]{\text{(1) } NaOC_2H_5} CH_3(CH_2)_3CH(CO_2C_2H_5)_2 \xrightarrow[\text{(2) } H^+, \text{ cold}]{\text{(1) } OH^-, H_2O, \text{ heat}}$

(b) $CH_2(CO_2C_2H_5)_2 \xrightarrow[\text{(2) } NaOC_2H_5, \text{ then } CH_3I]{\text{(1) } NaOC_2H_5, \text{ then } CH_3I} (CH_3)_2C(CO_2C_2H_5)_2 \xrightarrow[\text{heat}]{H_2O,\ H^+}$

(c) $CH_3\overset{O}{\overset{\|}{C}}CH_2CO_2C_2H_5 \xrightarrow[\text{(2) } CH_3CH_2CH_2Br]{\text{(1) } NaOC_2H_5} CH_3\overset{O}{\overset{\|}{C}}\underset{CH_2CH_2CH_3}{CH}CO_2C_2H_5 \xrightarrow[\text{(2) } CH_3I]{\text{(1) } NaOC_2H_5}$

$CH_3\overset{O}{\overset{\|}{C}}-\underset{CH_2CH_2CH_3}{\overset{CH_3}{C}}CO_2C_2H_5 \xrightarrow[\text{heat}]{H_2O,\ H^+}$

(d) $CH_2(CO_2C_2H_5)_2 \xrightarrow[\text{(2) } BrCH_2CO_2C_2H_5]{\text{(1) } NaOC_2H_5}$

(e) $CH_2(CN)_2 \xrightarrow[\text{(2) } Br(CH_2)_4Br]{\text{(1) } NaOC_2H_5} Br(CH_2)_4CH(CN)_2 \xrightarrow{NaOC_2H_5} BrCH_2(CH_2)_3\overset{\cdot\cdot}{C}(CN)_2 \longrightarrow$

(f) $\xrightarrow[\text{(2) } (CH_3)_2CHCH_2Br]{\text{(1) } NaOC_2H_5}$

16.9 (a) $\xrightarrow[\text{cold}]{Li^+\ {}^-N[CH(CH_3)_2]_2}$ $\xrightarrow[\text{allow to warm}]{C_6H_5CH_2Br}$ product

(b) [2-methylcyclopentanone] $\xrightarrow[\text{warm}]{\text{Na}^+ \ ^-\text{OC}_2\text{H}_5}$ [enolate anion] $\xrightarrow{\text{CH}_3\text{I}}$ product

(**a**) is under kinetic control; (**b**) is under thermodynamic control.

16.10 A cyclic amine has less steric hindrance around the nitrogen than an open-chain amine.

16.11 [cyclohexanone with CH₂C₆H₅ substituent] $=$O   +   [pyrrolidinium ion, $H-\overset{+}{N}-H$]

16.12 (a) $\text{CH}_3\text{CH}_2\overset{\text{O}}{\overset{\|}{\text{C}}}\text{CH}_2\text{CH}_3 + \text{HN}$[piperidine] $\overset{\text{H}^+}{\rightleftharpoons}$ [piperidine enamine] $\text{CH}_3\text{CH}_2\overset{}{\text{C}}=\text{CHCH}_3$ $\xrightarrow{\text{CH}_3\text{I}}$

[piperidinium enamine] $\text{CH}_3\text{CH}_2\overset{+}{\text{C}}\text{CH}(\text{CH}_3)_2$ $\xrightarrow{\text{H}_2\text{O, H}^+}$ $\text{CH}_3\text{CH}_2\overset{\text{O}}{\overset{\|}{\text{C}}}\text{CH}(\text{CH}_3)_2$

(b) [β-tetralone] $+$ HN[piperidine] $\overset{\text{H}^+}{\rightleftharpoons}$ [naphthalene-piperidine enamine] $\xrightarrow{\text{BrCH}_2\overset{\text{O}}{\overset{\|}{\text{C}}}\text{CH}_3}$

[iminium intermediate with $\text{CH}_2\overset{\text{O}}{\overset{\|}{\text{C}}}\text{CH}_3$ substituent] $\xrightarrow{\text{H}_2\text{O, H}^+}$ [tetralone with $\text{CH}_2\overset{\text{O}}{\overset{\|}{\text{C}}}\text{CH}_3$ substituent]

16.13 (**c**) and (**e**) because they contain α hydrogens

16.14 (a) $(\text{CH}_3)_2\overset{\text{OH}}{\underset{|}{\text{C}}}-\text{CH}_2\overset{\text{O}}{\overset{\|}{\text{C}}}\text{CH}_3$ **(b)** $(\text{CH}_3)_2\text{CHCH}_2\overset{\text{OH}}{\underset{|}{\text{CH}}}\text{CH}\overset{\text{O}}{\overset{\|}{\text{C}}}\text{H}$
$\underset{|}{\text{CH}(\text{CH}_3)_2}$

16.15 (a) $\text{C}_6\text{H}_5\text{CH}=\overset{}{\underset{|}{\text{C}}}\text{CHO}$ **(b)** $\text{C}_6\text{H}_5\text{CH}=\overset{\text{O}}{\overset{}{\underset{|}{\text{C}}}}\overset{\|}{\text{C}}\text{CH}_2\text{CH}_3$
$\underset{|}{\text{CH}_2\text{CH}_3}\underset{|}{\text{CH}_3}$

16.16 (a) $\text{C}_6\text{H}_5\text{CHO} + \text{CH}_3\text{CH}_2\text{CHO} \xrightarrow{\text{OH}^-}$

(b) $\text{CH}_3\overset{\text{O}}{\overset{\|}{\text{C}}}(\text{CH}_2)_4\overset{\text{O}}{\overset{\|}{\text{C}}}\text{CH}_3 \xrightarrow[\text{(2) H}^+, \text{heat}]{\text{(1) OH}^-}$

(c) [furan]$-\text{CHO} + \text{CH}_3\text{CHO} \xrightarrow{\text{OH}^-}$

(d) $\text{OHC}(\text{CH}_2)_5\text{CHO} \xrightarrow{\text{OH}^-}$

16.17 (a) $(\text{CH}_3\text{CH}_2)_2\text{C}=\text{O} + \overset{}{\underset{|}{\text{CH}_2}}\text{CN} \xrightarrow[\text{heat}]{\substack{\text{NH}_4\text{O}_2\text{CCH}_3 \\ \text{CH}_3\text{CO}_2\text{H}}}$
$\underset{|}{\text{CO}_2\text{C}_2\text{H}_5}$

(b) (cyclohexanone) =O + CH$_2$(CN)$_2$ $\xrightarrow[\text{heat}]{\substack{\text{NH}_4\text{O}_2\text{CCH}_3 \\ \text{CH}_3\text{CO}_2\text{H}}}$

16.18 (a) C$_6$H$_5$CH$_2$CCHCO$_2$C$_2$H$_5$ (with C=O above and C$_6$H$_5$ below)   **(b)** C$_6$H$_5$CH$_2$CCH$_2$C$_6$H$_5$ (with C=O above)

16.19 (a) (diester) $\xrightarrow{\text{NaOC}_2\text{H}_5}$ (cyclohexanone–CO$_2$C$_2$H$_5$) $\xrightarrow[\text{heat}]{\text{H}_2\text{O, OH}^-}$

(cyclohexanone–CO$_2^-$) $\xrightarrow{\text{H}^+}$ (cyclohexanone–CO$_2$H)

(b) (diester with H$_3$C, CH$_3$, CO$_2$CH$_3$, CO$_2$CH$_3$) $\xrightarrow{\text{NaOCH}_3}$ (cyclopentanone with H$_3$C, CH$_3$, =CO$_2$CH$_3$) $\xrightarrow{\text{H}^+}$ (cyclopentanone with H$_3$C, CH$_3$, CO$_2$CH$_3$)

16.20 (a) C$_2$H$_5$O$_2$CCCH$_2$CO$_2$C$_2$H$_5$ (with C=O above)   **(b)** (cyclopentanone =O with CH$_3$C=O)

16.21 (a) (structure with CH$_3$, CO$_2$C$_2$H$_5$, CO$_2$C$_2$H$_5$) $\xrightarrow{\text{NaOC}_2\text{H}_5}$ (cyclohexanone with CH$_3$, =O, CO$_2$C$_2$H$_5$) $\xrightarrow{\substack{(1)\ \text{H}_2\text{O, OH}^-,\ \text{heat} \\ (2)\ \text{H}^+}}$

(b) C$_6$H$_5$CO$_2$C$_2$H$_5$ + CH$_3$CO$_2$C$_2$H$_5$ $\xrightarrow{\text{NaOC}_2\text{H}_5}$ C$_6$H$_5$CCHCO$_2$C$_2$H$_5$ (with C=O above) $\xrightarrow{\substack{(1)\ \text{H}_2\text{O, OH}^-,\ \text{heat} \\ (2)\ \text{H}^+}}$

(c) C$_6$H$_5$CO$_2$C$_2$H$_5$ + CH$_2$(CN)$_2$ $\xrightarrow{\substack{(1)\ \text{NaOC}_2\text{H}_5 \\ (2)\ \text{H}^+}}$

(d) (CH$_3$)$_3$CCO$_2$C$_2$H$_5$ + CH$_3$CCH$_3$ (with C=O above) $\xrightarrow{\substack{(1)\ \text{NaOC}_2\text{H}_5 \\ (2)\ \text{H}^+}}$

16.22 CH$_3$CSCoA (with C=O above) $\xrightarrow{-\text{H}^+}$ $^-$:CH$_2$CSCoA (with C=O above) $\xrightarrow[\text{ester condensation}]{\text{CH}_3\text{CSCoA, } -\text{HSCoA}}$ CH$_3$C—CH$_2$CSCoA (with two C=O above) $\xrightarrow[-\text{HSCoA}]{\text{H}_2\text{O}}$

an enolate

CH$_3$C—CH$_2$CO$^-$ (with two C=O above) $\xrightarrow[-\text{CO}_2]{\text{H}^+}$ CH$_3$CCH$_3$ (with C=O above)

acetoacetate ion   acetone

16.23 (1,3-cyclohexanedione) $\rightleftharpoons$ (3-hydroxy-2-cyclohexenone)

16.24 (a) $CH_3CH(CO_2C_2H_5)_2 + CH_3CH_2CH=CHCO_2C_2H_5$ $\xrightarrow{NaOC_2H_5}$

$$CH_3CH_2\underset{\underset{CH_3\overset{|}{C}(CO_2C_2H_5)_2}{|}}{CH}CH_2CO_2C_2H_5 \quad \xrightarrow[\text{(2) H}^+]{\text{(1) H}_2O, \text{OH}^-, \text{heat}}$$

(b) $CH_2(CO_2C_2H_5)_2 + CH_3CH=CHCO_2C_2H_5$ $\xrightarrow[\text{(2) H}_2O, \text{H}^+]{\text{(1) NaOC}_2H_5}$ $CH_3\underset{\underset{CH(CO_2C_2H_5)_2}{|}}{CH}CH_2CO_2C_2H_5$ $\xrightarrow[\text{heat}]{\text{H}_2O, \text{H}^+}$

(c)

$+ \ CH_2(CO_2C_2H_5)_2$ $\xrightarrow[\text{(2) H}_2O, \text{H}^+]{\text{(1) NaOC}_2H_5}$

$\xrightarrow[\text{heat}]{\text{H}_2O, \text{H}^+}$

16.25

$\xrightarrow{\text{H}^+}$

an enolate

▼ Chapter 17

17.1 (a) 1°;   **(b)** 2°;   **(c)** salt of 3°;   **(d)** 1°;   **(e)** quaternary;   **(f)** 3°

17.2 (a) 1,4-butanediamine;   **(b)** 1,5-pentanediamine;
(c) (2-methylpropyl)amine, or isobutylamine;
(d) *N,N*-dimethylethylamine;   **(e)** *p*-chloro-*N*-methylaniline

17.3 (c) contains a chiral N, while **(d)** contains a chiral C. Each exists as a pair of enantiomers.

17.4 (a) $(CH_3)_2\underset{\underset{H}{|}}{N}H\text{-----}:N(CH_3)_2$

(b) $(CH_3)_2\underset{\underset{H}{|}}{N}H\text{-----}:N(CH_3)_2,$ $(CH_3)_2\underset{\underset{H}{|}}{N}H\text{-----}:\ddot{O}H,$ $(CH_3)_2\underset{\underset{H}{|}}{N}:\text{-----}HOH,$

$$H_2\ddot{O}:\text{-----}HOH$$

17.5

17.6 (a) (1) $CH_3CH_2CH_2Br$, (2) H_2O, OH^-
(b) (1) $CH_2=CHCH_2Br$, (2) H_2O, OH^-
(c) (1) $C_6H_5CH_2Br$, (2) H_2O, OH^-

17.7

17.8 (a) cyclohexanone $=O$ $\xrightarrow[\text{heat, pressure}]{\text{H}_2,\text{ Ni, NH}_3}$ cyclohexyl$-NH_2$

(b) $BrCH_2CH_2CH=CH_2$ $\xrightarrow{KCN}$ $NCCH_2CH_2CH=CH_2$ $\xrightarrow[\text{(2) H}_2\text{O, H}^+]{\text{(1) LiAlH}_4}$ product

(c) $C_6H_5CO_2H$ $\xrightarrow{SOCl_2}$ $C_6H_5\overset{O}{\overset{\|}{C}}Cl$ $\xrightarrow{2(CH_3)_2NH}$ $C_6H_5\overset{O}{\overset{\|}{C}}N(CH_3)_2$ $\xrightarrow[\text{(2) H}_2\text{O, H}^+]{\text{(1) LiAlH}_4}$ $C_6H_5CH_2N(CH_3)_2$

17.9 (a) $CH_3CH_2\overset{O}{\overset{\|}{C}}CH_3$ $\xrightarrow[\text{heat, pressure}]{\text{H}_2,\text{ Ni, excess NH}_3}$ $CH_3CH_2\overset{NH_2}{\overset{|}{C}H}CH_3$ ←

$\xrightarrow[\text{heat}]{\text{H}_2\text{O, OH}^-}$

(b) N^- $+ CH_3CH_2\overset{Br}{\overset{|}{C}H}CH_3$ $\longrightarrow$ $N\overset{CH_3}{\overset{|}{C}H}CH_2CH_3$ —

17.10 (a) (R)-$C_6H_5CH_2\overset{CH_3}{\overset{|}{C}H}NH_2$ **(b)** $H_2NCH_2CH_2CH_2NH_2$

17.11 (a) piperidine and **(b)** piperazine, based on electron withdrawal by N.

17.12 *p*-Methylaniline is a stronger base than aniline because the methyl group is *electron-releasing*. *p*-Nitroaniline is a weaker base than aniline because the nitro group is *electron-withdrawing*.

17.13 $HO\overset{CH_2CO_2H}{\underset{CH_2CO_2H}{\overset{|}{\underset{|}{C}}}}CO_2^-$ $\qquad$ $H_2\overset{+}{N}$ NH

17.14 $CH_3CH_2CH_2CH_2NH_2$

$\swarrow$ NaNO$_2$, HCl $\qquad$ $\downarrow$ NaNO$_2$, HCl $\xrightarrow{-N_2}$ $CH_3CH_2CH_2CH_2Cl$

$CH_3CH_2CH_2CH_2-\overset{+}{N}\equiv N$ $\xrightarrow{-N_2}$ $CH_3CH_2CH_2CH_2\overset{+}{O}H_2$ $\xrightarrow{-H^+}$ $CH_3CH_2CH_2CH_2OH$

$\uparrow$ Cl$^-$ or H$_2$O

$CH_3CH_2\overset{H}{\overset{|}{C}}HCH_2-\overset{+}{N}\equiv N$

$\swarrow -N_2$ $\qquad$ $\xrightarrow{-H^+}$ $CH_3CH=CHCH_3 + CH_3CH_2CH=CH_2$

$CH_3CH_2\overset{+}{C}HCH_3$ $\xrightarrow{Cl^-}$ $CH_3CH_2CHClCH_3$

$\xrightarrow[-H^+]{\text{H}_2\text{O}}$ $CH_3CH_2\overset{}{\underset{OH}{\overset{}{C}H}}CH_3$

17.15 The benzenediazonium ion is resonance-stabilized.

17.16 (a) $CH_3CH_2CH_2CH_2N(CH_3)_2 + CH_2=CH_2 + H_2O$
(b) $C_6H_5CH=CH_2 + (CH_3)_2NCH_2CH_3 + H_2O$

17.17 (a) $-CH_3$ OH$^-$; ; $+ (CH_3)_3N$

(b) [structure: cyclohexyl-$\overset{+}{N}(CH_3)_3$ OH^-]; [structure: cyclohexene] $+$ $(CH_3)_3N$

17.18 [structure: coniine — piperidine ring with $CH_2CH_2CH_3$ and N—H]

coniine

[structure: ring with vinyl group, N bearing H_3C and CH_3 and $CH_2CH_2CH_3$]

$C_{10}H_{21}N$

▼ Chapter 18

18.1 [resonance structures of phenanthrene] ⟷ [structure] ⟷ [structure] ⟷

[structure] ⟷ [structure]

Four of the five resonance structures show a C9–C10 double bond and two benzenoid rings.

18.2 The nitro group withdraws electron density by the inductive effect. Because the oxidizing agent attacks the other, less positive ring in Reaction 1, we would say that the attack is electrophilic.

The amino group donates electrons to the ring by resonance. Therefore, its ring is more negative. Again, this is the ring attacked by the electrophilic agent.

18.3 [structure: 9,10-dihydrophenanthrene]

18.4 (a) [naphthalene structure with NO_2 and CH_3 substituents]

(b) [naphthalene structure with CH_3 and NO_2]

(c) [naphthalene with NO_2 and SO_3H] $+$ [naphthalene with SO_3H and NO_2]

(d) [naphthalene with NO_2 and NO_2] $+$ [naphthalene with NO_2 NO_2]

18.5 (a) [naphthalene] $\xrightarrow[\text{H}_2\text{SO}_4]{\text{HNO}_3}$ [naphthalene-NO_2] $\xrightarrow[\text{(2) OH}^-]{\text{(1) Fe, HCl}}$ [naphthalene-NH_2] $\xrightarrow[\text{0}^\circ]{\substack{\text{NaNO}_2 \\ \text{HCl}}}$

[naphthalene-N_2^+ Cl^-] $\xrightarrow[\substack{\text{KCN} \\ \text{heat}}]{\text{CuCN}}$ product

(b)

(c)

18.6

18.7

The 5-intermediate has three resonance structures with one aromatic ring, while the 6-intermediate has only two.

18.8 not aromatic

18.9 (a)   **(b)**

(c)   **(d)**

▼ Chapter 19

19.1

19.2

(1) 2 CH₃MgI, ether
(2) H₂O, H⁺

$-H_2O$

19.3

H₂O, H⁺

$-H_2O$

19.4 Under acidic conditions, dehydration and rearrangements could occur.

19.5 (a) $\dfrac{120 - 20}{120 + 20} \times 100\% = 71\%$

(b) $\dfrac{50 - 50}{50 + 50} \times 100\% = 0$

19.6

OH⁻
$-H_2O$

H₂O
$-OH^-$

(S)

racemic

19.7

+ HO₂CCHCH₂OH

C₆H₅

tropine   tropic acid

19.8

$C_2H_5O_2C$ ... CH_2 ... $NHCO_2C_2H_5$ →[Na] ... →[Michael] ...

... →[proton transfer] ...

... →[Dieckmann] ... →[$-:OC_2H_5$] ...

... →[$-H^+$] ...

B   anion of B

▼ **Chapter 20**

20.1

π_6^* _____

π_5^* _____

π_4^* ⬇ _____

π_3 ⬆ _____

π_2 ⬆⬇ _____

π_1 ⬆⬇ _____

20.2 The double bonds in 1,4-pentadiene are not conjugated; thus, their π orbitals are independent of each other.

first C=C: π_2^* _____ second C=C: π_2^* _____

    π_1 ⬆⬇ π_1 ⬆⬇

20.3 [6 + 4]

20.4 **(a)** $2\ C_6H_5CH{=}CHCO_2H$ →[$h\nu$]

(b) $\begin{array}{l} CH_2CH{=}CH_2 \\ | \\ CH_2CH{=}CH_2 \end{array}$ →[$h\nu$]

In **(a)**, either *cis-* or *trans-*cinnamic acid (3-phenylpropenoic acid) could be used. Other stereoisomers (and structural isomers) of the product might be obtained as by-products.

20.5 A photo-induced [4 + 2] cycloaddition cannot occur when *either* the dienophile or the diene is excited.

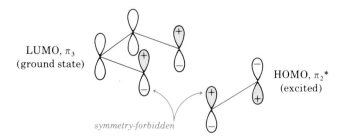

LUMO, π_3
(ground state)

HOMO, π_2^*
(excited)

symmetry-forbidden

20.6 A [6 + 4] cycloaddition reaction is thermally induced because this reaction path is symmetry-allowed.

π_3, HOMO of the triene

π_3^*, LUMO of the diene

20.7

symmetry-forbidden

20.8 **(a)**

trans

(b)

(3E,5Z,7Z)

20.9 **(a)** [3,3];   **(b)** [1,5]

20.10 **(a)** can proceed by a [1,5] sigmatropic shift, and **(b)** is a [3,3] sigmatropic shift. Both proceed readily.

20.11 Previtamin D contains a $(4n + 2)$ π system. The thermal reaction proceeds by disrotatory motion, and the photo-induced reaction proceeds by conrotatory motion.

thermal:

photochemical:

▼ ▬▬▬ Chapter 21

21.1 Tyrian purple   **21.2** **(c), (b), (a)**   **21.3** **(b)** 446 nm;   **(c)** 416 nm

21.4 The entire ring system of coronene has 24 pi electrons ($4n$, and not $4n + 2$). Therefore, not all the pi electrons are involved in the aromatic pi cloud. It is thought that only the peripheral carbons are part of the aromatic system.

21.5 A compound that appears violet absorbs at a shorter wavelength (about 570 nm) than one that appears blue-green (about 650 nm), and therefore is the compound with less delocalization. The structure in **(b)**, with only two —N(CH$_3$)$_2$ groups, has less delocalization and is the violet-colored compound. The structure in **(a)**, therefore, is the blue-green compound.

21.6 (a) $[CH_4]^{\dot{+}}$, ion-radical;   (b) $[CH_3]^+$, cation;
(c) $[CH_2CH_2]^{\dot{+}}$, ion-radical;   (d) $[CH_2CH_2]^+$, cation

21.7 (a) $[CH_4]^{\dot{+}}$;   (b) $[CH_3CH_2CH_3]^{\dot{+}}$

21.8 (a) 16;   (b) 43;   (c) 32;   (d) 18

21.9 (a) 30;   (b) 98;   (c) 46;   (d) 172

21.10 *B* contains Br, and *C* contains Cl.

21.11 (a)

$$\left[\begin{matrix} CH_3CHCH_2CH_2CH_3 \\ | \\ CH_3 \end{matrix}\right]^{\dot{+}} \quad \xrightarrow{-[CH_3]^{\cdot}} \quad \left[\begin{matrix} CHCH_2CH_2CH_3 \\ | \\ CH_3 \end{matrix}\right]^{+}$$

$$\xrightarrow{-[CH_3CH_2CH_2]^{\cdot}} \quad \left[\begin{matrix} CH_3CH \\ | \\ CH_3 \end{matrix}\right]^{+}$$

$[CH_3]^+$ and $[CH_3CH_2CH_2]^+$ might also be observed.

(b) $\left[\begin{matrix} CH_3 \\ | \\ CH_3CCH_3 \\ | \\ CH_3 \end{matrix}\right]^{\dot{+}} \xrightarrow{-[CH_3]^{\cdot}} [(CH_3)_3C]^+$

(c) $[CH_2{=}CHCH_2CH_2CH_3]^{\dot{+}} \xrightarrow{-[CH_3CH_2]^{\cdot}} [CH_2{=}CHCH_2]^+$

21.12 $CH_3OCH_2CH_2CH_3$ **21.13** (a) 58;   (b) 44;   (c) 60;   (d) 88 and 102

▼ Chapter 22

22.1 (a) aldotetrose;   (b) ketopentose or pentulose;   (c) aldohexose

22.2 (a) (*R*) and D;   (b) (*S*) and L;   (c) (3*S*,4*R*) and D

22.3 L-(+)- and *meso*-tartaric acid

22.4 (a)

```
      CHO
HO ── H
 H ── OH
HO ── H
HO ── H
     CH₂OH
```

(b)

```
      CHO
 H ── OH
 H ── OH
HO ── H
HO ── H
     CH₂OH
```

the enantiomer of D-arabinose

22.5

```
      CHO
 H ── OH
HO ── H
HO ── H
     CH₂OH
```

L-arabinose

22.6 L-idose

22.7

all groups axial

22.8 α:

β:

22.9 (a)

$$\text{CO}_2\text{H}$$
H—OH
$$\text{CH}_2\text{OH}$$

(b)   (c) none

(d)

22.10 allose and galactose   **22.11** $CH_3\overset{\text{O}}{\overset{\|}{C}}H + H\overset{\text{O}}{\overset{\|}{C}}CH_2OCH_3$

22.12

$$\text{CHO}$$
H—OH
HO—H
H—OH
H—OH
$$\text{CH}_2\text{OH}$$

$\xrightarrow{\text{HIO}_4}$ 5 $H\overset{\text{O}}{\overset{\|}{C}}OH$ + $H\overset{\text{O}}{\overset{\|}{C}}H$

to HCHO

22.13 (a)   (b)

22.14

$$\text{CO}_2\text{H}$$
H—OCH₃
$$\text{CO}_2\text{H}$$

+

$$\text{CO}_2\text{H}$$
H—OCH₃
CH₃O—H
$$\text{CO}_2\text{H}$$

+ H—OCH₃

$$\text{CO}_2\text{H}$$
H—OCH₃
$$\text{CH}_2\text{OCH}_3$$

+ CH_2OCH_3

$$\text{CO}_2\text{H}$$

22.15 (a)   (b)

22.16 (a), (b), and (e):

(c)   **(d)**

22.17 β-Cellobiose is more stable because each substituent on each ring is equatorial.

22.18 (a) **OH**;   **(b)** no reaction;   **(c)** no reaction;

(d)

22.19 (a) OH + CH₃CO₂H   **(b)** + CH₃CO₂⁻

▼ Chapter 23

23.1 $\underset{\begin{array}{c}|\\CHO_2C(CH_2)_7CH=CH(CH_2)_7CH_3\\|\end{array}}{\overset{\begin{array}{c}CH_2O_2C(CH_2)_7CH=CH(CH_2)_7CH_3\\|\end{array}}{CH_2O_2C(CH_2)_7CH=CH(CH_2)_7CH_3}}$ $\xrightarrow[\text{heat, pressure}]{\text{excess H}_2,\text{ catalyst}}$ $\underset{\begin{array}{c}|\\CH_2OH\end{array}}{\overset{\begin{array}{c}CH_2OH\\|\end{array}}{CHOH}}$ + 3 CH₃(CH₂)₁₆CH₂OH $\xrightarrow{\text{H}_2\text{SO}_4}$

CH₃(CH₂)₁₆CH₂OSO₃H $\xrightarrow{\text{NaOH}}$ CH₃(CH₂)₁₆CH₂OSO₃Na

23.2 *trans*-CH₃(CH₂)₁₂CH=$\underset{\begin{array}{c}|\\NH_2\end{array}}{\overset{\begin{array}{c}OH\\|\end{array}}{CHCHCHCH_2OH}}$

23.3 (a)

(b)

(c)

23.4 3β-Cholestanol is more stable because its OH is equatorial.

▼ **Chapter 24**

24.1

$$H_2N \overset{CO_2H}{\underset{CH_2SH}{\overset{|}{\underset{|}{C}}} H} \quad (R)$$

L-cysteine

The —CH₂SH group has higher priority than does the —CO₂H group because S has a higher atomic number than does O.

24.2 $(CH_3)_2CHCHO \xrightarrow[HCN]{CN^-} (CH_3)_2CHCHCN \xrightarrow{SOCl_2}$

with OH above the CHCN carbon, labeled A

$(CH_3)_2CHCHCN$ with Cl, labeled B $\xrightarrow{NH_3}$ $(CH_3)_2CHCHCN$ with NH₂, labeled C $\xrightarrow[H_2O]{OH^-}$ $(CH_3)_2CHCHCO_2^-$ with NH₂, labeled D

24.3 **(a)** acidic; **(b)** near-neutral; **(c)** near-neutral; **(d)** basic; **(e)** near-neutral

24.4 $HO_2CCH_2CH_2CHCO_2^- \ Na^+$ with NH₂ substituent

24.5 **(a)** pyrrolidine ring with N^+H_2 and —CO₂H Cl⁻ , pyrrolidine ring with NH and —CO₂⁻ Na⁺

(b) $HO-C_6H_4-CH_2CHCO_2H \ Cl^-$ with NH_3^+ , $^-O-C_6H_4-CH_2CHCO_2^- \ 2Na^+$ with NH₂

24.6 Like the nitrogen in pyrrole (Section 18.9), the nitrogen in tryptophan has no unshared bonding electrons; therefore, tryptophan is a neutral amino acid.

24.7 The other amino acids contain primary amino groups, which can form the imine group in Ruhemann's purple. Proline contains a secondary amino group, which cannot form an imine.

24.8 ala-gly-phe, ala-phe-gly, gly-phe-ala, gly-ala-phe, phe-ala-gly, phe-gly-ala

24.9

24.10 (a) Trypsin cleaves the amide bond (to the right) of lys and arg.

$$\text{trp–met–gly–gly–phe–arg–phe–trp–val–lys–ala–gly–ser} \xrightarrow[\text{H}_2\text{O}]{\text{trypsin}}$$

$$\text{trp–met–gly–gly–phe–arg} + \text{phe–trp–val–lys} + \text{ala–gly–ser}$$

(b) Chymotrypsin cleaves the amide bond (to the right) of tyr, phe, and trp.

$$\text{trp–met–gly–gly–phe–arg–phe–trp–val–lys–ala–gly–ser} \xrightarrow[\text{H}_2\text{O}]{\text{chymotrypsin}}$$

$$2 \text{ trp} + \text{met–gly–gly–phe} + \text{arg–phe} + \text{val–lys–ala–gly–ser}$$

(c) Pepsin cleaves the amide bond (to the left) of tyr, phe, and trp.

$$\text{trp–met–gly–gly–phe–arg–phe–trp–val–lys–ala–gly–ser} \xrightarrow[\text{H}_2\text{O}]{\text{pepsin}}$$

$$\text{trp–met–gly–gly} + \text{phe–arg} + \text{phe} + \text{trp–val–lys–ala–gly–ser}$$

24.11

24.12 The enzyme catalyzes: **(a)** transfer of an acetyl group from one molecule to another; **(b)** conversion of phenylalanine to tyrosine by hydroxylation of the benzene ring; **(c)** dehydrogenation of an ester or a salt of pyruvic acid.

▼ Chapter 25

25.1

25.2

25.3 Cytosine is more basic because it contains a free amino group that is not involved in an amide group.

— *amide nitrogen less basic*

25.4

thymine   guanine

only one hydrogen bond

Index

Correlation Chart for Group Assignments in Infrared Spectra

Typical Chemical Shifts in Nmr Spectra

Group	δ Value (ppm)
Proton on sp^3 Carbon:	
RCH$_3$	0.8 – 1.2
R$_2$CH$_2$	1.1 – 1.5
R$_3$CH	~1.5
ArCH$_3$	2.2 – 2.5
R$_2$NCH$_3$	2.2 – 2.6
R$_2$CHOR	3.2 – 4.3
R$_2$CHCl	3.5 – 3.7
R$_2$CHCR=CR$_2$	~1.7
$\overset{\displaystyle O}{\overset{\|}{RCCH_2R}}$	2.0 – 2.7
Proton on sp or sp^2 Carbon:	
R$_2$C=CHR	4.9 – 5.9
ArH	6.0 – 8.0
RCHO	9.4 – 10.4
RC≡CH	2.3 – 2.9
Proton on N or O:	
R$_2$NH	2 – 4
ROH	1 – 6
ArOH	6 – 8
RCO$_2$H	10 – 12